Growth Control
in Woody Plants

This is a volume in the

PHYSIOLOGICAL ECOLOGY series
Edited by Harold A. Mooney

A complete list of books in this series appears at the end of the volume.

Growth Control
in Woody Plants

Theodore T. Kozlowski
Department of Environmental Science, Policy and Management
College of Natural Resources
University of California, Berkeley
Berkeley, California

Stephen G. Pallardy
School of Natural Resources
University of Missouri
Columbia, Missouri

Academic Press

San Diego   London   Boston   New York   Sydney   Tokyo   Toronto

Copyright © 1997 by ACADEMIC PRESS

Academic Press, Inc.
525 B Street, Suite 1900, San Diego, California 92101-4495, USA
http://www.apnet.com

Academic Press Limited
24-28 Oval Road, London NW1 7DX, UK
http://www.hbuk.co.uk/ap/

Library of Congress Cataloging-in-Publication Data

Kozlowski, T. T. (Theodore Thomas), date.
   Growth control in woody plants / by Theodore T. Kozlowski, Stephen
  G. Pallardy.
        p.   cm. -- (Physiological ecology series)
   Includes bibliographical references and index.
   ISBN 0-12-424210-3 (alk. paper)
   1. Growth (Plants) 2. Woody plants--Development. 3. Woody
plants--Ecophysiology. I. Pallardy, Stephen G. II. Title.
III. Series: Physiological ecology.
QK731.K67   1997
582.1'50431--dc20   96-41837
                                    CIP

Printed and bound by CPI Group (UK) Ltd, Croydon, CR0 4YY
Transferred to Digital Print 2011

This book is dedicated to the memory of
Paul Jackson Kramer (1904–1995),
with whom we were privileged to work

Contents

8. Cultural Practices and Reproductive Growth

9. Biotechnology

Preface

This book deals with the physiological, environmental, and cultural regulation of growth of woody plants. It expands and updates major portions of two books: *Physiology of Woody Plants* by P. J. Kramer and T. T. Kozlowski (Academic Press, 1979), a second edition of which was recently published (Academic Press, 1997), and *The Physiological Ecology of Woody Plants* by T. T. Kozlowski, P. J. Kramer, and S. G. Pallardy (Academic Press, 1991). Since those books were published, intensive research has filled in gaps in our knowledge and altered some of our views about control of growth of woody plants. We therefore considered it important to update the available information on how physiological processes and environmental factors regulate plant growth and how management practices may alter the rate of growth and influence the quality of harvested products of woody plants.

This book was written for use as a text by students and as a reference for researchers and growers. The subject matter is interdisciplinary in scope and should be of interest to a wide range of scientists, including agroforesters, agronomists, arborists, biotechnologists, botanists, entomologists, foresters, horticulturists, plant breeders, plant ecologists, plant geneticists, landscape architects, plant pathologists, plant physiologists, and soil scientists. It should be of particular interest to those who grow woody plants for production of food and fiber.

The first chapter emphasizes the complexity of regulation of growth. Several aspects of genetic control of growth are addressed, including causes of genetic variation and opportunities for exploitation of genetic variation to increase production of wood, fruits, and seeds. Attention also is given to the difficulty of quantifying the effects of specific environmental stresses on growth.

The second chapter describes seed germination and seedling growth, environmental and physiological regulation of germination, physiology of young seedlings, and production of seedlings in nurseries. The third and fourth chapters address physiological regulation of vegetative and reproductive growth, with emphasis on carbohydrate, mineral, water, and hormone relations (as well as respiration) in regulation of growth. The fifth chapter describes the impact of abiotic environmental factors (light, water, temperature, soil fertility, salinity, pollution, wind, and fire) and biotic factors (insects and diseases) on vegetative growth. The sixth chapter addresses the effects of light intensity and day length, temperature, water, soil fertility, salinity, and environmental pollution on such aspects of reproductive growth as floral induction, bud dormancy, flowering, pollen formation, growth of pollen tubes, fertilization, and growth of fruits, cones, and seeds.

The seventh chapter analyzes the effects on vegetative growth of site preparation, drainage of soil, herbicides, irrigation, correction of mineral deficiencies, thinning of forest stands, pruning, use of growth regulators, and integrated pest management. The eighth chapter addresses effects of cultural management practices on reproductive growth. These include arrangement and spacing of trees, grafting, use of fertilizers, irrigation, thinning of forest stands, pruning, scoring or girdling of stems and branches, application of growth regulators, storage of harvested fruits, and prevention of freezing and chilling injury. The ninth chapter deals with the scope, techniques, accomplishments, and future potential of biotechnology.

We have defined some important botanical terms. For readers who are not familiar with some other terms we have used, we recommend that they consult the *Academic Press Dictionary of Science and Technology* (1992), edited by C. Morris.

Although we review literature on cultural practices that may influence growth of woody plants, we do not make any recommendations for use of specific management practices, experimental procedures and equipment, and materials. Selection of appropriate management practices and experimental procedures will depend on the objectives of individual growers, plant species and genotype, availability of management resources, and local conditions that are known only to each grower. However, we hope that an understanding of growth control in physiological, ecological, and management contexts will help growers to adopt management practices that will be appropriate for their situations.

A summary and list of general references have been added to the end of each chapter. References cited in the text are listed in the bibliography. We have selected references from a voluminous body of literature to make this book comprehensive, authoritative, and up to date. On controversial issues we often have presented contrasting views and have based our interpretations on the weight and quality of available research data. We caution readers that as new information becomes available we may revise some of our conclusions. We hope that readers also will modify their views when additional research provides justification for doing so.

We have used common names in the text for most well-known species of plants and Latin names for less common ones. A list of scientific and common names of plants cited is given following the text. Names of North American woody plants are based largely on E. L. Little's *Check List of Native and Naturalized Trees of the United States* (Agriculture Handbook No. 41, U.S. Forest Service, Washington, DC, 1979). However, to facilitate use of the common name index for a diverse audience, we decided not to use the rules of compounding and hyphenating recommended by Little. Those rules, while reducing taxonomic ambiguity in common names, often result in awkward construction and unusual placement within an alphabetical index. Names of plants other than North American species are from various sources.

We express our appreciation to many people who variously contributed to this volume. Much stimulation came from our graduate students, visiting investigators in our laboratories, and collaborators in many countries with whom we have worked and exchanged information.

Individual chapters were read by W. J. Libby and B. McCown. Their comments and suggestions are greatly appreciated. However, the chapters have been revised since they read them and they should not be held responsible for errors that may occur. We also sincerely acknowledge the able technical assistance of Julie Rhoads.

T. T. Kozlowski
S. G. Pallardy

1 _____

Introduction

_____

Woody plants are highly integrated organisms whose growth and development are regulated by interactions of their heredity and environment as they influence the availability of resources (carbohydrates, hormones, water, and mineral nutrients) at meristematic sites.

Genetic Control of Growth

Hereditary control of plant growth is contained largely in the DNA of cell chromosomes which through messenger RNA regulates the kinds of proteins and enzymes synthesized, which in turn control cell structure and plant responses. DNA also is present in mitochondria and chloroplasts, organelles that are important in producing essential proteins [e.g., large Rubisco (ribulose bisphosphate carboxylase/oxygenase) subunits]. Advances in molecular biology have demonstrated that genetic control of plant growth and development is very complex, with large numbers of genes being differentially expressed in various organs. In tobacco, for example, about 70,000 structural genes are expressed through the life cycle of the plant, but 6,000 of the genes transcribed in stem tissues are not transcribed in other parts of the plant (Kamaly and Goldberg, 1980, cited in van Loon and Bruinsma, 1992). Similarly, the gene coding for the small subunit of Rubisco is expressed in aerial parts of the plant but not in roots (Fluhr *et al.*, 1986, cited in Hutchison and Greenwood, 1991). The role of gene expression in plant growth and development is discussed further in Chapter 9.

Several causes of genetic variation in plants have been identified. The long-term and ultimate source of variation is mutation, which may involve a change in a single gene by alteration of DNA molecules or a loss or addition of one or more chromosomes or parts of chromosomes. The main short-term source of variation is genetic recombination, in which different combinations of genes are created in each generation of plants by segregation, assortment, and recombination of alleles and chromosomes. The level and distribution of genetic variation within a population are affected

1

by the breeding system, with inbreeding increasing total variation but reducing it within families, whereas outcrossing concentrates it within families. Genetic variation is increased by migration of alleles ("gene flow") through dispersal of seeds and pollen. Genetic variations within and among plant populations differ appreciably with their mating systems and gamete dispersal mechanisms. Plants with high potential for gene movement (e.g., wind-pollinated species with winged seeds) show less genetic variation among populations than do species with low mobility.

Genetic variation has been effectively exploited to meet growing demands for wood. Large increases in wood production have been achieved by combinations of parent selection, clonal selection, and interspecific hybridization (Zobel and Talbert, 1984; Stettler *et al.*, 1988; Lugo *et al.*, 1988). Much attention has been given throughout the world to planting genetically improved, fast-growing forest trees in close spacings along with using intensive cultural practices. Such trees commonly are harvested at short rotations, usually at 3 to 10 years. Early yields of wood from short-rotation plantations are suitable for timber and pulp, and the yields often are at least double those from conventional plantations (Zobel and van Buijtenen, 1989). This is particularly true for pines that are suitable for selective breeding (Kozlowski and Greathouse, 1970; Burley and Styles, 1976). In Brazil, closely spaced, high-yielding *Eucalyptus* species are grown for charcoal to operate steel mills, for pulp, and for fuelwood. Under optimal conditions yields are 40 to 50 tons ha^{-1} year $^{-1}$ of dry matter, but the average is closer to 25 tons (Eldridge, 1976; Hall and de Groot, 1987). Sachs *et al.* (1988) obtained data on potential yield of *Eucalyptus* trees on good sites near Davis, California. Productivity of biomass of selected river red gum (*Eucalyptus camaldulensis*) in an intensively managed plantation (1,100 trees ha^{-1}) was 25 to 27 tons $acre^{-1}$ for the third and fourth year after planting. Because the rate of growth did not increase after the third year, a three-year rotation was indicated.

Genetic variations in growth and development of woody plants differ among species, populations within a species, and individual plants. Genetic variations are particularly evident in species with large population sizes and extensive ranges. This is because the number of new mutations arising in each generation increases with population size, the rate of genetic drift varies inversely with population size, and environmental variation usually increases with the geographic range of a species (Mitton, 1995). There is a voluminous literature on genetic variation in woody plants (e.g., Wright, 1976; Burley, 1976; Zobel and van Buijtenen, 1989; Adams *et al.*, 1992; Mitton, 1995). Here we cite only a few examples.

In Douglas fir, genetic variations were reported in biomass partitioning, wood density, crown width, stem diameter growth, branch diameter and length, and needle size (St. Clair, 1994a,b). The variability in biomass partitioning and wood density indicated that genetic gains may be expected from selection and breeding of desirable genotypes. Some of the crown structure traits showed promise as potential ideotype traits. Large trees that grew vigorously in their growing space had tall, narrow crowns, large leaf areas, and preferential partitioning of carbohydrates to leaves over branches.

Much attention has been given to genetic variation in poplars, largely for three reasons (Dickmann and Stuart, 1983): (1) poplars produce more biomass in short-rotation intensive culture than most other deciduous or evergreen species; (2) wide genetic variability among poplars offers opportunity for genetic improvement and high yield of wood; and (3) there is a strong demand for use of poplar for pulp, lumber, and energy through combustion or conversion to alcohol or other fuels. Large increases in productivity of poplars have been made by combinations of interspecific hybridization, parent selection, and clonal selection (Weber *et al.*, 1985; Stettler *et al.*, 1988).

Genetic variations have been shown among poplars in photosynthesis and enzymatic traits (Weber and Stettler, 1981; Ceulemans *et al.*, 1987; Rhodenbaugh and Pallardy, 1993); partitioning of photosynthate (Bongarten and Teskey, 1987); stomatal size and frequency (Siwecki and Kozlowski, 1973; Pallardy and Kozlowski, 1979a; Kimmerer and Kozlowski, 1981); phenological characteristics such as timing of seasonal growth cessation and bud set (Weber *et al.*, 1985; Milne *et al.*, 1992; Ceulemans *et al.*, 1992); stem and crown form (Nelson *et al.*, 1981; Dickmann, 1985; Kärki and Tigerstedt, 1985); growth rate (Phelps *et al.*, 1982; Dunlap *et al.*, 1992; Heilman and Stettler, 1985; Rogers *et al.*, 1989; Milne *et al.*, 1992); wood properties such as specific gravity, fiber length, energy content, and chemical composition (Sastry and Anderson, 1980; Reddy and Jokela, 1982); rooting capacity (Pallardy and Kozlowski, 1979b; Ying and Bagley, 1977); and response to environmental stresses such as drought (Pallardy and Kozlowski, 1979c; Mazzoleni and Dickmann, 1988) and air pollution (Kimmerer and Kozlowski, 1981).

Characteristics associated with superior growth of poplar hybrids include favorable leaf expansion rates, leaf cell size and number, leaf mesophyll structure, stomatal conductance, stomatal responses to drought, canopy architecture, leaf retention, and resistance to disease (Stettler *et al.*, 1988). Enzymatic and stomatal frequency traits are associated with variations in growth rates of poplars (Weber and Stettler, 1981; Ceulemans *et al.*, 1984, 1987).

In contrast to the high genetic variability in Douglas fir and poplars, variability in red spruce is lower than that of most north-temperate woody species. Genetic variability is particularly low in discontinuous southern populations of red spruce, presumably because of genetic drift followed by inbreeding (Hawley and DeHayes, 1994).

Exploitation of genetic material is crucial for growing fruit trees and forest trees in seed orchards. Improvement of fruit trees has been documented since the time of recorded history. The best available wild edible species have been collected (e.g., blueberry, grape, plum, and walnut), or barely edible plants have been improved by selecting individuals with larger than average fruits and of better quality (e.g., apple and pear). A major objective in fruit-tree breeding is selection of early-flowering genotypes that produce high fruit yields (Alston and Spiegel-Roy, 1985; Sedgley and Griffin, 1989). Selection and breeding programs also have been directed toward producing plants with low chilling requirements, improved reproductive capacity in areas of low spring temperatures, enough cold hardiness to extend growing of fruit trees to colder regions, pest and disease resistance, and high fruit quality. Much attention has been given to selection of new rootstocks for controlling fruit size and production (Alston and Spiegel-Roy, 1985; see also Chapter 8).

For production of forest tree seeds it is essential to upgrade the genetic quality of planting stock by planting new seed orchards as improved clones become available. Desirable traits of introduced clones are high fecundity, synchrony of male and female flowering, and production of large seeds (Sedgley and Griffin, 1989). Seed orchards usually are designed to include large numbers of clones (25 to 30), each with similar opportunity for interbreeding. Many clones usually have been planted because (1) at the time of orchard establishment information often is lacking about the genetic quality of the clone and (2) high levels of outcrossing are maintained by clonal diversity despite variations among clones in seed yield (Griffin, 1982, 1984).

Environmental Regulation of Growth

Woody plants are subjected to multiple abiotic and biotic stresses. The important abiotic stresses include extremes in light intensity, drought, flooding, temperature extremes, pollution, wind, low soil fertility, and fire. Among the major biotic stresses are plant competition, attacks by insects and pathogens, some activities of humans, and herbivory (Kozlowski *et al.*, 1991).

Environmental stresses do not alter tree growth directly but rather indirectly by influencing the rates of and balances among physiological processes such as photosynthesis, respiration, assimilation (conversion of food into new protoplasm, cell walls, and other substances), hormone synthesis, absorption of water and minerals, translocation of compounds important in regulation of growth (e.g., carbohydrates, hormones, water, and minerals), and other processes and physicochemical conditions (see Chapter 1 of Kozlowski and Pallardy, 1997).

The centralized system of plant responses to stress can be triggered by a wide range of stresses (Chapin, 1991). However, the effects of individual environmental stresses on plant growth and development are complex and difficult to quantify for a variety of reasons. Whereas some stresses affect plants more or less continually, others exert strong effects more randomly. For example, because of shading by dominant trees in a forest stand, the leaves of understory trees are continually exposed to shading stress. Because leaves overlap, those in the interiors of crowns of dominant trees also undergo persistent shading.

Another difficulty is that the importance of specific environmental stresses on physiological processes and tree growth changes over time. An abruptly imposed severe stress, such as an insect attack or pollution episode, may suddenly dominate over other milder stresses that previously were the chief inhibitors of growth. If the soil is fully charged with water early in the growing season, a given amount of precipitation has little effect on growth, whereas half that amount after a drought later in the season generally stimulates growth. In Wisconsin, for example, the correlation of cambial growth of northern pin oak trees with temperature decreased in late summer as soil moisture was progressively depleted and growth was increasingly limited by water deficits (Kozlowski *et al.*, 1962). Another complexity is that correlation between the intensity of a specific environmental stress and degree of growth inhibition may suggest that this stress was actually controlling growth. Yet this stress factor may only be correlated with some other factor or factors that are more important in controlling growth but were not included in the analysis (Kozlowski *et al.*, 1991).

Still another problem is that growth responses to an environmental stress (or a decrease in its intensity) may not be apparent for a long time. Such lag responses are well illustrated by effects of environmental stresses on shoot growth and alleviation of stress on cambial growth. It also is clear from dendrochronological studies that severe stress events can impose long-term limitations on growth of trees for many years even when subsequent environmental conditions are favorable for growth (Jenkins and Pallardy, 1995).

Both short- and long-term lag responses of shoot growth to environmental changes have been shown. Rates of shoot growth lag behind changes in air temperature during the day (Luxmoore *et al.*, 1995). The length of shoots of some species is predetermined during bud formation. In these species, buds form during one year and expand into shoots in the following year (see Chapter 3 of Kozlowski and Pallardy, 1997). No matter how favorable the environment is during the year of bud expansion, the shoots expand for only a few weeks early in the summer. If the environment is favorable for growth during the year of bud formation, large buds form that will produce long shoots with many leaves in the next year (Table 1.1). When late-summer temperatures were low, only small buds formed on Norway spruce trees and expanded into relatively short shoots in the next growing season (Heide, 1974). In comparison, species that exhibit free growth or recurrently flushing shoot growth (see Chapter 3 of Kozlowski and Pallardy, 1997) generally expand their shoots late into the summer. In such species, shoot growth is affected much more by the environmental regime during the year of shoot expansion than is shoot growth of species exhibiting fixed growth.

Long lag responses to environmental changes also are shown by diameter growth responses of residual trees in thinned stands. The released trees respond to greater availability of light, water, and mineral nutrients by increasing the rate of photosynthesis of existing leaves and by producing more leaves. The greater leaf areas and availability of resources are accompanied by increased production of photosynthate and hormonal growth regulators, followed by their downward transport in the stem where they stimulate cambial growth. All these sequential changes require time, however, and the effect of stand thinning on diameter growth in the lower stem of the residual trees may not be apparent for a year or more (Fig. 1.1). Furthermore, dense stands of stagnated trees may not respond to thinning for an even longer time, and sometimes not at all. Worrall *et al.* (1985)

Table 1.1   Effect of Bud Size on Shoot Growth of 8-Year-Old Red Pine Trees[a,b]

	Bud diameter (mm)	Bud length (mm)	Shoot length (mm)
Terminal leader	8.2 ± 0.7	38.0 ± 2.8	742.0 ± 26.7
Whorl 1 shoots	5.9 ± 0.1	27.3 ± 0.7	484.8 ± 11.0
Whorl 2 shoots	5.5 ± 0.1	22.9 ± 0.8	403.2 ± 13.0
Whorl 3 shoots	4.5 ± 0.2	16.6 ± 0.9	271.4 ± 19.1
Whorl 4 shoots	3.8 ± 0.3	12.5 ± 1.0	132.1 ± 20.6
Whorl 5 shoots	3.7 ± 0.3	9.9 ± 0.8	65.2 ± 16.0
Whorl 6 shoots	3.3 ± 0.4	8.6 ± 1.4	74.4 ± 31.5

[a]From Kozlowski *et al.* (1973).
[b]Data are means and standard errors of bud diameters and lengths before initiation of shoot expansion (March 20, 1970) and final shoot lengths (August 19, 1970) at different stem locations.

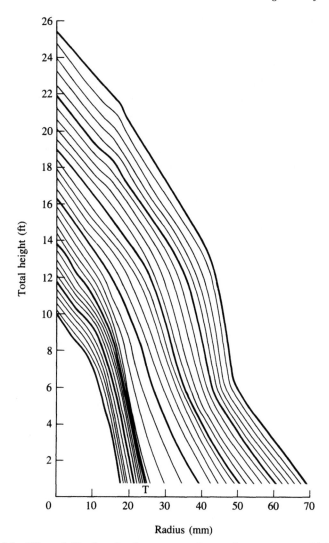

Figure 1.1   Effect of thinning closely grown ponderosa pine trees on cambial growth of residual trees. Before thinning the annual xylem increment was greatest in the upper stem. After thinning (T), the annual increment became greater in the lower stem, but there was a long lag response in the lower stem. (After Myers, 1963. Reprinted from *Forest Science*, **9**:394–404. Published by the Society of American Foresters, 5400 Grosvenor Lane, Bethesda, MD 20814-2198. Not for further reproduction.)

attributed the lack of height growth response of stagnated lodgepole pine trees after stand thinning to their higher than normal use of photosynthate in root growth and/or respiration.

Both physiological and growth responses to changes in environmental conditions vary with the environmental regime in which the plants were previously grown. For example, stomatal sensitivity to changes in light intensity and humidity differs with fertility of the soil in which the plants were grown (Davies and Kozlowski, 1974). Exposing plants to a high or low temperature regime affects the subsequent rate of photosynthesis at another temperature (Pearcy, 1977). The effect of a given pollution dosage on plants is influenced not only by prevailing environmental conditions but also by environmental regimes before and after the pollution episode (Kozlowski and Constantinidou, 1986b). By influencing plant metabolism, the environmental conditions following a pollution episode often regulate plant responses to various pollutants (Norby and Kozlowski, 1981b). A difficulty in evaluating the effects of environmental stresses on plant growth is that the effects of multiple stresses may or may not be additive. For example, as the CO_2 concentration of the air increases, growth is stimulated even though limitations by N supply and water absorption become more severe (Norby *et al.*, 1986a). In comparison, under a regime of both higher temperatures and CO_2 concentrations, growth may not change appreciably, or it may increase or even decrease, depending on interactions of C, N, and water supply (Pastor and Post, 1988). The complexity of the impact of multiple environmental stresses on plant growth is further shown by effects of combined air pollutants, which may be synergistic, additive, or antagonistic (Kozlowski and Constantinidou, 1986b). Mechanisms of synergism and antagonism are related to direct reactivity between pollutants, effects of individual or combined pollutants on photosynthesis and stomatal aperture, competition for reaction sites, changes in sensitivity of reaction sites to pollutants, and various combinations of these (Heagle and Johnston, 1979).

Either a deficit or excess of soil water predisposes woody plants to certain diseases (Schoeneweiss, 1978a). Outbreaks of stem cankers, diebacks, and declines often follow loss of vigor of trees after exposure to drought, soil inundation, mineral deficiency, or air pollution (Kozlowski, 1985b). Physiological changes in plants that are induced by environmental stresses often are prerequisites for attacks by various insects (Kozlowski *et al.*, 1991, pp. 25–28). Fungus infections may predispose plants to further infections by the same or different pathogenic organisms (Bell, 1982).

The effects of environmental stresses on plant growth and survival also vary greatly with plant vigor, and particularly with the amounts of stored carbohydrates and mineral nutrients in plants at the time of stress imposition. Low carbohydrate reserves in stems and roots often are associated with impaired growth of shoots and roots, as well as susceptibility to pathogens (Wargo and Montgomery, 1983; Gregory *et al.*, 1986). Trees with low carbohydrate reserves may die when subjected to environmental stresses because they lack sufficient reserves to heal injuries or maintain physiological processes at levels needed to sustain life (Waring, 1987). Death of balsam fir trees in the harsh climate at high altitudes in New Hampshire was preceded by depletion of carbohydrate reserves (Sprugel, 1976). Furthermore, defoliation often is followed by death of stressed trees (with low carbohydrate reserves), whereas unstressed trees are more likely to withstand defoliation. As much as 70 to 80% defoliation was required to decrease growth of vigorous poplar trees by 20% (Bassman *et al.*, 1982).

Yet another complication in relating growth of woody plants to their environmental regimes is that the changes in plant morphology that occur over time interact with the physiological processes that regulate growth. Plant morphology determines both the pattern of acquisition of resources and the condition of the internal environment of plants. As woody plants grow, their structure changes as a result of (1) progression from juvenility to maturation, (2) changes in growth rates in different parts of the crown, and (3) plastic changes associated with acclimation to a changing environment (e.g., sun and shade leaves) and variations in structure as a response to injury (Ford, 1992). All of these structural changes alter plant responses to their environment.

Woody plants undergo both structural and physiological changes as they progress from juvenility to adulthood and finally to a senescent state. After seeds germinate, young woody plants remain for several years in a juvenile condition during which they normally do not flower (see Chapter 4 of Kozlowski and Pallardy, 1997). The juvenile stage may differ from the adult stage in growth rate, leaf shape and structure, phyllotaxy, ease of rooting of cuttings, leaf retention, stem anatomy, and thorniness. The duration of juvenility varies greatly among species.

Although various species of adult trees age at different rates, they exhibit several common symptoms of aging. As a tree increases in size and builds up a complex system of branches it shows a decrease in metabolism, gradual reduction in growth of vegetative and reproductive tissues, loss of apical dominance, increase in dead branches, slow wound healing, and increased susceptibility to injury from certain insects and diseases and from unfavorable environmental conditions.

It is especially difficult, by interpreting data from field experiments, to assess the contributions of individual environmental factors to rates of physiological processes and plant growth. The use of controlled-environment facilities often helps in elucidating some of the mechanisms by which multiple stresses influence plant growth. Factors such as light intensity, temperature, and humidity are so interdependent that a change in one alters the influence of the others. It often is more informative to study impacts of environmental interactions in controlled-environment chambers or rooms than in the field (Kramer, 1978; Kozlowski, 1983; Kozlowski and Huxley, 1983). Examples are studies of interactions of light intensity and humidity (Davies and Kozlowski, 1974; Pallardy and Kozlowski, 1979a), light intensity and temperature (Pereira and Kozlowski, 1977a), light intensity and CO_2 (Tolley and Strain, 1984a), day length and temperature (Kramer, 1957), air pollution and temperature (Norby and Kozlowski, 1981a,b; Shanklin and Kozlowski, 1984), soil aeration and air pollution (Norby and Kozlowski, 1983; Shanklin and Kozlowski, 1985), soil aeration and temperature (Tsukahara and Kozlowski, 1986), soil fertility and air pollution (Noland and Kozlowski, 1979), water supply and CO_2 (Tolley and Strain, 1984b), water supply and soil fertility (McMurtrie *et al.*, 1990), and water supply and N supply (Walters and Reich, 1989; Liu and Dickmann, 1992a,b; Green and Mitchell, 1992). In controlled-environment facilities, plant growth can be studied under programmed diurnal and seasonal changes in climate. An important advantage of experiments in controlled environments over field experiments is that the data obtained are characterized by low variability and high reproducibility.

Beneficial Effects of Environmental Stresses

Not all environmental stresses are harmful to the growth of woody plants. Slowly increasing stresses often allow plants to adjust physiologically to stresses that would be harmful if they were rapidly imposed. The beneficial responses of plants to environmental stresses include those that occur during growth and development (e.g., adaptations to drought and freezing) and those evident in harvested products (e.g., fruit quality) (Grierson *et al.*, 1982).

Mild water deficits often have a variety of beneficial effects on plants. Plants previously subjected to water stress usually show less injury from transplanting and drought than plants not previously stressed. Hence, nursery managers often harden seedlings by gradually exposing them to full sun and decreasing irrigation. Seedlings of acacia and eucalyptus that had been repeatedly stressed by imposed drought had better control of water loss and were more drought tolerant than seedlings that were not previously stressed (Clemens and Jones, 1978).

A number of other benefits of mild water stress have been demonstrated. Water stress may decrease injury from air pollution because it induces stomatal closure and stomata are the principal pathways by which air pollutants enter leaves (e.g., Norby and Kozlowski, 1982). Moderate water stress increases the rubber content of guayule plants enough to increase rubber yield, even though the total yield of plant material is decreased (Wadleigh *et al.*, 1946). The oil content of olives also may be increased by water deficits (Evenari, 1960). The quality of apples, pears, peaches, and plums sometimes is improved by mild water deficits, even though fruit size is reduced (Richards and Wadleigh, 1952). The wood quality of water-stressed trees may be increased because of a higher proportion of latewood to earlywood (hence, more dense wood). Water deficits in some tropical trees (e.g., coffee, cacao) are necessary prerequisites for consistent flowering (Alvim, 1977; Maestri and Barros, 1977). Water stress also may increase cold hardiness (Chen *et al.*, 1975, 1977; Yelenosky, 1979). Cold hardiness also is induced by exposure of plants to low temperatures above freezing. For example, citrus trees readily hardened to cold when exposed to temperatures between 15.6 and 4.4°C (Yelenosky, 1976). A classic example of beneficial effects of stress on harvested products is maintenance of quality and prolongation of the market life of fruits by storing them in controlled atmospheres. The beneficial effects of controlled-atmosphere storage are mediated through physiological changes in fruits induced by abnormal levels of temperature, humidity, ethylene, CO_2, and O_2 (see Chapter 8).

Summary

Growth of woody plants is an integrated response controlled by interactions of heredity and environment, operating through the physiological processes of the plant. Hereditary control of plant growth is contained largely in the DNA of chromosomes which through messenger RNA regulates the kinds of proteins and enzymes that are synthesized (which in turn control cell structure and plant response). The causes of genetic variation in plant growth include mutation, genetic recombination, and gene migration. Genetic variation among woody plants has been exploited to greatly increase wood production. Large increases in productivity have been obtained by parent selection, clonal selection, and interspecific hybridization of pines, eucalypts, poplars, and other species. Exploitation of genetic material also has long been important for growing fruit trees as well as forest trees for seed orchards.

It is difficult to quantify the effect of specific environmental stresses on plant growth for several reasons: (1) some stresses affect plants continuously and others do so randomly, (2) the importance of individual environmental stresses varies with time, (3) correlation of changes in an environmental stress and plant growth does not necessarily involve a cause and effect relationship, (4) growth changes in response to an environmental stress may not be evident for a long time (until physiological processes are sufficiently altered to change the rate of growth), (5) growth responses to environmental changes are influenced by preconditioning environmental regimes, (6) the effects of environmental stresses on growth vary appreciably with plant vigor, and (7) changes in plant morphology that occur over time interact with the physiological processes that regulate growth. Controlled-environment facilities can be advantageously used to elucidate mechanisms by which environmental stresses affect the growth of plants.

Not all environmental stresses are harmful to plants. For example, mild water stress may decrease subsequent injury from transplanting and drought, decrease injury from gaseous air pollutants, increase the rubber content of guayule and the oil content of olives, improve the quality of edible fruits, increase wood quality, stimulate flowering in some tropical trees, and induce cold hardiness. Exposure of plants to low temperatures above freezing also may induce cold hardiness. Storage of fruits in controlled atmospheres maintains quality and prolongs the marketability of fruits.

General References

Ahuja, M. R., and Libby, W. J., eds. (1993). "Clonal Forestry, Volume 1: Genetics and Biotechnology; and Volume 2: Conservation and Application." Springer-Verlag, Berlin and New York.

Burley, J., and Styles, B. T., eds. (1976). "Tropical Trees: Variations, Breeding and Conservation." Academic Press, London.

Callaway, R. M. (1995). Positive interactions among plants. *Bot. Rev.* **61**, 306–349.

Cannell, M. G. R., and Jackson, J. E., eds. (1985). "Attributes of Trees as Crop Plants." Institute of Terrestrial Ecology, Huntingdon, England.

Cherry, J. H., ed. (1989). "Environmental Stress in Plants." Springer-Verlag, Berlin and New York.

Dickmann, D. I., and Stuart, K. W. (1983). "The Culture of Poplars." Department of Forestry, Michigan State University, East Lansing.

Faust, M. (1989). "Physiology of Temperate Zone Fruit Trees." Wiley, New York.

Fitter, A. H., and Hay, R. K. M. (1987). "Environmental Physiology of Plants." Academic Press, London.

Gates, D. M. (1980). "Biophysical Plant Ecology." Springer-Verlag, New York.

Jones, H. G., Flowers, T. J., and Jones, M. B., eds. (1989). "Plants Under Stress." Cambridge Univ. Press, Cambridge.

Katterman, F., ed. (1990). "Environmental Injury to Plants." Academic Press, San Diego.

Kozlowski, T. T. (1979). "Tree Growth and Environmental Stresses." Univ. of Washington Press, Seattle.

Kozlowski, T. T. (1995). The physiological ecology of forest stands. *In* "Encyclopedia of Environmental Biology" (W. A. Nierenberg, ed.), Vol. 3, pp. 81–91. Academic Press, San Diego.

Kozlowski, T. T., and Pallardy, S. G. (1997). "Physiology of Woody Plants." 2nd Ed. Academic Press, San Diego.

Kozlowski, T. T., Kramer, P. J., and Pallardy, S. G. (1991). "The Physiological Ecology of Woody Plants." Academic Press, San Diego.

Larcher, W. (1995). "Physiological Plant Ecology," 3rd Ed. Springer-Verlag, Berlin and Heidelberg.

McKersie, B. D., and Leshem, Y. Y. (1994). "Stress and Stress Coping in Cultivated Plants." Kluwer, Dordrecht, The Netherlands.

Mooney, H. A., Winner, W. E., and Pell, E. J., eds. (1991). "Response of Plants to Multiple Stresses." Academic Press, San Diego.

Mulkey, S. S., Chazdon, R. L., and Smith, A. P., eds. (1996). "Tropical Forest Plant Ecophysiology." Chapman and Hall, New York.

Raghavendra, A. S., ed. (1991). "Physiology of Trees." Wiley, New York.

Sachs, M. M., and Ho, T.-H. D. (1986). Alteration of gene expression during environmental stress in plants. *Annu. Rev. Plant Physiol.* **37**, 363–376.

Zobel, B. J., and Talbert, J. (1984). "Applied Forest Tree Improvement." Wiley, New York.

Zobel, B., and van Buijtenen, J. P. (1989). "Wood Variation: Its Causes and Control." Springer-Verlag, Berlin and New York.

Zobel, B. J., and Jett, J. B. (1995). "Genetics of Wood Production." Sprinter-Verlag, Berlin and New York.

Zobel, B. J., van Wyck, G., and Stahl, P. (1987). "Growing Exotic Trees." Wiley, New York.

2

Seed Germination and Seedling Growth

Introduction

Woody plants undergo their greatest mortality risk when they are in the ungerminated embryo stage of seed development and in the cotyledon stage of seedling development (Kozlowski, 1979, 1995). Hence, natural regeneration of many communities of woody plants depends on environmental conditions that are suitable for continuously maintaining germinating seeds and young seedlings in a physiologically efficient state. Even temporary mild environmental stresses typically induce physiological dysfunctions and drastic growth inhibition at these critical times in the lives of woody plants and lead to mortality of seeds and/or seedlings (Borger and Kozlowski, 1972a–d; Kozlowski *et al.*, 1991; Kozlowski, 1996).

The essential structure in a seed is the embryo, and most concern in handling and storing seeds is with providing conditions that will keep the embryo alive and ready to resume growth when the seed is planted. The resumption of growth of the embryo and its development into a new, independent seedling involve most of the important processes included in the realm of plant physiology such as respiration; absorption of water; conversion of foods into soluble forms; synthesis of enzymes and hormones; nitrogen and phosphorus metabolism; translocation of carbohydrates, hormones, water, and minerals to meristematic regions; and conversion of foods into plant tissues.

Seed Structure

Seeds of woody plants differ widely in size, shape, color, and structure. They range in size from those that are barely visible, such as those of sourwood and rhododendron, to those of coconut, which may have a fresh weight of about 9 kg. Surfaces of seed coats vary from highly polished to roughened. Seed appendages may include wings, arils, spines, tubercles, and hairs.

A true seed is a fertilized mature ovule that has an embryo, stored food material (rarely missing), and a protective coat or coats. In practice, however, the term "seed" is not always restricted to this definition. Rather, a seed often is perceived in a functional sense as a unit of dissemination, a disseminule. In this sense, the term seed is applied to dry, one-seeded or rarely two- to several-seeded fruits as well as to true seeds. For example, dry one-seeded fruits such as the samaras of elm and the nuts and husks of beech are generally, although incorrectly, referred to as seeds.

The embryo is a miniature plant comprising one or more cotyledons (first leaves), a plumule (embryonic bud), hypocotyl (stem portion), and radicle (rudimentary root) (Fig. 2.1). The size of the embryo varies greatly in seeds of different species. In some species the embryo is a rudimentary structure; in others it almost fills the seed (Fig. 2.2). The embryos of bamboo and palm seeds have only one cotyledon and therefore are classi-

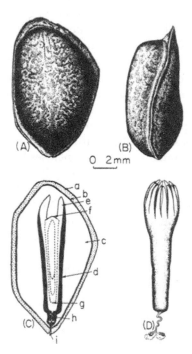

Figure 2.1 Structure of mature seed of sugar pine (A). (B) Exterior view of two planes. (C) Longitudinal section: a, seed coat; b, nucellus; c, endosperm; d, embryo cavity; e, cotyledons; f, plumule; g, radicle; h, suspensor; i, micropyle. (D) Embryo. From Anonymous (1948).

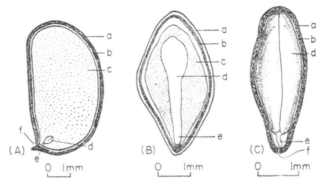

Figure 2.2 Variations in seed structure. (A) *Aralia*, with large endosperm and small embryo. (B) Hemlock, with large embryo surrounded by megagametophyte. (C) Shadbush, with no endosperm and embryo almost filling the seed cavity. a, Outer seed coat; b, inner seed coat; c, endosperm in A and C, and gametophyte in B; d, cotyledon; e, radicle; f, micropyle. From Anonymous (1948).

fied as monocotyledons. Embryos of most woody angiosperms are dicotyledonous, having two cotyledons, whereas gymnosperms have from two to as many as 18 cotyledons, depending on the species.

The food supply in seeds may be stored in the cotyledons or in tissue surrounding the embryo, which in angiosperms is endosperm. True endosperm is triploid in chromosome number, having been formed after union of the diploid fusion nucleus and a sperm, and is the principal food storage tissue of seeds of many dicotyledonous species. In seeds of certain plants, such as ailanthus, some food is stored in the endosperm and some in cotyledons. In gymnosperm seeds, food is stored primarily in the megagametophyte (female gametophyte) that encloses the embryo. The megagametophyte is haploid in chromosome number and different in origin from the endosperm, although it serves the same function.

The seed coats, which protect the embryo from desiccation and attacks by pests (Mohamed-Yasseen *et al.*, 1994), usually consist of an outer hard coat, the testa, and a thin, membranous inner coat. However, considerable variation occurs in seed coat characteristics. For example, in poplar and willow the testa is very soft, whereas in hawthorn, holly, and most legumes it is very hard. In elm, both the inner and outer seed coats are membranous. The simple seed coats of gymnosperms vary from hard in pines to soft in firs.

Seed Composition

Seeds contain variable quantities of foods in the form of carbohydrates, fats, and proteins. Seed starches are stored as granules in the endosperm, megagametophyte, and cotyledons. Lipids occur in fat bodies of seeds as fats and oils, depending on the proportions of saturated and unsaturated fatty acids present. Three-fourths or more of the seed proteins are stored in protein bodies, organelles 1 to 20 μm in diameter that contain a proteinaceous matrix and are bound by a single membrane. In beech cotyledons, the protein bodies originate by gradual subdivisions of vacuoles in which reserve proteins are deposited (Collada *et al.*, 1993). In addition to storing proteins, the protein bodies may contain mineral nutrients and crystals (White and Lott, 1983). Some seed proteins are distributed in nuclei, mitochondria, proplastids, microsomes, and cytosol (Ching, 1972).

Starch is the principal and most widespread storage carbohydrate in angiosperm seeds, but seeds of many gymnosperms store little starch (Ching, 1972; Kovac and Kregar, 1989b). Sucrose is the major stored sugar, but other sugars are found in varying amounts. Trehalose occurs in beech seeds; stachyose and raffinose are in honey locust seeds. Mannans (unbranched polymers of mannose) (see Chapter 7 of Kozlowski and Pallardy, 1997) occur in significant amounts in seeds of date palm and coffee (Matheson, 1984). Hemicellulose occurs in some seeds and is metabolized during germination, as in persimmon, coffee, and date palm (Crocker and Barton, 1953; De Mason and Thomson, 1981).

The seeds of many species store some or most reserves as fatty acids, with oleic, linoleic, and linolenic acids most common. Other fatty acids occur as glycerides. Among these are acetic, butyric, palmitic, stearic, lauric, and myristic acids. Other lipid compounds include esters of alcohols, sterols, phospholipids and glycolipids, tocopherols, and squalene (Mayer and Poljakoff-Mayber, 1989).

Although some seed proteins, such as enzyme proteins and nucleoproteins, are metabolically active, a large proportion is inactive. In addition to proteins, the nitrogenous material in seeds includes free amino acids and amides (see Chapter 9 of Kozlowski and Pallardy, 1997). Other seed constituents include variable quantities of minerals, phosphorus-containing compounds (phosphates, nucleotides, phospholipids, nucleoproteins), nucleic acids, alkaloids, organic acids, phytosterols, pigments, phenolic compounds, vitamins, and hormonal growth regulators (e.g., auxins, gibberellins, cytokinins, abscisic acid (ABA), and ethylene).

Higgins (1984) divided seed proteins into storage proteins and "housekeeping" proteins, the latter essential in maintaining normal cell metabolism. The storage proteins consist of relatively few species of proteins, whereas housekeeping proteins are composed of small amounts of many protein species. Storage proteins tend to be high in asparagine, glutamine, and arginine or proline. However, the composition of seed proteins varies among plant species. Whereas seeds of chestnut and beech accumulated globulin, those of oaks accumulated glutelin (Collada *et al.*, 1988, 1993). In *Hopea odorata* and *Dipterocarpus alatus,* both members of the Dipterocarpaceae, the amino compounds in seeds varied somewhat. In both species glutamic acid and glutamine were the major amino compounds present (Huang and Villanueva, 1993). *Hopea* also contained large amounts of aspartic acid, asparagine, serine, threonine, arginine, and alanine. Seeds of *Dipterocarpus* contained high levels of alanine, arginine, and threonine.

The proportions of carbohydrates, fats, and proteins vary greatly in seeds of different species of plants, with carbohydrates or lipids usually predominating. Seeds of white oak, silver maple, horse chestnut, and American chestnut have high carbohydrate contents (Table 2.1). Examples of seeds with very high lipid contents include those of English walnut, butternut, pecan, coconut, oil palm, and macadamia (Table 2.2). Seeds of eastern white pine and longleaf pine have high lipid and protein contents but are low in carbohydrates (Table 2.1). There often is appreciable variation in the composition of seeds of different species in the same genus. For example, seeds of sugar maple have high carbohydrate contents, but lipids and proteins predominate in seeds of box elder.

Although the chemical composition of seeds is genetically determined, the relative amounts of seed constituents are influenced by the environmental regime at the seed source and nutrition of the parent tree. For example, Durzan and Chalupa (1968) showed considerable variation in the chemical composition of the embryos and megagametophytes of jack pine seeds collected from different geographical sources. The climate at the seed source influenced the degree to which metabolism of carbon and nitrogen compounds proceeded before and during incipient germination.

Table 2.1   Carbohydrate, Fat, and Protein Contents of Some Tree Seeds[a]

	Percentage of air-dried seed mass		
Species	Carbohydrates	Fats	Proteins
Silver maple	62.0	4.0	27.5
Horse chestnut	68.0	5.0	7.0
Chestnut	42.0	3.0	4.0
Pedunculate oak	47.0	3.0	3.0
White oak	58.4	6.8	7.4
Northern red oak	34.5	22.5	—
Tung	5.0	21.0	62.0
Eastern white pine	4.8	35.4	30.2
Longleaf pine	4.5	31.7	35.2

[a]From Mayer and Poljakoff-Mayber (1963) and Woody Plant Seed Manual (Anonymous, 1948).

In angiosperm seeds, carbohydrates are stored in the endosperm, coty-ledons, or both. In gymnosperm seeds, the reserves are stored primarily in the megagametophyte and cotyledons. In seeds of apple, ash, and dog-wood, carbohydrates are stored primarily in the endosperm. In ailanthus, some reserves are stored in the endosperm and some in the cotyledons. The embryonic cotyledons of box elder and black locust store large amounts of carbohydrates. Those of ailanthus, green ash, and flowering dogwood store only small amounts. Species with seeds lacking endosperm (e.g., beech and black locust) have cotyledons adapted for both storage and photosynthesis (Marshall and Kozlowski, 1977). In Brazil nut, which has very small cotyledons, the embryonic axis is the major storage organ (Bewley and Black, 1978).

Patterns of Seed Germination

Seed germination may be considered to be resumption of embryo growth resulting in seed coat rupture and emergence of the young plant. Growth of the embryo requires both cell division and elongation, cell division occurring first in some species and cell elongation in others. For example,

Table 2.2 **Lipid and Protein Contents of Several Species of Woody Plants with Seeds High In Lipids**

Species	Major storage organ	Lipid (%)	Protein (%)
Queensland nut (*Macadamia ternifolia*)	Cotyledon	75–79	9
Brazil nut (*Bertholletia excelsa*)	Radicle/hypocotyl	65–70	17–18
Coconut (*Cocos nucifera*)	Endosperm	63–72	18–21
Hazelnut (*Corylus avellana*)	Cotyledon	60–68	18–20
Pecan (*Carya illinoensis*)	Cotyledon	65	12
Jojoba (*Simmondsia chinensis*)	Cotyledon	41–57	20–38
Oil Palm (*Elaeis guineensis*)	Endosperm	48–50	9–19
Italian stone pine (*Pinus pinea*)	Megagametophyte	45–48	34–35
Douglas fir (*Pseudotsuga menziesii*)	Megagametophyte	36	32
Cotton (*Gossypium hirsutum*)	Cotyledon	15–33	25–39
Tung (*Aleurites fordii*)	Endosperm	16–33	25
Olive (*Olea europaea*)	Endosperm/cotyledon	12–28	—

[a]The range of percent composition of unimbibed seed weight is given to include values reported in the literature. Modified from Trelease and Doman (1984).

cell division preceded cell elongation in embryo growth in Japanese black pine seeds (Goo, 1952). In cherry laurel, however, cell division and cell elongation began more or less simultaneously in embryonic organs (Pollock and Olney, 1959). Reserve foods in the seed sustain the growing embryo until the cotyledons and/or leaves expand to provide a photosynthetic system and roots develop to absorb water and minerals, thereby making the young plant physiologically self-sufficient. Seedlings of English walnut depended on reserve carbohydrates for respiratory substrates and growth for the first 21 days after the seeds were sown (Maillard *et al.*, 1994). Photosynthesis was confirmed by day 22. At day 29 current photosynthetic products were used for 25 and 30% of the respiration of the root and shoot, respectively. Current photosynthate was incorporated into the shoot beginning on day 32 and into the taproot after 40 days. After 43 days the contribution of reserve carbohydrates to plant growth were negligible.

As the embryo resumes growth during seed germination, the radicle elongates and penetrates the soil. In some woody plants—including most gymnosperms, beech, dogwood, black locust, ash, and most species of maple—the cotyledons are pushed out of the ground by the elongating hypocotyl (epigeous germination). In other species—including oak, walnut, buckeye, and rubber—the cotyledons remain underground while the epicotyl grows upward and develops foliage leaves (hypogeous germination) (Figs. 2.3 and 2.4).

Whereas all embryo cells divide during early seed germination, as seedlings develop the division of cells becomes localized in shoot and root apices. Important events following seed germination include sequential formation of leaves, nodes, and internodes from apical meristems. Shoots may originate from apical meristems in leaf axils, providing the young plant with a system of branches. The root apical meristem forms a taproot or primary root. Often branch roots or secondary roots originate at new apical meristems in the pericycle of the taproot.

Environmental Control of Seed Germination

Rapid germination of seeds usually is very desirable because the shorter the time required, the less opportunity there is for injury by insects, fungi, or unfavorable weather conditions or for seeds to be eaten by birds or rodents. Rapid seed germination is not always preferred, however, as in temperate regions with harsh winters and for opportunistic species that capitalize on seed banks (Kozlowski *et al.*, 1991, pp. 78–79). Among the most important environmental factors controlling seed germination are water, temperature, light, oxygen, various chemicals and interactions among them. Envi-

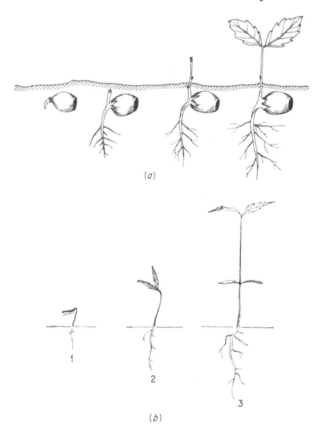

Figure 2.3 Stages in germination of seeds. (a) Germination of white oak acorns in which cotyledons remain below ground (hypogeous). (b) Germination of red maple in which cotyledons are pushed out of the ground (epigeous).

ronmental conditions required for germination commonly are more critical than conditions necessary for subsequent growth of seedlings.

Because plants with heavy seeds generally have large food reserves, they often survive postgermination environmental stresses better than plants with small seeds and limited food reserves. For example, heavy-seeded species such as water oak and sebastian bush produced large, sturdy seedlings that were relatively resistant to drought, flooding, herbivory, and damping-off (Jones *et al.*, 1994). In contrast, light-seeded species such as elm, red maple, and deciduous holly produced fragile seedlings that were very vulnerable to the same stresses. In Australian deserts, the larger seeded species can emerge from greater soil depths, have higher seedling

Figure 2.4   Epigeous germination of pine seed. Photo courtesy of St. Regis Paper Co.

growth rates, and show lower mortality than seedlings of small-seeded species (Buckley, 1982).

Water Supply

Nondormant seeds must imbibe a certain amount of water before they resume the physiological processes involved in germination. For example, seed respiration increases greatly with an increase in hydration above some critical level (Kozlowski and Gentile, 1959). The absolute amounts of water required to initiate germination are relatively small, usually not more than two or three times the weight of the seed (Koller, 1972). Following germination, however, a sustained supply of a large amount of water is needed by growing seedlings, and this requirement becomes greater as leaves grow and transpiration increases.

Practically all viable seeds, except those with impermeable seed coats or impermeable layers under the seed coats, can absorb enough water for germination from the soil at field capacity. In progressively drier soil, both the rate of germination and final germination percentage decrease. For example, Kaufmann (1969) found that germination of citrus seeds was greatly inhibited at soil water potentials of -0.23 and -0.47 MPa. After 31 days, 88% of the seeds germinated in soil at field capacity, whereas at −0.23 and −0.47 MPa only 23 and 3%, respectively, had germinated.

There is considerable variation among species in the effects of soil moisture availability on seed germination. As the soil dried from field capacity, germination of Japanese red pine and hinoki cypress seeds decreased, with hinoki cypress seeds being more sensitive to drying. For example, seed germination of hinoki cypress decreased by about 6% for each 0.1 MPa decrease in soil water potential. The corresponding decrease was about 2.5% for Japanese red pine (Satoo, 1966).

The influence of soil moisture stress on germination often depends on the prevailing temperature. For example, at temperatures near 38°C, moisture stress was critical for germination of mesquite seeds. However, at 29°C, embryo growth was largely influenced by temperature, and soil moisture stress was not a limiting factor until later in the germination process (Scifres and Brock, 1969). Soil water potential and temperature interacted in regulating germination of *Eucalyptus delegatensis* seeds (Battaglia, 1993). Seeds germinating at optimum temperature were less sensitive to moisture stress. Interactions of water supply and temperature on emergence of five species of temperate-zone trees and seed germination of four species of tropical trees are shown in Figs. 2.5 and 2.6.

Flooding

Activation of the biochemical processes necessary for seed germination depends on availability of water and oxygen. Unfortunately, the supply of oxygen to respiratory enzymes of seeds is reduced when the large soil pores are filled with water. In general, submersion of air-dry seeds in water for short periods stimulates germination, whereas prolonged soaking causes loss of seed viability (Kozlowski, 1984b). Seeds of green ash and box elder lost viability when they were submerged in water for more than short periods (Hosner, 1957, 1962). There are exceptions, however, and seeds of bald cypress and tupelo gum may retain viability when submerged for a long time. Seeds of these species do not germinate under water but do so when the floodwaters recede. In comparison, seeds of eastern cottonwood, black willow, and American sycamore often germinate under water.

Sometimes the flood tolerance of seeds and young seedlings varies greatly. Seeds of both *Parkia pendula* and *P. discolor* will germinate after submersion for as long as 7 months. However, seedlings of *P. pendula* can survive only a few weeks of submersion, whereas those of *P. discolor* can tolerate as much as 7 months of inundation. This difference emphasizes that classifying species as flood tolerant or intolerant can be misleading if the classification is based on response at only one developmental stage (Scarano and Crawford, 1992).

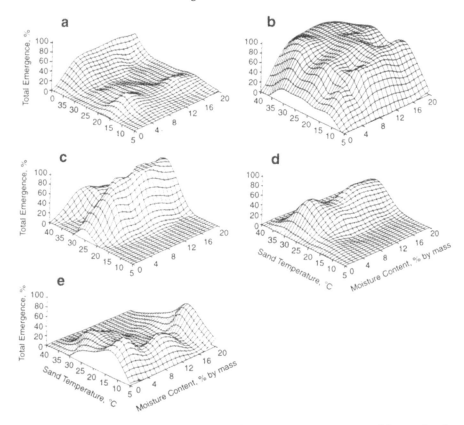

Figure 2.5 Interactions of soil moisture and temperature on emergence of five species of temperate-zone trees. a, *Fraxinus americana*; b, *Gleditsia triacanthos*; c, *Morus rubra*; d, *Platanus occidentalis*; e, *Prunus serotina*. From Burton and Bazzaz (1991).

Temperature

After seed dormancy is broken by low temperatures, much higher temperatures are needed to induce rapid germination (see below). Nondormant seeds also can germinate at low temperatures, but much longer times are required.

Germination of any lot of nondormant seeds can occur over a temperature range within which there is an optimum at which the highest percentage of germination is obtained in the shortest time. Minimum, optimum, and maximum temperatures for seed germination vary widely among seeds of different species and generally are lower for temperate-zone species than for tropical species. Seeds of tropical plants germinate best at temperatures between 15 and 30°C; those of temperate-zone plants, between 8 and 25°C; and those of alpine plants, between 5 and 30°C (Gates, 1993).

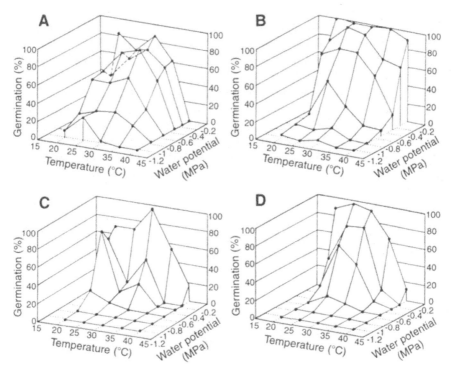

Figure 2.6 Interactions of water potential and temperature on germination of four tropical species. (A) *Combretum apiculatum*, (B) *Colophospermum mopane*, (C) *Acacia karroo*, and (D) *Acacia tortilis*. From Choinski and Tuohy (1991).

Germination temperatures also vary considerably with seed source. Germination temperatures between 24 and 30°C were best for ponderosa pine seed sources east of the Rocky Mountains, whereas temperatures of 35°C or higher were optimal for Pacific Northwest sources (Callaham, 1964). Optimal germination temperatures often are different for seeds obtained from different plants of the same species.

Seeds of many species germinate equally well over a rather wide temperature range. For example, seeds of lodgepole pine germinated at about the same rate at 20 as at 30°C (Critchfield, 1957), and germination of jack pine seeds did not vary appreciably at 15, 21, or 27°C under continuous light (Ackerman and Farrar, 1965). Kaufmann and Eckard (1977) found that total emergence of seedlings of Englemann spruce and lodgepole pine was similar at 16 and 25°C, but it was reduced at 35°C. However, at 16°C, spruce seedlings emerged 2 days sooner than pine seedlings. Other species have a rather narrow temperature range that may vary somewhat with preconditioning of seeds. For example, Norway maple seeds germinated best at temperatures between 5 and 10°C. Box elder seeds had 67% germination as temperature alternated between 10 and 25°C, but only 12% when temperatures were alternated between 20 and 25°C (Roe, 1941).

Although seeds of many species will germinate at a constant temperature, seed germination of most species requires or is increased by diurnal temperature fluctuations (Hatano and Asakawa, 1964). If seeds of Manchurian ash were first exposed to moist, low-temperature treatment and then placed at a constant temperature of 25°C, only a few germinated, and at a constant temperature of 8°C seed germination was greatly delayed. In comparison, alternating the temperature between 8°C for 20 hr and 25°C for 4 hr each day greatly accelerated germination. Germination of Japanese red pine seeds also was accelerated by diurnal thermoperiodicity. Maximum germination of *Phellodendron wilsonii* seeds was possible only with alternating exposure to temperatures of 35°C (8 hr in light) and 10°C (in darkness). This effect was demonstrated for stored seeds, fresh seeds, and even dry seeds (Lin *et al.*, 1994). A regime of alternating temperatures between 20 and 30°C was adopted by the International Rules for Seed Testing to test the germination of seeds of many species of woody plants.

Radiation

Whereas most seeds of temperate-zone species germinate as well in the dark as in the light, those of some species require light for germination. Seeds of these species will germinate at very low illuminance, those of spruce requiring only 0.08 lux; birch, 1 lux; and pine, 5 lux. Seeds of a few species require up to 100 lux for germination (Jones, 1961). Germination of seeds of some species (e.g., Fraser fir) is stimulated by light even though there is no absolute requirement (Henry and Blazich, 1990). The degree of stimulation of germination by light is influenced by several environmental factors, with stratification (moist prechilling) and temperature being very important. Seeds of a number of tropical species require light for germination. These include seeds of *Cecropia obtusifolia*, *C. peltata*, *Trema micrantha*, *T. orientalis*, and *Piper auritum* (Whitmore, 1983; Orozco-Segovia *et al.*, 1993). In Ghana, the seeds of 96 species germinated readily only in full light, and seeds of 25 other species germinated in the shade (Hall and Swaine, 1980). The light requirement for germination may vary over time. Seeds of some species may become more sensitive, and those of other species less sensitive, over time (Vazquez-Yanes and Orozco-Segovia, 1994).

Seedlings that develop from large seeds usually tolerate heavy shading better than seedlings emerging from small seeds. An increase in seed size from 0.03 to 30 mg among a variety of species was associated with increased seedling survival in heavy shade from 10–15 days to 30–40 days (Leishman and Westoby, 1994). The larger seeds provided a greater initial energy reserve, which may be advantageous in habitats in which canopy gaps are regularly created. Seedlings from large seeds showed greater early height growth. This may be an advantage for regeneration in habitats with a steep light gradient, as for seedlings emerging from seeds that germinate below litter.

Day Length   For seeds of the majority of light-sensitive species of woody plants, the most rapid and greatest total germination occurs in daily light periods of 8 to 12 hr. Interrupting the dark period with a short light flash or increasing the temperature usually has the same effect as extending the duration of exposure to light. In eastern hemlock, 8- or 12-hr days produced maximum seed germination, with no added response by increasing day length to 14 or 20 hr (Olson *et al.*, 1959). Eucalyptus seeds germinated well in 8-hr days and those of birch in 20-hr days. Seeds of Douglas fir, however, germinated in continuous light or 16-hr days, but not in 8-hr days (Jones, 1961).

Wavelength   Germination of seeds of a number of species of herbaceous and woody plants is sensitive to the wavelength of light. Examples of woody angiosperms showing such sensitivity are hairy birch, Manchurian ash, and American elm. Sensitive gymnosperms include species of *Abies*, *Picea*, and *Pinus* (e.g., Japanese black pine, eastern white pine, longleaf pine, and Virginia pine) (Hatano and Asakawa, 1964; Koller, 1972).

The germination response to wavelength is controlled by the phytochrome pigment system. Red light promotes germination and far-red light inhibits it. If seeds are exposed to red (650 nm) and far-red light (730 nm), their capacity to germinate depends on the irradiation that is given last, with the influence of red light in promoting germination being nullified if it is followed by far-red light. If, however, far-red light is followed by red light, germination is stimulated (Table 2.3). In forests, the litter may inhibit seed germination because of the low red to far-red ratio of the light transmitted by the litter layer (Vazquez-Yanes *et al.*, 1990). It should also be noted that the red–far red ratio of light beneath a canopy already is low relative to direct sunlight.

The red light requirement for promoting germination often is not rigid, varying with temperature or duration of water uptake by seeds. Toole *et al.* (1961) noted, for example, that germination of Virginia pine seeds occurred faster in seeds promoted with red light after a 20-day period of water imbibition at 5°C than in seeds given a 1-day period of imbibition. Greater germination of seeds was promoted by red light when they had previously absorbed water at 5°C rather than at 25°C.

Both metabolic activity and mitosis in embryos are stimulated by red light and inhibited by far-red light in light-sensitive seeds. Significant increases in respiration of Scotch pine seeds were induced by red light after imbibition for 24 hr, whereas mitotic activity was stimulated after 36 hr of

Table 2.3 Influence of Alternating Red
and Far-Red Irradiation on Germination
of Virginia Pine Seeds[a]

Character of irradiation[b]	Germination (%)
Dark control	4
R	92
R + FR	4
R + FR + R	94
R + FR + R + FR	3
R + FR + R + FR + R	93

[a]From Toole *et al.* (1961).
[b]R, red; FR, far red.

imbibition. Radicles did not emerge until after more than 48 hr of imbibition (Nyman, 1961).

Phytochrome, a widely distributed light-receptive, protein–pigment complex, is involved in the red-far red phenomena described above. Light acts on phytochrome to change it from an inactive form, with maximum absorption in the red part of the spectrum (660 nm, P_r), to the active form, with maximum absorption in the far-red (730 nm, P_{fr}) (Fig. 2.7). In mature seeds some P_{fr} often is present, but during imbibition it changes to the inactive P_r form. Seeds conditioned to germinate through the activity of red light can revert to a nongerminating state by exposure to far-red radiation, which changes P_{fr} back to P_r (Taylorson and Hendricks, 1976). In the dark, P_{fr} may gradually revert to P_r, or it may be destroyed by denaturation or enzyme-catalyzed conversion to an inactive form (Hart, 1988).

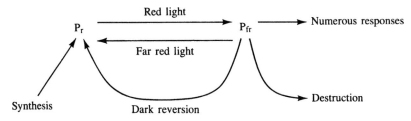

Figure 2.7 Photoconversion of phytochrome by red and far-red light into far-red absorbing (P_{fr}) and red-absorbing (P_r) forms, respectively. P_{fr} may gradually revert to P_r in the dark, or it may be destroyed. From *Plant Physiology*, 2nd Ed., by Frank B. Salisbury and Cleon W. Ross. © 1978 by Wadsworth Publishing Company, Inc. Reprinted by permission of the publisher.

Oxygen

As intense respiration is characteristic of the early phase of seed germination, it is not surprising that oxygen supply affects germination (see Chapter 6 of Kozlowski and Pallardy, 1997). Seeds usually require higher oxygen concentrations for germination than seedlings need for growth. The relatively high oxygen requirements of seeds of some species are the result of their seed coats acting as barriers to diffusion of oxygen into seeds. Removal of coats of red pine seeds or exposure of intact seeds to high oxygen concentrations greatly accelerated the rate of oxygen uptake. Removal of seed coats, followed by exposure of the decoated seeds to high oxygen concentrations, accelerated respiration even more (Kozlowski and Gentile, 1959).

Oxygen plays a role as the primary electron acceptor in respiration. In some species it also may be involved in inactivation of one or more inhibitors. Germination of isolated embryos of European white birch and European silver birch was prevented by aqueous extracts of seeds, but such inhibition could be decreased by exposure to light (Black and Wareing, 1959). It appeared that the intact seed coat prevented germination in the dark by reducing the oxygen supply below a critical level. Nevertheless, embryos without seed coats germinated in low concentrations of oxygen; hence, the embryo appeared to have a high oxygen requirement only when the seed coat was present.

As mentioned, soaking seeds for a few hours hastens germination, but prolonged soaking induces injury and loss in viability of many seeds, presumably because of the reduced concentration and availability of dissolved oxygen in comparison with that of air. Unless aseptic conditions are maintained, long soaking periods also may favor activities of harmful microorganisms. However, seeds of bottomland species, such as tupelo gum and bald cypress, have low oxygen requirements and can endure prolonged inundation without loss of viability. Hosner (1957) did not find any appreciable effect on germination of six bottomland species from soaking seeds for as long as 32 days.

Seedbeds

Because of wide differences among seedbeds in physical characteristics, temperature, and availability of water and mineral nutrients, establishment of plants varies greatly in different seedbeds (Winget and Kozlowski, 1965a). Mineral soil usually is a good seedbed because of its high infiltration capacity, adequate aeration, and capacity to establish close hydraulic contact between soil particles and seeds. Because of its high water-holding capacity, sphagnum moss often is a suitable seedbed for germination, but it may subsequently smother small seedlings. Decayed wood also is an excellent natural seedbed for seeds of some plants, probably because of its capacity for retention of water (Place, 1955).

Litter may or may not be a good seedbed depending on the plant species, amount and type of litter, and prevailing environmental conditions. Both seed germination and growth of young seedlings are influenced by litter. By altering the microenvironment of the topsoil, litter intercepts light and rain and affects the surface structure of the seedbed, hence regulating transfer of heat and water between the soil and aboveground atmosphere. Litter influences plant community structure directly (e.g., by effects on seed germination and seedling establishment) and indirectly by influencing availability of light, water, and mineral nutrients (Facelli and Pickett, 1991b).

Facelli and Pickett (1991a) predicted that the indirect effects of litter on the structure of old-field plant communities were important when (1) the tree population was not limited by dispersal or physical stress, (2) interspecific competition was more important than other biotic interactions in regulating establishment and growth of trees, (3) establishment of herbaceous plants was affected more by litter than was that of trees, and (4) plasticity of herbaceous plants was too slow to rapidly compensate for the lower density by increasing the size of individual plants.

In deserts, litter is considered beneficial because it increases the availability of soil water by inhibiting evaporation from the soil. In many forests, however, litter is a less suitable seedbed than mineral soil because it warms slowly, inhibits penetration by roots, prevents seeds from establishing hydraulic contact with the mineral soil, dries rapidly, and shades small seedlings.

The litter mat often prevents seeds and seedling roots from reaching the soil. In California, the roots of blue oak seedlings often failed to reach the soil when seeds germinated on top of a thick layer of litter (Borchert *et al.*, 1989). Tao *et al.* (1987) showed a strong inhibitory effect of litter on seed germination and establishment of both gymnosperms (*Picea jezoensis, Larix dahurica*) and angiosperms (*Populus davidiana, Betula costata, B. dahurica, B. platyphylla*) in Korean pine forests. *Tilia amurensis*, with very large and heavy seeds, was an exception. Its strong radicle readily penetrated the litter layer into the moist substrate. Matted litter consisting of leaf litter held together by fungal mycelium has a greater inhibitory effect on seedling establishment than loose litter.

Litter often retards seedling establishment by decreasing availability of resources. For example, litter may prevent germination of seeds that respond positively to light (Sydes and Grime, 1981a,b). Litter also may decrease availability of water to young seedlings. Shortly after seeds germinated in litter the rootlets of yellow birch seedlings tended to grow horizontally over the leaf mat (Winget and Kozlowski, 1965a). The major rootlets often were completely exposed to air, and only the secondary rootlets penetrated the leaf mat. The seedling stems often were prostrate or nearly so, with only a few millimeters of their stem tips oriented vertically. Occasionally the levering action of a rootlet against the surface of the leaf mat overturned seedlings, hence exposing the roots, followed by desiccation of these seedlings.

The effects of plant litter on germination and emergence of various species of plants may be quite different. Litter reduced and delayed final emergence of yellow birch seedlings, and the emerging seedlings had reduced root–shoot ratios and longer hypocotyls than control seedlings. By comparison, litter did not affect emergence of staghorn sumac seedlings but altered biomass allocation (Peterson and Facelli, 1992). The biomasses of stems, leaves, and roots of staghorn sumac seedlings were progressively reduced by increasing amounts of litter, and by leaf litter relative to needle litter. The different responses of these species to the presence of litter were attributed to differences in seed size (birch mean seed size, 1.0 mg; staghorn sumac seed size, 8.5 mg) and germination cues (birch seed requires light for germination; staghorn sumac seed requires chemical or heat scarification).

Chemicals

A variety of applied chemicals including certain herbicides (Sasaki *et al.*, 1968; Sasaki and Kozlowski, 1968d; Kozlowski and Sasaki, 1970), fungicides (Kozlowski, 1986a), insecticides (Olofinboba and Kozlowski, 1982), growth retardants, fertilizers, and soil salts sometimes inhibit plant establishment by direct suppression of seed germination, toxicity to young seedlings, or both (Kozlowski and Sasaki, 1970). The phytotoxicity of soil-applied chemicals varies greatly with the specific compound applied, dosage, plant species, environmental conditions, and manner of application. Toxicity often is low if the chemical is applied to the soil surface, intermediate if incorporated in the soil, and greatest if maintained in direct contact in solution or suspension with plant tissues (Kozlowski and Torrie, 1965). The high absolute toxicity of many chemicals often is masked because soil-applied compounds are variously lost by evaporation, leaching, microbial or chemical decomposition, and irreversible adsorption in the soil (Kozlowski *et al.*, 1967a,b).

Some herbicides are resistant to microbial action, and their persistence in the soil may affect tree growth long after weeds have been eliminated. For example, 2,3,6-TBA (2,3,6-trichlorobenzoic acid) may persist in the soil for several years. Direct contact of red pine seeds with 2,3,6-TBA did not suppress germination, but continuous contact of the herbicide with recently emerged seedlings caused abnormal plant development and often induced death of the seedlings (Kozlowski, 1986b).

Whereas some herbicides kill seedlings, others cause abnormal developmental changes such as curling, shriveling, or fusion of cotyledons, and chlorosis, distortion, and growth inhibition of various foliar appendages (Wu *et al.*, 1971). The primary mechanisms of herbicide toxicity are diverse and involve interference with plant processes as well as direct injury to cells and tissues.

Chemical growth retardants and inhibitors have been useful in producing compact plants and reducing the costs of storing and shipping nursery stock. However, some of these compounds may be toxic to seedlings in the cotyledon stage of development (Kozlowski, 1985d).

Soil salinity affects emergence of seedlings by decreasing the osmotic potential of the soil solution and by being toxic to the embryo and seedling. Salinity variously delayed and inhibited seedling emergence, reduced growth of shoots and roots, and altered the mineral concentration of several young citrus rootstocks (Zekri, 1993). Addition of 50 mol m^{-3} NaCl to a nutrient solution delayed seedling emergence by 3 to 5 days, except for Troyer citrange. Final emergence was reduced by less than 30% in Carrizo citrange, Troyer citrange, and Swingle citrange, and it was reduced over 65% in Ridge pineapple, Cleopatra mandarin, and Rough lemon. The biomasses of both shoot and root were reduced more than half by a 50 mol m^{-3} treatment. Clemens *et al.* (1983) showed considerable variation in response of germination of seeds of several *Casuarina* species to salinity. For example, *Casuarina stricta* seed showed a 40% decrease in germination when treated with 20 m*M* NaCl; germination of seeds of *C. distyla*, *C. inophloia*, *C. littoralis*, *C. luehmannii*, and *C. torulosa* was not affected.

Allelopathy

Seed germination and plant growth sometimes are inhibited by a variety of naturally occurring chemicals (allelochems) produced by plants (Rice, 1984; Hytönen, 1992). Toxic chemicals are released by both roots and shoots of some plants by volatilization, leaching, exudation from roots, and decay of plant tissues. Allelochems include phenolic acids, coumarins, quinones, terpenes, essential oils, alkaloids, and organic cyanides (Rice, 1974).

Perhaps the best known allelopathic chemical is juglone (Fig. 2.8), produced by plants in the genus *Juglans*, which inhibits growth of neighboring plants. Several genera of plants that grow in red pine stands are potentially allelopathic. These include *Prunus* (Brown, 1967), *Aster*, and *Solidago* (Fisher *et al.*, 1979). Water extracts of these plants did not inhibit germination of red pine seeds but adversely affected root and shoot growth of young seedlings (Table 2.4). Zackrisson and Nilsson (1992) presented convincing evidence of allelopathic effects of crowberry on regeneration of Scotch pine in Sweden. High dosages of allelopathic compounds occurred in forest soil when ground ice was present. At other times these toxins were detoxified by microorganisms.

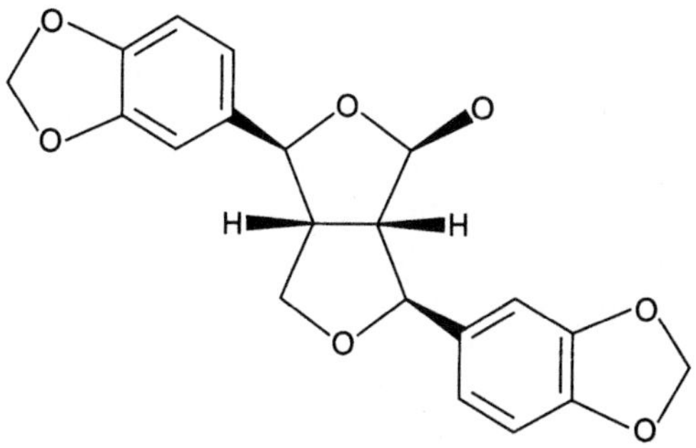

Figure 2.8 Structure of juglone, an alleopathic compound produced by black walnut.

Nilsson (1994) studied allelopathic effects of crowberry and root competition on growth of Scotch pine seedlings. Allelopathic effects were reduced by spreading activated charcoal on the soil to absorb the toxins that were leached from crowberry leaves and litter. Belowground competition was reduced by growing individual Scotch pine seedlings in plastic tubes. These experiments showed that two different growth-inhibiting mechanisms of crowberry were operating and that both belowground competi-

Table 2.4 Effect of Water Extracts of Leaves from Six Species and Distilled Water Control on Dry Weights of Roots and Shoots and Root–Shoot Ratios of Red Pine Seedlings after Seven Weeks of Treatment[a]

Effect	Root		Shoot		Root–shoot ratio	
	Weight[b] (mg)	Percentage of control	Weight[b] (mg)	Percentage of control	Ratio[b]	Percentage of control
Control	185.3 ± 18.6	100.0	412.9 ± 43.5	100.0	0.46 ± 0.03	100.0
Aster	139.1 ± 8.6[c]	75.1	246.4 ± 24.1[d]	59.7	0.62 ± 0.08	135.2
Lonicera	109.2 ± 10.2[d]	58.9	163.5 ± 18.4[e]	39.6	0.68 ± 0.03[e]	149.2
Prunus	143.1 ± 14.9	77.2	240.5 ± 29.0[d]	58.2	0.63 ± 0.05[c]	138.1
Rubus	126.7 ± 8.3[d]	68.4	207.6 ± 12.9[e]	50.3	0.62 ± 0.03[e]	134.5
Solanum	142.8 ± 14.2	77.1	274.5 ± 34.3[c]	66.5	0.55 ± 0.04	120.5
Solidago	124.7 ± 12.6[c]	67.3	204.2 ± 27.0[e]	49.4	0.65 ± 0.06[c]	143.1

[a]From *Plant Soil* **57** (1980), 363–374. Allelopathic potential of ground cover species on *Pinus resinosa* seedlings. Norby, R. J., and Kozlowski, T. T. Table 3, with kind permission from Kluwer Academic Publishers.
[b]Mean of 10 seedlings ± standard error.
[c]Significantly different from control at $p < 0.05$.
[d]Significantly different from control at $p < 0.01$.
[e]Significantly different from control at $p < 0.001$.

tion and allelopathy arrested growth of Scotch pine seedlings. Mahall and Callaway (1992) reduced the allelopathic influence of roots of creosote bush on adjacent plants by absorbing toxins in the soil with activated carbon.

Although many laboratory studies showed allelopathic effects of plant extracts, Lerner and Evenari (1961) advised caution in extrapolating the results of such experiments to field situations. Some studies used leaching or extraction procedures that gave little insight into natural release of allelochemicals (May and Ash, 1990). Hence, concentrations of allelopathic chemicals in many studies were higher than those to which plants in the field are subjected. Accumulation of allelochems in the field is modified by soil moisture and soil type. Furthermore, the field allelochems often are destroyed by soil microflora. Inderjit and Dakshini (1995) concluded that many laboratory assays of allelopathy had little relevance to field situations. This discrepancy typically results because of differences between laboratory tests and natural conditions, lack of standardized experimental methods, and lack of adequate controls.

Failure of regeneration of black cherry trees in old fields and forests was attributed to allelopathic interference by hay-scented fern (Horsley, 1977a,b). However, when black cherry seedlings were grown in soils from areas with and without hay-scented fern, height growth of black cherry seedlings did not differ, indicating that allelopathic chemicals from hay-scented fern were not accumulating in the field. Often the sites on which allelopathy is indicated are poorly drained, which may prevent leaching away or decomposition of allelochems.

Physiology of Seed Germination

The essential events in seed germination are resumption of growth by the embryo and its development into an independent seedling. Many changes are set in motion as germination begins, including the following: (1) seed hydration, (2) increased respiration, (3) enzyme turnover, (4) increase in adenosine phosphate, (5) increase in nucleic acids, (6) digestion of stored foods and transport of the soluble products to the embryo where cellular components are synthesized, (7) increase in cell division and enlargement, and (8) differentiation of cells into tissues and organs. The exact order of the early changes is not clear, and there is considerable overlap; however, with few exceptions, absorption of water is a necessary first step. Increase of hydration is associated with cell enlargement and cell division in the growing points as well as release of hormones that stimulate enzyme formation and activity. Although an increase in fresh weight of the seed accompanies imbibition, there is an early loss in dry weight due to oxidation of substrates and to some leakage of metabolites. After the root emerges and begins to absorb minerals and the cotyledons or leaves become photosynthetically active, the dry weight of the young seedling begins to increase until it regains and then surpasses the original seed weight.

Hydration

Water must be imbibed by seeds to increase protoplasmic hydration and set in motion a chain of metabolic events associated with germination. Imbibition of water softens hard seed coats, and the swelling of the imbibing embryo bursts the seed coat, permitting emergence of the radicle.

Movement of water from the soil into the seed depends on water relations of both the seed and the surrounding soil (Hadas, 1982; Bewley and Black, 1985). There is a large difference between the water potential (Ψ) of the dry, planted seed and moist soil, largely because of the low Ψ of the dry seed coat, cell walls, and storage components of the seed. As the seed imbibes water, this gradient decreases because the Ψ of the seed increases (becomes less negative) and the Ψ of the surrounding soil decreases. Hence, the rate of water uptake by the seed declines. Subsequent water uptake by the seed is appreciably influenced by the Ψ of the soil surrounding the seed, hydraulic conductivity of the soil, soil bulk density as affected by compaction, and extent of contact of the seed with soil particles.

Uptake of water by dry seeds typically occurs in three phases (Figure 2.9). An initial imbibitional phase of rapid water uptake is followed in order by low or negligible uptake and a phase of rapid uptake that culminates with radicle emergence (Hadas, 1982). The imbibitional phase of water uptake occurs in dormant, nondormant, viable, and nonviable seeds and involves physical hydration of the cell matrix, including walls and cytoplasmic colloids. As the seed imbibes water, a moving wetting front forms so there is a marked difference in the water content of the wetted cells and the adjacent, unhydrated cells.

In the lag phase of water uptake, the matrix components are no longer important in determining seed water status, and the Ψ of the seed is a function of its osmotic potential and the pressure potential that results from the pressure of the swollen contents on the cell wall. During this phase the seeds become metabolically very active.

The third phase of rapid water uptake occurs only in germinating seeds. At first this increase in water uptake is associated with volume changes in cells of the radicle as it elongates. Subsequently, water uptake from the surroundings is regulated by a decrease in the osmotic potential associated with osmotically active compounds formed during hydrolysis of stored food reserves (Bewley and Black, 1985).

The amount of water absorbed by various parts of the seed during germination varies greatly. In sugar pine, the embryo, which was the smallest seed component, absorbed the most water as a percentage of its dry weight.

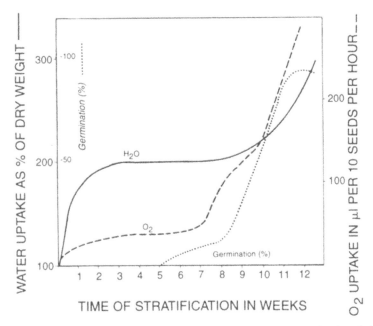

Figure 2.9 Water and oxygen uptake by apple seeds during stratification in relation to germination. From Lewak and Rudnicki (1977); modified from Duczmal (1963).

The megagametophyte absorbed less water and the seed coat the least (Stanley, 1958). The rate and pattern of water uptake by embryos are influenced by dormancy-breaking treatments. For example, in embryos of sugar pine seeds, water uptake increased as stratification time under moist conditions was increased. The rate and amount of water imbibed by seeds also vary with the nature of the seed coat, size and chemical composition of the seed, and water temperature.

Respiration

Respiration of seeds involves oxidative breakdown of organic constituents, primarily sugars, starches, fatty acids, and triglycerides, to provide energy in the form of ATP, which drives anabolic aspects of germination. The ATP is required for synthesizing new cellular constituents in seedlings and for forming protein-synthesizing machinery in producing enzymes for degradation and conversion of storage compounds. Respiration also provides much reducing power in the form of NADH. Hence, respiratory products play an important role in seed germination (Kozlowski, 1992).

Oxygen uptake of dry seeds is extremely low, but it increases greatly as water is imbibed and varies appreciably during three or four phases (Figure

2.9). In phase I, the imbibition phase, a burst of respiration is associated with activation of mitochondrial enzymes. In phase II, the rate of respiration is stabilized as seed hydration has slowed and preexisting enzymes are activated. Phase III is characterized by a second respiratory burst that is associated with activity of newly synthesized mitochondria and respiratory enzymes. Oxygen availability to the embryo also may be increased by rupture of the seed coat. In phase IV, which occurs only in depleted storage tissues, the rate of respiration decreases as respiratory substrates are exhausted (Bewley and Black, 1982).

When dormant sugar pine seeds imbibed water at 5°C they showed a rapid increase in O_2 uptake, ATP, and moisture content during the first 4 days (Murphy and Noland, 1982). A plateau phase followed until 60 days, after which a second marked increase occurred in all three features as dormancy was broken. The pattern was different when dormant seeds imbibed water and were maintained at 25°C. Water was absorbed much faster during the first 4 days, and a second increase in O_2 uptake did not occur after 60 days because the seeds did not break dormancy.

A high energy charge (EC) is important for termination of seed dormancy because of the role of ATP in respiration (see Chapter 6 of Kozlowski and Pallardy, 1997). The energy charge is the sum of the mole fraction of ATP plus one-half the mole fraction of ADP divided by the sum of the mole fractions of ATP, ADP, and AMP:

$$\text{Energy charge} = \frac{[\text{ATP}] + \frac{1}{2}[\text{ADP}]}{[\text{ATP}] + [\text{ADP}] + [\text{AMP}]} \tag{2.1}$$

The EC increased greatly during stratification of sugar maple seeds (Table 2.5), and, even at high absolute ATP levels, the embryonic axes elongated in response to gibberellic acid and kinetin only when the energy charge was near 0.8 or greater. The EC increased during the first 2 weeks of stratification of northern red oak acorns when germination was low (Hopper *et al.*, 1985). Thereafter the EC decreased before increasing again at 8 weeks after stratification. A second rise in the EC coincided with increased seed germinability. During the 28-day germination period, fluctuations in EC were correlated with growth of roots and shoots. Hopper *et al.* (1985) suggested that the EC may be useful for indexing the degree of seed dormancy and seedling vigor.

The rate and duration of each phase of respiration may be expected to differ among seeds of different species and genotypes because of variations in seed morphology, amounts of carbohydrate reserves, seed coat permeability, rates of imbibition, and metabolic rates (Bewley and Black, 1985).

Table 2.5 Effect of Stratification at 5°C on ATP, ADP, AMP, and Energy Charge in Sugar Maple Seeds[a]

Stratification period (days)	Adenylates (nmol/g dry weight)			Energy charge[b]
	ATP	ADP	AMP	
0	46.9 ± 12.6	180.2 ± 34.6	690.8 ± 31.1	0.15
1	137.9 ± 13.4	163.3 ± 24.7	194.9 ± 15.0	0.44
5	238.9 ± 27.9	119.4 ± 14.1	57.3 ± 14.1	0.72
10	327.2 ± 31.7	98.1 ± 17.0	60.2 ± 14.0	0.78
20	354.3 ± 63.2	99.8 ± 21.0	63.6 ± 18.3	0.78
40	416.3 ± 48.8	92.1 ± 11.5	63.2 ± 12.2	0.80
75	456.3 ± 70.7	91.4 ± 29.3	56.8 ± 11.8	0.83

[a]From Simmonds and Dumbroff (1974).
[b]Energy charge = ATP + $\frac{1}{2}$ADP/(ATP + ADP + AMP).

Enzyme Turnover

Even though some enzymes are active in dormant seeds, there generally is a dramatic increase in enzyme synthesis and activity after germination processes are set in motion. Changes in enzyme activities during and following germination have been studied in many woody plants. Examples include pines (Noland and Murphy, 1984, 1986; Gifford *et al.*, 1989; Groome *et al.*, 1991), spruce (Gifford and Tolley, 1989), tamarack (Pitel and Cheliak 1986), elm (Olsen and Huang, 1988), and hazel (Gosling and Ross, 1981; Li and Ross, 1990). The increased activity of some enzymes is the result of their conversion from an inactive to active form. However, many new enzymes also are synthesized after seeds imbibe water. Concurrently, some existing enzymes are degraded; hence, enzyme turnover is very great. The types of enzymes vary greatly among seeds and are related to the types of substrates present. As germination proceeds, the activity of enzymes involved in digestion of starch, lipids, proteins, hemicelluloses, and phosphates rapidly increases. Most of the enzymes involved in carbohydrate conversions become active because of synthesis, but some are mobilized by activation or release. The lipid bodies in seeds contain or acquire lipases that are involved in conversion of lipids to fatty acids and glycerol. Seeds also contain many proteolytic enzymes that are synthesized in the protein bodies or originate elsewhere in the cell and are translocated into the protein bodies (Mayer and Poljakoff-Mayber, 1989).

Phosphorus Metabolism

Phosphorus-containing compounds in seeds include nucleotides, nucleic acids, phospholids, phosphate esters of sugars, and phytin. There is considerable interest in the phosphate metabolism of germinating seeds because of the relationship between phosphorus and energy transfer in metabolism. When cherry embryos were chilled to break dormancy, phosphate was translocated from the cotyledons and appeared in the embryo axis as sugar phosphates, high-energy nucleotides, and nucleic acids. In unchilled embryos, however, only inorganic phosphates accumulated in cells (Olney and Pollock, 1960). Such experiments have been interpreted as showing that the breaking of seed dormancy is accompanied by phosphate metabolism and an increase in available energy to the embryo.

Metabolism of Food Reserves

Germination of seeds involves a drastic reversal of metabolic processes in the food storage tissues. Cells that initially synthesized insoluble starch, protein, and lipids during seed development suddenly begin to hydrolyze these materials. During seed development there is transport into the storage tissues, but during germination the soluble products of hydrolysis are translocated out of storage tissues to the meristematic regions of the seedling. This reversal of translocation involves considerable activation and deactivation of enzymes and conversion of "sinks" into "sources" (Dure, 1975). As a result the dry weight of storage tissues is reduced. For example, the dry weight of the megagametophyte of ponderosa pine seeds decreased greatly as it was depleted of reserves during germination (Fig. 2.10).

Carbohydrates　During the initial stages of germination, insoluble starch and reserve sugars are converted to soluble sugars, and the activities of amylases and phosphorylases increase. The soluble sugars are translocated from endosperm to cotyledon tissues to growing parts of the embryo. Cells of cotyledon tissues from dormant black oak acorns contain many starch granules, whereas cotyledons of germinating acorns contain few starch granules (Fig. 2.11). In seeds of *Araucaria araucana*, starch degradation is initiated by α-amylase and phosphorylase in the megagametophyte (Cardemil and Varner, 1984). The cotyledons of *Araucaria* can take up sucrose as well as glucose, fructose, and maltose (Lazada and Cardemil, 1990).

Lipids　Reserve fats in seeds are first hydrolyzed to glycerol and fatty acids by the action of lipases. These enzymes generally are not substrate specific and may hydrolyze different triglycerides to the same extent. Some of the hydrolyzed fatty acids are reused in synthesis of phospholipids and glycolipids that are needed as constituents of organelles and membranes. Most of the fatty acids, however, are converted to acetyl coenzyme A and then to sugar by reversal of the glycolytic pathway. In some species there is little or no conversion of lipids to carbohydrates. In oil palm, for example, most seed lipids are consumed by respiration during seed germination (Oo and Stumpf, 1983a,b). In seeds of jojoba wax esters serve as food reserves during germination (Moreau and Huang, 1977).

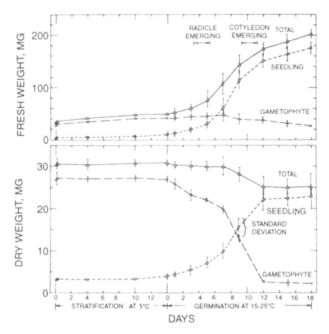

Figure 2.10   Changes in fresh and dry weight in megagametophyte, seedling, and total of both seeds and embryo of ponderosa pine seedlings during stratification and germination. From Ching and Ching (1972).

Compositional changes of Douglas fir seeds during germination are shown in Fig. 2.12. Lipids, the major food reserves, originally made up 48 and 55% of the dry weight of the gametophyte and embryo, respectively.

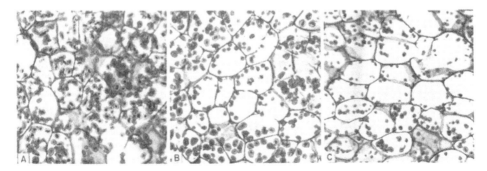

Figure 2.11   Starch distribution in cotyledons of black oak acorns: (A) cells of dormant acorns, (B) cells of stratified acorns, and (C) cells of germinated acorns. From Vozzo and Young (1975), *Bot. Gaz.*, © University of Chicago Press.

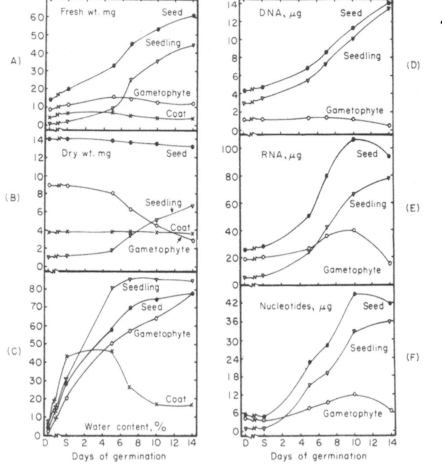

Figure 2.12 (A–R) Changes in weight and composition of embryo, megagametophyte, and whole seed of Douglas fir during germination. D, Air-dried seed; S, stratified seed. From Ching (1966).

During early germination, the lipids in the gametophyte decreased greatly and were used for embryo development. The dry weight of the embryo increased by 600%, and that of the megagametophyte decreased by 70%.

In oily seeds a concomitant increase in activities of lipase, isocitrate lyase, and fructose 1,6-biphosphatase indicates metabolism of stored lipids by gluconeogenesis. In hazel seeds held at 20°C, total lipase and isocitrate lyase activities were low. In seeds stratified at 5°C, total lipase and isocitrate lyase activity increased greatly in both the embryonic axis and cotyledons

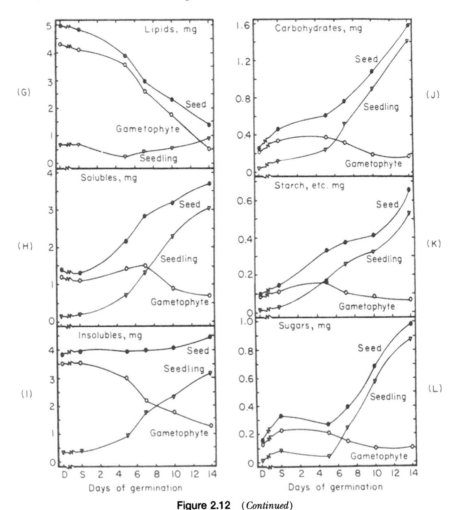

Figure 2.12   (*Continued*)

(Fig. 2.13). However, isocitrate lyase activity increased earlier in the embryonic axis; in the cotyledons the increase occurred only after 3 weeks. Similar increases in isocitrate lyase activity during cold stratification were reported for seeds of Norway maple (Davies and Pinfield, 1980) and sugar pine (Noland and Murphy, 1984).

Proteins   Reserve proteins in seeds frequently are broken down by proteolytic enzymes to soluble nitrogen compounds which are then used in metabolism and growth of various parts of the seedling. Insoluble crystal-

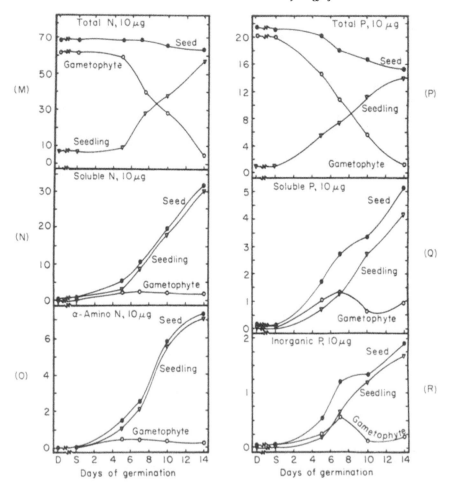

Figure 2.12 (*Continued*)

loid proteins were the major storage reserves in mature lodgepole pine seeds (Lammer and Gifford, 1989). These proteins were concentrated mainly in the megagametophyte, with smaller amounts in the embryonic axis. During germination these reserves were hydrolyzed to amino acids and transported to the developing embryonic axis. Hydrolysis of storage proteins in the megagametophyte began within 2 days of imbibition and was 90% complete within 5 days. In European silver fir there was about twice as much protein in the embryo of the resting seed as there was in the megagametophyte (Kovac and Kregar, 1989a). Soluble and insoluble pro-

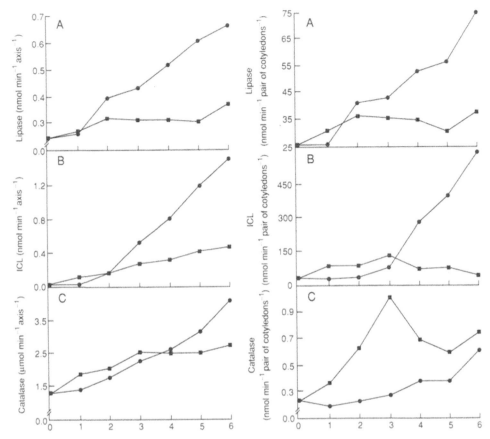

Figure 2.13 Changes in total activities of lipase (A), isocitrate lyase (ICL) (B), and catalase (C) from embryonic axes (left) and cotyledons (right) of hazel during cold stratification at 5°C (●) and warm incubation at 20°C (■). From Li and Ross (1990).

teins in the embryo supported embryo metabolism before the proteins stored in the megagametophyte were mobilized. The major proteins in the megagametophyte were hydrolyzed only after radicle protrusion occurred.

The germination pattern and protein metabolism during seed germination often differ somewhat among related species, as between *Hopea odorata* and *Dipterocarpus alatus,* both members of the Dipterocarpaceae (Huang and Villanueva, 1993). *Hopea* seeds began to germinate from the first day of imbibition and reached maximum germination rate by the fourth day. *Dipterocarpus* seeds germinated more slowly and had only a 50% rate of germination by the eighth day. The differences in the germination patterns of these species correlated well with the metabolic evolution of protein and amine compounds. In *Hopea*, polyamines increased during the first 3 days of germination and were maximal by the third day, one day before germination occurred. In *Dipterocarpus*, polyamines were maximal by the sixth day, and maximum germination occurred on the seventh day.

Protein degradation and increase in proteolytic activity generally begin after a lag of 1 day or more. In some, but not all cases, proteolytic activity appears to be regulated by hormones that form in the embryo or embryonic axis (Mayer and Marbach, 1981). Decrease in reserve proteins is associated with increases in amino acids and amides, and it is followed by synthesis of new proteins in the growing parts of the embryo (Mayer and Poljakoff-Mayber, 1989). Little soluble nitrogen accumulates at storage sites because rapid synthesis of new proteins in the developing embryo consumes the available nitrogen compounds.

Some of the proteases and peptidases are present in dry seeds, and others appear during germination. Generally the enzymes with proteolytic activity that develop in the cotyledons, endosperm, or megagametophyte are similar to those found in other plant tissues.

Seed Longevity and Aging

One of the most interesting characteristics of seeds is their wide variability in length of life, ranging from a few days to several decades or even centuries. It was reported, for example, that seeds of the arctic tundra lupine that had been buried for at least 10,000 years in frozen silt germinated readily in the laboratory (Porsild *et al.*, 1967). Seeds of albizia retained viability for at least 150 years and those of *Hovea* sp. for 105 years (Osborne, 1980). Seeds of some forest trees of the temperate zone can be stored for a long time. For example, viability of slash and shortleaf pine seeds was 66 and 25%, respectively, after 50 years of cold storage (Barnett and Vozzo, 1986). There was some loss of vigor during the storage period but probably not enough to influence the genetic makeup of the next generation of plants.

Longevity of seeds in the soil has important ecological implications, particularly with respect to weed control and plant succession. Seeds of many species may remain viable in the soil for 50 years or longer. Oosting and Humphreys (1940) concluded that seeds of weed species in North Carolina soils retained viability for several decades. Livingston and Allessio (1968) reached similar conclusions in Massachusetts. Hill and Stevens (1981) reported that seeds survived for 30 to 50 years under forests in Wales and Scotland. Fivaz (1931) concluded that *Ribes* seeds could retain viability in the soil for as long as 70 years.

Short-lived seeds of woody plants include those high in water content (e.g., those of *Taxus, Populus, Ulmus, Salix, Quercus, Carya, Betula,* and *Ae-*

sculus). Mature seeds of many tropical species deteriorate rapidly. Tropical genera whose seeds characteristically have a short life span include *Theobroma, Coffea, Cinchona, Erythroxylum, Litchi, Montezuma, Macadamia, Hevea, Thea,* and *Cocos.* However, by temperature and humidity adjustments during storage, the life of many tropical seeds can be prolonged from a few weeks or months to at least a year.

With respect to desiccation, seeds have been classified as orthodox or recalcitrant (Roberts, 1973). Recalcitrant seeds cannot be dried below some relatively high moisture content without rapidly losing vitality. Even when fully hydrated, recalcitrant seeds usually lose viability in a few weeks or months (E. H. Roberts *et al.,* 1984). Unlike recalcitrant seeds orthodox seeds can be dried to low moisture contents (usually to 5% and often to 2% on a fresh weight basis) without injury. The seeds of most herbaceous crop plants are orthodox, whereas recalcitrant seeds are mainly fleshy seeds produced by woody plants of both the temperate zone (e.g., chestnut, hazel, horse chestnut, oak, walnut) and the tropics (e.g., avocado, cacao, mango, rubber) and forest trees of the families Aracauriaceae and Dipterocarpaceae.

For a given genotype, the deterioration of orthodox seeds in storage is a function of time, moisture content, and temperature. Over a relatively narrow range of temperature and moisture contents, there is an approximately linear relation between storage temperature, moisture content, and the logarithm of viability period (Roberts, 1972). These considerations led to the following rules for storage of orthodox seeds (Murray, 1984):

1. For each 1% decrease in seed moisture content, the storage life of the seed is doubled.
2. For each 5.6°C decrease in storage temperature, the life of the stored seed is doubled.
3. The sum of the storage temperature (°F) and relative humidity (%) should not be greater than 100, with not more than half of the sum contributed by the temperature.

For detailed information on the storage conditions needed to prolong the life of seeds of many woody plants, the reader is referred to the chapter by Harrington in Volume 3 of *Seed Biology* (Kozlowski, 1972c) and the book by Young and Young (1992).

Considerable attention has been paid to the vigor of different seed lots. High seed vigor commonly is associated with rapid seed germination and seedling growth under field conditions. Expression of seed vigor is influenced by heredity; seed development, harvest, and storage conditions; and the seed germination environment. In general, high seed vigor is associated with high anabolic enzyme activity, respiration, ATP pool size, and synthesis of proteins, RNA, and DNA (Ching, 1982).

The vigor of aging seeds declines progressively, and the seeds change imperceptibly from one stage of deterioration to the next. Initial symptoms of aging include a decrease in capacity to germinate as well as an increase in susceptibility to attacks by microorganisms. As seed deterioration continues, the emerging radicles are progressively shorter, and cotyledons do not emerge from the seed coat. Finally, the seed dies.

As seeds age they undergo several structural, biochemical, and genetic alterations. These include reduction in capacity for synthesis of proteins, lipids, and RNA; injury to membranes and chromosomes; and decreased synthesis of labile enzymes and repair systems (Fig. 2.14) (Osborne, 1980, 1982; Priestley, 1986).

Because of worldwide concern for conservation of germplasm, much attention has been given to aging-induced genetic changes in seeds and their progeny. Induction of genetic variation in aging seeds includes changes in DNA, cytoplasm, RNA, and chromosomes. Most commonly, the chromosome aberrations in roots of plants produced from old seeds involve breakage of chromosomes. Because they are easily detected, chromosome aberrations have received primary attention. However, pollen abortion mutations and chlorophyll mutations also are associated with seed aging (Roos, 1982).

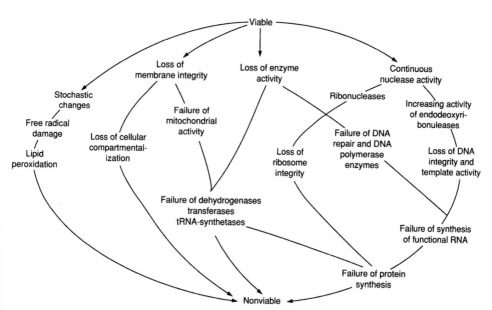

Figure 2.14 Changes during progressive senescence of dry seeds in storage. Reprinted with permission from Osborne, D. J. (1980). Senescence in seeds. *In* "Senescence in Plants" (K. V. Thimann, ed.), pp. 13–37. Copyright CRC Press, Boca Raton, Florida.

Seeds of Italian stone pine that were stored for more than 1 year lost capacity for germination and for increasing their protein and nucleic acid levels during germination (DeCastro and Martinez-Honduvilla, 1982). Although most synthetic events are initiated within minutes after dry embryos are rehydrated, initial DNA replication occurs rather late during metabolic reactivation. In old seeds, activation of DNA synthesis is further delayed, and phenotypic abnormalities and chromosomal aberrations at first mitosis become more numerous. After prolonged storage of dry seeds, the capacity to synthesize protein after imbibition may fail completely.

The lack of an effective DNA repair system in old seeds is associated with loss of membrane integrity and with leakage of cell contents. Decrease in germination capacity after many years of storage of Italian stone pine seeds was correlated with increased loss of reducing sugars (DeCastro and Martinez-Honduvilla, 1984). The greater leakage of solutes from old than young seeds implies that the integrity of the plasmalemma and/or the tonoplast is lost during aging. Ultrastructural studies provide additional convincing evidence of disruption of membrane integrity in old seeds (Priestley, 1986). Phospholipid degradation and peroxidation of unsaturated fatty acids, followed by membrane destruction, played an important role in aging of Norway maple seeds (Pukacka and Kuiper, 1988).

Seed Dormancy

Mature seeds of some species of woody plants of the temperate zone germinate rapidly if planted under favorable environmental conditions, but seeds of most species exhibit some degree of dormancy. Hence, they do not germinate promptly even when placed under the most favorable environmental conditions. The seeds of most tropical species of woody plants have little or no dormancy, but some remain dormant for years (Alexandre, 1980). In Malaysia, all the seed of two-thirds of the woody species germinated within 3 months after the seeds were shed (Ng, 1980).

Seeds that fail to germinate when shed are said to exhibit primary dormancy. Those that can germinate readily when shed and exposed to favorable environmental conditions may subsequently develop a dormant state, called secondary dormancy, if exposed to unfavorable conditions. Secondary dormancy often is induced by specific environmental conditions such as very high or low temperature, burial in soil, anaerobic conditions, or soaking in water (Khan and Samimy, 1982; Mayer and Poljakoff-Mayber, 1989).

Seed dormancy may be advantageous or disadvantageous. The prolonged chilling requirement for breaking dormancy of seeds of many temperate-zone plants prevents germination until spring. This tends to promote plant survival because earlier germination would subject the young plants to danger of injury by freezing. The seeds of some wild plants may remain dormant in the soil for many years, and the period of germination of such seeds may be spread over years. Such dormancy provides for establishment and survival of a species, even though the earliest emerging seedlings or even all of those in a given year are killed by drought or frost. Another advantage of seed dormancy involves the restriction of seed germination in hot and dry regions to the short, wet period of the year. In seeds of some desert plants, inhibitors in the seed coat prevent germination. However, the inhibitors are leached out by sufficient rain to wet the soil thoroughly, ensuring that the seeds germinate only when there is enough soil moisture for establishment (Wareing, 1963). In contrast, seed dormancy often is a nuisance to nursery operators who wish to have large quantities of seeds germinate promptly in order to produce large and uniform crops of seedlings. The causes of seed dormancy and methods of breaking it are therefore of both physiological and practical importance.

Causes of Dormancy

Knowledge of the causes of seed dormancy often makes it possible to intelligently apply appropriate treatments to overcome the dormant condition of individual seed lots. Seed dormancy results from a number of causes, including (1) immaturity of the embryo, (2) impermeability of seed coats, (3) mechanical resistance of seed coats to growth of the embryo, (4) metabolic blocks within the embryo, (5) combinations of (1) to (4), and (6) secondary dormancy (Villiers, 1972).

Seed Coat Dormancy Dormancy of seeds often is associated with seed coat characteristics and is due to the impermeability of the coat to water and/or gases, mechanical prevention of radicle extension, prevention of inhibitory compounds from leaving the embryo, or supplying of inhibitors to the embryo (Kelly *et al.*, 1992). Seed coat dormancy is especially common in the seeds of legumes (Morrison *et al.*, 1992) including those of black locust, honey locust, and redbud. Seeds of many nonlegumes, such as those of red cedar, basswood, eastern white pine, and apple, also have impermeable seed coats. Similarly, the seeds of some tropical species do not germinate promptly because they have impermeable and hard seed coats. Examples are *Podocarpus* spp., *Intsia palembanica, Parkia javanica, Sindora coriacea,* and *Dialinum maingayi* (Sasaki, 1980a,b). In some species (e.g., Scotch pine, Norway spruce) imbibition of seeds is regulated largely by impermeable layers surrounding the megagametophyte rather than by the seed coat proper (Tillman-Sutela and Kauppi, 1995a,b).

The nature of seed coat impermeability varies somewhat among species. In honey locust and black locust, the seed coat is a barrier to both imbibition of water and oxygen uptake. In eastern white pine and green ash, however, the seed coat is permeable to water but restricts gas exchange. In apple seeds, the seed coat impedes oxygen uptake, thereby making the supply inadequate for the high respiration rate necessary for germination of the embryo (Visser, 1954). Contributing to seed coat impermeability are compounds such as pectin, suberin, cutin, and mucilage.

Seed coats of some species are said to be mechanically resistant, preventing the embryo from further development once it becomes fully grown in the seed. However, Villiers (1972) stated that many such reported cases probably were traceable to other factors such as physiological dormancy of the embryo.

Embryo Dormancy   Sometimes the embryo is immature and requires a period of "afterripening" (storage under favorable conditions) to reach a certain stage of development before germination occurs. Examples are seeds of viburnum, holly, and ginkgo. However, the most common type of seed dormancy is one in which morphologically mature embryos are unable to resume growth and germinate. Such dormancy is common in apple, lilac, oak, chestnut, dogwood, hickory, pear, and sycamore, among the angiosperms. Gymnosperms that show embryo dormancy (see Chapter 3 of Kozlowski and Pallardy, 1997) include some pines, bald cypress, Douglas fir, hemlock, juniper, larch, spruce, and fir.

Physiological embryo dormancy, like bud dormancy, appears to develop in two stages in which mild and reversible dormancy passes progressively into deep-seated dormancy, which cannot be reversed by the same environmental conditions that induced it. Seeds of some species of *Fraxinus*, which normally become dormant, will germinate at once if they are harvested and planted before going through a drying phase.

Sometimes the failure of seeds to germinate is traceable to more than one specific type of dormancy. In some species of *Rosa*, seed germination is prevented by the mechanical restriction of a thick pericarp on embryo expansion, as well as by dormancy resulting from growth inhibitors in the achene (Jackson and Blundell, 1963). Such "double dormancy" has also been reported in seeds of hawthorn, junipers, yew, basswood, dogwood, osage orange, yellowwood, witch hazel, digger pine, Swiss stone pine, and whitebark pine. Dormancy of holly seeds has been traced to both immature embryos and hard seed coats. Delayed germination of acorns of water oak was attributed to at least three factors: (1) mechanical strength of the pericarp, (2) chemical inhibition by the pericarp, and (3) slowly increasing capacity to imbibe water (Peterson, 1983).

Some idea of the complexity of the seed dormancy mechanism in certain species may be gained from the analysis by Villiers (1975) of seed dormancy of European ash (Figure 2.15). This species produces dry, single-seeded, dormant fruits, with germination normally being postponed until at least the second season after the fruit is dispersed. The embryo is morphologically complete, but it must grow to about twice its original size on reimbibition of water before germination will occur. However, expansion of the immature embryo is inhibited by the fruit coat, which interferes with oxygen uptake. Hence, growth of the embryo is further inhibited until the outer layers of the seed coat begin to decay. Even the fully grown embryo is dormant and cannot emerge from the seed until it is chilled for several months. Therefore, the need for embryo development, as well as restriction of gas exchange, inhibit germination in the first spring. The chilling requirement is met in the second winter, and germination finally occurs in the second spring after the seed is produced. Germination can be delayed even longer by slow decomposition of the fruit coat and inadequate chilling during the second winter.

In western white pine, the papery inner layer of the seed coat and physiological conditions of the megagametophyte–embryo accounted for dormancy (Hoff, 1987). The hard seed coat contributed in a minor way to dormancy. Seven percent of the unstratified seeds were not dormant, 9% were dormant because of the hard seed coat, 34% were dormant because of the inner seed coat, and 50% were dormant because of physiological conditions of the megagametophyte–embryo.

In the genus *Acer*, seed dormancy usually is ascribed to constraints of the seed coat in *A. davidii, A. ginnala, A. negundo, A. pseudoplatanus,* and *A. velutinum,* and to embryo dormancy in *A. cappadocicum, A. hersii, A. platanoides, A. saccharum,* and *A. tataricum.* Nevertheless, variable degrees of both types of dormancy occur in all these *Acer* species. Embryo dormancy

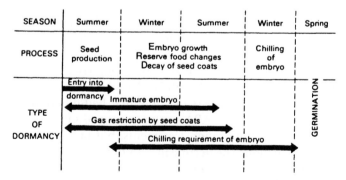

Figure 2.15 Complexity of dormancy mechanisms of seeds of European ash. From Villiers (1975).

and seed coat dormancy apparently represent consecutive developmental stages, with the former changing to the latter as embryo dormancy is progressively weakened. The differences between the two types of seed dormancy in *Acer* appear to be quantitative rather than qualitative. A common mechanism appears to regulate dormancy in all these species of *Acer* irrespective of the category to which a particular species is assigned (Pinfield and Dungey, 1985; Pinfield *et al.*, 1987; Pinfield and Stutchbury, 1990).

Hormones and Seed Dormancy

For a long time it was presumed that all embryo dormancy was controlled by hormonal growth regulators. Many investigators concluded that the onset of embryo dormancy was regulated by accumulation of growth inhibitors, particularly abscisic acid (ABA), and that breaking of dormancy was mediated by a shift in the balance toward hormonal growth promoters (e.g., gibberellins, cytokinins) that overcame effects of the inhibitors (Taylorson and Hendricks, 1977). It now appears that the mechanism of control of seed dormancy is more complex and cannot be adequately explained solely by changes in endogenous growth hormones.

A consistent role of ABA in regulating seed dormancy has not been established. In some species ABA decreases during the dormancy-breaking chilling of seeds (Sondheimer *et al.*, 1968; Webb *et al.*, 1973a; Orunfleh, 1991). In seeds of other species, the changes in ABA are similar under both conditions that maintain dormancy and those that prevent it (Khan and Samimy, 1982).

The degree of dormancy in apple seeds was correlated with their ABA concentration (Bouyon and Bulard, 1986). During chilling of apple seeds, ABA was transported from the cotyledons to the axis (Singh and Browning, 1991). Nevertheless, the level of dormancy decreased somewhat, possibly because the sensitivity of the axis declined as a result of chilling.

Several studies showed that a decline in ABA during stratification did not account for loss of seed dormancy. For example, ABA levels in peach and apple seeds decreased substantially after 3 weeks of chilling. However, ABA levels fell equally during low- and high-temperature stratification, but only the low-temperature treatment broke dormancy (Bonomy and Dennis, 1977; Balboa-Zavala and Dennis, 1977). The ABA content of dormant, unchilled seeds of three cultivars of apple that had different chilling requirements were similar. The ABA content dropped rapidly and similarly under 20°C and 5°C stratification treatments, but only cold stratification accelerated germination. The decline in ABA during stratification was attributed largely to leaching from the seed coat and nucellar membranes. The ABA content of the embryo remained almost constant. The radicle in intact seeds stratified at 5°C began growing 20 to 30 days after ABA in the seed coats and nucellar membranes had nearly disappeared. Radicles of embryos of unchilled seeds did not grow, even though ABA had leached from them as well (Subbaiah and Powell, 1992). Although the amount of ABA in embryos of sugar pine seeds decreased during stratification, no ABA was lost during the first 30 days, during which time germination capacity increased from 5 to 20%. Furthermore, the ABA content decreased three times as fast at 25°C as at 5°C, but seed dormancy was not broken at the higher temperature (Murphy and Noland, 1980a,b, 1981b).

Various lines of evidence have been advanced to show that seed dormancy is controlled by gibberellins. For example, application of exogenous gibberellin was associated with large increases in germination capacity in seeds of a number of species, suggesting that gibberellins accumulated in seeds as a response to environmental factors that initiated the causal events leading to loss of dormancy. In addition, low-temperature stratification of seeds was followed by increases in several gibberellins as well as loss of dormancy (Isaia and Boulard, 1978; Powell, 1988). When excised embryos of seeds with a chilling requirement are placed in an environment favorable for germination, they form a rosette of leaves, but internode expansion is negligible. Such dwarf seedlings, which resemble those of some gibberellin-deficient plants, can develop into more normal seedlings by exposure to chilling temperatures or applied gibberellins (Black, 1980/1981).

Considerable other evidence casts doubt on the importance of gibberellins alone in controlling seed dormancy. Exogenous gibberellins induced germination of partially stratified seeds of northern red oak (Vogt, 1970, 1974). However, the applied gibberellins did not completely replace the effects of stratification. For example, a minimum of 4 weeks of stratification was necessary to effectively stimulate germination with exogenous gibberellins (Vogt, 1974). Although gibberellins increased during stratification of seeds of Norway maple, their levels were not correlated with germination capacity (Pinfield and Davies, 1978). When peach seeds were chilled, gibberellins increased, but the increase appeared to reflect normal developmental changes more than release from dormancy (Gianfagna and Rachmiel, 1986). In hazel seeds, gibberellins increased appreciably after previously chilled seeds were exposed to warm temperatures (Ross and Bradbeer, 1968, 1971).

In general, gibberellin levels remain relatively low during seed stratification or show a transitory increase followed by a subsequent decrease. Hence, it now appears doubtful that gibberellins play an important role in the breaking of seed dormancy. Rather, they are more likely to be involved with the metabolism associated with germination and growth (Ross, 1984).

Some evidence for possible involvement of cytokinins in breaking seed dormancy comes from studies showing that exogenous applications are associated with release from dormancy (Black, 1980/1981). In addition, chilling of seeds of some species is associated not only with breaking of dormancy, but also with an increase in cytokinins (Taylor and Wareing, 1979; Kopecky *et al.*, 1975). Other studies show little change in cytokinins of seeds exposed to chilling temperatures. Examples are seeds of sycamore maple (Webb *et al.*, 1973b) and Norway maple (Pinfield and Davies, 1978). In the latter species, only negligible changes in cytokinins occur during the chilling period; large increases only accompany actual germination. During the first 20 days of stratification of sycamore maple fruits at 5°C, the amounts of cytokinins increased but subsequently decreased to levels lower than those in embryos of freshly harvested fruits. The lower amounts persisted during the remainder of the 60-day stratification period. In embryos from fruits stored at 17°C (which lacked capacity to germinate), amounts of both free and bound cytokinins were very low throughout the 60-day period (Julin-Tegelman and Pinfield, 1982). The rise in cytokinins of chilled seeds of apple and sugar maple is transient, and evidence that chilling stimulates the capacity of seeds to subsequently produce cytokinins is lacking (Black, 1980/1981).

Auxins apparently play a negligible role in release from seed dormancy. The amount of indoleacetic acid (IAA) in embryos of Scotch pine and sycamore maple decreased during early stages of stratification at low temperature but subsequently increased (Sandberg *et al.*, 1981; Tillberg and Pinfield, 1981). A decline in IAA during the first 2 weeks of stratification of rose achenes at 4°C apparently was not involved in dormancy release because a similar decrease in IAA was found in achenes stratified at 17°C which did not break dormancy (Tillberg, 1984).

Although exogenous growth promoters often are associated with the breaking of seed dormancy, the weight of evidence now indicates that other endogenous factors are more directly involved in controlling seed dormancy. Chilling of seeds also is associated with activation of a promotive force for germination, but endogenous factors other than hormone levels appear to be causally involved in the dormancy-breaking mechanism. These may include enzymatic changes and an increase in ATPase activity, possibly indicating a change in membrane permeability (Powell, 1988).

Some investigators associated the breaking of seed dormancy with changes in protein metabolism. In dormant embryos the cotyledons inhibit growth of the embryonic axis. Hence, at least a portion of the promotive effect of seed stratification is removal of the inhibitory effects of the cotyledons. In sugar pine seeds, the capacities for protein synthesis of both the embryo and megagametophyte were considered essential components of increased germination with loss of seed dormancy (Noland and Murphy, 1986). In both the embryo and megagametophyte of loblolly pine seeds, some proteins decreased greatly during stratification. In the megagametophyte, a more diverse set of proteins increased during stratification (Schneider and Gifford, 1994). Low-temperature stratification increased soluble proteins, amino acids, and several enzymes in American basswood seeds (Pitel *et al.*, 1989). These changes were closely correlated with rates of seed germination. Stratification of sugar maple seeds increased the capacity of the embryo for protein synthesis (Hance and Bevington, 1992). The release of embryos from dormancy was accompanied by changes in the levels of specific proteins. The data suggested that some of these proteins were associated with correlative inhibition of axis growth by the cotyledons.

The potential for embryo growth appears to be regulated by complex interactions of hormonal and other endogenous factors in tissue surrounding the embryo and the embryo itself. Both the nature and intensity of these interactions are appreciably influenced by environmental factors such as light, temperature, and water supply (Khan and Samimy, 1982; Battaglia, 1993).

Breaking of Seed Dormancy

Seed dormancy often can be broken by treatments that stimulate embryo metabolism, increase the permeability of seed coats to oxygen and water, and/or reduce the mechanical resistance to growth of the embryo. The efficiency of specific treatments varies greatly with the degree and kind of seed dormancy present. In some species, seed dormancy can be easily broken by any of several treatments, whereas seeds of other species respond only to a single, specific treatment. Seed dormancy of certain species sometimes cannot be broken by any of the methods commonly used.

Afterripening   For seeds having only a moderate degree of embryo dormancy, a period of afterripening in dry storage may be all that is necessary. During afterripening, the capacity for germination extends to a wider range of environmental conditions. The temperature range for germination, which for freshly harvested seeds may have been narrow and restricted to very low or very high temperatures, gradually widens. Specific requirements for alternating temperatures or specific light regimes are gradually dissipated.

The rate of afterripening varies from hours to years in seeds of different species. Even in a genetically uniform population of seeds that afterripen in the same environmental regime, the individual seeds are not in the same afterripening stage.

A number of physical and chemical changes often occur in seeds during dry storage. At least some of these result from changes in seed coats that alter their tensile strength or increase their permeability to water and gases. In other cases, changes occur in the embryo or surrounding tissues. For seeds with immature embryos, further morphological changes take place. In some species of *Fraxinus*, the morphologically complete embryos must increase in size before germination can occur. Biochemical changes during afterripening include decreases in amount of stored lipids, carbohydrates, and proteins; increases in metabolic activity; and changes in hormonal balance to favor growth promoters over inhibitors. Zarska-Maciejewska and Lewak (1983) distinguished three phases of metabolic activities that reached maxima at different times of afterripening of apple seeds. During the first phase, the primary cause of dormancy was removed. During the second phase, the highest rate of metabolism occurred. There was a massive hydrolysis of proteins (which was not directly influenced by temperature) and an increase in gibberellins and hydrolytic enzymes. The third phase was characterized by initiation of germination.

Stratification Embryo dormancy is commonly broken by stratification (storing moist seeds at low temperatures, usually between 1 and 5°C, for periods ranging from 30 to 120 days). Stratification of tree seeds is an old practice that can be traced back to the seventeenth century (Evelyn, 1670). In addition to breaking embryo dormancy, stratification also may alleviate other components of dormancy. For example, chilling of the embryo for 60 days removed embryo dormancy of apple seeds, but chilling of the whole seed required a much longer time (Bewley and Black, 1985).

Stratification, the most commonly used method in forestry and horticulture for breaking seed dormancy, attempts to simulate natural outdoor winter conditions to which seeds are exposed. The term "stratification" implies placing seeds in layers of moisture-retaining media such as peat moss, sawdust, or sand. However, low temperature and moisture are the essential features of this practice. Hence, seeds are considered to be "stratified" when they are moistened, put in plastic bags, and stored at low temperatures. The plastic bags should be opened periodically for aeration, even if the plastic is thin. After such treatment, the seeds can be planted; if allowed to desiccate, however, the seeds may lapse into a very deep state of secondary dormancy.

When freshly removed from fruits, the seeds of sour orange exhibited low O_2 uptake. However, when the seeds imbibed water and were stored at 4°C for 4 weeks, O_2 uptake increased throughout the storage period. Removal of the testa and tegumen was followed by increases in O_2 uptake of 69 and 529%, respectively (Edwards and Mumford, 1985). When measured at 5°C, the respiration rate of apple seeds and isolated embryos during stratification was very low. Mitochondria isolated from embryos also showed low respiratory activity. During germination, a rapid rise occurred in respiration of isolated embryos and mitochondria (Bogatek and Rychter, 1984).

The duration of moist stratification needed to break dormancy varies with species and, for a given species, is related to the winter climate at the seed source. For example, seeds of sweet gum from southern Mississippi germinated faster after 32 days of stratification than seeds from northern Mississippi after 64 days of stratification (Wilcox, 1968). Provenance differences in stratification requirements also have been shown for seeds of Douglas fir (Allen, 1960) and eastern white pine (Fowler and Dwight, 1964).

Early growth of seedlings depends not only on the stratification temperature but also on the duration of exposure of seeds to low temperature. Stratification of peach seeds at 0 to 6°C (especially 2°C) produced seedlings with more abnormal growth than stratification at 8 to 10°C at intermediate durations of exposure (30 to 50 days) (Frisby and Seeley, 1993). Abnormal (epinastic) growth was reduced following longer stratification at 4 to 14°C.

Chemicals The use of various chemicals to accelerate germination has attracted attention from time to time, but it has not been widely adopted as a practical method. Embryo dormancy of seeds often has been broken by application of gibberellic acid or cytokinins. In species with relatively mild embryo dormancy (e.g., loblolly pine, slash pine, Japanese red pine, subalpine larch, and western larch), oxidizing agents such as hydrogen peroxide stimulate respiration and may accelerate germination (Kozlowski, 1971a). However, use of hydrogen peroxide has practical limitations in inducing germination of some seeds with dormant embryos. For example, exposure of eastern white pine seeds to strong hydrogen peroxide stimulated germination of some seeds but killed others, with the result that final germination percentages of various seed lots were reduced (Kozlowski and Torrie, 1964).

Scarification Dormancy of seeds with impermeable seed coats often can be broken by soaking seeds in concentrated sulfuric acid for periods of 15 to 60 min. Permeability of seed coats in such seeds can also be increased by puncturing the seed coats or scratching the seeds with abrasives. For additional information on methods of breaking dormancy in seeds, see Young and Young (1992).

Heat The dormancy of seeds with hard seed coats sometimes can be broken by heat. An important natural agent for breaking dormancy of seeds of some species with hard seed coats is heat from fire or insolation. The hard seeds of the leguminous shrub *Acacia pulchella* required exposure to temperatures of 55 to 60°C for maximum germination (Portlock *et al.*, 1990). The duration of heating, depth of seed burial, and soil moisture content influenced the germination response to heating.

There apparently is considerable variation in the effect of temperature on breaking dormancy of seeds with hard seed coats. Seeds of *Macroptilium atropurpureum* were very sensitive to fluctuations in temperature (Moreno-Casasola *et al.*, 1994). Seeds of *Crotalaria incana* and *Indigo suffruticosa* showed a cumulative temperature effect on seedcoat softening. By comparison, the duration of exposure to temperature fluctuations did not have a marked effect on softening of seed coats in *Acacia farnesiana, A. macracantha,* or *Mimosa chaetocarpa.* Large fluctuations in temperature were more important. Seeds of *Chamaecrista chamaecristoides* required both large fluctuations in temperature and long exposure for breaking seed coat dormancy.

Physiology of Young Seedlings

In both gymnosperm and angiosperm seedlings, the cotyledons play a paramount role in seedling growth and development. The cotyledons are important in storage of foods and mineral nutrients, in photosynthesis, and in transfer to the growing axis of substances needed for growth.

Hypogeous Seedlings

In oaks, walnut, buckeye, and rubber, the cotyledons remain below ground and serve primarily as storage organs. Foods stored in the seed sustain the growing embryo until the leaves expand to provide a photosynthetic system and roots develop to absorb water and minerals, thereby making the young plant physiologically self-sufficient.

The carbohydrate sources of some hypogeous species often vary with environmental conditions. For example, white oak seedlings in a shaded understory exhibited only one annual flush of shoot growth in the spring (Reich *et al.*, 1980). Hence, all current photosynthate used in growth was provided by one set of leaves. In contrast, white oak seedlings grown under favorable environmental conditions exhibited four annual flushes of shoot growth that alternated with periods of root growth. During development of northern red oak seedlings, carbohydrate sources shifted from cotyledon reserves to photosynthesis of leaves of three sequential growth flushes (Hanson *et al.*, 1988a,b). The leaves of the first flush were the major source of total photosynthate, through the second flush of growth. The first-flush leaves contributed 80 to 100% of total photosynthate until the second-flush leaves were fully expanded. The second-flush leaves then became the major carbohydrate source until the third-flush leaves matured. The proportional reduction in the photosynthetic contribution of the first- and second-flush leaves as the seedlings developed was attributed largely to an increase in leaf area and, to a lesser extent, to a change in the rate of photosynthesis of new and existing leaves.

Epigeous Seedlings

The physiology of epigeously germinating seedlings varies somewhat among species. The epigeous cotyledons of some species store appreciable amounts of carbohydrates. Those of other species, such as pines, accumulate only small amounts, but they become photosynthetically active shortly after they emerge from the ground.

Following seed germination, a pine seedling is a system of competing carbohydrate sinks. Seedling development is an integrated continuum, with the site of synthesis of carbohydrates shifting during ontogeny from cotyledons to primary needles to secondary needles. There is a close dependency of growth of one class of foliar appendages on the capacity of the preceding class to synthesize compounds needed for growth. Hence, development of primary needles requires metabolites from cotyledons, and development of secondary needles depends on metabolites from primary needles. The young seedling in the cotyledon stage appears to be operating at threshold levels of growth requirements and is especially sensitive to environmental stresses. There is a strong influence of shoot environment early in ontogeny on initiation of all but the early formed primary needle primordia and on expansion of all primary needles, including those formed early (Sasaki and Kozlowski, 1968a, 1969, 1970). Initiation of a few primary needles depends on availability of current photosynthate of cotyledons. Low temperature or low light intensities during the cotyledon stage prevent initiation of most of the normal complement of primary needles of red pine (Tables 2.6 and 2.7). However, when seedlings are placed in a favorable environment, following prolonged exposure to low temperature or low light intensity, primordia of primary needles form readily and subsequently expand (Fig. 2.16).

The physiology of epigeous broad-leaved seedlings varies somewhat among species, depending on cotyledon function and patterns of leaf production. Species with seeds that lack endosperm (exalbuminous seeds) have cotyledons adapted for both storage and photosynthesis; examples are beech and black locust. Cotyledons of species with endosperm act not only as storage organs, but also as transfer organs in absorbing reserves from the endosperm and transferring them to growing axes. Endosperm was not essential for limited early growth of green ash seedlings (Marshall and Kozlowski, 1976b). However, cotyledons in contact with the endosperm exhibited higher rates of elongation and dry weight increase as well as slower depletion of proteins and lipids than cotyledons that were separated from the endosperm. The rate of growth of intact seedlings was much greater than in seedlings grown from excised embryos (Table 2.8).

The early development of epigeous broad-leaved seedlings occurs in

Table 2.6 Effect of Temperature on Development of Cotyledons and Primary Needles of Red Pine[a,b]

	Temperature (°C)				
	10	15	20	25	30
Average number of cotyledons per plant	6.54 ± 0.09	6.52 ± 0.09	6.46 ± 0.08	6.60 ± 0.09	6.36 ± 0.07
Average number of primary needles per plant	2.74 ± 0.38	9.06 ± 0.44	16.74 ± 0.39	16.62 ± 0.50	17.58 ± 0.66
Average length of cotyledons per plant (mm)	16.14 ± 0.35	16.76 ± 0.44	18.22 ± 0.41	18.08 ± 0.43	17.46 ± 0.49
Average length of primary needles per plant (mm)	2.57 ± 0.41	7.34 ± 0.45	21.10 ± 0.77	23.46 ± 0.96	21.02 ± 0.98
Average dry weight of cotyledons per plant (mg)	4.11	4.00	4.07	2.62	2.76
Average dry weight of primary needles per plant (mg)	0.34	2.13	9.01	7.10	8.21
Average dry weight of cotyledons plus primary needles (mg)	4.45	6.13	13.08	9.72	10.97

[a]From Kozlowski and Borger (1971).

[b]Values are means ± SEM ($n = 50$). Seedlings were grown from seed for 32 days at 23–25°C and a light intensity of 750 lux (1.6×10^3 ergs/cm^2/sec); they were subsequently grown for 44 days at indicated temperatures and a light intensity of 13,000 lux (2.6×10^4 ergs/cm^2/sec). At 10°C, 27 of 50 plants had expanded primary needles; at 15°C, 48 of 50 plants had expanded primary needles.

Table 2.7　**Effect of Light Intensity on Initiation and Expansion of Primary Needles of Red Pine[a,b]**

Treatment	Average number of visible primary needles after:				
	32 days	39 days	46 days	53 days	60 days
Dark, 25°C	0	0	0	0	0
750 lux (2.6 × 10³ ergs/cm²/sec), 23–25°C	0	0	0	0	0
6480 lux (1.3 × 10⁴ ergs/cm²/ sec), 25°C	3.9 ± 0.24	7.6 ± 0.24	12.3 ± 0.30	17.3 ± 0.39	21.7 ± 0.48
13,000 lux (2.6 × 10⁴ ergs/cm²/ sec), 25°C	7.4 ± 0.14	11.8 ± 0.32	16.3 ± 0.38	21.6 ± 0.43	25.3 ± 0.52

[a]From Kozlowski and Borger (1971).
[b]Values are means ± SEM. Time intervals indicate number of days after sowing.

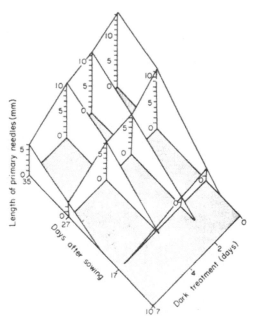

Figure 2.16　Effects of reexposure of red pine seedlings to light, after various periods of dark treatment, on elongation of primary needles. From Sasaki and Kozlowski (1970).

Table 2.8 Growth and Compositional Changes of Ten-Day-Old Green Ash Cotyledons and Seedlings with Endosperm Removed at Two-Day Intervals[a]

Seedling age (days)	Endosperm	Cotyledon length (cm)	Cotyledon dry weight (mg/cotyledon)	Intact seedling dry weight (mg)	Cotyledon protein (mg/cotyledon)	Cotyledon lipids (mg/cotyledon)	Cotyledon total nonstructural carbohydrates (mg/cotyledon)	Cotyledon chlorophyll (mg/cotyledon)
0	±	0.60	1.1	3.2	0.40	0.45	0.05	0.0
2	+	0.90 NS[b]	1.4	4.5 NS	0.31[d]	0.42 NS	0.16 NS	0.5 NS
2	−	0.85[c]	1.1[c]	3.4	0.19[d]	0.42 NS	0.13	0.5
4	+	1.25 NS	2.2[d]	6.9[c]	0.24[d]	0.60[c]	0.34 NS	2.5[c]
4	−	1.15 NS	1.1	3.5	0.13[d]	0.35[c]	0.28 NS	4.8[c]
6	+	1.90	2.3[d]	8.0[d]	0.20[c]	0.70[d]	0.54[c]	18.5[d]
6	−	1.45[c]	1.1[d]	3.7	0.14[c]	0.24[d]	0.25[c]	7.8
8	+	2.10	2.5[d]	8.9[d]	0.28[d]	0.50[d]	0.62[c]	22.0[d]
8	−	1.50[d]	1.2	4.0[d]	0.17[d]	0.14[d]	0.31[c]	9.2[d]
10	+	2.20	2.7[d]	10.3[d]	0.22[c]	0.53[d]	0.69[c]	21.0[d]
10	−	1.50[d]	1.4[d]	5.1	0.14[c]	0.20[d]	0.42[c]	10.5[d]

[a]From Marshall and Kozlowski (1976b). Importance of endosperm for nutrition of *Fraxinus pennsylvanica* seedlings. *J. Exp. Bot.* **27**, 572–574, by permission of Oxford University Press.

[b]NS, Not significantly different.

[c]Significant at $p \leq 0.05$.

[d]Significant at $p \leq 0.01$.

several stages. Li *et al.* (1985) divided the early development of apple seedlings into three stages. In the first stage, from the time the cotyledons emerged above ground, spread, and developed, cotyledon photosynthate was translocated mainly to the stem below the cotyledon and to the roots. In the second stage, from the first stage until the seedlings had developed leaves, cotyledon photosynthate moved largely to the shoot apex and immature leaves. In the final stage, as the cotyledons senesced, the function of supplying photosynthate to the root system was assumed by the first true leaf.

During their development, the cotyledons of some species progress sequentially through storage, transition, photosynthetic, and senescent stages. This progression requires drastic changes in enzyme activity, as mentioned earlier. In the storage phase the cotyledons are filled with reserves (e.g., carbohydrates, proteins, lipids) that are depleted during germination and used for early seedling growth. The kinds of reserves stored by cotyledons vary among species. For example, box elder cotyledons store primarily proteins, whereas those of ailanthus store large amounts of lipids. Embryonic cotyledons of box elder and black locust contain appreciable stored carbohydrates and those of ailanthus and green ash only small amounts (Fig. 2.17).

Cotyledons also store mineral elements in varying amounts. Embryonic cotyledons of exalbuminous seeds of red maple and black locust stored large amounts of nutrients that were translocated to rapidly developing axes during early seedling growth. In contrast, the small, thin cotyledons of green ash stored small amounts of mineral nutrients (Marshall and Kozlowski, 1975). Silver birch seedlings rapidly depleted the mineral nutrients in the seed and depended on an external nutrient supply by the time the first pair of leaves was produced (Newton and Pigott, 1991a).

The transition stage of cotyledon development is characterized by cotyledon expansion, depletion and utilization of cotyledon reserves by meristematic regions, differentiation of guard cells, emergence of cotyledons from seed coats, and synthesis of chlorophyll. The cotyledons emerge earlier from seed coats in species with exalbuminous seeds than in species with cotyledons embedded in endosperm (Marshall and Kozlowski, 1977).

The very important photosynthetic stage begins with attainment of positive net photosynthesis and ends with cotyledon senescence. Several studies have demonstrated an important relationship between photosynthesis of epigeous cotyledons and early seedling growth (Sasaki and Kozlowski, 1968a, 1970; Marshall and Kozlowski, 1974a,b, 1976a). The importance for seedling development of healthy, photosynthetically active cotyledons has been shown by the inhibition of seedling growth following shading or removal of cotyledons, or after application of inhibitors of photosynthesis to cotyledons (Table 2.9).

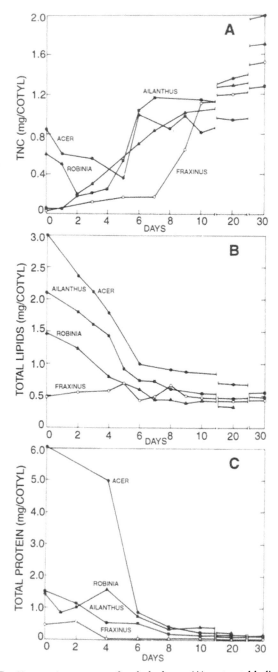

Figure 2.17   Changes in amounts of carbohydrates (A), extractable lipids (B), and total protein (C) per cotyledon during early seedling growth of four species of woody angiosperms. From Marshall and Kozlowski (1976c).

Table 2.9 Dry Weights (mg) of Various Parts from Black Locust Seedlings with Cotyledons Removed at Different Seedling Ages[a]

Plant part	Seedling age (days)	Seedling age at cotyledon removal (days)					
		2	4	6	8	10	Control
Roots	14	2.1	2.2	3.2	4.4	7.1	22.7
	21	5.6	9.8	7.8	8.7	14.2	45.4
	28	18.5	19.4	12.1	20.6	27.7	59.5
Hypocotyls	14	1.0	1.4	1.7	1.8	1.9	3.0
	21	1.8	1.9	2.1	2.1	2.9	4.9
	28	2.6	3.2	2.3	2.8	3.2	7.6
Foliage	14	2.6	2.9	4.2	5.0	4.4	11.4
	21	8.0	11.0	9.7	8.6	14.0	22.6
	28	16.0	18.0	14.6	24.8	23.7	71.2

[a]From Marshall and Kozlowski (1976a).

Both the rate of cotyledon photosynthesis and duration of the photosynthetic stage vary widely among angiosperm species. The rate of photosynthesis of cotyledons was much lower in American elm than in black locust, ailanthus, or green ash. High rates of cotyledon photosynthesis were maintained for a longer time in ailanthus and in green ash than in American elm or black locust (Figure 2.18). In cacao the epigeous cotyledons played an important role in storing of fats, amino acids, and carbohydrates. These reserves sustained the growing seedling until the foliage leaves emerged and began to export carbohydrates. The cotyledons were relatively unimportant photosynthetic organs (Olofinboba, 1975). Kitajima (1992) reported wide variations in cotyledon photosynthesis among 10 tropical tree species. Both gross and net photosynthesis were (per unit of dry weight) negatively correlated with cotyledon thickness. Whereas thin cotyledons had very high rates of photosynthesis, cotyledons thicker than 1 mm had photosynthetic rates just high enough to balance respiration.

The beginning of the senescent stage is characterized by yellowing of the cotyledons and continues until their abscission. The functional life span of cotyledons varies greatly among species and usually is shorter in angiosperms than in gymnosperms (DeCarli *et al.*, 1987). Epigeous cotyledons were relatively short-lived in honey locust, mulberry, and black willow (about 1 month); longer lived in cottonwood, hop hornbeam, and sycamore (about 2 months); still longer lived in Hercules'-club, hackberry, and blue beech (3 months); and unusually long-lived in white ash and magnolia (6 months or more) (Maisenhelder, 1969). It is difficult to

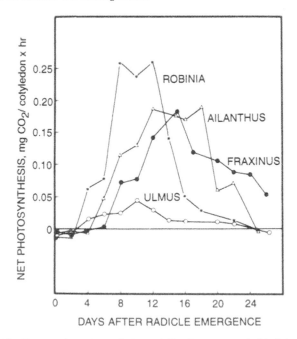

Figure 2.18 Photosynthesis in cotyledons of ailanthus, green ash, black locust, and American elm at various times after germination. From Marshall and Kozlowski (1976a).

establish precisely the life span of cotyledons of different species because cotyledon longevity is strongly influenced by environmental stresses. For example, drought can induce rapid senescence and abscission of mature cotyledons.

Production of Seedlings in Nurseries

Growth and survival of seedlings planted in the field are influenced by seedling quality, which varies appreciably with seed quality, nursery practices, and handling of nursery stock during lifting from the nursery and in outplanting.

Seedling Quality

The quality of seedlings reflects integration of many physiological and morphological characteristics. Ritchie (1984) divided components of seedling quality into "material attributes" (e.g., dormancy status, water relations, nutrition, and morphology) and "performance attributes" (e.g., root growth potential, hardiness to frost, and resistance to stress). Material attributes are more easily measured than performance attributes but, considered individually, they often have relatively low predictive value for seedling success unless they fall outside some normal range. Nevertheless, seedling size often has been a useful index of seedling success. For example, survival of grade 1 (root collar diameter >4.7 mm) loblolly pine seedlings was higher than that of grade 2 seedlings (root collar diameter 3.2 to 4.7 mm). Volume production of grade 1 seedlings was 17.5% higher than that of grade 2 seedlings (South *et al.*, 1985). On some sites, however, short seedlings may grow better than tall seedlings. For example, on very dry sites short seedlings, with small transpiring leaf areas, sometimes outperform tall seedlings.

Nursery Practices

It is important that nursery practices are favorable for maintaining physiological processes of seedlings that will result in high capacity for growth and survival after outplanting. The loss of many of the small absorbing roots during lifting and handling of nursery stock often leads to dehydration of transplanted trees (Kozlowski, 1975, 1976a; Kozlowski and Davies, 1975a,b). Hence, important requirements for survival of transplanted trees are a high root–shoot ratio and rapid growth of roots into a large volume of soil in order to maintain high rates of absorption of water and mineral nutrients. Reserve foods also are essential because photosynthesis of outplanted seedlings may not return to normal for several weeks.

Production of high-quality seedlings requires close attention to all phases of nursery management. These include preparation of nursery beds, soil management, planting procedures, control of seedling density, use of fertilizers, irrigation, and pest control. Sometimes they also may include root pruning and inoculation with mycorrhizal fungi (Chapter 7). For discussion of nursery practices, see Chapter 7 and books by Duryea and Brown (1984), Duryea and Landis (1984), Duryea and Dougherty (1991), and van den Driessche (1991a,b).

Storage of Seedlings

It often is necessary to place seedlings in cold storage because readiness of planting sites does not coincide with the time of lifting seedlings at the nursery. Furthermore, if the planting season must be extended, planting stock in cold storage can be kept dormant for a longer time than if left in nursery beds. If lifted, handled, and stored properly, seedlings from cold storage may grow better and survive as long or longer than recently lifted seedlings (Hocking and Nyland, 1971). Seedlings should be physiologically dormant when lifted for storage. If seedlings are to be stored over winter, their lifting from the nursery should be delayed as long as possible.

Storage Conditions   The quality of planting stock can be greatly lowered by improper storage. Seedlings should not be allowed to dehydrate either before or during cold storage. Dehydration of seedlings during storage may greatly lower the root growth potential (RGP) (Colombo, 1990; Deans *et al.*, 1990). To prevent dehydration of planting stock most storage areas are maintained at a relative humidity of 85% or higher. Often seedlings are wrapped in plastic films. While slowing dehydration, this practice increases molding, which can be counteracted by low temperature.

Temperatures of most seedling storages are maintained at either 1 to 2°C above freezing or 2 to 4°C below freezing (Camm *et al.*, 1994). However, satisfactory storage temperatures range from 3°C to −6°C. In northwestern parts of the United States and in Canada conifer seedlings are routinely stored at temperatures below freezing to reduce losses of reserve carbohydrates by respiration and to inhibit growth of molds (Simpson, 1990; Omi *et al.*, 1991). In British Columbia most conifers can be satisfactorily stored at −2°C (Van Eerden and Gates, 1990). Unfortunately, however, frozen seedlings usually thaw slowly before they are planted, often resulting in appreciable respiratory depletion of carbohydrates. Furthermore, molds sometimes develop during the periods of thawing.

Although many cold storages are maintained in darkness, a daily photoperiod during storage may improve growth and increase survival of outplanted seedlings that had been lifted in the autumn. Cold hardiness also may be increased by a daily photoperiod in storage. Camm *et al.* (1994) suggested that it might be useful to combine early lifting of seedlings with cold storage in the light.

Lifting Date   The date of lifting of seedlings for immediate planting or for cold storage is important because it influences the RGP of outplanted seedlings. Usually the RGP increases from a low value in the autumn to a high value in the early spring and then decreases.

Seedlings of Douglas fir, western hemlock, black spruce, and jack pine that were lifted from the nursery in September or October had less vigor than those lifted after mid-November (Camm *et al.*, 1994). Ponderosa pine seedlings lifted early in autumn also had lower RGP than those lifted in November. Stone *et al.* (1962) showed that mortality of transplanted conifers and RGP were inversely related. Hocking and Nyland (1971) suggested that to minimize adverse effects on seedling vigor lifting from the nursery should be delayed as long as possible.

In some regions (e.g., interior Washington State, British Columbia) the ground freezes and seedlings cannot be lifted in midwinter when their RGP is highest. Hence, seedlings have been lifted in the autumn and planted after overwinter storage, or they have been lifted in the spring and planted immediately or after a short period of storage. Ritchie *et al.* (1985)

showed that autumn lifting of lodgepole pine and interior spruce (*Picea glauca* × *P. engelmannii*) in British Columbia, beginning November 1 and with overwinter storage, was preferable to spring lifting and planting. Seedlings that were lifted in the spring had low RGP, low resistance to stress, low frost hardiness, and poor storage performance.

Duration of Storage Seedlings have been cold-stored from a few days in the southern United States to as long as 8 months in northern states and Canada. A moderate period of cold storage does not significantly influence mechanisms of seedling development. However, under certain conditions prolonged storage may be harmful because of depletion of reserve carbohydrates by respiration and a decrease in RGP. In a forest nursery in Scotland, the amounts of reserve carbohydrates in bulked needles, stems, and roots of unlifted Sitka spruce and Douglas fir seedlings were relatively constant (100 to 150 mg g^{-1}) from September to April. In comparison, carbohydrates were depleted by respiration in intact seedlings in cold storage at a rate of 0.4 to 0.6 mg g^{-1} day^{-1}, until only 40 to 50 mg g^{-1} of carbohydrate remained (Cannell *et al.*, 1990). Ritchie (1982) noted a decrease in RGP after storage of Douglas fir seedlings for 6 months. Dry weight losses of seedlings in cold storage often are associated with reduced survival of outplanted seedlings. The harmful effects of prolonged storage are more serious for seedlings lifted early than for late-lifted seedlings. Seedlings lifted in December and planted the next May had the benefit of a short storage time and development of high resistance to environmental stresses (Camm *et al.*, 1994).

Handling of Planting Stock

All of the skill devoted to maintaining appropriate nursery regimes can be negated by improper handling of nursery stock during lifting from the nursery and outplanting after storage. Hocking and Nyland (1971) recommended that stock should be lifted only when the leaves are dry. Roots should be kept moist, and the air temperature should be cool in the sorting and packing sheds. Injured, weak, diseased, and undersized plants should be culled.

Exposure of bare-rooted trees for even short periods during the outplanting process may severely inhibit growth and decrease survival. Even after seedlings are planted in the field, there often is excessive loss of water at a time when the roots grow too slowly to keep up with transpirational losses (Kozlowski *et al.*, 1991). Trees that survive transplanting may show growth inhibition for a long time. For example, height growth of white spruce transplants was reduced by about half in the first year after outplanting. In some cases, growth was inhibited for at least 10 years (Mullin, 1964).

Growth Potential of Outplanted Seedlings

Growth of planted seedlings varies greatly among species and genotypes. Many conifer seedlings grow much more slowly than seedlings of broad-leaved trees as shown by comparisons of dry weight increment and relative growth rates (Jarvis and Jarvis, 1963). Marked differences in productivity were attributed to species variations in photosynthesis, leaf area index, leaf longevity, and partitioning of photosynthate within the plant (Bond, 1986).

The very low rate of growth of conifer seedlings confers a competitive disadvantage on them when they are grown together with broad-leaved trees, shrubs, or herbaceous plants (Kozlowski *et al.*, 1991). Grasses restrict growth of conifer seedlings, often as a prelude to mortality of the conifers. For example, competing grasses severely reduced growth of 3 year old pines. After two growing seasons, shoot length, root collar diameter, and needle length were reduced by 40, 54, and 20%, respectively, compared to seedlings growing without such competition. Growth reduction was associated with multiple environmental stresses induced by the presence of grasses (Caldwell *et al.*, 1995). In competitive situations, conifers often grow best where the environmental conditions are poor for growth of broad-leaved seedlings. Hence, conifers often are excluded from sites where supplies of light, water, and mineral nutrients are high (Bond, 1989; Woodward, 1995).

Summary

Seeds are fertilized ovules that contain an embryo, stored food, and a protective coat. The embryo consists of cotyledons, an embryonic bud, a stem portion, and a rudimentary root. Mature seeds contain various amounts of carbohydrates, fats, proteins, enzymes, minerals, phosphorus-containing compounds, nucleic acids, alkaloids, organic acids, sterols, pigments, phenolic compounds, and hormones.

Seed germination involves resumption of embryo growth resulting in seed coat rupture and emergence of the young plant. Important environmental factors controlling seed germination are water supply, temperature, light (including light intensity, day length, and wavelength), oxygen, salinity, and plant litter. Because of wide differences among seedbeds in physical characteristics, temperature, and availability of water and mineral nutrients, seed germination and seedling establishment vary greatly in different seedbeds. A variety of naturally occurring chemicals (allelochems) and applied chemicals, including pesticides, herbicides, and growth retardants, may arrest plant establishment by suppression of seed germination, toxicity to young seedlings, or both.

Major events in germination include (1) seed hydration, (2) increase in respiration, (3) enzyme turnover, (4) increase in adenosine phosphates,

(5) increase in nucleic acids, (6) digestion of stored foods and transport of the soluble products to the embryo, (7) increase in cell division and enlargement, and (8) differentiation of cells into tissues and organs. Mature seeds of many species exhibit some degree of dormancy which may be associated with (1) immaturity of the embryo, (2) impermeability of seed coats, (3) resistance of seed coats to embryo growth, and (4) metabolic blocks within the embryo, or various combinations of these. For a long time it was presumed that embryo dormancy was regulated by balances of hormonal growth inhibitors and promoters. Current evidence indicates that control of seed dormancy is regulated by complex interactions of hormones and other endogenous factors in seeds. Seed dormancy can be broken by stratification of moist seeds at low temperature or high temperature, by afterripening in dry storage, or by mechanical and chemical treatment of seed coats to increase their permeability.

Seeds age progressively and exhibit reduced germination capacity and potential for synthesizing compounds needed for growth. Genetic changes in seeds and their progeny also are associated with seed aging.

In both gymnosperm and angiosperm seedlings, the cotyledons play an important role in growth and development of seedlings. Cotyledons are important in storage of foods and mineral nutrients, in photosynthesis, and in transfer to meristematic regions of substances needed for growth. In hypogeous seedlings, the cotyledons remain below ground and serve primarily as storage organs. In epigeous seedlings, the cotyledons emerge above ground. During their development, the cotyledons progress through storage, transition, photosynthetic, and senescent stages.

Growth and survival of outplanted seedlings are affected by nursery practices. Production of high-quality seedlings requires close attention to all phases of nursery management, including preparation of nursery beds, soil management, planting procedures, use of fertilizers, irrigation, root pruning, inoculation with mycorrhizal fungi, and pest control. Seedling quality also is influenced by time of lifting, duration of cold storage of seedlings, storage temperature and humidity, as well as handling of seedlings during lifting and planting.

Most outplanted conifer seedlings grow more slowly than seedlings of broad-leaved trees. Hence, conifer seedlings often are at a disadvantage in competitive situations. Conifers often grow best when the environmental regime is poor for growth of broad-leaved seedlings.

General References

Bass, L. N. (1980). Seed viability during long-term storage. *Hortic. Rev.* **2**, 117–141.

Bewley, J. D. (1994). "Seeds: Physiology of Development and Germination," 2nd Ed. Plenum, New York.

Bradbeer, J. W. (1988). "Seed Dormancy and Germination." Chapman & Hall, New York.

Copeland, D., and McDonald, M. B. (1995). "Principles of Seed Science and Technology." Chapman & Hall, London.

Duryea, M. L., and Brown, G. N., eds. (1984). "Seedling Physiology and Reforestation Success." Martinus Nijhoff/Dr. W. Junk Publishers, Dordrecht, The Netherlands.

Duryea, M. L., and Dougherty, P. M., eds. (1991). "Forest Regeneration Manual." Kluwer, Dordrecht, Boston, London.

Duryea, M. L., and Landis, T. D. (1984). "Forest Nursery Manual: Production of Bareroot Seedlings." Junk, The Hague.

Fenner, M., ed. (1992). "The Ecology of Regeneration in Plant Communities." Univ. of Arizona Press, Tucson.

Khan, A. A., ed. (1977). "The Physiology and Biochemistry of Seed Dormancy and Germination." North-Holland, Amsterdam.

Khan, A. A., ed. (1982). "The Physiology and Biochemistry of Seed Development, Dormancy and Germination." Elsevier Biomedical Press, Amsterdam.

Kigel, J., and Galili, G., eds. (1995). "Seed Development and Germination." Dekker, New York.

Kozlowski, T. T., ed. (1972). "Seed Biology," Vols. 1–3. Academic Press, New York.

Leck, M. A., Parker, V. T., and Simpson, R. L., eds. (1989). "Ecology of Soil Seed Banks." Academic Press, San Diego.

McDonald, A. J. S., Ericsson, T., and Ingestad, T. (1991). Growth and nutrition of tree seedlings. *In* "Physiology of Trees" (A. S. Raghavendra, ed.), pp. 199–220. Wiley, New York.

McDonald, M. B., Jr., and Nelson, C. J., eds. (1986). "Physiology of Seed Deterioration." Crop Sci. Soc. Am., Special Publ. No. 11, Madison, Wisconsin.

Mayer, A. M., and Poljakoff-Mayber, A. (1989). "The Germination of Seeds," 4th Ed. Pergamon, Oxford.

Murray, D. R. (1984). "Seed Physiology," Vols. 1 and 2. Academic Press, Sydney.

Priestley, D. A. (1986). "Seed Aging: Implications for Seed Storage and Persistence in the Soil." Comstock, Ithaca, New York, and London.

Putnam, A. R., and Tang, C. S. (1986). "The Science of Allelopathy." Wiley, New York.

Rice, E. L. (1984). "Allelopathy," 2nd Ed. Academic Press, New York.

Taylorson, R. B. (1989). "Recent Advances in the Development and Germination of Seeds." Plenum, New York.

Young, V. A., and Young, C. G. (1992). "Seeds of Woody Plants in North America." Dioscorides Press, Portland, Oregon.

3

Physiological Regulation of Vegetative Growth

Introduction

The major internal requirements for growth of woody plants include a supply of carbohydrates, nitrogen, and about a dozen mineral elements, enough water to maintain cell turgor in meristematic tissues and their derivatives, and hormonal growth regulators that aid in integrating physiological processes. The carbohydrates translocated to meristematic regions are converted to cellulose, lignin, pectic compounds, and lipids in the cell walls, and the amino acids and amides are incorporated into the protein framework and enzymes of new protoplasm. Small amounts of lipids are incorporated in cell membranes, and large amounts go into the suberin, cutin, and waxy coatings of leaves, stems, and fruits.

Transport of adequate supplies of metabolites to meristematic tissues where they are used as respiratory substrates and building materials is as important for plant growth as the physiological processes that produce them. In adult trees rapid transport of compounds required for growth occurs simultaneously upward and downward for distances of up to 100 m. The partitioning of metabolites among roots and shoots, cambia, and developing flowers, fruits, cones, and seeds must be closely regulated. An understanding of the tissues involved and factors affecting translocation is basic, not only for an appreciation of the physiology of woody plants, but also for understanding how certain pathogens, insects, and injuries such as girdling damage trees.

The control of vegetative growth involves close interdependency between roots and shoots. Roots depend on leaves for photosynthate and hormonal growth regulators, including auxins and gibberellins. Shoots, in turn, depend on roots for supplying water, mineral nutrients, and certain hormones such as cytokinins. Roots also play an essential role in N metabolism. For example, shoot growth often requires N compounds supplied by the roots (see Chapter 9 in Kozlowski and Pallardy, 1997).

The size of the root system relative to that of the shoot system is important in regulating plant growth. A small root system inhibits growth by limiting the supply of water, mineral nutrients, and certain hormones to the crown. Reduction in the size of the shoot system limits root growth by decreasing the availability of carbohydrates and shoot-produced hormones to the roots.

The root–shoot ratio, which varies among species and decreases with tree age, typically conforms to the allometric growth formula

$$y = bx^k \tag{3.1}$$

where y is the dry weight of the root, x is the dry weight of the shoot, and b and k are constants.

We now discuss separately the importance of carbohydrates, respiration, mineral nutrients, water, and endogenous hormones in regulating vegetative growth. It should be remembered, however, that important interactions occur among the various internal resources in regulating growth.

Carbohydrate Relations

As emphasized in Chapter 7 of Kozlowski and Pallardy (1997), carbohydrates are the most important constituents of woody plants. They comprise about three-fourths of the dry weight of woody plants and are the primary energy storage compounds and the organic substances from which most other organic compounds are synthesized.

Carbohydrate Transport

The amount of carbohydrates translocated in woody plants exceeds that of all other solutes combined. Carbohydrate transport may occur downward, upward, and laterally. When radioactive sucrose is applied to the cambial region, it moves along the rays to the inner sapwood (Höll, 1975). Sugar uptake from xylem vessels by contact ray cells also occurs and is followed by transport along the rays to the phloem (Van Bel, 1990). Lateral movement of sugar occurs almost entirely through the ray symplast (Sauter and Kloth, 1986). Norway spruce needles transported carbohydrates downward in the shoot axis as well as laterally along the rays (Fig. 3.1). Sugars were transported via the leaf trace phloem, then to the base of the stem in the sieve cells of the latest increment of secondary phloem. On the way down sugars moved radially from the sieve cells into phloem parenchyma cells, the cambium, rays, the inner periderm, and some cells of the pith and cortex, including the epithelial cells around resin ducts (Langenfeld-Heyser, 1987).

Most carbohydrates are transported from sources (suppliers) to sinks (utilization sites) during the growing season. However, some carbohydrates may be loaded into the phloem and transported basipetally along the shoot

CORTEX

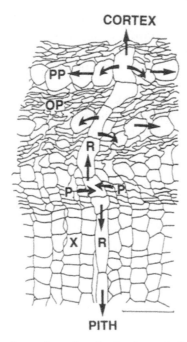

PITH

Figure 3.1   Possible pathway of translocation in winter of photosynthate in the vascular cylinder of a 6-year-old shoot of Norway spruce (see arrows). Simultaneously with long-distance translocation in the latest increment of secondary phloem (P), assimilates are unloaded into the rays (R). Radial translocation may occur either toward the cortex via the phloem rays or toward the pith via the xylem rays. Assimilates may be unloaded into the tangential bands of phloem parenchyma (PP) for temporary storage. OP, Obliterated phloem; X, xylem. From *Trees*, Phloem transport in *Picer abies* (L.) Karst. in mid-winter. I. Micro-autoradiographic studies on ^{14}C-assimilate translocation in shoots. Blechschmidt-Schneider, S., **4**, 179–186, Fig. 11 (1990). Copyright Springer-Verlag.

axis even when meristematic activity is not evident. In Germany, Norway spruce needles were photosynthetically active in January and transported appreciable amounts of carbohydrates out of the needles (Blechschmidt-Schneider, 1990).

Most carbohydrate transport occurs through the phloem. This is shown by stem-girdling experiments, analysis of phloem exudates, and experiments with radioactive tracers. However, some carbohydrates move into sinks from the xylem sap. Much evidence for translocation of carbohydrates in the xylem has been obtained in studies on sap flow in sugar maple (see Chapter 11 in Kozlowski and Pallardy, 1997). In this species, sucrose contents are associated with positive pressure in the xylem or bleeding sap (Cortes and Sinclair, 1985; Johnson *et al.*, 1987).

The flowers of some species, which are produced in late winter and early spring when the sieve tubes are not functional or only beginning to differentiate, receive some carbohydrates from the xylem. For example, the xylem pathway is important in supplying carbohydrates to growing catkins of poplars (Sauter, 1980, 1981) and European white birch (Sauter and Ambrosius, 1986).

In most woody plants, carbohydrates are transported largely or entirely as sucrose; in a few plant families, however, raffinose, stachyose, and verbascose are important constituents of xylem sap. The sugar alcohol sorbitol is found in apple and cherry, and mannitol in ash (see Chapter 7 of Kozlowski and Pallardy, 1997). Reducing sugars do not occur in the phloem sap.

Mechanism of Phloem Translocation   This section briefly addresses aspects of the anatomy of phloem tissues that are germane to an understanding of translocation mechanisms. This is followed by a discussion of the driving forces in translocation of sugars in the phloem.

Sieve Tubes of Angiosperms   In angiosperms, a number of conducting elements of the phloem, the sieve tube members, are joined end to end to form long sieve tubes. The term "sieve" refers to the end walls that are pierced by clusters of pores through which the protoplasts of adjacent sieve tube members are interconnected. Sieve plates, parts of the wall bearing sieve areas with large pores, generally occur on the end walls of sieve tube members (Fig. 3.2).

Protoplasts of young sieve tube members have all the cellular contents that normally occur in living plant cells. During final stages of their differentiation, however, the sieve tube members lose many of their cellular components, including the tonoplast, microtubules, ribosomes, and dictyosomes. In some species the nucleus disappears completely during differentiation; in others remnants of the degenerate nucleus persist for a long time (Evert, 1977). The sieve tube members of many species contain a proteinaceous substance called P-protein or slime (Fig. 3.3), which may occur as filaments distributed throughout the entire cell lumen or may occupy an entirely parietal position.

Small amounts, if any, of callose are present in sieve areas of mature conducting sieve tubes. As sieve tubes begin to senesce, the amount of callose increases, and when sieve tubes finally die and become inactive, their sieve plate areas may have very large amounts of callose. However, callose eventually disappears from old, inactive cells. The walls of sieve tube members commonly are regarded as primary. Sieve tube members are almost invariably accompanied by parenchymatous cells, the companion cells, with which the sieve tubes are structurally and functionally related.

Sieve Cells of Gymnosperms   The conducting elements of gymnosperm phloem, called sieve cells, are similar in shape and size to tracheids. During

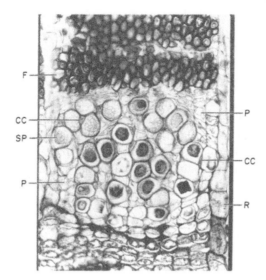

Figure 3.2 Cross section of secondary phloem of black locust. Some of the sieve tube members are sectioned in or near the plane of the sieve plate. The dense bodies in some sieve tube members are slime (P-protein) plugs. P, Parenchyma cell; R, ray; CC, companion cell; F, fibers; SP, sieve plate. Magnification: ×375. Photo courtesy of R. F. Evert.

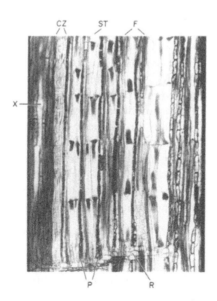

Figure 3.3 Radial section showing portion of secondary xylem, cambial zone, and secondary phloem of American elm. Note the numerous slime (P-protein) plugs lodged against the sieve plates of the individual sieve tube members. The vertical series of sieve tube members constitute the sieve tubes. CZ, Cambial zone; F, phloem fibers; P, phloem parenchyma cells; R, ray; ST, sieve tubes; X, xylem. Magnification: ×138. Photo courtesy of R. F. Evert.

differentiation, the sieve cells become disorganized and lose many of their contents. Sieve cells lack P-protein and have only primary walls, except in the Pinaceae where secondary walls also are present. In gymnosperms, the counterparts of the companion cells of angiosperms are the albuminous cells (also called Strasburger cells).

Longevity of Conducting Elements   In most species of angiosperms, the sieve elements are short-lived and function in translocation only during the season in which they are formed. In a few genera, however, they remain alive and functional for more than 1 year. For example, the sieve tubes of grape and yellow poplar remain functional for at least two seasons, the first summer of activity being followed by an inactive period during the winter. The sieve tubes of basswood may remain functional for 5 to 10 years. The dormant, inactive sieve tubes have heavy masses of callose on the sieve plates and sieve areas. When sieve tubes become active in the spring, the callose gradually dissolves. Phloem reactivation begins first near the cambium. The reactivating phloem can be distinguished from dormant phloem by the thin callose masses and conspicuous P-protein in the reactivating phloem (Esau, 1965). When sieve elements die, their associated companion cells or albuminous cells also die, emphasizing a close relationship between them.

In a number of species of angiosperms and gymnosperms, translocation of carbohydrates can occur very early in the year because the first functional sieve elements are those that formed last during the previous growing season and overwintered as undifferentiated cambial derivatives. These reach maturity early in the spring and are capable of transporting carbohydrates before the cambium is reactivated. Such overwintering sieve elements do not become inactive in translocation until after the newly formed sieve tubes are present and functional. In various evergreen angiosperms and in gymnosperms, some of the phloem remains active in carbohydrate transport for parts of two seasons (Esau, 1965). In some angiosperms, such as box elder and black locust, inactive sieve elements and associated cells collapse, but in others such as walnut, yellow poplar, and basswood they remain unchanged in structure (Tucker and Evert, 1969). The sieve tubes of woody monocotyledons must remain functional for the lifetime of the plants.

Driving Forces in Translocation   The most important translocated compounds such as sucrose accumulate in the phloem parenchyma cells of the leaves and subsequently are transported into the sieve tubes of the phloem in concentrations exceeding those in the surrounding tissues. This is explained by "phloem loading" or secretion by active transport into the phloem by expenditure of metabolic energy. The participation of metabolic energy in phloem loading is shown by reduced sugar transport in the phloem following application of inhibitors of metabolism to source leaves (Geiger and Savonick, 1975).

The rate of sugar transport in the sieve tubes varies with the size of the pool of available carbohydrates and strength of variously located sinks. Rates of translocation of ^{14}C-photosynthate have been reported as 9.6 to 13.2 cm hr^{-1} in black spruce, 6 to 12 cm hr^{-1} in jack pine (Thompson *et al.*, 1979), and 10 to 20 cm hr^{-1} in European larch (Schneider and Schmitz, 1989).

Over the years, several models have been proposed to explain the driving force for movement of sugars in the phloem. These models comprise versions of two basic mechanisms: (1) a metabolic mechanism in the sieve tubes that provides the driving force for carbohydrate transport independent of the water and (2) a primarily passive flow of carbohydrates along a concentration gradient. Several investigators who favored a metabolic mechanism contended that sieve tubes contained blockages and also had filamentous structures in their lumens. However, such structures apparently were artifacts of tissue fixation methods. The structure of sieve tubes, lack of correlation between sugar translocation and petiole metabolism, and demonstration of a physical basis of inhibition of sugar transport in the phloem provide compelling arguments against metabolic mechanisms of long-distance transport of carbohydrates (Giaquinta, 1980).

The pressure flow hypothesis (also called the mass flow hypothesis) is the most widely supported mechanism of phloem transport of carbohydrates (Thorne and Giaquinta, 1984). This involves a passive, bulk flow of sugar solution in the sieve tubes under the influence of a concentration gradient. Pressure flow requires that (1) sieve tubes possess differentially permeable membranes and function as osmotic systems, (2) sieve tubes permit longitudinal symplastic flow, (3) a loading mechanism exists for solutes at the source end and an unloading mechanism at the sink end, and (4) there is a gradient of decreasing turgor pressure from source to sink to act as a driving force for sugar transport.

Active loading creates a driving force by increasing the osmotic concentration in the conducting elements. Water entry follows, increasing the turgor pressure and driving phloem transport. The concentration gradient is maintained by regulated loading of solutes into the sieve element–companion cell complex in source regions and regulated unloading in sink regions (Gifford *et al.*, 1984). Unloading of sugars to a sink may occur with or against a concentration gradient. In many plants, unloading of sugar down a concentration gradient to consumer cells (sinks) occurs without additional expenditure of energy by sieve cells or accessory cells, with the driving force generated by use of solutes by the consumer cells (Geiger and Fondy, 1980).

Annual Carbohydrate Cycles

Woody plants use large amounts of carbohydrates in metabolism and growth (see Chapter 7 in Kozlowski and Pallardy, 1997). Annual trends in use and accumulation of carbohydrates differ among species and genotypes in accordance with their growth characteristics. In many deciduous trees of the temperate zone, the amounts of reserve carbohydrates of stems and branches decrease rapidly during early summer growth, reach a minimum later in the summer, subsequently increase to a peak in the autumn, and decline slowly during the winter (Fig. 3.4). Such patterns have been described for many species including sugar maple (Jones and Bradlee, 1933), gray birch (Gibbs, 1940), apple (Hansen and Grauslund, 1973), and peach (Stassen *et al.*, 1981). However, because of differences among species in hereditary patterns of growth and variations in environmental conditions within and between years, annual carbohydrate cycles often deviate somewhat from the basic pattern described above.

The seasonal duration of leaf retention varies appreciably among species and genotypes, hence influencing accumulation of carbohydrates late in the growing season. In Illinois, black alder retained its leaves and photosynthesized for a month longer than white basswood (Neave *et al.*, 1989). In Wisconsin, hybrid poplars retained green leaves for 2 to 6 weeks longer than native poplar trees did; hence, the hybrids accumulated carbohydrates later into the autumn (Nelson *et al.*, 1982).

In recurrently flushing species (see Chapter 3 in Kozlowski and Pallardy, 1997), some carbohydrates are depleted with each growth flush and replenished between flushes. In cacao, for example, soluble carbohydrates in the mature leaves, stems, and roots decreased during the period of expansion of the flush leaves but accumulated again after the leaves had expanded (Sleigh *et al.*, 1984). Northern red oak seedlings depleted carbohydrates with each of three sequential flushes of shoot growth (Hanson *et al.*, 1988a). Depletion of twig carbohydrates during shoot growth flushes of *Antiaris africana* in Nigeria also has been demonstrated (Olofinboba, 1969).

The seasonal pattern of starch accumulation differs from that for total carbohydrates. Two seasonal maxima of starch have been found in many species, one in the spring and one in late summer or early autumn. Starch accumulates in early spring before budbreak occurs and subsequently decreases as reserves are utilized during the early-season growth flush. As growth slows late in the summer and autumn starch accumulates again. In poplars, starch in the ray parenchyma cells was converted to sugars at the beginning of the dormant season (Witt and Sauter, 1994). The winter decline in starch reflects its conversion to sucrose at low temperatures. The seasonal cycle of starch storage in roots differs from that in stems because the starch in roots is not converted to sucrose during the winter (Cottignies, 1986; Dickson, 1991).

Annual carbohydrate cycles of evergreens differ appreciably from those of deciduous trees (Fig. 3.5). Evergreens accumulate some carbohydrates before seasonal growth starts, accumulate carbohydrates later into the year,

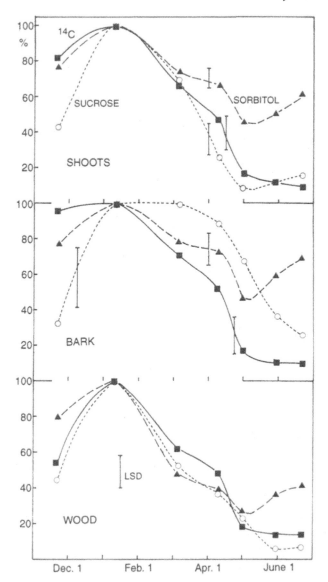

Figure 3.4   Seasonal variations in relative concentrations of soluble ^{14}C (■), sorbitol (▲), and sucrose (○) of shoots, bark, and wood of branches plus trunk of young apple trees. Bars indicate least significant difference (LSD $p = 0.5$). From Hansen and Grauslund (1973).

and show smaller seasonal variations in carbohydrate reserves. Photosynthesis in evergreens begins before shoot, stem, and root growth start. As a result, carbohydrates often accumulate in needles and branches before

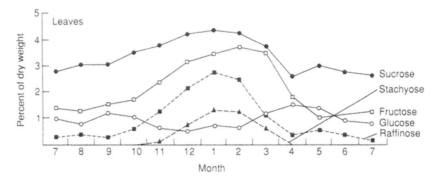

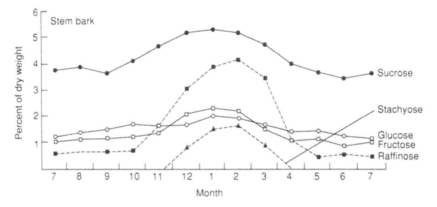

Figure 3.5 Seasonal variations in various carbohydrates in leaves and stem bark of mugo pine in southern Germany. From Jeremias (1964).

budbreak occurs as shown for red pine (Pomeroy *et al.*, 1970) and Norway spruce (Senser and Beck, 1979). The early-season increase in starch in the roots of Scotch pine seedlings coincided with that in the needles, indicating that current photosynthesis was the source of sugars that were converted to starch in both.

Differences in seasonal variations in carbohydrate contents of evergreen and deciduous trees are shown in Fig. 3.6. In California, the carbohydrate contents of branches varied much more in the drought deciduous California buckeye than in the evergreen California live oak (Mooney and Hays, 1973).

Source–Sink Relations

Each woody plant is a highly integrated system of competing carbohydrate sinks. Both the direction and rate of carbohydrate transport are regulated by the placement of vascular connections and the relative strengths of carbohydrate sinks as well as their proximity to stored and to currently produced carbohydrates (Kozlowski, 1992). Hence, an understanding of the direction and rate of carbohydrate transport requires consideration of the rate of growth taking place at various times and the parts of the plant in which meristematic activity is occurring.

Transport of carbohydrates and their use in growth are not uniform in time or space because different parts of trees grow at different rates and at

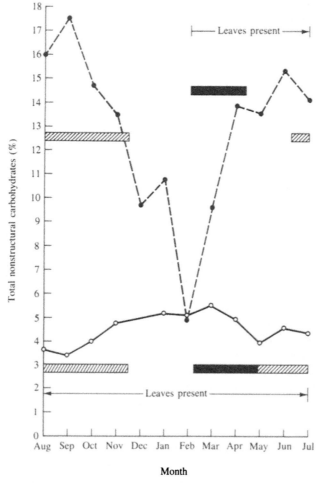

Figure 3.6 Seasonal cycles of total nonstructural carbohydrates in twigs of California buckeye (drought-deciduous, dashed line) and California live oak (evergreen, solid line). Hatched bars, Time of fruiting; solid bars, time of stem growth. From Mooney and Hays (1973).

different times of the year. In many temperate-zone trees, annual root growth starts before shoot growth does, and cambial growth in the lower stem begins later than growth of roots or shoots (see Chapter 3 in Kozlowski and Pallardy, 1997). Hence, the times and rates of carbohydrate transport to supply different meristems may be expected to vary appreciably.

Intense competition within trees for available carbohydrates is shown by alterations in the direction of carbohydrate movement as growth rates of different tissues and organs change (Kozlowski, 1992). In very young apple seedlings, carbohydrates from the cotyledons are first transported to the lower stem and root system, then to the shoot apex, and finally to the root system again as the cotyledons senesce and the first true leaf begins to provide photosynthate for the growing seedling (Li *et al.*, 1985). Competition for carbohydrates is dramatically shown by monopolizing of the available carbohydrates by fruits during years of heavy bearing, resulting in little transport to vegetative tissues and greatly reduced vegetative growth (Kozlowski *et al.*, 1991).

Shoot Growth

Both carbohydrate reserves and current photosynthate are used in shoot growth, the proportion of each varying with species, genotype, type of shoot, and shoot location on the tree.

The sink strength of the many buds on a tree differs widely because individual buds may fail to open, develop into long or short shoots, produce flowers, or die (see Chapter 3 in Kozlowski and Pallardy, 1997). Buds in the upper crown of a tree are more vigorous and are stronger carbohydrate sinks than those in the lower crown. In red pine, many small buds on the lower branches did not open and expand into shoots (Kozlowksi *et al.*, 1973).

Leaf Growth and Metabolism   Carbohydrates are used in both maintenance respiration and growth respiration of leaves. High respiration rates during early leaf growth are correlated with rapid synthesis of chlorophyll, proteins, and protoplasm. The requirement for respiratory energy decreases as the photosynthetic system matures (Dickmann, 1971).

Young expanding leaves import carbohydrates and use them, as well as their own photosynthate, in metabolism and growth. The rate of carbohydrate import increases progressively, even after a leaf becomes photosynthetically active, and reaches a maximum by the time a leaf is 20 to 30% expanded. Transport of carbohydrates into a growing leaf then slows gradually until the leaf begins to export carbohydrates well before it is fully expanded (Kozlowski, 1992). In Norway spruce needles, cell wall thickening was an indicator of needle maturation and masked the beginning of the transition from sink to source properties (Hampp *et al.*, 1994). In both angiosperms and gymnosperms, the transition of leaves from carbohydrate sinks to sources is related to changes in activities of sucrose-metabolizing enzymes. Both invertase and sucrose synthase (which catalyze breakdown of sucrose) decrease, and sucrose phosphate synthase (which catalyzes formation of sucrose) increases (Hampp *et al.*, 1994).

Different parts of leaves mature at different times. Hence, growing leaves often import carbohydrates to immature regions of the blade while at the same time transporting carbohydrates to other parts of the shoot (Dickson and Isebrands, 1991). Most simple leaves mature first at the tip. The photosynthate produced in the tip region is transported out of the leaf through the midvein and petiole. In comparison, the simple leaves of some species (e.g., northern red oak) develop from the base to the tip, and the young tip of the blade imports photosynthate after transport to the base has stopped. Maturation of compound leaves also progresses from the base to the tip (Larson and Dickson, 1986). At certain stages of their development, mature leaflets transport carbohydrates to distal developing leaflets on the same rachis as well as out of the leaf to other carbohydrate sinks.

The conversion of photosynthate to specific chemical compounds changes drastically during leaf development. In young cottonwood leaves, more than half of the recently produced photosynthate was incorporated into proteins and pigments; only 10% was in sugars. As the leaves grew, the percentage of photosynthate that was incorporated into sugars increased linearly. By the time the leaves reached maturity, more than half of the photosynthate was in the sugar fraction (Dickson and Shive, 1982).

Most of the C used by leaves for protein synthesis is derived from photosynthesis, whereas imported sucrose is preferentially used in synthesizing structural carbohydrates. As leaves continue to grow, the proportional use of carbohydrates from these two sources changes, with the amount of current photosynthate used by growing leaves eventually greatly exceeding the amount imported from other sources. In addition to exporting substantial amounts of carbohydrates, leaves that have achieved their maximum surface area use carbohydrates for maturation. For example, fully expanded leaves of northern red oak not only exported carbohydrates but also used some in blade thickening (Dickson *et al.*, 1990).

Variations in Use of Carbohydrates   Both deciduous trees and evergreens use stored carbohydrates, as well as current photosynthate, for shoot elongation. Because deciduous trees lack foliage when shoots begin to expand, their growth depends more on reserve carbohydrates than does that of evergreens. It has been estimated that as much as two-thirds of the carbohydrates used in early-season growth of shoots and flowers of apple trees is supplied by reserves (Hansen, 1977). Later in the season, deciduous trees use current photosynthate for shoot growth.

In species exhibiting free growth of shoots (see Chapter 3 of Kozlowski and Pallardy, 1997) such as poplar and apple, leaves in widely different stages of development occur on the same shoot. The direction of net translocation of carbohydrates from different leaves along a shoot changes as the shoot elongates and adds new leaves. The very young leaves near the shoot apex import carbohydrates from the mature leaves below them. When leaves are partly expanded they both import and export carbohydrates, but the lower fully expanded leaves export but do not import carbohydrates. As an individual leaf expands and the shoot adds new leaves above it, the pattern of carbohydrate export shifts from upward only to bidirectional and eventually, when it is overtopped by several well-expanded leaves, to a predominantly downward direction. As the leaf finally senesces, the rates of both photosynthesis and carbohydrate export decline (Larson and Gordon, 1969; Hansen, 1967b).

As new leaves developed at the base of elongating shoots of bigtooth aspen trees, they initially imported and used stored carbohydrates (Donnelly, 1974). The first-formed leaves began to export photosynthate around June 1 (Fig. 3.7). Initially transport was primarily upward to the shoot tip, but within 2 weeks these leaves were transporting carbohydrates downward toward the main stem. The translocation pattern from leaves located midway on the shoot was similar, but these leaves continued to export photosynthate to the stem tip until early July. The upper, last-

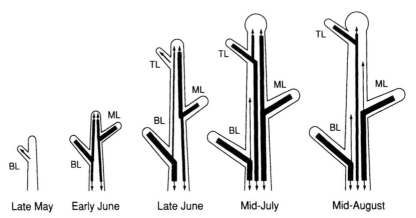

Figure 3.7 Seasonal changes in relative proportion of ^{14}C-labeled photosynthate translocated to the stem tip and lower stem from tip leaves (TL), middle leaves (ML), and basal leaves (BL) on shoots of bigtooth aspen. From Donnelly (1974).

formed leaves did not begin to export significant amounts of photosynthate until late July. In a 15-leaf poplar shoot, the top 5 leaves were expanding and importing photosynthate, the middle 5 leaves were exporting both upward and downward in the shoot, and the bottom 5 leaves were transporting photosynthate to the lower stem and roots (Dickson, 1989). In rapidly growing grape shoots, a young leaf at first imported carbohydrates from the leaves below it. When it was about half-expanded, the leaf began to export carbohydrates to the new unexpanded leaves above it (Figs. 3.8 and 3.9). Such strictly upward transport continued for only 1 or 2 days. Subsequently, some of the photosynthate also moved downward to the fruit. This pattern lasted for only 2 or 3 days, and thereafter translocation was entirely downward. The former function of the leaf of supplying the growing shoot tips had now been assumed by the young leaves located closer to the shoot tips (Hale and Weaver, 1962).

The presence of growing fruits often modifies the pattern of carbohydrate export from leaves. Fruits are powerful carbohydrate sinks and often monopolize available carbohydrates even to the extent of causing a reversal of translocation from wholly downward to bidirectional. When this change occurs, some of the photosynthate is exported upward into fruits from leaves below and downward from leaves higher up on the branch (see Chapter 7). Both Quinlan (1965) and Hansen (1967b) showed such a translocation pattern in apple shoots.

Although most evergreens can carry on photosynthesis throughout the

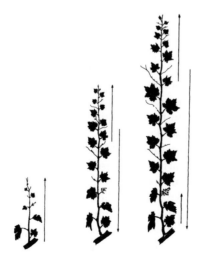

Figure 3.8 Three stages of development of grape shoots showing main direction of translocation of photosynthate from various leaves. From Hale and Weaver. (1962).

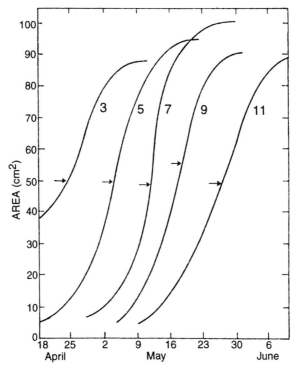

Figure 3.9 Growth curves of five leaves (numbers 3, 5, 7, 9, and 11) of grape showing the date (indicated by arrows) on which the leaves began exporting carbohydrates. From Hale and Weaver (1962).

year, they also use some carbohydrate reserves, in addition to current photosynthate, for shoot elongation. In pines, carbohydrates stored in the old needles as well as in twigs are used in shoot growth (Kozlowski and Winget, 1964a; Kozlowski and Clausen, 1965; Clausen and Kozlowski, 1967a). When red pine seedlings were exposed to $^{14}CO_2$ in late August, after annual shoot elongation stopped, some labeled photosynthate was stored in twigs, stems, and roots. During the next growing season some of the carbohydrates that had been stored in twigs were used in shoot elongation (Olofinboba and Kozlowski, 1973). When the phloem of red pine shoots was severed in early April just below the terminal buds (hence preventing upward movement of reserve carbohydrates into the buds), the buds did not open and expand into shoots (Kozlowski and Winget, 1964a). Significant decreases in the dry weight of old needles of red pine and balsam fir trees as shoots begin to elongate also suggest that translocation of carbohydrates from the old needles occurs and supplements the current photosynthate used during early-season shoot elongation (Clausen and Kozlowski, 1967a, 1970; Loach and Little, 1973).

In addition to using some stored carbohydrates for early-season shoot elongation, northern pines also use current photosynthate supplied at first by the old needles and then by current-year needles after they elongate. In early May the one-year-old needles of red pine were contributing most of their current photosynthate to the expanding buds. Two- and three-year-old needles supplied smaller amounts. The supply of current photosynthate to the new shoots from the three age classes of old needles decreased late in the growing season as proportionally more carbohydrates were synthesized by mature current-year needles (Dickmann and Kozlowski, 1968). Similarly, the old needles of eastern white pine seedlings supplied large amounts of carbohydrates to the expanding new shoots. By mid-July, however, the expanded current-year needles had replaced the old needles as the primary carbohydrate sources for shoot elongation (Ursino *et al.,* 1968).

The amount of carbohydrates used in shoot growth varies among species and genotypes in accordance with their hereditary patterns of growth. Species exhibiting fixed growth (see Chapter 3 in Kozlowski and Pallardy, 1997), which complete shoot expansion in a small proportion of the frost-free season, generally use smaller amounts of carbohydrates for shoot growth than species that exhibit free growth or recurrently flushing growth, with shoots elongating during a large part of the summer (Kozlowski *et al.,* 1991). Some tropical pines use very large amounts of carbohydrates for shoot growth because their shoots grow rapidly and more or less continuously throughout the year (Kozlowski and Greathouse, 1970). Wide genotypic differences in use of carbohydrates for shoot growth are traceable to variations in time of bud opening, rates of shoot growth, and seasonal duration of shoot growth (Kozlowski, 1992).

Cambial Growth

Large amounts of carbohydrates are required for cambial growth. A major use is for diameter increase of branches, stems, and large perennial roots by periclinal division of fusiform initials and of xylem and phloem mother cells, followed by their division and differentiation into xylem and phloem cells. More carbohydrates are used in production of xylem than phloem increments because (1) more xylem than phloem cells are produced annually, (2) production of xylem cells occurs for a longer part of the growing season than does production of phloem cells, and (3) more carbohydrates are used in wall thickening of xylem cells than of phloem cells. As plants increase in size, carbohydrates also are used in expansion of the cambial sheath by anticlinal division of fusiform cambial cells. During expansion of the cambial sheath there is overproduction of cambial initials, followed by death of many of these cells. The surviving initials are those with most ray contacts, suggesting that severe competition occurs for available carbohydrates. Small amounts of carbohydrates are used for production of phellem (cork) and phelloderm tissues by division of the phellogen (cork cambium) (Kozlowski, 1992).

Xylem and Phloem Cell Wall Formation   The walls of most mature xylem cells consist of a thin primary wall and a thick secondary wall (see Chapter 3 in Kozlowski and Pallardy, 1997). Phloem fibers also develop secondary walls, whereas sieve elements, companion cells, and most parenchyma cells of the phloem do not.

Large amounts of carbohydrates are used in formation of secondary walls which includes deposition of cellulose, hemicellulose, and lignin as well as small amounts of proteins and lipids. Cellulose is deposited in cell walls by apposition (see Chapter 3 in Kozlowski and Pallardy, 1997). The cellulose microfibrils apparently are formed at the outer surface of the plasmalemma (Goodwin and Mercer, 1983). Lignin is deposited as a matrix component with hemicellulose in the spaces between cellulose microfibrils (Torrey *et al.*, 1971; Higuchi, 1985). Lignification begins at cell corners in the middle lamella area and progresses through the primary and secondary wall. Mature gymnosperm tracheids and angiosperm vessels are heavily lignified, but libriform fibers of angiosperms contain little lignin. Although the lignin concentration is higher in the middle lamella than in the secondary walls of gymnosperm tracheids, most of the lignin is in the secondary wall, which comprises a higher proportion of the tissue volume.

Formation of cell walls generally involves conversion of monosaccharides to their sugar nucleotide derivatives, followed by their use in synthetic polymerization reactions (Hori and Elbein, 1985). The formation and secretion of cell wall polysaccharides begin with synthesis of membrane-associated transferases in the rough endoplasmic reticulum. The membranes that contain these enzymes move to the dictyosomes of the Golgi apparatus. In the case of matrix polysaccharides, nucleotide sugar translocators transfer the substance for polymerization to the dictyosomes, and polymerization is initiated by glucosyl transferases. Secretory vesicles that contain the newly synthesized polymers bud off from the dictyosomes, move to the plasma membrane, and fuse with it, releasing their contents to the cell wall space (Delmer and Stone, 1988).

The mechanism of polymerization of lignin precursors is not well understood, largely because the specific enzymes involved in such polymerization are uncertain. For a long time the view was widely held that peroxidase was the only cell-wall-associated oxidative enzyme involved in the polymerization step of lignin synthesis. It now appears that laccase, which is bound to lignifying walls and can polymerize lignin precursors (O'Malley *et al.*, 1993), is also involved in lignification (Savidge and Udagama-Randeniya, 1992; Bao *et al.*, 1993).

Variations in Use of Carbohydrates Both reserve carbohydrates and currently produced carbohydrates are used in cambial growth. However, more stored carbohydrates are used for early-season cambial growth of deciduous trees than of evergreens. In many deciduous trees, cambial activity begins below the buds before the leaves emerge and photosynthesis begins; hence, early-season cambial growth depends on carbohydrate reserves. Use of reserve carbohydrates is especially important for phloem production, which in some deciduous species precedes xylem production by several weeks (see Chapter 3 in Kozlowski and Pallardy, 1997). Early-season cambial growth is accompanied by depletion of reserve carbohydrates from storage tissues. As the season progresses, there is a shift toward greater use of current photosynthate in cambial growth. This is shown by decreases in cambial growth during periods of low photosynthesis (Kozlowski, 1982a,b; Creber and Chaloner, 1984) and during diversion of photosynthate to growing reproductive tissues (Kozlowski, 1992).

Although conifers use both reserve and currently produced carbohydrates in cambial growth, the early-season xylem increment depends largely on current photosynthate. When red pine or jack pine seedlings were exposed to $^{14}CO_2$ late in the growing season, only a small proportion of the total uptake was recovered from the trees at the end of winter dormancy, indicating that a large proportion of the reserve carbohydrates was consumed in respiration (Gordon and Larson, 1970; Glerum, 1980). Exposure of conifers to $^{14}CO_2$ while the cambium is active is followed by rapid incorporation of ^{14}C into structural compounds of the newly formed tracheids, further emphasizing the importance of current photosynthate for xylem production (Rangenekar *et al.*, 1969; Dickmann and Kozlowski, 1970a).

Temporal and Spatial Variations in Carbohydrate Use In many species of the cool temperate zone, the use of carbohydrates for xylem production often lasts for 2 to 4 months in the summer and continues for a longer time in evergreens than in broad-leaved trees (Winget and Kozlowski, 1965b). Moreover, even during that short time, the rate at which carbohydrates are used in cambial growth changes because of periodic competition of cambial meristems with stronger carbohydrate sinks and effects of recurrent climatic stresses, insects, and diseases. In contrast, in many tropical trees, carbohydrates are used in cambial growth during much of the year and sometimes even throughout the year (Fahn *et al.*, 1981). However, tropical trees that show cambial increments during each month of the year sometimes produce multiple rings of xylem, emphasizing uneven use of carbohydrates in cambial growth.

Transport and use of carbohydrates for cambial growth at different stem heights also vary. The use of carbohydrates in different parts of the stem varies with the times of initiation and ending of annual xylem production as well as with the number of xylem cells produced. As the dormant season ends, a wave of xylem production moves rapidly from the stem apex toward the base of the stem in ring-porous species. In comparison, in diffuse-porous species, basipetal movement of the cambial growth wave is much slower, and several weeks may elapse between initiation of xylem production at the stem apex and that in the lower stem (Kozlowski, 1971b).

The amounts of carbohydrates used in cambial growth at different stem heights vary with the size of the crown. In dominant trees with large crowns, the annual xylem increment narrows progressively below the crown but thickens near the stem base. In suppressed (overtopped) trees with small crowns, the annual xylem rings are thinner than in dominant trees, the stem height at which the annual xylem increment is thickest is higher, and the thickness of the xylem increment narrows rapidly below this height (see Chapter 3 of Kozlowski and Pallardy, 1997). In severely suppressed trees, there may not be any deposition of xylem in the lower stem (Watzig and Fisher, 1987; Kozlowski *et al.*, 1991).

The use of carbohydrates in xylem production sometimes is very uneven around the stem circumference. This is shown by limited xylem production on certain sides of stems of overmature, heavily defoliated, leaning, or suppressed trees (Kozlowski, 1971b; Kellogg and Barber, 1981).

In conifers, the branch whorls of the upper crown are the primary sources of current photosynthate for cambial growth. In average red pine trees the branches in the upper third of the crown supply most of the carbohydrates used in cambial growth. Such relatively unshaded branches have high photosynthetic capacity as well as a short translocation path to the main stem. The relative importance of carbohydrates produced by other branches varies with environmental conditions, tree age, competition, and other factors (Larson, 1969). In open-grown trees the retention of branches in the lower stem provides an important source of carbohydrates for cambial growth near the base of the stem. Under severe competition among trees in closed stands, natural pruning of lower branches or their lack of vigor confines the carbohydrate source for cambial growth to branches located in the upper stem.

The carbohydrate supply for cambial growth from individual branches decreases as they become increasingly suppressed. In gymnosperms, for example, as the lower branches are overlaid by more and more branch whorls, they become increasingly shaded and their photosynthetic activity declines. Such lower, suppressed branches have less carbohydrates available for export and for cambial growth than do the branches higher up on the stem.

There is a changing seasonal pattern in the importance of different-aged needles as carbohydrate sources for cambial growth. In young red pine trees, most of the earlywood in older stem internodes was produced while the shoots were elongating and during the early stages of needle expansion. The new needles did not contribute an appreciable amount of photosynthate for cambial growth at this time. Rather, carbohydrates exported from the old needles (mostly 1 year old) were used in earlywood production. When the new needles were almost fully expanded, they began to provide large amounts of carbohydrates for xylem production. At the same time, a change occurred in the direction of translocation of carbohydrates from the old needles, with transport now occurring primarily to the roots and moving upward to the new branches. The increased carbohydrate supply following maturation of the new needles was correlated with thickening of tracheid walls, indicative of latewood development. Toward the end of the growing season, the new needles supplied carbohydrates to the new buds and to cambial growth in upper stem internodes, and the old needles supplied carbohydrates for cambial growth of lower stem internodes. Late in the season, when the cambium was no longer a significant carbohydrate sink, all age classes of needles supplied carbohydrates to the roots, and reserve carbohydrates accumulated in parenchyma tissues throughout the tree (Gordon and Larson, 1968; Larson and Gordon, 1969).

Root Growth

Both the long-lived perennial roots and the short-lived fine roots are strong carbohydrate sinks. Carbohydrates are used in metabolism, elongation, and thickening of roots and in formation and subsequent growth of new roots. The rate and seasonal duration of carbohydrate use for root growth vary with species, genotype, age and size of plants, type of root, site, and environmental conditions (Kozlowski, 1992).

The amounts of carbohydrates allocated to root growth vary greatly on different sites. Belowground net primary production in Douglas fir forests ranged from 23 to 46% of total net primary production, depending on silvicultural treatment. Production of coarse roots comprised 37 to 54% of total belowground net primary production. Small roots made up 9 to 29% and very fine roots 29 to 60% of belowground net primary production (Gower *et al.*, 1992). Estimates of the C content of fine root production on 43 forest sites ranged from 25 to 820 g m^{-2} year^{-1}. More photosynthate was used in root respiration than in growth (Nadelhoffer and Raich, 1992). Some studies showed that more carbohydrates were used in growth of fine roots than in growth of large perennial roots. This difference may be inferred because, although the fine roots may comprise less than 1% of the biomass of a tree at any given time, their short lives and recurrent replacement (see Chapter 3 in Kozlowski and Pallardy, 1997) could account for up to two-thirds of biomass production (Grier *et al.*, 1981). The sink strength of fine roots was stronger on poor than on good sites (Pregitzer *et al.*, 1990; Nguyen *et al.*, 1990).

Both reserve carbohydrates and current photosynthate are used for root growth. In addition to their role in maintaining root respiration, reserve carbohydrates are important for early-season growth of existing roots. This is emphasized by the beginning of root growth of deciduous trees before the leaves unfold and become photosynthetically active. When Sitka spruce transplants were transferred from a nursery in Scotland to a growth chamber, they produced new roots even when the stems were girdled so the roots did not receive carbohydrates from the shoots. This further emphasized that stored carbohydrates were used in root growth. However, currently produced shoot metabolites also were required because the amount of root growth was greatly reduced by girdling of stems (Philipson, 1988).

Growth of perennial roots depends on current photosynthate in the latter part of the growing season. The roots of sugar maple were depleted of starch during shoot growth in May (Wargo, 1979). Starch began to reaccumulate in late June, and by late July the roots were refilled with starch. Continued cambial growth of roots did not deplete their starch reserves, indicating that they were using current photosynthate in metabolism and growth.

Root regeneration in young seedlings utilizes largely current photosynthate. Douglas fir seedlings that had been girdled or defoliated did not regenerate roots even though they had substantial carbohydrate reserves (Ritchie and Dunlap, 1980). Over short periods, the growth of roots of first-year silver maple seedlings was correlated with their photosynthetic rates (Richardson, 1953). When the rate of photosynthesis was lowered by shading, root growth was reduced within a day. In comparison, root growth of second-year seedlings depended more on reserve carbohydrates (Richardson, 1956a,b).

Much interest has been shown in the root growth potential (RGP) of tree seedlings because successful growth of transplanted trees depends on rapid regeneration of roots (Kozlowski *et al.*, 1991). Some investigators associated high RGP with abundant carbohydrate reserves (e.g., Stone and Jenkinson, 1970). Other studies did not associate higher RGP with amounts of stored carbohydrates. Douglas fir seedlings that had been girdled or defoliated did not regenerate roots even though they had substantial carbohydrate reserves (Ritchie, 1982).

Rooting of Cuttings   The use of carbohydrates in rooting is shown by depletion of carbohydrates from the basal ends of cuttings when photosynthesis or downward translocation of carbohydrates is inhibited. When starch is present it is the primary source of carbohydrates for initiation and development of root primordia.

Reserve carbohydrates were used in rooting of cuttings of conifers that had been stored over winter. The carbohydrate contents and rooting were linearly correlated (Behrens, 1987). Seasonal variations in rooting also were correlated with carbohydrate concentrations in the bases of cuttings of olive (del Rio *et al.*, 1991).

Over a range of intermediate amounts of carbohydrates in cuttings, it often is difficult to determine the causal role of carbohydrates in rooting. Despite the essentiality of carbohydrate for rooting of cuttings, they alone do not appear to control rooting. Often the carbohydrate contents of cuttings and rooting capacity are poorly correlated, as in poplar (Okoro and Grace, 1976), mesquite (De Souza and Felker, 1986), and obeche (Leakey and Coutts, 1989).

Some minimal amount of carbohydrate is necessary for rooting of cuttings, and when carbohydrate contents are below this threshold level, the growth and development of roots stop. When stock plants are depleted of reserve carbohydrates and cuttings from these plants are planted under conditions that suppress photosynthesis, the energy charge usually is too low for rooting to occur (Veierskov, 1988). For optimal rooting, the carbohydrate content of cuttings must be adequate to supply energy throughout the period of rooting during photosynthesis-limiting conditions. This has been shown by inhibiting photosynthesis by shading, defoliating cuttings, or use of genetic mutants (Veierskov, 1988).

Removal of all leaves of *Terminalia spinosa* cuttings prevented formation of any adventitious roots, emphasizing the importance of photosynthesis for rooting. In this easy-to-root species, variations in leaf area had little effect on the final rooting percentage. This contrasted with *Triplochiton scleroxylon* and *Khaya ivorensis*, in which optimum leaf areas for rooting were 50 cm^2 and 10 to 30 cm^2, respectively (Newton *et al.*, 1992). Studies with leafless hardwood cuttings (Vieitez *et al.*, 1980) and etiolated stock plants (Pal and Nanda, 1981) show that rooting is inhibited when the initial carbohydrate content is low.

As emphasized by Haissig (1986), determination of the regulatory role of carbohydrates in rooting of cuttings is difficult because (1) quantitative correlations between carbohydrate contents of cuttings and rooting capacity may not reflect a cause and effect relation; (2) whereas light intensity influences the carbohydrate content of cuttings, it also may induce effects that influence rooting capacity independently of carbohydrate levels; (3) data from studies in which carbohydrates are supplied to cuttings may be incorrectly interpreted; (4) assumptions that the only roles of carbohydrates are as sources of energy and C skeletons may be incorrect; and (5) variations in the physiology of stock plants may cause changes in the amounts and types of compounds available for metabolism. The complexity of the regulation of rooting of cuttings is discussed in pages 156 to 161.

Mycorrhizae   Mycorrhizal fungi are strong carbohydrate sinks, using large amounts of carbohydrates for metabolism and growth of the fungal biomass. The strong sink strength of mycorrhizal fungi is related largely to high respiration rates associated with their large amounts of cytoplasm and mitochondria as well as very active enzyme systems (Barnard and Jorgensen, 1977; Smith and Gianinazzi-Pearson, 1988).

The strong sink strength of mycorrhizae is shown by rapid basipetal translocation of ^{14}C-photosynthate to the cortex, Hartig net, and mantle of mycorrhizal hyphae surrounding the roots (Cox *et al.*, 1975; Bauer *et al.*, 1991). Whereas shoots of nonmycorrhizal red pine plants exported only 5% of their photosynthate to the roots, those of mycorrhizal plants exported 54% (Nelson, 1964). It has been estimated that 6 to 10% more photosynthate is used by mycorrhizal roots than by nonmycorrhizal roots (Snellgrove *et al.*, 1982; Koch and Johnson, 1984).

Developing Forest Stands

Net primary productivity increases with age of forest stands to a maximum near the time of canopy closure and generally declines thereafter (Kozlowski *et al.*, 1991). Such changes are correlated with progressive changes in carbohydrate production and partitioning as well as in respiration. As a stand grows, production of photosynthate typically increases to a maximum at canopy closure and stabilizes or declines slightly thereafter. The decrease in aboveground productivity after canopy closure is correlated with increases in maintenance respiration and a proportional increase in allocation of carbohydrates to fine roots and mycorrhizae (Vogt *et al.*, 1982; Cannell, 1989; Luxmoore *et al.*, 1995).

Growth Respiration

Respiration is required for an increase in plant biomass in the form of cellulose, hemicellulose, lignin, protein, and other compounds derived from photosynthetically fixed carbon (see Chapter 6 in Kozlowski and Pallardy, 1997).

Direct relationships between two measures of respiration (metabolic heat rate and CO_2 production rate) and growth have been shown. Metabolic heat rates of five clones of coast redwood began to increase in early May, peaked in mid-July, and declined to a low level in September. Height growth and average metabolic heat rate of all clones over a 6-week period were highly correlated. The fastest growing clone increased 39 cm in height and had an average metabolic heat rate of 22.8 μW mg^{-1}. Trees with a metabolic heat rate below 10 μW mg^{-1} grew negligibly or not at all. This inhibition of growth suggested that approximately 10 μW mg^{-1} was the maintenance respiration rate (Anekonda *et al.*, 1993). In another study of 192 genotypes of coast redwood, several growth traits (height, basal diameter, and stem volume) were highly correlated with metabolic heat rates and rates of CO_2 production (Anekonda *et al.*, 1994). The data suggested that respiration rates of juvenile trees could be used to predict their longer term growth rates.

Mineral Relations

Mineral nutrients participate in many essential functions in plants, especially as constituents of plant tissues, osmotic regulators, controllers of membrane permeability, coenzymes, and activators or inhibitors of enzyme systems (see Chapter 10 in Kozlowski and Pallardy, 1997). Mineral deficiencies occur commonly and often limit growth and development of plants.

The annual mineral requirements of woody plants are met in part by internal redistribution, in part by return from plants to the soil followed by reabsorption, and in part by uptake from soil reserves (Attiwill, 1995).

Mineral Transport

Practically all movement of mineral nutrients from roots to shoots occurs in the xylem sap, carried by mass flow in the transpiration stream. There also is considerable lateral movement of minerals between the xylem and phloem, and minerals diffuse out of the xylem during ascent, presumably because of their use in meristematic regions along the way. Figure 3.10 demonstrates such movement. The cells of stem tips, cambial regions, and other metabolically active regions all accumulate minerals.

Upward Movement Data on the amounts and proportions of various minerals in the xylem sap are given in Table 3.1. The rate of transport of minerals is controlled chiefly by the rate of transpiration and therefore is much slower in defoliated trees than in those with leaves. Not all mineral elements move upward in the xylem at the same rate, with movement of some ions apparently being slowed by adsorption on the walls of xylem elements. Examples of slow-moving ions are Ca^{2+}, Fe^{2+}, and Zn^{2+}. Also, some kinds of plants apparently retain more of certain ions in their roots than others (Ehlig, 1960). As pointed out in Chapter 10 of Kozlowski and Pallardy (1997), cells are highly selective with respect to ion accumulation and root systems are fairly selective, but at least traces of almost every ion in the root environment leak through the ion barriers and occur in the free space or apoplast of leaves.

Translocation of Foliar-Absorbed Nutrients The rate of movement of nutrients deposited on leaves from the atmosphere or applied in foliar sprays

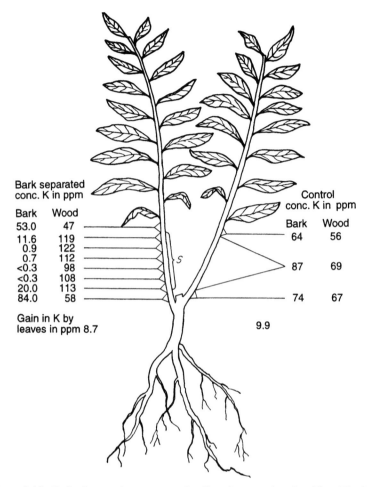

Bark separated
conc. K in ppm

Bark	Wood
53.0	47
11.6	119
0.9	122
0.7	112
<0.3	98
<0.3	108
20.0	113
84.0	58

Gain in K by
leaves in ppm 8.7

Control
conc. K in ppm

Bark	Wood
64	56
87	69
74	67

9.9

Figure 3.10 Path of upward movement of radioactive potassium in willow. The bark of the left-hand branch was separated from the wood by a layer of paraffined paper, whereas the right-hand branch was left intact. Very little potassium was found in the bark where it was separated from the wood, but large amounts were present where it was in normal contact. This experiment indicates that upward movement occurred chiefly in the xylem but that lateral movement from wood to bark also occurred. From Stout and Hoagland (1939).

varies greatly with the mobility of different ions. Whereas ions of P, K, Na, S, Cl, and Rb are highly mobile, ions of Zn, Cu, Mn, Fe, and Mo are partly mobile, and ions of Mg, Ca, Sr, and B are considered immobile (Bukovac and Wittwer, 1957, 1959). Chelated forms of mineral nutrients are widely used for foliar applications because they are more mobile than inorganic sources.

Translocation of P and N varies with the form in which they are applied. Urea N was transported more rapidly than were other forms. However, translocation of urea N was less effective when applied to the foliage than to the soil (Forshey, 1963). Translocation of N also varies with leaf development and is especially high when N sprays are applied shortly before or during leaf abscission.

Calcium ion is not readily translocated out of leaves. Among the factors that favor export of Ca^{2+} from leaves are leaf injury, high humidity, high Ca^{2+} concentration in applied sprays, presence of other divalent cations in the spray, Ca^{2+} chelation, and a volume of spray solution greater than the volume transpired during uptake by leaves (Hanger, 1979). Translocation of foliar-applied nutrients is discussed in more detail by Bukovac and Wittwer (1959) and Swietlik and Faust (1984).

Resorption of Minerals Trees typically accumulate mineral nutrients in amounts that exceed their current requirements for growth. Mobile nutrients, especially N, P, and K, are stored in leaves, branches, stems, and roots and later used in growth at all stages of development from seedlings to adult trees (Luxmoore *et al.*, 1981; see Chapters 9 and 10 in Kozlowski and Pallardy, 1997). Resorption, storage, and remobilization of mineral nutrients permit woody plants to be relatively independent of the pool of soil nutrients during periods of high mineral requirements. During the early summer flush of growth, when requirements for mineral nutrients are high, the nutrients stored in plants probably account for higher growth

Table 3.1 Concentrations of Solutes in Sap of Xylem and Phloem of Red Ash[a,b]

	Xylem sap (mg/ml)	Phloem sap (mg/ml)
Dry matter	1.4	220
Sucrose	0.128	140
Potassium	0.177	2.1
Magnesium	0.0115	0.077
Sodium	0.004	0.037
Calcium	0.018	0.049
Phosphorus	0.005	0.052
pH	4.9–5.0	7.5

[a]From van Die and Willemse (1975).

[b]Sieve tube exudate and xylem sap were collected for analysis on October 10.

rates than would be possible if only the nutrients in the soil pool were used. Resorption may also decrease fluctuations in the annual growth of plants, which depends on mineral nutrients resorbed during the prior growing season as well as uptake from the soil during the current season (Ryan and Bormann, 1982). By mobilizing leaf N and P and storing them in nearby twigs and branches where they are readily available for foliar expansion the following spring, bald cypress trees conserved these nutrients, which were not readily supplied from external sources (Dierberg and Brezonik, 1983).

The leaves are important storage organs for mineral nutrients. For example, approximately 65% of the annually absorbed N in aboveground tissues was stored in the needles of Monterey pine; only 15 and 20% were stored in the branches and stems, respectively (Beets and Pollock, 1987a,b). The amount of internal retranslocation of mineral nutrients is regulated by the rates of nutrient absorption from the soil and by plant growth. Because retranslocation is driven more by shoot growth than by soil nutrient supply, shoots continually compete for stored minerals (Nambiar and Fife, 1991).

A large portion of the mineral nutrients in leaves moves back into twigs and branches during a short period of leaf senescence before the leaves are shed (Fig. 3.11). Reports of autumnal N translocation out of leaves of species that do not fix N range from 40 to 80% of their leaf N for apple, birch, poplar, willow, and basswood. During a 3- to 4-week period, 16% of the dry matter, 52% of the N, 27% of the P, and 36% of the K were depleted from senescing apple leaves (Oland, 1963). Retranslocation of mineral nutrients from leaves to twigs in a chestnut oak forest exceeded the amounts lost from trees by leaching or those recycled in the litter. The backflux of mineral nutrients from the leaves to the twigs of bald cypress trees exceeded the amounts lost in litterfall and throughfall (Dierberg *et al.*, 1986). Resorption of N, P, and Cu by senescing leaves of trembling aspen averaged 43, 51, and 10%, respectively (Killingbeck *et al.*, 1990). In contrast, Al, B, Ca, Fe, Mg, Mn, and Zn concentrations either did not change in leaves or increased during leaf senescence. Resorption of N, P, and Cu was strongly influenced by timing of leaf abscission and was lowest in plants that lost their leaves earliest and in leaves that senesced earliest on individual plants. In Scotch pine stands in Finland, between 70 and 85% of the N, P, and K and between 30 and 60% of the Mg were retranslocated from senescing needles (Fig. 3.12).

It has been suggested that species adapted to infertile sites resorb proportionally more minerals from leaves and use nutrients more efficiently than species growing on nutrient-rich soils (Vitousek, 1982). For example, resorption of N varied from 4.6% for alder on nutrient-rich sites to 76.7% for pine on nutrient-poor sites (Stachurski and Zimka, 1975). However, other studies did not find such differences. For example, a constant proportion of N was reabsorbed from senescing loblolly pine needles regard-

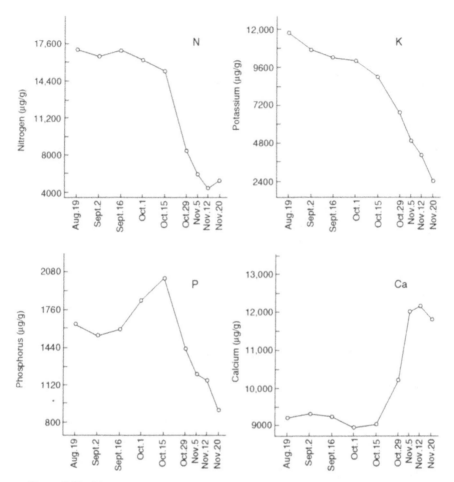

Figure 3.11 Mean concentration of N, K, P, and Ca expressed as micrograms per gram during leaf senescence of chestnut oak. From Ostman and Weaver (1982).

less of the level of available N (Birk and Vitousek, 1986). Del Arco *et al.* (1991) also showed that site fertility was not related to efficiency of mineral resorption from leaves. As emphasized by Birk and Vitousek (1986), it is difficult to make meaningful comparisons of various studies on resorption efficiency because some investigators reported mineral retranslocation on a concentration basis, others on a per leaf basis, and still others on a whole canopy basis.

Large variations have been shown among species in nutrient resorption, and only a few examples are given here. In a chestnut oak forest, 78.5% of the N was resorbed from the leaves (Ostman and Weaver, 1982). Other

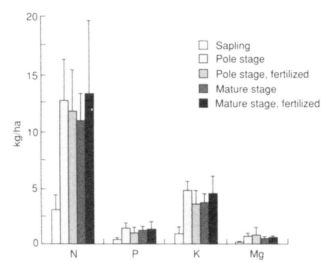

Figure 3.12 Mean resorption of nutrients (kg ha^{-1}) during needle senescence in Scotch pine stands in 1983 to 1987. Vertical bars represent standard deviations of the mean of 5 years. From Helmisaari (1992).

estimates for resorption of N were 39% in loblolly pine (Switzer and Nelson, 1972), 33% in northern hardwoods (Bormann *et al.*, 1977), and 61% in cottonwood (Baker and Blackman, 1977). Resorption of K amounted to 63% in chestnut oak (Ostmann and Weaver, 1982), 22% in loblolly pine (Switzer and Nelson, 1972) and 66% in bald cypress (Schlesinger, 1978).

Negi and Singh (1993) studied seasonal changes in the N content of leaves of 26 species of evergreen and deciduous trees of the central Himalaya. The foliar N contents of all species decreased continuously during their leaf life span, at first because of dilution caused by increased leaf growth and subsequently by resorption as the leaves senesced. Resorption of N mass varied from 33 to 75%, but trees with N concentrations below 2.5% in mature leaves did not retranslocate more than 60% of their N. The period during which N concentration of leaves declined because of resorption varied between deciduous and evergreen trees. Most resorption usually occurred in the autumn in deciduous trees and early summer in evergreens. Much of the N decrease in leaves of evergreens occurred as new leaves formed, which presumably received N from the senescing leaves.

Autumnal fluxes in leaf and bark N pools of N-fixing alder trees were substantially lower than in other winter-deciduous broad-leaved trees (Côté and Dawson, 1986). In the final stages of leaf senescence, the total leaf N concentration was approximately halved in eastern cottonwood and white basswood but decreased much more slowly in black alder.

Leaves in the upper crown of European beech trees were much more efficient in translocation of N than leaves in the lower crown, possibly reflecting the longer period of senescence of sun leaves over shade leaves. The higher leaf temperatures in the upper crown also may promote hydrolysis of organic N and P compounds and their translocation out of leaves (Staaf and Stjernquist, 1986).

Recirculation of Minerals Although the xylem sap is the primary source of minerals and N for growing regions in shoots, recirculation or recycling in the phloem also plays an important role in mineral nutrition (Fig. 3.13). Thus growing leaves may receive minerals from both the xylem and phloem. Continual movement of minerals into leaves in the transpiration stream tends to cause accumulation of ions as the season progresses. This sometimes causes injury to leaves of plants that are too heavily fertilized or are growing in saline soil. Elements such as Ca, Mg, and Na in saline soil usually accumulate as the growing season progresses, but the concentrations of such mobile elements as N, P, and K often decrease, when ex-

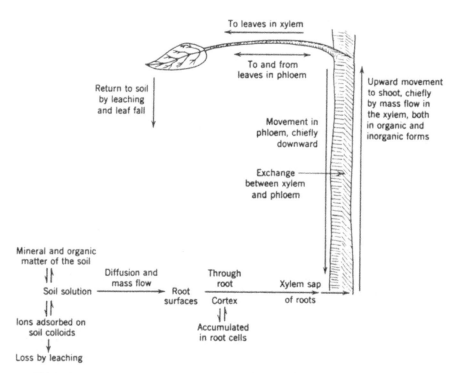

Figure 3.13 Circulation of minerals within a tree and their return to the soil by leaching and leaf fall. From Kramer (1969). Copyright 1969; used with permission of McGraw-Hill Book Company.

pressed as percentage of dry weight (see Fig. 3.11). There is a general tendency for the more mobile elements to move out of older tissues, which are less active metabolically, to young leaves, growing fruits, and other metabolically active regions (Williams, 1955).

In addition to resorbed minerals, small amounts of ions are lost from the free space or apoplast of healthy leaves by the leaching action of rain (see Chapters 9 and 10 of Kozlowski and Pallardy, 1997). Some plants lose salt through secretion from salt glands, and some is lost in the guttation water forced out of hydathodes by root pressure. Experiments with radioactive tracers such as ^{32}P show that, when supplied to the leaves, P often moves down to the roots in the phloem, where some leaks out and some is transferred to the xylem and moves upward (Kramer and Kozlowski, 1979).

The mobility of various ions in the phloem varies widely. Whereas ions of K, Rb, Na, Mg, P, S, and Cl are highly mobile, ions of Fe, Mn, Zn, Cu and Mo are intermediate, and ions of Ca, Li, Sr, Ba, and B are relatively immobile (Bukovac and Wittwer, 1957). However, the mobility of ions in the intermediate group varies with species, stage of plant growth, and amount of the element in the plant. For example, Cu, Zn, and S are mobile only when the concentration is high, and mobility is low in plants deficient in these elements. The concentration of immobile elements may be adequate in old leaves while young leaves in the same plant show deficiency symptoms.

Conversely, freely mobile elements such as N, P, and K often move out of older leaves, causing deficiency symptoms, while the young leaves do not show such symptoms. This situation must be taken into account when leaves are sampled to test for deficiency symptoms. Recirculation of minerals is discussed further in Chapters 9 and 10 of Kozlowski and Pallardy (1997).

Mineral Requirements

Woody plants use both stored and recently absorbed minerals for growth. During the first 6 weeks of growth of Sitka spruce in the spring, stored N was used for growth of new foliage. Subsequent growth depended on absorption of N by roots (Millard and Proe, 1992).

Achieving the optimal level of mineral nutrition in plants is complicated because at least 12 essential elements are involved, a variety of environmental factors influence availability and absorption of mineral nutrients, uptake and use of mineral nutrients vary among species and genotypes, and nutrients interact (May and Pritts, 1993).

Nutrient interactions are well known. Interactions occur between macronutrients, micronutrients, and various combinations. Both temperature and Fe influence P-related Zn disorders (Adams, 1980). Furthermore, P interacts with B. Boron is required for root elongation which influences absorption of P (Pollard *et al.*, 1977). However, P also affects uptake of B (Bingham and Garber, 1960). Although applied nutrients had important effects on yield of strawberries, the responses varied with levels of other mineral nutrients and with soil pH (May and Pritts, 1993). At a soil acidity of pH 5.5, yield varied linearly with B and quadratically with P. At a soil acidity of pH 6.5, P interacted with both B and Zn.

The amounts of mineral nutrients needed by growing plants vary greatly with the stage of plant development and growth rate. During seed germination and very early seedling development, when growth depends on minerals stored in seeds, the effects of fertilizers are limited. During the subsequent exponential growth period, seedling growth is very responsive to fertilizers. When the exponential period of growth stops and the relative growth rate (RGR) decreases because of self-shading and plant aging, the RGR is not increased by higher external concentrations of nutrients. At very high external or internal concentrations of nutrients, RGR may decline precipitously (Figs. 3.14 and 3.15).

As emphasized by Ågren (1985b), investigators of plant nutrition often overlooked the fact that mineral nutrients in the plant, and not necessarily the amount of fertilizer added, dominated control of plant growth. Up to an optimum, plant growth depends on the internal concentration of mineral nutrients. If mineral nutrients are supplied at a constant rate during the exponential period of growth, the internal nutrient concentration will

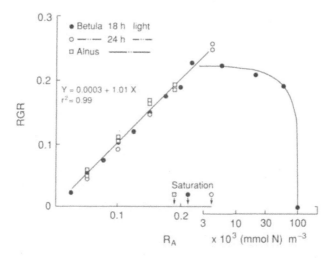

Figure 3.14   Relation between external N supply and relative growth rate (RGR). The relative addition rate (R_A) of N is in the suboptimum and optimum range; the external concentration is in the supraoptimum range. The linear relation within the suboptimum and optimum range is independent of species or day length. From Ingestad (1982).

decrease with time, and growth consequently will decline. Ideally, therefore, as plants increase in size, the supply of mineral nutrients should be progressively increased to maintain an optimal internal nutrient concentration, either by increasing the fertilizer dosage or by applying fertilizer more often.

Ingestad and Lund (1979, 1986) introduced the concept of "relative addition rate" (R_A) of mineral nutrients. The parameter R_A was expressed as the amount of nutrients to be added per unit of time in relation to the amount of nutrients in the plant. During the exponential period of growth, strong linear relations were shown for many plant species between (1) relative growth rate (RGR) and R_A (Fig. 3.14), (2) internal nutrient concentration and R_A, and (3) RGR and internal nutrient concentration (Fig. 3.15; Ingestad, 1980, 1981, 1982, 1987).

Ingestad (1988) developed a fertilization model that incorporated two concepts: (1) nutrient flux density in the soil (amount of nutrient available per unit of soil and unit of time) and (2) nutrient productivity (growth rate per unit of nutrient in the plant). Nutrient productivity was used to calculate the nutrient uptake rate necessary to maintain an optimal internal nutrient content and which, therefore, was necessary to be matched by the nutrient flux density.

Figure 3.14 emphasizes the positive effects of higher external concentrations of nutrients on plant growth because of increased R_A as well as negative effects of high concentrations. The growth of red pine seedlings

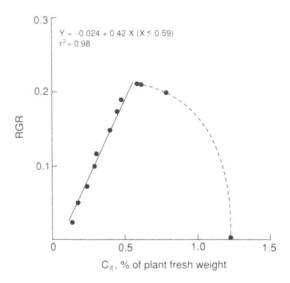

Figure 3.15 Relation between internal N content (C_{it}) and relative growth rate (RGR) of silver birch. From Ingestad (1982).

in containers was increased more by fertilizer application at an exponential rate than a constantly increasing rate (Timmer and Armstrong, 1987). Superior seedlings were grown by supplying fertilizer by the R_A technique, with only 25% of the fertilizer dose conventionally used for producing seedlings in containers. The appropriate R_A rates will vary for species with different growth rates. For example, to achieve maximum growth rates, nutrients were added at a R_A of 19% day^{-1} for seedlings of *Populus simonii* and 27% day^{-1} for *Paulownia tomentosa* (Jia and Ingestad, 1984).

Shoot Growth

Deficiencies of any of the essential mineral elements inhibit shoot growth, especially leaf growth, even before visible symptoms are evident. The reduced leaf area of mineral-deficient plants results from fewer leaves, smaller leaves, and more leaf shedding than in plants with adequate mineral supplies. The leaves of mineral-deficient plants tend to be chlorotic and sometimes have dead areas at the tips and margins or between the veins. Sometimes the shoots develop in tufts or rosettes and also may show dieback (Fig. 3.16; Kozlowski and Pallardy, 1997).

The most commonly observed symptom of mineral deficiency is chlorosis, caused by interference with chlorophyll synthesis. It varies in the pattern, the extent to which young and old leaves are affected, and the severity, according to the species, the specific deficient element, and the degree of deficiency. Chlorosis most often is associated with lack of N, but it also is caused by deficiencies of Fe, Mn, Mg, K, and other elements. Furthermore, numerous unfavorable environmental factors induce chlorosis, including an excess or deficiency of water, unfavorable temperature, air pollution, and an excess of minerals. Chlorosis is also caused by genetic factors, which produce plants ranging from albinos, totally devoid of chlorophyll, to those showing various degrees of striping or mottling of leaves. The wide variety of factors producing chlorosis suggests that it is caused by general disturbances of metabolism as well as by deficiency of a specific mineral element.

One of the most troublesome and common types of chlorosis is found in a wide variety of woody plants growing on alkaline and calcareous soil (usually with Ca contents > 20%) (Marschner, 1986). This type of chlorosis generally is caused by unavailability of Fe at high pH, but sometimes by Mn deficiency. The most severe lime-induced chlorosis occurs on fine textured, poorly aerated, cold soils where conditions are unfavorable for mineral uptake. Drought also can increase chlorosis caused by Fe deficiency. In leaves of Fe-deficient angiosperms the midrib and smaller veins remain green while the interveinal areas become pale green, yellow, or even white. The youngest leaves usually are most severely affected. In conifers, the young needles are pale green or yellow, and if the deficiency is severe, the needles often turn brown and are shed. Interpretation of Fe chlorosis has been confused by statements that the concentration of Fe in chlorotic tis-

Figure 3.16   Boron deficiency in peach showing bunched, distorted, and chlorotic leaves. From Childers (1961).

sue was as high or higher than in healthy tissue. Many of the high Fe contents reported are caused by surface contamination with Fe and are greatly reduced if the leaf surfaces are washed before analysis (Stone, 1968).

Sensitivity to lime-induced chlorosis varies among species. In general among deciduous fruit trees, peach and pear are most sensitive, with sweet cherry, plum, apricot, apple, and sour cherry showing progressively less sensitivity. There also are wide differences in susceptibility of different cultivars, rootstocks, and ecotypes. Some grape cultivars are susceptible to lime-induced chlorosis, but the symptoms can be overcome by grafting to chlorosis-resistant rootstocks (Bavaresco *et al.*, 1993). When calcareous ecotypes of the eucalypt manna gum were subjected to Fe stress, they developed chlorosis later, grew new roots faster, and removed more Fe from solution when compared with an acid ecotype (Ladiges, 1977).

Chlorosis also may occur in trees growing on acid soils. Van Dijk and Bienfait (1993) reported an unusual type of chlorosis in Scotch pine trees growing on acidic soils in the Netherlands. The current-year needles were yellowish, and the discoloration was most pronounced at the bases of needles. The Fe content of chlorotic needles was approximately half that of green needles in the same stands or of trees in healthy stands.

Cambial Growth and Productivity

In mineral-deficient plants, cambial growth is reduced in accordance with lowered downward transport of carbohydrates and hormonal growth regulators. This is emphasized by increases in diameter growth following application of fertilizers to trees and shrubs on infertile sites (see pp. 471–476 of Kozlowski *et al.*, 1991).

Many studies show that addition of fertilizers to infertile sites increases shoot growth (especially foliage mass), which is followed by increased cambial growth. In 20-year-old Douglas fir trees, N fertilization increased leaf size, the number of leaves per shoot, and the length and number of branches (Brix and Ebell, 1969). On N-deficient sites, the leaf area index (LAI) increased up to 60% following N fertilization (Vose and Allen, 1988). The increases in leaf growth of loblolly pine were greatest in the middle and lower parts of the crown (Fig. 3.17), probably because shading in these positions inhibited production and survival of foliage.

The importance of leaf surface area to stem growth is emphasized by high correlations between LAI and productivity. For example, strong positive correlations between LAI and productivity were found for fertilized Douglas fir (Binkley and Reid, 1984) and jack pine trees (Magnussen *et al.*, 1986).

The increases in productivity following fertilization generally are delayed in accordance with the rate of increase in leaf mass. Although heavy fertilization increased the needle mass of Douglas fir trees by 90% after 7 years, needle production did not peak until 2 to 3 years after fertilization, and maximum foliage mass was not recorded until 4 to 7 years after treatment (Brix, 1981).

Productivity of forest stands is positively correlated with LAI only up to some critical value. At higher LAI values, net productivity often decreases, largely because of tree mortality and reduced photosynthetic efficiency of the lower leaves. For example, net primary productivity of a Douglas fir forest peaked at about half the maximum LAI, and it decreased at higher values as growth losses from tree mortality exceeded the small increments in growth of the remaining trees (Waring, 1983).

Root Growth

Growth of roots is inhibited as downward translocation of carbohydrates and hormonal growth regulators is reduced in mineral-deficient plants.

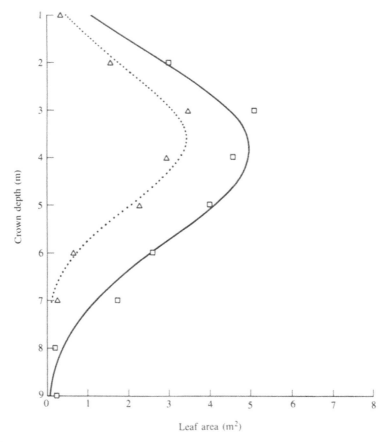

Figure 3.17   Effect of N fertilization on foliage area of crowns of young fertilized (□) and unfertilized (△) loblolly pine trees. Note that most of the increase in leaf area occurred in the middle and lower part of the crown. From Vose (1988). Reprinted from *Forest Science* **34**: 564–573. Published by the Society of American Foresters, 5400 Grosvenor Lane, Bethesda, MD 20814-2198. Not for further reproduction.

This is emphasized by variations in the extent, depth, and density of rooting of trees with different amounts of foliage. For example, the length and branching of roots of Douglas fir trees of different crown classes varied in the following order: dominant > intermediate > suppressed trees. Total root length of dominant trees often was greater than in much older intermediate trees (McMinn, 1963).

The importance of mineral supply to root growth is emphasized by stimulation of root growth by fertilizers. When only part of a root system is fertilized, growth of nonwoody and woody roots usually is stimulated on the side of the root system to which mineral nutrients are added (Coutts and Philipson, 1976). Hence, the asymmetric growth of root systems in fertile "patches" is attributable to a limited area of the crown receiving the benefits of mineral nutrients and subsequently returning carbohydrates and hormones to that part of the root system in fertile soil. Local application of N fertilizer stimulated growth of Sitka spruce roots more than application of P, and addition of K did not promote root growth. Adding N plus P had a greater stimulatory effect than adding either alone (Philipson and Coutts, 1977).

Proliferation, longevity, and mortality of fine roots depend on availability of water and mineral nutrients. Production of fine roots of broad-leaved trees was much greater in response to added water plus N when compared with water alone. The fine roots produced in response to added water plus N also lived longer (Pregitzer *et al.*, 1993).

Paradoxically, mineral deficiencies not only decrease root growth but also increase the proportional translocation of carbohydrates to the root system, hence increasing the root- shoot ratio. For example, a higher proportion of the total net primary productivity (TNPP) was allocated to fine roots in stands on infertile sites compared to fertile sites (Keyes and Grier, 1981). On fertile sites, the fine roots of hybrid poplars turned over more slowly, comprised a smaller portion of the root biomass, and were weaker carbohydrate sinks than the fine roots of poplars on mineral-deficient sites (Pregitzer *et al.*, 1990; Nguyen *et al.*, 1990).

Formation of mycorrhizae is suppressed in very fertile soils. In general, mycorrhizal development varies inversely with available N, P, K, and Ca. Fertilizer applications reduced mycorrhizal infections in pedunculate oak and silver birch seedlings (Fig. 3.18), with N having a greater effect than P

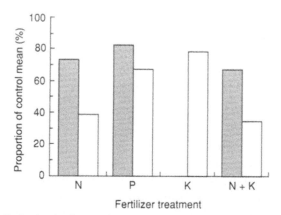

Figure 3.18 Reduction in the number of mycorrhizal root tips following application of fertilizers to seedlings of silver birch. Shaded bars denote field experiment, and open bars represent pot experiments. Only significant ($p < 0.05$) results are shown. From Newton and Pigott (1991b).

or K. The effects of fertilizers were dramatic, with mycorrhizal infections sometimes reduced by as much as 60%.

The response of mycorrhizae to a high nutrient level may not be evident for some time. For example, reduced mycorrhizal development in green ash seedlings after fertilizer application was not apparent until the second growing season (Douds and Chaney, 1986). As the light intensity is increased, the higher is the fertility level that allows for a given development of mycorrhizae. This relation applies over a wide range of nutrient availability but breaks down under extreme mineral deficiency or excess.

Water Deficits

Growth of plants is inhibited by too little water (Kozlowski, 1971a,b, 1972a,b, 1977, 1982a; Kramer and Boyer, 1995) as well as too much water (Kozlowski, 1984a,b; 1996). The impact of water deficits on growth may be considered at several scales including cellular, tissue, and organ levels of organization.

Cellular and Tissue Growth

In general, cell enlargement is more sensitive than cell division to dehydration. Cell growth begins when loosening processes cause walls to relax. This, in turn, causes turgor (and consequently Ψ) within cells to be reduced and induces a water potential gradient between the cell and its external environment (Cosgrove, 1993a,b). Water movement into cells in response to this water potential gradient increases cell volume. For sustained growth, solutes must continuously be imported into growing cells so that the osmotic potential (Ψ_π) does not increase in the face of osmotic dilution by inflowing water. Solute import thus assures maintenance of the water potential gradient that will sustain inward water movement.

Volume growth of cells can be described by an equation that considers both cell turgor and cell wall properties:

$$\frac{dV}{dt}\frac{1}{V} = m(\Psi_p - Y) \tag{3.2}$$

where $dV/dt\ 1/V$ is the rate of volume growth per unit of original volume, m is wall extensibility, Ψ_p is turgor potential (or pressure), and Y is a minimum turgor potential below which no growth occurs (the "yield threshold") (Lockhart, 1965; Boyer, 1985). Both m and Y depend on the physical properties of cell walls, but the biochemical basis of these parameters is not understood. Although represented as constants, these parameters are variable and reflect the dynamic metabolic activity in growing tissues (Passioura and Fry, 1992; Passioura, 1994). Pritchard (1994) suggested that m could be the result of enzyme cleavage of load-bearing bonds (or "tethers") within cell walls; Y might be associated with the tension necessary to "open" cleavage sites in the proper conformation for enzyme action.

Wall loosening may be related to the activity of specific wall enzymes that break the hemicellulosic tethers that connect cellulose microfibrils. One such enzyme, xyloglucan endotransglycosylase (XET), is present in walls of growing cells and has been hypothesized as such a wall-loosening enzyme because xyloglucan cleavage and chain transfer can be detected in growing cell walls (Smith and Fry, 1991; Tomos and Pritchard, 1994). McQueen-Mason *et al.* (1992) reported that a crude protein fraction extracted from the growing region of cucumber hypocotyls and containing no XET activity could induce wall extension activity. This fraction did contain two active proteins which the authors named "expansins." Although XET presumably promotes wall loosening by cleavage of covalent bonds, expansins show no hydrolytic activity. Rather, they appear to disrupt noncovalent hydrogen bonds between cellulose microfibrils and their hemicellulose coats, thereby functioning as "biochemical grease" to promote wall loosening (McQueen-Mason and Cosgrove, 1995). Much more research must be done before a clear picture of the biochemical basis of changes in wall physical properties can be obtained.

Equation 3.2 assumes that growing cells are in direct contact with the water supply and hence that there is no resistance to water flow to the cells. In growing leaves, this is clearly not the case, as expanding cells may be several cells removed from the xylem water supply. In this case, tissue growth can be described by an equation that is similar to Eq. 3.2 (Boyer, 1985):

$$\frac{dV}{dt}\frac{1}{V} = L(\Psi_0 - \Psi_w) \tag{3.3}$$

where $dV/dt\ 1/V$ is the relative volume of water entering the tissues per unit time ($\approx$ rate of volume growth), L is the hydraulic conductivity of tissues between growing cells and xylem water source, and Ψ_0 and Ψ_w are the water potentials of the xylem water source and elongating tissues, respectively. When the growth rate is constant, Eqs. 3.2 and 3.3 can be set equal, and rearrangement provides a combined growth equation that characterizes the relationship between tissue growth and controlling factors:

$$\frac{dV}{dt}\frac{1}{V} = \frac{mL}{m + L}(\Psi_0 - \Psi_\pi - Y) \tag{3.4}$$

where $(\Psi_0 - \Psi_\pi - Y)$ represents the "net osmotic force" for growth (i.e., the maximum possible turgor potential of the growing tissue reduced by the yield threshold turgor potential).

Thus, water deficits may reduce growth of enlarging tissues by affecting some combination of (1) Ψ of the xylem water supply, (2) osmotic potential (and hence turgor) of the growing tissue, (3) the yield threshold Y, (4) wall extensibility m, and (5) the hydraulic conductivity of the pathway through which water must flow to the growing tissues L. Sudden reduction in the Ψ of water supply, or other environmental changes such as dark–light transitions or root chilling (Pardossi *et al.*, 1994), may reduce both turgor and growth in the short term, but prolonged drought often leads to solute accumulation and osmotic adjustment in growing regions. As a result, turgor often is maintained, or at least does not drop below the yield threshold, in dehydrated growing tissues (Michelana and Boyer, 1982; Roden *et al.*, 1990; Nonami and Boyer, 1990; Schultz and Matthews, 1993; Chazen and Neumann, 1994). There also is evidence that water stress influences wall extensibility, causing cell wall "hardening." For example, growing leaves of three poplar clones did not lose turgor when irrigation water was withheld; however, cell wall extensibility was considerably reduced (Fig. 3.19), as was the rate of leaf extension (Roden *et al.*, 1990). Similarly, Nonami and Boyer (1990) observed substantial reductions in m in water-stressed stem tissue of soybean seedlings; however, they also reported that hydraulic conductivity of the tissues was reduced. There has been little study of the dynamic changes in parameters of Eq. 3.3 in water-stressed plants. Nonami and Boyer (1990) asserted that the initial reduction in growth of soybean seedling stems was attributable to collapse of the

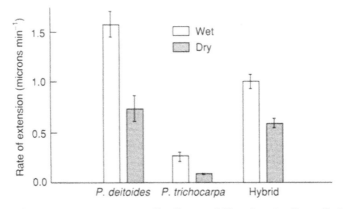

Figure 3.19 Effect of irrigation on cell wall extensibility of poplar. From Roden *et al.* (1990).

water potential gradient between xylem and growing cells. Subsequent responses included relatively rapid reductions in m and $(\Psi_p - Y)$ and slower declines in L. All of these parameters, except L, recovered at least partially during prolonged water stress.

Hormonal growth regulators also play an important role in controlling growth during droughts and may have opposite effects in different organs (see pp. 121–123). Thus, growth inhibition under water stress appears to be a complex phenomenon, subject to both direct biophysical and indirect metabolic control.

Shoot Growth

Inhibition of shoot growth by internal water deficits varies among species and genotypes, severity and time of drought, shoot location on the plant, and with different aspects of shoot growth such as bud formation, height growth and elongation of branches, leaf expansion, and leaf abscission.

Bud Formation and Internode Elongation Shoot growth is decreased by inhibitory effects of water deficits on both bud formation and bud elongation (Kozlowski, 1968a, 1972a). In species with fixed growth (e.g., northern pines; see Chapter 3 of Kozlowski and Pallardy, 1997), water deficits during the year of bud formation usually control ultimate shoot length more than water deficits during the next year because buds expand into shoots during the late spring and early summer before droughts normally develop. Large buds often produce long shoots; small buds generally produce shorter shoots (Kozlowski *et al.*, 1973). Irrigation of young red pine trees in late summer in Ontario, Canada, caused formation of large buds which produced long shoots in the subsequent year (Clements, 1970). Late-summer droughts resulted in formation of small buds which, in turn, expanded into short shoots. Irrigation in the spring of the year following bud formation had little effect on ultimate shoot length. The importance of water deficits during bud formation on shoot growth in the following summer also is shown by high correlations between shoot length and the amount of rain during the previous year (Kozlowski, 1971a).

In certain species with preformed shoots (fixed growth), environmental conditions during the year of bud expansion $(n + 1)$ have some influence on shoot growth. For example, in green ash, not all the putative leaf primordia (formed in year n) that overwintered in the bud developed into foliage leaves in year $n + 1$. Instead, the last-formed primordia differentiated into scales for the next terminal bud, appreciably reducing the leaf surface and shoot growth for that year (Remphrey, 1989).

In contrast to their effects on species with fixed growth of shoots, summer droughts greatly reduce current-year shoot elongation in species that exhibit free growth or recurrently flushing growth. These two groups continue to expand their shoots during much of the summer, whereas species with fixed growth normally complete shoot expansion in 2 to 6 weeks (Kozlowski, 1971a, 1982a; see Chapter 3 in Kozlowski and Pallardy, 1997). In recurrently flushing species, a midsummer drought cannot inhibit expansion of the bud in the first growth flush (which preceded the drought), but it may decrease the number of shoot primordia that are forming in the new bud and will expand in a late-season flush of shoot growth. Height growth of red pine (with fixed growth) was not significantly reduced by late-summer droughts (Lotan and Zahner, 1963). In contrast, a late-summer drought reduced the number of growth flushes of loblolly pine and greatly reduced its height growth (Fig. 3.20).

There are some exceptions to these generalizations. For example, species with fixed growth that normally expand their shoots for only a few weeks early in the growing season sometimes produce abnormal late-season shoots by opening and expansion of buds that normally would not open until the following year (see Chapter 3 in Kozlowski and Pallardy, 1997). The expansion of such late-season shoots may be reduced by severe late-summer droughts. Still another complication is that some species (e.g., larch) produce both long shoots and short shoots on the same tree. The short shoots, which show negligible internode elongation, comprise preformed needles that expand rapidly early in the summer. The long shoots, in contrast, continue to increase in length late into the summer and to form and expand new needles. Hence, a late-summer drought may be expected to reduce growth of long shoots but to have little effect on short shoots which completed expansion before a drought occurred (Clausen and Kozlowski, 1967b, 1970).

Water deficits generally affect growth of variously located shoots differently. For example, in recurrently flushing southern pines (e.g., loblolly pine), late-summer droughts generally decrease elongation of shoots in the upper crown more than those in the lower crown. This is because shoots in the upper crown exhibit more seasonal growth flushes, and elongate later into the summer, than shoots in the lower crown (Kozlowski, 1971a).

Leaf Growth   Both the number and size of leaves are reduced by water deficits. The number of needle fascicles of red pine trees was controlled to a large extent by water supply during the previous growing season (Garrett and Zahner, 1973).

Both extensibility of cell walls and turgor are important for leaf growth. As mentioned previously, Cosgrove (1993a,b) considered wall yielding as the major limiting process for growth of cells. Many studies suggest that water deficits also can directly inhibit leaf expansion by lowering cell turgor, because reduction in cell enlargement occurs too fast to be mediated by metabolism. Furthermore, irrigation of mildly stressed plants often is

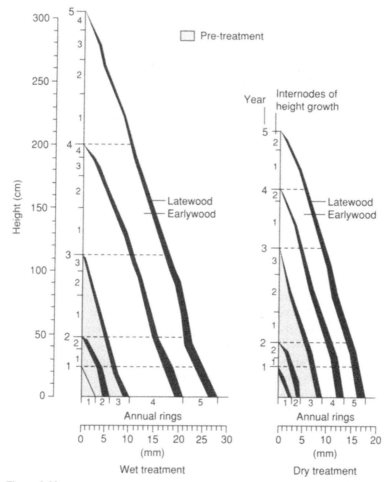

Figure 3.20 Effect of severe water deficits on height growth and cambial growth of loblolly pine seedlings. Wet treatment trees were irrigated throughout the year by surface watering of the soil to field capacity. Dry treatment trees received no irrigation water from May 1 to the end of October. From Zahner (1962). Reprinted from *Forest Science* **8**, published by the Society of American Foresters, 5400 Grosvenor Lane, Bethesda, MD 20814-2198. Not for further reproduction.

followed by resumption of leaf expansion within seconds (Hsiao, 1973). However, growing leaves of plants exposed to prolonged water stress exhibit reduced Ψ_π because of solute accumulation. Hence, turgor may quickly be regained although growth remains inhibited. For example, Schultz and Matthews (1993) found that when water was withheld from grape plants, the leaves accumulated sufficient solutes to maintain Ψ_p at or above the level of well-watered plants; nevertheless, leaf growth was completely inhibited. Reduced cell wall extensibility was found in water-stressed leaves, and it may have been partly responsible for growth inhibition. Several studies have associated shoot and leaf growth inhibition with accumulation of abscisic acid (ABA) (Watts *et al.*, 1981; van Volkenburgh and Davies, 1983; Creelman *et al.*, 1990; Saab *et al.*, 1990; Zhang and Davies, 1990). Interestingly, ABA may *maintain* root growth at a low level under water stress (see below).

The importance of cell turgor for leaf expansion varies with leaf age. Turgor is greater more often than the yield threshold in maturing cells compared to rapidly growing immature cells (Van Volkenburgh and Cleland, 1986). In Tasmanian blue gum, the rate of leaf growth of small juvenile leaves with active cell division was low. After leaf size exceeded 5 cm^2, the rate of leaf growth accelerated rapidly, because of the high extensibility of the cell walls (permitting rapid cell expansion) and a low yield threshold (permitting growth to continue even at low turgor) (Metcalfe *et al.*, 1990).

Normal diurnal changes in leaf dehydration do not appreciably reduce final leaf size, but dehydration of leaves for long periods results in small leaves (Boyer, 1976a,b). When Delicious apple trees were irrigated daily over a 30-week period, they had 14 times the leaf area that was produced by trees irrigated at 3-week intervals. This difference reflected an increased number of leaves as well as greater expansion of leaves of trees that were irrigated daily (Chapman, 1973). The needles of irrigated red pine trees were 40% longer than those of unirrigated trees (Lotan and Zahner, 1963), and growth of loblolly pine needles was proportional to their Ψ (Miller, 1965).

Leaf Shedding   Much of the reduction in leaf area as a result of drought is caused by premature leaf shedding. In some species, the leaves merely wither and die before they are shed. In others, leaf shedding as a result of drought involves true abscission and is associated with hormonal changes (see Chapter 3 in Kozlowski and Pallardy, 1997).

The extent of leaf shedding during drought varies among species and genotypes. Whereas California buckeye often sheds all its leaves and yellow poplar sheds some during drought, the leaves of dogwood usually wilt and die rather than abscise. Black walnut leaves senesce and are shed in response to mild droughts (Davies and Kozlowski, 1977; Parker and Pallardy, 1985); white oak is much less sensitive. In black walnut trees that shed more than 80% of their leaves, abscission increased from 8.1% at a predawn leaf Ψ of -1 to -1.5 MPa to nearly 100% at a Ψ of -3.5 MPa (Fig. 3.21). Leaf shedding occurred first in the older, basal leaves and progressed to the

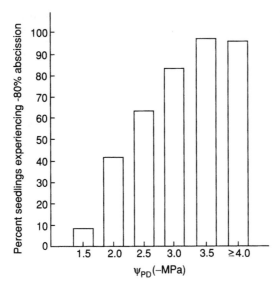

Figure 3.21 Effect of increasing water deficits, given as more negative predawn water potential (Ψ_{PD}) on abscission of black walnut leaves. Modified from Parker and Pallardy (1985).

younger leaves toward the apex. In Missouri, some leaf abscission of black walnut trees in response to drought occurred by July 3, and by September 25 most of the leaves had fallen. By comparison, only a few leaves of white oak began to change color but were not shed by September 25 (Ginter-Whitehouse *et al.*, 1983). In another study, post oak and white oak seedlings showed very little water-stress-induced leaf shedding even under severe drought conditions. In contrast, sugar maple and especially black walnut seedlings showed substantial leaf abscission during drought (Table 3.2). Before they were shed, the leaves of droughted English walnut trees lost 87% of the maximum hydraulic conductivity, whereas stems lost only 14%. The data indicated that drought-induced leaf shedding was preceded by cavitation in leaf petioles before cavitation occurred in stems (Tyree *et al.*, 1993). Some genotypic variations in drought-induced leaf shedding also have been reported (Kelliher and Tauer, 1980; Parker and Pallardy, 1985).

Shedding of needles of a developing loblolly pine stand was influenced by stand density (hence, partly by availability of water) and by weather (Hennessey *et al.*, 1992). In a young stand, the amount of annual needle loss was correlated with stem basal area. After crown closure occurred, the effects of basal area were minimal, and variations in weather predominated in influencing shedding of needles. Annual needle fall was up to 29% greater, and needle shedding occurred as much as 2 months earlier, in dry

Table 3.2 Effect of Drought Severity (as Indicated by Predawn Leaf Water Potential) on Leaf Abscission and Replacement of Leaf Area of Four Species of Woody Plants[a]

	Predawn Ψ_1 (MPa)[b]			
	-0.1	-1	-2	-3
Percent abscission				
Quercus stellata	9.50a	9.50b	9.60c	11.30c
Quercus alba	13.54a	13.54b	13.60bc	14.58c
Acer saccharum	13.91a	13.99b	16.36b	35.79b
Juglans nigra	13.55a	34.75a	62.33a	85.34a
New leaf area (cm^2)				
Quercus stellata	12.58a	12.76a	12.96b	13.15b
Quercus alba	3.08b	4.55a	6.18b	7.81b
Acer saccharum	6.00ab	6.00a	6.67b	17.47b
Juglans nigra	11.20ab	27.58a	45.78a	63.98a

[a]From Pallardy and Rhoads (1993).
[b]Values within a column not followed by the same letter are significantly different ($p \leq 0.01$).

than in wet years. Needle shedding was influenced by weather of the previous year as well as the current year.

Annual needle shedding in Monterey pine stands was linearly correlated with total foliage biomass and stand basal area (Raison *et al.*, 1992). Duration of retention of needles varied from 4 years in stands not undergoing water stress to approximately 2 years in stands exposed to drought. Most of the shed needles were more than 2 years old. In most years, needle shedding in stands undergoing drought peaked in the summer or autumn. In wet years or in irrigated stands, peak needle fall was delayed by 2 to 6 years.

Species that shed only some of their leaves during summer droughts include such broad-leaved evergreens as species of *Eucalyptus* and *Citrus* and many tropical and subtropical evergreens. In dry tropical areas, drought initiates defoliation in many trees. Most of the trees are considered facultatively deciduous. They retain their leaves until water stress increases sufficiently to induce abscission. In dry years the leaves are shed early. Some trees in dry tropical areas are obligately deciduous. Their leaf shedding is regulated more by day length than by water stress (Addicott, 1991). Trees of tropical rain forests commonly shed most or all of their leaves even during a mild drought. The timing of leaf shedding of various tropical species is correlated with periods of drought. In fact, water supply determines whether some species are classified as deciduous or evergreen

(Kozlowski, 1992). In Singapore rubber trees retained their leaves for 13.3 months, but in Ceara, Brazil, and South Malaysia leaf retention averaged 10 months because of differences in amounts of rainfall (Addicott, 1982).

In some tropical species new leaves emerge even before a drought ends. In West Africa, for example, many trees shed their leaves early in the dry season but produce a new leaf crop before the dry season ends. *Acacia albida* in the Sudan bears a crop of leaves during the dry season but is leafless during the rainy season (Radwanski and Wickens, 1967).

Desiccation and shedding of leaves are induced not only by lack of soil moisture (Pook *et al.*, 1966) but also by hot, dry winds that have been variously called foehns, chinooks, siroccos, levantos, and Santa Anas (Kozlowski, 1976a).

Cambial Growth

Water deficits inhibit several aspects of cambial growth, including division of fusiform cambial initials and xylem and phloem mother cells as well as increase in size and differentiation of cambial derivatives. The number of xylem cells produced, seasonal duration of xylem production, time of initiation of latewood, and duration of latewood production are influenced by tissue hydration.

The effects of water deficits on cambial growth may be direct as well as indirect. Internal water deficits inhibit cambial growth directly because high turgor of cells is necessary for cell enlargement. Increasing water deficits in Scotch pine stems decreased incorporation of glucose units into the walls of differentiating tracheids (Whitmore and Zahner, 1967). After 4 weeks of drought, the lumen diameters of recently produced tracheids of Monterey pine were small. Furthermore, rehydration and restoration of cell expansion did not occur for several days after water was resupplied (Sheriff and Whitehead, 1984). However, cell division continued when cell expansion was decreased to a minimum, emphasizing the greater sensitivity of cell expansion to water stress (Kramer, 1983).

Water deficits often interact with hormonal growth regulators in influencing cambial growth, as shown in Figs. 3.22 and 3.23. Increasing water deficits (decreasing Ψ) reduced the effects of growth hormones on production and enlargement of cambial derivatives of European ash. The greatest inhibitory effect was attributed to water deficits, with even mild water stress reducing both division and expansion of cells. The effects of hormonal growth regulators on cambial growth are discussed in more detail in pp. 138–151.

In many regions, water deficits often have only a minor role in initiating seasonal cambial activity, probably because water stress is not severe early in the growing season. However, after seasonal cambial growth is initiated, water deficits play a major role in controlling subsequent growth. As inter-

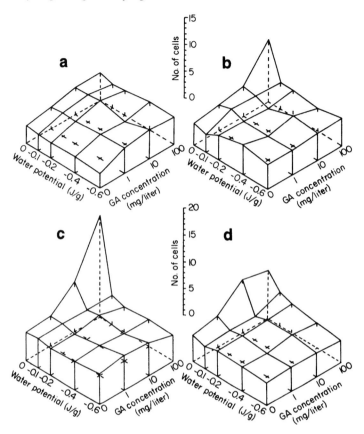

Figure 3.22 Interactions between water potential and IAA and GA in their effects on production of cambial derivatives in European ash. (a) Zero IAA; (b) 1 mg/liter IAA; (c) 10 mg/liter IAA; (d) 100 mg/liter IAA. From Doley and Leyton (1968).

nal water deficits increase, production of xylem cells slows or stops (Fig. 3.24), and if internal water deficits are decreased by rain or irrigation, xylem production often resumes and accelerates. During a dry summer in Arkansas, diameter growth of loblolly pine stopped by August but resumed during a rainy autumn, with one-third of the annual xylem increment being produced in September and October (Zahner, 1958, 1968). When several sequential droughts occur during the growing season, multiple rings of xylem often form in woody plants. The seasonal xylem increment is greatly increased and sometimes may even be doubled by irrigation.

In addition to decreasing the width of the annual xylem increment, water deficits induce early formation of latewood, and continued deficits shorten the period of latewood formation. Red pine trees released by stand

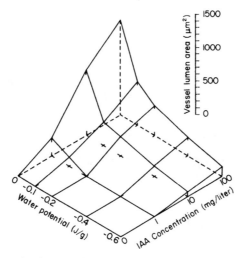

Figure 3.23 Interaction between IAA concentration and water potential in their effects on cross-sectional areas of lumens in vessels of European ash. From Doley and Leyton (1968).

thinning began to form latewood 2 weeks later than trees in an unthinned stand. The longer periods of latewood formation in the released trees were attributed to their less severe water deficits (Zahner and Oliver, 1962). As shown in Fig. 3.25, small amounts of latewood formed in loblolly pine trees during years of rather severe drought.

Root Growth

Water deficits commonly inhibit initiation, elongation, branching, and cambial growth of roots. Inhibition of root growth during a drought is associated with increased plant water deficits (Fig. 3.26). Both the number and elongation of roots of black walnut decreased rapidly as predawn xylem Ψ of leaves decreased from 0 to -0.5 MPa, and root growth became negligible as Ψ changed from -0.5 to -1.0 MPa (Fig. 3.27).

At the tissue level, water deficits cause a shortening of the growth zone in root tips (Spollen and Sharp, 1991). Any growth that may occur generally becomes restricted to the first few apical millimeters of roots. As is the case with leaf growth, osmotic adjustment occurs in growing regions of roots when they are subjected to water stress. Hence, turgor may be sustained across the entire region, although at a lower level than in unstressed roots (Spollen *et al.*, 1993; Pritchard and Tomos, 1993). Growing tissues of roots are quite close to the water supply of the plant, and neither the hydraulic conductivity nor the gradient in Ψ between water source and growing tissue appear to limit growth under water stress (Pritchard, 1994).

It is interesting to compare the relative sensitivities of root growth and

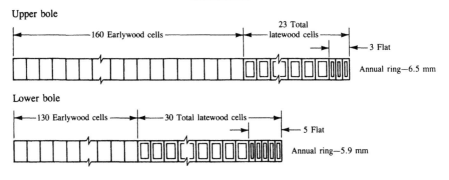

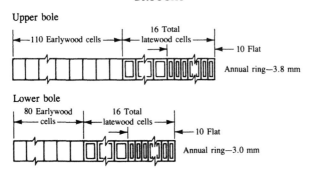

Figure 3.24   Difference in width of tracheids and in proportion of latewood in xylem rings of red pine trees grown under irrigation and under drought conditions. Note differences in xylem production in the upper and lower bole. From Zahner (1968).

shoot growth to water deficits. In general, root growth appears less sensitive to water deficits than shoot growth. Root growth under water stress is still reduced but proportionally less than that in shoots (Sharp and Davies, 1979). At the whole-plant level, this pattern may be partly responsible for altered root–shoot ratios observed in environments characterized by frequent periods of moderate water stress. For example, Keyes and Grier (1981) observed that belowground allocation of biomass of Douglas fir stands was substantially greater on drought-prone, nutrient-poor sites than on better quality sites. It should also be noted that severe water stress will cause drastic reduction or complete cessation of root growth, thereby decreasing or eliminating the tendency toward higher root–shoot ratios (Sharp and Davies, 1979; Rhodenbaugh and Pallardy, 1993; Pallardy and Rhoads, 1993).

At the tissue level, elongation of shoot tissues also is inhibited at higher

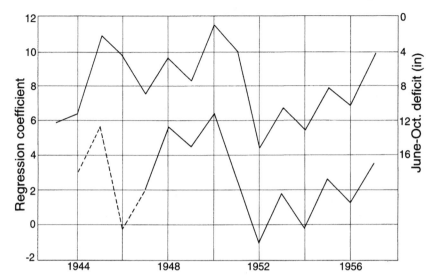

Figure 3.25 Relation between latewood formation and summer water deficit. Upper curve represents soil water deficits, and lower curve represents regression coefficients for latewood percentage on internodes from the apex. From Smith and Wilsie (1961).

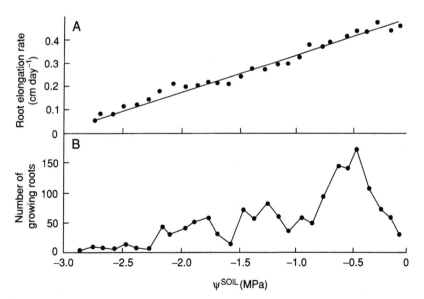

Figure 3.26 Effect of soil Ψ on root elongation (A) and number of growing roots (B) of white oak growing at temperatures higher than 17°C. From Teskey and Hinckley (1981).

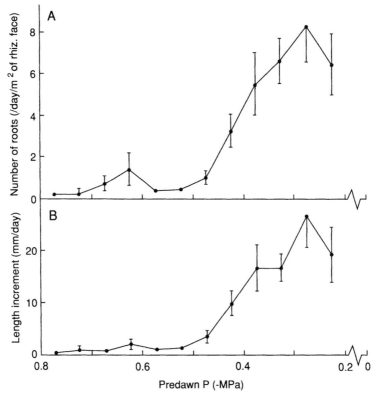

Figure 3.27   Relation between predawn xylem pressure potential (*P*) and number of growing roots (A) and total length increment of roots of black walnut (B). From Kuhns *et al.* (1985).

Ψ than in root tissues (Westgate and Boyer, 1985). This differential response is not well understood, but it may involve variations in the response of root and shoot growth to ABA that accumulates in water-stressed tissues. Whereas ABA tends to inhibit shoot growth, it may induce changes in roots that maintain some level of growth in water-stressed roots over that which otherwise would occur (Saab *et al.*, 1990; Wu *et al.*, 1994; Sharp *et al.*, 1994). ABA-linked influences on cleaving enzymes (such as XET, see above) that loosen cell walls have been hypothesized, and some evidence supports this idea (Wu *et al.*, 1994). Inhibition of ethylene synthesis by ABA also may promote sustained growth in water-stressed roots (Spollen *et al.*, 1993).

At the root system level, gross changes in morphology accompany water deficits. For example, large, branched root systems developed in loblolly pine seedlings growing in well-watered soil, and small, relatively unbranched root systems developed in seedlings undergoing periodic soil drying cycles (Kozlowski, 1949). Root growth often decreases as the soil dries even when the soil temperature is increasing (Richards and Cockcroft, 1975). However, early in the season soil temperature may be more important than soil moisture in controlling root growth. For example, at soil temperatures below 17°C, temperature was the dominating soil factor controlling root growth of white oak seedlings; at temperatures above 17°C the soil Ψ was more important. At soil temperatures above 17°C, elongation of roots was linearly related to soil Ψ (Fig. 3.26).

On arid sites, penetration of soil by roots generally is restricted to the depth to which the soil is wetted. Not only is root elongation stopped in dry soil, but roots also tend to become suberized to their tips, reducing their capacity for absorbing water and minerals. Plants subjected to severe droughts may not regain their full capacity to absorb water for a few days after the soil is rewetted (Kramer, 1983). As the soil dries, at least some of the decreased growth of roots is due to the physical resistance of the soil to penetration by roots.

Both the amount and type of mycorrhizae formed are responsive to soil moisture supply. When several species of fungi invade a host, they often respond differently to soil moisture stress. For example, a white fungus, the major mycorrhiza formed in Virginia pine when soil moisture content was high, was absent after the soil dried. In contrast, a black fungus (*Cenococcum granifome*) mycorrhiza formed when soil moisture availability was low (Worley and Hacskaylo, 1959).

Water Excess

Temporary or continuous flooding of soils is very common and often is associated with inhibition of vegetative growth, injury, and death of unadapted woody plants. The effects of flooding vary with species and genotype, age of plants, condition of the floodwater (moving or stagnant), and time and duration of flooding. Whereas flooding during the growing season reduces growth of plants, flooding during the dormant season often does not (Kozlowski, 1984a–c, 1985a; Kozlowski *et al.*, 1991).

Mechanisms of Growth Inhibition and Injury

The mechanisms by which flooding of soil affects plants are complex and involve altered carbohydrate, mineral, and hormone relations. Soil inundation is followed by rapid stomatal closure, leading to early reduction of photosynthesis (Pereira and Kozlowski, 1977b; Kozlowski and Pallardy, 1979; Sena Gomes and Kozlowski, 1980a, 1986). Following an initial decrease in photosynthesis by stomatal closure and reduced CO_2 absorption, the low rate of photosynthesis is associated with decreased activity of carboxylation enzymes, loss of chlorophyll, and reduced leaf area (which is traceable to inhibition of leaf formation and expansion as well as premature leaf abscission) (Kozlowski and Pallardy, 1984).

By restricting root growth, flooding decreases absorption of mineral nutrients (see Chapter 10 in Kozlowski and Pallardy, 1997). Furthermore, energy release by anaerobic respiration of roots often is too low for adequate uptake of mineral nutrients. In addition, injury to membranes of root cells causes loss of ions by leaching (Rosen and Carlson, 1984).

Flooding alters synthesis, destruction, and translocation of hormonal growth regulators. Levels of auxins, ethylene, and ABA are increased; those of gibberellins and cytokinins are decreased. The effects of these changes on root growth are discussed in pp. 154 to 161.

Flooded soils contain many potentially phytotoxic compounds, including sulfides, CO_2, soluble Fe and Mn, ethanol, acetaldehyde, and cyanogenic compounds. Although it often has been claimed that the adverse effects of flooding are caused by excess ethanol in plant tissues, this seems unlikely because ethanol is readily eliminated from plants. Furthermore, when ethanol was supplied to nutrient solutions at concentrations 100 times greater than those reported in flooded soils, plants were not injured (Jackson *et al.*, 1982). Although toxic compounds such as reduced Fe, fatty acids, and ethylene inhibited root growth of conifer seedlings, the inhibitory effects of O_2 deficiency were much more substantial (Sanderson and Armstrong, 1980a,b).

Hormone Relations

Each of the major classes of plant growth hormones is variously involved in regulation of dormancy and growth of shoots, stems, and roots (see Chapter 13 in Kozlowski and Pallardy, 1997).

Hormone Transport

All the classes of plant hormones except ethylene are found in both xylem and phloem and apparently move for long distances in plants. Labeled growth regulators move out of leaves at similar rates as other organic compounds. Inactive auxin conjugates, which move in the vascular system, may be "activated" by enzymatic hydrolysis at unloading zones (Rubery, 1988). Auxins, gibberellins, ABA, and cytokinins all move out of roots in the xylem. Thus, there is no question about the occurrence of long-distance transport of hormonal growth regulators. However, the physiological significance of long-distance transport is not clear because plant hormones can be synthesized in all the organs of plants. It therefore appears possible that *in situ* synthesis of hormones might explain most of the results attributed to transport from roots or other distant tissues. Further study of hormone sources and sinks is needed in order to evaluate the importance of long-distance hormone transport.

In addition to moving rapidly in the vascular system, endogenous hormones move slowly through ground tissues. In shoots, basipetal auxin transport occurs in several tissues depending on the season (Lachaud and Bonnemain, 1982, 1984). In mid-April, auxin was translocated by almost all cells of the young primary shoots of European beech. The capacity for auxin transport was lost in cortical parenchyma by the end of April and by May in pith parenchyma. During the entire period of cambial activity, auxin was transported and retained mostly by cells of the cambial zone and its recent derivatives. In September (before the onset of dormancy) and in February (at the end of dormancy), auxins were translocated primarily in the phloem. When the cambium was reactivated in the 1-year-old shoots, auxin again was translocated in cells of the cambial zone and differentiating cambial derivatives. Auxin transport is strongly polar, and its downward movement from stem tips and young leaves often produces a gradient of decreasing concentrations from stem tips to roots. Polar auxin transport is blocked by such inhibitors as TIBA (2,3,5-triiodobenzoic acid) and NPA (*N*-1-naphthylphthalamic acid) as well as by anaerobiosis and metabolic poisons such as cyanide and dinitrophenol.

Transport of cytokinins from roots to shoots has been demonstrated in many woody plants, including *Citrus* (Hutton and Van Staden, 1983), *Malus*, and *Populus* (Van Staden and Davey, 1979). In Douglas fir, cytokinins are carried in the xylem sap and accumulate in vegetative shoots before and during budbreak (Doumas *et al.*, 1986).

Both the path and rate of translocation of applied growth-retarding chemicals vary greatly. Maleic hydrazide (MH) moves very rapidly. When MH is injected into stems, it may be translocated throughout the plants within a day. In contrast, the triazoles move rather slowly, but their distribution can be facilitated by low-pressure injection (Domir and Roberts, 1983). Twenty-seven days after [^{14}C]paclobutrazol was injected into stems of saplings, only 23% of the ^{14}C activity had been translocated to the apical shoots. Labeled paclobutrazol was found in both the xylem and phloem, but ^{14}C activity in the roots did not increase (Sterrett, 1985). Paclobutrazol absorbed by roots was transported primarily in the xylem and accumulated in the leaves. Within a day, 26% of the paclobutrazol was in the leaves; after 28 days, 55% was in the leaves. When paclobutrazol was applied to the foliage, none was translocated to the stem or roots (Wang *et al.*, 1986). Uniconazole labeled with ^{14}C and injected into the rootstock of 1-year-old apple trees also moved very slowly. After 4 months, only 16% of the recoverable ^{14}C activity was translocated into the new shoots, and most remained in the rootstock. Nevertheless, shoot growth was inhibited (Sterrett, 1990).

Flurprimidol is not very mobile in plants. Thirty-five days after flur-primidol was injected into 1-year-old apple trees, 10% had moved into the new shoots, 1.5% had moved into the scion phloem, and 80% remained near the injection site. There was no detectable downward movement. The compound appeared to be transported more readily through the xylem than the phloem, similarly to paclobutrazol transport (Sterrett and Twor-koski, 1987).

Bud Dormancy

Induction of endodormancy (see Chapter 3 in Kozlowski and Pallardy, 1997) involves initiation by environmental signals and sequential molecular, cellular, and morphological changes. With completion of the chilling requirement and progression of buds from endodormancy to ecodormancy, a number of changes take place (Lang, 1989). Both fresh and dry weights of buds increase. Free sugars, amino acids, and organic acids increase, while total sugars and starch decrease. Total proteins decrease, but with increased postdormancy activity, total proteins, RNA, and DNA increase. Activities of enzymes and respiration rates also increase, but whether the rate of respiration regulates dormancy is unclear. Treating buds with thiourea breaks bud dormancy but does not increase respiration. Furthermore, inhibiting respiration by limiting the O_2 supply induced budbreak in peach (Erez *et al.*, 1980).

For a long time it was believed that endodormancy of buds was largely controlled by a balance between growth-inhibiting hormones (e.g., ABA) and growth-promoting hormones (e.g., gibberellins, cytokinins). In sycamore maple, most growth inhibitors were found in buds and leaves when the buds were not expanding, and the smallest amounts occurred when the shoot apex was expanding (Phillips and Wareing, 1958). When the buds began to expand, there was a decrease in inhibitors, an increase in growth promoters, or both. In sycamore maple, the release of bud dormancy was better correlated with an increase in gibberellin content of buds than with a decrease in growth inhibitors (Eagles and Wareing, 1964).

As more information accumulated about the complex nature and control of the sequential phases of bud dormancy, it became necessary to distinguish between the possible role of endogenous hormones in breaking dormancy and in influencing bud growth after dormancy was broken. In early studies this distinction often was not made. It now appears that the mechanisms controlling induction and release of bud dormancy are more complex than envisioned earlier and involve interactions among several growth hormones and other compounds as well.

Several studies cast doubt on the view that hormones alone control the induction and breaking of bud dormancy. It has been claimed that ABA is responsible for bud dormancy and that breaking of dormancy by chilling causes a decrease in ABA. In some species, the capacity for bud opening increases progressively downward on the plant axis (Powell, 1987). In broad-leaved trees, there often is less ABA in buds located in the lower crown than in those in the upper crown. Furthermore, the onset of dehydration of buds at the beginning of exposure to short days, with a simultaneous increase in ABA, suggests that ABA might be involved in induction of growth cessation and development of bud dormancy (Rinne *et al.*, 1994). However, the amount of ABA in buds declined about equally when plants were exposed to warm or cold temperatures (Mielke and Dennis, 1978; Barros and Neill, 1988). Furthermore, in apple trees a decrease in ABA began well before the buds were exposed to chilling (Fig. 3.28). The amount of endogenous ABA was similar in bay willow seedlings grown under short days or long days (Johansen *et al.*, 1986). Also, no correlation was evident between the ABA content of buds and short-day-induced cessation of shoot growth of sycamore maple (Phillips *et al.*, 1980). Martin (1991) concluded that ABA has no direct effect in controlling bud dormancy.

Gibberellins often have been implicated in dormancy release because their concentration in expanding buds often increases. In general GA-like activity increases prior to or concurrently with budbreak, and activity decreases as bud set is approached. Determining the role of gibberellins in

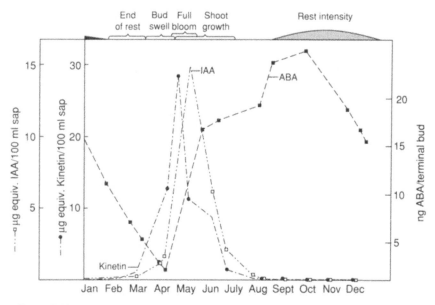

Figure 3.28 Changes in growth hormones in apple trees during the annual growth cycle. From Seeley and Powell (1981). *J. Am. Soc. Hortic. Sci.* **106**, 405–409.

release of dormancy is complicated because most of the gibberellic acids (GAs) in plants are inactive. Hence, the total amounts of GAs in buds during changes in dormancy release may be misleading.

Walser *et al.* (1981) concluded that GA applications could substitute for chilling in breaking dormancy. However, both Leike (1967) and Paiva and Robitaille (1978b) observed that GA induced budbreak only after the buds had been chilled. It thus appears unlikely that GA can substitute for chilling in breaking endodormancy. As emphasized by Leike (1967), an increased concentration of GA in expanding buds may not be the cause, but rather the result of release of dormancy by chilling.

A role of cytokinins in dormancy release has been postulated because of increasing levels during bud growth and stimulation of bud growth by exogenous cytokinins. However, the weight of evidence does not support a primary role for cytokinins in release of buds from dormancy.

Although cytokinins in buds increase during dormancy release (Domanski and Kozlowski, 1968), the increase is small (Wood, 1983; Young, 1989). Cytokinins apparently stimulate bud growth after release from dormancy but do not appear to directly control dormancy release. This conclusion is consistent with studies that show that applied cytokinins, at concentrations much higher than endogenous levels, induce growth of buds after they have been partially chilled (Shaltout and Unrath, 1983).

Ethylene does not appear to have an important regulatory role in release of dormancy of buds that have a chilling requirement. Stimulation of bud growth by exposure to ethylene was attributed to effects of ethylene after partial or full release from dormancy by chilling (Paiva and Robitaille, 1978a; Wang *et al.*, 1985). Ethylene probably was not involved in breaking dormancy of apple buds because ethylene production did not increase in buds in which dormancy was broken by removal of bud scales (Lin and Powell, 1981).

The release of bud dormancy is a sequential process, with each phase characterized by different metabolic and hormonal effects as shown for English walnut (Dathe *et al.*, 1982). Current evidence indicates that control of bud dormancy is very complex and involves interactive effects of growth hormones and other compounds. Some of these interactions are still not fully understood and much more research will be needed before we fully understand the mechanisms involved. The attractive model of Saure (1985) suggests that release of bud dormancy may occur in the following sequential phases, as it apparently does in seeds (Zarska-Maciejewska and Lewak, 1983):

Phase I. Removal of the primary cause of dormancy occurs in phase I, largely involving the enzymes that are most active at chilling temperatures. These changes cause transition from endodormancy to ecodormancy.

Phase II. The catabolic phase II is characterized by high metabolic activity, increasing GA activity which promotes synthesis and/or activation of hydrolytic enzymes (e.g., amylases, proteases, ribonuclease), and development of a translocation system, causing mobilization of reserve carbohydrates essential for growth. During this phase, bud swelling indicates a transition from endodormancy to ecodormancy.

Phase III. Phase III is characterized by additional increases in gibberellin and cytokinin activity, further mobilization of reserve carbohydrates, increasing auxin activity, and complete release from dormancy. The end of this phase is characterized by budbreak and rapid shoot growth.

Shoot Growth

Most elongation of shoots results from cell expansion, with gibberellins playing a dominant role in the hormonal complex regulating shoot expansion. Evidence for this conclusion comes from an increase in shoot elongation following application of gibberellins, the counteracting effects of applied gibberellins on those of endogenous growth inhibitors or applied growth retardants, and correlation of shoot extension with endogenous gibberellin levels.

A number of investigators have shown that exogenous gibberellins increased height growth and internode elongation in a variety of woody angiosperms and gymnosperms. The effects of exogenous GA vary in detail and depend on the specific gibberellin applied; concentration, frequency, and method of application; and on the species and age of trees.

Cell elongation following GA treatment requires recognition of GA by a receptor molecule, interaction of the activated receptor with the cell, and production of a wall-loosening factor or inhibition of a wall-stiffening factor (Jones, 1983). Applying GA_3 to the shoot apex of bay willow in the spring delayed growth cessation and shoot tip abortion. Both growth cessation and shoot tip abortion were induced by short days or application of growth retardants, and these effects were antagonized by GA_3 (Juntilla, 1976). Gibberellin-like activity in the apical part of bay willow shoots decreased prior to cessation of shoot growth, indicating that low GA activity in the shoot may be a prerequisite for cessation of shoot growth (Juntilla, 1982). The effects of specific gibberellins on shoot growth vary appreciably. For example, exogenous GA_{20} and GA_1 stimulated shoot elongation of bay willow under short-day conditions and could substitute for transfer of plants to long-day conditions, whereas GA_{53}, GA_{44}, and GA_{19} were inactive. Both GA_{20} and GA_1 were active even when applied at concentrations one-thousandth of the concentration of applied GA_{19} (Juntilla and Jensen, 1988).

Interactions between day length and gibberellins on shoot elongation have been well documented, with short-day-induced growth cessation of many deciduous species prevented by GA_3.

Gibberellins may be synthesized in partially expanded leaves as well as in other organs. The inhibition of internode extension following removal of the young leaves of apple was partially or entirely eliminated by exogenous GA_3 (Powell, 1972). Removal of the young leaf just as it was unfolding inhibited elongation not only of the internode below it but also of two internodes above it. Applied GA_3 completely replaced the leaf with regard to expansion of the internode below, but it only partially substituted for the leaf in controlling growth of internodes above it. This suggested that the leaf supplied essential growth-controlling factors in addition to gibberellins.

In contrast to the stimulation of shoot growth in angiosperms by short-term treatments with GA, early experiments showed negligible or only slightly stimulatory effects of gibberellins on shoot growth of conifers. In these studies, however, the lack of stimulation of shoot growth by applied gibberellins may have been due to insufficient amounts of GA, application of an inappropriate GA, poor absorption of GA, or overriding inhibitory factors (Ross *et al.*, 1983). Several lines of evidence now show that endogenous gibberellins are very important in regulating shoot growth of conifers (Dunberg, 1976; Juntilla, 1991). Spray treatments of GA applied monthly for 9 months increased height growth (observed for 18 months) of loblolly pine seedlings (Roberts *et al.*, 1963). Both soil drenches (applied several times weekly) and foliar sprays (applied weekly for 3 months) of GA_3 increased height growth of balsam fir seedlings. The increase was associated with a change in the distribution of photosynthate rather than in the rate of photosynthesis (Little and Loach, 1975). Ross *et al.* (1983) cited studies that showed that exogenous gibberellins (including GA_3 and $GA_{4/7}$) stimulated shoot growth in 26 species of conifers.

If auxins play an important role in shoot elongation, they should show a relation to the seasonal timing of growth. Ample evidence exists of such a relation for broad-leaved trees. For example, the auxin content increased in the spring as apple shoots began to expand. Auxin levels declined during the growing season, and the decrease was followed by slowing of shoot growth (Hatcher, 1959).

Auxin-induced loosening of cell walls dominates the rate of cell expansion, with increase in cell size occurring as a result of a turgor-driven extension of cell walls. The major steps in auxin action on cell extension are shown in Fig. 3.29. The initial site of auxin action is on or within the cytoplasm. Subsequent sequential steps include auxin uptake, its movement to a receptor, a hormone–receptor interaction such as release of a secondary messenger, transport of the secondary messenger to a site of action, and eventual action of the secondary messenger on the growth process, probably by inducing cell wall loosening (Evans, 1985).

Although applied auxins increased elongation of excised stem sections

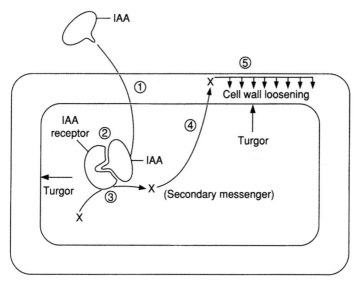

Figure 3.29   Major steps in initial action of auxin in cell elongation: (1) uptake of auxin, (2) binding of auxin to a receptor, (3) action of the auxin–receptor complex to produce or release a secondary messenger (x), (4) movement of the secondary messenger to the cell wall, and (5) action of the secondary messenger on the wall to induce wall loosening, thereby allowing for cell expansion with prevailing cell turgor. Reprinted with permission from Evans, M. L. (1985). The action of auxin on plant cell elongation. *Crit. Rev. Plant Sci.* **2**, 317–366. Copyright CRC Press, Boca Raton, Florida.

of some gymnosperms (Terry *et al.*, 1982), auxins appear to be less important in intact trees. This is shown by negligible or inhibitory effects of applied auxins on shoot elongation (Heidmann, 1982; Ross *et al.*, 1983). There are some exceptions, however, and Dunberg (1976) found that the auxin content of Norway spruce shoots was highly correlated with their rates of elongation. Some evidence for auxin involvement in regulation of stem elongation of gymnosperms comes from inhibition of height growth following application of triiodobenzoic acid (TIBA), an inhibitor of polar auxin transport (Little, 1970).

Abscisic acid (ABA) does not appear to have a major role in regulating shoot growth when the potential for growth is high. When apple seedlings were treated with ABA in January in a greenhouse (low growth rate), shoot growth stopped and terminal buds formed. However, a similar treatment in June (high growth rate) had little effect on shoot elongation (Powell, 1982).

Some investigators have implicated polyamines in regulation of shoot growth. For example, Rey *et al.* (1994) found correlations of high spermidine and spermine levels with rapid shoot growth and leaf expansion of hazel trees. They also reported that low spermidine and spermine levels, together with increasing amounts of putrescine, may be associated with induction of bud dormancy.

Leaf Senescence and Abscission

Because leaf senescence typically precedes natural leaf shedding, it is considered to be a necessary signal for abscission. Senescence of leaves is associated with several biochemical and structural changes including proteolysis, loss of chlorophyll, and degradation of nucleic acids (with loss of RNA occurring faster than loss of DNA). Nevertheless, synthesis of RNA and proteins continues during the decrease in nucleic acids. Losses of chloroplast structure and integrity are dominant features of leaf senescence. Granal thylakoids swell and are dismantled, but the chloroplast envelope retains its integrity until late stages of senescence. The rate of photosynthesis declines progressively during senescence, the result at least in part of loss of key photosynthetic enzymes. During senescence, the levels of some enzymes decrease (e.g., pectin methylesterase) and those of others increase (e.g., proteases, RNase, and other hydrolases) (Lauriere, 1983). The rate of respiration declines gradually (Rhodes, 1980) or is stable during early senescence and subsequently exhibits a climacteric-like increase followed by a marked decline (Roberts *et al.*, 1985). Deterioration of membranes is an early feature of senescence and results in increased cell permeability, loss of ionic gradients, and decreased activity of important membrane-associated enzymes. Generally, development of senescence also is accompanied by export of carbohydrates, amino acids, and minerals from leaves. Minerals also may be lost by leakage from leaf cells.

As summarized by Addicott (1982), abscission results from the activity of hydrolytic enzymes that are synthesized by sequential reactions from DNA to hydrolases (Fig. 3.30). The energy for such synthesis is provided by respiration. Synthesis of cell walls continues at a low level during the life of the abscission zone (shown by the lower chain of reactions from DNA to synthetases in Fig. 3.30). The growth hormones regulate enzyme synthesis, with delay of abcission by IAA associated with stimulation of production of synthetase. In contrast, ABA stimulates production of hydrolase. Either a decrease in IAA or an increase in ABA is followed by synthesis of hydrolase, leading to breakdown of pectins and hence to abscission. The rise in ethylene level in the abscission zone usually is correlated with increased synthesis of hydrolase.

The onset and rate of abscission of leaves are regulated by hormonal messages involving interactions of several endogenous hormones. The hormones that influence abscission may form in the abscission zone, but the most important effects are induced by hormones translocated to the abscission zone from the subtended leaf.

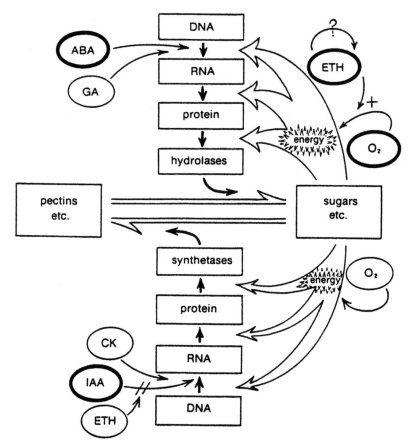

Figure 3.30 Sequential events in the abscission zone that lead to abscission of leaves. The dominating hormones are given heavy emphasis. From Addicott (1982).

The role of individual hormones within the hormonal complex that controls abscission varies appreciably. Auxins, cytokinins, and gibberellins delay abscission; ABA and ethylene commonly accelerate abscission. Growing leaves with high levels of auxins, cytokinins, and gibberellins do not abscise, but when levels of these hormones decrease, the leaves begin to senesce and eventually abscise.

Auxin appears to dominate regulation of abscission. As long as the supply of auxin reaching the abscission zone from the leaf blade is high, abscission does not occur. The capacity of senescent leaves to produce auxins declines and eventually is followed by abscission (Addicott, 1991). The amount of ABA increases greatly in senescing leaves and those undergoing water stress (when the amounts of auxin, gibberellins, and cytokinins are decreasing). Endogenous ethylene is considered a powerful accelerator but not an initiator of abscission (Addicott, 1991). Production of ethylene follows rather than precedes abscission. Sometimes abscission is initiated when essentially no ethylene is present. The stimulation of abscission by ethylene has been attributed to its capacity to accelerate degradation of IAA, block IAA transport, and increase synthesis of ABA.

Numerous studies show that application of auxins, gibberellins, or cytokinins to leaves that are still green in the autumn retards their senescence. For example, Osborne and Hallaway (1960) treated leaves of *Prunus serrulata* with auxins. The treated leaves not only retained their green color but also maintained high rates of photosynthesis and high protein levels. Brian *et al.* (1959) treated leaves of European ash, sweet cherry, and sycamore maple with GA_3 in the early autumn. The treated leaves remained green and did not abscise by November 21, whereas all untreated leaves were shed by that date. Thimann (1980) emphasized that cytokinins prevent or at least greatly delay proteolysis and loss of chlorophyll. The effect of cytokinins on attached leaves usually is less marked than when the hormone is applied to detached leaves. This probably is because attached leaves receive a supply of endogenous cytokinins from the roots.

Evidence that exogenous ethylene can initiate leaf abscission is impressive. A concentration as low as 0.1 μl/liter in the air can induce leaf abscission. The oldest leaves are shed first, indicating their greater sensitivity to ethylene. Abscission of both leaves and fruits can be stimulated by applications of Ethrel (2-chloroethylphosphonic acid), also called ethephon.

Several compounds other than hormonal growth regulators have been implicated in regulating leaf senescence. For example, promotion of senescence has been attributed to several amino acids (e.g., serine, arginine, cysteine, lysine, and ornithine) as well as fatty acids. Some dibasic compounds, especially the polyamines, spermine, spermidine, and putrescine, inhibit senescence (Brown *et al.*, 1991).

Cambial Growth

A voluminous literature shows that each of the major plant growth hormones is present in the cambial region and may be variously involved in regulation of production, growth, and differentiation of cambial derivatives (Little and Savidge, 1987). As emphasized in Chapter 13 of Kozlowski and Pallardy (1997), the effects of hormones are exerted by interactions among them as well as through second messengers which may or may not be plant hormones.

Auxins apparently exert a predominant role in regulating cambial activity (including mitosis of fusiform initials and differentiation of cambial derivatives) as shown by the following lines of evidence: (1) Both basipetal transport of auxin and cambial growth are reduced by disbudding of branches (Hejnowicz and Tomaszewski, 1969), defoliation (Kulman, 1971), or phloem-blockage (Evert and Kozlowski, 1967; Evert *et al.*, 1972). (2) Application of exogenous auxin to disbudded, defoliated, or phloem-severed plants increases production of xylem and phloem in gymnosperms (Larson, 1962b; Little and Wareing, 1981; Little and Savidge, 1987) and angiosperms (Digby and Wareing, 1966; Doley and Leyton, 1968; Zakrzewski, 1983). (3) Production of cambial derivatives often is correlated with the amount of exogenous auxin up to an optimal level (Little and Bonga, 1974; Sheriff, 1983). (4) Exogenous auxin promotes enlargement of cambial derivatives and thickening of their walls (Larson, 1960, 1962b; Gordon, 1968; Porandowski *et al.*, 1982; Sheriff, 1983).

The importance of auxin in regulating different aspects of cambial growth varies appreciably. For example, auxin promotes tracheid enlargement, but its effect on wall thickening is less obvious. In broad-leaved trees a stimulatory action of auxin on the relative number of differentiated vessels was shown by Digby and Wareing (1966) and Zakrzewski (1983). Both production of large-diameter earlywood vessels and rapid differentiation of vessels have been attributed to IAA (Aloni and Zimmermann, 1983; Aloni, 1987). Because exogenous IAA has little effect on cambia that are more than 1 year old, it is not clear if IAA regulates cambial growth directly or does so indirectly through hormone-directed transport.

Evidence for involvement of gibberellins in control of cambial growth is inconsistent. Exogenous GA_3 stimulated cambial growth and affected the anatomy of cambial derivatives in several broad-leaved trees (Wareing *et al.*, 1964; Savidge and Wareing, 1981). However, exogenous gibberellins did not stimulate cambial activity in brittle willow (Robards *et al.*, 1969), apple (Pieniazek *et al.*, 1970), or beech (Lachaud, 1983).

Gibberellins stimulated cambial growth in certain conifers (Little and Loach, 1975; Ross *et al.*, 1983). Little and Savidge (1987) considered the role of gibberellins in regulation of cambial growth of conifers as unclear. Wang *et al.* (1992) reported that $GA_{4/7}$ increased tracheid production in Scotch pine seedlings, with the effect mediated through an increase in IAA in the cambial region.

A few studies showed that exogenous cytokinins, alone or together with IAA, accelerated production of xylem and phloem or ray tissues (Casperson, 1968; Philipson and Coutts, 1980a; Zakrzewski, 1983). In other studies, exogenous cytokinins did not stimulate cambial growth (Zajaczkowski, 1973; Wodzicki and Zajaczkowski, 1974; Little and Bonga, 1974).

Although ABA appears to be involved in control of cambial growth, there is some uncertainty about its precise role. Some evidence indicates that ABA inhibits cambial activity (Little and Eidt, 1968; Jenkins, 1974; Little, 1975). The formation of latewood has been attributed to increased ABA in the cambial zone (Jenkins and Shepherd, 1974; Wodzicki and Wodzicki, 1980). Late in the growing season, ABA caused a decrease in tracheid enlargement in Monterey pine (Pharis *et al.*, 1981). Nevertheless, actual involvement of ABA in the earlywood–latewood transition has been questioned (Little and Wareing, 1981; Savidge and Wareing, 1984).

Several studies suggest that ethylene plays a role in the hormonal complex that regulates the amounts of xylem and phloem cells produced and the anatomy of cambial derivatives. Involvement of ethylene in regulating cambial growth is emphasized by the following evidence: (1) Application of Ethrel to stems accelerates xylem and phloem production in several species of trees, including Monterey pine (Barber, 1979), eastern white pine (Brown and Leopold, 1973; Telewski and Jaffe, 1986b), Aleppo pine (Yamamoto and Kozlowski, 1987b), Japanese red pine (Yamamoto and Kozlowski, 1987c), apple (Robitaille and Leopold, 1974), brittle willow (Phelps *et al.*, 1980), American elm (Yamamoto *et al.*, 1987a), and Norway maple (Yamamoto and Kozlowski, 1987f). (2) Exogenous ethylene stimulates growth of bark tissues more than xylem tissues (Yamamoto and Kozlowski, 1987a–c; Yamamoto *et al.*, 1987a). (3) Treatment of cuttings of Norway spruce seedlings with Ethrel influences incorporation of carbohydrates into cell walls (Ingemarsson *et al.*, 1991). High concentrations of Ethrel inhibited expansion of xylem cells while stimulating incorporation of cellulose in cell walls. Addition of small amounts of Ethrel, which slightly stimulated ethylene emission, led to increases in the size of xylem cells, amounts of phloem tissue, and intercellular spaces in the cortex (Ingemarsson *et al.*, 1991). The activity of several enzymes involved in lignin synthesis was stimulated in plants treated with ethylene (Roberts and Miller, 1983).

Eklund and Little (1995) reported that ethylene evolution was not specifically associated with IAA-induced tracheid production in balsam fir. Ethylene did not mimic the promoting effect of IAA on tracheid production. Ethylene could promote production of tracheids but only when its application was followed by unphysiologically high concentrations in the cambial region which, in turn, induced accumulation of IAA. We interpret the weight of evidence to show that ethylene has a synergistic effect on xylogenesis in the presence of auxin.

Cambial Sensitivity to Growth Hormones   The effects of hormone levels on cambial growth vary with seasonal changes in cambial sensitivity to the presence of hormones. Some evidence shows that the actual onset of cambial activity is not controlled by changes in IAA level alone. Once cambial cells regain the capacity to respond to a stimulator, auxin levels then appear to regulate cambial activity. The capacity of cambial cells to respond to IAA is restored before new cambial derivatives are cut off by the cambium

(Little and Savidge, 1987; Lachaud, 1989). During the period of intense cambial activity, auxin levels control the intensity of mitosis, but not the duration of cambial growth (Little and Savidge, 1987; Lachaud, 1989).

Tracheid production was stimulated less by application of IAA to Scotch pine shoots late in the growing season than it was by earlier application (Wodzicki and Wodzicki, 1981). Little (1981) found similar rates of auxin transport in trees with resting or active cambia. Additionally, the auxin concentration at the beginning of seasonal cambial activity often is not very high, and it may not decline markedly during the growing season. Such observations indicate that the sensitivity of cambial cells to IAA varies during the year and that an autumnal resting state reflects a lack of capacity of cambial cells to respond to IAA. The resting state presumably changes to a quiescent state in the winter, and, after adequate chilling, cambial cells regain their capacity to respond to IAA. Creber and Chaloner (1984) emphasized that woody plants of the temperate zone develop an endogenous rhythm that is locked into the annual temperature cycle and photoperiodic cycle. This rhythm is characterized by cellular changes that reflect variable responses to endogenous growth regulators which stimulate cambial activity during the part of the year that is most favorable for growth.

Responses of stem tissues to ABA also vary appreciably during the season. An ABA-mediated inhibition of cambial growth more likely reflects a change in cambial sensitivity to ABA rather than an increase in ABA (Lachaud, 1989).

Apical Control of Cambial Growth Evidence for a dominant regulatory role of apically produced growth hormones on cambial growth comes from correlations between bud growth in the spring and initiation of xylem production below buds, basipetal migration of the cambial growth wave, arrested cambial activity in defoliated or disbudded trees or below phloem-blocking stem girdles, and initiation of cambial growth with exogenous hormones. Neither the roots nor stem tissues near the base of the stem can supply enough growth hormones to sustain normal cambial growth.

Initiation of Cambial Growth After bud dormancy is broken by adequate chilling, apically produced growth-promoting hormones move down the branches and stem and provide the stimulus for initiation of seasonal cambial growth. Indoleacetic acid produced in developing leaves and expanding shoots is transported basipetally at a rate of about 1 cm hr^{-1} (Little and Savidge, 1987; Savidge, 1988).

Whereas the cambium in the part of a pruned branch above the uppermost bud remains inactive, xylem and phloem cells are produced by cambial activity below the same bud. Also, disbudding of shoots during the dormant season greatly impedes cambial growth in the same shoots during the next growing season. Disbudding of Scotch pine cuttings during the dormant season inhibited tracheid production, whereas lateral application of IAA to disbudded cuttings promoted tracheid production (Little *et al.*, 1990). Similar responses were found in lodgepole pine (Savidge and Wareing, 1984) and balsam fir (Sundberg and Little, 1987).

The basipetal spread of initiation of seasonal cambial activity occurs faster in ring-porous than in diffuse-porous trees. Digby and Wareing (1966) concluded that expanding buds were primary sources of auxins in diffuse-porous trees. A gradient of auxin below the buds was correlated with the downward spread of cambial activity. In ring-porous trees, however, it was postulated that an auxin precursor was present before the buds opened and was converted to auxin throughout the tree axis near the time the buds opened, hence stimulating rapid resumption of cambial growth throughout the stem.

Defoliation of trees by insects or other agents may be expected to decrease the xylem increment to an appreciable extent by inhibiting synthesis and downward translocation of IAA (Sundberg and Little, 1987). In conifers, removal or defoliation of current-year shoots decreases tracheid production more by decreasing the IAA level than by reducing the supply of photosynthate. This is because the young leaves use most of their photosynthate for their own growth and metabolism. The 1-year-old needles are the major source of photosynthate for cambial growth of branches and stems of pines (Kozlowski, 1992).

Production of Cambial Derivatives   The rate of production of cambial derivatives is influenced by the level of IAA, as shown in balsam fir (Little and Bonga, 1974), Sitka spruce (Little and Wareing, 1981), and Monterey pine (Sheriff, 1983). Regulation of division of cambial cells by apically produced hormones also is shown by a reduction in both auxin content and xylem increment below phloem blockages in branches and stems.

Wilson (1968) reported an increase in the number of xylem and phloem cells produced above phloem-severing stem girdles of eastern white pine trees, apparently because of a higher than normal mitotic index (percentage of cambial zone cells in mitosis) and because of the longer duration of mitotic activity. Below the phloem blockage, cell division and expansion stopped within a few weeks after the trees were girdled. Wodzicki and Wodzicki (1973) found marked reductions in both auxin content and cambial growth below phloem blockages in Scotch pine trees. In the following year, both auxin levels and cambial activity were negligible below the phloem block. Applications of IAA to disbudded cuttings and attached shoots of Scotch pine were followed by an increase in the internal concentration of free IAA to a level similar to that in cuttings with opening buds (Sundberg and Little, 1991). The increase in the internal level was positively correlated with tracheid production (Fig. 3.31). In intact cuttings,

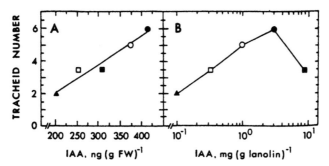

Figure 3.31   Relationship after 21 days between the number of lignified tracheids in Scotch pine shoots and the sum of the internal IAA concentrations measured on days 7, 14, and 21 (A) and the concentration of applied IAA (B) in 1-year-old cuttings that were debudded and treated apically with 0.11 (▲), 0.33 (□), 1 (○), 3 (●), or 9 (■) mg IAA g^{-1} lanolin. From Sundberg and Little (1991).

the amount of IAA was not optimal for tracheid production because laterally applied IAA increased tracheid production, as also was shown for intact seedlings of balsam fir (Little and Wareing, 1981) and Sitka spruce (Denne and Wilson, 1977; Philipson and Coutts, 1980a).

Differentiation of Cambial Derivatives   Variations in the sizes of cambial derivatives and in thickening of their walls are regulated to an appreciable extent by apically produced growth hormones. In gymnosperms, earlywood tracheids form when internal control mechanisms favor radial expansion over wall thickening. In red pine, radial expansion of tracheids usually occurs along the entire stem during the period of shoot growth as long as the needles are elongating. Later in the season, reduction in tracheid diameter resulting in latewood production begins at the base of the stem and subsequently occurs upward in the stem and outward in the growth ring (see Chapter 2 in Kozlowski and Pallardy, 1997).

The transition from earlywood to latewood has been correlated with a decrease in IAA. In red pine trees, large-diameter earlywood tracheids formed during the period of shoot elongation and high auxin synthesis, whereas narrow diameter, latewood cells formed after shoot elongation stopped and auxin synthesis was reduced (Larson, 1962a,b, 1963c, 1964).

When Larson (1962a) exposed red pine trees to short days, growth of needles was first reduced and later stopped, and this change was followed by production of small-diameter tracheids. In trees exposed to long days, the needles continued to elongate, and large-diameter tracheids were produced. Eventually, however, both needle growth and production of large-diameter cells stopped.

As seen in Fig. 3.32, the concentration of ^{14}C-labeled IAA was very high

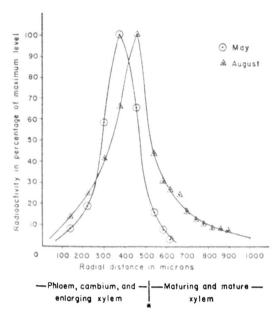

Figure 3.32   Radial distribution of radioactivity derived from exogenously applied ^{14}C-IAA in the tissues of decapitated pine stems after a 2-day labeling period. The vertical line marked with an asterisk (*) below the horizontal axis marks the transition from enlarging to maturing xylem. From Nix and Wodzicki (1974).

in the cambial zone during the growing season and declined greatly toward the maturing xylem of shortleaf pine stems. This distribution of auxin varied seasonally in relation to the transition to latewood formation (Nix and Wodzicki, 1974). As IAA in balsam fir stems declined during the growing season, radial expansion of cells in the inner periphery of the cambial zone also decreased (Sundberg *et al.*, 1987).

   Although some hormonal growth promoters such as auxins are produced by the cambium itself after it is activated by an apical stimulus (Sheldrake, 1971), normal development of xylem and phloem depends on a continuous supply of growth-regulating hormones from the shoots. Exogenous IAA promoted expansion of cambial derivatives as well as thickening of their secondary walls (Denne and Wilson, 1977; Savidge and Wareing, 1982). In Monterey pine seedlings, tracheid wall thickness increased approximately linearly as exogenous IAA concentration was increased up to 20 mg liter^{-1} (Sheriff, 1983). Evert and Kozlowski (1967) and Evert *et al.* (1972) showed that severing the phloem of trembling aspen and sugar maple stems during the dormant season, or at various times during the growing season, affected subsequent production and differentiation of xylem and phloem below the phloem block. When the phloem was interrupted during the dormant season, xylem differentiation during the next growing season did not occur below the blockage. If the phloem was severed shortly after seasonal cambial activity began, relatively few xylem elements were produced during the same season, and these were short and without normally thickened walls. The most conspicuous effect of phloem interruption was curtailment of secondary wall formation in both the xylem and phloem. When phloem blocks were applied in midseason, the first part of the annual xylem increment had cells of normal length and wall thickening.

Usually the diameters of vessels of angiosperms increase from the leaves to the roots. The basipetal increase in vessel diameter is associated with a decrease in the number of vessels per unit of stem cross sectional area (Zimmermann and Potter, 1982). At a given stem height, the size of vascular elements increases from the inner growth ring, through a number of annual growth rings, until a constant size is attained.

Assuming that differentiation of vascular elements is controlled to a considerable extent by basipetal auxin flow, Aloni and Zimmermann (1983) explained changes in vessel size and frequency along the stem axis by the following: (1) auxin concentration decreases progressively down the tree axis, (2) impeding auxin transport causes local increases in auxin, (3) the distance from the source of auxin to differentiating vessels determines the amount of auxin they receive, (4) the higher the amount of auxin, the more rapid the rate of differentiation (hence, the duration of vessel differentiation increases from shoots to roots), (5) vessel diameters are determined by the rate of differentiation, with rapid differentiation resulting in narrow vessels and slow differentiation in wide vessels (i.e., basipetal decrease in auxin is associated with a progressive increase in vessel diameters), and (6) the higher the auxin concentration, the higher is the frequency of vessel elements. Hence, the frequency of vessels decreases from shoots to roots. When Aloni and Zimmermann (1983) blocked the downward flow of auxin in red maple stems, the vessels above the blockage (where auxin concentration was high) were narrow and vessel frequency was increased.

Formation of Rays The formation and development of rays appear to be regulated by hormones, with ethylene prominently involved. Lev-Yadun and Aloni (1995) suggested that ethylene originating in the xylem flows outward, disturbing radial auxin flow, and regulates both the initiation of new rays and the expansion of existing rays. The formation of new rays following stem wounding (when large amounts of ethylene accumulate) suggests that ethylene is involved in the conversion of fusiform initials to ray initials. The importance of ethylene in regulating ray development also was emphasized by Yamamoto *et al.* (1987a), who found that application of Ethrel to stems of American elm seedlings was followed by formation of abnormally wide rays composed of unusually large ray cells.

Formation of Resin Ducts   Formation of resin ducts appears to be regulated by growth hormones. Ethephon (Ethrel), which increases production of ethylene, when applied to stems of Aleppo pine and Japanese red pine seedlings, stimulated production of resin ducts in the xylem (Yamamoto and Kozlowski, 1987b,c). It appeared that when internal ethylene reached a critical level it caused a change in cambial activity from production of mostly tracheids and a few resin ducts to a higher proportion of the latter. Resin ducts also may form in response to exogenous auxin. However, resin ducts did not form until several weeks after auxins were applied to Aleppo pine stems (Fahn and Zamski, 1970), but they formed shortly after ethephon was applied (Yamamoto and Kozlowski, 1987b). Hence, the effects of auxin on formation of resin ducts may involve auxin-induced ethylene formation (Fahn, 1988). Oleoresin production in red pine stems was greatly stimulated by applied ethephon, indicating that ethylene may be a predominating hormonal factor in regulation of differentiation of traumatic resin ducts as well as oleoresin synthesis (Wolter and Zinkel, 1984).

Reaction Wood   A consequence of leaning of trees is a redistribution of the amount and nature of cambial growth on the upper and lower sides of the stem and formation of abnormal "reaction" wood. Reaction wood, which is called compression wood in gymnosperms because it occurs on the lower (leeward) side and tension wood in angiosperms because it occurs on the upper side of leaning stems, is well documented in trees on persistently windy sites (Savill, 1983). Reaction wood has several commercially undesirable characteristics such as a tendency toward shrinking, warping, weakness, and brittleness which adversely affect its utilization.

In gymnosperms, the compression wood forms preferentially on the lower side of inclined stems and branches (Fig. 3.33), but sometimes it forms in various amounts on opposite sides of stems. Formation of compression wood usually involves increased xylem production on the lower sides of leaning stems and growth inhibition on the upper side. Compression wood functions in the righting of inclined stems and branches, apparently by expanding during and after cell differentiation. As soon as vertical orientation is attained, compression wood stops forming (Westing, 1965, 1968; Scurfield, 1973).

The structure of compression wood differs greatly from that of normal wood, with compression wood having tracheids that are more nearly rounded in cross section and with large intercellular spaces (Fig. 3.34). Compression wood tracheids are very thick walled, so no ready distinction can be made between earlywood and latewood. The inner layer (S_3) of the

COMPRESSION
WOOD

Figure 3.33 Eccentric cambial growth and compression wood in ponderosa pine. Photo courtesy of U.S. Forest Service, Forest Products Laboratory.

secondary wall of compression wood tracheids usually does not form, or it develops poorly. The inner layer (the middle S_2 layer in normal wood) of compression wood is very thick. Because of more anticlinal divisions in compression wood, the tracheids generally are much shorter than in normal wood. Sometimes a higher frequency and larger size of ray cells are noted in compression wood, reflecting a higher overall rate of growth. Timell (1973) did not find significant ultrastructural differences between tracheids of normal wood and compression wood. Tracheids in compression wood have more cellulose and lignin than those in normal wood. Also, the cellulose of compression wood is less crystalline than that of normal wood.

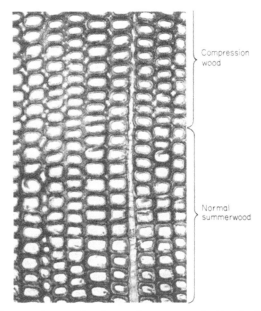

Figure 3.34   Compression wood and normal summerwood of longleaf pine. Photo courtesy of U.S. Forest Service, Forest Products Laboratory.

Tension wood forms characteristically on the upper sides of leaning angiosperm trees. Often, but not always, formation of tension wood is associated with eccentric cambial growth. When both increased xylem production and tension wood occur, their locations usually coincide, but sometimes they occur on opposite sides of branches or leaning stems. In S-shaped or recurved stems, the eccentric growth may occur on different sides at various stem heights. In deciduous angiosperms, tension wood is most developed in earlywood and does not extend through all the latewood. In some evergreen angiosperms, such as *Eucalyptus* spp., however, tension wood often extends throughout both the earlywood and latewood (Wardrop and Dadswell, 1955). A characteristic feature of tension wood is the preponderance of gelatinous fibers, which can be identified microscopically by the gelatinous appearance of their secondary walls (Fig. 3.35).

Tension wood is not so easily identified (except microscopically) as is compression wood. It has fewer and smaller vessels than normal wood and proportionally more thick-walled fibers. Ray and axial parenchyma cells are unchanged. Vessels of tension wood usually are well lignified, but sometimes lignification is reduced. Reduced lignification of the G layer (see below) is a common feature.

The cell wall layer designated as S_2 or S_3 in normal wood is replaced in ten-

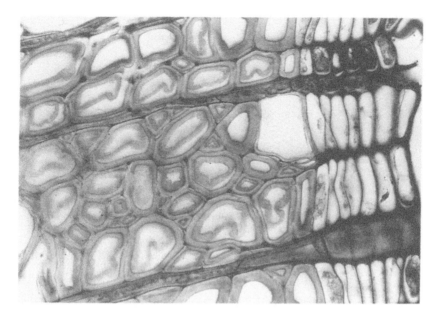

Figure 3.35 Transverse section of poplar stem showing cambial zone cells (right) and subjacent tension wood cells (left). Photo courtesy of Graeme Berlyn, Yale University, New Haven, Connecticut.

sion wood by an unlignified, often convoluted layer, designated the G layer. When the S_2 layer is thus replaced, the new designation is $S_2(G)$. Sometimes the G layer is produced in addition to the S_1, S_2, and S_3 layers of normal xylem (Fig. 3.36). As the G layer consists almost wholly of cellulose, it has a highly ordered parallel molecular orientation. Both the thickness and form of the G layer vary conspicuously among species as well as among trees within a species. In some genera, such as *Acacia*, the G layer is convoluted, and in *Eucalyptus gigantea* it may almost fill the cell lumen (Wardrop, 1961).

Formation of reaction wood is a geotropic phenomenon that is associated with internal redistribution of hormonal growth regulators. For example, compression wood forms toward the focus of gravitational attraction under conditions of a reversed gravitational field. If trees are grown upside down, compression wood forms on the morphologically upper sides of branches, which physically are the lower sides. When trees are grown on a revolving table, the reaction wood forms on the outer side of the stem.

Several investigators concluded that increased cambial growth on the leeward sides of tilted stems occurs because of a high auxin gradient that causes mobilization of food. Many investigators attributed formation of compression wood to high auxin levels (for review, see Timell, 1986, Vol.

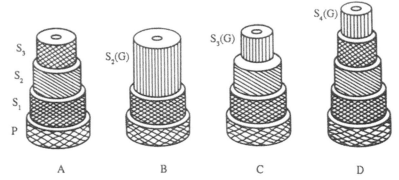

Figure 3.36 Organization of cell walls in wood fibers. (A) Normal wood fiber of structure
$P + S_1 + S_2 + S_3$; (B) tension wood fiber of structure $P + S_1 + S_2(G)$; (C) tension wood fiber
of structure $P + S_1 + S_2 + S_3(G)$; (D) tension wood fiber of structure $P + S_1 + S_2 + S_3 + S_4(G)$. P, Primary wall; S_1, outer layer of secondary wall; S_2, middle layer of secondary wall; S_3,
inner layer of secondary wall; S_4, gelatinous layer; G, unlignified layer. After Wardrop (1964).

2). Induction of compression wood often follows application of IAA to
buds as well as to stems. Compression wood that was induced by applied
IAA could not be distinguished from naturally occurring compression
wood by appearance or physical properties (Larson, 1969). Furthermore,
application of inhibitors of auxin transport to inclined stems stops formation of compression wood (Phelps *et al.*, 1977; Yamaguchi *et al.*, 1983).

Formation of tension wood is associated with auxin deficiency. This is
shown by lower concentrations of auxin on the upper sides of tilted stems
(Leach and Wareing, 1967), inhibition of tension wood formation by auxins applied to the upper sides of tilted stems, and induction of tension
wood by applied auxin antagonists such as TIBA (2,3,5-triiodobenzoic
acid) or DNP (2,4-dinitrophenol) (Morey, 1973). The few and small vessels
in tension wood also are associated with auxin deficiency. Decreases in
vessel frequency from the leaves to the roots are correlated with a gradient
of decreasing auxin concentration (Zimmermann and Potter, 1982).

An important role has been ascribed to ethylene in production of reaction wood, including compression wood in gymnosperms (Barker, 1979)
and tension wood in angiosperms (Nelson and Hillis, 1978). Brown and
Leopold (1973) stimulated ethylene production in pine stems bent in
loops and suggested that ethylene played a regulatory role in formation of
compression wood. This view also was supported following detection of
endogenous ACC (1-aminocyclopropane–1-carboxylic acid), an auxin precursor, in compression-wood-associated cambium but not in the opposite-
wood-associated cambium of lodgepole pine (Savidge *et al.*, 1983).

Other studies questioned whether ethylene has a direct role in formation of compression wood. Flooding of soil stimulated ethylene production by Aleppo pine and induced formation of abnormal xylem with round, thick-walled tracheids and surrounded by intercellular spaces (Yamamoto *et al.*, 1987b). However, these tracheids developed an S_3 wall layer which is not present in well-developed compression wood (Côté and Day, 1965). Application of Ethrel to Aleppo pine seedlings induced formation of slightly abnormal tracheids but without the essential features of those in compression wood (Yamamoto and Kozlowski, 1987b). Tilting of Japanese red pine seedlings induced compression wood to form, whereas application of Ethrel to stems of upright plants did not. Furthermore, application of Ethrel to tilted seedlings counteracted the formation of compression wood even though the ethylene contents of stems increased appreciably (Yamamoto and Kozlowski, 1987c).

Ethylene also has been associated with formation of tension wood (Nelson and Hillis, 1978). However, the following lines of evidence indicate that ethylene did not directly induce formation of tension wood in Norway maple seedlings: (1) ethylene production of stems was stimulated more by flooding of soil than by tilting of stems, but only tilting induced formation of tension wood, (2) the increase in ethylene of tilted seedlings was as great or greater on the lower side than on the upper side of the stem, but tension wood formed only on the upper side, (3) application of Ethrel to upright stems did not induce formation of tension wood, and (4) application of Ethrel to tilted stems inhibited formation of tension wood (Yamamoto and Kozlowski, 1987f). Application of Ethrel to upright stems of American elm seedlings increased their ethylene contents but did not induce formation of tension wood (Yamamoto and Kozlowski, 1987a).

We interpret the weight of evidence as showing that formation of reaction wood is regulated by hormonal interactions, with auxin playing a primary role. Because ethylene regulates auxin levels by influencing auxin synthesis, it may have an indirect role in the formation of reaction wood. However, our emphasis on auxin and ethylene should not obscure the extreme complexity of regulation of various aspects of cambial growth and involvement of other compounds in the formation of reaction wood.

Hormonal Interactions

We emphasize that specific hormones interact with other hormones and with nonhormonal substances as well in regulating cambial growth (see Chapter 13 in Kozlowski and Pallardy, 1997). The abnormal cambial growth below a phloem blockage is associated with a deficiency of both hormonal growth regulators and carbohydrates. In some species cell division to produce xylem is curtailed more rapidly than cell wall thickening below stem girdles, suggesting that different control systems are involved in various phases of cambial activity. In eastern white pine stems, in which the phloem was severed after seasonal xylem production began, tracheid production below the phloem blockage persisted for only a few weeks, whereas thickening of tracheid walls continued for several months, although at a reduced rate (Wilson, 1968). The data suggested that hormone deficiency below the phloem blockage controlled cell division primarily, whereas carbohydrates, which were present as reserves below the stem girdles and were depleted slowly, had a greater effect than hormonal supplies on thickening of tracheid walls. Nevertheless, hormonal growth regulators have an important role in building of secondary walls by influencing incorporation of sucrose into the walls of cambial derivatives (Gordon and Larson, 1970; Kozlowski, 1992).

That auxin alone does not regulate cambial growth is shown by growth responses to exogenous IAA of plants of different age and degree of foliation. In defoliated stem segments of lodgepole pine, applied IAA promoted division of cambial cells and differentiation of tracheids along the entire lengths of treated shoots of 1-year-old stems but not older stem segments (Savidge and Wareing, 1981; Savidge, 1988). Mature conifer needles produced a tracheid differentiation factor which was necessary, in addition to IAA, for normal cambial growth (Savidge and Burnett, 1993). When Denne and Wilson (1977) applied IAA to disbudded spruce shoots, both tracheid diameters and wall thickness were similar to those of controls. However, when intact plants were treated with IAA, both tracheid diameters and wall thickness increased, suggesting that IAA acted synergistically with some compound translocated from the buds. That IAA alone does not regulate cambial growth also is shown by the variations in cambial response to IAA applied at different times in the growing season. Tracheid production was stimulated less by application of IAA to Scotch pine shoots late in the growing season than by earlier application (Wodzicki *et al.*, 1982).

Whereas exogenous IAA or α-naphthaleneacetic acid (NAA) initiated cambial growth in disbudded Scotch pine shoots, GA or cytokinins did not. However, normal xylem was produced only by applied IAA (or NAA) plus GA and cytokinins (Hejnowicz and Tomaszewski, 1969). Synergistic effects of IAA plus GA were shown on cambial growth of sycamore maple and Lombardy poplar (Wareing, 1958). When GA_3 alone was applied to disbudded poplar shoots some abnormal xylem formed. Fully differentiated xylem formed following applications of low GA_3 and high IAA concentrations. Phloem production was stimulated by high GA_3 and low IAA concentrations (Fig. 3.37). Similarly, IAA and GA_3 synergistically influenced cambial growth of English oak (Zakrzewski, 1983).

Differentiation of various cell types in American elm xylem was correlated with the concentration of Ethrel applied (Yamamoto *et al.*, 1987a). However, endogenous ethylene probably acted by enhancing the effect of a

A

Mean xylem width (epu)

B

Mean phloem width (epu)

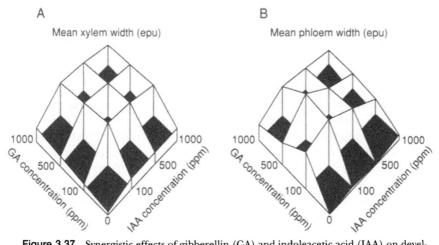

Figure 3.37 Synergistic effects of gibberellin (GA) and indoleacetic acid (IAA) on development of xylem (A) and phloem (B) in *Populus*. The growth regulator concentrations shown were those in lanolin and do not indicate concentrations in the tissues. Xylem width and phloem width are given in eyepiece units (epu). From Wareing *et al.* (1964).

low auxin content in the lower stem. Several compounds are involved in regulating anatomical changes that follow applications of Ethrel to plants. Endogenous auxins, cytokinins, ABA, and gibberellin stimulate ethylene production and modify its influence on plants (Fuchs and Lieberman, 1968; Abeles, 1973).

The Ethrel-induced formation of resin ducts in the xylem of Japanese red pine (Yamamoto and Kozlowski, 1987c) probably was not caused by ethylene alone because such ducts form in response to exogenous auxins, and auxins stimulate ethylene production in plants (Fahn and Zamski, 1970; Abeles, 1973). Cytokinins also stimulate ethylene synthesis, whereas ABA inhibits it (Reid and Bradford, 1984).

Internal control of lignification is complex, with auxins, cytokinins, ethylene, and other compounds variously involved (Wardrop, 1981; Miller *et al.*, 1985). When present together in optimal amounts, auxins and cytokinins influenced lignification of differentiating tracheary elements in tissue cultures (Fukuda and Komamine, 1985; Savidge and Wareing, 1981). Xylem differentiation in intact plants (Rhodes and Wooltorton, 1973) and tissue cultures (Kuboi and Yamada, 1978) was correlated with activity of phenylalanine ammonia-lyase (PAL), which is regulated by both auxins and cytokinins (Haddon and Northcote, 1975; Kuboi and Yamada, 1978). Formation of cell walls in Norway spruce tracheids was regulated by interactions of IAA, Ca^{2+}, and ethylene (Eklund, 1991). Cellulose deposition was progressively stimulated by IAA as the amount of Ca^{2+} increased. By comparison, IAA promoted lignification and deposition of noncellulosic polysaccharides only in the absence of Ca^{2+}. With addition of Ca^{2+} the IAA effect disappeared. Similarly, IAA promoted deposition of noncellulosic polysaccharides only in the absence of ethylene. At increasing levels of ethylene, the effect of IAA on noncellulosic polysaccharides disappeared and IAA promoted deposition of cellulose.

Interactions between inhibitors and growth promoters also appear to be involved in control of cambial growth. For example, high concentrations of inhibitors extracted from European larch tissues reduced the effect of auxin on xylem differentiation. A slight synergistic action was demonstrated between IAA and inhibitors at low concentrations of both substances (Wodzicki, 1965). Cronshaw and Morey (1965) also demonstrated that exogenously applied growth inhibitors modified the auxin stimulation of cambial activity and xylem differentiation.

Substances other than growth hormones apparently also influence cambial growth. For example, a high sucrose to auxin ratio promotes phloem differentiation, whereas other combinations of these stimulate xylem differentiation (Wetmore and Rier, 1963). Robards *et al.* (1969) applied IAA, NAA, 2,4-D (2,4-dichlorophenoxyacetic acid), GA, FAP (6-furfurylaminopurine), myoinositol, and sucrose singly and in mixtures to the apical ends of disbudded willow stems. Each of these substances had some effect on xylem differentiation. Production and differentiation of xylem cells were increased most when IAA, GA, and FAP were applied together, but the response was greater when the mixture was augmented by inositol or sucrose. Wodzicki and Zajaczkowski (1974) applied auxins, together with various vitamins and substances known to regulate cell metabolism, to decapitated stems of young Scotch pine trees. Several compounds, particularly inositol, vitamin A, and pyridoxine, acted synergistically with auxin in regulating differentiation of cambial derivatives. High levels of putrescine in the cambial region of Norway spruce shoots suggested that polyamines may also be involved in regulation of cambial growth (Königshofer, 1991).

Root Growth

Growth of roots depends to a large degree on a continuous supply of apically produced hormonal growth regulators. Defoliation, phloem blockage, and insect or fungus injury to leaves decrease the downward flow of growth hormones and inhibit root growth, even when the amounts of nonstructural carbohydrates in roots are high (Fig. 3.38). Following defoliation of seedlings of mountain ash (*Eucalyptus regnans*), root elongation stopped in 2 to 4 days (Wilson and Bachelard, 1975).

Mycorrhizae   Over the years there has been vigorous debate about the relative importance of excess carbohydrates and hormonal growth regula-

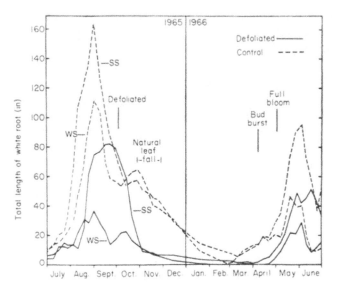

Figure 3.38 Effects of defoliation on root development of Worcester Pearmain apple trees from July, 1965, to June, 1966. SS, Strong shoot growth; WS, weak shoot growth. From Head (1969).

tors in controlling formation of mycorrhizae. Whereas Björkman (1970) contended that the concentration of carbohydrates in roots was the primary regulator, Slankis (1973, 1974) concluded that auxin of fungal origin was more important.

The carbohydrate theory was supported by the following major lines of evidence (Björkman, 1970): (1) mycorrhizal fungi depend entirely on carbohydrates supplied by the host for their energy, (2) mycorrhizae do not become established until sugar in the roots of the host plants accumulates to a threshold concentration, and (3) plants with adequate N have lower carbohydrate levels than plants showing N deficiency. Björkman (1970) emphasized that mycorrhizae did not form when the levels of carbohydrates were low in high-N plants. Hence, he reasoned that the regulatory function of soil N was mediated through soluble carbohydrates in the host plant.

The importance of auxin in regulating mycorrhizal development was supported by the following observations (Slankis, 1973, 1974): (1) structures resembling mycorrhizae are induced by fungal exudates or high concentrations of applied auxins, (2) when the supply of exogenous auxin is terminated, formation of mycorrhizae stops and root elongation and formation of root hairs follow, (3) high levels of exogenous N inhibit formation of mycorrhizae, and (4) high concentrations of auxins are present in mycorrhizal roots whereas nonmycorrhizal roots contain only traces.

Nylund (1988) reviewed the evidence for both the carbohydrate and auxin theories and concluded that neither had conclusively been proved experimentally. As early as 1973 Hacskaylo postulated that mycorrhizal development was a complex process and was not controlled by carbohydrate supply or auxin alone. He envisioned a control mechanism that incorporated sequential involvement of different regulatory compounds. Hacskaylo (1973) suggested that root exudates stimulated spore germination and growth of hyphae. In roots with low carbohydrate concentrations, an unidentified "M factor" was inoperative, and hence infection did not occur. When the fungus contacted a root, subsequent root growth was altered by an exuded metabolite or metabolites which might be auxin or some other growth hormone or hormones. Subsequently, more carbohydrates were translocated to the roots in accordance with their strong sink strength or some effect of growth hormones. Uptake of mineral nutrients was increased, thereby increasing photosynthesis and auxin production by the host. Nylund and Unestam (1982) concluded that development of mycorrhizae is regulated by interactions among carbohydrates, auxins, and enzymes. They divided mycorrhizal formation into several sequential stages, including (1) stimulation of fungal growth by root metabolites, (2) formation of a hyphal envelope on the root, (3) intercellular penetration by single hyphae, (4) change in fungal morphology with production of labyrinthic tissue leading to formation of a Hartig net, and (5) extension of labyrinthic tissue to form a mantle.

Models of interactive effects of several regulatory compounds on growth of mycorrhizae are supported by a growing mass of data (Nylund, 1988). Because much evidence shows that growth of plants often is influenced more by factors controlling the use of carbohydrates than by carbohydrate supplies alone, and that growth is regulated by hormonal interactions, it seems unlikely that carbohydrate or auxin levels alone control mycorrhizal initiation and development. For example, auxin effects are mediated by various cofactors and at least partly by ethylene (Rupp and Mudge, 1985).

Rooting of Cuttings   Physiological control of rooting of cuttings involves subtle relations among carbohydrates, N compounds, enzyme activity, and growth hormones (Haissig, 1974a–c, 1982a,b). Normal metabolism must be greatly altered to regenerate a root system when preparation of cuttings eliminates absorption of water and mineral nutrients from the soil. Hence, both leaf-produced carbohydrates and growth hormones play an important role in root induction and development. Ordinarily mineral nutrition is less crucial because root formation usually is not influenced by a deficiency or excess of mineral nutrients (except for N and rarely for B). Water deficits in cuttings are common and influence root initiation through their effects on metabolism of carbohydrates and hormones (Gaspar and Coumans, 1987).

Root formation at the ends of cuttings requires that certain cells have a capacity to sequentially dedifferentiate, become meristematic, and differentiate into root initials. Control of rooting of cuttings may involve somewhat different physiological controls in at least six successive developmental stages (Gaspar and Hofinger, 1988): (1) induction (the period preceding cell division), (2) transverse first division of pericycle cells, (3) longitudinal first division of daughter cells, (4) continued cell divisions without increase in gross volume of the meristematic cluster of cells, (5) volume increase of the cell cluster by cell expansion, and (6) protrusion of roots. Variations in control of these phases can be inferred from known differences of sensitivity to IAA at different stages of plant development (Imaseki, 1985).

Hormonal growth regulators play an important role in stimulating rooting of cuttings. This is shown by increased rooting following treatment of cuttings with synthetic hormones. An important regulatory role of auxin in the hormonal complex that regulates initiation of roots on cuttings is shown by several lines of evidence. The effects of young leaves and active buds, which are sources of auxin and generally stimulate rooting, often can be replaced with exogenous auxin. In many species of woody plants, correlations are high between the content of endogenous free auxin and rooting capacity when the auxin level is high at the time the cuttings are made. The free auxin content increases in the bases of cuttings sometime prior to rooting (Gaspar and Hofinger, 1988). The IAA content at the base of Scotch pine cuttings that had been treated with indolebutyric acid (IBA) was three times as high as in untreated cuttings (Dunberg *et al.*, 1981). In cuttings of *Cotinus coggyria* taken in the spring (which rooted readily), the IAA content was approximately 10 times as high as in autumn cuttings (which rooted poorly) (Blakesley *et al.*, 1985). The failure of cuttings of broad-leaved trees to root was associated with low auxin levels (Smith and Wareing, 1972a,b). Inhibition of rooting by TIBA (triiodobenzoic acid), which inhibits auxin transport and action, also supports the essentiality of auxin for root initiation (Gaspar and Coumans, 1987).

Other studies questioned the regulatory role of auxin alone in rooting of cuttings. For example, the amounts of IAA in cuttings of *Rhododendron ponticum* (easy to root) were similar and did not vary appreciably during the season (Wu and Barnes, 1981). High rooting capacity is associated with high levels of cofactors that act synergistically with auxins. Considerable evidence shows that auxins are oxidized or conjugated with other compounds before they exert a role in inducing formation of root primordia. Auxins readily form conjugates with amino acids and sugar alcohols

(Cohen and Bandurski, 1982). Because rapid decreases in IAA precede root initiation and coincide with formation of IAA conjugates (Haissig, 1986; Gaspar and Hofinger, 1988), several investigators questioned that unmodified auxins induce initiation of root primordia.

The best rooting of rose and apple cuttings followed treatment of the cut ends with IAA. Before rooting occurred, however, IAA increased but then rapidly disappeared as IAA conjugates formed (Collet and Le, 1987). Rooting of jack pine cuttings was stimulated more by synthetic phenyl esters, phenyl thioesters, and phenyl amides of IAA and IBA than by the free acids (Haissig, 1979, 1983).

Although auxins stimulate development of root primordia, they alone apparently do not induce root initiation (Haissig, 1974b; Kling *et al.*, 1988). For example, IAA alone induces root primordia only in predisposed cells, and the predisposition is not caused by IAA alone. Induction of initiation of root primordia requires auxin synergists that are synthesized in young leaves and translocated to the bases of cuttings (Haissig, 1974a). A variety of unrelated compounds may function as auxin synergists. Examples are indole-, phenylacetic-, and phenylbutyric acids, α- and β naphthol, sodium bisulfite, phenols, and steroid hormones (Gaspar and Coumans, 1987). In most cases, sterols alone do not promote or only slightly promote rooting, but they act synergistically with auxin. Interference by sterols of auxin metabolism was indicated when vitamin D_2 stimulated root formation by altering peroxidase activity during the inductive phase of rooting (Moncusin and Gaspar, 1983).

Rooting also may be stimulated by leaf-produced compounds that are not auxin synergists. An extract from the leaves of Amur maple, that was not synergistic with IAA, strongly stimulated root initiation in the difficult-to-root species *Acer griseum* (Kling *et al.*, 1988). Some investigators reported that exogenous ethylene stimulated adventitious root formation on cuttings (Kawase, 1971; Dua *et al.*, 1983; Gonzalez *et al.*, 1991). Production of adventitious roots by flooded plants has been attributed to high endogenous ethylene levels (Kawase, 1972).

Under some conditions, ethylene has a contributory role in the sequence of events that lead to adventitious root formation, but it does not appear to have a primary, direct effect. This is shown by the inability of some investigators to induce rooting with applied ethylene (Shanks, 1969; Schier, 1975) and by the failure of environmentally stressed plants with high ethylene levels to initiate adventitious roots. Stimulation of rooting by ethylene application usually is more successful in intact plants than in cuttings, in herbaceous plants than in woody plants, and in plants with preformed root primordia than in plants lacking such primordia (Mudge, 1988).

Stimulation of ethylene production is a common plant response to drought, temperature extremes, insect attack, air pollution, and disease. However, high endogenous ethylene levels generally do not induce production of adventitious roots in plants exposed to such stresses. In flooded plants, adventitious rooting often is unrelated to levels of endogenous ethylene. For example, more ethylene was produced by flooded Tasmanian blue gum seedlings than by Murray red gum seedlings, yet Murray red gum produced more adventitious roots (Tang and Kozlowski, 1984b). Flooding of bur oak seedlings increased ethylene production by more than 500%, yet only a few adventitious roots formed (Tang and Kozlowski, 1982a). Flooding also greatly stimulated ethylene production in paper birch seedlings, but adventitious roots did not form on submerged stems (Tang and Kozlowski, 1982c). Such observations suggest that compounds other than ethylene are required for production of adventitious roots. Ethylene contents of stems of flooded trees often are as high or higher above the water line than below it, but adventitious roots form only on the submerged portions of stems (Tang and Kozlowski, 1984a,b). Ethylene was less important than auxins transported from the shoots in regulating formation of adventitious roots of flooded box elder seedlings. Blocking of downward auxin transport with NPA (*N*-1-naphthylphthalamic acid) reduced formation of adventitious roots in flooded seedlings, even when large amounts of ethylene were present in the lower stem. Application of NAA to the stems, below the height at which NPA had been applied, restimulated production of adventitious roots (Fig. 3.39). These observations were consistent with a voluminous literature showing the essentiality of auxins for development of adventitious roots.

Exogenous gibberellins generally inhibit formation of adventitious roots but under certain conditions may stimulate root formation. Exposure of stock plants and cuttings to low irradiance or short days may reduce the inhibitory effect of gibberellins on rooting. Auxins also counteract the inhibitory effects of gibberellins on root formation (Hansen, 1988).

The effects of gibberellins on rooting vary with the stage of root development. For example, GA_3 was inhibitory during the phase that preceded formation of organized root primordia in Monterey pine. In contrast, it stimulated rooting when applied at the time of the first observable stage of root initiation. When applied after meristemoids were established, it inhibited root formation (Smith and Thorpe 1975a,b). This observation was consistent with inhibition by GA_3 of cell division in established root primordia of brittle willow noted by Haissig (1986).

Unlike auxins, applied cytokinins usually inhibit root formation on cuttings (Van Staden and Harty, 1988). Exogenous cytokinins inhibited rooting of cuttings of red maple and Murray red gum (Bachelard and Stowe, 1963). However, the effects of cytokinins vary with plant species, genotype, cytokinin dosage, the specific cytokinin applied, and the time of applica-

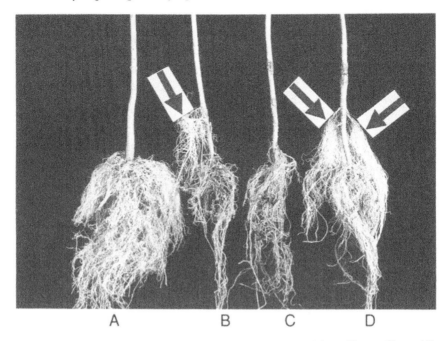

Figure 3.39   Effects of various treatments on root growth of box elder seedlings. (A) Unflooded seedling. (B) Flooded, lanolin-treated seedling. (C) Flooded, NPA-treated seedling. (D) Flooded, NPA plus NAA-treated seedling. In (D), NPA was applied to the stem at 4 cm above the water level and NAA at 3 cm above the water level. Adventitious roots (arrows) formed in B and D but not in A or C. Reprinted from *Environ. Exp. Bot.* **27**, Yamamoto and Kozlowski, Effects of flooding of soil and application of NPA and NAA to stems on growth and stem anatomy of *Acer negundo* seedlings, 329–340. Copyright 1987, with kind permission from Elsevier Science Ltd., The Boulevard, Langford Lane, Kidlington OX5 1GB, UK.

tion. Kinetin has a strong inhibitory effect on root initiation if applied during the preinitiative stage of root development, but this effect is lessened or disappears if applied after meristemoids are established.

Rooting was inhibited less by the synthetic cytokinins kinetin and ribosylkinetin than by natural cytokinins (Forsyth and Van Staden, 1986), and cytokinin ribosides inhibited rooting less than the free bases (Wightman *et al.*, 1980). Whereas root formation in myrobalan, sweet cherry, peach × almond, and quince was stimulated in shoot cultures maintained without exogenous auxin, kinetin and zeatin did not stimulate root formation (Nemeth, 1979). The auxin–cytokinin balance also controls root initiation, with high cytokinin–auxin ratios favoring shoot growth over root growth.

Complexity of Control of Rooting of Cuttings   The weight of evidence indicates that control of rooting of cuttings is complex and involves regulatory influences of several different compounds (Haissig, 1986; Davies and Hartmann, 1988). Development of roots on cuttings has the following prerequisites: (1) availability and receptivity of parenchyma cells for regeneration of meristematic regions and (2) modifications of a complex of phenolic compounds, growth promoters, inhibitors, cofactors, auxins, enzyme systems, and carbohydrate accumulation and partitioning as well as N compounds for synthesis of proteins and nucleic acids (Haissig, 1974b,c, 1982a,b; Davies and Hartmann, 1988). A deficiency of appropriate enzymes or enzyme activators, separation of reactants by cellular compartmentation, or a deficiency of cofactors may account for failure of cuttings to form adventitious roots (Haissig, 1972, 1974a).

Summary

The major internal requirements for growth of woody plants include adequate supplies of carbohydrates, mineral nutrients, water, and hormonal growth regulators at meristematic sites.

Most carbohydrates are transported from sources to sinks (utilization sites) during the growing season. Carbohydrates move downward, upward, and laterally in woody plants. The pressure flow ("mass flow") hypothesis is the most widely supported mechanism for carbohydrate transport. This involves a passive, bulk flow of sugar solution in the sieve tubes of the phloem under a concentration gradient.

Both stored and currently produced carbohydrates are used in metabolism and growth. Reserve carbohydrates are more important for early shoot growth in deciduous species than in evergreens. Species with fixed growth use less carbohydrates for shoot elongation than species with free or recurrently flushing growth patterns. Both reserve and currently produced carbohydrates are used in production of xylem and phloem mother cells, their division and differentiation into xylem and phloem cells, and expansion of the cambial sheath. Large amounts of carbohydrates are used in formation of secondary walls of xylem cells including synthesis and deposition of cellulose, hemicellulose, and lignin. Carbohydrates are used in metabolism, elongation, and thickening of roots, and in formation and growth of new roots.

Respiration is required for growth as shown by direct correlation between CO_2 production or metabolic heat rate and growth.

Almost all transport of mineral nutrients from roots to shoots occurs in the xylem sap, caused by mass flow in the transpiration stream. Minerals also diffuse out of the xylem during their ascent. The rate of transport from leaves of foliar-absorbed minerals varies greatly with the mobility of different ions. A large portion of the mineral nutrients in leaves moves back into twigs and branches before the leaves are shed.

Deficiencies of mineral nutrients are common and inhibit growth. Leaf growth often is reduced before visible symptoms of deficiencies are evident. Reduction in leaf area results from fewer and smaller leaves and increased leaf shedding. Mineral deficiency also induces chlorosis, injury to leaves, and shoot dieback. Cambial growth of mineral-deficient plants is reduced as a result of lowered availability of carbohydrates and hormonal growth regulators. The decreased availability of carbohydrates results more from a reduced leaf area than from lowered photosynthetic efficiency. Mineral deficiencies not only inhibit root growth but also increase proportional translocation of carbohydrates to the root system, hence increasing the root-shoot ratio.

Water deficits directly and indirectly inhibit growth. Both extensibility of cell walls and turgor are important for growth of cells. In general, cell enlargement is more sensitive than cell division to dehydration. Growth of cells begins when loosening processes cause their walls to relax. This causes a decrease in cell turgor and induces a water potential gradient between the cell and its external environment. Water moves into the cell in response to this water potential gradient, thereby increasing cell volume.

Shoot growth is decreased by effects of water deficits on bud formation, leaf growth, internode elongation, and premature leaf shedding. In species exhibiting fixed growth of shoots, water deficits during the year of bud formation control shoot length more than do water deficits during the year of bud opening and expansion into a shoot. In contrast, summer droughts during shoot expansion have greater effects on shoot length of species with free or recurrently flushing growth patterns. Water deficits inhibit diameter growth of stems and branches by decreasing the size of xylem and phloem increments. Water deficits generally have a minor role in regulating initiation of cambial activity but a major role in controlling subsequent cambial growth. Water deficits induce early formation of latewood, and continued drought shortens the period of latewood formation. Water deficits inhibit initiation, elongation, and cambial growth of roots. On arid sites, penetration of soil by roots usually is restricted to the depth to which the soil is wetted.

Flooding of soil leads to inhibition of vegetative growth, injury, and death of many woody plants. Adverse effects on shoot growth include inhibition of leaf formation and expansion, as well as reduced internode elongation, and premature chlorosis, leaf senescence, and abscission. Soil inundation not only reduces cambial growth but often alters stem anatomy. The small root systems of flooded plants result from inhibition of formation and growth of roots as well as root decay. The complex mechanisms by which water excess affects plant growth involve altered carbohydrate, mineral, and hormone relations.

All the major classes of plant hormones except ethylene are found in both xylem and phloem and move for long distances. In addition to moving rapidly in the vascular system, hormones move slowly through ground tissues.

Interactions among hormonal growth regulators influence several aspects of growth. Release of endodormancy of buds is a sequential process, with each phase being characterized by different metabolic and hormonal effects. Both gibberellins and auxins regulate shoot growth. Auxins, cytokinins, and gibberellins delay leaf senescence and abscission; ethylene and ABA accelerate abscission. Several amino acids and free fatty acids also promote senescence, whereas polyamines inhibit senescence.

Each of the major plant growth hormones is present in the cambial region and may be involved in controlling production and differentiation of cambial derivatives. Regulation is exerted by interactions among hormones and through second messengers which may not be hormones. Auxins play a predominant role in controlling mitosis of fusiform cambial initials and differentiation of cambial derivatives. Evidence for the role of gibberellins in regulating cambial growth is inconsistent. There is uncertainty about the precise role of ABA in regulating cambial growth. Ethylene appears to have a synergistic effect on cambial growth in the presence of auxin and cytokinins. The effects of hormones on cambial growth vary with seasonal changes in cambial sensitivity to the presence of hormones. Seasonal initiation of cambial growth does not appear to be controlled by IAA levels. When cambial cells subsequently regain capacity to respond to a stimulator, auxin levels appear to regulate cambial activity. The sizes of cambial derivatives and thickening of their walls are regulated by apically produced growth hormones. Formation of reaction wood is a geotropic phenomenon that involves internal redistribution of hormonal growth regulators. Formation of compression wood in conifers has been linked to high auxin levels, and formation of tension wood in broad-leaved trees has been attributed to auxin deficiency. Because ethylene regulates auxin synthesis, it may have an indirect role in the formation of reaction wood.

Growth of roots depends on a continuous supply of apically produced hormonal growth regulators. Control of rooting of cuttings is complex and involves subtle relations among carbohydrates, N compounds, enzyme activity, growth hormones, and cofactors.

General References

Addicott, F. T. (1982). "Abscission." Univ. of California Press, Berkeley.

Addicott, F. T. (1991). Abscission: Shedding of parts. *In* "Physiology of Trees" (A. S. Raghavendra, ed.), pp. 273–300. Wiley, New York.

Baker, N. R., Davies, W. J., and Ong, C. K., eds. (1985). "Control of Leaf Growth." Cambridge Univ. Press, Cambridge.

Brett, C., and Waldron, K. (1990). "Physiology and Biochemistry of Plant Cell Walls." Unwin Hyman, London.

Cannell, M. G. R. (1989). Physiological basis of wood production: A review. *Scand. J. For. Res.* **4**, 459–490.

Cutler, H. G., Yokota, T., and Adam, G., eds. (1991). "Brassinosteroids." ACS Symp. Ser. No. 474. American Chemistry Society, Washington, D.C.

Davies, P. J., ed. (1988). "Plant Hormones and Their Role in Plant Growth and Development." Kluwer, Dordrecht and Boston.

Davies, T. D., Haissig, B. E., and Sankhla, N., eds. (1988). "Adventitious Root Formation in Cuttings." Dioscorides Press, Portland, Oregon.

Faust, M. (1989). "Physiology of Temperate Zone Fruit Trees." Wiley, New York.

Hansen, A. D. (1982). Metabolic responses of mesophytes to plant water deficits. *Annu. Rev. Plant Physiol.* **33**, 163–203.

Harley, J. L., and Russell, R. S., eds. (1979). "The Root Soil Interface." Academic Press, London.

Harley, J. L., and Smith, S. E. (1983). "Mycorrhizal Symbiosis." Academic Press, London.

Jackson, M. B., ed. (1986). "New Root Formation in Plants and Cuttings." Martinus Nijhoff, Dordrecht, The Netherlands.

Jones, H. G., Flowers, T. J., and Jones, M. B., eds. (1989). "Plants Under Stress." Cambridge Univ. Press, Cambridge.

Jordan, C. F. (1995). Nutrient cycling in tropical forests. *In* "Encyclopedia of Environmental Biology" (W. A. Nierenberg, ed.), Vol. 2, pp. 641–654. Academic Press, San Diego.

Kossuth, S. V., and Ross, S. D., eds. (1987). Hormonal control of tree growth. *Plant Growth Regul.* **6**, 1–215.

Kozlowski, T. T., ed. (1968–1983). "Water Deficits and Plant Growth," Vols. 1–7. Academic Press, New York.

Kozlowski, T. T., ed. (1984). "Flooding and Plant Growth." Academic Press, New York.

Kozlowski, T. T. (1992). Carbohydrate sources and sinks in woody plants. *Bot. Rev.* **58**, 107–222.

Kozlowski, T. T., and Pallardy, S. G. (1997). "Physiology of Woody Plants," 2nd Ed. Academic Press, San Diego.

Kozlowski, T. T., Kramer, P. J., and Pallardy, S. G. (1991). "The Physiological Ecology of Woody Plants." Academic Press, San Diego.

Kramer, P. J., and Boyer, J. S. (1995). "Water Relations of Plants and Soils." Academic Press, San Diego.

Landsberg, J. J. (1986). "Physiological Ecology of Forest Production." Academic Press, London.

Larcher, W. (1995). "Physiological Plant Ecology," 3rd Ed. Springer-Verlag, Berlin, Heidelberg, and New York.

Larson, P. R. (1994). "The Vascular Cambium: Development and Structure." Springer-Verlag, New York.

Little, C. H. A., and Pharis, R. P. (1995). Hormonal control of radial and longitudinal growth in the tree stem. *In* "Plant Stems" (B. L. Gartner, ed.), pp. 281–319. Academic Press, San Diego.

Marschner, H. (1995). "The Mineral Nutrition of Higher Plants," 2nd Ed. Academic Press, London.

Mitchell, C. P., Ford-Robertson, J. B., Hinckley, T., and Sennerby-Forsse, L. (1992). "Ecophysiology of Short Rotation Forest Crops." Elsevier Applied Science, London and New York.

Nooden, L. D., and Leopold, A. C., eds. (1980). "Senescence and Aging in Plants." Academic Press, San Diego.

Oliveira, C. H., and Priestley, C. A. (1988). Carbohydrate reserves in deciduous fruit trees. *Hortic. Rev.* **10**, 403–430.

Pollock, C. J., Farrar, J. F., and Gordon, A. J., eds. (1992). "Carbon Partitioning within and between Organisms." Bioscientific Pub., Oxford.

Schulze, E.-D., ed. (1994). "Flux Control in Biological Systems." Academic Press, San Diego.

Smith, W. K., and Hinckley, T. M., eds. (1995). "Resource Physiology of Conifers: Acquisition, Allocation, and Utilization." Academic Press, San Diego.

Thimann, K. V. (1980). "Senescence in Plants." CRC Press, Boca Raton, Florida.

Timell, T. E. (1986). "Compression Wood," Vols. 1–3. Springer-Verlag, Berlin and New York.

Wardlaw, I. F. (1990). The control of carbon partitioning in plants. *New Phytol.* **116**, 341–381.

4

Physiological Regulation of Reproductive Growth

Introduction

Physiological regulation of reproductive growth is considered more complex than regulation of vegetative growth. This is because production of fruits, cones, and seeds requires successful completion of each one of a sequential series of distinct stages including (1) initiation of floral primordia, (2) flowering, (3) pollination, (4) fertilization, (5) fruit set, (6) growth of the embryo, and (7) growth of fruits and cones and growth of seeds (see Chapter 4 of Kozlowski and Pallardy, 1997). The internal resources needed to complete all phases of the reproductive process are similar to those for vegetative growth and include carbohydrates, mineral nutrients, water, and endogenous hormones.

Carbohydrate Relations

Reproductive structures are very strong sinks, as shown by rapid transport of large amounts of reserve carbohydrates and current photosynthate into growing flowers and fruits. As growing fruits mobilize carbohydrates, the rates of both photosynthesis and leaf respiration increase. The rate of photosynthesis of a fruiting apple tree was 2 to 2.5 times higher than in a nonfruiting tree (Wibbe *et al.*, 1993). Despite the strong sink strength of reproductive organs, it is unlikely that carbohydrate supplies alone usually control reproductive growth. Nevertheless, there is some evidence that a threshold of reserve carbohydrates is important for flower initiation of some alternate-bearing trees. In Wilking mandarin, nonbearing trees contained a total of 13.26 kg starch and 10.66 kg soluble sugars in vegetative tissues; in comparison, bearing trees contained 2.95 kg starch and 6.75 kg soluble sugars. Presumably, much of the reserve carbohydrate pool would be transported and used in initiation and growth of the next year's fruit crop. Removal of fruits by midsummer was followed by differentiation of flower buds in the following year and was positively correlated with accumulation of carbohydrates (Goldschmidt and Golomb, 1982).

Fruits compete with shoots for carbohydrates, but rapidly growing fruits are much stronger carbohydrate sinks (Kozlowski, 1992). The amounts of hormonal growth promoters are higher in fruits than in shoots, thus conferring a hormone-directed translocation advantage to the fruits. Even when trees are undergoing environmental stress, fruits continue to import carbohydrates, leading to deficiencies in vegetative tissues (Sedgley, 1990). When irrigation was reduced in high-density peach orchards during the pit hardening stage of fruit growth, shoot growth stopped but fruit growth continued during the period of stress. Subsequently, the fruits continued to grow as competition with shoots for water and carbohydrates was reduced. The yield of fruits was increased up to 30% by withholding water to decrease shoot growth (Chalmers *et al.*, 1981).

The strong sink strength of reproductive structures is shown by lower amounts of nonstructural carbohydrates in branches, stems, and roots of trees bearing fruits or cones than in nonbearing trees. Examples of this difference were shown in apple (Hansen, 1967a), avocado (Scholefield *et al.*, 1985), pecan (Wood and McMeans, 1981), peach (Ryugo and Davis, 1959), prune (Hansen *et al.*, 1982), and citrus trees (Goldschmidt and Golomb, 1982). In sweet cherry trees, the small amounts of nonstructural carbohydrates in shoots during the 4 to 6 weeks before harvest were attributed to the strong sink strength of the fruits (Roper *et al.*, 1988).

The amounts of carbohydrates used in metabolism and growth of reproductive structures vary with plant species and genotype, tree vigor and age, time, and developmental patterns of flowers, fruits, and cones (Kozlowski, 1992). Because trees in the juvenile stage of development do not flower, only their vegetative tissues are carbohydrate sinks. The length of the nonflowering period varies from as short as 3 to 4 years in some species to more than 20 years in others. However, it is difficult to identify a specific age of initial and maximal flowering for any species because reproductive growth is greatly influenced by plant competition, tree vigor, and environmental conditions. Even after woody plants reach the adult stage at which they are capable of flowering, they often do not flower and produce fruit each year thereafter (Kozlowski *et al.*, 1991).

Carbohydrate Sources

The carbohydrates used in fruit growth generally are supplied from fairly local sources (Kozlowski, 1992). In apple, peach, pecan, and catalpa, for example, fruit growth depends largely on the carbohydrates synthesized by leaves on the same branch that bears the fruit (Stephenson, 1980; Kozlowski, 1992). Nevertheless, during periods of very rapid fruit growth, some carbohydrates may be transported to shoots from considerably longer distances. Much of the time the carbohydrate requirements of growing kiwi fruits were supplied by two or three nearby leaves. When these could not supply carbohydrates fast enough, the fruits imported carbohydrates over a distance of seven nodes (approximately 1 m) (Lai *et al.*, 1989). The spur leaves of sweet cherry trees do not have the capacity to supply all the carbohydrates used in fruit growth. In addition to carbohydrates provided by leaves on branches on which the fruit is borne, photosynthate for fruit growth is supplied by leaves acropetal to the fruit as well as from reserves and leaves basipetal to the fruit (Roper *et al.*, 1987). In grape, the leaves and fruits are close together. However, following defoliation, photosynthate moved readily from a leaf to a fruit cluster over a distance of 3.8 m (Meynhardt and Malan, 1963). When the distance between the source leaves and apples was increased, fruit growth was not appreciably influenced, further emphasizing long-distance transport of carbohydrates from leaves to these fruits (Hansen, 1977).

A number of deciduous trees flower before their leaves are fully expanded. Examples are willow, birch, maple, pecan, peach, and cherry. In sweet cherry, flowering precedes leaf expansion, and fruits are ready to be harvested in 65 to 75 days after bloom (Roper and Kennedy, 1986). In this species large amounts of reserve carbohydrates are used in flowering. In other fruit trees (e.g., apple), the leaves on fruiting spurs are well expanded and capable of exporting carbohydrates before the flowers emerge. Hence, reserve carbohydrates are used in metabolism and growth of apple flowers only during their early development (Hansen, 1971). In addition to using stored carbohydrates, reproductive structures use large amounts of current photosynthate as shown for many species of angiosperms and gymnosperms.

Variations in Sink Strength

The rate at which carbohydrates are transported into reproductive structures differs seasonally as well as with species and genotype, age of trees, tree vigor, and growth patterns of reproductive structures (Kozlowski, 1992).

The capacity of fruits to mobilize carbohydrates varies seasonally in accordance with differences in their rates of growth. For example, the fruits of catalpa accumulated less than 10% of their final dry weight during the first 5 weeks of growth and the remaining 90% during the following 4 weeks (Stephenson, 1980). The rate at which coffee fruits imported carbohydrates increased rapidly during the first 6 weeks of fruit growth. The rate then slowed but increased rapidly again before and during fruit ripening (Cannell, 1971a,b).

In rapidly growing grape shoots, carbohydrate translocation followed the pattern described for shoots exhibiting free growth, except that it was modified by development of a fruit cluster. In early stages of development of grape shoots, when photosynthate moved into the shoot tip from the first leaf to start exporting, the flower cluster was a weak sink and remained so until after full bloom. Following the set of berries, about 10 days after full bloom, the berry cluster became the dominant sink. At that time, the shoot tip became a somewhat weaker sink as the rate of shoot growth decreased. When fruit development started, large amounts of carbohydrates moved into the fruit cluster from leaves above as well as from leaves below the cluster (Hale and Weaver, 1962). Grape berries are particularly strong carbohydrate sinks during ripening. Under nonstressful conditions, carbohydrate reserves were not translocated to berries (Candolfi-Vasconcelos *et al.*, 1994). In defoliated vines, however, stored carbohydrates in the lower parts of the plants were transported to the fruits. In the 3 weeks following inception of berry ripening (veraison), 12% of labeled C reserves were translocated to the fruits of defoliated plants compared to 1.6% found in berries of intact vines. Transport of carbohydrates from the trunk and roots to the berries was highest during the middle of the ripening period, when 32% of the labeled C was found in the fruit compared to 0.7% in the undefoliated plants.

In elongating fruiting pecan shoots, carbohydrate translocation from different leaves varied with differential development along the shoot axis and with varying strengths of fruit sinks (Fig. 4.1). Beginning with bud opening and continuing through leaf expansion, carbohydrates were not exported from leaves. The immature leaves were partially self-supporting, but they also depended on carbohydrates stored during the previous season. When a leaf was fully expanded, it exported carbohydrates both upward and downward until the leaf above it matured (Fig. 4.1, pattern II), after which translocation of carbohydrates from the lower leaf was only downward (Fig. 4.1, pattern III). This pattern was maintained as leaves higher up on the stem matured sequentially, and eventually all the leaves were exporting carbohydrates.

During early development of the pecan shoot, a high percentage of the leaves exported carbohydrates both upward and downward, indicating that both immature leaves and pistillate flowers were strong sinks for current photosynthate. The proportion of leaves that exported carbohydrates in both directions decreased to a low level during fruit enlargement. During rapid kernel development, however, carbohydrate translocation from basal leaves again occurred in an upward as well as a downward direction. After the fruit matured there was another shift, and all leaves translocated carbohydrates only downward. During a non- or weak-fruiting year, and after all the leaves on a branch were fully expanded, downward transport of carbohydrates predominated (Davis and Sparks, 1974).

Coffee fruits, which have a double sigmoid growth pattern, have two

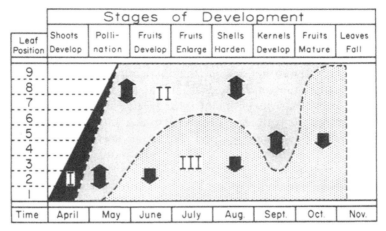

Figure 4.1   Direction of translocation of photosynthate from pecan leaves at various times during the growing season: pattern I, assimilation but no export; pattern II, bidirectional translocation; pattern III, basipetal translocation. From Davis and Sparks (1974).

major periods of high carbohydrate requirements. Much of the final dry weight of fruits is accumulated early, and hence their carbohydrate requirements increase rapidly during the first 6 weeks to reach a peak at 12 to 14 weeks after flowering. At that time, carbohydrates may be transported from rather distant parts of the tree into heavily fruiting branches. The carbohydrate requirement of fruits then decreases but increases again before and during ripening. The dry weight of fruits increases greatly during the latter part of the endosperm filling stage (at 25 to 29 weeks), indicating a high carbohydrate requirement at that time (Cannell, 1971a,b).

During the critical early stages of reproductive development in avocado there was enough current photosynthate from old leaves to support growth of both developing fruitlets and new leaves (Finazzo *et al.*, 1994). Even during rapid growth of fruitlets, the old leaves supplied enough carbohydrates for downward transport to branches as well as upward transport to the fruitlets and young leaves at the shoot apex (Fig. 4.2).

At certain stages of their development, the reproductive structures of gymnosperms are very strong carbohydrate sinks. In their first year of development, the conelets of red pine grew little and consequently were relatively weak carbohydrate sinks. In the second year of development, however, the rapidly growing cones utilized more carbohydrates than the growing shoots, with which they competed. Reproductive structures (first-year conelets and second-year cones) obtained almost three times as much carbohydrate from the old needles as was mobilized by shoots (Dickmann and Kozlowski, 1970a, 1973).

There was a changing seasonal pattern in sources of carbohydrates for growth of second-year cones in red pine. Early in the season, carbohydrate reserves were important because the cones began to increase in dry weight during mid-April in Wisconsin, when temperatures were cool and the rate of photosynthesis was low. By early May the cones were importing currently produced carbohydrates. The 1-year-old needles (those expanded during the previous growing season) were the most important source of photosynthate for early- and mid-season growth of cones, with 2- and 3-year-old needles being progressively less important. Translocation of carbohydrates to cones from the currently expanding needles became important only late in the growing season after these needles were mature. Before that time the new needles retained most of the carbohydrates they synthesized. Photosynthesis by the green cones was relatively unimportant (Dickmann and Kozlowski, 1970b).

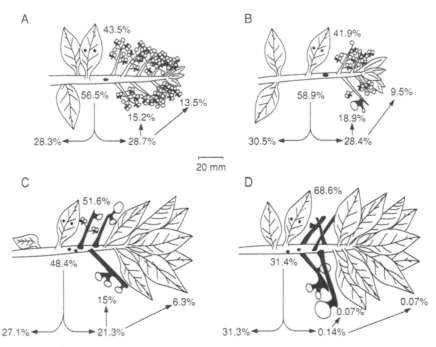

Figure 4.2 Distribution of ^{14}C-photosynthate within an avocado branch from the source (indicated by black spots) when the midvein lengths of the young leaves at the shoot apex were (A) <20 mm (stage 1), (B) 20–30 mm (stage 2), (C) 30–50 mm (stage 3), and (D) >50 mm (stage 4). The arrows indicate the direction of assimilate flow. From Finazzo *et al.* (1994).

Mineral Relations

Phloem transport is important for accumulation of mineral nutrients during fruit growth (except for phloem-immobile elements such as Ca). Minerals are recycled from the xylem to the phloem in the leaves and then transported to developing fruits together with carbohydrates (Ho, 1992).

Much evidence shows that mineral nutrients regulate reproductive growth, largely by maintaining a sufficiently large leaf surface that produces the metabolites and growth hormones necessary for each stage of reproductive growth. Mineral deficiencies may adversely influence each sequential stage of reproductive growth. Some idea of the importance of mineral nutrients may be gained from studies of stimulation of reproductive growth by fertilizers (Chapter 8).

Where mineral deficiencies exist the initiation of floral primordia is increased following addition of appropriate fertilizers (Chapter 8), with the greatest effects generally apparent after addition of N (Govind and Prasad, 1982). Examples are apple (DeLap, 1967; Dennis, 1979) and lime (Arora and Yamadigni, 1986).

Deficiencies of macronutrients, especially N, reduce the percentage of flowers that set fruits. Both soil application and foliar application of N fertilizers increased fruit set of apple (Hill-Cottingham and Williams, 1967; Oland, 1963). Where micronutrient deficiencies exist, appropriate fertilizers increase fruit set. Examples are application of fertilizer containing Zn to sweet lime (Arora and Yamadigni, 1986); application of Mn plus Zn to cranberries (DeMoranville and Deubert, 1987); and application of B to prune (Chaplin *et al.*, 1977), cherry (Westwood and Stevens, 1979), and apple trees (Bramlage and Thompson, 1962).

Deficiencies of certain mineral nutrients commonly inhibit reproductive growth more than they arrest vegetative growth. For example, reproductive development of grapes was inhibited more than vegetative growth by P deficiency. Initiation, differentiation, and maintenance of reproductive primordia were especially sensitive to P supply throughout the season, with maintenance of fruit clusters being more sensitive than initiation of clusters (Skinner and Matthews, 1989). Both the rate of fruit growth and size of fruits at harvest are controlled by relations between the numbers of fruits and leaf surface. Application of N fertilizers may decrease the size of harvested fruits if floral initiation and fruit set are increased proportionally more than leaf growth. However, if the added N does not appreciably increase floral initiation and fruit set, it usually greatly increases the size of the mature fruits.

Localized deficiency of Ca seems to be associated with a wide variety of fruit disorders, including bitter pit, cork spot, Jonathan spot, internal breakdown, low-temperature breakdown, senescent breakdown, water core, and cracking of apples; cork spot of pears; cracking of prunes and cherries; soft nose of mango; and soft spot of avocado (Shear, 1975). Calcium appears to have an important role in maintaining membrane integrity, lowering the rate of respiration, and postponing ultrastructural changes in apple cells (Sharples, 1980). Fruit quality seems to be related to the ratio of Ca to other elements, with the ratios of N, K, and Mg to Ca being related to the occurrence of bitter pit of apple. According to Bramlage *et al.* (1980), the effects of N, P, K, Mg, and B are exerted primarily through interactions with Ca. Johnson *et al.* (1983) stated that although spraying apple fruits with Ca before harvest improves storage quality, it is ineffective if the background concentration in the fruit is very low.

The supply of P seldom affects fruit quality at picking, but P deficiency is associated with rapid deterioration of fruit in storage in areas where this element is deficient. Injury from K deficiency is rare, but high levels sometimes are associated with bitter pit, scald, and storage breakdown of apples. Although B deficiency is common, an excess of B also can cause physiological disorders in fruits.

Water Deficits

Dehydration of woody plants usually, but not always, inhibits reproductive growth. The effects of water deficits may be expected to vary appreciably with species and genotype, severity and duration of the water deficit, stage of reproductive growth, and weather.

Floral Initiation

Severe water deficits during the period of flower bud initiation often reduce the number of flower buds formed, as shown for olive (Hartmann and Panetsos, 1961), apricot (Brown, 1953; Jackson, 1969), and grape (Smart and Coombe, 1983). However, short periods of water deficits at certain stages of the reproductive cycle sometimes increase floral initiation, with the response often being associated with suppression of vegetative growth (Menzel, 1983; see also Chapters 5 and 8). For example, moderate water deficits for as short as 2 weeks induced production of flowers of Tahiti lime, with the stimulatory response being greater as the duration of the water deficit was increased (Table 4.1).

In many tropical trees, especially those growing in seasonal climates, floral initiation and anthesis are not continuous but are separated by a prolonged period of bud inactivity (Borchert, 1983). Anthesis is regulated to a large extent by changes in the water balance of trees. Bud opening commonly is induced by temporary rehydration of trees after leaf fall, by small amounts of precipitation during the dry season, or by onset of the wet

Table 4.1 Effect of Moderate Water Stress over Time on Leaf Xylem Pressure Potential and Flower Induction in Tahiti Lime[a]

Duration of water stress (weeks)	Leaf xylem pressure potential (MPa)		Shoots/plant	Shoot type (%)			Flowers/plant	Flowering shoots (%)[b]
	Predawn	Midday		Vegetative	Mixed	Generative		
Control	-0.34 ± 0.08[c]	-1.48 ± 0.15	4.50[d] ± 1.9	100.0	0	0	0	0
2	-0.90 ± 0.42	-2.25 ± 0.08	6.25 ± 2.2	68.0	16.0	16.0	3.0 ± 0.82	32.0
3	-1.62 ± 0.82	-2.21 ± 0.25	8.00 ± 2.6	46.9	21.9	31.2	5.0 ± 2.16	53.1
4	-0.87 ± 0.09	-2.89 ± 0.23	9.75 ± 3.0	43.6	20.5	35.9	9.0 ± 2.16	56.4
5	-2.89 ± 0.62	-2.83 ± 0.19	9.75 ± 1.5	10.3	56.4	33.3	21.0 ± 8.04	89.7

[a]From Southwick and Davenport (1986).

season. In tropical deciduous forests, flowering of many species occurs during the dry season and is triggered by leaf fall and associated rehydration of previously water-stressed trees. For example, in wet forests of Brazil and Costa Rica, the flowers of *Erythrina* and *Tabebuia* opened immediately after drought-induced leaf fall occurred (Alvim and Alvim, 1978; Reich and Borchert, 1984). During droughts in Costa Rica, forest trees shed many leaves, and a high solute content and high water potential of stems was necessary for budbreak and flowering (Borchert, 1994). The onset of flowering followed rehydration of trees by the first rainfall of about 25 mm. After each rain there was a succession of flowering of many trees and shrubs (Fig. 4.3). Such rehydration of trees even triggered anthesis of species that usually bloom 6 to 8 weeks later (Opler *et al.*, 1976).

The capacity of buds to open on rehydration varies with the stage of bud development. In coffee trees only the buds that had progressed to the "open white cluster" stage of development opened in response to rehydration following a drought (Crisosto *et al.*, 1992). Flower opening was stimulated by rehydration after a period of water deficit if the predawn leaf Ψ declined below -0.8 MPa, or when the water deficit was less severe but more prolonged. Bud opening in response to rehydration also depends on

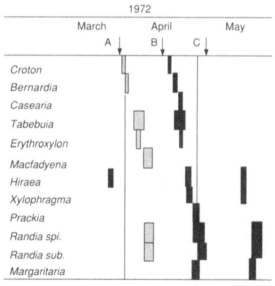

Figure 4.3 Induction of flowering by small amounts of rain during the second half of the dry season at two sites in Costa Rica, Camelco Ranch (solid bars) and Hacienda La Pacifica (shaded bars). Dates of rainfall are indicated by arrows. From Opler *et al.* (1976), with permission of Blackwell Science Ltd.

the extent of bud dormancy. For example, buds of *Lonchocarpus* and *Lysiloma* did not open until 4 weeks after the onset of heavy rains (Frankie *et al.*, 1974).

Fruit Set

Water deficits often reduce fruit set. For example, water stress after flowering decreased fruit set in apple trees to about one-third of that in unstressed (irrigated) trees. Furthermore, irrigated trees shed fewer fruits (Powell, 1974). Water deficits also decreased fruit set in blueberries (Davies and Buchanan, 1979). Drastic reductions in fruit set in grape were associated with hot, dry winds.

Growth of Fruits

The size of fruits generally is reduced by plant water deficits before and during the period of fruit enlargement. Severe water stress at each of five stages of reproductive growth reduced yield of grapes. Water deficits during the first 3 weeks after flowering decreased yield most, primarily because of reduced fruit set. Thereafter, losses in yield were associated with small berry size and, following water stress after the fruit softening stage, with failure of fruits to mature (Hardie and Considine, 1976). Irrigation of apple trees from mid-June to harvest had little effect on fruit development. In contrast, reducing plant water stress by irrigation during the previous 3 months not only stimulated fruit growth but also influenced bud morphogenesis in the following year (Powell, 1976; see Chapter 8). In comparison, growth of peach fruits was reduced by water deficits during the final (cell expansion) stage of growth but not during the early (cell division) stage. This difference emphasized that cell enlargement was more sensitive than cell division to water deficits (Li *et al.*, 1989). In arid regions, water deficits were considered more important before fruit set than after fruit set for growth of citrus fruits (Hilgeman and Reuther, 1967).

Shrinking and Swelling of Fruits and Cones

Reproductive organs progressively increase in size by division and expansion of cells. Superimposed on such growth changes are reversible changes in size associated with shrinkage and expansion as reproductive structures dehydrate and rehydrate (Kozlowski, 1972a). Such shrinking and swelling may complicate studies of growth of fruits and cones.

Diurnal and seasonal shrinkage and swelling of fruits and cones have been reported for many species of trees, including red maple, oak, citrus, hazelnut, black walnut, apple, peach, pear, avocado, cherry (Fig. 4.4), grape, red pine, jack pine, and white spruce (Kozlowski, 1972a,b).

During the night when the stomata are closed and the transpiration rate is low or negligible, differences in Ψ among plant organs are small. During

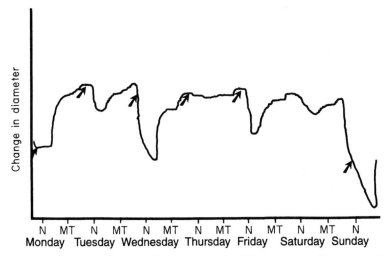

Figure 4.4   Diurnal changes in diameters of fruits of Montmorency cherry in a mid-stage of fruit development. Arrows indicate times of irrigation. From Kozlowski (1968b).

the day, high transpirational losses are followed by transport of water from fruits to leaves while absorption of soil water by roots is not adequate to supply the leaves. Hence, fruits tend to shrink during the day.

Some investigators suggested that fruit shrinkage resulted from transpiration of fruits and that transport of water from fruits to other tissues was small (Jones and Higgs, 1982). However, when attached apples were wrapped with aluminum foil to prevent transpiration, they shrank appreciably during the day. Transpiration of water from the fruits accounted for only 20 to 35% of the fruit shrinkage. Water transport from fruits explained most of the shrinkage. This observation was reinforced by a higher correlation of fruit shrinkage with leaf Ψ than with the vapor pressure deficit (VPD) of the air (Tromp, 1984a).

Diurnal shrinkage of fruits usually lags behind leaf shrinkage, indicating withdrawal of water from fruits. For example, fruits of calamondin orange began to shrink about an hour after the leaves began shrinking (Fig. 4.5). Apparently, transpirational water loss created a Ψ gradient from the fruit to the leaves, and hence water was withdrawn from the fruit along this gradient. Fruits continued to shrink for several hours after leaf shrinkage stopped. The length of time between the beginning of leaf expansion and fruit expansion depended on environmental conditions influencing transpiration of leaves. When the VPD was high during the day, the lag in initiation of fruit expansion over leaf expansion was greater than when the VPD was low (Chaney and Kozlowski, 1971).

Young fruits usually shrink less than old fruits. For example, Shamouti

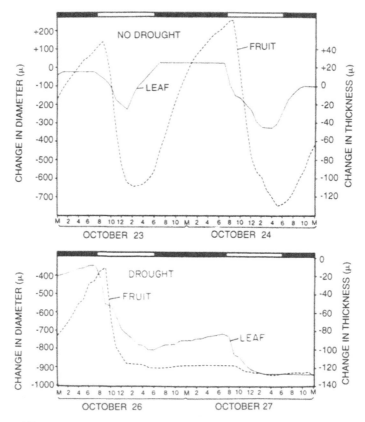

Figure 4.5 Changes in fruit diameter and leaf thickness of calamondin orange during a period of daily irrigation (no drought) and a period during which irrigation was discontinued (drought). The photoperiod is shown at top with darkened portions representing periods from sunset to sunrise. M, Midnight; 12, noon. From Chaney and Kozlowski (1971).

orange fruits did not shrink during the day until early June, by which time the fruits were not as large as ripe olives. Thereafter, the fruits began to shrink during the day and swell at night (Rokach, 1953). The failure of young fruits to shrink during the day often is associated with low transpiration rates of leaves. During early growth of Montmorency cherry trees, the leaves were not fully expanded, transpiration was low, and fruits did not shrink during the day (Kozlowski, 1965). Japanese pear fruits showed little diurnal shrinkage during early growth. During subsequent rapid growth, diurnal fruit shrinkage was appreciable on clear days, but not on rainy days or when the trees were sprinkled with water (Endo, 1973; Endo and Ogasawara, 1975). Daytime shrinkage of grapes was greater before ripening began than it was during ripening (Greenspan *et al.*, 1994).

Both diurnal and seasonal shrinkage of cones have been well documented (Chaney and Kozlowski, 1969a,b; Kozlowski, 1972b). Young cones usually increase in diameter more or less continuously. During mid-development they often shrink during the day and swell during the night. For example, the diameters of young jack pine and white spruce cones increased intermittently. They expanded primarily during the night while the VPD of the air was decreasing or low. In a mid-stage of development, they shrank during the day and expanded at night (Fig. 4.6). In contrast, mature cones commonly dehydrate and show nearly continuous shrinkage. Such patterns were shown for both white spruce cones which ripen in 1 year (Fig. 4.7) and jack pine cones which ripen in 2 years. From resumption of growth to early July, second-year red pine cones increased in diameter in a stepwise pattern, with no significant daily shrinkage. Between mid-July and mid-August, the growth of cones was negligible, but diurnal shrinkage and expansion were appreciable (Dickmann and Kozlowski, 1969a).

Shedding of Flowers and Fruits

Abscission of flowers and fruits generally is increased by plant water deficits, as shown for apple (Landsberg and Jones, 1981), citrus (Kriedemann and Barrs, 1981), peach (Proebsting and Middleton, 1980), and nut trees (Sparks, 1989). Misting or application of antitranspirants shortly after bloom may appreciably decrease June drop of citrus fruits during a drought and hence increase yield (Kriedemann and Barrs, 1981). Extensive abortion of pecan fruits during a drought was reduced by irrigation (Table 4.2). Drought induced fruit abortion but not defoliation, indicating greater sensitivity of fruit abortion than leaf abscission to water deficits (Sparks, 1989).

Fruit Quality

Plant water deficits have important impacts on fruit quality, with specific effects varying among species and genotypes as well as with the severity and duration of water stress. Usually the proportion of small fruits is increased by drought. Water deficits during early growth of grapes led to small berries at harvest. Continuous water deficits during fruit development delayed ripening and adversely affected the color and appearance of grapes (Lavee and Nir, 1986).

Water deficits during the growing season may induce skin cracking and russet in apples, which reduce storage life by increasing water loss of stored fruits or promoting infection by rot fungi. Other disorders of apples associated with water deficits include drought spot, bitter pit, and scald (Landsberg and Jones, 1981). Peaches exposed to severe drought for several weeks before harvest had tough, leathery textures. Drought also prolonged matu-

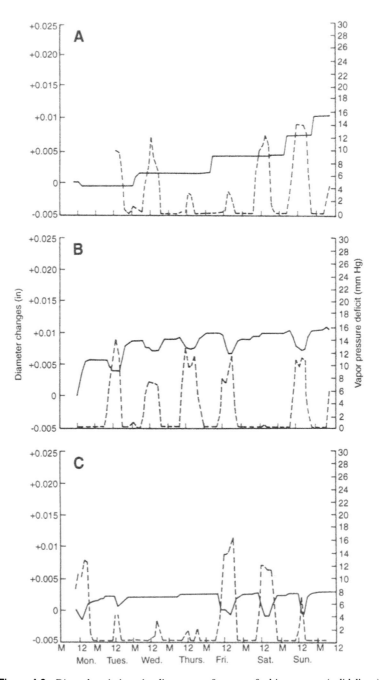

Figure 4.6 Diurnal variations in diameters of cones of white spruce (solid lines) and changes in vapor pressure deficit (dashed lines) at various times during the growing season: (A) June 12–19, (B) June 19–26, and (C) June 26–July 3. From Chaney and Kozlowski (1969b).

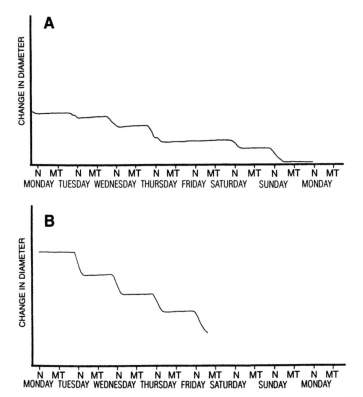

Figure 4.7   Progressive late-season shrinkage of maturing white spruce cones: (A) August 29–September 5 and (B) September 5–9. From Chaney and Kozlowski (1969b).

ration of pears (Veihmeyer and Hendrickson, 1952). Severe water stress also may increase sunburn and shrivel of grapes.

In citrus there does not appear to be an important advantage in maintaining a high soil moisture content throughout the entire period of fruit development. After fruits are set, moderate moisture stress results in higher soluble solids, thinner peels, and higher yields, probably in association with reduced vegetative growth. Nevertheless, Kriedemann and Barrs (1981) advised caution in withholding irrigation because very severe water stress not only reduces yield but also may lower fruit quality. For example, fruits of grapefruit trees that were exposed to severe drought in the summer had high levels of titratable acidity in the following winter, as well as lowered palatability (Levy *et al.*, 1978).

Beneficial Effects of Water Deficits

Not all effects of plant water deficits are harmful to reproductive growth. As mentioned, for example, by inhibiting vegetative growth water deficits

Table 4.2   Effect of Irrigation during Drought on Fruit
Abortion and Fruit Growth in Three Pecan Cultivars[a]

Cultivar and irrigation status	Fruit abortion (no./m² soil surface)	Fruit weight (g/fruit)
Caddo		
Nonirrigated	47.3	0.33
Irrigated	1.3	0.77
Western		
Nonirrigated	79.4	0.62
Irrigated	2.8	1.02
Cape Fear		
Nonirrigated	44.0	0.58
Irrigated	7.3	0.83

[a]From Sparks (1989). *HortScience* **24**, 78–79.

sometimes favor formation of flower buds. Withholding irrigation during the period of rapid shoot growth may stimulate reproductive growth while arresting shoot growth (Chapter 8). In California and Italy, irrigation sometimes is withheld in the summer to cause wilting of lemon trees. In the water-stressed trees, flowering is induced at an abnormal time. By this treatment, the fruits ripen in the spring when the price of lemons is high (Barbera *et al.*, 1985).

Air humidity influences pollen dispersal by regulating evaporation of water from the exothecium surrounding the anthers, hence affecting their opening. Low air humidity caused desiccation of anther cells of conifers and promoted opening of anthers; high humidity impeded opening and pollen dispersal (Sarvas, 1962).

Dehydration and shrinkage of certain fruits and cones induce their opening and facilitate seed dispersal. Progressive drying of mature cones precedes their opening. The cones of black spruce, white spruce, and tamarack dehydrated from a moisture content near 400% (dry weight basis) to less than 40% in September (Clausen and Kozlowski, 1965). The opening of cones on drying is related to the greater shrinkage of ventral (abaxial) tissues than dorsal (adaxial) tissues in the cone scales (Harlow *et al.*, 1964). After cones open as a result of dehydration, they may close and reopen again in response to changes in the relative humidity of the air (Allen and Wardrop, 1964).

The quality of wine sometimes is improved by mild droughts. For example, mild water stress during the second active period of growth of wine grapes accelerated ripening. This probably was related to increased sugar accumulation, stimulated by an increase in abscisic acid (ABA) and prevention of sugar dilution by rapid water movement into berries at that stage (Lavee and Nir, 1986).

Flooding

Inundation of soil often has adverse effects on reproductive growth. For example, flooding may induce abscission of flowers and fruits (Addicott, 1982). Reduced mineral absorption and altered metabolism of flooded plants (Kozlowski and Pallardy, 1984) impair development of flower buds and render them susceptible to abscission. The greatly increased production of ethylene in flooded plants contributes to abscission of flowers and young fruits. The effects of flooding on reproductive growth are discussed further in Chapter 6.

Hormone Relations

Reproductive growth of woody plants is regulated by a series of hormonal signals that influence initiation, growth, and development of flowers, fruits, cones, and seeds (Fig. 4.8). In most angiosperm species, growth of the ovary is negligible during anthesis. Pollen is a rich source of growth hormones; following pollen deposition on the stigma, growth of the ovary is stimulated. Transition from a flower to a growing fruit (fruit set) is accompanied by wilting or abscission of petals and stamens (see Chapter 4 of Kozlowski and Pallardy, 1997). Following fertilization and fusion of the

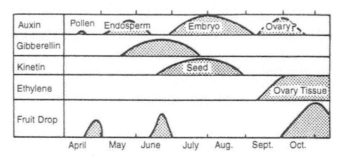

Figure 4.8 Appearance and disappearance of various hormones during growth of a pome fruit. From Thimann (1965).

endosperm nuclei, developing seeds produce hormones that control growth and differentiation of the young fruits. Seed hormones also accelerate translocation of carbohydrates and mineral nutrients into growing fruits (Coombe, 1976). The effects of individual hormones may vary appreciably during different phases of reproductive growth.

Floral Initiation

The induction of floral primordia has long been thought to have a hormonal basis. Sachs suggested as early as 1887 that floral induction is caused by a chemical substance produced in the leaves. Chailakhyan enunciated the concept of a specific flowering hormone and called it "florigen" many years ago (see Chailakhyan, 1975), and extensive research has been conducted to isolate such a hormone. Most attempts at isolating florigen have involved direct extraction of leaves and apices by organic solvents. These generally have produced negative results or some positive results that subsequently were questioned because of low reproducibility. Early investigators usually assumed that florigen was a single, organic substance of low molecular weight. The inability to isolate and characterize florigen may indicate that a complex of substances is involved and that flowering is induced by several regulatory compounds in proper balance. Although there is considerable evidence that flowering is induced by hormonal factors, the specific nature of the factors remains an enigma.

Much evidence shows that gibberellins play an important role in promoting flowering in conifers (Ross and Pharis, 1985; Owens, 1991; Chapter 8). In angiosperm trees, floral induction often was inhibited by exogenous gibberellins and promoted by cytokinins, implying regulation by a hormonal balance (Hoad, 1984). Stimulatory effects of cytokinins on floral induction are shown by strong correlations between amounts of endogenous cytokinins in shoots of apple trees and return bloom (Grochowska *et al.,* 1983).

Abbott (1984) showed a high correlation between regrowth of roots following root pruning in May and return bloom. If reduced root production in an "on" year (during which the supply of root-produced cytokinins to shoots is limited) is associated with a high level of gibberellic acid (GA) in spurs, the hormonal balance presumably does not favor floral induction. During an "off" year, however, in which both root growth and cytokinin production are favored, together with negligible amounts of seed-produced gibberellins, return bloom is favored (Hoad, 1984). Polyamines also have been implicated in promotion of floral initiation (Edwards, 1986).

As noted previously, much evidence shows that gibberellins play an important role in promoting flowering in conifers. Many environmental factors and cultural practices that promote flowering also inhibit shoot elongation. This observation is consistent with the nutrient-diversion hypothesis which emphasizes that a higher nutrient concentration is required for buds to differentiate reproductively rather than vegetatively. According to Sachs and Hackett (1977), elongating shoots are strong carbohydrate sinks, and factors that inhibit shoot elongation tend to favor flowering. However, promotion of flowering in conifers often is associated with *increased* shoot elongation (Ross and Pharis, 1985). Treatment of Monterey pine with $GA_{4/7}$ caused a reallocation of carbohydrates to potentially reproductive tissues (Ross *et al.*, 1984). In comparison, GA_3 induced similar translocation of carbohydrates into long shoots but, unlike $GA_{4/7}$, did not promote flowering. Hence, it appeared that the benefits of diversion of carbohydrates and increased growth of long-shoot primordia during early development were secondary to a direct, morphogenetic effect on differentiation of cone buds. Owens *et al.* (1985, 1986) also demonstrated that $GA_{4/7}$ promoted cone initiation in conifers without significantly altering the size or mitotic activity of vegetative buds. Pharis *et al.* (1987) concluded that $GA_{4/7}$ had a direct morphogenetic role in promoting flowering that was independent of the diversion of assimilates or effects on the vigor and mitotic activity of potentially reproductive primordia.

Pollination

Fruit growth is stimulated following pollination and growth of pollen tubes. Large amounts of auxins and gibberellins that are synthesized during early growth of pollen tubes stimulate fruit growth. The correlation between pollination intensity and fruit production may result from effects of the hormones produced by pollen tubes and seeds. It often is difficult to separate the effects of the hormones from these two sources.

Fruit Set

Hormonal growth regulators are thought to control fruit set, perhaps by increasing transport of carbohydrates and mineral nutrients into developing ovaries or fruits. In some species, fruit set is stimulated by the auxin present in pollen, and further growth depends on auxins and other hormones produced by growing seeds. The effects of the pollen and seeds can be enhanced by exogenous hormones (Chapter 8).

It is unlikely that the fundamental physiology of various fruit types differs among species and cultivars. Fruit set probably requires adequate amounts of more than one hormone. The extent to which an applied hormone induces fruit set apparently depends on variations among species and cultivars in deficiencies of specific hormones in unpollinated flowers. It should not be inferred, however, that a specific exogenous hormone that induces fruit set is the limiting one.

Fruit Growth

Increase in the size of fruits is highly correlated with availability of growth hormones. The presence of seeds is important in fruit growth because they produce a sequence of different hormones during their development in the fruit. A major function of fertilized ovules or seeds in fruit growth is to synthesize hormones that initiate and maintain a metabolic gradient along which carbohydrates and hormones are translocated from other parts of the plant.

Seeds and fleshy tissues of fruits contain high levels of hormones that generally correlate well with rates of cell division and growth rates of fruits. The concentration of free indoleacetic acid (IAA) in fleshy tissues of apple fruits decreased slowly during early June, and the concentration of auxin conjugates decreased rapidly. Nevertheless, the IAA content on a per fruit basis increased, emphasizing that the decrease in concentration was a dilution effect during rapid fruit growth (Treharne *et al.*, 1985).

The sizes and shapes of some fruits, including grape, pear, apple, and blueberry, are highly correlated with seed number. Apples with few seeds often fail to achieve full size and tend to abscise early. Fruits in which seeds are unevenly distributed usually grow irregularly and often are misshapen. An explanation for selective maturation of many-seeded fruits is that more seeds produce more growth hormones, and hence the fruit is a stronger sink for carbohydrates (Stephenson, 1981).

The importance of auxin for fruit growth is emphasized by the high correlation between fruit size and seed development (seeds are rich sources of auxins) as well as an increase in fruit size in response to exogenous auxin. The achenes in strawberry fruits provide the auxins that are essential for growth of the receptacle (Archbold and Dennis, 1985). In peach fruits, IAA concentrations rise to high levels during rapid enlargement of mesocarp cells (Miller *et al.*, 1987). The precise mechanism by which auxin stimulates growth of fruits is not settled, but there is some evidence that auxin increases the level of invertase, which hydrolyzes sucrose to hexoses, thereby increasing import of more sucrose (Lee, 1988).

Fruit growth is not controlled by auxin alone. Both phloem unloading and sucrose uptake by sinks are promoted by ABA. Cytokinins and gibberellins increase cell division or cell enlargement. More cells and larger cells increase the number of sites available for deposition of carbohydrates, hence increasing the rate of dry weight increment of fruits.

Extracts of young seeds of almond, apricot, and plum show gibberellin-like activity. Gibberellin activity in the seeds, endocarp, and mesocarp of apricot fruits was well correlated with growth rates in these tissues for the first 60 days after anthesis (Fig. 4.9). Gibberellin activity increased following anthesis and reached a maximum 20 days after anthesis. Most of the gibberellin activity was in the seeds and least in the endocarp. Thereafter,

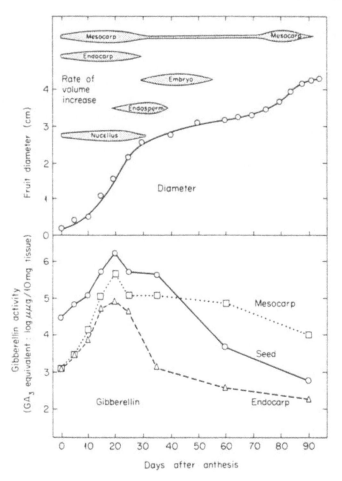

Figure 4.9 Concentrations of gibberellin-like substances in methanol extracts of seed, endocarp, and mesocarp of apricots compared with the rate of fruit growth during development. From Jackson and Coombe (1966). Copyright 1966 by the American Association for the Advancement of Science.

the concentration of gibberellins declined in all three tissues, with the most rapid decrease occurring in the endocarp and the slowest in the mesocarp (Jackson and Coombe, 1966).

The importance of cytokinins in fruit growth also is well documented (Stevens and Westwood, 1984). Cytokinins in immature seeds and fruitlets, together with auxins, regulate cell division in fruits. The presence of cytokinins has been shown indirectly by stimulation of cell division in tissue explants following application of extracts of seeds and fruits of apple, plum, peach, pear, and quince. Natural cytokinins can be replaced by synthetic cytokinins. In this way tissues of various edible fruits have been subcultured successfully for many years.

Fruit Ripening

All major categories of plant hormones are variously involved in regulating fruit ripening, with ethylene playing a dominant role. Whereas ethylene and ABA induce ripening, auxins, gibberellins, and cytokinins wholly or partly retard ripening.

Evidence showing that ethylene functions naturally in ripening of climacteric fruits is very strong. Ethylene production is one of the earliest indicators of ripening, and application of ethylene to fruits accelerates ripening. Furthermore, removal of ethylene as it forms or inhibition of ethylene synthesis retards ripening (Tucker and Grierson, 1987). Fruit ripening induced by ethylene treatment is similar biochemically to natural ripening. The volatile emanations of ethylene from some fruits trigger ripening in adjacent fruits. For example, ethylene produced by orange fruits causes premature ripening of bananas. It has long been known that incomplete combustion of organic fuels releases ethylene which accelerates fruit ripening. The Chinese burned incense in rooms to accelerate ripening of pears. Ethylene in smoke produced by kerosene stoves has been used in railroad cars and packing houses to induce ripening of oranges.

The rate of ethylene production varies widely among different fruits, but the endogenous level needed to induce ripening (0.1 to 1 ppm) is similar for a large number of fruits. Ethylene appears to induce ripening in these fruits as long as they are in a receptive state. As fruits grow their sensitivity to ethylene progressively increases. Hence, the concentration of ethylene and the duration of exposure required to induce ripening decrease as fruit maturation progresses.

In contrast to many fruits, avocados do not ripen on the tree. Furthermore, exposure of attached avocados to ethylene does not induce ripening. Hence, it is unlikely that endogenous levels of ethylene regulate the onset of ripening in avocado fruits.

Two control systems for ethylene biosynthesis in fruits have been demonstrated: system 1, which is common to both nonclimacteric and climacteric fruits, produces basal ethylene as well as the ethylene produced when tissues are wounded; and system 2, which is unique to climacteric fruits, accounts for the autocatalytic ethylene production that accompanies ripening.

Yang (1987) presented a useful model showing the role of system 1 and system 2 ethylene in maturation and ripening of climacteric fruits (Fig. 4.10). The model assumes that binding of ethylene to two kinds of receptors is associated with biochemical changes leading to ripening. System 1 receptor, which is present in both immature and mature fruits, promotes

Immature fruits

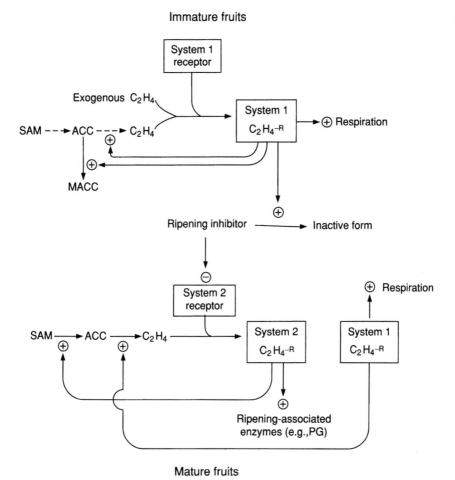

Figure 4.10 Model showing the sequence of ethylene action on the regulation of ethylene biosynthesis in fruit maturation and ripening. The symbols ⊕ and ⊖ indicate the metabolic processes that are positive and negatively regulated by the specific ethylene–receptor complex or the ripening inhibitor. SAM, *S*-adenosylmethionine; ACC, 1-aminocyclopropane-1-carboxylic acid; MACC, *N*-malonyl-ACC; PG, polygalacturonase; R, receptor. From Yang (1987).

development of ethylene-forming enzyme (EFE). System 2 receptor develops or becomes functional in mature fruits and regulates development of 1-aminocyclopropane-1-carboxylic acid (ACC) synthase, the rate-limiting enzyme in ethylene synthesis. The small amount of system 1 ethylene in preclimacteric fruit, together with a system 1 receptor, destroys a "ripening inhibitor." Inactivation of the inhibitor is followed by development of system 2 receptor. Subsequently, an ethylene–system 2 receptor complex forms that stimulates formation of ACC synthase. A previously existing ethylene–system 1 complex regulates development of EFE. The presence of both ACC synthase and EFE results in autocatalytic production of system 2 ethylene and subsequent synthesis of the specific enzymes that regulate ripening. Hence, during fruit development there is a transition from immature tissue that lacks system 2 receptor to mature climacteric tissue in which system 2 receptor is formed.

In a variety of fruits, including bananas, pears, and grapes, auxin pretreatments delay ethylene-induced ripening. Endogenous auxin appears to be a ripening resistance factor that must be depleted to a critical level before ethylene can trigger the ripening process. Under certain conditions, however, applied auxins have been shown to enhance fruit ripening by increasing ethylene production. The effects of auxins are not easily quantified because they not only stimulate ethylene production but also alter the sensitivity of tissues to auxin.

Abscisic acid induces ripening in both climacteric and nonclimacteric fruits. This is shown by three major lines of evidence: (1) endogenous ABA increases before or during fruit ripening, as in citrus (Rasmussen, 1975), peach (Looney *et al.*, 1974), and cherry (Davison *et al.*, 1976); (2) treatments that accelerate senescence cause an increase in endogenous ABA; and (3) exogenous ABA accelerates fruit ripening. Treatment of grapes with Ethrel (ethephon) during a mid-stage of development promoted ripening and advanced the onset of increase in ABA (Coombe and Hales, 1973).

Both gibberellins and cytokinins regulate chloroplast senescence but do not appear to play a dominant role in influencing other aspects of ripening (Rhodes, 1980). Gibberellins block the capacity of ethylene to induce ripening. Yellowing of green bananas was delayed by applied gibberellins, but other characteristics of ripening were not affected (Vendrell, 1970). Similarly, kinetin applied to banana slices before they were exposed to ethylene greatly retarded yellowing of the peel but did not influence development of the respiratory climacteric (Wade and Brady, 1971).

During senescence of fruits, the mechanism controlling aerobic respiration breaks down, enzyme activity causes solubilization of cell wall pectins, cells begin to separate, and softening and physical breakdown occur. These changes also create conditions favorable for invasion by fungi and bacteria. Polygalacturonase (PG) activity was not detectable in the preclimacteric stage of avocado fruit, increased during the climacteric, and continued to increase during the postclimacteric stage to three times the amount at the edible soft stage (Awad and Young, 1979). Cellulase activity was low in preclimacteric fruit, started to increase as respiration accelerated, and eventually reached a level twice that at the edible stage. Cellulase activity started to increase 3 days before PG activity was detected. Increased production of ethylene followed the increase in respiration and cellulase activity. The rapid increase in cell wall depolymerizing enzymes, the rise in respiration rate, and ethylene production were closely correlated.

Abscission

Shedding of reproductive structures is regulated by interactions among auxins, gibberellins, ABA, and ethylene, with auxins and ethylene being most important. The interactive mechanisms commonly are dominated by changes in the distribution of auxin and in ethylene production (Brady and Spiers, 1991).

Auxins almost always inhibit abscission, whereas ethylene and ABA promote it. Gibberellins and cytokinins sometimes stimulate abscission and at other times inhibit it, but they do not appear to be strongly involved in regulating abscission. Ethylene and ABA often interact to augment promotion of abscission by each (Addicott, 1983).

The importance of auxin in preventing abscission is demonstrated by changes with age in the tendency of cotton fruits to abscise. Large floral buds rarely abscise, but after anthesis the young fruits show increased capacity for abscission. A high concentration of IAA in floral buds before anthesis is correlated with high resistance to abscission. A low concentration of IAA at anthesis and for several days thereafter apparently promotes abscission of young cotton bolls (Guinn and Brummett, 1988).

There is considerable evidence that ethylene is present in flowers and young fruits in amounts that accelerate abscission. After weakening of the abscission zone is initiated, a continual supply of ethylene is necessary to continue the weakening effect. The promotive effect of ethylene on abscission is counteracted by the auxin level in the abscission zone. Ethylene in the range of 0.1 to 100 μl liter^{-1} reduces the inhibitory effect of applied IAA. Auxin in the range of 10^{-5} to 10^{-2} M progressively inhibits the accelerating effect of ethylene on abscission (Sexton *et al.*, 1985). Application of inhibitors of ethylene synthesis such as aminoethoxyvinylglycine (AVG) slows the natural rate of abscission as shown for apple (J. A. Roberts *et al.*, 1984), avocado (Davenport and Manners, 1982), and citrus fruits (Sipes and Einset, 1982), and raspberry (Bourdon and Sexton, 1989, 1990).

Ethylene is a natural regulator of flower abscission. Stead and Moore (1983) showed that petal fall was correlated with elevated rates of ethylene formation. Halevy *et al.* (1984) concluded that increased sensitivity to ethylene is critical in shedding of corollas. The response of flower petals to ethylene depends on the physiological age of the tissue, with the more mature petals senescing in response to lower ethylene levels. Pollination of flowers often triggers rapid senescence of petals by stimulating ethylene production in styles, ovaries, receptacles, and petals (Borochev and Woodson, 1989).

Ethylene can accelerate abscission of reproductive structures indirectly by changing the concentrations of other hormones that also influence abscission. Ethylene also apparently accelerates abscission directly as shown by its very rapid effects. For example, reproductive structures often are shed within 2 hr after plants are exposed to ethylene (Sexton *et al.*, 1983). Increase in cell wall hydrolase activity has been attributed to direct effects of ethylene. Reducing endogenous ethylene levels by application of AVG is followed by reduction in production of hydrolases (Sexton *et al.*, 1985). Abscission at the junction of the peach fruit and peduncle is induced by activities of both cellulase and polygalacturonase, which induce degradation of both the middle lamella and primary cell wall (Zanchin *et al.*, 1995).

Considerable evidence shows that ABA plays an important role in stimulating abscission of reproductive organs. Increased ABA has been positively correlated with abscission of citrus flowers; young fruits of citrus, persimmon, cotton, peach, and pear; and mature fruits of citrus, plum, peach, and grape. Furthermore, applied ABA induces or accelerates abscission of flowers of grape and cotton as well as fruits of apple, cherry, peach, and grape. ABA acts indirectly by reducing the abscission-inhibiting levels of IAA and sometimes increasing the amount of ethylene. Such changes augment the direct abscission-promoting effect of ABA (Addicott, 1983).

The acceleration of abscission by environmental stresses commonly is mediated by hormonal effects. For example, water deficits accelerate shedding of cotton bolls by increasing ethylene production and ABA contents of bolls and their abscission zones and by decreasing the auxin content of the abscission zone. Indoleacetic acid either delays or prevents abscission, partly because it blocks synthesis of an abscission-causing cellulase (Guinn and Brummett, 1988).

Summary

Carbohydrates, mineral nutrients, water, and hormonal growth regulators play important roles in physiological regulation of reproductive growth. Reproductive structures are strong carbohydrate sinks as shown by rapid transport of large amounts of reserve carbohydrates and current photosynthate into rapidly growing flowers, fruits, and cones. The amounts of carbohydrates used in metabolism and growth of reproductive structures vary with species and genotype, tree vigor and age, time, and developmental patterns of flowers, fruits, and cones. Most reserve carbohydrates and current photosynthate used in fruit growth are supplied by local sources, especially nearby leaves. However, during very rapid fruit growth some carbohydrates may be obtained from more distant sources.

With adequate supplies of mineral nutrients woody plants maintain a leaf surface area capable of producing the metabolites and growth hormones necessary for all phases of reproductive growth. The importance of mineral nutrients in reproductive growth is emphasized by stimulation by fertilizers of floral induction, fruit set, and growth of flowers, fruits, and cones.

Plant water deficits usually reduce floral initiation, fruit set, fruit enlargement, and fruit quality and increase abscission of reproductive structures. The effects of water deficits on reproductive growth vary with species and genotype, severity and duration of the deficit, stage of reproductive growth, soil type, and weather. Some effects of water deficits are beneficial. Short periods of water stress at critical stages favor formation of flower buds. Water deficits also may release dormancy of flower buds of some species and improve the taste of wine. Low air humidity accelerates pollen dispersal by increasing evaporation of water from anthers. Dehydration of fruits and cones facilitates seed dispersal. Shrinking and swelling of reproductive structures as they dehydrate and rehydrate often complicate studies of growth of fruits and cones. Flooding of soil may induce abscission of flowers and fruits by reducing mineral absorption and altering plant metabolism as well as by increasing ethylene production. Some fruits may crack when trees are flooded.

Reproductive growth is regulated by a series of hormonal signals and interactions. Hormones produced by pollen tubes and seeds stimulate fruit set and growth. An important function of fertilized ovules or seeds is to synthesize hormones that maintain a metabolic gradient along which carbohydrates and hormones are transported into fruits from other organs and tissues. Whereas ethylene and ABA induce fruit ripening, auxins, gibberellins, and cytokinins retard ripening. Shedding of reproductive structures is regulated by several growth hormones but is dominated by ethylene, which promotes abscission, and by auxin, which prevents abscission. Acceleration of abscission by environmental stresses commonly is mediated by hormonal influences. In addition to the regulatory effects of all major classes of growth hormones (auxins, gibberellins, cytokinins, abscisic acid, and ethylene), polyamines also play a regulatory role in reproductive growth.

General References

Abeles, F. W., Morgan, P. W., and Saltveit, M. E., Jr. (1992). "Ethylene in Plant Biology," 2nd Ed. Academic Press, San Diego.

Addicott, F. T. (1982). "Abscission." Univ. of California Press, Berkeley.

Atkinson, D., Jackson, J. E., Sharples, R. O., and Waller, W. M., eds. (1980). "Mineral Nutrition of Fruit Trees." Butterworth, London.

Brady, C. J. (1987). Fruit ripening. *Annu. Rev. Plant Physiol.* **38**, 155–170.

Burd, M. (1994). Bateman's principle and plant reproduction: The role of pollen limitation in fruit and seed set. *Bot. Rev.* **60**, 83–139.

Davies, P. J. (1988). "Plant Hormones and Their Role in Plant Growth and Development." Kluwer, Dordrecht and Boston.

Faust, M. (1989). "Physiology of Temperate Zone Fruit Trees." Wiley, New York.

Friend, J., and Rhodes, M. J. C., eds. (1981). "Recent Advances in the Biochemistry of Fruits and Vegetables." Academic Press, London.

Greyson, R. I. (1994). "The Development of Flowers." Oxford Univ. Press, New York.

Halevy, A. J. (1985). "Handbook of Flowering," Vols. 1–5. CRC Press, Boca Raton, Florida.

Kay, S. J. (1991). "Postharvest Physiology of Perishable Plant Products." Van Nostrand-Reinhold, New York.

Kozlowski, T. T., ed. (1973). "Shedding of Plant Parts." Academic Press, New York.

Kozlowski, T. T. (1981). "Water Deficits and Plant Growth," Vols. 6 and 7. Academic Press, New York.

Kozlowski, T. T. (1992). Carbohydrate sources and sinks in woody plants. *Bot. Rev.* **58**, 107–222.

Kozlowski, T. T., and Pallardy, S. G. (1997). "Physiology of Woody Plants," 2nd Ed. Academic Press, San Diego.

Kozlowski, T. T., Kramer, P. J., and Pallardy, S. G. (1991). "The Physiological Ecology of Woody Plants." Academic Press, San Diego.

Mattoo, A. K., and Suttle, J. C., eds. (1991). "The Plant Hormone Ethylene." CRC Press, Boca Raton, Florida.

Monselise, S. P., ed. (1986). "CRC Handbook of Fruit Set and Development," Vols. 1–5. CRC Press, Boca Raton, Florida.

Roberts, J. A., and Tucker, G. A., eds. (1985). "Ethylene and Plant Development." Butterworth, London.

Sedgley, M. (1990). Flowering of deciduous perennial fruit crops. *Hortic. Rev.* **12**, 223–264.

Sedgley, M., and Griffin, A. R. (1989). "Sexual Reproduction of Tree Crops." Academic Press, London and New York.

Seymour, G. B., Taylor, J. E., and Tucker, G. A., eds. (1993). "Biochemistry of Fruit Ripening." Chapman & Hall, London, Glasgow, and New York.

Tucker, G. A., and Grierson, D. (1987). Fruit ripening. *In* "The Biochemistry of Plants" (D. D. Davies, ed.), Vol. 12, pp. 265–318. Academic Press, New York.

Weichmann, J. (1986). The effect of controlled-atmosphere storage on the sensory and nutritional quality of fruits and vegetables. *Hortic. Rev.* **8**, 101–127.

Wright, C. J., ed. (1989). "Manipulation of Fruiting." Butterworth, London.

5

Environmental Regulation of Vegetative Growth

Introduction

Environmental factors are stressful when they are in limited supply or in excess. Within these extremes, woody plants are exposed to various degrees of stress. The farther away the environment is from the optimum state the more stressful it is. Because woody plants rarely grow in continuously optimal environments they are under some degree of stress very much of the time.

As emphasized in Chapter 1, woody plants are exposed to many abiotic and biotic stresses of varying intensity and duration. Hence, their growth is an integrated response to the effects of multiple environmental stresses. However, the importance of individual stress factors varies as their intensity and/or duration changes. Sudden imposition of a very severe stress, or increasing the severity of a mild stress, can predominate over other environmental stresses that previously regulated growth more strongly through their interactive effects on physiological processes in plants.

Growth responses of woody plants typically lag behind changes in weather. For example, there was a delay of 1 and 2 days, respectively, in the influence of temperature and solar radiation on shoot elongation in Sitka spruce. After these delays, the growth response to a given stimulus continued for more than 1 day (Ford *et al.*, 1987b). As emphasized in Chapter 1, there often are very long lag responses of cambial growth to changes in weather (Kozlowski *et al.*, 1962).

The effects of specific weather variables on growth are not constant during the growing season. Over the first half of the growing season about 80% of the variation in shoot elongation of Sitka spruce could be explained by growth responses to daily air temperature and solar radiation (Ford *et al.*, 1987a). As the season progressed, the dominance of both of these factors decreased.

With the foregoing considerations in mind, we present an overview of the effects of the important environmental factors on vegetative growth and survival of individual woody plants and plant communities. We emphasize the interactive effects of environmental factors and allude briefly to the intermediation of physiological processes in growth control by various environmental factors.

Light

The term "light" refers to the portion of radiant energy that is visible to the human eye and represents only a small part of the total spectrum of radiation to which vegetation is exposed, as shown in Fig. 5.1. At the short wavelength end of the spectrum is ultraviolet radiation, and at the long wavelength end is infrared radiation, or heat. Radiation in the region from approximately 400 to 700 nm is used in photosynthesis, and the region from 660 to 730 nm has important qualitative, photomorphogenic effects on plant growth. Physiologically, the effects of radiation on plants can be grouped in two categories: the high-energy (>1,000 lux) photochemical

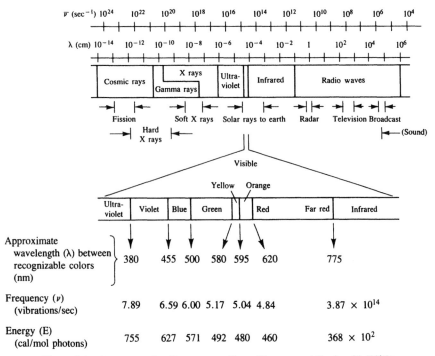

Figure 5.1   Spectrum of radiant energy. From Kramer and Kozlowski (1979).

effects, operating through the process of photosynthesis, and the low energy (<100 lux) effects. The latter include such photomorphogenic effects as etiolation and responses to photoperiod and the directive effects on stem and leaf orientation termed phototropism. Physically, the effects of light on plant growth can be treated as depending on irradiance or intensity, photoperiod or daily duration of illumination, and wavelength or quality.

Light Intensity

Light intensity is the most important attribute of solar radiation in nature because of its role in controlling photosynthesis. Where conditions favor high rates of photosynthesis, increase in biomass typically increases linearly with the amount of intercepted light (Landsberg, 1986). Photosynthesis of many trees and canopies is limited by light intensities lower than one-third or one-fourth of full sun, and a large number require full sunlight for maximum photosynthesis (see Chapter 5 of Kozlowski and Pallardy, 1997). It is difficult to quantify light conditions in forest and plant canopies because they differ with several factors, including (1) clumping of leaves, (2) gaps in forests, (3) variation in leaf orientation angles, (4) penumbra (partial shadows between regions of complete shadow and complete illumination), (5) topography, (6) seasonal variations in phenology of trees, and (7) daily and seasonal movements of the sun (Baldocchi and Collineau, 1994).

In forests and orchards, the amount of light available to the lower canopy is low, largely because of shading by neighboring trees as shown for yellow poplar and beech forests (Table 5.1; Figs. 5.2 and 5.3). Much less light penetrates to the forest floor in tropical forests than in temperate forests. Penetration of light in tropical forests is effectively blocked by the extensive leaf development in their multiple-storied canopies (Terborgh, 1985). The light intensity that penetrates a forest canopy may approximate 50 to 80% of full sunlight in leafless deciduous forests of the temperate zone, 5 to 25% in deciduous forests in full leaf, 10 to 15% in open, even-aged pine stands, and less than 1% in some tropical forests (Spurr and Barnes, 1980; Baldocchi and Collineau, 1994). Of course, the amount of light transmitted to the forest floor decreases as the number of trees per unit area increases.

The amount of solar radiation that penetrates a deciduous forest varies greatly during the season (Fig. 5.3). For example, the amount of light reaching the forest floor of a 50-year-old tulip poplar forest in Tennessee was greatest in the spring before the leaves expanded. The amount was lowest early in the autumn when the forest was in full leaf and the sun was low. After the leaves were shed later in the autumn, the light intensity on the forest floor decreased again with the winter decline in intensity.

Table 5.1 Amount of Radiation Received within a Yellow Poplar Forest throughout a Year[a]

Phenoseason	Duration (days)	Total radiation received [langleys (percentage of yearly total)]			
		Above canopy (32 m) (%)	Upper canopy (16 m) (%)	Midcanopy (3 m) (%)	Forest floor (0 m) (%)
Winter leafless	91	13,300 (11.5%)	5,400 (17.5%)	2,400 (16.9%)	1,500 (16.5%)
Spring leafless	55	15,000 (13.0%)	7,600 (24.6%)	4,900 (34.5%)	4,100 (45.0%)
Spring leafing	30	10,200 (8.8%)	4,000 (12.9%)	2,500 (17.6%)	1,800 (19.8%)
Summer leafing	26	11,700 (10.1%)	2,400 (7.8%)	800 (5.6%)	300 (3.3%)
Summer full-leaf	67	31,500 (27.2%)	5,100 (16.5%)	1,500 (10.6%)	700 (7.7%)
Autumn full-leaf	57	21,100 (18.2%)	3,300 (10.7%)	1,000 (7.0%)	200 (2.2%)
Autumn partial-leaf	39	12,900 (11.2%)	3,100 (10.0%)	1,100 (7.8%)	500 (5.5%)
Photosynthetically active period total		74,500 (64.4%)	14,800 (47.9%)	5,800 (40.8%)	3,000 (33.0%)
Dormant period total		41,200 (35.6%)	16,100 (52.1%)	8,400 (59.2%)	6,100 (67.0%)
Yearly total		115,700	30,900	14,200	9,100

[a]From Hutchinson and Matt (1977).

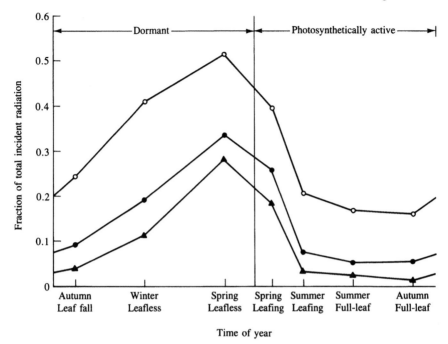

Figure 5.2 Annual course of daily fractional penetration of solar radiation in a yellow poplar forest at 16 m elevation (○), 3 m elevation (●), and 0 m elevation (▲). From Hutchinson and Matt (1977).

Light Interception   The amount of solar radiation absorbed by a tree canopy varies with canopy shape, leaf area, leaf distribution, position of the sun, and fractions of sunlight that are direct and diffuse (Kozlowski *et al.*, 1991).

The light intensity decreases with increasing depth of tree crowns. However, the rate of decrease from the exterior toward the center of the crown varies greatly among trees. In some trees, the crown is so dense that essentially no light reaches the interior and leaves do not form in the heavily shaded portions. In trees with more open crowns, leafy branches may extend to the crown interior, which is penetrated by diffuse light. Most leaves in the crown interior do not develop under a continuous low light intensity. Rather, they are exposed to longer periods of shading than those in the crown periphery. Hence, a tree canopy includes a gradient of leaves ranging from those exposed to long periods of high light intensity and mixtures of intermediate and low light intensity to those exposed to a minimal period of high light intensity and longer periods of intermediate and low light intensities (DeJong and Doyle, 1985).

The conical shapes of crowns of many conifers and the layered arrangement of their whorls of branches, separated by relatively long stem por-

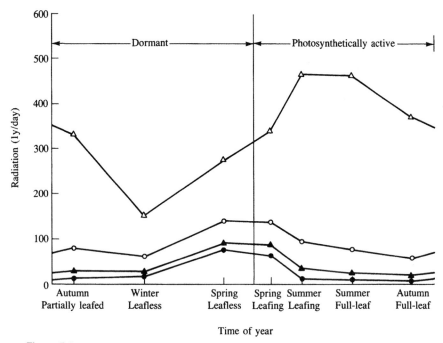

Figure 5.3 Annual regimes of total radiation in a yellow poplar forest above canopy (32 m, △), in the upper canopy (16 m, ○), in the midcanopy (3 m, ▲), and at the forest floor (0 m, ●). From Hutchinson and Matt (1977).

tions, account for efficient interception of light by the entire crown (Oker-Blom and Kellomaki, 1983; Smith and Brewer, 1994). In addition, high interception of direct-beam sunlight, with only small losses due to penumbral light spreading, is associated with conical crowns when the leafy branch tips are exposed to sunlight (Smith *et al.*, 1989; Schoettle and Smith, 1991).

Exposure of conifer needles to light varies not only with crown height but also along individual shoots. For example, in lodgepole pine the daily integral of photosynthetically active radiation decreased along the needle-bearing part of a shoot from the oldest needles and beyond (Fig. 5.4). The decline in light interception was much greater in the upper crown than in the lower crown. However, light interception at the first leafless position of the shoot was similar throughout the crown (Schoettle and Smith, 1991).

Penetration of light into the canopy is decreased by clumping of foliage (Margolis *et al.*, 1995). Hence, in modeling light penetration into tree crowns a distinction often is made between random models, based on the assumption that leaves are randomly dispersed in the canopy (or at least their horizontal location is random), and nonrandom models that consider different foliage groupings. In a Scotch pine stand, the grouping of foliage decreased absorption of light by the canopy. The grouping effect was especially important in stands with few trees but also in the upper parts of dense stands (Oker-Blom and Kellomaki, 1983).

The kinds and arrangements of leaves differ considerably among species of plants, with correspondingly variable effects on mutual shading. The leaves of vines on walls usually form a mosaic in which each has maximum exposure to light, and the same situation often exists with leaves of the surface of a tree crown. By comparison, the needles of pines are borne in fascicles, causing mutual shading. When the needles of loblolly pine seedlings were spread out flat and fully exposed to light in a cuvette, they were light-saturated at approximately the same intensity as leaves of deciduous trees. However, normally displayed needles of pine seedlings required a light intensity three times higher for saturation because of mutual shading (Kramer and Clark, 1947).

Up to 90% of the photosynthetic photon flux density (PPFD) intercepted by citrus trees is absorbed by leaves in the outside meter of the tree crown. In some citrus trees, the light intensity in the crown interior may be less than 2% of the light intensity outside the crown (Monselise, 1951). In hedged peach trees, most absorption of light took place in the outer 25 cm of the crown (Kappel *et al.*, 1983). Penetration of light into the crowns of apple trees decreased rapidly as leaves expanded during April and May. By

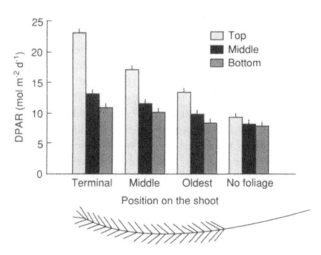

Figure 5.4 Variation in exposure to light at four points along the shoots in each crown third of lodgepole pine trees. The data are daily integrals of photosynthetically active radiation (DPAR). From Schoettle and Smith (1991).

mid-May, 70% of the total seasonal decrease in PPFD had already taken place (Porpiglia and Barden, 1980).

Light intensities in tree crowns also differ among plant varieties. For example, the crown interiors of citrus trees with dense crowns, such as Marsh seedless grapefruit and Clementine mandarin, are much darker than those of Shamouti orange (Monselise, 1951). It is difficult, however, to assess the long-term physiological effects of such comparisons because sunflecks influence the light regime within tree crowns and on the forest floor. In many forest stands up to 80% of the total irradiance at the forest floor may consist of sunflecks which, by influencing photosynthesis, are essential for growth of many understory plants (see Chapter 5 of Kozlowski and Pallardy, 1997). The smearing of sunflecks by penumbras reduces the light energy below the midcrown of many forest stands (Baldocchi and Collineau, 1994). Sunflecks may temporarily increase photosynthesis and offset respiratory losses of carbohydrates, thereby maintaining some shade-adapted plants that otherwise might not survive (see Chapter 5 of Kozlowski and Pallardy, 1997).

Crown Characteristics   Light intensity affects crown size and density by influencing branching, bud formation, shoot expansion, and leaf distribution and structure.

Branching   Trees growing in the open typically have larger crowns and more branches than those growing in the shade, as shown by differences in bifurcation or branching ratios (the ratio of numbers of distal to proximal branches). Open-grown sugar maple saplings not only had fuller crowns and more branches than trees grown in the shade (Steingraber *et al.*, 1979), but the well-lit leaders of sugar maple also had more profuse branching than shaded branches lower down on the stem (Steingraber, 1982). The branching ratios for flowering dogwood and red maple were higher in open fields than in closed canopy forests (Pickett and Kempf, 1980). The branching of some shrubs also is suppressed by shading, as in *Lindera benzoin*, a common understory shrub (Veres and Pickett, 1982).

The time of initial branching varies with tree age and degree of shading. Young sugar maple trees exhibited strong apical dominance and increased in height but did not branch for approximately the first 5 years (Bonser and Aarssen, 1994). Thereafter, the switch to branch formation was delayed in the shade of a closed stand when compared to branching of open-grown trees.

The effect of light on branching is dramatically shown in trees exposed to uneven lighting around the crown periphery. Unidirectional lighting eventually results in development of a strongly one-sided crown, with more branches on the well-lit side of the crown (Fisher and Hibbs, 1982).

Shoot Growth   Light intensity influences bud formation as well as expansion of internodes and leaves. These responses involve effects of light intensity on various plant processes such as chlorophyll synthesis, photosynthesis, hormone synthesis, stomatal opening, and transpiration. The specific effects of changes in light intensity are not easily assessed because increases in light intensity often are accompanied by higher temperatures which, in turn, influence each of the physiological processes already mentioned, and respiration as well.

As discussed in Chapter 3 of Kozlowski and Pallardy (1997), the ultimate length of shoots in some species is correlated with the size of the bud and the number of shoot primordia that are present in the winter bud. The low light intensities received by the lower and inner branches of trees inhibit bud development and thereby impose a limitation on shoot growth. In understory striped maple trees, bud development was controlled by the light intensity reaching the leaves. All buds contained a pair of preformed early leaves as well as a pair of rudimentary leaf primordia. At low light intensities these rudimentary primordia formed bud scales; at higher light intensities they formed leaves (Wilson and Fischer, 1977). After buds form, their expansion into shoots also is influenced by light intensity. For example, shading of tamarack shoots at various times during the growing season inhibited their elongation (Clausen and Kozlowski, 1967b).

In addition to influencing the formation and expansion of shoot primordia, natural shading often induces death of the lower branches on tree stems. The degree of such natural pruning in response to the low light intensities within dense tree stands varies among species (see Chapter 3 of Kozlowski and Pallardy, 1997). A common response to opening of stands by thinning or by pruning branches and suddenly exposing tree trunks to increased light intensities is stimulation of epicormic shoot formation.

Species vary greatly in shoot growth response to light intensity. Logan (1965, 1966a,b) compared the shoot growth of seedlings of several species of angiosperms and gymnosperms in full light and in shelters admitting about 45, 25, or 13% of full sun. In general, heavy shading reduced shoot growth of the conifers more than that of the broad-leaved species (Table 5.2). Height growth of white birch, yellow birch, silver maple, and sugar maple actually increased at the greatest level of shading. However, the dry weight increments of both shoots and foliage of all four species decreased under heavy shading. Sugar maple showed a greater percent reduction in shoot weight under heavy shade than did the other three species (Table 5.2). Shading also had only a limited effect on shoot elongation of basswood seedlings. In comparison, American elm seedlings showed an increase in height and in stem weight as light intensity was increased up to 45% of full light (Logan, 1966b). Jackson and Palmer (1977a,b) exposed

Table 5.2 Effect of Shading for 5 Years on Shoot Growth Characteristics of Angiosperm and Gymnosperm Seedlings[a,b]

	Height (cm) at light intensity of				Dry weight of shoots (g) at light intensity of				Dry weight of foliage (g) at light intensity of			
	13%	25%	45%	100%	13%	25%	45%	100%	13%	25%	45%	100%
Angiosperms												
White birch	180.3	203.2	226.1	152.4	90.7	156.7	318.5	243.4	39.1	54.9	97.2	96.6
Yellow birch	152.4	195.6	213.4	149.9	88.1	154.8	237.6	236.2	36.5	51.1	72.3	95.0
Silver maple	111.8	121.9	119.4	76.2	34.2	44.2	54.5	50.7	18.5	27.5	34.0	50.4
Sugar maple	149.9	144.8	182.9	139.7	56.4	76.6	171.2	235.0	32.7	43.3	77.7	154.6
Gymnosperms[c]												
Tamarack	76.2	111.8	172.7	170.2	4.2	10.2	30.3	40.7	—	—	—	—
Jack pine	40.6	73.7	99.1	111.8	2.7	16.2	32.1	52.7	—	—	—	—
White pine	27.9	38.1	55.9	55.9	2.6	5.4	17.3	22.7	5.4	13.6	28.4	31.3
Red pine	15.2	30.5	38.1	40.6	1.4	4.4	12.1	23.5	1.7	10.1	22.3	38.5

[a]From Logan (1965, 1966a).
[b]Lines connect treatments in which species showed no significant differences.
[c]Data on dry weight of shoots are given after 4 years.

Cox's Orange Pippin apple trees to 100, 37, 25, or 11% of full daylight, in 1970 or 1971, or both years. Shading inhibited shoot growth in both years.

Low light intensities greatly decreased shoot growth of tamarack, jack pine, eastern white pine, and red pine seedlings (Logan, 1966a). By the fourth year the seedlings of all four species at the two lowest light intensities (13 and 25% of full light) were smaller than those grown in 45% or 100% of full light. Five-year-old jack pine plants required full light for maximum height growth, but height growth of the other species was not greatly different at the two highest light intensities. Nevertheless, shoot dry weight of each of the four species was reduced by each shading treatment. Shading depressed shoot growth more in red pine than in eastern white pine.

Sun and Shade Leaves There is considerable difference in the structure and physiology of leaves grown in the sun and those grown in the shade. This applies to the shaded leaves produced within a tree crown compared to those on the periphery of the same crown as well as to leaves of entire plants growing either in shade or in full sun. Shade-grown leaves of broad-leaved trees, which are larger and thinner than those of sun-grown leaves (Table 5.3; see Chapter 2 of Kozlowski and Pallardy, 1997), provide more

Table 5.3 Photosynthetic and Anatomical Characteristics of Leaves of Black Walnut Seedlings Grown under Several Shading Regimes[a]

Shading treatment[b]	Light transmission (% of PAR)[c]	Stomatal density (no. mm^{-2})	Palisade layer ratio[d]	Leaf thickness (μm)	Quantum efficiency (mol CO_2 fixed per mol PAR absorbed)
Control	100.0	290.0a	1.00a	141.3a	0.023ab
GL1	50.0 20.2	293.1a	1.44b	108.3b	0.023a
ND1	15.7	264.0b	1.25ab	104.9b	0.026ab
GL2	20.9 8.3	209.2c	1.85c	89.9c	0.030bc
ND2	3.3	187.0d	1.93c	88.2c	0.033c

[a]From Dean *et al.* (1982).

[b]The GL treatments consisted of green celluloid film with holes to simulate a canopy that had sunflecks. For GL1 and GL2, upper value indicates the sum of 100% transmission through holes and transmission through remaining shaded area; lower value indicates transmission through shade material only (GL1, one layer; GL2, two layers). ND1 and ND2 represent treatments with two levels of shading by neutral density shade cloth. Mean values within a column followed by the same letter are not significantly different ($p \leq 0.05$).

[c]PAR, Photosynthetically active radiation (400–700 nm).

[d]Palisade layer ratio equals the cross-sectional area of palisade layer of control leaves divided by that of leaves of other treatments.

efficient light harvesting per unit of dry weight invested; hence, shade leaves confer a greater potential for plant growth per unit of leaf dry weight. However, several studies have shown that physiological processes of woody plants may or may not be related to their leaf anatomical adaptations to shading. Studies of seedlings in tropical evergreen forests showed that closely related species that occurred within the same forest type, but differed in successional status and crown stratum of occupation, had physiological processes that were not related to their anatomical adaptations (Strauss-Debenedetti, 1989). Other studies of closely related species that occurred within the same forest type but were of the same successional and crown status showed a close relation between anatomical adaptations and efficiency of physiological processes of seedlings (Ashton and Berlyn, 1992). A close relationship also was shown between anatomical plasticity of leaves and efficiency of physiological processes to shading for three temperate-zone oak species (black oak, northern red oak, and scarlet oak) (Ashton and Berlyn, 1994).

Shade Tolerance   Trees and shrubs vary widely in their capacity to grow in the shade, and this difference often becomes a decisive factor in their success during competition. In fact, climax plant communities usually are composed of shade-tolerant species because their seedlings are able to become established in the shade.

A number of investigations focused on shade tolerance of saplings because the capacity of juvenile stages to survive and grow in low-light environments is important in plant succession (Kozlowski *et al.*, 1991). In a wet tropical forest, 92% of the competing trees were expected to die before growing to a diameter of 4 cm (Clark and Clark, 1992). Kobe *et al.* (1995) emphasized that shade tolerance of saplings in a northern hardwood forest involved a trade-off between growth in high light and survival in low light. More shade-tolerant species grew slowly in high-light regimes and were less likely to die in low-light conditions than were less shade-tolerant species.

Forest trees often have been assigned tolerance ratings in several classes representing various degrees of capacity to endure shade (Table 5.4). Shade tolerance varies with the age of trees and with environmental conditions. Trees tend to show the greatest shade tolerance in their youth, and those on good sites and in the southern parts of their range are more tolerant to shade than those on poor sites or in the northern part of their range (Baker, 1950; Wuenscher and Kozlowski, 1971). It is nearly impossible to accurately compare the shade tolerances of species that are native to different regions.

The mechanism controlling shade tolerance is complex and not fully understood. It is unclear whether the traits of sun or shade plants are adaptations to light intensity, to factors associated with light intensity, or to the synergistic effects of both. Nevertheless, silvicultural systems and recom-

Table 5.4 Relative Shade Tolerance of Forest Trees[a]

Gymnosperms	Angiosperms
Very tolerant	
Eastern hemlock	American beech
Balsam fir	Sugar maple
Western hemlock	Flowering dogwood
Western red cedar	American holly
Alpine fir	American hop hornbeam
Tolerant	
Red spruce	Red maple
Black spruce	Silver maple
White spruce	Basswood
Sitka spruce	Buckeye
White fir	Tan oak
Redwood	Bigleaf maple
Intermediate	
Eastern white pine	Yellow birch
Slash pine	White oak
Western white pine	Red oak
Sugar pine	Black oak
Douglas fir	White ash
Giant sequoia	American elm
Intolerant	
Red pine	Yellow poplar
Shortleaf pine	Paper birch
Loblolly pine	Sweet gum
Ponderosa pine	Black cherry
Lodgepole pine	Hickories
Noble fir	Black walnut
Very intolerant	
Longleaf pine	Quaking aspen
Jack pine	Gray birch
Tamarack	Willows
Digger pine	Cottonwood
Western larch	Black locust
Whitebark pine	

[a]Adapted from Baker (1950).

mendations for planting ornamental woody plants on the basis of their shade tolerance ratings have been very successful.

As emphasized by Givnish (1988), persistence of plants in an understory is associated with a wide variety of leaf and canopy traits that differ with light intensity. Such traits include leaf size, stomatal conductance, leaf absorbance, leaf N content, mesophyll photosynthetic capacity, chlorophyll/protein and chlorophyll/Rubisco ratios, internal leaf architecture, and leaf area index (Table 5.5). An important adaptation of many shade-tolerant plants is their capacity to maintain adequate photosynthesis in shade, but drought tolerance and disease resistance also are important (Fenner, 1987; Givnish, 1988). Except for cases in extremely harsh environments, physiological characters alone may not explain ecological success. For example, photosynthetic capacity, stomatal responses, or annual C gain did not explain successional position and competitive capacity of woody plants growing in favorable environments. Instead, growth often was closely related to partitioning of photosynthate (Küppers, 1994). Leaf longevity was important in determining shade tolerance of tropical trees in Panama (King, 1994). Both shade-intolerant and shade-tolerant species showed increased biomass and height growth with increasing light intensity, but the relative performance of several species varied across a light gradient. Shade-intolerant species grew as fast in the shade as shade-tolerant species, but the shorter leaf life span of the former required higher production rates to maintain a given leaf area, largely excluding shade-intolerant species from shaded understory sites.

Mulkey *et al.* (1993) showed that plants may grow in the shade because of different combinations of adaptive characters. They identified at least two different suites of characters that enabled closely related tropical shrubs to grow together in the shade. *Psychotria limonensis* and *P. marginata* exhibited high leaf longevity (~3 years), high plant survivorship, low leaf N content, and high leaf mass per unit area. In comparison, *Psychotria furcata* had low leaf survivorship (~1 year), high plant mortality, low leaf mass per unit area, high leaf N content, the highest leaf area to total plant mass, the lowest levels of self-shading, dark respiration, and light compensation, and the highest stem diameter growth rates. This combination of characters accounted for high whole-plant carbon gain and high leaf and population turnover in *Psychotria furcata.*

In forests, understory plants are subjected not only to shade but also to deficiencies of water and mineral nutrients, which contribute to reduced plant growth and to mortality (Kozlowski, 1949; Kozlowski *et al.*, 1991). In California, availability of soil moisture was more important than light intensity in controlling growth of understory white fir saplings (Conard and Radosevich, 1982). When soil moisture was increased and plants were shaded, height growth was increased by 200% in 4 years. On one site,

Table 5.5 Characteristic Differences between Plants Adapted or Acclimated to Sunny or Shady Extremes in Irradiance Level[a]

Trait	Sun	Shade
Leaf level		
Photosynthetic light response		
Light-saturated rate	High	Low
Compensation irradiance	High	Low
Saturation irradiance	High	Low
Biochemistry		
N, Rubisco, and soluble protein content/mass	High	Slightly lower
Chlorophyll *a*/chlorophyll *b* ratio	High	Low
Chlorophyll/soluble protein ratio	Low	High
Anatomy and ultrastructure		
Chloroplast size	Small	Large
Thylakoid/grana ratio	Low	High
Morphology		
Leaf mass/area	High	Low
Leaf thickness	High	Low
Stomatal size	Small	Large
Stomatal density	High	Low
Palisade/spongy mesophyll ratio	High	Low
Mesophyll cell surface/leaf area ratio	High	Low
Leaf orientation	Erect	Horizontal
Iridescence, lens-shaped epidermal cells	None	Rare
Reddish leaf undersides	Very rare	Infrequent
Canopy level		
Leaf area index	High to low	Low
Phyllotaxis	Spiral	Distichous
Twig orientation	Erect	± Horizontal
Asymmetric leaf bases	Very rare	Infrequent
Plant level		
Fractional allocation to leaves	Low	High
Fractional allocation to roots	High	Low
Reproductive effort	High	Low

[a]Derived from Boardman (1977), Björkman (1981), Bazzaz *et al.* (1987), and Givnish (1987).

growth increased by 140 to 160% when soil moisture was increased but shade was not provided.

Cambial Growth   The effects of light intensity on cambial growth are complex and mediated chiefly through export from leaves of carbohydrates and hormonal growth regulators to stems and roots. As mentioned previously, the light during one year often influences the formation of leaf primordia and the number of leaves produced during the subsequent year. In addition, the light regime influences the expansion and anatomy of leaves and thereby controls their potential production of assimilates and growth hormones, hence regulating cambial growth.

The increased diameter growth of dominant over suppressed trees, and of residual trees following thinning of a stand, is an integrated response to an overall improved environmental regime characterized by greater availability of water and minerals as well as better illumination (Chapter 7). Perhaps cambial growth in variously shaded branches of a tree is a more direct indicator than growth in the main stem of the importance of light on cambial growth.

The branches of many trees, especially gymnosperms, undergo successive suppression. Many pines, for example, annually produce a whorl of branches at the bases of the terminal leader and each branch leader. Consequently, each whorl of branches from the apex downward becomes progressively more shaded after each flush of shoot growth.

Forward and Nolan (1961) studied the pattern of xylem increment at nodes of branches of different ages in 25- to 30-year-old red pine trees. In a wide range of environments, annual xylem production increased for some distance from the branch apex downward and then it decreased. This pattern was a constant feature of all upper branches. In trees that had been grown in the open, suppressed, or suppressed and later released, the upper branches and parts of lower ones that were formed while near the tree apex showed a pattern of xylem increment similar to that of the main stem. As a branch was progressively suppressed during aging, there was a redistribution of its increment, and growth of the xylem sheath was restricted in the lowermost whorls at all internodes. Hence, even in open-grown trees the lower branches were suppressed.

The lower, suppressed branches of many trees may not form annual rings of wood for several years before the branches die. For example, in 35- to 77-year-old Douglas fir trees Reukema (1959) noted that there often were 9 to 10 fewer xylem rings in the base of the branch than in the main stem at the point of attachment. Labyak and Schumacher (1954) observed that a branch of loblolly pine in the lower half of the crown with fewer than three branchlets, or one in the lower fourth of the crown with fewer than five branchlets, did not contribute to xylem increment in the main stem. Such observations indicate that suppressed lower branches often are unable to supply carbohydrates and growth hormones for growth of the main stem. Obviously the pruning of such lower "negative" branches will not detract from diameter increment of the main stem. Figure 5.5 shows the volume contributed to growth of the main stem by branches at various stem heights.

Root Growth   The extent and density of root systems are influenced by light intensity as it affects availability of carbohydrates and hormonal growth regulators to roots. Hence, the total length and branching of root systems vary with crown class in the following order: dominant > intermediate > suppressed trees (Table 5.6). Root growth also varies among species in response to different light regimes. For example, heavy shading reduced root growth of both oaks and pines, but growth of pine roots was reduced much more (Table 5.7). In trees grown in either light or shade, there were more stored carbohydrates in oak than in pine. Also, amounts of carbohydrate reserves were greater in light-grown pine than in shade-grown pine, but in oak there were no significant differences among plants grown in the light or shade. This finding was consistent with the observation of Kramer and Decker (1944) that oaks were very efficient in carrying on photosynthesis at low light intensities whereas pines were not.

Photoperiod (Day Length)

At the equator the days are of equal length (about 12 hr) during the entire year. As the distance from the equator increases, the days become progressively longer in the summer and shorter in the winter (Fig. 5.6). Boston, Massachusetts, for example, has about 15 hr of daylight on the longest

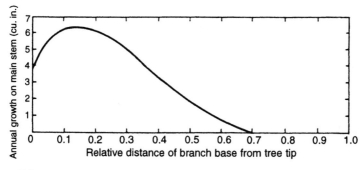

Figure 5.5   Volume of main stem growth contributed by branches at various heights on a loblolly pine 30 years old and 15 m in height. A live crown ratio (percentage of length of stem with live branches) of 0.4 produced the maximum volume of clear bole, but a live crown ratio of only 0.3 produced 89% as much clear bole. From Labyak and Schumacher (1954).

Table 5.6   Root Development in 25-Year-Old Douglas Fir Trees[a]

	Crown class		
Parameter	Dominant	Intermediate	Suppressed
Average height (m)	20.0	10.0	5.0
Average diameter at 1.5 m (cm)	15.7	8.1	5.1
Average length of roots 1 cm diameter (cm)			
Primary	1129	484	55
Secondary	887	257	51
Tertiary	156	10	—
Quaternary	—	—	—
Total and range	2172 (1747–2561)	751 (530–971)	106

[a]From McMinn (1963).

day of the year and about 9 hr on the shortest day. In Arctic regions daylight is practically continuous at midsummer.

Many plant responses, including seed germination, bud dormancy, stem elongation, leaf growth, leaf shedding, and development of cold hardiness, are synchronized by day length. Howe *et al.* (1995) concluded that two aspects of photoperiod are important in regulating plant growth: (1) the critical photoperiod (the longest photoperiod that elicits a short-day response) and (2) photoperiodic sensitivity (change in response per unit change in photoperiod). For each trait analyzed (e.g., time of bud set,

Table 5.7   Effect of Shading on Growth of Loblolly Pine and Overcup Oak[a]

		Dry weight (g)		Root–shoot
Treatment	Height (cm)	Roots	Shoots	ratio
Loblolly pine				
Full light	42	25.2	20.1	1.25
Shade	35	6.1	7.2	0.84
Overcup oak				
Full light	59	44.1	21.1	2.01
Shade	66	38.7	20.1	1.92

[a]The plants were 3 years old when harvested and had been grown for two growing seasons in full light or shade. From Kozlowski (1949).

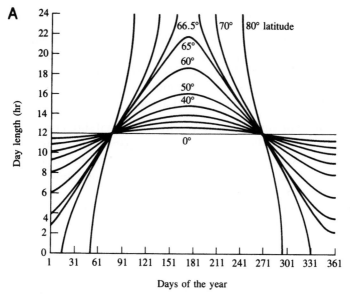

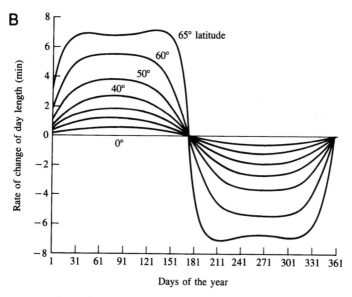

Figure 5.6 Annual cycle of day length (A) and rates of change in day length (B) at various latitudes. From *Plant Physiology*, 2nd Ed., by Frank B. Salisbury and Cleon W. Ross. © 1978 by Wadsworth Publishing Company Inc. Reprinted by permission of the publisher.

branching) a northern ecotype of black cottonwood had a longer critical photoperiod and greater photoperiodic sensitivity than did southern ecotypes (Howe *et al.*, 1995). Ecotypic differences in photoperiodic sensitivity indicated that variations in photoperiodic responses were not adequately described by the critical photoperiod alone.

Shoot Growth The timing and duration of seasonal shoot growth are influenced by day length. Short days stop shoot expansion of many woody plants of the temperate zone and induce development of a dormant state (Fig. 5.7), whereas long days delay or prevent dormancy; however, these responses vary among species.

As shoot elongation slows under short-day conditions, successively shorter internodes are produced, shoot elongation stops, and resting buds form. Such responses have been shown for black locust, yellow poplar, red maple, horse chestnut, sycamore maple (Vince-Prue, 1975; Kozlowski *et al.*, 1991), and Douglas fir (MacDonald and Owens, 1993a,b).

In Douglas fir, bud development involves initiation and differentiation of a bud scale complex and a preformed shoot within the complex. MacDonald and Owens (1993a) demonstrated a strong influence of short days on cessation of shoot growth of Douglas fir seedlings, leading to early bud development. The development of buds involved two stages, initiation of bud scales and initiation of leaves, separated by a transitional stage. The change from leaf formation to initiation of bud scales was faster, fewer bud scales were formed, and development of the bud scale complex was faster under abrupt short days than under gradual short days. The transitional phase was shorter (hence, leaf initiation began earlier) under abrupt short days. More leaf primordia were initiated under abrupt short days than under gradual short days. In another study MacDonald and Owens (1993b) showed that within the first week of exposure to short days, the shoot apices of growing Douglas fir seedlings ended leaf initiation, began and completed initiation of bud scales, and enlarged but initiated no or few new leaf primordia (transitional phase), and then began to initiate new leaves.

Cessation of vegetative growth by short days can be brought about in different ways. For example, in poplar and dogwood grown under short days, the leaf primordia developed into scales instead of leaves; in sumac and lilac the terminal meristem died and abscised (Nitsch, 1957). Generally, the induction of dormancy by short days leads to formation of a terminal resting bud in monopodial species and to abscission of the shoot apex in sympodial ones. Some species, such as yellow poplar, red gum, beech, and some pines which are induced to a dormant state under short days, will resume growth if placed under long-day conditions. That these are true photoperiodic responses is shown by shoot expansion of even leafless dormant trees of these species when exposed to long days, whereas they are maintained in a dormant state under short days.

Figure 5.7 Height growth of slash pine after 15 months under 12-, 14-, and 16-hr day lengths (left to right). From Photocontrol of growth and dormancy in woody plants. Downs, R. J., *In* "Tree Growth" (T. T. Kozlowski, ed.). Copyright © 1962 Ronald Press. Reprinted by permission of John Wiley & Sons, Inc.

Exposure to short days has been used experimentally to stop shoot elongation of many species of woody plants. Nevertheless, short days are not the primary factors that induce dormancy of species that exhibit only a single flush of annual shoot growth, because shoot elongation of these species generally stops before the days begin to shorten. Under natural conditions shoot growth is more likely to be inhibited by the shortening days of late summer in species that exhibit free growth (see Chapter 3 of Kozlowski and Pallardy, 1997) or recurrently flushing growth (Kozlowski *et al.*, 1991).

Exposure of dormant buds to long days may or may not break their dormancy, depending on the species. Bud dormancy of European beech was broken by continuous light regardless of the degree of dormancy. However, the buds of Scotch pine and oaks could be induced to grow into shoots by long days or continuous light if they were in a quiescent or relatively mild state of dormancy, but not when they were in a deep state of dormancy (Wareing, 1956). Long days variously interact with low temperatures in breaking bud dormancy. Both chilling and long days were required for breaking of dormancy in European beech buds. In Norway, nonchilled beech buds sampled in mid-October required long days for bud opening (Heide, 1993b). Buds that were chilled until March still needed exposure to long days for normal bud opening. Buds sampled in November and December did not open regardless of day length but opened only after they were exposed to substantial chilling.

Shoot growth of a number of species is prolonged by long days. Shoot elongation of yellow poplar and sweet gum seedlings continued through the winter in plants exposed to 16-hr days in a greenhouse (Kramer, 1936). In comparison, shoot growth of Scotch pine and sycamore maple eventually stopped even in continuous light (Wareing, 1956). In controlled environments, shoot growth was greatly increased by long days in such tropical species as afara, cacao, coffee, and Carib pine. In afara seedlings, the effects of day length on shoot growth were detected after only 3 days of treatment (Longman and Jenik, 1974). Shoots of some species that normally grow in flushes (e.g., some oaks) respond to long days by producing additional flushes rather than by growing continuously. Shoot growth of certain species, such as European mountain ash, European ash, and apple, usually was unresponsive to day length (Wareing, 1956; Vince-Prue, 1975).

Leaf Abscission   Shedding of leaves is variously influenced by day length. Leaves of tulip poplar and smooth sumac were shed early when exposed to short days. In comparison, the leaves of black locust, sycamore maple, European white birch, red gum, and white oak were retained under short days if the temperature did not drop (Kramer, 1936; Wareing, 1954). In Sapporo, Japan, the shoots of poplar trees usually stop growing in September and shed their leaves in mid- to late October. However, trees close to street lights often retain their leaves to mid-November (Sakai and Larcher, 1987). Day length often interacts with temperature in influencing leaf shedding. For example, at 15°C and short days, black locust plants entered dormancy and shed their leaves; at 21 or 27°C and short days they became dormant but retained their leaves (Nitsch, 1962).

Cambial Growth   In many species of trees cambial growth is linked to shoot expansion and ceases soon after shoot elongation stops. This relationship appears to be the result of stimulation of cell division in the cambium by hormones translocated from the growing shoots. Hence, in some species the effect of photoperiod on shoot growth also influences the seasonal duration of cambial growth. There are other species, however, with a very short period of shoot expansion and a much longer duration of cambial growth. Day length also influences cambial activity in this group. For example, duration of cambial activity in Scotch pine, which annually has a short period of shoot elongation, was greater under 15-hr than under 10-hr photoperiods. Cambial growth could be prolonged in the autumn by supplementing the natural photoperiod with artificial light (Wareing, 1951). Wareing and Roberts (1956) showed that photoperiod also had an important effect in controlling cambial growth of black locust seedlings. Plants were first exposed to short-day conditions in order to cause shoot extension to stop. If they were further exposed to short days for several additional weeks, cambial growth ceased. If, however, they were exposed to long days after shoot extension stopped, cambial activity usually resumed. Experiments in which the stem was girdled to block phloem transport demonstrated that the cambial stimulus, which originated in the leaves when they were exposed to long days, moved only downward in the stem.

In addition to regulating the seasonal duration of cambial growth, day length influences the specific gravity of wood by controlling the proportion of large-diameter to small-diameter cells in the annual xylem ring. Waisel and Fahn (1965) found that photoperiod affected cambial growth of black locust, with the types of cambial derivatives formed being influenced more than the width of the xylem increment.

Frost Hardening   Short-day induction of frost hardening has been demonstrated for many species of trees including Norway spruce, Monterey pine, black spruce, and white spruce. Hardening of shoots of white spruce was influenced by day length, whereas hardening of roots responded only to temperature (Bigras and D'Aoust, 1993). Short-day treatments often have been used to control height growth and induce cold hardiness of seedlings grown in containers. However, short-day treatment also induces early budbreak, hence increasing the risks of exposure of plants to spring frosts.

Site of Photoperiodic Perception   The locus of photoperiodic sensitivity varies among species and may be in very young leaf primordia, resting buds, partially expanded leaves, or mature leaves. In beech the photoperiodic perception arises in young leaf primordia in the buds. When dormant buds were covered and twigs exposed to continuous illumination, dormancy was not broken. In contrast, continuous illumination of buds promoted their growth, even in the absence of bud scales, emphasizing a direct effect of light on the primordial tissue. Approximately 0.7% of the incident light was transmitted through the bud scales of leaf primordia, and that was enough for a periodic response (Wareing, 1953). In comparison, the primary site of photoperiodic perception in sycamore maple and black locust was in the mature leaves. The mechanism of perception of day-length-sensitive plants involves phytochrome, which can convert differences in photoperiod into metabolic reactions that control growth (Schopfer, 1977).

Photoperiodic Adaptation   Often the beginning and cessation of seasonal shoot growth become genetically fixed in relation to the local photoperiod. Such adaptations probably evolve in response to a specific length of growing season. For example, when Pauley and Perry (1954) collected poplar clones from a wide latitudinal range and grew them in the intermediate day length of Weston, Massachusetts, the clones from high latitudes and longer days ceased height growth earlier in the season than those from lower latitudes and shorter days. Vaartaja (1959) also showed that photoperiodic ecotypes (ecological variants adapted to local conditions) occur very commonly in species with wide north-to-south ranges in the northern hemisphere. Such ecotypic variations have important practical significance and should be carefully considered when selecting sources of seeds for planting. Seeds from northern long-day races should not be planted in southern latitudes with short summer days because the trees developing from such seeds are likely to stop growing early and be small. However, such seeds might be selected for sites with short growing seasons at high elevations in southern latitudes. Ecotypes with long growing seasons in a particular latitude should be avoided as seed sources for habitats with short seasons at the same latitude, because the trees that develop from these seeds may be subject to frost damage. For the same reason, such seeds should not be moved northward and planted in a long-day environment.

Light Quality or Wavelength

The light that reaches tree crowns is transmitted, reflected, and absorbed (Table 5.8). As sunlight filters through tree crowns its quality is altered. For example, in passing through leaves daylight is depleted of red, blue, and, to some extent, green wavelengths, largely because of absorption by chlorophyll and, to a lesser extent, light scattering. Essentially no far-red radiation

Table 5.8 Reflectivity, Transmissivity, and Absorptivity as a Function of Wavelength for a Green Leaf and for Light of Normal Incidence[a]

	Wavelength (nm)						
	400	450	500	550	670–680	740–750	1,000
Reflectivity (%)	10	8	9	21	9	49	40
Transmissivity (%)	3	3	6	17	4	47	40
Absorptivity (%)	87	89	85	62	87	4	20

[a]From Federer and Tanner (1966).

is absorbed by leaves, and a substantial amount is transmitted. Hence, the proportion of far-red radiation on the forest floor is increased (the red–far red ratio, R:FR, is decreased). The decrease in R:FR is higher under closed broad-leaved forest stands than under evergreen coniferous forests. For example, R:FR under a sugar maple forest in the northeastern United States varied between 0.08 and 0.11 (in daylight, R:FR was 1.1); in an eastern white pine forest R:FR varied between 0.25 and 0.26 (Federer and Tanner, 1966). Of course, the light quality also may be expected to vary under a tree canopy as a result of sunflecks and canopy gaps. Nevertheless, R:FR often has proved to be a reliable index of the degree of shading (Kozlowski *et al.*, 1991).

There has been much interest in light quality because of its effects on plant growth. Light quality has important effects in seed germination and subsequent plant growth (Chapter 2). The greatest increase in plant dry weight usually occurs in the full spectrum of sunlight. Expansion of leaf blades is prevented by darkness, retarded in green light, intermediate in blue light, and greatest in white light. There is considerable variation in growth responses of different species to changes in the R:FR ratio. Monterey pine was most responsive to a change in the R:FR ratio; New Zealand kauri was intermediate; and *Dacrydium cupressinum* showed the least response (Warrington *et al.*, 1988). Whereas seedlings of *Khaya senegalensis* (shade tolerant) were relatively insensitive to the R:FR ratio, those of *Terminalia ivorensis* (shade intolerant) were sensitive. When the R:FR ratio was low, both the specific leaf area and relative growth rate (RGR) of *Terminalia* were increased (Kwesiga and Grace, 1986). The primary pigment involved in plant responses to light quality is phytochrome (Hart, 1988). In sour cherry, however, phytochrome controlled leaf expansion while stem elongation was influenced more by a blue-absorbing pigment that acted independently of phytochrome (Ballardi *et al.*, 1992).

Under natural conditions, growth of woody plants usually is influenced less by changes in light quality than by changes in light intensity. On the other hand, light quality is very important for growth of plants under artificial illumination. Excessive stem elongation tends to occur under incandescent lamps, but stem elongation of some species is below normal under cool white or daylight-type fluorescent lamps. A few incandescent lamps added to banks of fluorescent tubes provide the far-red wavelengths missing from fluorescent lights, and this combination usually produces more satisfactory growth.

Water Supply

Distribution and growth of woody plants depend more on water supply than on any other factor of the environment (see Chapters 11 and 12 of Kozlowski and Pallardy, 1997). Both species composition and productivity are sensitive to too little or too much water, with the former condition predominating. Aridity characterizes nearly one-third of the world's land area. Over most of the remaining land area, growth of plants is variously reduced by periodic droughts. The staggering amounts of such losses in plant productivity usually are not realized because relatively few reliable data are available to show how much more growth would occur if woody plants growing under natural conditions had favorable water supplies throughout the growing season.

Typically, in the southern part of the United States forest soils are recharged with water to field capacity only for a short time after a rain. Most light rains during the growing season are intercepted by tree crowns or recharge only the surface soil layers to field capacity. Hence, soil water deficits during the growing season are both prolonged and severe (Kozlowski, 1969).

Species Composition and Plant Productivity

The species composition of forests depends on the annual amount of precipitation, its variability between years, its seasonal distribution, and the ratio of precipitation to potential evaporation. Water supply also affects biomass production. For example, annual net primary production (NPP) varies from as much as 3000 g m^{-2} in wet regions to 250 to 1000 g m^{-2} in semiarid regions, and only 25 to 400 g m^{-2} in arid regions (Fischer and Turner, 1978).

In the central United States annual rainfall of at least 380 mm is needed to support open woodlands, 500 mm for open forests, and 640 mm for closed forests (Spurr and Barnes, 1980). Some idea of the dramatic effect of moisture gradients on species composition can be gained by comparing vegetation types in corresponding summer dry, winter wet, or Mediterranean-type climates of Chile and California. The dominant vegetation types are similar at about the same position along a moisture gradient varying from about 1600 to 80 mm of annual rainfall (Mooney and Dunn, 1970). Evergreen forest predominates at the wet end of the gradient. Toward the dry end of the gradient, the forest is replaced in order by dense evergreen shrubs, an open drought-deciduous shrub community, and an open community composed of drought-deciduous shrubs and many succulents (Mooney *et al.*, 1970).

In the tropics where temperature varies little during the year, vegetation types vary from semidesert to rain forests. As the amount of rainfall increases, the number of woody species increases until deciduous forests predominate. Evergreen forests occur where there is even more rain and a shorter dry season. The wettest portions of the equatorial zone, without a distinct dry season, support lush rain forests.

As shown in Table 5.9, dry tropical forests are floristically and structurally simpler than wet tropical forests. Net primary productivity of dry forests averages one-half to three-fourths of that of wet forests. Net primary productivity is correlated with the amount of annual precipitation and duration of the wet season, but it is modified by temperature, soil, and topography (Murphy and Lyugo, 1986).

The species composition of forests depends not only on total rainfall but also on its seasonal distribution. In general, broad-leaved deciduous forests of the temperate zone are restricted to regions in which the rainfall varies between 50 and 200 cm annually and is evenly distributed throughout the year. Conifer forests and sclerophyll evergreen hardwoods grow in regions with rainy winters but prolonged summer droughts (Hinckley *et al.*, 1981). Seasonal distribution of rainfall also determines the species composition of tropical forests. For example, at Cherrapungi, India, despite high annual rainfall (11,610 mm), mesophytes do not grow because almost all the rain falls in eight consecutive months (the other four months receive a total of <100 mm). In other areas, where rainfall is more evenly distributed throughout the year, much less rain than falls at Cherrapungi can support rain forest (Daubenmire, 1978). Species diversity is very high in tropical rain forests of Borneo where monthly rainfall exceeds 100 mm throughout the year. Diversity is lower for Amazonian or African tropical forests that have a few relatively dry months, and it is still lower for seasonal tropical forests and savanna woodlands in which there is a dry season of several months (Iwasa *et al.*, 1993).

The effect of the amount and seasonal distribution of rainfall on species composition and productivity is modified by the ratio of precipitation to evaporation (P/E ratio). This is because evapotranspiration is a measure of simultaneous availability of water and solar energy, the most critical resources for photosynthesis. Currie and Paquin (1987) calculated that 87%

Table 5.9   Characteristics of Tropical and Subtropical Dry Forests Compared with Tropical and Subtropical Wet and Rain Forests[a]

Structural characteristics	Dry forest[b]	Wet forest[c]
Number of tree species	35–90	50–200
Canopy height (m)	10–40	20–84
Number of canopy strata	1–3	3 or more
Leaf area index	3–7	5–8
Basal area of trees (m^2 ha^{-1})	17–40	20–75
Biomass (tons ha^{-1})		
Stems and branches	28–266	209–1163
Leaves	2–7	7–10
Roots	10–45	11–135
Total	78–320	269–1186
Net primary productivity (tons ha^{-1} $year^{-1}$)		
Above ground	6–16	10–22
Roots	2–5	3–6
Total	8–21	13–28
Diameter growth (mm $year^{-1}$)	1–2	2–5 or more
Growth periodicity	1–2 annual pulses	Continuous or intermittent

[a]Adapted from Murphy and Lyugo (1986). Reproduced, with permission, from the *Annual Review of Ecology and Systematics*, Volume 17, © 1986, by Annual Reviews Inc.
[b]Annual rainfall 500–2000 mm; strongly seasonal; annual ratio of potential evaporation to precipitation (PET/P) > 1.
[c]Annual rainfall >2000 mm; little or moderate seasonality; annual PET/P normally < 1.

of the large-scale variability in richness of plant species in North America was regulated by climatic conditions and topography, with 76% explained by evapotranspiration. Rosenzweig (1968) found actual evapotranspiration (precipitation minus runoff and percolation) to be an excellent indicator of aboveground productivity of mature plant communities from forests to deserts.

In addition to regulating species composition and productivity of communities of woody plants, water deficits variously affect shoot growth, cambial growth, and root growth of woody plants. This is discussed in Chapter 3.

Shrinking and Swelling of Stems

Measurements of cambial growth by determining increase in stem diameter sometimes are complicated by shrinking and swelling of stems. Cambial tissues are under water stress of varying intensity almost daily because of the high tensile forces that develop in the adjacent mature xylem (Zahner, 1968). As daily transpiration of mountain ash (*Eucalyptus regnans*) began, water was withdrawn from developing cells (with the cells nearest vessels affected most), with water stress in developing xylem tissues being greater than in phloem tissues. The xylem-associated tissues contained about twice as much water in the early morning as in the afternoon (Stewart *et al.*, 1973).

Diurnal shrinkage and swelling of tree stems occur commonly, even in plants growing in well-watered soil, because absorption of water during the day does not keep up with transpirational water loss (Fig. 5.8). At night, absorption of water exceeds transpirational water loss; hence, tree stems rehydrate and swell. Transmission of water deficits from the leaves to the lower stem occurs reasonably quickly. For example, the daily lag of minimum stem diameter behind maximum leaf water deficit was 1 to 2 hr for sugar maple and white birch, and 3 to 4 hr for red spruce and balsam fir (Pereira and Kozlowski, 1976, 1978).

During extended droughts the cambial sheath undergoes continuous water stress as shown by stem shrinkage for periods of days to months. This has been well documented for trees of the temperate zone (Kozlowski, 1958, 1962, 1967a, 1968a–c; Kozlowski *et al.*, 1962; Bormann and Kozlowski, 1962; Ogigirigi *et al.*, 1970; Braekke and Kozlowski, 1975; Chalmers *et al.*, 1983; Brough *et al.*, 1986) and tropics (Dawkins, 1956; Doley, 1981).

The amount of stem shrinkage varies with species and genotype, soil moisture supply, leaf area, and atmospheric conditions that affect transpi-

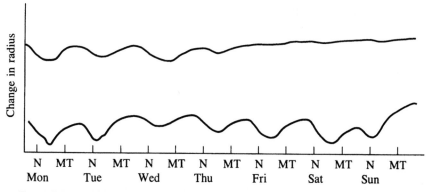

Figure 5.8 Dendrograph traces showing shrinkage of red pine stems in the afternoon followed by swelling at night. The upper trace is for the week of July 10 to 17 in Wisconsin, and the lower trace is for August 21 to 28. The upper trace shows no daily shrinkage and swelling of the stem during the latter part of the week when cloudy and rainy weather prevailed. From Kozlowski (1968a).

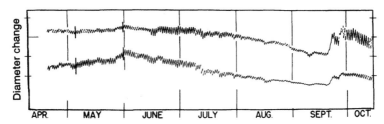

Figure 5.9   Diurnal and seasonal changes of stem diameters of two Douglas fir trees in southwestern Colorado. Adapted from Fritts *et al.* (1965). Reproduced by permission of the Society for American Archeology from *American Antiquity*, Vol. 31.

ration. Shrinkage also is influenced by pathogens that induce formation of gums and tyloses which block the water-conducting vessels (Kozlowski *et al.*, 1962).

Reversible changes in stem diameter as a result of hydration changes often are small. Sometimes, however, they are appreciable and mask the amount of diameter change resulting from growth through cell division and enlargement. For example, the amount of stem shrinkage of several broad-leaved trees during a severe drought in New Jersey exceeded total diameter growth up to that time. By mid-August, the diameters of some trees were smaller than they were before the growing season began. The trees remained in a dehydrated condition until the soil was recharged by rain in December. Then the trees rapidly rehydrated and their stems expanded (Buell *et al.*, 1961). Stems of Douglas fir trees in southern Colorado shrank progressively for more than 3 months, except for slight daily rehydration and expansion during the night (Fig. 5.9). After a rain in September, the trees rapidly rehydrated and their stems swelled accordingly (Fritts *et al.*, 1965). Swelling of tree stems when rain follows a drought occurs rather rapidly (Kozlowski and Peterson, 1962). For example, the stems of droughted trembling aspen trees rehydrated and swelled measurably within 30 min after a cloudburst (Kozlowski and Winget, 1964b).

Flooding

Continuous or temporary flooding of forest ecosystems, which occurs commonly as a result of overflowing of rivers, storms, construction of dams, and overirrigation, leads to soil anaerobiosis and consequent ecological effects. A substantial fraction of the volume of well-drained soils consists of air. To remain healthy and provide woody plants with adequate amounts of water, mineral nutrients, and certain hormonal growth regulators, roots require a constant supply of O_2. In well-drained soils both consumption of soil O_2

and production of CO_2 by roots are counteracted by gas exchange between the soil and air. As much as 2.5 to 17 liters of O_2 per square meter of land area may diffuse into the soil each day as O_2 is depleted by respiration of roots and microorganisms (Russell, 1973). When a soil is flooded, the water occupies the previously gas-filled pores, and gas exchange between the soil and air is then restricted to molecular diffusion in the soil water. Such diffusion, however, is very slow, and the supply of O_2 to roots is limited. Almost all of the remaining O_2 in a flooded soil is consumed by microorganisms within a few hours.

Effects of Flooding

The poor soil aeration in flooded soils is accompanied by a number of soil and plant changes that adversely influence growth (Kozlowski *et al.*, 1991). Alterations in soil structure include breakdown of aggregates, deflocculation of clay, and destruction of cementing agents. Flooding also reduces the soil redox potential, increases the pH of acid soils (largely because of a change of Fe^{3+} to Fe^{2+}), and decreases the pH of alkaline soil (mainly because of CO_2 accumulation which eventually forms H_2CO_3). Soil inundation also decreases the rate of decomposition of organic matter, sometimes by as much as half. Whereas the organic matter in unflooded soil is decomposed by a variety of aerobic organisms (including bacteria, fungi, and soil fauna), decomposition in flooded soil is accomplished only by anaerobic bacteria.

A variety of toxic compounds accumulate in flooded soils. Some of these (e.g., ethanol, acetaldehyde, and cyanogenic compounds) are produced by the roots. Compounds produced by anaerobic bacteria in flooded soil include gases (N_2, CO_2, methane, and hydrogen), hydrocarbons, alcohols, carbonyls, volatile fatty acids, nonvolatile acids, phenolic acids, and volatile S compounds (Ponnamperuma, 1984).

Flooding injury has been linked to altered respiration, with several possible causes of injury. Some investigators claimed that the end products of glycolysis (e.g., CO_2, ethanol, and/or lactic acid) accumulated to toxic levels in flooded tissues (Crawford, 1967, 1978). Others suggested that depletion of carbohydrates in flooded plants leads to starvation. Still another view is that energy consumption through utilization of ATP exceeds the capacity of glycolysis to synthesize ATP. Hence, the energy charge of flooded tissues decreases to levels that do not maintain metabolic control (McKersie and Leshem, 1994).

In some plants the harmful effects of flooding are linked to blocking of water transport. Bubbling of CO_2 through a nutrient solution in which *Euphorbia* roots were immersed induced wilting and stem degradation (Erstad and Gislerod, 1994). Wilting was attributed to reduced water transport associated with occlusion of vessels by protoplasm from collapsed adjacent parenchyma cells. In contrast, water transport in *Pelargonium* cuttings was not affected by similar treatments. Andersen *et al.* (1984) noted that root anaerobiosis led to xylem blockage, reduced leaf Ψ, and caused wilting of *Pyrus communis*. Both *Pyrus betulaefolia* and *P. calleryana* were less affected by root anaerobiosis.

Growth of Plants

Flooding of soil during the growing season affects woody plants at all stages of development. Typical responses of unadapted plants include inhibition of seed germination, decrease in growth, changes in plant morphology and anatomy, and often death of plants. Because the effects of flooding are mediated by physiological processes, reduction in growth may not be obvious for some time after flooding is initiated.

Inundation of soil adversely affects shoot growth of many species (Fig. 5.10) by inhibiting formation and expansion of leaves, reducing internode elongation, and inducing chlorosis, leaf senescence, and abscission. Inhibition of leaf formation and expansion of flooded seedlings has been shown for paper birch (Tang and Kozlowski, 1982c), American elm (Newsome *et al.*, 1982), and Japanese larch (Tsukahara and Kozlowski, 1984). Whereas leaves of unflooded seedlings of American sycamore expanded fully in 110 days, those of flooded seedlings expanded by only 10% (Tsukahara *et al.*, 1985). The leaves of many species of flooded plants turn yellow, senesce, and are shed early (Tang and Kozlowski, 1982c).

Flooding may increase or decrease diameter growth of woody plants. Usually, however, early increases in diameter of flooded trees are followed by decreased diameter growth. Cambial growth of most flood-intolerant trees is reduced by prolonged flooding during the growing season. Examples include seedlings of jack pine, red pine (Tang and Kozlowski, 1983), Japanese larch (Tsukahara and Kozlowski, 1984), box elder (Yamamoto and Kozlowski, 1987e), Japanese cryptomeria (Yamamoto and Kozlowski, 1987d), and rubber (Sena Gomes and Kozlowski, 1988). Flooding also may alter the anatomy of xylem and phloem tissues. For example, the stems of flooded Aleppo pine seedlings contained more parenchyma tissue (abundant and enlarged ray cells, numerous resin ducts, and more parenchyma cells) than did stems of unflooded seedlings (Yamamoto *et al.*, 1987b). Sometimes flooding increases stem diameters, especially in seedlings, by increasing bark production rather than by increasing xylem increment (Fig. 5.11). In contrast, diameter growth of flooded Manchurian ash seedlings increased because of greater numbers and sizes of wood fibers rather than bark increment (Yamamoto *et al.*, 1995). Flooding also may alter the

Figure 5.10   Effect of flooding of soil on shoot growth of seedlings of red gum (*Eucalyptus camaldulensis*; top) and blue gum (*E. globulus*; bottom). Unflooded seedlings are at left, flooded seedlings at right. Photo by A. R. Sena Gomes.

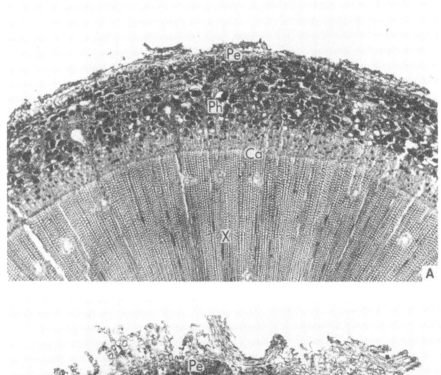

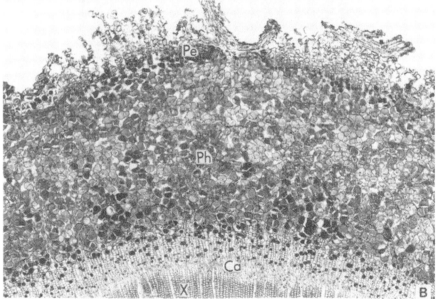

Figure 5.11 Effect of flooding on growth of bark tissues of Aleppo pine: (A) unflooded seedling; (B) flooded seedling. Ca, Cambium; Pe, periderm; Ph, phloem; X, xylem. From Yamamoto *et al.* (1987b).

shapes of tracheids in gymnosperms. For example, the tracheids of flooded Aleppo pine and Japanese red pine seedlings were abnormally short, round, thick-walled, and surrounded by intercellular spaces (Fig. 5.12).

Both formation and growth of roots are greatly reduced by prolonged soil inundation (Fig. 5.13). Examples include roots of seedlings of bur oak (Tang and Kozlowski, 1982a), American sycamore (Tang and Kozlowski, 1982b), paper birch (Tang and Kozlowski, 1982c), American elm (Newsome *et al.*, 1982), Aleppo pine (Sena Gomes and Kozlowski, 1980c; Yamamoto *et al.*, 1987b), Japanese cryptomeria (Yamamoto and Kozlowski, 1987d), cacao (Sena Gomes and Kozlowski, 1986), and rubber (Sena Gomes and Kozlowski, 1988).

Flooding of soil also is followed by death and decay of roots, largely by increased activity of *Phytophthora* fungi. These organisms are stimulated by the low vigor of flooded plants and attraction of fungal zoospores to root exudates, including sugars, amino acids, and ethanol (Stolzy and Sojka, 1984). Woody roots are much more tolerant of flooding than nonwoody

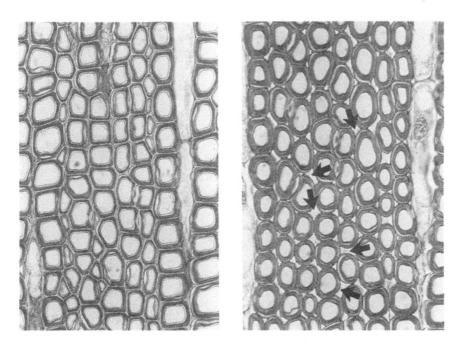

Figure 5.12 Effect of flooding on xylem anatomy of Aleppo pine: (Left) unflooded seedling; (right) flooded seedling. Note rounded tracheids with thick walls and extensive intercellular spaces (arrowheads) in the xylem of the flooded seedling. From Yamamoto *et al.* (1987b).

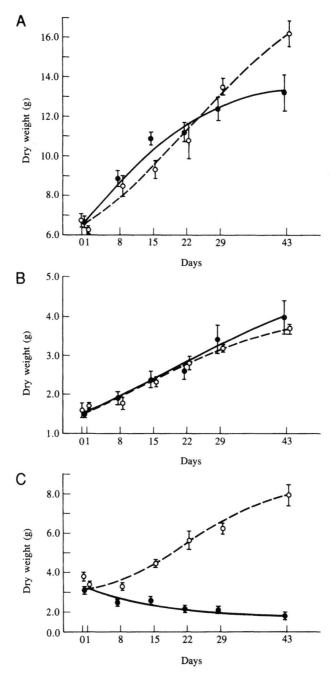

Figure 5.13   Effects of flooding on dry weight changes of (A) needles, (B) stems, and (C) roots of unflooded (○), and flooded (●) Aleppo pine seedlings. From Yamamoto *et al.* (1987b).

roots. For example, all the woody roots of rooted cuttings of lodgepole pine survived flooding, but all the nonwoody roots were killed (Coutts, 1982).

Factors Influencing Plant Responses to Flooding

Responses to flooding vary with species and genotype, age of plants, condition of the floodwater, and time and duration of flooding (Kozlowski, 1984d).

Species and Genotype   Although broad-leaved forest trees generally tolerate flooding better than conifers do, there are wide variations in response within each group. Willows, tupelo gum, black gum, and mangroves are considered flood tolerant; yellow poplar, flowering dogwood, and sweet gum are not. Seedlings of silver maple and buttonbush tolerated flooding better than those of hackberry (Hosner, 1958, 1960). Among conifers, bald cypress and tamarack tolerate flooding well, whereas eastern white pine and red pine do not. Sensitivity to flooding often varies greatly among closely related species, as among species of eucalypts. Seedlings of river birch are much more flood tolerant than those of paper birch (Norby and Kozlowski, 1983). Flood tolerance of Japanese alder and Japanese birch seedlings, both species in the Betulaceae, varied greatly. Height growth of alder was influenced only slightly by flooding; however, growth of birch was severely depressed, and many plants were killed by flooding (Terazawa and Kikuzawa, 1994).

The flood tolerance of fruit trees varies greatly. Rowe and Beardsell (1973) ranked fruit trees in the following order of flood tolerance: quince, extremely tolerant; pear, very tolerant; apple, tolerant; citrus and plum, intermediately tolerant; cherry, intermediately sensitive; apricot, peach, and almond, sensitive; and olive, very sensitive. Such ratings are complicated by differences in flood tolerance among rootstocks on which fruit trees are grown (Table 5.10). For example, Red Delicious apple scions on MM.111 and seedling rootstocks were more sensitive to flooding than those on M.27 or MM.106 rootstocks. Scions on M.26 rootstocks were least affected (Rom and Brown, 1979). Rowe and Beardsell (1973) compiled the following rating of sensitivity of apple rootstocks to flooding of soil: fairly resistant, M.2, M.3, M.6, M.7, M.13, M.14, M.15, M.16, Crab C, and Jonathan; moderately sensitive, M.4, M.9, and M.26; very sensitive, M.2, MM.104, and MM.109; and extremely sensitive, M.779, M.789, M.793, and Northern Spy.

Age of Plants   Mature trees tolerate flooding better than seedlings or overmature trees. The flood tolerance of a plant may vary with its stage of development. For example, 7 months of submersion did not reduce subsequent germination of *Parkia pendula* seeds (Scarano and Crawford, 1992). However, 1-month-old seedlings grew only during the first 3 weeks of flooding. By the fourth week, leaflets were shed. Returning these seedlings to unflooded conditions during the fifth week did not lead to recovery, and

Table 5.10   Sensitivity of *Prunus* spp. Rootstocks to Waterlogging[a]

Sensitivity to waterlogging	Rootstock	Species
Resistant	Damas de Toulouse	*P. domestica*
	Damas GF1869	*P. domestica*
	GF8-1	*P. cerasifera* cv. Marianna
	S2544-2	
Moderately resistant	GF31 hybrid	*P. cerasifera* × *P. salicina*
	Myrob B	*P. cerasifera*
	P936	*P. cerasifera*
	P938	*P. cerasifera*
	P855	*P. cerasifera*
	P34	*P. cerasifera*
	St. Julian A	*P. domestica*
	St. Julian GF355-2	*P. domestica*
	Brompton	*P. domestica*
	Ciruelo 43	*P. domestica*
Moderately sensitive	S37	*P. salicina*
	S2540	
	S2541	
	S300	
Sensitive	S2514	
	S2508	
	S763	
	S2538	
Extremely sensitive	Apricot	*P. armeniaca*
	St. Lucie 39	*P. mahaleb*
	Cherry	*P. avium*
	Peach	*P. persica*
	GF305	*P. davidiana*

[a]After Rowe and Beardsell (1973); from Kozlowski (1984b), by permission of Academic Press.

all the plants subsequently died. Submersion of whole seedlings led to their death by the third week. By comparison, both seeds and seedlings of *Parkia discolor* tolerated 7 months of submersion. These differences indicate that classifications of species as flood tolerant or intolerant can be misleading if they disregard the tolerance of different developmental stages.

Condition of the Floodwater   Flooded trees are injured much more by stagnant water (which contains less O_2) than by moving water. In fact, even the most flood-tolerant species (e.g., bald cypress) may be injured by standing water. Flooding of bald cypress seedlings with stagnant water substantially decreased the rate of height growth, needle initiation and growth, and dry weight increase of needles and roots (Table 5.11). Height growth and dry

Table 5.11 Effects of Flooding of Soil with Standing Water for 8 or 14 Weeks on Growth of Bald Cypress Seedlings[a]

Weeks	Treatment	Number of leaves	Height (cm)	Leaf area (cm^2)	Dry weight (g)				Root–shoot ratio
					Leaves	Stems	Roots	Whole plant	
0	—	19.0	31.0	115	0.58	0.24	0.37	1.19	0.45
8	Unflooded	42.9	76.0	903	5.12	3.46	4.71	7.84	0.55
	Flooded	40.6	62.6	410	2.43	2.94	2.15	7.52	0.40
14	Unflooded	49.1	84.3	988	6.00	7.05	10.06	23.32	0.78
	Flooded	41.0	59.5	437	2.50	4.17	4.69	11.35	0.70

[a]Reprinted from *Environ. Pollut. (Ser. A)* **38**, Shanklin, J., and Kozlowski, T. T., Effect of flooding of soil on growth and subsequent responses of *Taxodium distichum* seedlings to SO$_2$. 199–212. Copyright 1985, with kind permission from Elsevier Science Ltd., The Boulevard, Langford Lane, Kidlington OX5 1GB, UK.

weight increment of swamp tupelo and tupelo gum seedlings growing in moving water were several times greater than in seedlings growing in standing water (Hook *et al.*, 1970b).

Time and Duration of Flooding   Flooding during the growing season is much more harmful than flooding during the dormant season. The greater inhibition of growth and more extensive injury by flooding during the growing season are associated with the high O_2 requirement of growing roots.

Whereas some species of trees are killed by less than 1 month of flooding, others can survive continuous flooding for at least two growing seasons. Inundation of soil for a few weeks or more during the growing season retards growth of most woody plants. Sometimes, however, temporary flooding with moving water may increase growth of some flood-tolerant species such as tupelo gum (Klawitter, 1964).

Adaptations to Flooding

A number of wetland species have various morphological and physiological adaptations that enable their roots to maintain more normal physiological processes. Some species absorb O_2 through stomatal pores or lenticels from which it moves downward and diffuses out of the roots to the rhizosphere. Such O_2 transport benefits plants by oxidizing reduced soil compounds such as toxic ferrous and manganous ions (Opik, 1980). Entry of O_2 through leaves and downward transport are well known in willows (Armstrong, 1968) and lodgepole pine (Philipson and Coutts, 1978, 1980b) and through lenticels of twigs, stems, and roots of several species of woody plants (Hook, 1984).

Many woody plants form hypertrophied lenticels on submerged portions of stems. Examples are black willow, cottonwood, green ash, bur oak, American sycamore, and American elm (Kozlowski, 1984b; Tang and Kozlowski, 1984b; Angeles *et al.*, 1986), ponderosa pine, red pine, jack pine, Virginia pine, and tamarack (Hahn *et al.*, 1920). The flood-induced hypertrophied lenticels not only facilitate stem aeration but also release potentially toxic compounds such as ethanol, acetaldehyde, and ethylene. Gas exchange is facilitated more by flood-induced lenticels than by lenticels of unflooded plants (Hook, 1984). Hypertrophy of lenticels in flooded plants is modified by the temperature of the floodwater. For example, abundant lenticel hypertrophy of mango occurred at 30°C but not at 15°C (Larson *et al.*, 1991).

Aeration of some flooded plants also is improved by formation of aerenchyma tissue with large intercellular spaces, through which O_2 diffuses rapidly. Examples include mangroves (Gill and Tomlinson, 1975), lodgepole pine (Coutts and Philipson, 1978b), loblolly pine (Hook, 1984), and pond pine (Topa and McLeod, 1986).

Trees of swamp habitats or those subject to tidal flooding, such as mangroves, often have specialized root systems ("pneumatophores") that are involved in gas exchange (see Chapter 6 in Kozlowski and Pallardy, 1997). For example, black mangrove produces air roots that protrude from the mud around the base of the stem. Enough air diffuses in through lenticels on the vertical roots when the tide rises to maintain aerobic conditions in the root system without appreciably reducing respiration during tidal inundation (Curran *et al.*, 1986). Some nonlenticular gas exchange also occurs through breaks near pneumatophore tips.

A number of flood-tolerant species initiate adventitious roots on the submerged portion of the stem, on the original root system, or both (Fig. 5.14). When the original roots of flooded plants die back to major secondary roots or the primary roots, new roots emerge from these points.

There has been some disagreement about the physiological importance of flood-induced adventitious roots. Gill (1975) and Tripepi and Mitchell (1984) questioned the assertion that adventitious roots were important in flood tolerance. However, the following lines of evidence indicate that such roots do confer flood tolerance. (1) Flood tolerance and formation of

Figure 5.14 Extensive formation of adventitious roots on the stem of a previously flooded American sycamore seedling. The horizontal line shows the height to which the seedling was flooded. From Kozlowski (1986c).

adventitious roots are correlated. For example, flood-intolerant species (e.g., Aleppo pine, jack pine, red pine, and paper birch) produced few or no adventitious roots when flooded (Sena Gomes and Kozlowski, 1980c; Tang and Kozlowski, 1982c, 1983). In contrast, flood-tolerant species such as melaleuca (Sena Gomes and Kozlowski, 1980b), green ash, river birch (Kozlowski, 1984b; Tang and Kozlowski, 1984b), and American sycamore (Kozlowski, 1986c) produced many adventitious roots when flooded. (2) Absorption of water and mineral nutrients is increased by adventitious roots of flooded plants. Transpiration and absorption of water by green ash seedlings were increased as much as 900% in seedlings with adventitious roots on submerged portions of stems over seedlings from which such roots had been excised. Increased production of adventitious roots of flooded seedlings also was correlated with eventual reopening of stomata that had closed shortly after the soil was flooded (Sena Gomes and Kozlowski, 1980a; Kozlowski, 1985a). Furthermore, when flood-induced adventitious roots were excised from the submerged parts of stems of American sycamore seedlings, subsequent growth was reduced. (3) Flood-induced adventitious roots are important in oxidizing the rhizosphere and detoxifying soil toxins (Hook *et al.*, 1970a; Hook and Brown, 1973; Hook, 1984). (4) Flood-induced adventitious roots increase the supply of root-synthesized hormones (e.g., gibberellins, cytokinins) to the shoots (Reid and Bradford, 1984).

Temperature

Together with water supply temperature is a dominating factor in the distribution and growth of woody plants. Temperature affects growth and development of woody plants directly by inducing injury and indirectly by influencing physiological processes. Most plant processes are highly temperature dependent. When temperatures are low the available energy often is inadequate for maintaining the biochemical processes essential for plant growth. Low temperatures also decrease permeability of membranes and increase protoplasmic viscosity. When temperatures become excessively high, molecular activity may become so rapid that enzymes controlling metabolic processes are denatured or inactivated. Inhibition of growth at high temperatures also may be associated with excessive transpiration (see Chapter 12 of Kozlowski and Pallardy, 1997) and high rates of respiration that deplete carbohydrate pools (see Chapter 6 of Kozlowski and Pallardy, 1997) which otherwise would be available for growth.

Effects of Low Temperatures

Low temperatures not only injure woody plants but also limit their distribution and regulate their growth. Much injury to plants in mild climates is caused by low temperatures following periods of mild weather that permitted growth to begin and caused loss of cold hardiness. In addition to exhibiting freezing injury, many tropical and subtropical plants are susceptible to "chilling injury" when exposed to temperatures a few degrees above freezing.

Plant Distribution   Climatic zones, which are determined to a large extent by variations in temperature related to latitude and altitude, support plant communities that are determined by natural selection for temperature regimes (Grace, 1987). The dominant effect of temperature is emphasized by higher tree lines on southern than on northern exposures of mountains, and by shifting of tree lines on higher elevations during long warm periods. Changes in the latitudinal limits of the boreal forest in North America are correlated with long-term changes in the boundaries of the Arctic air mass (Bryson and Murray, 1977). Temperature, radiation, rainfall, and wind speed are correlated with altitude, but fluctuations in altitudinal tree lines are correlated better with temperature regimes than with other environmental variables (Grace, 1989). Although the altitude of the tree line varies geographically, it coincides fairly well with the 10°C isotherm for mean summer air temperature (Friend and Woodward, 1990).

At increasing altitudes trees become progressively shorter and more deformed, and finally they disappear. As the tree line is approached, the short trees become multibranched, having many secondary branches below the main stem. At the tree line, woody plants are low, prostrate shrubs without a dominant main stem (Grant and Mitton, 1977). There is some evidence that certain species at the tree line are dwarf genotypes. Examples are genotypes of heather, Swiss stone pine, and mugo pine (Grant and Hunter, 1962; Grant and Mitton, 1977). However, the dwarfing of most trees at high elevations occurs without an obvious genetic component.

The harsh climate and short growing season at high elevations affect growth of woody plants in several ways: (1) inhibiting formation of leaf primordia by controlling cell division, (2) arresting leaf growth, internode expansion, cambial growth, and root growth; (3) decreasing seed production and dispersal; and (4) inducing mortality. These responses are associated with adverse effects of low temperature on carbohydrate, water, mineral, and hormone relations.

Low temperatures interact with strong winds near tree lines. High wind speeds break young shoots and often cause severe abrasion of leaves by ice particles and sand. The branches of trees exposed to strong winds commonly develop only on the leeward sides of trees. Near tree lines, individual trees or clumps of trees often are confined to valleys and lee sides of protecting rocks or ridges. Where wind velocities are very high, the tree line is lowered, but in well-protected areas some trees may survive at several hundred feet above the tree line.

Trees at high elevations are very prone to winter desiccation, in part because the short growing season prevents full development of cuticles on leaves (Baig and Tranquillini, 1980). Hence, when the soil is frozen and atmospheric conditions favor transpiration, leaves may become severely dehydrated. As elevation increased from 732 to 1402 m in New Hampshire, the thickness of the cuticle of balsam fir needles decreased from 3.01 to 2.2 μm and transpiration increased by 59% (De Lucia and Berlyn, 1984). In Scotland, desiccation of buds of mountain ash and European white birch trees increased progressively with altitude during the coldest part of the winter (Barclay and Crawford, 1982). The needles on windward sides of Engelmann spruce and alpine fir trees at high elevations had lower water contents and lower water potentials and were less viable than the leeward needles (Hadley and Smith, 1983). Snow cover prevented cuticle abrasion, resulting in higher needle water contents than in the severely abraded and dehydrated needles above the snow line (Hadley and Smith, 1987).

Success of short vegetation at high elevations where tall vegetation fails to grow is favored because short vegetation commonly is warmer than the air. Dwarf vegetation is aerodynamically smoother than tall vegetation and dissipates heat more slowly. Short vegetation is especially likely to be warmer than tall vegetation on sunny days when the wind speed is low (Wilson *et al.*, 1987; Grace *et al.*, 1989). The higher leaf temperatures of dwarf vegetation have important effects on rates of photosynthesis. At high altitudes the optimal temperature for photosynthesis at high irradiance usually varies from 15 to 25°C, but air temperature is as much as 10°C lower than the optimum. On days of high irradiance, the leaf temperatures of small trees commonly increase to the optimal range for photosynthesis (Friend and Woodward, 1990).

Growth of Plants   The effects of temperature on distribution and growth of woody plants have been studied with a variety of objectives in mind; see Kozlowski *et al.* (1991) for a review. Some studies showed that temperature regimes influence productivity of forests. For example, aboveground net primary productivity decreased by 80% from tropical to boreal pine forests (Gower *et al.*, 1994). In England a 1°C rise in temperature may increase plant productivity by approximately 10% if other factors such as water supply and soil fertility are not limiting (Grace, 1988). Most investigations dealt with effects of air temperatures on growth of whole plants or various plant parts (Kozlowski *et al.*, 1991), but some determined the effects of soil temperatures on growth (Kuhns *et al.*, 1985; Lyr and Garbe, 1995). Certain studies emphasized the importance of cardinal temperatures (the minimum temperature below which a physiological process, including growth, is not measurable; the maximum above which it is not measurable; and the optimum at which it proceeds most rapidly). Other studies documented the importance of day versus night temperatures on growth. Still others evaluated effects of very high or very low temperatures on physiology, growth, and distribution of woody plants.

Many attempts have been made to correlate heat sums or degree-days with growth of woody plants. Hellmers (1962) defined heat sum as total daily degree-hours. The temperature was the numerical value above 0°C times the number of hours that plants were at that temperature during a 24-hr day. Cleary and Waring (1969) developed a heat sum index that related the effects of field temperatures to distribution of Douglas fir trees. However, their procedure did not account for interactions of temperature with other environmental factors. Thomson and Moncrief (1982) developed a model, based on degree-day accumulation, for predicting dates of seasonal initiation of shoot expansion from data on dates of bud opening and weather records. Unfortunately, several heat sum studies assumed a constant relationship between temperature and growth during all stages of plant growth and during the day and night. The effects of temperature on vegetative growth are discussed in more detail in the following sections.

Shoot Growth   Temperature influences several aspects of shoot growth, including bud development, release of bud dormancy, expansion of leaves and internodes, and induction of bud dormancy. Both the temperature of the year of bud formation and temperature of the year of bud expansion into a shoot may influence the amount of shoot growth (Kozlowski, 1983).

Bud Dormancy   In the temperate zone, bud dormancy is most commonly induced by low temperatures and broken by winter cold. Freezing temperatures are not required for breaking of dormancy, and bud temperatures between 2 and 7°C are most effective.

The amount of chilling needed for release of bud dormancy varies with plant species, genotype, location of buds on a plant, site, and even with weather of the previous season. In Norway the vegetative buds of *Betula pubescens*, *B. pendula*, and *Prunus padus* were released from dormancy in December; those of *Populus tremuloides* in January; and those of *Alnus incana* and *A. glutinosa* in February (Heide, 1993a). More than 2,000 hr of chilling were required for breaking dormancy of buds of sugar maple seedlings in southern Canada; much less chilling was needed for seedlings from more southerly sources (Kriebel and Wang, 1962; Taylor and Dumbroff, 1975). When tested in Gainesville, Florida, seedlings of red maple provenances from cold regions required up to several weeks of chilling to release bud dormancy, whereas seedlings from warm regions required little or no chilling (Perry and Wang, 1960). Chilling of loblolly pine seedlings in storage was as effective as natural chilling for breaking dormancy (Carlson, 1985).

The chilling requirements of buds vary with their location on the plant

and type of buds. Terminal vegetative buds have a lower chilling requirement than lateral vegetative or flower buds. There also may be differences
in chilling requirements of buds within each of these groups. For example,
the chilling requirement of terminal buds of spur shoots of apple trees was
lower than for terminal buds of long shoots (Latimer and Robitaille, 1981).
In warm regions with limited amounts of chilling, the terminal buds may
open long before the lateral buds do, resulting in stronger apical dominance than typically occurs in a cold climate, where chilling induces opening of variously located buds at about the same time.

The effects of low temperature on dormancy release vary with the depth
of bud dormancy. When buds are in a state of deep dormancy, the range of
low temperatures needed to break dormancy is narrow. In comparison,
during late dormancy, when buds have been exposed to some chilling, the
temperature range for promotion of budbreak widens.

Interruption of a long period of chilling by exposure of plants to high
temperatures often negates the dormancy-breaking effect of low temperature. For example, 8-hr interruptions at 24°C with 16 hr at 6°C completely
negated the chilling effect in peach (Erez *et al.*, 1979). The high-temperature interruption had to exceed 4 hr to have a negating effect (Couvillon
and Erez, 1985b). Exposure of apple trees to 15 or 20°C during the first or
second 500 hr of chilling (in a 1500-hr, 5°C chilling period) reduced budbreak and elongation of both shoots and roots (Young, 1992).

Shoot Expansion   After bud dormancy is broken by low temperature,
exposure to warm temperature for a critical period is required for normal
shoot expansion. For example, after bud dormancy of sugar maple seedlings was broken by cold, exposure to at least 140 hr of warm temperatures
was necessary for shoot elongation to begin (Taylor and Dumbroff, 1975).
The rate of leaf unfolding of two hibiscus cultivars was curvilinear, increasing to a maximum at approximately 30°C (Karlsson *et al.*, 1991). In general,
a decrease in the unfolding rate of leaves may be expected above the
optimum temperature for leaf unfolding. In most deciduous trees of the
temperate zone, leaf emergence generally depends on the cumulative thermal sum (degree-hours or degree-days) to which buds are exposed after a
prerequisite cold period (Lechowicz, 1984).

Temperature influences both initiation and expansion of conifer needles. The optimum temperature for needle formation in black spruce and
white spruce seedlings was 25°C; needle initiation was reduced at both
lower and higher temperatures (Pollard and Logan, 1977).

Temperature also regulates the rate and seasonal duration of internode
elongation, as emphasized by reduction of internode elongation at progressively higher altitudes (Fig. 5.15).

In species exhibiting fixed growth of shoots (see Chapter 3 of Kozlowski

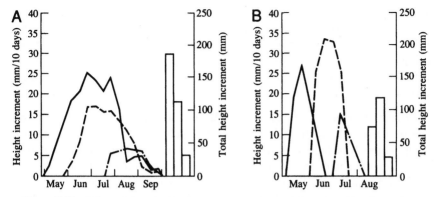

Figure 5.15 Effect of altitude on the amount and seasonal duration of height growth of young European larch (A) and Norway spruce (B) in Austria. Trees were studied at three altitudes: 700 m (solid line), 1300 m (dashed line), 1950 m (dot–dash line). Adapted from Tranquillini and Unterholzner (1968) and Oberarzbacher (1977); from Kozlowski (1983), by permission of Westview Press.

and Pallardy, 1997), the amount of annual shoot elongation often is correlated with the size of the winter bud (Kozlowski *et al.*, 1973), with large buds producing long shoots. Hence, the amount of shoot growth often is better correlated with the temperature regime of the year in which buds form than with temperature of the next year when buds expand into shoots. For example, environmental conditions during bud development influenced the subsequent year's growth of shoots of red pine (Olofinboba and Kozlowski, 1973).

Cambial Growth Temperature regulates cambial growth by influencing the time of seasonal initiation of division of fusiform cambial cells as well as the subsequent rate and duration of production of xylem and phloem tissues (Kozlowski *et al.*, 1991). Responses of cambial growth to temperature changes often are rather rapid. Early in the growing season in Wisconsin there was a 1-day lag effect of temperature on daily radial growth of northern pin oak trees (Kozlowski *et al.*, 1962). Fritts (1959) found that the maximum temperature of the preceding day influenced diameter growth of white oak and sugar maple trees in Indiana.

The effects of latitude and altitude on seasonal duration of cambial growth are correlated with air temperature. Cambial growth of some temperate-zone species in cold regions may continue for only a few weeks in the summer, resulting in small xylem increments. In contrast, in moist and hot tropical regions cambial growth of certain species may be more or less continuous throughout the year (Kozlowski *et al.*, 1991).

The predominant effect of temperature on cambial growth is illustrated by variations in growth of trees at different altitudes. Yearly diameter growth of Norway spruce trees in the Austrian alps was 6 mm at low altitudes and 3 mm near timberline (Fig. 5.16). The reduced cambial growth at high altitudes was associated with a short growing season, late seasonal initiation of cambial growth, and low cambial activity during the growing period (Holzer, 1973).

Thermal Shrinkage and Expansion of Stems   Studies of effects of temperature on cambial growth, as determined from changes in stem diameter, are complicated by effects of temperature on stem shrinkage and expansion. Most thermal shrinkage of stems and branches with lowering of temperature, and expansion on warming, occur during the winter (Kozlowski, 1965). In Wisconsin, tree stems contracted greatly as the air temperature dropped rapidly to below freezing values (Winget and Kozlowski, 1965b). During very cold winter periods, stem diameters decreased up to three times as much (and in some cases five times as much) as they increased as a result of cambial growth during the entire previous summer. Thermal shrinkage and expansion of tree stems occur very rapidly as the tempera-

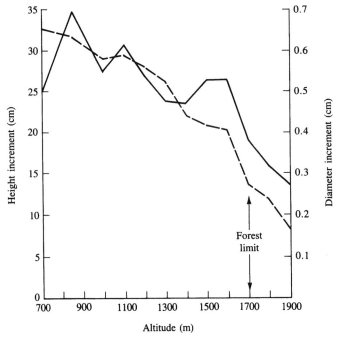

Figure 5.16   Effect of altitude on height (dashed line) and diameter (solid line) growth of mature (70- to 140-year-old) Norway spruce trees in the Austrian alps. After Holzer (1973); from Kozlowski (1983), by permission of Westview Press.

ture changes. In North Carolina, stems of shortleaf pine shrank after a cold spell and expanded almost immediately after a freeze ended (Byram and Doolittle, 1950). Most daily stem shrinkage of white ash trees in winter occurred near 6:30 A.M. when the daily temperature was lowest; the largest increases in stem diameter were recorded near noon when the highest temperature was recorded (Small and Monk, 1959). Most of the thermal shrinkage and expansion of stems is localized in the bark. After transfer from a warm (20°C) to a cold (−20°C) regime, twigs with bark shrank much more than those from which the bark had been removed (McCracken and Kozlowski, 1965).

Root Growth   In much of the temperate zone, roots begin to grow shortly after the soil becomes free of frost. Hence, root growth begins earlier in woody plants growing at low than at high altitudes. In the southeastern United States some growth of tree roots occurs in every month of the year.

The optimum temperature for root growth varies with several factors, including species and genotype, stage of plant development, and availability of soil moisture and O_2. For temperate-zone woody plants the minimum temperature for root growth is between 0 and 5°C, with the optimum being between 20 and 25°C. In Missouri, growth of black walnut roots began at a soil temperature of 4°C, increased slowly as the soil warmed to 13°C, and accelerated more rapidly to a maximum at 17 to 19°C. The number of growing roots peaked at 21°C (Kuhns *et al.*, 1985). Species of northern origins had lower root temperature optima than those of more southern origins (Lyr and Garbe, 1995).

The growth of roots of seedlings in the cotyledon stage of development is particularly sensitive to temperature, with root growth being reduced more than shoot growth by temperature extremes (Kozlowski and Borger, 1971). Over a 7-week period, the dry weight increment of red pine seedlings in the cotyledon stage was decreased by temperatures above or below 20°C. At 10°C, root growth was decreased by 61% and shoot growth by 26%. At 30°C root growth was reduced by 56% and shoot growth by 35% (Kozlowski, 1967b).

The effect of soil temperature on root growth is strongly modified by soil moisture supply. At soil temperatures below 17°C, temperature predominated in influencing root elongation of white oak trees in Missouri. However, at temperatures above 17°C, soil water supply became the more dominant regulatory factor (Teskey and Hinckley, 1981).

Soil temperature often affects growth of plants by regulating development and physiological efficiency of mycorrhizal fungi (Borges and Chaney, 1989). The optimum temperature for growth of different species of mycorrhizal fungi varies between 8 and 27°C. The form of the growth curve may show a sharp or indistinct optimum depending on the species of fungus (Harley and Smith, 1983). Colonization of roots by mycorrhizae is mediated by effects of temperature on growth of the fungal mycelium, spore production and germination, composition of root exudates, and growth and maturation of root tissues.

Colonization of Monterey pine roots by four mycorrhizal fungi was less at a soil temperature of 16°C than at 25°C (Table 5.12). In some tests there was no colonization at 16°C. Infection of green ash roots by *Glomus macrocarpum* or *G. fasciculatum* was higher at 25°C than at 15 or 35°C (Borges and Chaney, 1989). As emphasized by Hacskaylo *et al.* (1965), growth of mycorrhizal fungi at different temperatures is greatly influenced by the laboratory media used. Hence, strong correlations may not be evident between growth of mycorrhizal fungi on synthetic media and in the different environment of the rhizosphere (Hacskaylo *et al.*, 1965; Bowen and Theodorou, 1973).

Chilling Injury   Vegetative and reproductive tissues of many tropical and subtropical plants, and a few temperate-zone plants, are injured by exposure to temperatures below approximately 15°C but above freezing. Such

Table 5.12   Colonization of Monterey Pine Roots and Growth on Melin-Norkrans Agar by Strains of Mycorrhizal Fungi at Different Temperatures[a]

Fungus	Soil temperature (°C)	Root length (mm)[b]	Length colonized (mm)[b]		Colony diameter in Melin-Norkrans medium (mm)[c]
			Seedlings	Fibers	
Suillus luteus No. 1	16	73.4	6.2	4.5	8.9
	20	77.6	15.2	7.4	70.9
	25	82.9	24.9	6.9	63.8
S. luteus No. 3	16	85.6	18.6	5.1	37.7
	20	91.7	29.4	5.1	60.7
	25	92.9	31.8	9.1	61.3
S. granulatus No. 8	16	75.8	16.0	6.1	7.0
	20	90.7	28.0	6.5	41.0
	25	87.1	26.3	8.1	44.8
Rhizopogon luteolus No. A[d]	16	41.9	4.1	4.4	40.0
	25	60.1	27.3	3.2	82.0
	LSD: *Suillus*		5.1	($p = 0.05$)	
			6.7	($p = 0.01$)	
	R. luteolus		7.4	($p = 0.01$)	

[a]From Theodorou and Bowen (1971). *Aust. J. Bot.* **19**, 13–20.
[b]Growth in soil for 4 weeks.
[c]Growth for 28 days, except for *R. luteolus* which was for 11 days.
[d]This experiment was carried out at a different time from that with the species of *Suillus*.

chilling injury often is a prelude to reduced growth and sometimes death of plants.

Chilling injury is expressed in a wide range of visible symptoms and metabolic disturbances. Chill-induced changes include reduced photosynthesis (see Chapter 5, Kozlowski and Pallardy, 1997), altered protein synthesis, necrotic lesions (e.g., discoloration or pitting of foliage), increased susceptibility to decay organisms, changes in membrane permeability, and ultrastructural changes. Chilling of callus tissue of red-osier dogwood led to disruption of the tonoplast within 24 hr. Additional chilling disrupted cell organelles, including nuclei, mitochondria, and protoplastids (Niki *et al.*, 1978). An increase in ion leakage, and in K^+ leakage specifically, was a sensitive indicator of chilling injury in grapefruit callus tissue (Forney and Peterson, 1990). Unfortunately, measurements of ion leakage are destructive and time consuming. Cohen *et al.* (1994) found that weight loss (water loss) of grapefruits and lemons was a useful nondestructive indicator of chilling injury before visible injury symptoms were evident. Weight loss of fruits was associated with development of large cracks around the stomata. Such cracks were sites for fungus infection.

Symptoms of chilling injury vary among species and genotypes, various tissues, and with the chilling temperature and duration of exposure. In general, chilling injury increases as the temperature is lowered or as the duration of exposure to a chilling temperature is increased. Chilling and light intensity often interact to injure plants. For example, injury to the photosynthetic apparatus by chilling is accentuated by high light intensity (Powles *et al.*, 1983; see Chapter 5 of Kozlowski and Pallardy, 1997).

Chilling-sensitive plants may show chilling injury at any stage of development except the dry seed stage (Morris, 1982). Dry seeds of chilling-sensitive species can be stored for long periods at chilling or freezing temperatures without adverse effects (Wolk and Herner, 1982).

Both germinating seeds and young seedlings of tropical origin are susceptible to chilling injury. Sensitivity to chilling is evident as soon as seeds begin to imbibe water. Some species also show increased sensitivity a few days later. When cotton seeds were chilled as soon as they began to imbibe water, the tips of their radicles aborted. However, if the seeds were chilled 24 hr after imbibition started, the root cortex was injured (Christiansen, 1964). Symptoms of chilling injury in seedlings include wilting, desiccation, bleaching, necrosis of leaves, decreased growth, and often death of plants (Wolk and Herner, 1982).

Chilling injury occurs in two stages. The most widely accepted explanation for chilling injury is that an almost instantaneous change in the physical state of plant membranes (the primary response) induces secondary dysfunctions that lead to visible injury. These secondary responses include changes in membrane-associated enzyme activities and membrane permeability (see Chapter 8 in Kozlowski and Pallardy, 1997) and consequent changes in ATP levels, leakage of solutes, loss of compartmentalization, and ionic imbalances. Whereas the primary response is reversible, the secondary responses may not be. The secondary events are reversible in the short term if the chilling stress is removed. If the chilling stress continues, however, the process becomes irreversible. In fact, after this stage is reached, increasing the temperature above the chilling level often accentuates injury (Lynch, 1990).

Freezing Injury   Freezing injury to vegetative tissues includes killing back of shoots, injury to the cambium, frost cracks, and injury to roots that often leads to plant mortality. Freezing injury is variously expressed as discoloration, dieback, rupture of tissues, malformations following injury to meristematic or incompletely differentiated tissues, delay in plant development, and plant mortality (Sakai and Larcher, 1987).

Injury to Shoots   Cold hardiness of shoots disappears rapidly in the spring before buds open. Hence, both buds and young leaves often are killed by freezing in the spring. Some shoots are injured more than others, reflecting differences in rates of dehardening. Lateral buds, which may expand before terminal buds do, are injured most (Cannell and Sheppard, 1982). Freezing injury to late-season growth flushes is well known (Kozlowski, 1971a).

The sequence of events associated with winter injury to conifer foliage has been well documented (Adams *et al.*, 1991; Perkins *et al.*, 1991). Winter injury to red spruce trees, which was confined largely to the most recent year's foliage, began with chlorosis and was followed by color changes from yellow-orange to orange-red, and finally to red-brown pigmentation in needles (Friedland *et al.*, 1984). The affected needles abscised in the next growing season. The mesophyll was injured most. The cytoplasm lost its granular texture, and the integrity of thylakoid, tonoplast, and plasma membranes was sequentially lost. Desiccation of cells was associated with disruption of cellular membranes. When the integrity of the plasma membrane was lost, disintegrating cell constituents were released, causing discoloration of cells and tissues.

Injury to Stems   Abnormal "frost rings" sometimes result from freezing injury to the cambium. Frost rings are made up of an inner part comprising frost-killed cells, usually xylem mother cells and differentiating cambial derivatives, and an outer layer of abnormal xylem cells that were produced after the frost occurred. Frost rings of gymnosperms typically have abnormal, underlignified and collapsed tracheids as well as traumatic parenchyma cells. The rays usually are expanded and laterally displaced (Fig. 5.17).

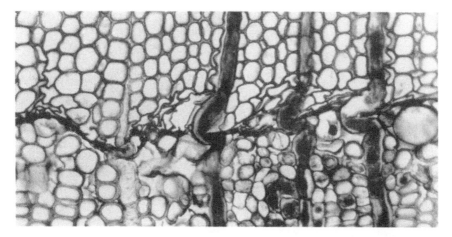

Figure 5.17 Frost ring in a conifer. The effects of frost are indicated by collapsed tracheids, displacement of rays, and excessive production of parenchyma tissue. Photo courtesy of U.S. Forest Service.

During periods of freezing the stems and branches of trees may develop frost cracks (Fig. 5.18). Cracking of stems on very cold nights sometimes produces a loud noise resembling a gunshot. Frost cracks generally form in only some of the trees of a stand, presumably because of differences in hydration among trees. Stems of trees with very wet wood are more likely to crack than dry stems.

Frost cracks form when wood shrinks more tangentially than radially. They are caused largely by freezing out of water from wood cell walls into cell lumens. The higher vapor pressure of some of the bound water in cell walls causes diffusion of water to the ice in cell lumens. Formation of frost cracks also is influenced by faster cooling of the outer wood than the inner wood, expansion of freezing water in cell lumens, and formation of lenses of ice in the wood (Kubler, 1983, 1987, 1988).

Injury to Roots Although roots are not as hardy as stems, roots often are injured less because they are protected by soil and snow cover from exposure to cold (Goulet, 1995). Nevertheless, many roots, especially the fine roots, are killed when the soil freezes. The amount of root mortality varies with species and rootstocks. Small trees often are injured by frost heaving when roots are frozen in a mass of ice and subsequently additional layers of ice form below. Some of the roots are broken as seedlings are lifted. Larger trees are less prone to frost heaving because the frozen surface soil layer slides upward along their stems and downward on thawing. In comparison, small seedlings tend to remain lifted on thawing because their lower roots

Figure 5.18   Frost cracks (arrows) on stems of 13-year-old Golden Delicious apple trees on East Malling VII rootstock. Extreme splitting from the main scaffold branches to the stock–scion union is visible in the tree at left. At right is a tree that has recovered from previous injury. After Simons (1970); from Kramer and Kozlowski (1979), by permission of Academic Press.

are more likely to be on a frozen layer. Injury or death of roots follows if roots are broken or the exposed roots are dehydrated. The amount of injury by frost heaving is reduced by snow cover, ground covers, mulches, use of fertilizers, irrigation, early seeding, and careful planting practices that induce large and deep root systems.

Mechanisms of Freezing Injury   Injury to plant tissues at subfreezing temperatures can be caused directly by intracellular freezing of cell contents or indirectly by tissue dehydration resulting from extracellular freezing. However, most freezing injury probably is caused by dehydration, and tolerance to freezing depends on the capacity of protoplasm to survive such dehydration (Levitt, 1980a; Sakai and Larcher, 1987). When frost-hardened plants are cooled slowly, ice forms first in the intercellular spaces; as the temperature decreases, water moves out of cells to these ice nuclei. As a result, the concentration of the cell sap is increased and its freezing point lowered. Rapid cooling of many cold-hardened plants often causes injury, possibly because water does not move out of cells fast enough to prevent intracellular freezing.

Freeze-induced dehydration injury results from destabilization of cell membranes. The plasma membrane is a primary site of injury because of its central role in cell responses during a freezing–thawing cycle. Maintenance of differential permeability of the plasma membrane is essential for plant survival. Changes in the ultrastructure of the plasma membrane may take different forms depending on the degree of cell dehydration. When cells are cooled to 0 to −5°C, little dehydration occurs, and the cells retain their capacity to respond osmotically on warming and thawing. Nevertheless, lysis occurs during osmotic expansion. When cells are cooled to temperatures below −5°C, they dehydrate more and lose their capacity to respond osmotically during thawing. Both forms of injury result from dehydration, but the mechanisms are different (Steponkus and Lynch, 1989). Injury may follow dehydration at the cellular level, involving transport of bulk water or water loss from membranes (Steponkus and Webb, 1992).

In many species of plants, freezing is prevented or postponed by supercooling (undercooling). A decreasing water content and increasing concentration of osmotically active compounds in the cell sap not only lower the freezing point but also decrease the threshold subcooling temperature and prolong supercooling. Hence, some tissues do not freeze for a long time, if at all, when exposed to very low temperatures.

Deep supercooling has been reported in more than 240 species of woody plants. The capacity for supercooling varies greatly among species as well as different organs and tissues of the same plant. In some woody plants, the hardy xylem ray parenchyma and axial parenchyma cells, floral tissues, phloem, and buds may supercool and do not show typical extracellular freezing. Supercooling occurs to lower temperatures in xylem parenchyma cells than in the bark, with xylem parenchyma cells of some deciduous trees supercooling to temperatures as low as −47°C (Quamme, 1985).

Many frost-sensitive plants are injured by low temperatures because of the presence of catalysts that limit supercooling of their tissues. Prominent among such catalysts are ice-nucleation-active (INA) bacteria. For example, *Pseudomonas syringae* and *Erwinia herbicola* are widespread on plant tissues and catalyze ice formation at temperatures as warm as −1°C. Freezing injury in many sensitive plants is a function of the number of INA bacteria on plant tissues and the nucleation frequency (activity) of bacterial cells (Lindow, 1983).

In some species, freezing is induced by ice nucleators other than bacteria. Ice nucleation activity in the wood of peach was nonbacterial in origin, as emphasized by the following lines of evidence (Gross *et al.*, 1988): (1) peach stems were INA-active beginning at near −2°C whether INA bacteria were detected or not, (2) suppression of INA bacteria in orchards with bactericides did not provide significant protection from freezing injury, (3) bacterial ice nuclei in the field were not active for extended periods, and (4) treatment of peach shoots with inhibitors of bacterial ice nucleation did not influence the ice nucleation temperature of shoots. Most other organic and inorganic materials such as dust particles nucleate ice only at temperatures below $-10°C$, and they do not limit supercooling of plant tissues above $-5°C$, the temperature at which most frost-sensitive plants are injured (Lindow, 1983).

Two distinct exotherms have been identified during cooling of stems of woody plants (Malone and Ashworth, 1991). The first or high exotherm, which corresponds to freezing of water in the xylem vessels and intercellular spaces, is not harmful. The second exotherm corresponds to freezing of a portion of the supercooled water. This exotherm, which generally occurs at or above $-40°C$, is correlated with injury to the xylem parenchyma cells. The number of cells killed during freezing is proportional to the amount of supercooled water that freezes. The xylem ray parenchyma cells of flowering dogwood showed two responses to freezing: (1) fragmentation of protoplasm with indistinguishable plasma membranes and damage to cell ultrastructure without intracellular ice formation and (2) formation of large intracellular ice crystals. It appeared that intracellular ice formation and possibly cavitation accounted for freezing injury in the xylem parenchyma cells (Ristic and Ashworth, 1993).

The effects of winter injury on subsequent plant growth vary with the severity of injury and duration of exposure to freezing temperatures. In New Hampshire a single winter injury event did not affect growth of subalpine red spruce 1 and 2 years later (Peart *et al.*, 1992). However, repeated winter injury in 3 successive years, or 30% or more needle damage, resulted in a 60% decrease in cambial growth and a 30% decrease in height growth of red spruce (Wilkinson, 1990). It has been suggested that winter injury may be responsible for growth decline and high mortality of red spruce trees in high-elevation forests of the northeastern United States (Friedland *et al.*, 1984; Johnson *et al.*, 1986, 1988).

Cold Hardiness   Many plants native to warm regions cannot be moved to cold regions because they do not develop enough cold hardiness, do not harden rapidly enough to survive early cold weather, deharden too rapidly, and are killed by freezing temperatures. The practical implications of lack of adequate cold hardiness of southern seed sources are of particular interest to foresters, horticulturists, fruit growers, and arborists.

Acclimation of woody plants to cold varies widely among species and genotypes as well as different organs and tissues. The cold hardiness of twigs of species collected in midwinter from various parts of North America differed greatly (Table 5.13). Four northern species—trembling aspen,

Table 5.13 Variations in Freezing Tolerance of North American Tree Species and Minimum Temperatures at Northern Limits of Natural Ranges or Artificial Plantings[a]

Relative hardiness classification	Representative species	Average minimum temperatures at northern limits of growth (°C)		Observed freezing resistance (°C)
		Natural range	Artificial plantings	
Tender evergreen species	*Quercus virginiana*	−3.9 to −6.7	−9 to −12	−7 to −8
Hardy evergreen species	*Magnolia grandiflora*	−9 to −12	−18 to −20	−15 to −20
Hardy deciduous species	*Liquidambar styraciflua*	−18 to −20	−26 to −29	−25 to −30
Very hardy deciduous species	*Ulmus americana*	−37 to −46	−40 to −43	−40 to −50
Extremely hardy deciduous species	*Betula papyrifera*	Below −46	Below −46	Below −80
	Populus deltoides	−32 to −34	−37 to −45	Below −80
	Salix nigra	−32 to −34	−37 to −45	Below −80

[a]After Sakai and Weiser (1973); from Kramer and Kozlowski (1979), by permission of Academic Press.

balsam poplar, white birch, and tamarack—resisted freezing to −80°C, and some survived temperatures of −196°C after prefreezing to −15°C. Conifers of the western United States such as ponderosa pine, western white pine, lodgepole pine, Jeffrey pine, blue spruce, Engelmann spruce, alpine fir, white fir, and western larch survived freezing at temperatures between −60 and −80°C. In comparison, species from the warmer southeastern parts of the United States, including slash pine, longleaf pine, live oak, and southern magnolia, survived temperatures only down to −15°C. Both the degree of cold hardiness and time of its annual occurrence often vary in closely related species. For example, maximum cold tolerance of Sydney blue gum was acquired in late June and a month later in brown barrel. The rate of dehardening also varied and was faster in brown barrel than in Sydney blue gum (Fig. 5.19). The cold hardiness of these species was correlated with their natural temperature regimes.

Intraspecific variation in cold hardiness has been demonstrated in many species of trees and shrubs. Climatic races that have adapted to warm climates cannot be moved northward because they are not sufficiently cold hardy. Examples are eastern white pine (Maronek and Flint, 1974) and northern red oak (Flint, 1972). Acclimation of northern provenances to cold often is correlated with their phenology, with seed sources that set

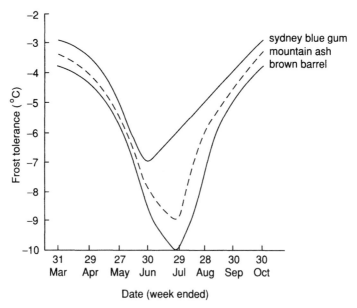

Figure 5.19 Seasonal changes in relative frost tolerance of Sydney blue gum, mountain ash, and brown barrel during 1976. From Menzies *et al.* (1981).

buds early being injured less by early frosts. For planting in cold regions it often is desirable to select planting stock for late bud opening (Dormling, 1982; O'Reilly and Parker, 1982).

Cold hardiness of different organs and tissues in the same plant typically varies greatly. In general, reproductive tissues are more sensitive than vegetative tissues to cold. Roots, young leaves, and cambial cells are particularly sensitive to cold.

Development of Cold Hardiness   Most woody plants of the temperate zone become dormant before they become cold hardy (Colombo *et al.*, 1989). Loss of cold hardiness of some species when brought indoors during the winter occurs only after they are released from dormancy. However, some hardy plants develop cold hardiness independently of bud dormancy (Sakai and Larcher, 1987). For example, pines of the southern United States become dormant but never acquire tolerance to very low temperatures. Glerum (1976) emphasized the importance of distinguishing between species that must become dormant in winter in order to achieve maximum cold hardiness and species that do not have such a requirement.

Cold hardiness of woody plants of the temperate zone develops in two stages. Stage 1 occurs at temperatures between 10 and 20°C in the autumn when carbohydrates and lipids accumulate. These compounds are substrates and energy sources for the metabolic changes that occur in the next stage. Stage 2, which is promoted by freezing temperatures, involves synthesis of proteins and membrane lipids as well as structural changes.

The rate of seasonal development of cold hardiness varies somewhat between species of warm climates and those of cold climates. Woody plants of mild climates (e.g., tropical pines, olive, eucalyptus) become cold hardy in response to progressively decreasing temperatures. Cold hardiness stops at stage 1, is acquired within a day or two, and can be rapidly lost without significant changes in metabolism or development. Cold hardening of these species is rather insensitive to photoperiod (Paton, 1982; Sakai and Larcher, 1987).

Initiation of cold hardiness in trees and shrubs of the cool temperate zone begins and continues slowly as their metabolism and growth decline, long before very low temperatures occur. In Finland cold hardiness of Scotch pine and Norway spruce increased in mid-September when the mean daily temperature decreased to within the range of 5 to 10°C (Repo, 1992). Such plants slowly become metabolically active in the late winter and early spring. In many species of this group, the effects of temperature on development of cold hardiness are modified by photoperiod. This is emphasized by winter injury to plants near street lamps (Kramer, 1937; Sakai and Larcher, 1987). In some species a combination of low temperature and short days induces cold hardiness. Examples are cloudberry (Kaurin *et al.*, 1982) and Scotch pine (Smit-Spinks *et al.*, 1985). Day length also may affect initiation of dehardening in the spring. For example, when days were less than 10 hr long, the onset of dehardening of Monterey pine was delayed by up to 10 days. Once the dehardening process started, however, temperature had the major effect in regulating cold hardiness (Greer and Stanley, 1985).

Development of cold hardiness of Monterey pine was influenced by photoperiod, night temperature, and severity of frosts (Greer and Warrington, 1982). Temperatures between −3 and −5°C predominated in controlling cold hardiness. Night temperatures as low as 1°C also had an influence, whereas temperatures above 5°C did not. A minimal photoperiod of less than 11 hr for up to 42 days was required for low-temperature hardening. Development of frost hardiness in seedlings of western red cedar and yellow cedar (*Chamaecyparis nootkatensis*) was primarily temperature dependent (Silim and Lavender, 1994). Exposure to low temperature induced and maintained cold hardiness; exposure to warm temperature resulted in loss of hardiness. In comparison, frost hardiness of white spruce seedlings was under more complex environmental control. Initiation of frost hardiness was regulated by day length, and maintenance of hardiness was related to depth of dormancy in apical buds. Conditions that led to bud set, while resulting in development of a moderate degree of cold hardiness, also predisposed plants to respond immediately to low temperature.

Biochemical Changes   During acclimation to cold, changes occur in activity of enzymes; hormone balance; and concentrations of sugars, proteins, amino acids, nucleic acids, and lipids. Cold hardiness has been variously attributed to such changes (Sakai and Larcher, 1987).

As plants begin to acclimate to cold in the autumn sugars accumulate directly from photosynthesis as their use in growth and respiration decline and also as a result of starch hydrolysis (Kozlowski, 1992). Sugars might increase cold hardiness by accumulating in vacuoles and decreasing the likelihood of formation of intracellular ice, by causing changes that increase tolerance of freeze-induced dehydration, and by diluting compounds such as electrolytes that in high concentrations may be toxic to membranes (Sakai and Larcher, 1987). The susceptibility of heavy-bearing pecan trees to winter cold was attributed to their low carbohydrate reserves (Wood, 1986).

Correlations between cold hardiness and soluble protein contents of plant tissues also have been shown (Li and Weiser, 1967; Brown and Bixby, 1975). Changes in soluble proteins may include an increase in water-binding proteins that could lower the amount of free cellular water, thereby decreasing formation of intracellular ice during acclimation to cold.

Development of cold hardiness is accompanied by increases in lipids, particularly phospholipids, in various organs and tissues, including, for example, the bark of poplar (Yoshida, 1974), shoots of apple trees (Ketchie and Burts, 1973), and chloroplasts of eastern white pine needles (de Yoe and Brown, 1979). The amount of membrane lipids of cold-hardened Norway spruce needles was almost twice as high as in unhardened needles (Senser, 1982). Phospholipids and galactolipids increased in needles of *Podocarpus* species at the time of maximum cold hardiness (Latsaque *et al.*, 1992). Steponkus *et al.* (1988, 1990) concluded that maintenance of stability of plasma membranes during development of cold hardiness was largely a consequence of changes in their lipid composition.

As emphasized by Kozlowski *et al.* (1991), biochemical changes in plant tissues exposed to decreasing temperatures are inevitable because (1) low temperature affects various enzyme-mediated processes differently, (2) slowing of growth is associated with decreased use of soluble carbohydrates and N compounds, and (3) reduction in the rate of respiration also decreases use of soluble carbohydrates. The accumulation of proteins in the autumn in the bark of trees of the temperate zone may not account for freezing tolerance (Guy, 1990). Hence, both sugars and proteins accumulate during cold hardening, but such accumulation may not be causally related to increase in cold hardiness.

Cytological Changes Exposure to cold induces cytological changes in plant cells. Changes in mesophyll cells may include clumping of chloroplasts, movement of chloroplasts into cell corners or around cell nuclei, splitting of large vacuoles into small ones, and reduction in chlorophyll content (Havranek and Tranquillini, 1995). Ultrastructural changes in cortical cells of apple trees in Japan were divided into two sets, one occurring during acclimation to cold (September to January) and another associated with deacclimation (February to May) (Kuroda and Sagisaka, 1993). During acclimation to cold, changes occurred in microvacuolation, and there was an increase in volume of the cytoplasm. The cells became temporarily rich in organelles involved in protein synthesis (vesicular endoplasmic reticulum, polysomes, dictyosomes, and vesicles). Plastids containing starch, protein–lipid bodies, and mitochondria also were abundant. From mid-November to March the plastids aggregated around the nucleus, and "plastid initials" formed by pinching off from mature plastids. Following development of maximum cold hardiness in January, starch disappeared from the plastids during deacclimation, and fusion of vacuoles began in late February. Starch granules reappeared in plastids, and organelles involved in protein synthesis became abundant. By mid-May when cold hardiness was low, most of the cells contained large vacuoles. The data suggested that seasonal replication of vacuoles and cytoplasm may be associated with the dehydration that occurs during freezing, and replacement of organelles may be linked to the metabolism required for development or loss of cold hardiness.

Winter Desiccation Injury Injury to evergreens during the winter often results from desiccation of leaves rather than from direct thermal effects. Usually the leaves turn brown and are shed the following spring, but the buds survive. Winter desiccation injury occurs when absorption of water cannot keep up with transpirational losses. In the temperate zone, transpiration often accelerates as the air temperature rises above freezing during sunny winter or spring days and increases the vapor pressure gradient between the leaves and surrounding air. Because the soil is cold or frozen, the roots cannot absorb water fast enough to replace transpirational losses, and the leaves dehydrate. In addition, when stems are chilled a few degrees below 0°C, water in the tracheids freezes, hence preventing water transport to the leaves even when some of the roots are in unfrozen soil (Havis, 1971). The extent of winter desiccation injury varies with many factors including plant species and genotype, soil water content, time and depth of freezing of soil, depth of snowfall, air humidity, and wind velocity (Kozlowski, 1976a, 1982a).

Winter desiccation is important in limiting the altitudinal and latitudinal ranges of conifers in forests of North America, Europe, and Japan (Kozlowski, 1982a; Sakai and Larcher, 1987). Winter drying injury, called "redbelt" when it occurs in horizontal bands on mountainsides, results when the soil is frozen or when warm winds follow belts. When lower branches of conifers are buried in snow they generally are not injured.

In general, temperate-zone conifers are more susceptible than boreal conifers to winter desiccation. Such cold-hardy species as Scotch pine and jack pine show less injury than red spruce, eastern hemlock, and eastern white pine. In the Adirondack mountains of New York State the order of desiccation injury among species was as follows: red spruce > eastern hemlock > eastern white pine > balsam fir (Curry and Church, 1952). In Sweden black spruce seedlings were injured more than those of Scotch pine or Norway spruce, and those of lodgepole pine were injured least (Christersson and von Fircks, 1988).

Symptoms of winter desiccation injury often resemble those of freezing injury, making it difficult to distinguish between them. Often the leaves of conifers are injured by both desiccation and freezing (Wardle, 1981).

Winter desiccation injury adversely affects tree growth, with the amount of growth reduction often being proportional to the amount of injury. In young lodgepole pine stands in the western United States winter desiccation injury ranged from 0 to 80% and averaged near 35% (Bella and Navratil, 1987). Reduction in height growth and bud size of young trees was proportional to the amount of injury. Growth was reduced up to two-thirds in trees in which 60% or more of the foliage showed necrosis. On intermediate-sized and mature trees, reduction of annual volume increment in the growing season following injury was variable and sometimes as much as 50%, with no additional growth reduction in subsequent years.

Effects of High Temperatures

Plant distribution and growth may be influenced both positively and negatively by high temperatures. Since the 1960s, there has been much concern that global warming ("the greenhouse effect"), that is, increases in greenhouse gases that prevent some outgoing long-wave radiation from being reradiated back to space, may raise global temperatures appreciably, with several possible ecological consequences. There also has been concern with heat injury to woody plants by exposure to much higher temperatures than could possibly be attributed to global warming.

Global Warming   The gases that may cause global warming include carbon dioxide (CO_2), nitrous oxide (N_2O), methane (CH_4), chlorofluorocarbons (CFCs), and a few other trace gases such as ammonia (NH_3) and sulfur dioxide (SO_2). More than 70% of the predicted greenhouse warming may be accounted for by CO_2 (49%), methane (18%), and N_2O (6%) (Denmead, 1991). The CO_2 concentration of the air is increasing steadily (see Chapter 5 of Kozlowski and Pallardy, 1997). More than 80% of the increase in CO_2 is due to combustion of fossil fuels and the rest to deforestation and biomass burning (Dickinson and Cicerone, 1986). Sources of methane include rice fields, wetlands, burning of biomass, ruminant animals, landfills, production of natural gas, and mining of coal. Wetlands are currently among the largest contributors to the global methane flux (Bridgham *et al.*, 1995). Sources of the additional amounts of N_2O have not been precisely quantified, but burning of biomass, combustion of fossil fuels, and emissions from soils undoubtedly are important. Some investigators estimate that soils may contribute as much as 90% of the total N_2O emissions. Increases in atmospheric N_2O concentrations have been attributed to greater use of N fertilizers and irrigation in agriculture, as well as increased N deposition in forest soils (Melillo *et al.*, 1989). The uncertainty of the magnitude of soil sources of N_2O reflects difficulties in measuring fluxes over large areas and over long periods as well as lack of reliable methods for extrapolating to large areas the data obtained locally (Denmead, 1991).

Much work with seedlings in controlled environments with natural photoperiods and light intensity and with temperature regimes following those out of doors showed that elevation of CO_2 above ambient levels increased growth variously in different species (Fig. 5.20; Tolley and Strain, 1984b; Strain and Cure, 1985; Sionit *et al.*, 1985; Norby *et al.*, 1986a). Sionit and Kramer (1986) summarized the results of many experiments dealing with effects of high CO_2 concentrations on seedlings as follows:

1. Dry weight, stem diameter, and height growth are increased by CO_2 concentrations in the range of 400 to 700 ppm.
2. Concentrations of CO_2 higher than 500 to 600 ppm induce little additional growth and sometimes cause a decrease.
3. A high concentration of CO_2 appears to induce morphological changes such as increased branching, leaf area, and leaf thickness in some species.
4. Seedlings of different species do not respond similarly to elevated CO_2 concentrations.
5. The competitive capacity of seedlings of various species is likely to be affected differently by increasing CO_2 concentration.

Elevated CO_2 generally promotes an increase in leaf area of most species, reflecting larger leaves and/or greater production of leaves. The increase in area of individual leaves may be the result of an increase in the number of cells or a higher rate of expansion of leaf cells (Ceulemans and Mousseau, 1994). Usually the increase in biomass with CO_2 enrichment is higher for broad-leaved trees than for conifers.

The additional foliage produced by plants in CO_2-enriched air may mag-

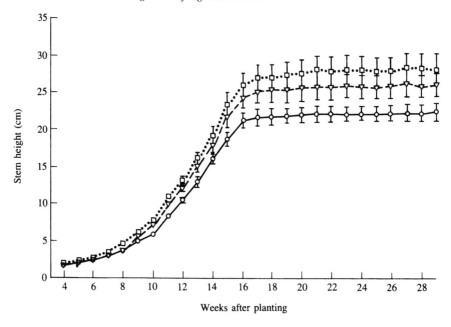

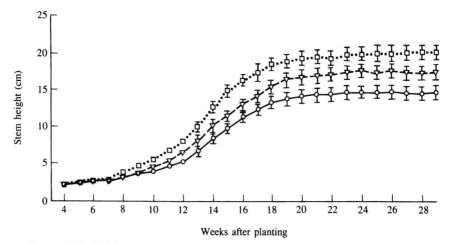

Figure 5.20 Height growth of seedlings of sweet gum (top) and loblolly pine (bottom) grown for 6 months with 350 (solid line), 500 (dashed line), and 650 μl liter^{-1} of CO_2 (dotted line) in an air-conditioned greenhouse with natural photoperiod and temperature simulating average outdoor temperatures. From Sionit *et al.* (1985).

nify the growth increase by higher photosynthetic efficiency per unit of leaf area. Whole-tree net photosynthesis of *Pinus eldarica* seedlings increased rapidly and reached a maximum at a CO_2 enrichment of 300 μmol mol^{-1}

above ambient, where it added nearly 50% to the basic growth stimulation provided by the direct photosynthesis-enhancing effect of elevated CO_2 (Garcia *et al.*, 1994).

The stimulation of plant growth in CO_2-enriched air is associated not only with increased photosynthesis (see Chapter 5 of Kozlowski and Pallardy, 1997), but also with improved mineral relations and water use efficiency (WUE; dry weight increment per liter of water used). Enrichment of CO_2 of the air increased the dry weight increment of N-deficient white oak seedlings by 85% (Norby *et al.* 1986a,b). High N use efficiency was shown by allocation of a greater proportion of the N pool to the fine roots and leaves. Absorption of other mineral nutrients increased with CO_2 concentrations; P and K uptake by plants in CO_2-enriched air increased proportionally with seedling growth. The increased absorption of P may have resulted from greater proliferation of fine roots and mycorrhizae. In shortleaf pine and white oak seedlings mycorrhizal density was greater at elevated than at ambient CO_2 levels. The greater mycorrhizal density and greatly increased first-year root growth of seedlings may facilitate subsequent absorption of mineral nutrients (O'Neill *et al.*, 1987a,b). When ambient air was augmented with CO_2, shortleaf pine seedlings allocated more carbohydrates to fine roots than did seedlings in ambient air. The increase was associated with greater production of fine roots and mycorrhizae (Norby *et al.*, 1987). Nutrient availability of N-fixing plants growing in infertile soils may be increased by enhanced nodulation and total nitrogenase activity as was found for black locust and European alder seedlings. These responses were attributed to increases in photosynthesis (Norby, 1987).

The increase in growth of white oak seedlings in CO_2-enriched air was associated not only with increased mineral uptake but also with an increase in water use efficiency, although total water use per plant was not affected. The increased WUE, a common plant response to high CO_2 levels, has been attributed to reduction in transpiration through stomatal closure while the rate of photosynthesis is either maintained or increased (Norby *et al.*, 1987).

Because short-term growth of CO_2-enriched seedlings grown in containers might not be indicative of long-term growth responses of trees in the field, Idso *et al.* (1991) grew young sour orange trees in the field with CO_2 enrichment at 300 ppm above ambient. After 2 years, the CO_2-enriched trees had 79% more leaves, more stem area (Fig. 5.21), 56% more primary branches, 70% more secondary branches, and 240% more tertiary branches. Furthermore, the CO_2-enriched trees produced some fourth-, fifth-, and sixth-order branches, whereas trees grown in ambient CO_2 did not. Over a 3.5-year period CO_2 enrichment also greatly stimulated root growth (Fig. 5.22). Idso and Kimball (1992b) calculated that doubling of the mean CO_2 concentration of 360 ppm would increase tree growth by 3.8

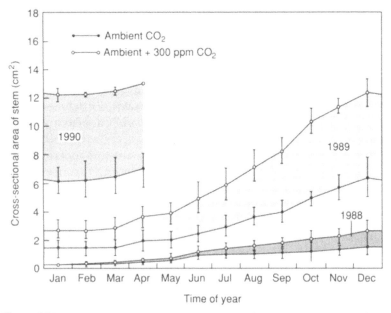

Figure 5.21 Effect of elevated CO_2 on cambial growth (as stem cross-sectional area) of sour orange trees. The trees were grown at ambient or ambient plus 300 ppm CO_2 for approximately 2.5 years. The error bars show standard deviation ranges. From Idso *et al.* (1991), with permission of Blackwell Science Ltd.

times. Exposure of Monterey pine seedlings to high CO_2 levels for 2 years increased biomass production and also increased wood density because of thickening of tracheid walls (Conroy *et al.*, 1990). These studies were performed on isolated trees, however, and growth stimulation of forests by CO_2 enrichment may be expected to be less dramatic because enhanced growth ultimately will lead to closed canopies and substantially more light limitation on photosynthesis and growth than for open-grown trees. Biomass after 4 yr was substantially greater in white oak saplings in open-top chambers with CO_2 enrichment than in control plants. However, there was no sustained effect of CO_2 on relative growth rate after the first year (Norby *et al.*, 1995). The data suggested that the increased growth rate could continue only as long as leaf area could increase, a condition that would not occur indefinitely in a forest.

Growth of various species under different environmental conditions differs widely in response to CO_2 (Ceulemans and Mousseau, 1994). In contrast to the data of Idso *et al.* (1991) showing a large increase in tree growth by CO_2 enrichment with optimal nutrient and water supplies, some studies conducted under less than optimal conditions showed little effect of increased CO_2 on tree growth. For example, growth of yellow poplar trees in CO_2-enriched air, but with no supplemental water or fertilizer, was not increased over three seasons (Norby *et al.*, 1992).

Aboveground biomass increase may vary among species because different species respond to increases in available carbohydrates in different ways (Ceulemans and Mousseau, 1994). Elevated CO_2 increased the biomass of sugar maple seedlings by 4.3 times, of sweet gum by 1.4 times, and American sycamore by 1.6 times after 69 days (Tschaplinski *et al.*, 1995). Several species maintained high photosynthetic rates during elevated CO_2 and increased the size or number of aboveground storage organs such as leaves (Kramer and Siunit, 1987; El Kohen *et al.*, 1993). In other species new root sinks were induced on storage organs (El Kohen *et al.*, 1993) or by increased turnover of fine roots (Norby *et al.*, 1992).

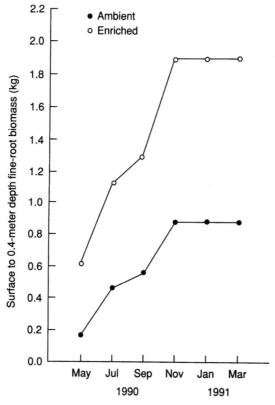

Figure 5.22   Total surface to 0.4-m-depth fine root biomass of the ambient-treatment and CO_2-enriched sour orange trees at six sampling dates. From Idso and Kimball (1992a), with permission of Blackwell Science Ltd.

Growth responses to elevated CO_2 may vary greatly in closely related species. For example, CO_2 enrichment increased height growth and biomass of *Eucalyptus tetrodonta* seedlings but not those of *E. miniata* (Duff *et al.*, 1994). With CO_2 enrichment, allocation of biomass to the main stem wood and main stem leaf mass increased, while allocation to branch wood and branch leaves declined in *E. tetrodonta*; allocation patterns of *E. miniata* were not altered. These two species exist as codominants in woodlands and open forests of northern Australia. The data indicated that the balance may shift to favor *E. tetrodonta* under CO_2 enrichment. DeLucia *et al.* (1994) studied growth of ponderosa pine trees in disjunct forest stands on similar substrates at high-elevation montane sites and low-elevation desert sites. Climatic variations between the sites were similar to those that would be expected from a doubling of atmospheric CO_2. The desert trees allocated proportionally more biomass to sapwood production and less to leaf production. In phytotron experiments a rise in temperature was followed by an increase in biomass of stems and a decrease in leaves. Hence, the shifts in partitioning of photosynthate as a result of temperature changes negated the effects on growth of the CO_2-driven increase in photosynthesis.

A number of investigators called attention to potential ecological effects of global warming. These might include changes in species composition of ecosystems, shifts in ranges of species and plant communities, and extinction of certain species of plants (Davis, 1983; Peters and Darling, 1985; Cohn, 1989; Houghton and Woodwell, 1989; Botkin, 1993). Changes very likely would reflect the impacts of combinations of direct and indirect effects; the direct effects would include those of increased CO_2 and changes in solar radiation, temperature, rainfall, humidity, and wind on physiological and phenological responses of trees. Secondary responses would involve plant–plant, plant–animal, and plant–microbe interactions, as well as feedback effects between changing ecosystems and variations in the atmosphere and climate (Gucinski *et al.*, 1995).

Kellomäki *et al.* (1995) suggested that climatic warming could increase the risk of frost damage to certain trees because of premature onset of growth during warm spells in the late winter and early spring. Concern has been expressed that if melting of glaciers and rise in sea levels occurred, some coastal areas would be inundated with salt water. Tree mortality would then be high, and even flood-tolerant species such as bald cypress would be at risk. Brubaker (1986) emphasized that the response times of different species to global warming would vary with the specific response measured, longevity of plants, rates of seed production and dispersal, methods of propagation, genetic diversity, phenotypic plasticity, and competitive potential.

There is much uncertainty about the long-term effects of global warming on forests. The effects will depend on how much the temperature will rise. Predictions of the actual amount of global warming that will occur are difficult to make. Some investigators question whether there will be significant warming (Roberts, 1989; Kerr, 1989). Many investigators believe that the climatic changes that are likely to occur from the increase in CO_2 concentration are no greater than the variations in weather during unfavorable seasons with present CO_2 concentrations (Idso, 1991; Idso *et al.*, 1991). The greatest amount of warming claimed for the earth over the course of the industrial revolution is only approximately 0.5°C, and very little of that resulted from increases in CO_2 and trace gases.

Uncertainty about the effects of global warming on forests also stems from the modifying effects of other environmental conditions on the influence of rising CO_2. Studies of tree rings have shown that increases in tree growth over the last few decades were correlated with increasing CO_2 (Graybill, 1986; Parker, 1986). Interpretation of such correlations is complicated by greater increases in growth than can be attributed to elevated CO_2 levels alone.

Ecosystems and species within them are likely to differ appreciably in response to increasing CO_2 depending on water supply, soil fertility, and temperature conditions. Natural ecosystems vary to the extent that their physiological functions are controlled by these factors (Mooney *et al.*, 1991a). Global warming probably will affect wet and dry tropical forests differently. The phenology of tropical rain forests will change little if, as expected, the water balance does not change much. The phenology of dry tropical forests is likely to be more sensitive to higher temperature and CO_2 levels. Higher temperature should result in more rapid water depletion, earlier leaf shedding, longer periods without leaves, and more strongly synchronized phenology (Reich, 1995). Process models have shown that whereas the responses in tropical and dry temperate forests would be affected by CO_2 levels, northern and moist temperate forests would reflect the effects of temperature or N availability (Melillo *et al.*, 1993).

General circulation models indicate that different species of trees will show individualistic responses to climatic change. For example, in New York State changes in temperature and water supply are likely to increase annual volume increment of red pine but decrease it in Norway spruce (Pan and Raynal, 1995). Pastor and Post (1988) concluded that different tolerances of species to drought will lead to segregation along moisture gradients in the landscape.

Presently the information on total responses of ecosystems to enhanced CO_2 is incomplete because the complex interactions involved have not been adequately evaluated (Mooney *et al.*, 1991a; Kerstiens *et al.*, 1995). Despite the fertilizing effects of CO_2 on individual plants, a proportional increase in global productivity of various ecosystems may not occur. Mooney and Koch (1994) described some difficulties in evaluating the web of interactions involved, including the following: (1) the proportion of C of tissues usually increases under increased CO_2 (this change influences both the decomposition of organic matter and the rate of tissue herbivory; the rate of microbial decomposition generally is reduced for tissues with high C/N ratios, but herbivory may be increased because herbivores need to consume more plant tissue to obtain a given amount of protein), (2) plants exposed to high CO_2 levels may adjust stomatal aperture so they lose less water for the same amount of carbon fixed (this response would influence both canopy energy and water flux), and (3) increased growth of ecosystem components would be greater below ground than above ground. Körner and Arnone (1992) reported that the major responses of a tropical ecosystem under elevated CO_2 were increased production of fine roots and a doubling of soil respiration. This study emphasized the difficulty in scaling responses from physiological baselines to entire ecosystems without accounting for various interactions among components.

Most research on effects of elevated CO_2 has dealt with aboveground plant responses. As pointed out by Norby (1994), more information is needed on (1) root functions in nutrient acquisition and mediation of whole-plant responses to CO_2, (2) carbon storage in roots as components of whole-plant storage, and (3) root turnover and the implications of carbon storage as soil organic matter. The relative importance of roots is likely to vary in different ecosystems. As emphasized by Mooney and Koch (1994), much more research is needed before we will fully understand the important interactions and feedbacks of various ecosystems to enhanced CO_2.

Heat Injury   Woody plants sometimes are injured by high temperatures below the thermal death point. Symptoms of heat injury include scorching of leaves and fruits, sunscald, leaf abscission, and even death of plants.

The bases of stems of tree seedlings begin to show damage close to the soil line when surface temperatures reach between 52 and 66°C. As the temperature within this range increases, the duration of exposure that causes death shortens progressively. Injured stems develop lesions, first on the south side of the stem, and the stem swells immediately above the lesion as a result of accumulation of photosynthate. Leaves are more resistant than stems to heat damage.

Direct heat injury to woody plants is relatively rare compared to the large amount of low-temperature injury. Perhaps the greatest danger of heat injury is to seedlings growing in soil exposed to the sun. Damage to seedling stems is caused by heat conducted from the surrounding soil, reradiation from the soil surface, and direct solar radiation (Helgerson, 1990).

Heat injury tends to be associated with soils with low heat capacity and conductivity. Seedlings growing in sandy and gravelly soils are especially vulnerable to heat injury. Heat damage occurs more on south, southeast, or southwest facing slopes where temperatures are higher than on north-facing slopes (Schramm, 1966).

A type of lesion caused by desiccation of the inner bark and cambium is known as sunscald or bark scorch. Sunscald typically occurs on south sides of stems of smooth-barked older trees that have been transplanted or exposed to the sun after a stand of trees has been thinned. The large lesions may extend for several feet up the stem. Sunscald degrades logs and provides an entrance for insects and fungi, which cause further deterioration.

Much attention has been given to effects of high temperatures on roots. The high ratio of surface area to volume of plant containers and the absorption of solar radiation by container walls often result in increases in soil temperatures of 4 to $10°C \, hr^{-1}$, and temperature maxima approaching $50°C$. Such high soil temperatures often injure roots and decrease their growth. For example, high temperatures sequentially inhibited growth, induced death of root tips and whole roots, and eventually led to death of shoots of five species of woody plants (Wong *et al.*, 1971). The critical temperature for injury to root cell membranes of two species of *Ilex* decreased linearly as the duration of exposure increased exponentially. The critical temperature for injury was higher for *Ilex vomitoria* 'Helleri' than for *Ilex vomitoria* 'Schellings' (Ingram, 1986).

Heat tolerance varies among plant species and genotypes. Seidel (1986) found small (2–4°C) but significant differences in heat tolerance of four conifers. Heat tolerance was highest in ponderosa pine seedlings, intermediate in Douglas fir and grand fir, and lowest in Engelmann spruce (Fig. 5.23). Thermotolerance varied among five *Betula* taxa. River birch maintained the highest rates of net photosynthesis, while the rate of paper birch was reduced the most (Ranney and Peet, 1994). In the same tree, heat tolerance decreases with greater meristematic activity and increases with maturity of tissues. For example, sensitivity to heat was greater in current-year shoots of black spruce than in the lignified first-year shoots (Koppenaal and Colombo, 1988).

Sensitivity of plants to heat varies with the time of year and with the poststress environment. Shirazi and Fuchigami (1993) exposed dormant red osier dogwood plants to a nonlethal heat stress (47°C) for 1 hr during either October, November, or December, followed by exposure to 0°C, 23°C, or natural conditions in Corvallis, Oregon. The plants that had been exposed to heat in October died within 9 weeks at 0°C, whereas those in other poststress environments survived and were not injured. For plants exposed to nonlethal heat stress during November or December, stem dieback occurred at 0°C after 12 and 15 weeks, respectively. Plants in the other poststress environmental regimes were not injured. The greater sensitivity of plants to nonlethal heat stress in October than in December may have been related to greater development of cold hardiness later in the year.

When previously cold-stored seedlings are transported to the field they may be injured if exposed to excessive heating. White spruce seedlings that were removed from cold storage ($-2°C$) and thawed at $5°C$ tolerated temperatures of $20°C$ for up to 4 days without adverse effects on growth. However, exposure to temperatures greater than $30°C$ for more than 48 hr resulted in severe physiological deterioration of the seedlings (Binder and Fielder, 1995).

Mechanisms of Heat Injury   Often it is difficult to distinguish between heat injury and desiccation injury caused by excessive water loss at high temperatures, or when soils are cold. The mechanism of heat injury is very complex. According to Levitt (1980a), the several types of primary indirect injuries (e.g., growth inhibition, starvation, toxicity, biochemical lesions, and protein breakdown) in response to moderately high temperatures for hours to days are metabolic and may be traceable to a single central mechanism. This mechanism may involve deficiency of an essential metabolite or an increase in a toxic compound or compounds. However, direct "heat-shock" injury, following exposure to high temperatures for seconds or minutes, is largely the result of damage to cell membranes, as shown by leakage of ions. Heat shock also changes some aspects of cell ultrastructure that may be linked to overall physiological changes. In some plants heat shock leads to dissociation of the endoplasmic reticulum (phospholipid membranes associated with ribosomes, polysome formation, and protein synthesis). Heat shock also may induce changes in nucleoli: affected nucleoli have a less granular appearance and lower heterochromatin content. Such changes are consistent with inhibition by heat shock of RNA processing and ribosome assembly (Brodl, 1990).

Levitt (1980a) concluded that variations in heat tolerance of plants are largely due to differences in protein thermostability. The rate of inactivation of enzymes increases rapidly at high temperatures, with most plant enzymes being inactivated above $70°C$. Enzyme inactivation by heat, which almost always results from denaturation of proteins, is influenced by pH, ionic strength, protein concentration, and the protective action of substrates and inhibitors (Dixon *et al.*, 1979).

Figure 5.23   Effect of temperature and duration of exposure on survival of 2- to 4-week-old seedlings of four species: (A) ponderosa pine, (B) Douglas fir, (C) grand fir, and (D) Engelmann spruce. The numbers on tops of bars indicate survival percentages. From Seidel (1986).

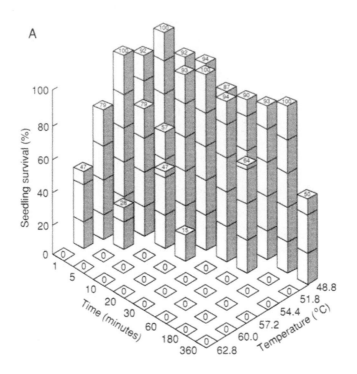

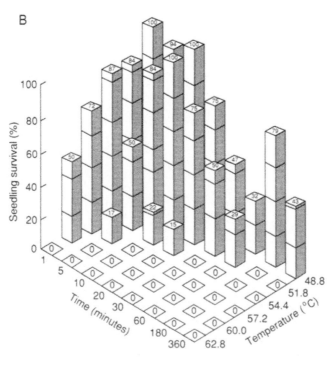

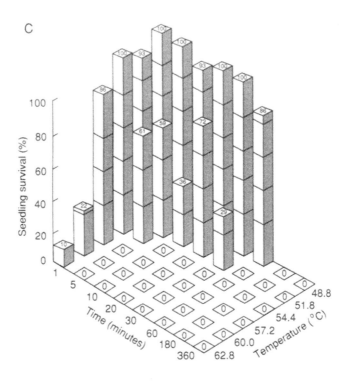

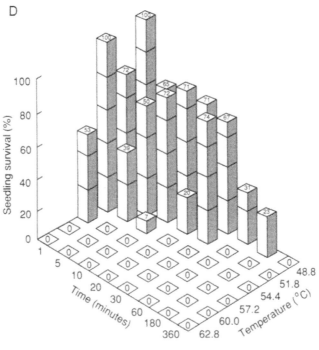

Figure 5.23 (*Continued*)

The mechanism of heat injury may vary with the rate of temperature increase. When the temperature rises rapidly, coagulation of protoplasm within plant cells usually accounts for disruption of protoplasmic structure. A gradual increase in temperature causes protein degradation, releasing ammonia which is toxic to cells (Katterman, 1991).

Plants often are injured by dehydration associated with high transpiration rates caused by high temperature. As the temperature rises, a rapid increase in transpiration occurs, largely because of an increase in the vapor pressure gradient between the leaf and surrounding air. Winds also reduce the boundary layer thickness, which influences transpiration. Hence, injury and shedding of leaves often are associated with hot, dry winds (Kozlowski, 1976a, 1982a, 1991).

Adaptation to Heat Stress Woody plants can adapt to heat as shown by increasing heat tolerance from low in the spring to high in the summer. Heat tolerance also is higher in unusually hot summers than in normal ones. The effect of preconditioning of plants by exposure to supraoptimal temperatures is cumulative, with heat tolerance being extended by repeated exposure to high temperatures. Such acquired tolerance varies among species and genotypes (Columbo *et al.*, 1992; Columbo and Timmer, 1992). For example, in jack pine and white spruce seedlings hardened at 38°C for 6 days the increased heat tolerance lasted for 14 and 10 days, respectively. The heat tolerance of similarly hardened black spruce seedlings persisted for only 4 days (Koppenaal *et al.*, 1991).

Induction of heat tolerance in seedlings of some species appears to be induced during seed germination at high temperatures. At temperatures between 25 and 35°C, 100% of the seeds of the heat-tolerant *Prosopis chilensis* germinated within 24 hr (Medina and Cardemil, 1993). At higher temperatures the rate of germination was reduced; at 50°C seeds did not germinate. After seed germination at 25°C the optimal temperature for seedling growth was 35°C, and the seedlings did not grow at 50°C. When seeds were germinated at 35°C, the optimal temperature for seedling growth was 40°C (some seedlings grew at 50°C), indicating that heat tolerance was induced during seed germination at 35°C. The induction of heat tolerance was correlated with synthesis of heat-shock proteins (see below).

Heat-Shock Proteins When plant tissues are rapidly exposed to temperatures 5 to 10°C above usual growth temperatures, synthesis of normal proteins decreases, and new polypeptides known as heat-shock proteins (HSPs) are produced. The heat-shock response can be detected within minutes of exposure of plants to high temperature. The absolute temperature rather than the magnitude of the temperature increase appears to be critical for synthesis of heat-shock proteins. Above the threshold temperature, the nature of the response depends on both the temperature and duration of exposure (Howarth and Ougham, 1993). If plant tissues that were exposed to heat are returned to their normal growing temperature, production of heat-shock proteins stops and synthesis of other proteins gradually occurs.

Each cell responds individually to heat shock, which may result from the following (Brodl, 1990): (1) changes in the oxidative/reductive environment of the cell, (2) changes in cellular ion levels, (3) presence of heat-denatured proteins in the cell, or (4) developmental induction. In addition to being triggered by high temperatures, heat-shock proteins may form in response to ethanol, arsenite, heavy metals, amino acid analogs, glucose starvation, drought, and wounding (Vierling, 1991).

Heat-shock proteins are believed by some investigators to play an important role in preventing accumulation of damaged proteins in cells during heat shock (Howarth and Ougham, 1993). Katterman (1991) suggested that they aid cells in surviving and recovering from environmental stress. Burke and Orzech (1988) stated that heat-shock proteins may protect enzymes from inactivation and nucleic acids from the cleavage induced by elevated levels of specific metals. In soybean not only is the rate of synthesis of heat-shock proteins correlated with development of tolerance of high temperature, but accumulation of these proteins parallels the acquired thermotolerance (Lin *et al.*, 1984; Vierling, 1991). Heavy metal tolerance appears to be increased by heat-shock proteins. In soybean seedlings accumulation of heat-shock proteins coincided with a decrease in solute leakage after heat stress (Lin *et al.*, 1985). Exposure of tomato plants to a short period of heat stress, prior to a heavy metal stress, prevented membrane damage and was associated with formation of cytoplasmic heat-shock proteins (Neumann *et al.*, 1994). As is the case with endogenous hormones, more than one mechanism may account for the protective effects of HSPs (McKersie and Leshem, 1994). One view is that HSPs aggregate into heat-shock granules (HSGs) (dense granular complexes comprising HSP and mRNA). The HSGs, which are transient sites for non-heat-shock mRNA, prevent its degradation during heat shock. As the temperature decreases following heat shock, the dispersed HSGs associate with polysomes which are active in protein synthesis.

Exposure of leaves to temperatures slightly below injuriously high temperatures may decrease subsequent denaturation of Photosystem II components at high temperatures. More research is needed to resolve the extent and nature of the protective role of heat-shock proteins.

Soil Fertility

Productivity of most woody plants is related to availability and absorption of mineral nutrients. Unfortunately, low soil fertility often limits plant growth (see Chapter 10 of Kozlowski and Pallardy, 1997). Mineral deficiencies are accentuated by flooding of soil, forest fires, whole-tree harvesting, and harvesting of trees on short rotations.

Trees growing on soils of low fertility typically are deficient in the internal resources necessary to sustain the physiological processes that regulate growth (Chapter 3). This is emphasized by increases in carbohydrates, foliar nitrogen, chlorophyll, and amino acids in trees receiving fertilizers over unfertilized trees (Billow *et al.*, 1994).

Deficiencies of mineral nutrients usually are chronic and often not obvious. However, some idea of the extent of low soil fertility can be gained from the voluminous literature that shows that fertilizers increase growth of forest, fruit, and shade trees (see Chapter 7). Efficient use of fertilizers requires information about which nutrient or nutrients are limiting plant growth and the extent of the deficiency of each nutrient. The nutritional status of woody plants can be diagnosed by plant tissue analysis, soil analysis, or determination of plant responses to fertilizers containing specific elements. Methods of determining the nutritional status of plants are discussed by Mead (1984) and Walker (1991). The importance of mineral nutrients for growth of woody plants is discussed further in Chapter 7.

Salinity

The global extent of saline soils ranges between 450×10^6 and 950×10^6 ha (Downton, 1984). Soils may be salty because of existing salt deposits, salt spray along seacoasts, runoff water from highways treated with deicing salts, and salts in irrigation waters.

Salinity adversely affects both the distribution and growth of many plants. Accumulation of excessive salts excludes trees and shrubs from vast areas of land. Only a few halophytes (plants that can complete their life cycles in saline environments), such as greasewood and saltbush, can tolerate highly saline soils. The harmful effects of the Cl^- anion and the Na^+ cation to broad-leaved trees differ, with Cl^- injury appearing earlier than Na^+ injury. Chloride injury typically appears as marginal chlorosis of leaves and is followed by midleaf scorching and tipburn. Eventually more than half the leaf tissue may become necrotic. Accumulation of Na^+ in leaves induces mottling (e.g., in avocado) or tipburn (e.g., in stone fruit species). Salt injury to leaves often is followed by leaf shedding and shoot dieback. Salt injury to conifers appears as necrosis of needle tips and spreads to the bases of needles.

Salinity typically inhibits growth of plants and predisposes them to adverse effects of other environmental stresses such as drought, flooding and low temperature. In nonhalophytes salinity typically inhibits shoot growth more than root growth. For example, salinity increased the root–shoot ratio in leucaena (Gorham *et al.*, 1988) and mesquite (Jarrell and Virginia, 1990). In contrast, the root–shoot ratio of some halophytes is decreased by salinity, reflecting increased allocation of carbohydrates to leaf growth.

Salt tolerance varies widely among plant species and genotypes. Honey locust, northern red oak, white oak, black locust, horse chestnut, Japanese black pine, Austrian pine, jack pine, white spruce, blue spruce, and yews are considered salt tolerant. In contrast, sugar maple, box elder, European beech, American beech, American hornbeam, red pine, eastern white pine, balsam fir, and eastern hemlock are rated as sensitive to salinity (Holmes, 1961; Dirr, 1976, 1978). Salt tolerance of fruit trees varies greatly (Chapter 6).

Salt tolerance may vary appreciably among closely related species. Of 16 species of *Eucalyptus* tested, *E. drepanophylla*, *E. argophloia*, *E. camaldulensis*, and *E. robusta* were most tolerant to salt; *E. cloeziana* and *E. pilularis* were least tolerant (Fig. 5.24). Although *Acacia* species native to Australia are considered highly salt tolerant, there is much variation among them (Craig *et al.*, 1990). Clemens *et al.* (1983) ranked salt tolerance of species of *Casuarina* as follows: tolerant, *C. equisetefolia*; moderately tolerant, *C. cunninghamiana*, *C. glauca*, and *C. littoralis*; moderately sensitive, *C. cristata*, *C. decaisneana*, *C. stricta*, and *C. torulosa*; and sensitive: *C. inophloia*. Whereas *C. equisetefolia* showed little growth reduction and no visible symptoms when treated with 150 m*M* NaCl, *C. inophloia* showed greatly reduced growth accompanied by chlorosis and/or death of shoot tips. Ball and Pidsley (1995) found wide variations in salt tolerance of two species of mangrove. *Sonneratia alba* grew maximally in salinities ranging from 5 to 50% seawater. In comparison, maximum growth of *S. lanceolata* occurred in salinities ranging from 0 to 5% seawater.

Mechanisms of Salinity Effects

Salinity influences water, carbohydrate, mineral, and hormonal relations of plants. Salts that enter the soil decrease the capacity of roots to absorb water by decreasing the Ψ gradient from the soil to the roots.

As emphasized in Chapter 5 of Kozlowski and Pallardy (1997), salinity lowers the rate of photosynthesis by affecting stomatal aperture and by nonstomatal effects as well. The adverse long-term effects of salinity involve lowering the rate of photosynthesis by changes in the photosynthetic process and decreasing total photosynthesis as a result of inhibition of leaf initiation and expansion as well as leaf shedding. Salinity also elevates the rate of respiration and increases the proportion of maintenance respiration over growth respiration (Waisel, 1991).

Salinity decreases mineral acquisition by plants through osmotic effects and, more commonly, by direct interactions of ions in the substrate. The major ions can influence absorption of mineral nutrients by competitive interactions or by influences on ion selectivity of membranes. Examples include Na^+-induced K^+ deficiency and Ca^{2+}-induced Mg^{2+} deficiency (Grattan and Grieve, 1992).

Salinity induces changes in endogenous hormones that are associated with early senescence of plants. Senescence is promoted by increased production of abscisic acid (ABA) and ethylene. A decrease in availability of cytokinins may inhibit growth of salt-stressed trees (Waisel, 1991).

The overall mechanism by which salinity reduces growth of woody plants is complex. The decreases in supplies of water, carbohydrates, mineral nutrients, and certain hormones in plants exposed to salinity do not necessarily indicate that each of these lowered resources is controlling growth. Although it often is claimed that growth of plants in saline soil is regulated by turgor, leaf turgor of salt-affected plants is not always reduced. Furthermore, the turgor of leaves of salt-sensitive varieties usually is higher than that of leaves of salt-tolerant relatives. As emphasized by Munns (1993), turgor is essential for growth of cells, but the rate of cell wall expansion may be controlled by rheological processes of the cell wall (e.g., "wall softening" and "strain hardening") and not by turgor alone (Chapter 3). Munns (1993) also emphasized that growth of plants is affected by salinity earlier than is their photosynthetic rate. This observation suggests that salinity influences photosynthesis per plant more because of a smaller leaf area than a lower rate of photosynthesis.

Munns (1993) presented a useful model of short- and long-term responses of plants to salinity. Growth is first inhibited by a decrease in soil Ψ. This water stress effect is regulated by inhibitory signals from the roots. Later a salt-specific effect appears as salt injury in the old leaves, which die because of a rapid increase in salt in the cell walls or cytoplasm. This two-phase growth response is shown in Fig. 5.25 by comparing genotypes that are sensitive, moderately tolerant, and tolerant to salinity. In the short term (phase 1) these genotypes respond similarly. In the long term (phase 2) they respond differently because of variations among them in the time required for salt to reach a maximum concentration in the vacuoles of mesophyll cells. Death of leaves begins during phase 1; the effect on leaf expansion is evident much later. The loss of leaves reduces the supply of photosynthate or hormones to meristematic tissues and thereby inhibits growth.

Adaptations to Salinity

Certain plants can adapt to saline environments by tolerating salt or avoiding it by excluding salt passively, extruding salt actively, or diluting the entering salt. Adaptation may depend largely on a single mechanism or on several, depending on the species.

A

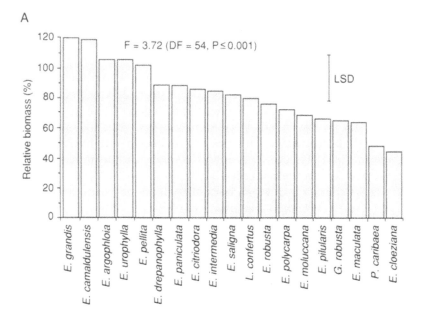

B

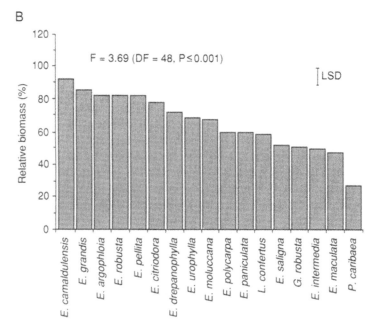

Figure 5.24   Effect of concentration of sodium chloride on growth (biomass) and mortality (indicated by absence on graphs) of 16 species of *Eucalyptus* and other woody plants (*Grevillea robusta, Lophostemnon confertus,* and *Pinus caribaea*): (A) 50 mM NaCl, (B) 100 mM NaCl, (C) 150 mM NaCl, and (D) 200 mM NaCl. From Sun and Dickinson (1993). *For. Ecol. Manage.* **60**, 1–4, with permission of Elsevier Science.

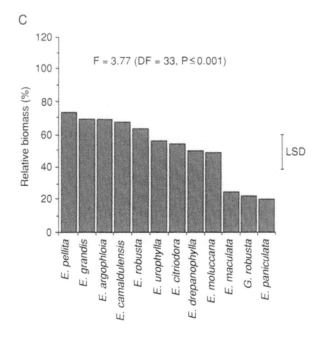

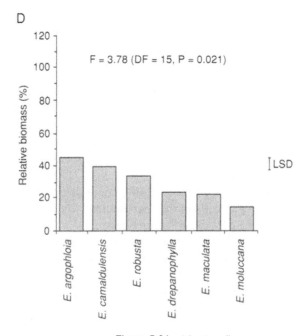

Figure 5.24   (*Continued*)

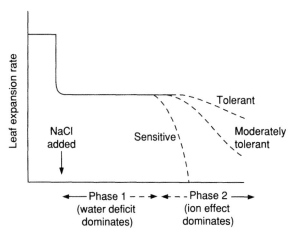

Figure 5.25   Model of short- and long-term responses of sensitive, moderately tolerant, and tolerant genotypes to salinity. From Munns (1993), with permission of Blackwell Science Ltd.

Some plants tolerate soil salinity because their root systems tend to prevent salt from reaching the shoots. For example, variable salt tolerance in citrus is associated with differences in the capacity of rootstocks to restrict accumulation of salt (Chapter 6; Mobayen and Milthorpe, 1980) and with low sensitivity of foliage to chloride and sodium ions (Nieves *et al.*, 1991).

Rapid osmotic adjustment may account for salt tolerance (McKersie and Leshem, 1994). For osmotic adjustment to occur the internal osmotic potential must be lowered either by absorption of solutes or synthesis of organic solutes (Fig. 5.26). Both mechanisms appear to be variously involved in different plants. Halophytes operate primarily by absorbing inorganic ions. Osmoregulation in salt-tolerant nonhalophytes results from synthesis and appropriate partitioning of organic solutes but with an unavoidable cost of photosynthate (Epstein, 1980).

Certain plants release excess salts by shedding some of their salt-laden leaves. In a few species (e.g., *Atriplex*) salt accumulates in bladders that grow on leaf surfaces. Eventually these bladders are shed or collapse and release salt solution (Waisel, 1991). Some halophytes release excess salts through salt glands. The secreted solution consists largely of NaCl, but many other organic and inorganic ions are present (Waisel, 1991). The rate of salt secretion through glands varies among plant species and with salt composition, age of tissues, plant water balance, temperature, light, and O_2 supply.

Some mangroves possess mechanisms for both salt tolerance and avoidance (Scholander, 1968). For example, Waisel *et al.* (1986) identified three mechanisms of salt resistance in *Avicennia marina*: (1) salt exclusion be-

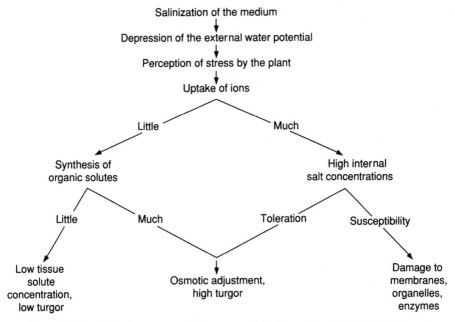

Salinization of the medium

↓

Depression of the external water potential

↓

Perception of stress by the plant

↓

Uptake of ions

Little   Much

Synthesis of
organic solutes

High internal
salt concentrations

Little   Much   Toleration   Susceptibility

Low tissue
solute
concentration,
low turgor

Osmotic adjustment,
high turgor

Damage to
membranes,
organelles,
enzymes

Figure 5.26 Cardinal responses of plants to salinity. From Epstein (1980). *In* "Genetic Engineering of Osmoregulation," with permission of Plenum Press.

cause of low permeability of roots to salts, which was by far the most important salt-rejecting mechanism, preventing absorption of 80% of the soil salts near the root surface, (2) salt tolerance (capacity to maintain normal metabolism when internal salt levels are high), and (3) salt evasion (release of some of the absorbed salt). Approximately 40% of the salt that reached the leaves was released through salt glands. *Bruguiera gymnorhiza* lacks salt glands but is a good salt excluder.

Progress in developing salt-tolerant trees has been slow for a variety of reasons including lack of financial support, lack of data on interactions of salinity and other environmental factors, incomplete understanding of plant responses to salinity, and lack of criteria for selection of salt-tolerant plants (Tal, 1985). Allen *et al.* (1994) expressed cautious optimism about prospects for improving the tolerance of woody plants to salinity by intensive screening and selection.

Soil Compaction

Many soils are compacted by pedestrian traffic, grazing animals, and heavy machinery (Greacen and Sands, 1980). Soil compaction in planting areas at new construction sites is a serious problem. Because of their loose, friable structure, forest soils are particularly prone to compaction (Froehlich *et al.*, 1985).

Compaction increases the bulk density and mechanical resistance of soil. For example, compaction increased the bulk density of soil at a campsite by 21% (Legg and Schneider, 1977). As bulk density is increased, total pore

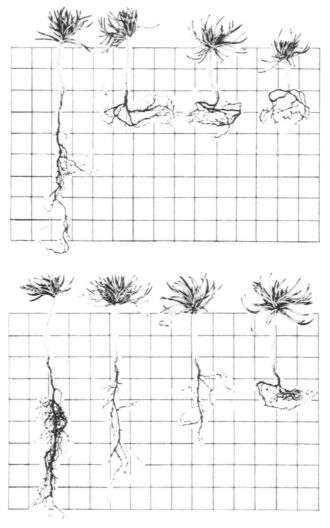

Figure 5.27   Effect of increasing soil compaction on growth of Austrian pine seedlings on a silt loam soil (top) and a sandy loam soil (bottom). Bulk densities of the soil from left to right are 1.2, 1.4, 1.6, and 1.8 g cm^{-3}. From Zisa *et al.* (1980).

space (especially pores greater than 50 μm) decreases while the proportion of micropores increases. The decrease in pore space inhibits diffusion of O_2 into and diffusion of CO_2 out of the soil. Soil compaction also is often associated with development of a soil crust that inhibits infiltration, hence increasing surface runoff of water.

Growth of plants in compacted soil is greatly reduced (Fig. 5.27) as shown for seedlings of Douglas fir (Heilman, 1981), ponderosa pine (Froehlich *et al.*, 1986), red maple, and black oak (Donnelly and Shane, 1986). The mechanisms of growth reduction are complex and are associated with changes in soil strength, infiltration and retention of water, and soil aeration. An individual root can penetrate only those soil pores with a diameter greater than that of the root. Root elongation and penetration into soils of high bulk density often are confined to cracks rather than occurring within the soil matrix (Patterson, 1976). Establishment of many loblolly pine seedlings was prevented because the roots could not penetrate the compacted soil (Foil and Ralston, 1967). Depletion of soil O_2 also reduces root growth. The growth of roots of apple trees was reduced when soil compaction lowered the soil air space to less than 15%, and roots essentially stopped growing when the air space decreased to 2% (Richards and Cockcroft, 1974). Methods of alleviating soil compaction are discussed by Rolf (1994).

Pollution

A wide variety of naturally occurring and man-made toxic chemicals in the air and soil adversely affect woody plants. The major environmental pollutants include sulfur dioxide (SO_2), ozone (O_3), fluorides, oxides of nitrogen (NO_x), peroxyacetylnitrate (PAN), and particulates such as cement kiln dusts, soot, lead particles, magnesium oxide, iron oxide, foundry dusts, and sulfuric acid aerosols. Pollutants are classified as primary or secondary. Primary pollutants such as SO_2 and hydrogen fluoride (HF) originate in a toxic form at the source. Secondary pollutants are generated by interactions between primary pollutants. Examples are O_3 and PAN that are formed by action of sunlight on products of fuel combustion.

Some plants give off chemical compounds that may inhibit seed germination and growth of other plants (Chapter 2; Rice 1984). Such allelopathic chemicals (allelochems) are variously released by plants to the soil by leaching, volatilization, excretion, exudation, and decay either directly or by the activity of microorganisms. In addition, a variety of commonly applied chemicals, including deicing salts, herbicides, fungicides, insecticides, and antitranspirants, when improperly used, also may act as pollutants (Kozlowski *et al.*, 1991).

The effects of pollution on woody plants range from negligible or subtle physiological changes to inhibition of growth, injury, leaf shedding, and death of plants, as well as changes in the structure and function of ecosystems (Smith, 1990; Kozlowski *et al.*, 1991).

Injury

Gaseous air pollutants may injure leaves after they are absorbed, mostly through stomatal pores. Whereas O_3 can enter the leaf interior only through stomatal pores, other gaseous pollutants (e.g., N_2O) can enter through the cuticle as well as the stomata. Most uptake of SO_2 occurs through the stomata; diffusion through the cuticle is much less important (Matyssek *et al.*, 1995). Pollution injury commonly is classified as acute, chronic, or hidden. Acute injury, which is severe and characterized by death of tissue, occurs after a short-term, high pollution dosage or when a very pollution-sensitive plant is exposed to a lower dosage. In broad-leaved trees both SO_2 and HF cause collapse of spongy mesophyll cells and those of the lower, stomata-bearing epidermis, followed by injury to palisade cells (Ormrod, 1978).

Ozone can cause a reduction in leaf area and in width of epidermal and mesophyll cells, as well as an increase in stomatal frequency (Matyssek *et al.*, 1995). As O_3 dose increased during expansion of silver birch leaves, the differentiation of leaves was increasingly altered (Günthardt-Georg *et al.*, 1993). Leaf area was reduced and leaf air space increased as groups of mesophyll cells collapsed. Finally, the epidermal cells collapsed. When subsidiary cells collapsed, the guard cells opened passively for a while and subsequently collapsed.

Chronic injury follows absorption of low amounts of pollutants over long periods of time. Such injury, characterized by slow development of chlorosis and early leaf senescence, may be associated with necrotic lesions. Sometimes the physiological activity of affected plants is impaired well before visible symptoms of injury are evident. Hence, investigators often refer to "hidden" or "physiological" pollution injury.

The visible symptoms of acute pollution injury often vary with the specific pollutant to which plants are exposed. For example, SO_2 injury to broad-leaved trees is characterized by necrotic leaf lesions while tissue around the leaf veins remains green (Fig. 5.28). Typical symptoms of O_3 injury are interveinal chlorosis together with flecking (initially bronze and subsequently necrotic spots which progressively increase in diameter) on the upper surface of the leaf. Eventually, foliar necrosis may be widespread over the leaf (McKersie and Leshem, 1994).

Symptoms of leaf injury are not as definite in gymnosperms as in angiosperms. In pines acute SO_2 injury is characterized by reddish-brown discoloration of needle tips. As pollution continues, the discoloration pro-

Figure 5.28   Injury to leaves of ash trees by SO_2. Photo courtesy of U.S. Department of Agriculture.

gresses toward the base of the needle. Acute O_3 injury is characterized by death of needle tips or whole needles. Eventually all except the current-year needles may be shed. The cumulative external O_3 dose that initiated leaf abscission in conifers was reported as 46 to 126 ppm hr^{-1} for slash pine (Hogsett *et al.*, 1985), 111 ppm hr^{-1} for Scotch pine (Skeffington and Roberts, 1985), 130 to 220 ppm hr^{-1} for loblolly pine (Stow *et al.*, 1992), and 400+ ppm hr^{-1} for ponderosa pine (Coyne and Bingham, 1982). Such apparent variations in sensitivity of different species to O_3 should be viewed with caution because of differences in experimental procedures of investigators, tree age, and environmental conditions.

Ascertaining the specific cause of tipburn of conifer needles often is difficult because several different pollutants, including SO_2, O_3, and fluoride, can cause tipburn. Furthermore, tipburn may be caused by some herbicides, deicing salts, excess fertilizers, and winter injury (Kozlowski, 1980a,b; Kozlowski and Constantinidou, 1986a).

Air pollutants inhibit wax formation and degrade surface waxes of leaves (see Chapter 8 of Kozlowski and Pallardy, 1997; Percy and Riding, 1978; Sauter *et al.*, 1987; Barnes *et al.*, 1988). Exposure of Scotch pine needles to SO_2 accelerated weathering of needle waxes and induced thickening of the wax tubes (Crossley and Fowler, 1986). Acid precipitation induced cuticular cracking, desiccation, and erosion of trichome surfaces of flowering dogwood leaves (D. A. Brown *et al.*, 1994).

Ultrastructural Changes

The effects of sulfur compounds, fluorides, and O_3 on the fine structure of leaf cells of broad-leaved trees and conifers are rather similar. Chloroplasts become disorganized early. Chloroplast membranes are stretched and rupture even before visible symptoms become evident. Subsequently the cell contents, including chloroplasts, aggregate into masses and eventually the cell walls collapse. The final disorganization of cells occurs only when macroscopic symptoms also are evident (Soikkeli and Karenlampi, 1984). In Norway spruce needles exposed to SO_2 for three seasons, ultrastructural changes were identified 2 years before visible injury occurred (Sutinen *et al.*, 1990). Changes included a decrease in chloroplast size, increase in density of the stroma, accumulation of ribosome-like granules, and a decrease in the size of starch grains. Eventually the cytoplasm disintegrated.

Growth of Plants

A voluminous literature shows that both air and soil pollutants, alone and in combination, variously alter growth of woody plants, particularly near point sources of pollution such as smelters. Sulfur dioxide reduced growth of seedlings of American elm (Constantinidou and Kozlowski, 1979a), paper birch, Murray red gum, blue gum, and red pine (Norby and Kozlowski, 1981a,b). More comprehensive reviews are presented by Mudd and Kozlowski (1975), Kozlowski and Constantinidou (1986a), Scholz *et al.* (1989), Schulze *et al.* (1989), and Smith (1990).

Pollutants typically decrease growth of leaves as well as expansion of internodes. The leaf area of trees often is reduced because pollutants inhibit leaf formation and expansion and accelerate leaf abscission. The rates of leaf expansion of American elm seedlings exposed to SO_2 were only 40% of those of control plants (Constantinidou and Kozlowski, 1979a). Leaf expansion also was arrested by O_3 (Jensen, 1981, 1982), fluorides (Ferlin *et al.*, 1982), and particulates (Darley, 1966). Premature defoliation is a common response to air pollution (Carlson, 1980; Usher and Williams, 1982; Reich and Lassoie, 1985). Even at low O_3 dosages, 20 to 50% of the leaves of silver birch seedlings were shed, while 40 to 70% of the remaining leaves showed advanced discoloration by the end of the growing season (Matyssek *et al.*, 1992). Mature ponderosa pine trees exposed to oxidants retained mostly the 1-year-old needles (Axelrod *et al.*, 1980).

Decreases in height growth of seedlings of several species of woody plants have been attributed to SO_2 (Suwannapinunt and Kozlowski, 1980; Norby and Kozlowski, 1981c; Tsukahara *et al.*, 1985, 1986), O_3 (Duchelle *et al.*, 1982), and particulates (Greszta *et al.*, 1982; McClenahen, 1983). Ozone damage and growth inhibition are highly correlated with photosynthetic reduction (see Chapter 5 of Kozlowski and Pallardy, 1997). Leaves exposed to O_3 may senesce prematurely; leaf carbon balance may be altered (resulting in decreased gain or export of carbon); synthesis and metabolism of carbon compounds within the leaf may be altered; and carbon may be diverted from storage compounds to repair processes (Friend and Tomlinson, 1992).

Pollutants inhibit cambial growth, often without visible symptoms of injury. For example, reduced diameter growth of tree seedlings has been attributed to SO_2 (Tsukahara *et al.*, 1985, 1986), oxidants (Ohmart and Williams, 1979), and fluorine (Treshow and Anderson, 1982). Treating the soil with Pb or Cd decreased diameter growth of American sycamore saplings by as much as 65% (Carlson and Bazzaz, 1977). Examples of species showing reduced diameter growth of mature trees near point sources of pollution include eastern white pine in southern Canada (Navratil and McLaughlin, 1979), Austrian pine in England (Gilbert, 1983), and Scotch pine in Finland (Havas and Huttunen, 1972). The reduction in growth of European aspen that had been exposed to low levels of O_3 for 2 years was attributed to decreases in photosynthesis, carboxylation efficiency, and water use efficiency (Matyssek *et al.*, 1993).

In addition to inhibiting wood production, pollutants may alter wood structure. For example, the amount of latewood in annual rings of Norway spruce was reduced by exposure to SO_2 (Keller, 1980). Several conifers that had been exposed to air pollution had short tracheids, and broad-leaved trees had short vessels, tracheids, and fibers (Grill *et al.*, 1979).

Although growth of both shoots and roots is reduced by pollution, root growth usually is reduced more. This is because the reduction of photosynthesis by pollution is followed by proportionally lower allocation of the carbohydrate pool to the roots than to growing shoots. Root–shoot ratios were decreased in European white birch seedlings that were exposed to SO_2 or a combination of SO_2 and NO_2 (Freer-Smith, 1984).

By reducing the flow of photosynthate from leaves to roots, air pollutants may inhibit mycorrhizal development (Kasana and Mansfield, 1986). Infection of northern red oak seedlings with mycorrhizal fungi was reduced by exposure to high concentrations of SO_2 (Reich *et al.*, 1985) and O_3 (Reich *et al.*, 1986).

Structure of Forests

Forests are affected both directly and indirectly by pollution, with the indirect effects being much more important. The indirect effects include changes in soil properties and in the capacity of trees to compete for light, water, and mineral nutrients or cope with insect pests and fungal pathogens (Taylor *et al.*, 1994).

The effects of pollutants on forest stands are many and mediated through individual plants to the ecosystem. Tree stands exposed to a pollution stress may respond only slightly or drastically depending to a large extent on the pollution load. High dosages of pollutants can induce severe stand degradation by setting in motion a retrogression characterized by reduction in structural complexity, biomass productivity, and species diversity (Whittaker, 1975). Bormann (1990) described the response of forests to increasingly severe pollution as a continuum of stages. At low levels of pollution, forests are a sink for some pollutants, and effects on ecosystem function often are negligible. As pollution increases, sensitive species become subtly and adversely influenced. For example, the rate of photosynthesis of sensitive species may decline and they become predisposed to insect or fungus attack. As the load of pollutants increases further, populations of sensitive species decrease. With still more pollution, a forest may be peeled off in layers, with trees, shrubs, and herbaceous plants dying in order. As the ecosystem is unable to repair itself by substituting tolerant for intolerant species, productivity is reduced. Runoff, erosion, and loss of mineral nutrients from the ecosystem are accelerated. Finally the ecosystem collapses because it is so severely injured that it cannot recover.

For a long time it was assumed that forest soils rarely contain an excess of mineral nutrients and tree growth should be stimulated by added N. However, some forests apparently are receiving too much N from polluted air (see Chapter 9 of Kozlowski and Pallardy, 1997). The high levels of N deposition may be a major threat to extensive areas of temperate forests (Aber *et al.*, 1989; Högberg and Johannison, 1993). Forests that are subjected to high N loading eventually reach a state of N saturation (when N inputs exceed the combined requirements of higher plants and microorganisms). Excess N may inhibit productivity of forest ecosystems and possibly contribute to forest decline. Effects of N saturation include increased leaching of nitrate and cations from soils, and possible increases in emissions of trace gases such as nitrous oxide that may alter atmospheric chemistry and contribute to global warming.

Under N saturation conditions changes in species composition of forest ecosystems gradually occur. For example, N deposition in Swedish forests is believed to be a major cause of increased growth of grasses, especially *Deschampsia flexuosa*, and decreased abundance of shrubs such as *Vaccinium myrtillus* and *V. vitis-idaea* (Rosen *et al.*, 1992). Before such changes in species composition are evident these plants accumulate N in specific amino acids. In N-enriched conditions *Deschampsia* had high levels of asparagine and low levels of arginine, compared with high levels of arginine and glutamine in the two *Vaccinium* species. Hence, N metabolism was similar in *Vaccinium* and Scotch pine but different in *Deschampsia* (Näsholm and Ericsson, 1990; Näsholm *et al.*, 1994).

Not all trees in a forest stand are similarly affected by pollution. In a mixed forest stand, growth of some species may even be stimulated if they are given a competitive advantage by a greater inhibitory effect of pollution on other species. Growth of other species may be inhibited by the effect of the pollutant and also by their lowered competitive potential (Kozlowski, 1985c). For example, in the mixed conifer forest of the Sierra Nevada, air pollution reduced cambial growth (basal area increment) of ponderosa pine but increased growth of white fir (Kercher *et al.*, 1980).

As emphasized by Taylor *et al.* (1994), the single-factor response methods that were used for many years in agricultural science have not been very useful in clarifying the effects of pollutants on forests. This conclusion is based on recognition of deposition of multiple pollutants on forests, longevity of forest trees, genetic variability and multiple age classes of forest stands, indirect and cumulative effects of pollutants, and concern with food chain transport. Hence, much emphasis is now directed toward (1) productivity studies of individual trees, forest stands, and communities, (2) biogeochemistry studies, and (3) modeling (Taylor *et al.*, 1994).

Acid Precipitation (Acid Rain)

Sulfur dioxide and nitrogen oxides, together with HCl and some other compounds, mix in the atmosphere with O_2 and water to form acidic solutions that are deposited on plants and soil as acid precipitation, snow, and fog. Although rain formed in a nonpolluting area should have an acidity of about pH 4.6, rains with much lower pH values have been detected in many parts of the world. For example, rain and snow sometimes are 5 to 30 times more acid than would be expected in an unpolluted atmosphere (Likens *et al.*, 1979).

Opinions among scientists are divided about the impact of acid precipitation on woody plants. Lucier and Haines (1990) summarized the following often-cited mechanisms by which acid deposition could adversely affect plant growth: (1) increased leaching of basic cations from the foliage and soil as a prelude to mineral deficiency or imbalance, (2) increased mobilization of Al, causing injury to roots, mineral deficiency, and decrease in drought tolerance of plants, (3) inhibition of biological processes in soils, inducing slowing of decomposition of organic matter, development of mineral deficiency, and injury to mycorrhizae, and (4) increased availability of

N, leading to N saturation. Acid rain has been shown to decrease frost tolerance of trees. For example, frost tolerance of red spruce seedlings was reduced following exposure to acid mist containing SO_4^{2-}, NH_4^+, NO_3^-, and H^+ ions (Sheppard, 1994). Frost hardiness was reduced through direct attacks of SO_4^{2-} ions on membrane proteins. Indirect effects were mediated through increased consumption of sugars (reducing the pool available for cryoprotection) and impairment of the photosynthetic mechanism.

A number of short-term studies (reviewed by Kozlowski *et al.*, 1991) called attention to adverse effects of acid precipitation on seed germination, injury to leaf cells, and inhibition of plant growth. In contrast, many other short-term studies showed that acid precipitation stimulated growth of woody plants as a result of N fertilization. Both the rate of photosynthesis and growth of eastern white pine and loblolly pine seedlings were increased by simulated acid precipitation (Reich *et al.*, 1987; Hanson *et al.*, 1988c). Slash pine seedlings irrigated weekly for 28 months with pH 3.3 solutions showed increased rates of photosynthesis. The increases were attributed to improved mineral nutrition (the irrigation treatments added the equivalent of 124 kg N ha^{-1} year^{-1}). The data were consistent with those of Schier (1985) who found that short-term growth of pitch pine seedlings was stimulated by irrigation with acid solution as a result of increased N availability and input of nutrient cations from acid-induced weathering of soil minerals or release of mineral nutrients from organic matter.

A body of evidence shows that the short-term beneficial effects of acid precipitation do not carry over to the long term. Initially the effects may be beneficial from increased N availability, if cations in the soil are in ample supply but N is limiting growth. Subsequently a nutritional imbalance may develop as acidic deposition accelerates leaching of cations from the soil. In the subarctic Bäck *et al.* (1994) sprayed 5-m-tall Scotch pine trees with acidified water over the period of 1985 to 1992. This treatment reduced shoot growth up to 40% and needle growth by 15%. The effects were evident after 6 years of treatment. Sulfate was much more toxic than nitrate. Schier (1985) found that the early beneficial effects of acid precipitation on plant growth did not last because soil cations were progressively depleted. During the short time that Schier monitored soil chemistry the concentration of Ca and Mg in the leachate from the A horizon of soil decreased by 50 and 73%, respectively. McLaughlin *et al.* (1993) showed in greenhouse experiments that acid deposition adversely affected the physiology of red spruce seedlings. Plant responses included a reduction in the rate of photosynthesis and an increase in respiration, with increased respiration contributing more than reduced photosynthesis to a lowered photosynthesis–respiration ratio. The increased rate of respiration was strongly correlated with loss of foliar Ca. These observations were consistent with a dominant role of increasing respiration of mature red spruce trees at upper elevations in the southern Appalachian Mountains, where trees frequently are exposed to very acidic clouds (McLaughlin *et al.*, 1990, 1991).

Oren *et al.* (1988) concluded that imbalanced mineral nutrition was more important than deficiency of a particular element in reducing the vigor of declining trees that had been exposed to acidic deposition for some time. Acid rain can leach mineral nutrients from leaves. For example, Mg^{2+}, K^+, Ca^{2+}, and Na^+ are washed from leaves. Furthermore, Al^{3+} ions and some heavy metals are released in soils in amounts that may be toxic to roots. The subsequent decrease of absorption of water and mineral nutrients induces even greater deficiencies of Mg^{2+}, K^+, and Ca^{2+} in plants (McKersie and Leshem, 1994). The nutrient relations of plants also may be altered because of adverse effects of acid precipitation on mycorrhizae. Simulated acid precipitation decreased mycorrhizal infection in eastern white pine and northern red oak seedlings (Reich *et al.*, 1986; Stroo *et al.*, 1988). Some harmful long-term effects may result from interactions between acid precipitation and other pollutants. For example, a combination of exposure to SO_2 and irrigation with simulated acid precipitation induced greater ultrastructural modifications in Scotch pine needles than did SO_2 alone (Saastamoinen and Halopainen, 1989). Hidy (1995) concluded that acidic deposition is not the primary cause of forest declines. Rather, it contributes to the processes that increase susceptibility of trees to naturally occurring stresses such as drought and disease.

Forest Declines Concern with the effects of air pollution has increased progressively because of its possible association with widespread synchronized mortality of trees (Plochmann, 1984; Schuett and Cowling, 1985; McLaughlin, 1985; Smith, 1995; Little, 1995). Examples of forest declines in Europe include mortality of stands of Norway spruce, silver fir, Scotch pine, European beech, and oaks (Huettl, 1989). It has been estimated that in some western European countries more than half of all the forest trees are in decline because of pollution (McKersie and Leshem, 1994). In North America forest declines include mortality of sugar maple (McLaughlin *et al.*, 1985), red spruce (McLaughlin *et al.*, 1987), white ash (Castello *et al.*, 1985), and oaks (Tainter *et al.*, 1984; Biocca *et al.*, 1993; Jenkins and Pallardy, 1995).

Typical symptoms of forest decline in conifers include chlorosis, loss of needles, bud mortality, growth inhibition, bark injury, resin exudation, and tree mortality. Although stands of all ages are affected, injury is more severe in old than in young stands. Chlorosis in Norway spruce exposed to pollution was associated with Mg deficiency. Tip yellowing started in the older needles of the lower and middle crown. Eventually whole needles became chlorotic and were shed. As new shoots began to grow in the spring, the previous year's needles turned yellow, probably because Mg was translocated from the old to new needles (Mies and Zoettl, 1985). Decline of broad-leaved trees is characterized by yellowing and early senescence of leaves, growth inhibition, dieback of branches beginning with those at the top of the crown and progressing downward, stem necrosis, bark peeling, injury to roots, and tree mortality. The effects of pollutants on forest declines are discussed further in pages 296 to 298.

Biodiversity   Natural variations among species and plant communities are reflected in biodiversity. Such diversity has its origin at the genetic level and is propagated to higher levels of organization through populations (Barker and Tingey, 1992).

There is much concern throughout the world with declining biodiversity as a result of habitat modification and release of pollutants into the air, water, and soil. In 1991 it was estimated that approximately 10,000 species were becoming extinct each year, with the rate of loss being the highest since the mass extinctions that occurred 65 million years ago. It is likely that the threat of pollution to biodiversity will increase proportionally with industrial growth.

The increasing loss of biodiversity inevitably leads to irreplaceable losses of species, populations, and ecosystems. The end results of disruption of ecological processes and decrease in biodiversity are current or potential losses of food, fiber, and medicinal plants. Although we may not know of any economic or human use for many unstudied plants, past experience warns us that many of them would likely prove to be valuable.

Air pollution has the potential of decreasing biodiversity if it (1) changes genetic diversity within populations, (2) decreases the reproductive potential of species, (3) reduces yields of crops or products of natural vegetation, or (4) adversely affects the structure and function of ecosystems (Barker and Tingey, 1992).

Mechanisms of Pollution Effects

Pollutants influence woody plants by complex mechanisms. Pollutants rapidly induce biochemical injuries in cells that are later expressed in many metabolic changes in whole plants. Hence, it is difficult to select a specific reaction that might be a unique determinant of an effect of air pollution. Furthermore, biochemical responses are modified by the types and dosages of pollutants as well as by environmental changes.

Air pollutants adversely affect photosynthesis (see Chapter 5 in Kozlowski and Pallardy, 1997), respiration, stomatal aperture, metabolic pools, transport of carbohydrates and growth hormones to meristematic regions, transpiration, membrane permeability, and enzymatic activity (Kozlowski and Constantinidou, 1986a). Both the structural and functional changes that are induced in plants by pollutants are directly related to impaired cellular metabolism. Changes in metabolite pools typically precede both visible symptoms and growth reduction. Visible injury occurs when the expenditure of energy and metabolites necessary to counteract the biochemical changes induced by pollutants exceed the input of energy and availability of reserve foods (Kozlowski and Constantinidou, 1986a).

The effects of specific pollutants on plant responses may vary. For example, O_3 increases membrane leakage and alters permeability to ions. By oxidizing unsaturated bonds in the fatty acids of membrane lipids, O_3 can alter plasmalemma structure, membrane functions, and cell metablism (Matyssek *et al.*, 1995). Increases in respiration of pine foliage on exposure to O_3 suggest that O_3-induced growth reduction is mediated to an appreciable extent by increased respiration (Edwards *et al.*, 1992). Plants exposed to O_3 also undergo several biochemical changes before visible injury occurs. These alterations include increases in activities of enzymes associated with defense mechanisms, a surge in ethylene production, changes in polyamine metabolism, and increases in activities of phenylpropanoid and flavonoid pathway enzymes. Activities of superoxide dismutase and peroxidases, which protect cells from oxidative injury caused by hydroxyl radicals, H_2O_2, and superoxides, also increase (Kangasjärvi *et al.*, 1994). Weak acids such as SO_2 and HF appear to act as poisons at specific metabolic sites. Numerous rapid changes in the chemical composition of plant membranes indicate that the plasma membranes may be among the initial sites of reaction to HF (Rakowski *et al.*, 1995). Fluorides also inhibit the activity of a wide variety of enzymes (Kozlowski and Constantinidou, 1986a).

Pollutants entering the soil, especially heavy metals, can injure roots and reduce their efficiency in absorbing water and mineral nutrients (Godbold and Hütterman, 1985; Smith, 1995). Soil pollutants also may acidify the soil and increase losses of nutrients from the rooted soil substrate (Ulrich, 1986). Although most Al is incorporated into sparingly soluble aluminosilicate soil minerals, acid precipitation increases Al solubility in many soils, thus aggravating Al toxicity in forests. Common symptoms of Al toxicity in roots include reduced root elongation and branching, thickening of roots, and necrosis of cells near root and shoot meristems. Aluminum controls growth primarily by inhibiting mitosis (Tepper *et al.*, 1989). Aluminum also affects synthesis of ATP, calmodulin function, cell wall synthesis, membrane permeability, and transport processes (Haug, 1983).

Mechanisms of tolerance of plants to toxic metals have proved elusive in part because of the failure to relate adaptive mechanisms (present only in tolerant phenotypes) to constitutive mechanisms (present in most phenotypes). This failure often has distorted the relative importance of a proposed tolerance mechanism (Meharg, 1994).

Protection of cells against toxic metals is complex and involves many processes including organometal synthesis, metal reduction, metal chelation, and ion transport. All of these may be constitutive to cell functions. The mechanisms of adaptation to highly contaminated environments may involve only one of these processes, as many adaptive mechanisms are under control of major genes (MacNair, 1993).

Adverse effects of pollutants also may result from their destruction of the stratospheric O_3 layer, resulting in increasing exposure to ultraviolet radiation, or by increasing temperature as a result of increasing CO_2 concentration (see pp. 256–269).

Pollution stresses commonly are superimposed on a variety of abiotic and biotic stresses, often with disastrous results. A simultaneous pollution stress exerted on both tree crowns and roots is more harmful than pollution stress imposed on only one or the other. Some of the pathways and mechanisms by which pollutants influence growth and survival of woody plants are summarized in Fig 5.29.

Factors Affecting Plant Responses to Pollution

There are many problems in quantifying the effects of pollution on woody plants because of the complexity of pollution stress, variations in growth characteristics of trees and communities of trees, and the impacts of many abiotic and biotic stresses that interact with the effects of pollutants. The responses of plants to pollution vary with species and genotype, types and combinations of pollutants, pollutant dosage, stage of plant development, plant response measured, and environmental regimes.

Species and Genotype   The tolerance of woody plants to pollution varies among species and genotypes, largely because of differences in absorption of pollutants by leaves or in biochemical tolerance of pollutants, but mostly the former. Variations in metabolism of pollutants into less toxic compounds and dilution of pollutants by rapid redistribution within plants also may be involved in mechanisms of tolerance.

Pollution tolerance commonly is closely correlated with stomatal conductance, which often is a function of stomatal size and frequency. Stomatal conductance varies widely among species as well as clones and cultivars (Siwecki and Kozlowski, 1973; Pallardy and Kozlowski, 1979c). White ash leaves with large stomata (high stomatal conductance) absorbed more SO_2 than leaves of sugar maple with small stomata (low stomatal conductance) (Jensen and Kozlowski, 1975). Seedlings of river birch were more sensitive than paper birch seedlings to SO_2, partly because of the higher stomatal conductance of the former (Norby and Kozlowski, 1983). Sensitivity of poplar clones to SO_2 was correlated with their stomatal conductance (Kimmerer and Kozlowski, 1981). Leaf conductances of 10 species of trees and

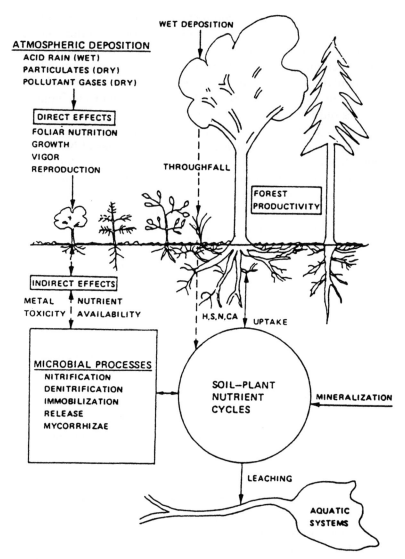

Figure 5.29 Sites of reaction of wet- and dry-deposited atmospheric pollutants on forest ecosystems. From McLaughlin (1985).

shrubs in California were good indicators of uptake and tolerance of SO_2 exposure (Winner *et al.*, 1982).

Genotypic variations in susceptibility to pollution are not always associated with differences in stomatal conductance. Kargiolaki *et al.* (1991) found wide differences in susceptibility of four poplar clones to $SO_2 + O_3$

but no differences in stomatal conductance. The differences in response were attributed to variations among clones in ethylene production. Lack of response of an eastern cottonwood clone to air pollution suggested that it possessed mechanisms which suppressed ethylene production. This clone may have contained or may have activated detoxifying compounds for both SO_2 and O_3 in its leaves or operated a more efficient mechanism than other clones did for scavenging free radicals.

Some investigators emphasized variations in biochemical tolerance of pollutants by woody plants. For example, certain clones of Norway spruce fixed more S in organic fractions and maintained higher buffering capacity after exposure to SO_2 than did SO_2-susceptible clones (Braun, 1977a,b). Activation of superoxide dismutase, catalase, and peroxidase (enzymes that could protect cells from O_2-induced oxidation of unsaturated fatty acids of membranes) may be a mitigating mechanism in some O_3-tolerant plants (Bennett *et al.*, 1984). Differences also have been shown in enzymatic patterns of trees that vary in pollution tolerance (Mejnartowicz, 1984).

Pollution Dosage   To establish legislation governing allowable emissions of pollutants, reliable data are needed on critical dosages of pollutants that injure plants and decrease their growth. However, determining precise dosages of pollutants that injure plants is complicated because plant responses differ with many factors including methods of evaluation of pollution effects, environmental conditions, the specific plant response measured, frequency of pollutant exposures, length of time between exposures, extent of fluctuations in concentrations of pollutants, time of day of exposures, and their sequence and pattern. Additionally, total flux of pollutants is influenced by characteristics of plant canopies and leaf boundary layers. The ambient dose (the dose to which a plant is exposed) is not a direct quantitative measure of the effective dose (the dose that induces a physiological response). The effective dose is a function of the rate at which the pollutant molecules arrive at mesophyll cells in the leaf. This rate is regulated largely by the gas-to-liquid pathway, which varies with the environmental regime and species. Only very few molecules of a gaseous pollutant react with leaves. In addition, only a portion of the total flux enters the leaf interior, with the rest adsorbed to the leaf surface. The effective dose may vary among gaseous pollutants even though the ambient dose is equivalent (Taylor *et al.*, 1982).

Data on effects of pollutant dosage on plants have been obtained mainly from two types of studies: (1) exposure of plants to known concentrations of pollutants in controlled-environment facilities for specified periods (Kozlowski and Constantinidou, 1986a; Smith, 1990) and (2) evaluation of plant responses at varying distances from point sources of pollution such as smelters and power-generating stations (Miller and McBride, 1975; Smith,

1990, pp. 485–499). Interpretation of data from such studies requires that the concentrations of pollutants be known and that there is reasonable control and characterization of environmental conditions. These requirements often are better met in controlled-environment studies than in studies conducted in the field.

Pollution dosages that induce foliar injury differ with the specific pollutant. The critical dosage of SO_2 for acute injury may vary from 70 pphm for 1 hr to 18 pphm for 8 hr of exposure (Linzon, 1978). Threshold concentrations of O_3 for deciduous trees of the eastern United States range from 20 to 30 pphm for 2 to 4 hr. Conifers are somewhat more sensitive (National Academy of Sciences, 1977b). The threshold dosage for NO_x was reported as 160 to 260 pphm for up to 48 hr. However, 2,000 pphm was critical for 1-hr exposures, and 100 pphm for 20-hr exposures (National Academy of Sciences, 1977a). Fluorides are much more toxic than other gaseous pollutants, with threshold concentrations 10 to 1000 times lower than those for SO_2, O_3, or NO_x (Weinstein, 1977).

The importance of pollution dosage on responses of woody plants is dramatically illustrated by decreasing plant mortality at increasing distances from point sources of pollution. Near an iron-sintering plant in southern Ontario, Canada, the number of plant species was 2 to 4 within 5 km of the pollution source and 43 species at distances greater than 16 km from the source (Fig. 5.30). In Biersdorf, Germany, there was a denuded zone in the immediate vicinity of an iron ore roasting furnace. This zone was surrounded by a transition zone with isolated clusters of grass. As the amount of pollutants decreased, species diversity increased until a mixed beech–oak forest was established (Guderian and Kueppers, 1980).

There has been considerable controversy about the relative importance of specific air pollutants that injure forest trees located far from the polluting source. Woodman (1987) concluded that O_3 was the only regionally dispersed air pollutant that injured and killed forest trees at long distances from the source. He reviewed evidence for O_3 injury to eastern white pine, ponderosa pine, Jeffrey pine, sugar pine, and incense cedar far from the pollution source. Duchelle *et al.* (1982) reported on long-distance oxidant-type injury to tulip poplar, green ash, hickories, black locust, table mountain pine, Virginia pine, pitch pine, and eastern hemlock. Although SO_2 has injured forests as much as 100 km from point sources, on a regional scale it is more important as a precursor pollutant that leads to formation of acid rain (Linzon, 1986).

Combinations of Pollutants   Air pollutants only rarely exist singly. Usually the environment contains a mixture of gaseous and particulate pollutants. When plants are injured or their growth altered by a combination of pollutants, it generally is difficult to quantify the importance of the individual

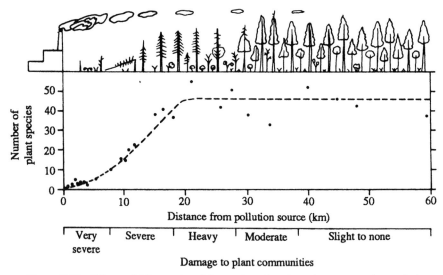

Figure 5.30 Effects of SO_2 produced by an iron-sintering plant on numbers of plant species. Reproduced from Whittaker (1975); after Gordon and Gorham (1963).

pollutants in the mixture. In general, plants are injured more by combinations of pollutants than by single pollutants (Fig. 5.31). However, the effects of combined pollutants on physiology and growth of plants may be synergistic, additive (Constantinidou and Kozlowski, 1979a), or antagonistic (Noble and Jensen, 1980). Antagonistic effects between acid rain and O_3 on height and biomass increment of loblolly pine were reported by McLaughlin *et al.* (1994). Among the factors influencing plant responses to pollutant mixtures are concentrations of each gas in the mixture, relative proportions of the gases, whether there is simultaneous or intermittent application of the combined pollutant stress, and the age and physiological condition of the plant (Reinert *et al.*, 1975; Constantinidou and Kozlowski, 1979a,b).

Stage of Plant Development The susceptibility of woody plants to pollution varies with their stage of development. Seedlings are more susceptible than old trees, with plants in the cotyledon stage being particularly sensitive. For example, red pine seedlings in the cotyledon stage that had been exposed to 1.3 ppm SO_2 for 15 min became chlorotic. The treatment also inhibited expansion of primary needles and caused death of needle tips (Constantinidou *et al.*, 1976). Conifer seedlings in the cotyledon and primary needle stages are more sensitive than those with well-developed secondary needles. However, immature secondary needles are quite sensitive to air pollutants (Treshow and Pack, 1970).

Responses of leaves of broad-leaved species to pollution also vary with their stage of development. The younger, fully expanded leaves and those near full expansion are most sensitive to SO_2. The small expanding leaves are least sensitive.

Response Parameters   The effects of pollutants on plants vary among specific plant responses, which fall into three broad categories: (1) injury, (2) growth, yield and development, and (3) physiological and biochemical responses. Pollution tolerance of species or genotypes often is ranked differently by various investigators who use dissimilar criteria for arriving at relative rankings.

Leaf injury has been variously assessed, for example, as percentages of leaves injured, tip necrosis in conifer needles, leaf necrosis in broad-leaved trees, and leaf shedding. A variety of growth responses have been evaluated, including height growth, internode elongation, leaf growth, cambial growth, root growth, increase in dry weight of plant parts or whole plants, and relative growth rate (RGR). Some investigators quantified several physiological and biochemical responses to pollution including stomatal aper-

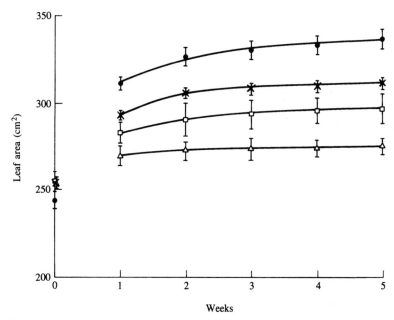

Figure 5.31   Effects of SO_2 ($\square$; 2 ppm for 6 hr), O_3 (X; 0.9 ppm for 5 hr), and SO_2–O_3 mixtures ($\triangle$; 2 ppm SO_2 and 0.9 ppm O_3 for 5 hr followed by 2 ppm SO_2 for 1 hr) on leaf growth of American elm seedlings, compared to control seedlings ($\bullet$). From Constantinidou and Kozlowski (1979a).

ture, membrane permeability, photosynthesis, accumulation of metabolites, and enzymatic activity.

Some plant responses are much more sensitive than others to pollution. For example, 6 weeks after treatment, SO_2 decreased leaf growth, dry weight increment, and relative growth rate but not height or diameter growth of bald cypress seedlings (Shanklin and Kozlowski, 1985). On the basis of needle injury, dry weight increment, RGR, and diameter growth, the tolerance to SO_2 varied in the following order: Japanese black pine > Japanese red pine > Japanese larch. On the basis of height growth response, however, the order of tolerance to SO_2 was Japanese red pine > Japanese black pine > Japanese larch (Tsukahara *et al.*, 1985).

The selection of response parameters to determine pollution impacts should be carefully considered by investigators. In general, data on pollution tolerance of different plants have been more consistent when based on leaf injury than on some aspect of growth or physiological change. Data on growth responses are complicated by variations in species growth characteristics, age of plants, experimental conditions, and lack of uniformity in methods of measuring growth. Changes in dry weight or biomass may be useful for assessing relatively long-term effects of pollutants. However, RGR often is a more useful measure of pollution tolerance of plants that vary in growth rate and size at the beginning of an experiment. Relative growth rate permits comparison of the effects of pollutants on the rate of growth independently of the size of the plants that are being compared (see Chapter 3 of Kozlowski and Pallardy, 1997). Measurement of the rates of physiological processes is complicated by the high sensitivity of physiological processes to stresses other than pollution and by the difficulty in relating rates of physiological processes to growth. For example, both high and low positive correlations as well as negative correlations between rates of photosynthesis and growth of trees have been demonstrated (Kozlowski *et al.*, 1991, p. 44). Still another problem is that of deciding which physiological process should be measured to determine pollution tolerance. These problems are discussed further by Kozlowski and Constantinidou (1986a) and Smith (1990, 1995).

Environmental Factors and Interactions   The effects of air pollutants on woody plants are influenced by prevailing environmental regimes as well as those occurring before and after a pollution episode. Prevailing environmental conditions influence responses of plants to gaseous air pollutants largely by controlling stomatal aperture. Plants absorb more gaseous air pollutants when the soil is wet than when it is dry because the stomata are more open. Stomata are good sensors of air humidity (Sena Gomes *et al.*, 1987). Seedlings of paper birch absorbed more SO_2 and were injured more when fumigated at high humidity than when fumigated at low humidity

(Norby and Kozlowski, 1982). Air temperature also influences plant responses to air pollution. For example, seedlings of paper birch, white ash, and red pine exposed to SO_2 at 30°C had more-open stomata and absorbed more of the pollutant than seedlings fumigated at 12°C (Norby and Kozlowski, 1981b).

Exposing plants to a prefumigation stress also affects their responses to pollution. Green ash seedlings that had been grown for 4 weeks at 15°C had fewer leaves and more-closed stomata than those grown at 25°C. When seedlings of both groups were exposed to SO_2 at 22°C, those preconditioned at 15°C absorbed less pollutant and were injured less (Shanklin and Kozlowski, 1984). Flooding of soil induced stomatal closure of river birch seedlings and reduced subsequent uptake of SO_2 (Norby and Kozlowski, 1983). Sulfur dioxide reduced growth more in unflooded bald cypress seedlings than in seedlings that had been flooded for 8 weeks prior to exposure to the pollutant (Shanklin and Kozlowski, 1985).

Postfumigation environmental regimes may affect responses to pollution by altering plant metabolism. For example, red pine seedlings fumigated with SO_2 formed new photosynthetically active needles at postfumigation temperatures of 22 or 32°C, hence counteracting the growth-inhibiting effects of the injured needles. In comparison, at a postfumigation temperature of 12°C the fumigated seedlings did not replace the injured needles (Norby and Kozlowski, 1981a).

Many environmental factors, including pollution, appear to be variously involved in inducing forest declines. However, pollution does not appear to be the primary cause of forest decline except in heavily polluted areas near point sources of pollution. Nevertheless, pollution can contribute to forest decline in areas in which climatic stresses initiate declines and other stresses contribute (Schäfer *et al.*, 1988). The weight of evidence indicates that the indirect effects of pollutants (in which pollution alters the capacity of a plant to compete for limited resources or withstand various environmental stresses) are more important than direct effects such as formation of necrotic lesions (Johnson and Taylor, 1989).

Forest declines generally appear to be responses to several sequential stress factors (Smith, 1995). Processes leading to decline of forest stands are set in motion by abiotic stresses, but eventual mortality typically is preceded by a biotic stress. The specific stresses involved vary from region to region. Manion (1981) considered tree declines to be caused by a succession of events, including contributions of predisposing factors (e.g., low site quality) that render trees more susceptible to inciting factors (e.g., drought, air pollution, insect attacks) and ending with contributing factors (e.g., root diseases, bark beetles) that precede death of trees.

Decline of red spruce in the United States has been attributed to several abiotic and biotic stresses (Smith, 1990). The major abiotic stresses include extremes of temperature and moisture supply, wind, soil conditions, and air pollution. Contributing biotic stresses may include foliar, branch, stem, and root pathogens as well as the spruce beetle (*Dendroctonus rufipennis*), spruce budworm (*Choristoneura fumiferana*), and swift moth (*Pharmacis mustelinus*). The specific causal factors vary with site, latitude, and altitude. In Germany decline of Norway spruce has been attributed to at least four sets of initiating factors on different sites. In the United States and Canada, sugar maple decline also has been attributed to different sets of initiating stresses (Smith, 1990). In the western United States, extensive mortality of conifers results from drought followed by infestations of bark beetles (Miller, 1989). In many cases infectious diseases (e.g., root diseases, needle casts, and rusts) weaken trees and render them more susceptible to bark beetle attack. Adverse growing conditions also interact with insect and disease complexes to hasten death of trees. In areas that are exposed to O_3 and acidic deposition, the superimposed air pollutants may contribute to forest declines. Interactions among pollutants and insects and diseases are rather complex. Pollutants may directly affect insects by influencing their rates of growth, fecundity, finding of hosts or mates, and mortality. Indirect effects of pollutants may be mediated through changes in host age, structure, distribution, vigor, and acceptance (Smith, 1990, 1995).

Both the types of insects present and their numbers are influenced by pollution. Insect populations may be increased or decreased. For example, SO_2 pollution reduced populations of white pine weevils (Linzon, 1966) but increased numbers of the European pine shoot moth (Sierpinski and Cholodny, 1977).

Much evidence shows that pollutants predispose trees to attacks by certain insects. Examples are gypsy moth infestations on white oak (Jeffords and Endress, 1984) and bark beetle attacks on ponderosa pine (Larsson *et al.*, 1983). The injury caused by bark beetles is associated with reduced oleoresin exudation pressures, lower rates of resin flow, and increases in crystallization of resin (Cobb *et al.*, 1968). Christiansen *et al.* (1987) correlated the severity of bark beetle attacks with availability of carbohydrates for defensive reactions by trees.

Pollutants also may influence insect attacks on vigorous trees by altering water relations and biochemical changes of trees that attract or repel insects. For example, production of terpenes by balsam fir trees exposed to SO_2 was followed by increased insect infestations (Renwick and Potter, 1981). In contrast, ethane produced by SO_2-injured tissues (Kimmerer and Kozlowski, 1982) may repel wood-boring beetles (Sumimoto *et al.*, 1975).

Pollutants may increase or decrease the severity of diseases depending on whether the host or pathogen is affected most. Pollutants also may change the habitat of the pathogen by affecting the structure and microenvironment of plant surfaces. By reducing tree vigor, air pollutants even at low concentrations may predispose trees to certain diseases. For example, injury by *Armillaria mellea* fungi was attributed to weakening of trees by air pollution (Sinclair, 1969). Trees weakened by SO_2 were more susceptible to pine needle blight (Chiba and Tanaka, 1968) and wood rots (Jancarik, 1961). Ozone-induced injury favored infection of pines by *Fomes annosus* (James *et al.*, 1980).

The severity of disease also may be reduced because of the greater toxicity of pollutants to certain pathogens than to the host. For example, rust fungi were absent on trees close to a smelter and abundant at greater distances (Scheffer and Hedgcock, 1955). The causal agent of white pine blister rust was almost absent to a distance of 40 km from an SO_2-emitting smelter in southern Canada (Linzon, 1978).

Wind

The effects of wind on woody plants can be both harmful and beneficial. Wind breaks stems and branches, topples and uproots trees, alters stem form, and injures leaves. It also erodes soils, affects physiological processes of plants, disperses air pollutants, and increases the risks of insect attack, disease, and fire (Kozlowski *et al.*, 1991; Coutts and Grace, 1995)). By aiding in dispersal of pollen and plant propagules, wind is beneficial to plant reproduction.

Susceptibility to wind varies among species, with conifers being affected more than broad-leaved trees by wind. Whereas eastern white pine and red pine stands were susceptible to wind damage at 15 years of age, stands of broad-leaved trees were not damaged until they were 20 years old and were not completely blown down until 80 to 100 years old (Foster, 1988). In mixed forest stands the susceptibility to strong winds varied as follows: eastern white pine > conifer plantations > eastern white pine–broad-leaved trees = eastern hemlock–broad-leaved trees–eastern white pine > broad-leaved trees (Foster and Boose, 1992).

The extent of wind damage within a stand is related to the canopy positions of different species. For example, fast-growing pioneer species of the overstory (eastern white pine, red pine, poplar, and paper birch) were injured much more by strong winds than were slow-growing species in lower canopy and understory positions (hickories, red maple, white oak, black oak, and eastern hemlock) (Foster, 1988).

When heavy rains precede strong winds and soften the soil, uprooting of trees is more likely to occur than if the soil is dry. On the other hand, stems of trees growing in dry soil are more likely to be broken by wind. Trees with diseased roots or stems are much more susceptible than healthy trees to windthrow or windbreak (Shaw and Taes, 1977). Trees in dense stands become very susceptible to windthrow after some of the trees are harvested. Open-grown trees with strongly tapered stems are particularly stable (Petty and Worrell, 1981). Trees without stem buttresses are more susceptible than buttressed trees to windthrow.

Injury to leaves of forest trees along windy seacoasts is well documented. In Northern Ireland, for example, wind damaged up to 46% of the leaf area of sycamore maple trees (Rushton and Toner, 1989). Leaf injury may be caused by tearing and shredding of leaves as well as by leaf dehydration. Tearing of leaves of palm, banana, apple, pear, and citrus leaves is well known (Waister, 1972). Wind caused tearing, browning, curling, necrosis, and abscission of leaves of cacao seedlings (Sena Gomes and Kozlowski, 1989). Injury to sycamore maple leaves by wind included formation of lesions, tearing of blades, collapse of epidermal and mesophyll cells, and disruption of leaf waxes (Wilson, 1980, 1984). Dehydration and shedding of leaves often are induced by hot, dry winds such as the Santa Anas of southern California that cause extreme damage to citrus trees.

Strong winds greatly modify tree size and form as shown by the dwarfed and often one-sided crowns of trees growing at high elevations or along windy seacoasts (Telewski, 1995; Fig. 5.32). Strong winds near upper altitudinal limits of tree growth often erode away the buds on the windward sides of trees. As a result, branches may develop only on the leeward sides.

Growth of Plants

Mild winds often decrease shoot and height growth of trees, increase diameter growth, and alter stem form. Wind (6 m sec^{-1}) dehydrated and injured the leaves of trembling aspen and also reduced leaf growth (Flückiger *et al.*, 1978). Wind also reduced needle elongation of loblolly pine (Telewski and Jaffe, 1981) and Fraser fir (Telewski and Jaffe, 1986a).

Stems of trees that sway in the wind are thicker, shorter, and more tapered than trees that are guyed to prevent them from swaying (Larson, 1965; Burton and Smith, 1972). Most cambial growth of conifers exposed to wind from one direction occurs on the leeward sides of stems. For example, the xylem increment was greater on the leeward sides than the windward sides of white spruce, eastern white pine, and lodgepole pine stems (Bannan and Bindra, 1970). Tilting of stems by wind also induces formation of abnormal reaction wood (e.g., compression wood on the leeward sides of conifers and tension wood on the windward sides of broad-leaved trees). Formation of reaction wood is discussed further in Chapter 3.

Windbreaks and Shelterbelts

Windbreaks are porous barriers that decrease wind velocity in sheltered areas on the leeward side (and to some extent on the windward side) of the barrier by deflecting the wind. Belts of trees as windbreaks can be of considerable practical value because they decrease soil erosion, reduce

Figure 5.32 Cluster of Engelmann spruce trees at the timberline (10,800 ft) in the Medicine Bow Mountains of Wyoming. A group of "flag" trees rise from a clump of stunted krummholz trees that never grow above the level of winter snow cover. The clump probably is a clone produced by layering. From Billings (1978). "Plants and the Ecosystem." © Wadsworth Publishing Company.

mechanical damage to plants, increase crop yield, control snow drifting, and improve cover and increase food supply for wildlife (Caborn, 1965; Baer, 1989). Windbreaks often are used to protect young trees in plantations and forest nurseries (Gloyne, 1976).

Shelterbelts generally increase crop yields on the leeward side, largely because of greater water supply and lowered plant water deficits associated with reduced evapotranspiration and added water from snowmelt. Evaporation on the leeward side of a shelterbelt often is decreased by as much as 20 to 30%. The trapped snows account for greater areas of unfrozen soil than in unprotected areas, hence providing better infiltration and decreased water runoff in the spring. Shelterbelts may increase crop yields from a few to several hundred percent depending on the crop, region, age and composition of the shelterbelt, and other factors.

Physiological Processes

The effects of wind on plant growth are complex and mediated by changes in water, food, and hormone relations (Kozlowski *et al.*, 1991; Coutts and Grace, 1995).

Water Relations   Although transpiration often is increased by low wind speeds as a result of reduction of the boundary layer surrounding the leaves, the desiccating effects of wind are modified by cooling of leaves, which decreases the leaf to air vapor pressure difference (Dixon and Grace, 1984).

Wind increased the rate of transpiration of alder and larch but decreased it in Norway spruce and Swiss stone pine, presumably because of differences in stomatal responses (Tranquillini, 1969). Wind speeds of 5.8 to 2.68 m sec^{-1} increased transpiration of white ash but decreased it in sugar maple. In white ash the high rate of transpiration caused by an increase in the vapor pressure gradient in wind was not reduced by stomatal closure. Unlike the small stomata of sugar maple, the large stomata of white ash were covered by a cuticular ledge. The exposed guard cells of sugar maple leaves dehydrated and closed faster than the covered guard cells of white ash (Davies *et al.*, 1974). The transpiration rate of cacao decreased as the wind speed increased up to 6 m sec^{-1}, apparently in response to lowering of the leaf to air vapor pressure gradient associated with cooling of the leaves (Sena Gomes and Kozlowski, 1989).

When plants are exposed to wind, their rate of transpiration typically increases rapidly and then gradually declines as the leaves dehydrate and stomata close (Fig. 5.33). In hinoki cypress such a gradual decline continued until the rate of transpiration was lower than it was in calm conditions (Satoo, 1962).

Food Relations   Strong winds often decrease total photosynthesis by causing injury and shedding of leaves. At low speeds wind regulates photosynthesis by affecting boundary layer thickness, leaf temperature, and stomatal aperture. An increased rate of photosynthesis of apple leaves resulted partly from reduction in thickness of the boundary layer (Avery, 1966). Changes in leaf temperature caused by wind affect both stomatal conductance and mesophyll conductance (Grace and Thompson, 1973). Strong winds also may decrease carbohydrate supplies by stimulating respiration (Todd *et al.*, 1972). Such increases in respiration have been attributed to the mechanical effects of wind (Jaffe, 1980).

Hormone Relations   Wind greatly modifies production and translocation of growth hormones, especially auxins and ethylene. Increase in diameter growth and formation of reaction wood in trees exposed to wind have been attributed to redistribution of hormonal growth regulators. For example, various investigators have linked the stimulation of cambial growth on the leeward sides of tilted conifers to a high auxin gradient that causes mobilization of carbohydrates. There also is some evidence that auxin may stimulate cambial growth in conifers by increasing ethylene production (Leopold *et al.*, 1972). Formation of tension wood in broad-leaved trees has been attributed to auxin deficiency (Chapter 3).

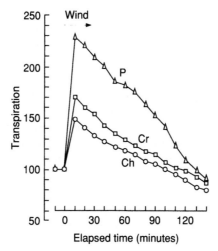

Figure 5.33 Change in transpiration with elapsed time after the beginning of the exposure to artificial wind (relative values, average of six determinations) in three species: P, *Pinus densiflora*; Cr, *Cryptomeria japonica*; Ch, *Chamaecyparis obtusa*. From Wind, transpiration, and tree growth. Satoo, T., *In* "Tree Growth" (T. T. Kozlowski, ed.). Copyright © 1962 Ronald Press. Reprinted by permission of John Wiley & Sons, Inc.

Fire

The effects of fire on woody plants range from catastrophic to beneficial. Fire influences growth and survival of plants and plant ecosystems by influencing physiological processes, destroying plants, and affecting site quality. The responses of plants to fire depend on the type and severity of the fire as well as on the plant species. More damage is caused to trees by crown fires than by ground fires. The most serious cause of injury by crown fires is reduction in photosynthetic surface by defoliation. This is especially serious in young forest stands that are more likely to be completely defoliated than older, taller trees. Ground fires often kill areas of bark and cambium at the bases of trees, thereby interfering with phloem translocation of carbohydrates. However, Jemison (1944) found that girdling of the bases of stems of deciduous trees did not have serious adverse effects on phloem translocation after the fire. New xylem and phloem tissues were laid down in newly oriented paths during the next growing season, providing new translocation paths around the fire wounds. Even though tree growth may not be greatly reduced by ground fires, wood quality often is lowered by fire scars.

The species composition of forests often is determined by fire. Because plant succession to a climatic climax stage is prevented by periodic fires, many forest ecosystems are maintained in a subfinal "fire climax" stage. In the northeastern United States trembling aspen, white birch, red pine, and jack pine are perpetuated by fire. Douglas fir is the principal fire climax species over much of the northwestern part of the United States.

Site Quality

Fire alters site quality, largely by burning organic matter and heating the soil surface. Severe fires reduce the capacity of soils to store water, cause soil compaction, accelerate soil erosion, and increase fluctuations in soil temperature. Forest fires result in losses of mineral nutrients from plants by volatilization, ash convection, and subsequent leaching. As much as 70% of the N and other nutrients may be lost during hot fires. Even though much N is lost by combustion, it usually is rather rapidly replaced by addition in rain, increased microbiological activity, and N fixation (Fig. 5.34).

The mineral nutrients dissolved in ash have several fates. They may be lost by surface runoff, leached into the soil and retained, or leached through the soil profile. Despite potential losses of nutrients by volatilization and runoff, availability of mineral nutrients to plants usually is higher after than before a fire (Christensen, 1985; Rashid, 1987). The increase may reflect direct production by burning, leaching into mineral soil, stimulation of microbial mineralization because of greater availability of soil moisture, and decreased absorption associated with root mortality.

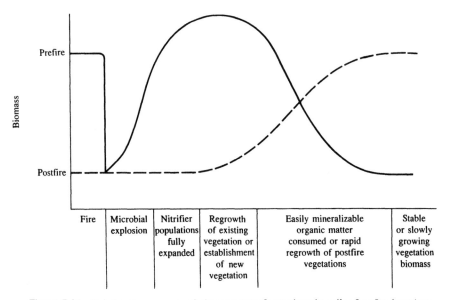

Figure 5.34 Relative time course of nitrogen transformations in soils after fire by microorganisms (solid line) and vegetation (dashed line). From Woodmansee and Wallach (1981).

Higher availability of mineral nutrients following fire may or may not improve site quality. If mineral nutrients leach below the depth of rooting or are washed off the surface, site quality is lowered. However, if soluble minerals are leached into the root zone and absorbed by roots, site quality is improved. Deterioration of a site by fire is more likely to occur on very coarse sands or heavy soils than on sandy to loamy soils (Spurr and Barnes, 1980).

Fire may decrease site quality by increasing soil erosion through elimination of vegetation and reduction in soil infiltration capacity. The amount of erosion varies greatly with slope and amount of plant cover. Soil erosion following fire on juniper lands was negligible on level sites and severe for more than 30 months on steep slopes. Seeding of steep slopes with grasses reduced erosion by as much as 90% (Wright *et al.*, 1982).

Insects and Diseases

Trees that have been weakened by fire are more susceptible than vigorous trees to attack by certain insects, especially wood-boring beetles. Insects attracted by heat and smoke preferentially attack fire-scarred trees, particularly those with fungal decay (Gara *et al.*, 1984). Almost every tree in a conifer forest may become infected with wood borers within a few weeks after a fire. Outbreaks of spruce budworm (*Choristoneura fumiferana*) started in mixed conifer–hardwood forests exposed to fire and subsequently spread into spruce–fir stands (Berryman, 1986).

Fires aid in controlling some plant diseases and spreading others. Seedling regeneration on ash beds often is successful following fire partly because fire kills fungi that cause seed decay, damping-off, and root rots. For example, fires reduce or control brown needle spot of longleaf pine, leaf-spot of blueberry, and *Nectria* cankers (Ahlgren, 1974). On the other hand, both surface roots injured by fire and fire scars on stems are entry points for disease fungi such as *Fomes* spp. that cause heart rot (Berry, 1969). Fire also spreads disease by stimulating formation of dense stands of host plants. An example is the rapid spread of powdery mildew in thick stands of blueberry (Demaree and Wilcox, 1947).

Adaptations to Fire

Fire tolerance of woody plants has been attributed to characteristics that increase survival after fire and those that accelerate regeneration of tree stands. Species with thick or corky bark (e.g., redwood, western larch, ponderosa pine, Douglas fir, longleaf pine, and bur oak) often escape injury by fire. In contrast, thin-barked species (e.g., alpine fir, Engelmann spruce, lodgepole pine, and eastern white pine) are more likely to be killed by ground fires.

Several species of forest trees survive fires by sprouting (Kozlowski *et al.*,

1991). Destruction of foliage by fire is followed by release of dormancy in buds that survive under the bark or by formation of adventitious buds that develop into shoots. Sprouting after fire is especially important for regeneration of some broad-leaved woody plants such as oaks. Most chaparral species produce vigorous stump sprouts after fire (Biswell, 1974). Sprouting also is important for survival of a few species of conifers after fire. An example is pitch pine.

Another adaptation to fire is the production of serotinous (late to open) cones. In lodgepole pine, jack pine, pitch pine, sand pine, and black spruce, cones containing viable seeds may remain closed on the tree for years. When the resinous material on serotinous cones is destroyed by fire, the cone scales open and the seeds are released. Where fires are common, release of large amounts of seeds from serotinous cones may determine the dominant species, as in lodgepole pine stands.

Some effects of fire (including removal of litter, exposure of mineral soil, and a rise in soil temperature) increase availability of mineral nutrients, and hence increase seed germination and seedling establishment (Weaver, 1974; Miyanishi and Kellman, 1986). Heat stimulated seed germination of California red fir and ponderosa pine (Wright, 1931) and a number of chaparral species (Biswell, 1974, 1989). Fire also stimulates seed germination of some species by improving the light regime (see Chapter 2).

Some woody plants survive fire by more than one adaptation. [For example, redwood has fire-resistant bark and prolific sprouting capacity; chaparral produces seeds at an early age, retains seed viability for decades, and has strong sprouting ability (Biswell, 1974, 1989)].

Insects and Diseases

Many insects and diseases inhibit growth of woody plants by altering the rates and balances of plant processes. An initial insect or fungus attack usually sets in motion a series of physiological dysfunctions. The impact of insect attacks or disease may or may not be drastic, depending on the extent to which various plant processes are disturbed and on the vigor of plants at the time of attack (Kozlowski, 1969).

Insects

Defoliating insects decrease the amounts of carbohydrates and hormonal regulators available for plant growth. Insects that feed on terminal or lateral buds cause deformation of trees. Phloem-boring insects interfere with transport of carbohydrates to meristematic regions and often inoculate the host with disease-causing microorganisms. Insects that feed on roots decrease absorption of water and mineral nutrients. Seed and cone insects often injure reproductive structures in forest seed production areas and in fruit orchards. In forests, however, seed and cone insects sometimes are beneficial because they may reduce the amounts of carbohydrates that are allocated to reproductive growth, hence diverting such carbohydrates to vegetative growth.

The effects of defoliation by insects vary with tree species, age of trees, weather, environmental preconditioning of trees, and presence of secondary insects and diseases. Defoliation of woody plants by insects usually inhibits growth much more than does removal of an equivalent amount of foliage by pruning of branches. This is because pruning of forest trees typically removes only the physiologically inefficient lower branches, whereas many insects remove foliage in the upper crown, which is more important in synthesizing carbohydrates and hormonal growth regulators (see Chapter 5 of Kozlowski and Pallardy, 1997). Furthermore, some insects preferentially defoliate trees that lack vigor and have small carbohydrate reserves (Kozlowski, 1973).

Evergreen conifers often are killed by a single complete defoliation before midsummer (before the buds form that will open in the following year). Defoliation after midsummer usually does not kill these trees. Recurrently flushing pines, such as loblolly and slash pines, can withstand a severe defoliation much better than pines that produce only one annual flush of shoot growth (e.g., red and eastern white pine). Deciduous conifers, such as larch, usually can be defoliated for several years before they are killed (Coulson and Witter, 1984).

Deciduous angiosperms are relatively tolerant of defoliation and may survive defoliation for a few years because they often contain large amounts of stored carbohydrates and replace destroyed foliage by new leaf-bearing shoots. Sometimes two severe defoliations in the same year may kill broad-leaved trees, but many can survive a single annual defoliation repeated for many years.

The impact of defoliation on xylem increment and its distribution along the stem is extremely variable. Partial defoliation may inhibit cambial growth in the same season, or there may be a lag in response depending on the plant species, severity of defoliation, time of its occurrence, and other factors (Kozlowski, 1971b).

Defoliators such as spruce budworm may destroy all current-year foliage without greatly affecting xylem increment at the base of grand fir trees during the first year of defoliation. However, xylem production in the upper stem usually is greatly reduced. After the first year of such defoliation, xylem production at the stem base, as well as at the top, is greatly reduced (Williams, 1967). Early-season defoliation of shortleaf pine by the first generation of the red-headed pine sawfly (*Neodiprion lecontei*) was followed by inhibition of xylem production during the same season, with recovery occurring during the next season. In comparison, late-season defoliation did not inhibit cambial growth appreciably during the same year but did so in the following year. The delayed effect of late-season defoliation resulted because most of the annual xylem increment was produced before the defoliation started (Benjamin, 1955).

Predisposition of Plants to Insect Attack   Physiological changes in woody plants that are induced by environmental stresses commonly are prerequisites for attack by certain insects. Both growth and survival of insects in stressed trees have been attributed to improved nutrition for the insect, lowered chemical defenses of plants, and a better environment for insect growth.

The age at which trees are attacked varies with specific insects. Most bark beetles favor mature or overmature trees, but some attack trees of specific age groups. *Dendroctonus brevicomis* attacks old trees that lack vigor; *Dendroctonus monticolae* attacks younger trees; and *Ips confusus* attacks young trees or tops of old trees (Rudinsky, 1962). Epidemics of spruce budworm (*Choristoneura fumiferana*) in old stands of balsam fir are associated with abundant production of staminate strobili (Kramer and Kozlowski, 1979).

A variety of environmental stresses predispose trees to attack by certain insects. Increased susceptibility of ponderosa pine to bark beetles, which is increased by drought, is associated with decreased flow of oleoresins. There is considerable evidence that attacks on trees by boring insects that live in the inner bark and outer wood are more severe in dry years than in wet years. Vité (1961) stated that infestation of ponderosa pine by beetles is much more severe in water-stressed trees than in well-watered trees. It also is reported that the extent and duration of water stress have an important effect on the susceptibility of loblolly pine to attack by the southern pine beetle. Apparently the high oleoresin pressures existing in unstressed trees discourage invasion by borers (see Chapter 8 in Kozlowski and Pallardy, 1997). A similar situation exists in white fir when engraver beetles establish themselves at a water stress of -2 MPa, but not at -1.5 MPa because of greater resin flow in the less-stressed trees (Ferrell, 1978).

Mattson and Haack (1987b) proposed a hypothetical framework that integrated influences of drought on host plants with necessary conditions for enhanced insect performance. The authors identified six general effects of drought that may increase the attractiveness of plants or improve the environment for insect development and reproduction: (1) elevation of temperatures during drought furnishes a more favorable environment for growth of insects, (2) water-stressed plants are more sensorially attractive or acceptable to insects, (3) water stress renders a plant more physiologically suitable to insects, (4) drought promotes the function of insect capacities for detoxification of deleterious chemicals, (5) drought may create conditions favorable to insect mutualistic microorganisms but not to natural enemies, and (6) drought may induce adaptive genetic changes in insects. The type of water stress imposed on a plant and its timing relative to the insect life cycle also may have significant impacts on plant susceptibility. Mopper and Whitham (1992) concluded that Colorado pinyon pine trees would be most prone to drought-related increases in attacks by pinyon sawfly if trees were subjected to sustained moderate water stress or if a drought was broken at the point of peak insect feeding. Under these conditions, the nutritional value of tissues would be enhanced and the amount of defensive compounds would be reduced. Sustained severe stress would reduce the suitability of pinyon pine trees for sawflies.

Attacks by some insects may physiologically predispose woody plants to attacks by other insects. For example, defoliation by the Douglas fir tussock moth (*Orygia pseudotsuga*) increased the susceptibility of grand fir to the fir engraver beetle (*Scolytus ventralis*) (Wright *et al.*, 1979, 1984). Plant diseases also may predispose trees to attack by bark beetles. After fires, outbreaks of the mountain pine beetle (*Dendroctonus ponderosae*) in lodgepole pine forests occurred mainly in old trees that were weakened by the fungus *Phaeolus schweinitzii* (Geiszler *et al.*, 1980).

Defense Mechanisms   Woody plants may be defended against insects by both physical and chemical mechanisms. Physical defenses may involve waxes, cutin, and suberin that make the surface slippery and camouflage taste. Cellulose, lignin, and Ca increase the toughness of leaves (Schowalter *et al.*, 1986). Woody plants also produce compounds containing terpenes, phenols, tannins, and other substances that are toxic or inhibitory to insects or interfere with their digestive processes. Plants also may respond to insect attack by producing chemicals that normally are not toxic but inhibit growth and development and reduce fecundity of insects. For example, trees may produce tannins that bind with proteins and carbohydrates, making plant tissues indigestible, or proteolytic enzyme inhibitors that interfere with digestion. Wounding caused by insect feeding may induce expression of genes coding for protein inhibitors of digestive enzymes of insects (Chapter 9). By tying up N, woody plants also may decrease its availability to feeding insects or produce very fibrous needles that are not readily eaten or digested (Berryman, 1986). It also has been claimed that defensive chemicals are synthesized in trees that are not actually attacked but that are located close to trees defoliated by insects (Rhoades, 1983; Baldwin and Schultz, 1983).

Although the nutritional quality of leaves influences the choice of food for defoliating insects, both chemical and structural defenses of plants are more important as determinants of palatability of plant tissues. For example, large amounts of saponins in young leaves of various species of trees deter colonization by the southern red mite (*Ologonychus ilicis*), fall webworm (*Hypantria cuvea*), and eastern tent caterpillar (*Malacasoma americanum*) (Potter and Kimmerer, 1989). Young leaves of jack pine and creosote bush contain high concentrations of resins or resin acids that have adverse effects on insects (All and Benjamin, 1975).

Rapid-growing species on good sites commonly undergo greater injury from defoliating insects and have lower amounts and different types of defensive chemicals than slow-growing species, especially those on poor sites. In a rain forest, for example, fast-growing species of trees were eaten as much as six times faster by insects as slow-growing species (Coley *et al.*, 1985). The concentration of defensive chemicals in leaves of slow-growing species on poor sites sometimes was twice as high as in fast-growing species on good sites (Feeny, 1976; Coley, 1983).

Both the rate of flow and chemical composition of monoterpenes play a complex role in both colonization of bark beetles and resistance of trees to attack by beetles. Monoterpenes may repel or attract insects and associated fungi depending on the rate of monoterpene flow, composition of the resin, and the specific insect and associated fungus involved (Johnson and Croteau, 1987).

Copious resin flow was associated with resistance of ponderosa pine trees to attacks by bark beetles (Vité, 1961). Abundant production of oleoresins by resin ducts repelled or "pitched out" invading bark beetles. The flow of oleoresins decreased during droughts and increased the susceptibility of trees to attacks. The success of bark beetles was favored by low oleoresin exudation pressure (OEP). Initial attacks occurred at random, but only beetles in trees with low OEP made successful attacks that subsequently led to mass attacks. Oleoresin exudation pressure decreased as water stress in trees increased. Hence, drought predisposed trees to attacks by bark beetles.

Some oleoresins also attract insects. Thinning of a Douglas fir stand increased the abundance of insect vectors of black stain root disease, including *Hylastes nigrinus*, *Pissodes fasciatus*, and *Steremnius carinata*. The increase was attributed to attraction of the insects to resin and the dead inner bark of slash (Harrington and Reukema, 1983; Witkosky *et al.*, 1986).

During colonization, bark beetles are sequentially exposed to oleoresins with different monoterpene compositions. Initially the primary resins present in cortical blisters and resin ducts may repel pioneer beetles. However, wounding by beetle attacks stimulates production of parenchyma tissue that secretes new resin. Subsequent invasion by fungi is followed by transport to the wound site of additional resin from the sapwood at some distance (Raffa and Berryman, 1983). The monoterpenes of grand fir that repel the fir engraver beetle (*Scolytus ventralis*) and are toxic to the associated fungus, *Trichosporium symbioticum*, occur primarily in induced resin rather than in the primary resin (Russell and Berryman, 1976).

Some monoterpenes attract bark beetles. Whereas limonene repels the western pine beetle, α-pinene acts as the aggregation hormone for the same insect (Raffa *et al.*, 1985). The importance of intraspecific variation in monoterpene composition to resistance of ponderosa pine to bark beetle attack is based on the following evidence (Smith, 1966a,b): (1) monoterpenes of ponderosa pine differ in their toxicity to *Dendroctonus brevicomis* beetles, (2) differences occur in monoterpene compositions of unattacked ponderosa pine trees and those killed by bark beetles, and (3) the monoterpene composition of ponderosa pine varies widely.

Diseases

Disease symptoms on woody plants differ greatly and may be expressed as color changes, necrosis, vein clearing, wilting, and leaf spots on leaves; atrophy, hypertrophy, and rotting of tissues; and dieback or abscission of plant parts. Malformations such as cankers, galls, intumescences, witches'-brooms, rosettes, fasciation, and leaf crinkling or rolling usually indicate a diseased condition. Such symptoms commonly are associated with inhibition of plant growth and mortality (Tattar, 1978).

Plant diseases strongly influence the structure and composition of forests, plant succession, and biodiversity. The effects are exerted largely through death of trees, which may occur at the small scale (gap phase) or the large scale (forest development). There is much evidence of the effects of pathogen-induced mortality on structure of forests. For example, white pine blister rust, caused by *Cronartium ribicola*, affects the distribution of eastern and western white pines. Chestnut blight, caused by *Cryphonectria parasitica*, eliminated chestnut trees in eastern North America. Root diseases, caused by *Armillaria* spp. and *Phellinus weirii* in the western United States, alter the composition of forests and lead to development of "root disease climaxes" (Van der Kamp, 1991). In Australia mortality of eucalypts, caused by root rots, alters open forests with a sclerophyllous understory to open forests with large gaps and a sedge-dominated ground flora (Weste, 1986). The model of Castello *et al.* (1995) integrates both a decline–disease spiral and gap phase dynamics of plants (Fig. 5.35).

The emphasis on pathogen-induced tree mortality should not obscure the fact that death of trees follows severe physiological dysfunctions that adversely affect carbohydrate, water, mineral, and hormonal relations of trees. Leaf diseases reduce photosynthesis through loss of photosynthetic tissue, damage to chloroplasts, inhibition of chlorophyll synthesis, altered metabolism, and combinations of these. In a number of vascular diseases resistance to water transport through infected xylem is much greater than

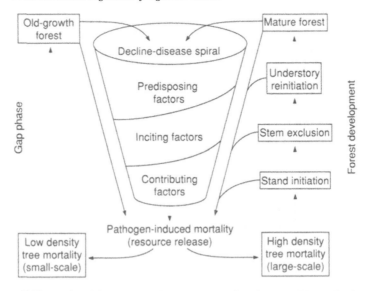

Figure 5.35   Model of the role of pathogens in mortality of trees and forest development. Certain pathogens are considered effective agents of mortality at some stages of forest development, others at many stages, and still others as predisposing, inciting, or contributing factors in tree mortality. The upward arrows indicate direction through forest development, and downward arrows indicate mortality. From Castello *et al.*, © 1995 American Institute of Biological Sciences.

through uninfected xylem. Obstructions in the xylem include the mycelium of fungal pathogens, cells of bacteria, materials that result from partial breakdown of host tissues, materials secreted by the pathogen or host, and structures formed by renewed growth of living cells in the xylem (Talboys, 1978). Gums and tyloses often obstruct the water-conducting vessels in the xylem. In such diseases as chestnut blight, Dutch elm disease, and oak wilt, injury is associated with embolized vessels (Zimmermann and McDonough, 1978). Diseases influence the mineral nutrition of plants by causing mineral deficiency at various sites in the plant, impaired utilization of minerals, or toxicity resulting from abnormal accumulation around the site of invasion. Nutrient deficiency of the host may occur as a result of immobilization in the soil, decreased absorption from the soil, utilization by the pathogen, impaired nutrient transport, accumulation in nutrient sinks around infection sites, or loss by exudation (Huber, 1978). Some symptoms of disease are associated with altered hormone relations. Deformation of peach leaves caused by *Taphrina deformans*, hypertrophy of crown gall, root deformation by clubroot, and galls produced on trees by *Cronartium* have been attributed to high auxin levels (Kramer and Kozlowski, 1979).

Predisposition of Plants to Disease   Environmental stresses often predispose woody plants to disease. Outbreaks of stem cankers, diebacks, declines, and some root rots follow loss of tree vigor as a result of environmental stresses. Probably the two most common predisposing stresses are drought and freezing, but flooding, low soil fertility, air pollution, and combinations of these also are important (Kozlowski, 1985b). Woody plants often are invaded by fungi regardless of prevailing stresses, but the invading organisms generally are not pathogenic until the host plant has been physiologically predisposed by stress (Schoeneweiss, 1978a; Houston, 1984).

Dieback of ash was induced by drought, followed by production of cankers caused by *Cytophoma pruinosa* and *Fusicoccum* sp. Although the fungi were present in the bark of unstressed trees, they induced cankers only after the trees were predisposed by drought (Silberborg and Ross, 1968). Susceptibility to disease is influenced by both the degree and duration of drought. When paper birch seedlings were exposed to drought their susceptibility to *Botryosphaeria dothidea* did not increase until the plant Ψ fell below -1.2 MPa, when susceptibility increased greatly and cankers formed within 4 days (Schoeneweiss, 1978b). In at least one case the increased production of a specific metabolite by plants under drought has been shown to be related to disease. Griffin *et al.* (1986) reported that the hyphal growth rate of *Hypoxylon mammatum*, a cause of canker disease of trembling aspen, a serious problem during drought, was stimulated by proline. As aspen produced more proline under drought, the enhanced growth of *H. mammatum* cankers in drought-stressed trees could be attributed to this increase. A decrease in antifungal compounds (e.g., catechol, salicin, and salicortin) also appeared to be important in drought-induced susceptibility of aspen to *H. mammatum* (Kruger and Manion, 1994).

Some leaf-infesting fungi such as powdery mildews are more troublesome in dry than in wet weather. Nevertheless, the spread of many fungi is increased by wet weather because leaf surfaces must remain wet for many hours to permit germination of spores of fungi such as *Venturia inaequalis*, which causes apple scab. The relationship of water stress to plant disease is discussed in detail by Schoeneweiss (1986).

Flooding of soil often increases root diseases because several pathogenic fungi grow vigorously in poorly aerated soil. Inundation of soil leads to death and decay of large portions of root systems, primarily because of increased activity of *Phytophthora* fungi which can tolerate low soil O_2 content (see pp. 229–231).

Freezing temperatures predispose trees to invasion by canker and dieback pathogens. For example, frost injury increased susceptibility of birch and spruce to *Nectria* (Gäumann, 1950) and of European white birch to *Botryosphaeria dothidea* (Crist and Schoeneweiss, 1975). Warm temperatures also may predispose woody plants to disease. Hence, woody plants that are adapted to cool climates often become more susceptible to disease when the temperature rises. As the temperature increased from 16 to 26°C, elm trees became increasingly susceptible to Dutch elm disease (Birkholz-Lambrecht *et al.*, 1977). At very high temperatures, the increased susceptibility of woody plants to certain pathogens is associated with adaptation of the pathogen to high temperature (Bell, 1981).

Defense Mechanisms   Woody plants can protect themselves from disease by establishing mechanical barriers to disease-causing organisms and by producing a variety of defensive chemicals. When host tissues are invaded by pathogenic fungi, the host often responds by differentiating barriers of cork cells. Certain nonpathogenic fungi can overcome such barriers and are followed by pathogenic microorganisms. When protective barriers are overcome, changes in the host are induced that compartmentalize the wounded tissue (Shigo, 1984).

Woody plants produce secondary compounds such as phenols, coumarins, tannins, and lignins that provide protection against disease-causing organisms (Feeny, 1976; see Chapter 7 of Kozlowski and Pallardy, 1997). Lignified cell walls impede progression of microorganisms as shown by slow biodegradation of highly lignified tissues (Walter, 1992). As indicated by Ride (1978), lignification may impede growth of fungi through plant tissues by (1) rendering cell walls resistant to mechanical penetration, (2) increasing resistance of cells to dissolution by fungal enzymes, (3) restricting diffusion of fungal toxins, (4) inactivating fungal membranes, enzymes, toxins, and elicitors by phenolic precursors of lignin, and (5) losing plasticity that is necessary for growth of the fungal tip when tissue becomes lignified.

Both the amount and diversity of defensive chemicals produced are higher in tropical than in temperate-zone plants, and they are higher in plants of wet than of arid regions (Langenheim, 1984). In some species of plants a large pool of defensive chemicals is maintained during much of the life of the plant. In other species defensive chemicals, called phytoalexins, are present in trace amounts and accumulate rapidly in response to infection (Creasy, 1985). Substances of pathogen origin known as elicitors regulate production of phytoalexins. Elicitors appear to be recognized by plant cells through interactions on plant plasma membranes. The elicitor–receptor interactions presumably generate signals that activate the nuclear genes involved in synthesis of phytoalexins. The details of the sequence of molecular events involved in plant defense reactions are discussed by Yoshikawa *et al.* (1993). Disease resistance does not appear to be conferred by a single phytoalexin but rather by interactions involving accumulation and detoxification of phytoalexins, growth of the pathogen, production of microbial toxins, and creation of barriers that resist injury by pathogens

(Kuc, 1976). Some species of woody plants resist pathogenic organisms through increased wood density, deposits of defensive chemicals in wood, and compartmentalization of tissues at wound sites (Loehle, 1988a,b).

Summary

The major environmental stresses that influence distribution and growth of woody plants include shading, water deficits, flooding of soil, temperature extremes, low soil fertility, salinity, soil compaction, pollution, wind, fire, insects, and diseases. Growth inhibition is an integrated response to multiple environmental stresses rather than to a single stress.

Light affects growth of plants through its intensity, quality, and duration. In stands of trees, the amount of light available to trees of the lower canopy is only a small fraction of the amount to which canopy trees are exposed. In many individual trees, the light intensity decreases rapidly with increasing depth of the tree crown.

Shading variously inhibits bud development, expansion of leaves and internodes, cambial growth, and root growth. The structure of leaves grown in the sun varies appreciably from that of leaves grown in the shade. Sun-grown leaves usually are smaller and thicker and have more stomata per unit area, longer palisade cells, often more than one layer of palisade cells, and more chlorophyll per unit of leaf area. The mechanism controlling shade tolerance of different species of plants is complex. An important adaptation is capacity to maintain high rates of photosynthesis in the shade, but drought tolerance and disease resistance also are involved.

Day length (photoperiod) affects many plant responses, including seed germination, bud dormancy, leaf growth, internode elongation, leaf shedding, and development of cold hardiness. Exposure to short days stops shoot elongation of many species. Nevertheless, short days are not the most important environmental factors that account for growth cessation of species in which shoot elongation stops well before the days begin to shorten.

Under natural conditions the growth of woody plants is influenced less by light quality than by light intensity or day length. However, light quality is important in regulating growth of plants under artificial lights. Excessive stem elongation may occur under incandescent lamps and is below normal under cool white or daylight-type fluorescent lamps.

Species composition and productivity of plant communities are sensitive to both water deficits and excesses. Although plants obtain most water from the soil, they may also absorb small amounts of liquid water, water vapor, and dew from the atmosphere. The water needs of woody plants are partly fulfilled by water stored in leaves, stems, roots, and fruits and cones, with stems being the most important water reservoirs.

The species composition of forests depends on the annual amount of precipitation, its variability between years, and the ratio of precipitation to evaporation. Broad-leaved deciduous forests of the temperate zone usually occur where annual rainfall varies between 50 and 200 cm and is evenly distributed during the year. Conifer forests and sclerophyll evergreen hardwoods are favored by rainy winters and long summer droughts. In the tropics, plant communities vary from deserts to lush rain forests depending on the amount and seasonal distribution of rainfall.

Flooding of soil variously inhibits seed germination, shoot growth, cambial growth, and root growth. Death and decay of roots in flooded soils occur largely as a result of increased activity of *Phytophthora* fungi. Soil changes associated with flooding include decreases in O_2 and increases in CO_2, replacement of aerobic by anaerobic microorganisms, and increased accumulation of potentially toxic compounds. Plant responses to flooding vary with plant species and genotype, age of plants, condition of the floodwater, and time and duration of flooding. Adaptations to flooding may include production of lenticels on submerged stems and roots as well as formation of aerenchyma tissues, pneumatophores, and adventitious roots.

Low temperatures may injure woody plants, limit their distribution, and regulate their growth. To a large extent low temperature determines the latitudinal and altitudinal limits of tree growth. Temperature also influences bud development, release and induction of bud dormancy, expansion of leaves and internodes, seasonal initiation and cessation of production of cambial derivatives, and root growth. Studies of effects of temperature on cambial growth are complicated by thermal shrinkage and expansion of stems.

Both vegetative and reproductive tissues of tropical and subtropical plants, and a few temperate-zone plants, are injured by temperatures below approximately 15°C but above freezing. Chilling injury is characterized by reduced photosynthesis, necrotic lesions, susceptibility to decay organisms, changes in membrane permeability, and ultrastructural changes. Sensitive plants may show chilling injury at all stages of development except the dry seed stage. The most widely accepted explanation for chilling injury is that a rapid change in the physical state of plant membranes induces several secondary changes that lead to visible injury.

Freezing temperatures injure woody plants as shown by killing back of shoots, frost cracks, frost rings, and injury to roots. Freezing injury can be caused by intracellular freezing of cell contents or by dehydration of tissues resulting from extracellular freezing, with the latter mechanism predominating. Freezing of plant tissues often is prevented or postponed by supercooling (undercooling). A decreased water content and increased concentration of osmotically active compounds in the cell sap lower the freezing point, decrease the threshold subcooling temperature, and prolong super-cooling.

Many plants that are native to warm regions cannot be moved to cold regions because they do not develop enough hardiness to cold. Acclimation to cold varies widely among species, genotypes, and different organs and tissues on the same plant. In general, reproductive tissues are more sensitive than vegetative tissues to cold. Cold hardiness of temperate-zone plants develops in two stages. The first stage occurs at temperatures between 10 and 20°C when carbohydrates and lipids accumulate. These compounds are energy sources for metabolic changes that occur in the second stage which is promoted by freezing temperatures.

Winter injury of evergreens often results from desiccation of leaves rather than from direct thermal effects. Winter desiccation injury occurs when the soil is cold or frozen and absorption of soil water is too slow to replace transpirational losses.

Growth of plants may be increased or decreased by high temperature. There is much concern that increases in greenhouse gases (e.g., CO_2, N_2O, methane, chlorofluorocarbons) in the atmosphere may raise global temperatures enough to induce changes in species ranges, composition, and plant productivity, melting of glaciers, and flooding of low-lying coasts with salt water.

The extent of the effects of global warming on woody plants is controversial and will depend on how much the temperature will increase. Much work with seedlings in controlled environments with natural photoperiods and light intensity and with temperature regimes like those out of doors shows that elevation of CO_2 above ambient levels increases growth of seedlings. The stimulation of growth is associated with increased rates of photosynthesis as well as improved mineral relations and water use efficiency. Some studies with isolated older trees also show growth stimulation by CO_2 enrichment of the air. Other studies show wide variations among species in response to elevated CO_2. In the long term, growth stimulation by increased CO_2 may be less dramatic in forests than in isolated trees because enhanced growth will ultimately lead to closed canopies in forests and more shading than in open-grown trees. Ecosystems very likely will differ in response to global warming depending on water supply, soil fertility, and temperature conditions.

Woody plants often are injured by high temperatures below the thermal death point. Heat injury includes scorching of leaves, sunscald, leaf abscission, and death of plants. High-temperature injury varies with the injurious temperature and duration of exposure. Injury by heat may involve starvation, toxicity, protein breakdown, and damage to cell membranes. Woody plants can adapt to heat as shown by seasonal variations in heat tolerance (low in the spring, high in the summer). When plant tissues are rapidly exposed to temperatures 5 to 10°C above normal growth temperatures, heat-shock proteins are produced. The heat-shock proteins may protect cells from heat injury.

In many parts of the world, the productivity of woody plants is reduced by low soil fertility, as emphasized by increases in growth following addition of fertilizers in nurseries, forest stands, and fruit orchards. The impacts of mineral deficiencies are accentuated by flooding, forest fires, whole-tree harvesting, and harvesting of trees on short rotations.

Salinity because of existing salt deposits, salt spray along seacoasts, use of deicing salts, and salt in irrigation water adversely affects the growth and distribution of woody plants. Accumulation of excessive salts excludes woody plants from vast areas of land. Salt tolerance varies greatly among plant species and genotypes. Salts decrease the capacity of roots to absorb water, lower the rate of photosynthesis by stomatal and nonstomatal inhibition, decrease total photosynthesis per plant as a result of small leaf area, decrease absorption of mineral nutrients, and injure leaves.

Compaction of soils by pedestrian traffic, grazing animals, and heavy machinery adversely affects growth of plants by decreasing soil aeration, developing soil crusts that inhibit water infiltration and increase surface runoff, and impeding penetration of soils by roots. Forest soils with their loose and friable structure are very susceptible to compaction.

Many gaseous pollutants (e.g., SO_2, O_2, fluorides, nitrogen oxides, peroxyacetylnitrates) and a variety of particulates adversely affect growth of woody plants. Plants also release a variety of chemicals that inhibit growth. The effects of pollutants on plant communities range from negligible to severe. High dosages of pollutants affect forest stands by reducing structural complexity, biomass productivity, and species diversity. However, in mixed forest stands the growth of some species may be stimulated if they are given a competitive advantage by a greater inhibitory effect of pollution on other species. The responses of plants to pollution vary with plant species and genotype, type and combinations of pollutants, pollutant dosage, stage of plant development, plant response measured, and interactions with a number of abiotic and biotic environmental stresses.

The effects of acid precipitation on growth of woody plants have been shown to be both harmful and beneficial. In some short-term studies, acid precipitation inhibited seed germination, injured leaf cells, inhibited seedling growth, and decreased frost tolerance. In other studies, acid precipitation stimulated growth of woody plants as a result of increased N availability. The beneficial short-term effects of acid precipitation may not carry over to the long term, to a considerable extent because of eventual nutritional imbalances of plants associated with accelerated leaching of cations from the soil and adverse effects on development of mycorrhizae.

Wind has both harmful and beneficial effects on plants. The harmful effects include breaking and uprooting of trees, injuring leaves, eroding soils, dispersing pollutants, and increasing risks of insect attack, disease, and fire. Beneficial effects of wind include dispersal of pollen and seeds. Belts of trees as windbreaks can be used to decrease soil erosion, reduce injury to plants, increase crop yields, and also improve cover and increase food supplies for wildlife.

The effects of fire on woody plants vary from catastrophic to beneficial. Fire affects plants by destroying them and by altering site quality by burning of organic matter and heating the soil surface. Plant succession to a climatic climax stage commonly is prevented by fire. Hence, many forest stands are maintained in a subfinal fire-climax stage. Fires predispose trees to attacks by certain insects. Fires also control some plant diseases but spread others. Adaptations of fire-resistant trees include thick bark, serotinous cones, and high capacity for sprouting after fire.

Insects have both harmful and beneficial (e.g., dispersing pollen) effects on woody plants. Defoliating insects decrease the amounts of carbohydrates and hormonal growth regulators available for growth. Phloem-boring insects interfere with transport of carbohydrates to meristematic regions and spread disease-causing organisms. Insects that feed on roots decrease absorption of water and minerals. Seed and cone insects injure reproductive structures. Growth and proliferation of insects in environmentally stressed trees have been attributed to improved nutrition for the insect, lowered chemical defenses of plants, and a better environment for insect growth. Environmental stresses commonly predispose plants to insect attack. Woody plants protect themselves against insects by producing compounds that are toxic to insects or interfere with their digestive processes.

Disease symptoms associated with reduced growth and mortality of woody plants include necrosis, leaf spots, atrophy, hypertrophy, dieback, leaf abscission, cankers, galls, and intumescences. A variety of environmental stresses predispose plants to disease. Woody plants can protect themselves from disease by creating mechanical barriers to disease-causing organisms and producing defensive chemicals such as phenols, coumarins, tannins, and lignins.

General References

Anderson, L. S., and Sinclair, F. L. (1993). Ecological interactions in agroforestry systems. *For. Abstr.* **54**, 489–523.

Baker, N. R., and Bowyer, J. R., eds. (1994). "Photoinhibition of Photosynthesis: From Molecular Mechanisms to the Field." Bios Scientific, Oxford.

Barker, J. R., and Tingey, D. T., eds. (1992). "Air Pollution and Effects on Biodiversity." Van Nostrand-Reinhold, New York.

Bernays, E. A. (1992). "Insect–Plant Interactions," Vol. 4. CRC Press, Boca Raton, Florida.

Berryman, A. A. (1986). "Forest Insects. Principles and Practice of Population Management." Plenum, New York and London.

Binzel, M. L., and Reuveni, M. (1994). Cellular mechanisms of salt tolerance in plant cells. *Hortic. Rev.* **16**, 33–69.

Botkin, D. B. (1993). "Forest Dynamics: An Ecological Model." Oxford Univ. Press, Oxford and New York.

Bowen, G. D., and Nambiar, E. K. S. (1984). "Nutrition of Plantation Forests." Academic Press, London.

Burns, R. M., and Honkala, B. H., eds. (1990). "Silvics of North America." U.S. Dept. of Agriculture Handbook No. 654, Washington, D.C.

Caldwell, M. M., and Pearcy, R. W., eds. (1994). "Exploitation of Environmental Heterogeneity by Plants: Ecological Processes Above- and Belowground." Academic Press, San Diego.

Ceulemans, R., and Mousseau, M. (1994). Effects of elevated atmospheric CO_2 on woody plants. *New Phytol.* **127**, 425–446.

Cherry, J. H., ed. (1989). "Environmental Stress in Plants." Springer-Verlag, Berlin and New York.

Christensen, N. L. (1995). Fire ecology. *In* "Encyclopedia of Environmental Biology" (W. A. Nierenberg, ed.), Vol. 2, pp. 21–32. Academic Press, San Diego.

Coulson, R. N., and Witter, J. A. (1984). "Forest Entomology: Ecology and Management." Wiley (Interscience), New York and Chichester.

Coutts, M. P., and Grace, J., eds. (1995). "Wind and Trees." Cambridge Univ. Press, Cambridge and New York.

Crawford, R. M. M. (1989). "Studies in Plant Survival." Blackwell, Oxford.

Fitter, A. H., and Hay, R. K. M. (1987). "Environmental Physiology of Plants." Academic Press, London and San Diego.

Goulet, F. (1995). Frost heaving of forest tree seedlings: A review. *New Forests* **9**, 67–94.

Graham, R. L., Turner, M. G., and Dale, V. H. (1990). How increasing CO_2 and climate change affect forests. *BioScience* **40**, 575–587.

Hart, J. W. (1988). "Light and Plant Growth." Unwin Hyman, London.

Heinrichs, E. A., ed. (1988). "Plant Stress–Insect Interactions." Wiley (Interscience), New York.

Hennessy, T., Dougherty, P., Kossuth, S., and Johnson, J., eds. (1986). "Stress Physiology and Forest Productivity." Martinus Nijhoff, Boston.

Horsfall, J., and Cowling, E., eds. (1977). "Plant Pathology—An Advanced Treatise," Vols. 1–3. Academic Press, New York.

Idso, E. A. (1989). "Carbon Dioxide and Global Change: Earth in Transition." IBR Press, Tempe, Arizona.

Jones, H. G. (1992). "Plants and Microclimate," 2nd Ed. Cambridge Univ. Press, Cambridge.

Jones, H. G., Flowers, T. J., and Jones, M. B., eds. (1989). "Plants Under Stress." Cambridge Univ. Press, Cambridge.

Katterman, F., ed. (1990). "Environmental Injury to Plants." Academic Press, San Diego.

Kimmins, J. P. (1987). "Forest Ecology." MacMillan, New York.

Kozlowski, T. T., ed. (1968–1983). "Water Deficits and Plant Growth." Vols. 1–7. Academic Press, New York.

Kozlowski, T. T., ed. (1984). "Flooding and Plant Growth." Academic Press, Orlando, Florida.

Kozlowski, T. T., and Ahlgren, C. E. (1974). "Fire and Ecosystems." Academic Press, New York.

Kozlowski, T. T., Kramer, P. J., and Pallardy, S. G. (1991). "The Physiological Ecology of Woody Plants." Academic Press, San Diego.

Kramer, P. J., and Boyer, J. S. (1995). "Water Relations of Plants and Soils." Academic Press, San Diego.

Lee, R. E., Jr., Warren, G. J., and Gusta, L. V., eds. (1995). "Biological Ice Nucleation and Its Applications." American Phytopathological Society, St. Paul, Minnesota.

Levitt, J. (1980). "Responses of Plants to Environmental Stresses," Vols. 1 and 2. Academic Press, New York.

Li, P. H., and Christersson, L., eds. (1992). "Advances in Plant Cold Hardiness." CRC Press, Boca Raton, Florida.

Lindquist, S., and Craig, E. A. (1988). The heat shock proteins. *Annu. Rev. Genet.* **22**, 631–677.

Little, C. E. (1995). "The Dying of the Trees." Viking Penguin, New York.

Long, S. P., and Woodward, F. I., eds. (1988). "Plants and Temperature." Dept. of Zoology, Univ. of Cambridge, Cambridge, England.

Lucier, A. A., and Haines, S. G., eds. (1990). "Mechanisms of Forest Responses to Acidic Deposition." Springer-Verlag, Berlin and New York.

McKersie, B. D., and Leshem, Y. Y. (1994). "Stress and Stress Coping in Cultivated Plants." Kluwer, Dordrecht, The Netherlands.

Manion, P. D. (1991). "Tree Disease Concepts." Prentice-Hall, Englewood Cliffs, New Jersey.

Manion, P. D., and Lachance, D., eds. (1992). "Forest Decline Concepts." American Phytopathological Society, St. Paul, Minnesota.

Mansfield, T., and Stoddart, J., eds. (1993). "Plant Adaptation to Environmental Stress." Chapman & Hall, New York.

Mitch, W. J., ed. (1994). "Global Wetlands: Old World and New." Elsevier, Amsterdam and New York.

Mitch, W. J., and Gosselink, J. G. (1994). "Wetlands," 2nd Ed. Van Nostrand-Reinhold, New York.

Mitchell, C. P., Ford-Robertson, J. B., Hinckley, T., and Sennerby-Forsse, L., eds. (1992). "Ecophysiology of Short Rotation Forest Crops." Elsevier Applied Science, London and New York.

Mooney, H. A., Winner, W. E., and Pell, E. J., eds. (1991). "Response of Plants to Multiple Stresses." Academic Press, San Diego.

Mudd, J. B., and Kozlowski, T. T., eds. (1973). "Responses of Plants to Air Pollution." Academic Press, New York.

Nover, L. (1991). "Heat Shock Response." CRC Press, Boca Raton, Florida.

Nover, L., Neumann, D., and Scharf, K.-D. (1989). "Heat Shock and Other Stress Response Symptoms of Plants." Springer-Verlag, Berlin.

Peters, R. L., and Lovejoy, T. E., eds. (1992). "Global Warming and Biological Diversity." Yale Univ. Press, New Haven, Connecticut.

Porter, J. R., and Lawlor, D. W. (1991). "Plant Growth: Interactions with Nutrition and Environment." Cambridge Univ. Press, Cambridge.

Rizvi, S. J. H., and Rizvi, V., eds. (1992). "Allelopathy: Basic and Applied Aspects." Chapman & Hall, London and New York.

Sakai, A., and Larcher, W. (1987). "Frost Survival of Plants." Springer-Verlag, Berlin and New York.

Schaffer, B., and Andersen, P. C., eds. (1994). "Handbook of Environmental Physiology of Fruit Crops." CRC Press, Boca Raton, Florida.

Scholz, R., Gregorius, H.-R., and Rudin, D., eds. (1989). "Genetic Effects of Air Pollutants in Forest Tree Populations." Springer-Verlag, New York.

Schulte-Hostede, S., Darrall, N. M., Blank, L. W., and Wellburn, A. R., eds. (1988). "Air Pollution and Plant Metabolism." Elsevier, London and New York.

Schulze, E.-D., Lange, O. L., and Oren, R., eds. (1989). "Forest Decline and Air Pollution." Springer-Verlag, Berlin.

Shannon, M. C., Grieve, C. M., and Francois, L. E. (1994). Whole-plant response to salinity. *In* "Plant–Environment Interactions" (R. E. Wilkinson, ed.), pp. 199–244. Dekker, New York.

Smith, W. H. (1990). "Air Pollution and Forests," 2nd Ed. Springer-Verlag, New York.

Smith, W. H. (1995). Air pollution and forests. *In* "Encyclopedia of Environmental Biology" (W. A. Nierenberg, ed.), Vol. 1, pp. 37–51. Academic Press, San Diego.

Smith, W. K., and Hinckley, T. M., eds. (1995a). "Ecophysiology of Coniferous Forests." Academic Press, San Diego.

Smith, W. K., and Hinckley, T. M., eds. (1995b). "Resource Physiology of Conifers." Academic Press, San Diego.

Solomon, A. M., and Shugart, H. H., eds. (1993). "Vegetation Dynamics and Global Change." Chapman & Hall, New York and London.

Staples, R. C., and Toenniessen, G. J., eds. (1984). "Salinity Tolerance in Plants; Strategies for Crop Improvement." Wiley, New York.

Taylor, G. E., Jr., Pitelka, L. F., and Clegg, M. T., eds. (1991). "Ecological Genetics and Air Pollution." Springer-Verlag, New York.

Taylor, G. E., Jr., Johnson, D. W., and Andersen, C. P. (1994). Air pollution and forest ecosystems: A regional to global perspective. *Ecol. Appl.* **4**, 662–689.

Teller, A., Mathy, P., and Jeffers, J. N. R., eds. (1992). "Responses of Forest Ecosystems to Environmental Changes." Elsevier, New York.

Trabaud, L., ed. (1987). "Fire Ecology." SPB, The Hague.

Treshow, M., ed. (1984). "Air Pollution and Plant Life." Wiley, New York and Chichester.

Udger, W. N., and Brown, K. (1995). "Land Use and the Causes of Global Warming." Wiley, Chichester and New York.

Ulrich, B., and Pankrath, J., eds. (1983). "Effects of Accumulation of Air Pollutants on Forest Ecosystems." Reidel, Dordrecht, The Netherlands.

Ungar, I. A. (1991). "Ecophysiology of Vascular Halophytes." CRC Press, Boca Raton, Florida.

van den Driessche, R., ed. (1991). "Mineral Nutrition of Conifer Seedlings." CRC Press, Boca Raton, Florida.

Vince-Prue, D. (1975). "Photoperiodism and Plants." McGraw-Hill, Maidenhead, England.

Wang, C. Y. (1990). "Chilling Injury of Horticultural Crops." CRC Press, Boca Raton, Florida.

Waring, R. H., and Schlesinger, W. H. (1985). "Forest Ecosystems: Concepts and Management." Academic Press, Orlando, Florida.

Whelan, R. J. (1995). "The Ecology of Fire." Cambridge Univ. Press, Cambridge.

Wild, A. (1989). "Russell's Soil Conditions and Plant Growth." Wiley, New York.

Wilkinson, R. E., ed. (1994). "Plant–Environment Interactions." Dekker, New York.

Woodward, F. I. (1987). "Climate and Plant Distribution." Cambridge Univ. Press, Cambridge.

6

Environmental Regulation of Reproductive Growth

Introduction

Several environmental factors interact to influence reproductive growth, both indirectly by altering physiological processes (especially carbohydrate and hormone relations) and directly, as by thermal injury. Because of close correlations between vegetative growth and reproductive growth in adult trees, any environmental factor that influences vegetative growth may eventually have some effect on reproductive growth. Among the most important of these environmental factors are light intensity, water supply, temperature, soil fertility, salinity, and pollution.

Light Intensity

Much interest has been shown in the effects of light intensity on reproductive growth because formation of flower buds, fruit set, and size and quality of fruits are reduced by shade. Furthermore, abscission of flowers and fruits is stimulated by shading. Information on the effects of light intensity on reproductive growth of various species and cultivars is essential to managers of fruit and nut orchards for making decisions about pruning of branches to control tree form, choosing rootstocks to control tree size, and spacing of trees; and to forest managers in determining methods and timing of stand thinning and pruning of branches, particularly in seed orchards.

Flowering as well as fruit initiation and growth are influenced by light intensity. In deciduous forests many shaded understory plants produce leaves and flowers when irradiance of the understory is maximal (Janzen, 1972). For example, phenologies of trees from eight widely distributed tropical forests indicated that production of leaves and flowers coincided with seasonal peaks of irradiance (Wright and van Schaik, 1994). Shading of trees reduced flower bud initiation in apple (Table 6.1) and sour cherry

**Table 6.1   Effect of Artificial Shading of Apple
Trees in 1970 on Number of Flower Buds
per Tree in 1971 [a]**

Light intensity of outer crown [b]	Number of flower buds
100	159
37	96
25	69
11	33

[a]From Pereira (1975).
[b]As percentage of full sunlight.

(Flore, 1981). Low light intensity within the canopy also inhibited initiation of pistillate flowers in kiwi (Grant and Ryugo, 1984). The shaded branches of walnut trees initiated staminate rather than pistillate flowers (Ryugo *et al.*, 1980, 1985), further emphasizing the strong effects of shade on reproductive growth.

Fruit growth is influenced by the local light climate within a tree crown, largely because the carbohydrates used by growing fruits come from leaves relatively nearby (Kozlowski, 1992). In many fruit trees only the outer crown receives enough light to produce fruits of high quality. In very large Red Delicious apple trees the light intensity decreased to 42% of full sunlight at a depth of 2 m from the top of the tree (Heinicke, 1963, 1964). The light intensity in the center of a 4 m high Cox apple tree decreased from 95 to 34% of full sunlight within 1 m (Jackson, 1970). Penetration of light into the crowns of apple trees decreased rapidly as the leaves expanded during April and May. By mid-May, 70% of the total seasonal decrease in light intensity already had taken place (Porpiglia and Barden, 1980). Up to 90% of the light intercepted by citrus trees is absorbed by leaves in the outside meter of the tree crown. In some citrus trees, the light intensity in the crown interior may be less than 2% of the light intensity outside the crown (Monselise, 1951). In hedged peach trees, most absorption of light took place in the outer 25 cm of the crown (Kappel and Flore, 1983).

Crop Yield

Because of mutual shading within tree crowns, variously located fruits on the same tree are exposed to different microclimates and, therefore, vary in characteristics determined by the light intensity that reaches the fruit surface. Many experiments have shown that yield of fruits is reduced by low light intensity. Shading after completion of blossoming (therefore not influencing the number of flowers or their pollination and fertilization) decreased the number of apple fruits at harvest (Pereira, 1975). Jackson and Palmer (1977a,b) shaded Cox's Orange Pippin apple trees during the postblossom growing season in 1970 and 1971. Shading reduced fruit yield, the number of fruits, and mean fruit weight in the year of shading (Table 6.2). Shading also reduced flower bud formation and decreased the percentage of flowers that set fruits in the following year. The effects of shade on crop yield in the year following shading were at least as great as they were in the year of shading. Jackson *et al.* (1977) showed that shading reduced the size of apple fruits through reduction of cell size and the number of cells per fruit. Fruits grown under shade also had less dry matter and starch per unit of fresh weight. Concentrations of N, P, K, Ca, and Mg were similar in fruits of the same size produced in the shade or in full light, but the concentrations of Ca, N, and P were higher in small fruits than in large fruits. Most effects of canopy shading on fruit size occur relatively early in the growing season. Canopy shading effects on the final size of apple fruits were evident at 1 to 5 weeks after bloom.

Increasing the relative growth rate (RGR) of apple fruits by 5 or 10% during the 1- to 5-week stage of fruit development increased the weight of fruits by 20 to 43%, respectively. In comparison, increasing RGR by 10% from the time of normal summer pruning increased final fruit size by only 3% (Lakso *et al.*, 1989). This emphasized the importance of rapid early growth of fruits in increasing partitioning of carbohydrates into the fruits during the entire season.

Crop yield often is appreciably reduced by abscission of reproductive structures as a result of shading. For example, shading reduced retention of fruitlets of Cox (Jackson and Palmer, 1977a,b) and Delicious apple trees (Schneider, 1978). Shading of entire apple trees at 20 days after bloom induced most fruit abscission. The effect of shading decreased progressively at 27 and 34 days and was negligible at 41 days after bloom

Table 6.2 Effect of Shading on Fruit Yield per Tree in 1970 of Cox's Orange Pippin Apple Trees[a]

| | Light intensity (% of full daylight) | | | |
	100	37	25	11
Yield (kg)	14.27	11.96	8.87	6.31
Number of fruits	131.50	111.30	93.20	83.10
Mean fruit weight (g)	124.50	113.30	102.20	80.20

[a]From Jackson and Palmer (1977b).

(Kondo and Takahashi, 1987). Abscission of the entire crop of young fruits of cotton was induced by 3 days of darkness (Vaughan and Bate, 1977).

Fruit Quality

The light environment influences various aspects of fruit quality including fruit size, color, and concentration of soluble solids. In apples the most positive relation between light intensity and fruit quality usually is for soluble solids and fruit size (Seeley *et al.*, 1980; Robinson *et al.*, 1983). Shading reduces several aspects of fruit quality more than it reduces fruit production per hectare (Tromp, 1984b; Palmer, 1989).

The effect of light on fruit color varies appreciably with plant species and variety. High light intensities are not required for high-coloring strains of Delicious apples but are essential for other strains of Delicious and McIntosh apples as well as for peaches. In comparison, only low light intensities are necessary for color development in sweet cherries (Patten and Proebsting, 1986).

The color of apples results from a blend of several pigments including anthocyanins, flavonols, carotenoids, and chlorophyll. The red color is produced by cyanidin glycosides copigmented with flavonols and other compounds (Lancaster, 1992).

Much attention has been given to anthocyanin synthesis in developing apples. Young apples (~2 mm in diameter) have a background green color with a temporary peak of red color that disappears early but reappears during fruit ripening. In Orange Pippin apples the amount of anthocyanin pigments tripled during the month of ripening; chlorophyll concentration decreased fourfold; and carotenoids increased fourfold (Knee, 1972). In Golden Delicious apples both chlorophyll and carotenoids decreased during ripening and subsequently increased to their original values by harvest time (Gorski and Creasy, 1977).

Production of anthocyanin pigments in many apple cultivars is influenced by interactions of several factors, including light intensity, temperature, N supply, rootstock, and such applied chemicals as ethephon and daminozide (Saure, 1990). However, light intensity is considered the most important environmental variable because anthocyanin synthesis is light dependent. Hence, apples kept in the dark or exposed to very low light intensities do not develop red color. Once anthocyanin synthesis is triggered by light, the pigment continues to accumulate in the dark. Suppression of red pigmentation by shading has been well documented in apples (Seeley *et al.*, 1980; Robinson *et al.*, 1983). In Delicious apples both the intensity of red color and soluble solids content were progressively higher as the light intensity was increased (Campbell and Marini, 1992).

Two separate components of the action of light on anthocyanin production have been identified (Saure, 1990). These include (1) a red–far

re light than triangular trees of the same height and basal diameter.
vertheless, the distribution of light in the former is poorer because the
s are saturated with light and the sides are unsaturated. Figure 6.2 shows
he typical values of light interception by hedgerow orchards of different
ometries. Tall, triangular section hedgerows intercept a higher propor-
n of the available light than low hedgerows do. The percentage of avail-
le direct light that is intercepted is controlled by the relation of the
dgerow to the sun and distance between trees. Hence, the influence of
w orientation on interception of light varies with time of day and season,
itude, and orchard geometry.

Apple trees attain their maximum leaf area by midsummer but usually do
t intercept more than 65 to 70% of the available light when leaves are
lly expanded, and several years are required to achieve this level (Table
3). The loss of light interception by apple orchards is reflected in de-
reased crop yield. For example, yields of apples increased linearly to about
0% of the available light. In young Granny Smith apple trees trained to
everal orchard systems and several tree densities and on several rootstocks,
ght interception in the fourth year varied from 10 to 40%. Light intercep-
ion and leaf area per hectare were highly correlated. Fruit yield over two
easons was positively correlated with leaf area per hectare and with the
mount of intercepted light (Fig. 6.3).

Photoperiod

Because floral initiation in many temperate-zone trees takes place in the
autumn it might be inferred that day length is involved in regulation of

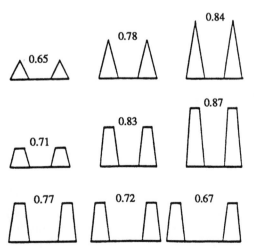

Figure 6.2 Effect of hedgerow geometry on the fraction of light intercept
apple trees in England. From Jackson (1985).

red reversible phytochrome-mediated response that is
day length and (2) formation of large amounts of an
posure to high fluence rates of visible and near-visible
750 nm).

Formation of anthocyanins in apples depends on a bal
the endogenous capacity of anthocyanin formation, base
level of phytochrome activity and probably also on photo
and (2) endogenous inhibition of anthocyanin formation
is controlled by gibberellin activity and ethylene [and/
(ABA)] activity. Hence, accumulation of anthocyanin pig
difference between a capacity for synthesis and an inhibit
(Fig. 6.1). Other environmental factors affect fruit color
ducing inhibition of anthocyanin formation, thus modifyir
light. Both soil and tree factors (e.g., N, dwarfing rootstocks,
affect anthocyanin formation by limiting the supply and tr
berellins from the roots to the shoots. Pruning of branches
reddening of apples by exposing them to more light. Howev
ing gibberellin activity, very severe pruning may inhibit antho
tion, despite exposure of fruits to high light intensities. A
ethephon may increase anthocyanin formation by reducing
gibberellins (Saure, 1990).

The efficiency of light capture by apple trees is greatly influe
arrangement and spacing of trees as well as by density of the ca
leaf characteristics (Jackson, 1985). The amount of light int
hedgerows depends on the proportion of the ground that is c
the height of the hedge. Trees with vertical sides and flat top

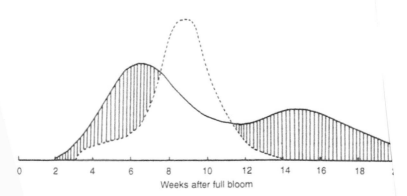

Figure 6.1 Model of regulation of anthocyanin formation in apples including caj
hocyanin formation (solid line), repression of anthocyanin formation (dashed lin
ocyanin formation (shaded area). From Saure (1990), with permission of Elsev

Table 6.3 Light Interception by Golden Delicious Apple Orchards of Different Ages[a]

Age of orchard (years)	Spacing (m)	Month	Light interception (%)
1	2.7 × 0.9	October	11
5	2.7 × 0.9	September	50
7	3.0 × 1.0	September	67
9	3.0 × 1.0	September	70

[a]From Jackson (1980).

flowering. However, with minor exceptions photoperiod does not appear to directly regulate flowering of woody plants, including fruit trees and forest trees (Piringer and Downs, 1959; Sedgley and Griffin, 1989). Flowers

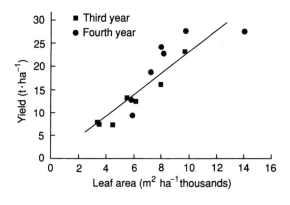

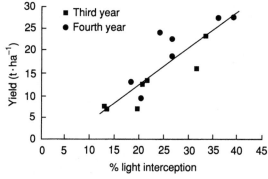

Figure 6.3 Relations among light interception, leaf area per hectare, and fruit production for third- and fourth-year Granny Smith apple trees. From Barritt *et al.* (1991). *Hort Science* **26**, 993–999.

of coffee are induced under a wide variety of day lengths that occur in the tropics (Cannell, 1972).

A few studies showed that day length regulated flowering in certain woody plants. For example, short days accelerated floral initiation in *Rhododendron* (Criley, 1969). Flowering of *Cupressus* showed a photoperiodic response but only when flowering was induced by gibberellins (Pharis *et al.*, 1970). Interactions of high temperature and short days regulated flowering of carib pine (Slee, 1977).

Water Supply

Either too little or too much water can variously influence each stage of reproductive growth. However, it usually is much more difficult to quantify the effect of water supply on reproductive growth than on vegetative growth. A major problem is that reproductive growth, especially of many forest trees, is both irregular and unpredictable. Another difficulty is that several stages of reproductive growth often are present on woody plants at the same time. Furthermore, drought often favors some aspects of reproductive growth (e.g., pollen shedding) while inhibiting others (e.g., enlargement of fruits and seeds) (Kozlowski, 1982a). In addition, various other environmental stresses, alone or in combination, may override or obscure the effects of water supply on reproductive growth. Still another problem is that measurements of effects of drought on reproductive growth are complicated by diurnal and seasonal shrinkage and swelling of fruits and cones as they dehydrate and rehydrate.

Despite these difficulties it is clear that adequate soil moisture is essential for high yields of fruits, cones, and seeds. Isolated heavy rains during the dry season are followed by rapid rehydration and mass flowering in tropical deciduous forests (Chapter 4). Flowering also is associated with reduction of water stress in trees by leaf shedding during a drought (Borchert, 1983, 1991; Reich and Borchert, 1982). The weight of evidence shows that prolonged droughts during most of the reproductive cycle reduce fruit yields. For example, a drought during which the soil Ψ decreased to -0.2 to -0.3 MPa reduced yields of different blueberry cultivars by 7 to 71% (Andersen *et al.*, 1979). Fruit yield of satsuma orange trees grown in pots was increased by half (for the same number of fruits per tree) when the soil Ψ ranged

---

Figure 6.4   Effects of waterlogging of soil on reproductive growth of apple. (A) Number of blossom clusters (open symbols) and fruit load (filled symbols), through the 1984 season (third year of treatment). Full bloom occurred on May 26 (day 0) and harvest on September 21 (day 118). (B) Reduction in yield of spring-, summer-, and autumn-waterlogged trees relative to controls. Means within a day separated by Duncan's multiple range test, $p = 5\%$. From Olien (1987). *J. Am. Soc. Hortic. Sci.* **112**, 209–214.

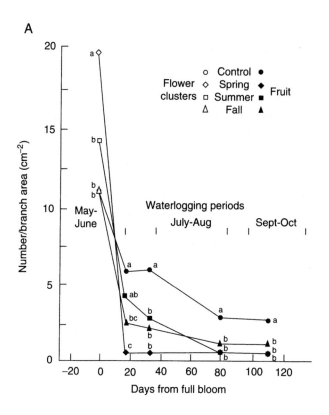

A

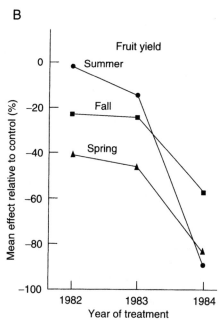

B

between -0.04 and -0.11 MPa, compared with -0.97 to -1.95 MPa in sparsely watered controls. These trees appeared to achieve full fruiting potential when the soil Ψ was maintained above -0.25 MPa (Suzuki and Kaneko, 1970). In general, high yields of grapes are obtained only with high soil moisture contents throughout the period of reproductive growth (Smart and Coombe, 1983). A large decrease in yield of grapes was attributed to a 22-day period of drought beginning at flowering (Hardie and Considine, 1976). Drought during early berry development was followed by a reduction in final berry size, even when the water deficit was alleviated. The small size of berries associated with drought has been attributed largely to fewer cells per berry (Coombe, 1960; Harris *et al.*, 1968). The importance of water supply to reproductive growth is dramatically illustrated by the beneficial effects of irrigation on yield of fruits, cones, and seeds (Chapter 8).

Flooding

Inundation of soil during the growing season decreases flowering, fruit set, crop yield, and fruit quality. The extent of inhibition of reproductive growth varies with species and genotype, age of plants, and time and duration of flooding.

Soil waterlogging for 6-week periods in the spring, summer, or autumn reduced apple yields (Fig. 6.4; Table 6.4), with yield being reduced most

Table 6.4   Effect of Flooding on Fruit Yield, Yield Efficiency, and Components of Yield over 1982–1984[a]

| Treatment | Yield (kg tree^{-1}) | Yield efficiency[b] (kg cm^{-2}) | Components of yield[c] | | |
			Number of fruit per tree	Fruit diameter (cm)	Fresh weight per fruit (g)
Control	12.5 a[d]	0.60 a	100 a	6.7 a	148 a
Spring flooding	6.0 b	0.35 b	59 a	6.5 ab	147 a
Summer flooding	9.4 ab	0.55 ab	90 a	6.5 ab	118 a
Fall flooding	8.4 ab	0.42 ab	86 a	6.4 b	120 a
Correlation with yield ($n = 48$)			$+0.92$	NS[e]	NS

[a]From Olien (1987). *J. Am. Soc. Hortic. Sci.* **112**, 209–214.
[b]Yield efficiency was calculated from yield per trunk cross-sectional area.
[c]Components of yield were determined from a subsample of 20 fruits collected at harvest.
[d]Means within a column separated by Duncan's multiple range test, $p = 5\%$.
[e]NS, Not significant.

(52%) by flooding in the spring (Olien, 1987). Flooded highbush blueberry plants had 61 to 77% fewer flower buds and 55 to 66% fewer flowers per bud when compared with unflooded plants. Fruit set was decreased by 45% and fruit abscission was increased by flooding. Fruit weight, fruit size, and percentage of soluble solids were reduced in flooded plants (Abbott and Gough, 1987a,b).

Some fruits, including apples, avocados, cherries, and grapes, may crack when trees are flooded or heavily irrigated after long periods of drought (Kaufmann, 1972). For example, cracking of apples was attributed to loss of skin elasticity and subsequent rapid water uptake by fruit cells following irrigation. High internal hydrostatic pressures in grapes sometimes cause cracking of the skin in areas of high rainfall and heavily irrigated vineyards (Lavee and Nir, 1986). Fruit cracking is a major factor limiting production of sweet cherries (Sekse, 1995).

Temperature

Yields and quality of fruits and seeds are variously influenced by temperature through its effects on floral initiation, release of bud dormancy, flowering, fruit set, growth, and ripening of fruits and cones. Both yield and quality of apple crops in England were regulated by temperature at three critical periods of reproductive growth (Pereira, 1975): (1) absence of late spring frosts (most important), (2) temperature at pollination, and (3) temperature in the month following full bloom. Warm weather in the first half of the month following bloom usually increased yield, whereas warm weather in the second half often decreased it.

Floral Induction

Low temperatures have been implicated in floral induction in woody plants, often in association with inhibition of vegetative growth (Jackson and Sweet, 1972). For example, low temperatures induced floral induction in sweet orange (Moss, 1969), litchi (Menzel, 1983), and macadamia (Stephenson and Gallagher, 1986).

The effects of temperature on floral induction are influenced by the time and duration of exposure to low temperature, prevailing light intensity, and age of plants. Transferring seedlings of crimson mallee from a heated greenhouse (24°C day temperature, 19°C night temperature) to a cold greenhouse (15/10°C) and back to the heated greenhouse induced formation of floral buds (NeSmith and Krewer, 1992). Production of floral buds was further stimulated when the seedlings were transferred to the cold greenhouse when the light intensity was high. Under low light intensities and short cold periods (0 to 5 weeks), the plants grew only vegetatively. Low temperatures reduced the rate of leaf initiation and plant growth, probably in association with redistribution of carbohydrates and mineral nutrients as shown for other species. Treatments or conditions that inhibit vegetative growth (e.g., stem girdling, root pruning, water deficits, and mineral deficiency) have been correlated with accelerated flowering (Chapter 8).

There is strong evidence that low temperatures do not induce floral initiation in all species. In olive, for example, floral initiation occurs by mid-autumn, and subsequent exposure of plants to low temperature releases the previously initiated floral buds from dormancy (Pinney and Polito, 1990). Rallo and Martin (1991) cited the following lines of evidence for lack of floral induction by chilling of olive plants: (1) the alternate bearing habit of olive trees implies that most floral buds in bearing ("on-year") plants are not induced during the previous growing season, (2) the presence of fruitlets or their seeds, as much as 5 months before the chilling periods, prevents flower induction and hence flowering in the following year, and (3) shading of plants before mid-autumn prevents flowering in the next year.

Bud Dormancy

Endodormancy (see Chapter 3 of Kozlowski and Pallardy, 1997) of flower buds commonly is broken long before the temperatures are high enough to permit flowering. However, buds sometimes open prematurely during a short period of warm weather, only to be killed later by freezing weather. This is a common problem with fruit trees, especially with peach and to a lesser degree apple; citrus fruits maturing during the winter in Florida often are injured by untimely freezes. In contrast, in mild climates, lack of sufficient cold weather to break bud dormancy results in failure of flowers to open properly. For example, the peach crop was a failure in most of the southeastern United States following the mild winter of 1931–1932. Failure of fruit trees to flower is a common problem when trees of the temperate zone are transplanted to subtropical climates.

When the amount of chilling is inadequate, peach flower buds near branch tips may develop into blossoms and basal buds may remain dormant. Small and deformed blossoms and low germination of pollen result from inadequate chilling. The stigma and style typically do not grow while the rest of the flower achieves normal size. Such abnormal flowers often do not set fruit (Kozlowski, 1983).

The chilling requirement (exposure to temperature between 5 and 10°C) for breaking dormancy of flower buds differs among species and cultivars. For example, it varies from 50 to 1700 hr for different species of *Prunus*; 150 to 650 hr for blueberry; 500 to 600 hr for pecan; 1500 hr for Delicious apple; to as much as 3,000 hr for some cultivars of pear (Austin *et al.*, 1982; Sedgley and Griffin, 1989). Although 1,000 hr of chilling will break dormancy of buds of most varieties of peach, the amount of chilling varies from near 50 to more than 1,100 hr for different varieties (Table 6.5).

Flowering

Each sequential event that occurs during flowering is regulated by temperature. These events include petal opening, anthesis, stigmatic secretion, pollen germination, growth of pollen tubes, maturation of embryo sacs, fusion of polar nuclei, double fertilization, petal fall, and ovary growth. Low temperatures sometimes slow floral development so much that initial fruit set is prevented.

The time of flowering in the spring depends not only on adequate exposure to cold but also on subsequent accumulation of heat. The base

Table 6.5 February 15th Chilling Requirements of Flower Buds of Peach Cultivars[a]

Cultivar	Hours of chilling required at 7.2°C (45°F)
Mayflower	1150
Raritan Rose, Dixired, Fairhaven	950
Sullivan Early Elberta, Trigem, Elberta, Dixired, Georgia Belle, Halehaven, Redhaven, Candor, Rio Oso Gem, Golden Jubilee, Shippers Late Red	850
Afterflow, July Elberta (Burbank), Hiland, Hiley, Redcap, Redskin	750
Maygold, Bonanza (dwarf), Suwannee, June Gold, Springtime	650
Flordaqueen	550
Bonita	500
Rochon	450
Flordahome (double flowers)	400
Early Amber	350
Jewel, White Knight No. 1, Sunred Nectarine, Flordasun	300
White Knight No. 2	250
Flordawon	200
Flordabelle	150
Okinawa	100
Ceylon	50 –100

[a]From Teskey and Shoemaker (1978).

temperature for heat accumulation may vary from 2.5°C in some peach cultivars to 4.5°C in almond and pear (Sedgley, 1990). The heat sum requirement for flowering also varies greatly, from approximately 680 hr in lingonberry to 3,423 hr in some pears and 8,900 hr in some almond cultivars. Applied gibberellins can partly substitute for the heat sum requirement.

By their effects on temperature, both latitude and altitude influence flowering dates. The average daily air temperature varies linearly with latitude, with a slope of -0.45 per degree Celsius between 10 and 55° latitude. Both leafing and blossoming occur earlier at lower latitudes. Flower development of apple trees starts 2.5 days earlier per degree Celsius at lower latitudes, according to studies conducted between 43 and 65° (Wagenmakers, 1991). The effect of temperature is shown by comparing dates of flowering of the same species in different parts of its range. Flowering dogwood blooms near Jacksonville, Florida, in mid-February and in Columbus, Ohio, in early May (Wyman, 1950). At East Malling, England, about three-fourths of the year-to-year variation in time of flowering of apple trees was accounted for by a linear regression of accumulated degree-days above 5.5°C from February 1 to April 15 (Jackson, 1975). Most pines showed a delay in flowering of about 5 days for each degree of latitude northward. Flowering of western white pine also was delayed by 5 days with each 300 m of increase in elevation (Bingham and Squillace, 1957). Kronenberg (1994) developed, tested, and adapted a model for flowering of lilac. According to the model, the flowering dates of lilac are determined by release of buds from dormancy by exposure for 400 hrs to a temperature below 7°C and subsequently to a temperature sum ranging from 340 degree-days in southern countries to 420 degree-days in northern countries, in a period with mean maximum day temperatures of at least 9.5°C.

There has been some disagreement about the relative importance of chilling and heat sums in determining the date of flowering of different species. Couvillon and Erez (1985a) concluded that the time of flowering of fruit trees depended more on the chilling requirement than on the heat sum. In contrast, Gianfagna and Mehlenbacher (1985) reported that the time of flowering of apple trees depended more on the heat sum than on chilling. Models for predicting bud opening, based on accumulation of chilling and heat units, are discussed by Kronenberg (1985), Anderson *et al.* (1986), and Sedgley (1990).

Pollen Formation, Germination, and Growth of Pollen Tubes

Temperature influences pollen formation, germination, and growth of pollen tubes. In mango formation of fertile pollen required temperatures in the range of 10 to 35°C (Issarakraisila and Considine, 1994). The developmental phase from meiosis to the prevacuolate microspore stage was the most temperature-sensitive stage of pollen development. The critical temperature for formation of fertile pollen was about 10°C. Trees exposed to night temperatures below 10°C during meiosis produced pollen grains with low viability (<50%). Temperature during the day usually was not a limiting factor for pollen development, providing it did not exceed 35°C.

The effect of temperature on pollen germination varies among species and genotypes. In English walnut the minimum temperature for pollen germination was near 14°C; the optimum temperature, near 28°C. Both the minimum and optimum temperatures for pollen germination were higher in black walnut than in English walnut (Fig. 6.5) (Luza *et al.*, 1987). Pollen germination at temperatures below 9°C was greater in almond than in peach; optimal temperatures for germination were 16°C for almond and 23°C for peach. Maximal pollen tube elongation occurred at temperatures 5 to 8°C higher than those for maximal pollen germination. Pollen tube elongation of the two species did not differ over a broad range of temperatures (Weinbaum *et al.*, 1984). Pollen tube growth in avocado increased with a rise in temperature; however, growth frequently was abnormal at high temperatures (33°C day/28°C night), and pollen tubes did not reach the ovary at low temperatures (17°C day/12°C night) (Sedgley, 1977).

In years of average summer temperatures in the Lake Region of England, pollen germination and growth of pollen tubes of littleleaf linden were arrested, resulting in negligible fruit production; hence regeneration of the species did not occur. However, if the temperature maxima were 3 to 8°C higher than the long-term average temperature, pollen germinated and pollen tubes grew readily from the stigma to the ovules, emphasizing that fruit production and hence regeneration, were regulated by temperature conditions (Pigott, 1992). Jefferies *et al.* (1982) derived a model for pollen tube growth of Victoria plum based on accumulated temperatures. Above a threshold temperature of 2.5°C, maximum growth of pollen tubes

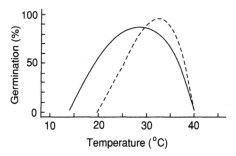

Figure 6.5 Effects of temperature on germination of pollen of English walnut (solid line) and black walnut trees (dashed line). From Luza *et al.* (1987).

was 0.34 mm per degree-day, and the tubes grew to half their final length at 16.6 degree-days above 2.5°C. The model indicated that, for fertilization to occur, between 40 and 50 degree-days above 2.5°C were required after pollination. This was equivalent to 16 to 20 days at 5°C, 5 to 7 days at 10°C, or 3 to 4 days at 15°C.

Fruit Set

Temperature strongly influences fruit set. The response is mediated by effects of temperature on flowering, stigma receptivity, ovule longevity, and growth of pollen tubes. Furthermore, extreme temperatures may induce seed abortion. In Manzanillo olive fruit set was completely inhibited at a constant temperature of 30°C (Cuevas *et al.*, 1994). All ovaries abscised within 2 weeks after bloom. The most favorable temperature for fruit set, near 25°C, was associated with most rapid growth of pollen tubes, and abundant and early fertilization. At 20°C, pollen tubes grew slowly, resulting in delayed and reduced fertilization.

Fruit Growth and Yield

By influencing the rate of growth and maturation of fruits, temperature influences the duration of the period between flowering and fruiting. Because low temperatures commonly arrest fruit growth, the time to harvest of a given cultivar is shorter in warm regions than in cool ones.

Fruit growers often use heat units (expressed in degree-days or degree-hours) to quantify the effect of temperature on fruit growth of different species and cultivars. Granny Smith apples require many heat units to mature fruits. Hence, they cannot be grown successfully at northern latitudes where the growing season is short and summers are cool (Ryugo, 1988).

Temperature has important effects on fruit quality, for example, in influencing the development of red color in apples. Each stage of apple ripening has an optimal temperature regime for anthocyanin formation. The decreasing temperatures of autumn are correlated with rapid formation of anthocyanin pigments. Low temperatures contribute to formation of red color by reducing gibberellic acid (GA) activity. However, at other stages of fruit development, higher temperatures also are required for proper development of higher quality apples (Saure, 1990).

Chilling Injury

Reproductive tissues of many tropical and subtropical plants, and a few temperate-zone plants, are injured after exposure to temperatures below approximately 15°C but above freezing. Injury of fruits by chilling is of special interest because many fruits are shipped and stored at low temperatures to extend their life. The storage life of peaches and nectarines is limited by chilling injury, which occurs in susceptible cultivars after 2 to 3 weeks of cool storage below 10°C (Lill *et al.*, 1989).

Symptoms of chilling injury in fruits vary widely (Kozlowski *et al.*, 1991) and may include surface lesions, surface and internal discoloration, breakdown of tissues, failure to ripen, accelerated senescence, increased susceptibility to decay organisms, and shortened storage and shelf life. Chilled peaches, plums, and nectarines develop a dry mealy texture, lose flavor and color, and show browning of the flesh. Often they fail to ripen. In peaches, changes in the wall structure of mesocarp cells accompanied the onset of mealiness and leatheriness due to chilling injury (Luza *et al.*, 1992). In leathery peaches, the mesocarp parenchyma cells collapsed and intercellular spaces increased. As internal breakdown progressed, dissolution of the middle lamella, thickening of the primary walls, and plasmolysis of mesocarp parenchyma cells followed. The large amount of cell wall thickening in leathery fruits indicated that some cell wall synthesis occurred.

Bananas are very sensitive to chilling. With mild chilling green bananas may not show any symptoms, but when they ripen the color of the peel often varies from dull yellow to grayish-yellow or smoky gray (Fig. 6.6).

Figure 6.6 Discoloration of peel and flesh of banana following chilling (top) compared with unchilled fruit (bottom). Photo courtesy of U.S. Department of Agriculture.

More severe symptoms include a dull yellow shine, browning of the skin, failure to ripen, loss of flavor, and increased susceptibility to mechanical injury. Chilling in the field often causes brown epidermal streaks and softening of the pulp during ripening as well as predisposition of bananas to storage rots and blemishes (Tai, 1977).

Among citrus fruits, grapefruits and limes are most sensitive to chilling. Grapefruits develop pits or depressions of the rind when stored at 4 to 5°C (Fig. 6.7). However, at 0 to 1°C the rind shows superficial brown staining. Chilled oranges develop small sunken spots in the rind and uniform browning over large areas. Lemons generally do not show external symptoms of chilling injury but exhibit "watery breakdown," a softening of the rind and flesh. Apples develop "soft scald," a brown discoloration as a result of death of patches of epidermis and cortex, and they also undergo low-temperature breakdown in the cortex.

The critical temperature at which chilling injury becomes apparent in tropical fruits is near 10 to 12°C, but species susceptibility varies with the region of origin and with plant variety. The lower temperature limit is near 12°C for many varieties of banana; between 8 and 12°C for citrus, avocado, and mango; and 0 to 4°C for temperate-zone fruits such as apples (Simmonds, 1982).

The critical duration of exposure at chilling temperature for induction of injury varies from a few hours to several weeks depending on the specific fruit and temperature. Fruits chilled for short periods may appear to be uninjured when removed from low-temperature regimes, but injury symptoms often appear in a few days at higher temperatures. The effects of chilling are cumulative, and low temperatures in the field before harvest and during transport of fruits can accentuate the effects of chilling in storage.

Chilling injury also is influenced by the degree of maturity of fruits. Bananas are most sensitive at the stage at which they normally are harvested. Immature, green grapefruits are very susceptible to chilling and become less susceptible as their color changes. Sensitivity of avocados is relatively high in the respiratory preclimacteric stage, increases to a maximum at the climacteric respiratory peak before the fruit is ripe, and decreases rapidly as the fruit ripens (Kosiyachinda and Young, 1976).

Freezing Injury

Severe winters often are disastrous to fruit and seed crops because of the high sensitivity of reproductive structures to frost. Freezing injury includes midwinter killing of dormant flower buds as well as spring and autumn injury to flowers and fruits. Freezing temperatures before budbreak resulted in a wide range of abnormalities in pecan flowers, whereas freezing

Figure 6.7 Surface pitting of grapefruit following chilling. Photo courtesy of U.S. Department of Agriculture.

temperatures during anthesis did not. Abnormal flowering depended on a critical stage of development of pistillate flowers within an 8- to 10-day interval before budbreak (Sparks, 1992).

Freezing injury may alter the bearing sequence of fruit trees. When early frost damage destroys the fruit crop of a biennially bearing tree, the normally bearing year becomes a vegetative one. The next year usually is one of heavy fruiting, and the biennial habit is continued thereafter (Davis, 1957).

In the temperate zone, the killing of flower buds by frost is very common. The flower primordia in winter buds may be 10 or 20°C more sensitive to frost than vegetative buds on the same tree. Flowers that have already expanded typically freeze at temperatures of −1 to −3°C. The flowers of winter-flowering species are exceptions and freeze at −5 to −15°C (Sakai and Larcher, 1987). Differences among species in frost resistance of reproductive structures are well known and are often related to times of flowering. Apricot trees bloom before peach trees, which bloom before apple trees. Hence, spring frost injury is likely to be greatest in apricot, intermediate in peach, and least in apple. There also is much variability among cultivars in damage by frost. In peach, for example, the cultivars Elberta, Sunhigh, Redskin, and Jubilee are injured more by frost than Veteran, Vedette, Early Red Free, and Sunrise (Teskey and Shoemaker, 1978).

When frost injures flowers or young fruits, the mature fruits may be abnormal and of low quality. For example, in apple trees injured by frost at the time of flowering, extensive discoloration of the apple fruit surface developed (Simons, 1959). Trees injured by frost 1 month after bloom had abnormally shaped fruits with large areas of dead cells (Simons and Lott, 1963).

Supercooling of Valencia orange fruits was limited to temperatures of $-4°C$ or higher as a result of the presence of ice-nucleating bacteria and a coexisting nucleating source (Constantinidou and Menkissoglu, 1992; Constantinidou *et al.*, 1991). Flower buds supercool by avoiding ice nucleation in the sensitive floral primordia. This may be accomplished through tissue dehydration by transport of water from floral primordia and vascular traces within the scales of floral buds. Risk of ice nucleation also is reduced by a high sucrose content of the primordium (Sedgley, 1990). Control of the ice-nucleation-active pool of citrus leaves reduced nucleation events by approximately 30% at a nucleation temperature of $-3°C$ or higher. Hence, treatments that promote supercooling and protect citrus tissues from frost injury by controlling the ice-nucleation-active bacterial pool probably will not be more than 30% successful (Constantinidou, 1992).

In some years, spring freezing is a major cause of mortality and abortion of gymnosperm conelets. Usually only part of the conelet crop is killed by freezing. Susceptibility to frost varies with the stage of conelet development, and injury is greater in ovulate than in staminate conelets. Ovulate conelets at the stage of maximum pollen receptivity are very vulnerable to frost damage, whereas in postpollination stages they generally are resistant. The ovulate conelets of shortleaf pine that were still covered with bud scales were only slightly damaged by frost or not at all, whereas those that had emerged from the bud scales were severely injured or killed. Staminate conelets were uninjured (Hutchinson and Bramlett, 1964). Variations in temperature within a tree crown or stand also may account for mortality of only part of the conelet crop. In addition to killing first-year conelets, freezing temperatures often reduce the number of sound seeds in the cones that survive. Some ovules are killed, although the cones are not destroyed, or pollen is rendered sterile, thus inducing subsequent ovule abortion.

Two separate freezing events have been identified in dormant flower buds of woody plants (Ashworth, 1990). The first occurs at a temperature a few degrees below 0°C and corresponds to the freezing of water in bud scales and tissues of the subtending axis. The second event (the "low-temperature exotherm") corresponds to freezing of a fraction of the supercooled water and is closely correlated to the temperature at which the buds are killed. The temperature of the second freezing event generally varies from -15 to $-30°C$ and depends on plant species and the degree of acclimation to cold. In buds with many florets, several low-temperature exotherms have been identified, each corresponding to the temperature at which individual florets were killed (George and Burke, 1977; Ishikawa and Sakai, 1985). In contrast to the pattern of freezing in buds of some species, in dormant flower buds the developing floral region did not freeze as a unit. Rather, the low-temperature exotherm corresponded to the lethal freezing of supercooled water in the anthers and parts of the pistil (Ashworth, 1990).

High-Temperature Injury

High air temperatures often injure fruits, including apples, avocados, cherries, grapes, peaches, and oranges. Common symptoms include burns and lesions, which vary somewhat among species. Injured orange fruits may have sunburned peels and dried flesh, and they may show increased granulation (Ketchie and Ballard, 1968). Prunes develop translucent areas in the mesocarp next to the pit cavity. These subsequently turn brown, and the injured cells generally disintegrate.

Soil Fertility

As emphasized in Chapter 10 of Kozlowski and Pallardy (1997), soil fertility is too low in many parts of the world for satisfactory growth of reproductive structures of forest and fruit trees. A balanced supply of mineral nutrients ensures normal growth of fruits and cones by regulating each phase of reproductive growth. The mineral requirements of reproductive structures are particularly high. For example, almost a third of the N absorbed by apple trees is used in fruit growth. In years of heavy fruiting very little N is stored in trees, and reserves often are depleted. A crop of apples was estimated to remove 80 kg ha^{-1} and a crop of oranges 200 kg ha^{-1} of minerals (Labanauskas and Handy, 1972).

The use of fertilizers in fruit orchards is a well-established and necessary practice. Fertilizer applications also are useful in seed orchards of forest trees (Chapter 8).

Salinity

By reducing flowering and fruit yield, salinity often inhibits reproductive growth. In a 10-year study salinity not only decreased yield of Valencia oranges but also delayed fruit maturity. The reduction in yield was due entirely to fewer fruits per tree (Francois and Clark, 1980). The adverse effects of salinity on reproductive growth often are preceded or accompanied by necrosis of leaf tips and margins, premature leaf shedding, and twig dieback (Chapter 5). Salinity effects are especially prominent in arid regions. Irrigation water usually contains more salts than are removed by harvesting of fruits. Hence, continuous irrigation of fruit orchards without leaching progressively increases salinization of the root zone (Shannon *et al.*, 1994).

Most fruit trees are sensitive to salinity. Pear, apple, orange, grapefruit, plum, apricot, peach, and lemon trees show low salt tolerance; olive and fig, moderate tolerance; and date palm, high salt tolerance. However, there are appreciable varietal differences in salt tolerance of a given species within each of these groups (McKersie and Leshem, 1994). Because of wide variations in tolerance of rootstocks to salinity, sensitivity to salt varies within a species. Differences in salt tolerance among citrus rootstocks were attributed to salt exclusion from the shoots by inhibition of salt absorption by the roots and compartmentalization within the root system. The scion had little effect on transport of salt from the roots to the shoots (Behboudian *et al.*, 1986). Differences in chloride concentrations of leaves of citrus on different rootstocks showed variations among rootstocks in root to shoot salt transport. After treatment with 100 mM NaCl for 6 weeks, the chloride concentration of leaves was 7 times higher for the rootstock Etrog citrus than for Rangpur lime (Walker and Douglas, 1982).

Responses to salinity often vary with prevailing environmental conditions. For example, soil salinity of irrigated fruit trees often is accompanied by flooding of the soil, which increases the harmful effects of salinity. The adverse effects of flooding of saline soils have been demonstrated where flooding is intermittent or while only a part of the root zone is flooded (West and Taylor, 1984).

Pollution

The yield and quality of flowers, fruits, cones, and seeds often are decreased by exposure to pollutants in the air and soil. For example, photochemical oxidants reduced yields of oranges and lemons by as much as half because of premature leaf and fruit drop (Thompson and Taylor, 1969). The yield of oranges also was reduced by exposure to NO_2 and peroxyacetylnitrate (PAN) (Thompson *et al.*, 1970).

Fluorides are very toxic to reproductive growth. For example, adverse effects on reproductive growth of raspberry and blueberry plants were demonstrated downwind from a phosphorus plant that emitted gaseous and particulate fluorides (Fig. 6.8). Most loss of reproductive potential was attributed to flower mortality (89% for blueberry and 78% for raspberry at the most polluted sites). Comparative values for the control site were 27% and 26%, respectively. More flowers were produced on raspberry plants at the most contaminated site. This response was associated with widespread production of suckers and lateral shoots after the death of shoot tips. In contrast, both flower and bud production in blueberry decreased with proximity to the F source.

Both cone and seed production as well as seed quality of gymnosperms may be adversely affected by air pollution. Seed production of several conifers was negligible near a smelter in Montana (Hedgcock, 1912). Cone production by ponderosa pine, lodgepole pine, western larch, and Douglas fir was reduced by SO_2 from smelters (Scheffer and Hedgcock, 1955). In the Urals SO_2 from a smelter reduced the weight of seed cones of Scotch pine by 32 to 50%, of pollen cones by 23 to 32%, and of seeds by 12 to 15% (Mamajev and Shkarlet, 1972). In France, air pollution not only reduced the size of Scotch pine cones and number of seeds per cone but also increased the rate of cone shedding (Roques *et al.*, 1980). In Ohio air pollution reduced the weight of red pine seeds, percentage of filled seeds, and seed germination capacity (Houston and Dochinger, 1977).

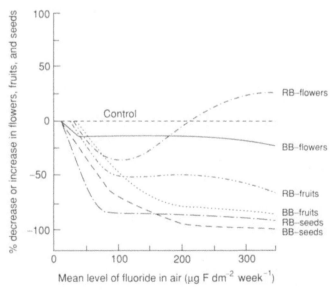

Figure 6.8 Percentage decrease (or increase) in numbers of healthy flowers, fruits, and seeds per plant with increasing fluoride levels. RB, Raspberry; BB, blueberry. From Staniforth and Sidhu (1984).

Air pollutants adversely affect several stages of reproductive growth (Fig. 6.9). The reduction in yield and quality of reproductive structures may be associated with effects of pollutants on injury to reproductive structures, mechanisms of flowering and fruiting, and abscission of both leaves and reproductive structures. In addition, air pollutants often are toxic to pollinating insects. Hence, they have important effects on insect-pollinated species such as maples, willows, basswood, and apple. Inhibitory effects on reproductive growth commonly result from a smaller carbohydrate pool or less partitioning of carbohydrates to reproductive than to vegetative organs. By inducing leaf abscission and lowering the photosynthetic capacity of leaves, air pollutants decrease the amounts of carbohydrates and hormonal growth regulators available for reproductive growth. Inhibition of reproductive growth in plants exposed to O_3 is mediated largely by reduced physiological efficiency of injured foliage.

Some pollutants lower fruit quality by injuring fruits. Whereas fluorides and SO_2 injure fleshy fruits, there is little evidence of direct injury by O_3 (Bonte, 1982). Examples of injuries to fleshy fruits by fluorides include lesions of apricots and peaches, deformed fruits and black, collapsed tissue around the calyx of pears and the stones of plums, and malformations at the apex of the receptacle of strawberry. Deformation of fruits by pollution

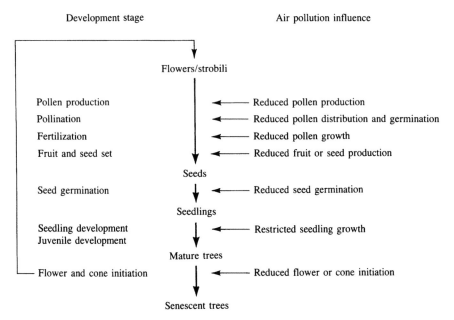

Figure 6.9 Effects of air pollutants on various developmental stages of woody plants. From Smith (1981). "Air Pollution and Forests," with the permission of Springer-Verlag.

often is associated with a decrease in the number of developed seeds. Injury to fleshy fruits commonly occurs at very low concentrations of F, usually well below the level required for leaf injury. Strawberry plants are particularly sensitive to F during anthesis and fertilization; at other times exposure to F affects only the leaves (Pack, 1972). In pear trees exposed to F, 98% of the fruits were deformed and seedless, whereas none were deformed in unexposed trees (Bonte *et al.*, 1980).

Pollen Germination and Elongation of Pollen Tubes

Reduction in seed production often has been traced to effects of individual and combined pollutants on germination of pollen and growth of pollen tubes (Masaru *et al.*, 1976; Wolters and Martens, 1987). Most studies have been conducted in laboratories and greenhouses, however, and caution is advised in extrapolating data obtained in such experiments to field situations because pollen germination and tube growth *in vitro* are much more sensitive to pollutants than *in vivo* because of chemical, physical, and temporal differences between the two conditions (Chaney and Strickland, 1984). The lesser response of pollen to air pollutants *in vivo* may reflect an additional buffering capacity of the stigmatic surface and protection of the pollen tube from air pollutants after it penetrates the style. Despite the greater impact of pollutants on pollen *in vitro*, considerable evidence shows that pollutants at ambient concentrations in the field adversely affect pollen germination and tube growth.

There are many examples of inhibitory effects of pollutants on pollen germination and growth of pollen tubes in both fruit trees and forest trees, and only a few will be cited. The effects of pollutants may be expected to vary with plant species, type of air pollutant, pollutant concentrations, exposure time, relative humidity, and other factors. Whereas dry pollen is quite tolerant to pollutants, pollen exposed to a pollutant in or on a germinating medium is very sensitive.

Fumigation with HF during anthesis inhibited pollen tube growth in sweet cherry and apricot. The effect was similar whether fumigation occurred before or after pollination. Hence, the effect was exerted on growth of the pollen tube and not on stigma receptivity or inactivation of pollen germination (Facteau *et al.*, 1973; Facteau and Rowe, 1981). Pollen tube elongation on the stigma of apricot was reduced more by exposure to a high HF concentration for a short time than by a low concentration for a longer period (Facteau and Rowe, 1977).

Both germination of pollen and growth of pollen tubes of red pine and eastern white pine from polluted areas were lower than those from unpolluted areas (Houston and Dochinger, 1977). Pollen germination and pollen tube growth of trembling aspen, red pine, Austrian pine, and Colorado blue spruce were reduced by SO_2 (Karnosky and Stairs, 1974). Sulfur dioxide inhibited germination of pollen of Austrian pine and Scotch pine more than that of mugo pine or European silver fir (Keller and Beda, 1984).

In vitro studies showed that pollen germination of 13 forest species was influenced at pH of 5.6 to 2.6. Germination at pH 3.0 was appreciably reduced in two-thirds of the species studied, with pollen of angiosperm species being more sensitive than pollen of gymnosperms. The 50% lethal dose (LD_{50}) for pollen germination of broad-leaved trees during 3-hr exposures was in the range of pH of 3.95 to 3.60, values comparable to those periodically measured in rainfall (Cox, 1983). *In vitro* experiments by Paoletti (1992) showed that a sodium dodecylbenzenesulfonate detergent or a growth medium of pH 4.0 to 5.0 inhibited pollen germination and pollen tube growth more in broad-leaved species than in conifers. In the broad-leaved trees, pollen germination and pollen tube elongation were equally sensitive to both the detergent and acidity. In the conifers, however, pollen tube elongation was more sensitive than pollen germination to the detergent and acidity. Both pollen germination and growth of pollen tubes of red maple were reduced by acid mist (pH 5.6 to 2.6) applied in the field (Table 6.6), with the inhibitory effects being less severe than on pollen germinated *in vitro* (Van Ryn *et al.*, 1986).

The effects of heavy metals on pollen germination are of particular concern because they often accumulate in soil and plant tissues in toxic amounts and have long residence times. Heavy metals inhibited pollen germination and tube growth of red pine in aqueous solutions (Chaney and Strickland, 1984). The most toxic ion on pollen germination was Cd^{2+}, followed by Cu^{2+}, Hg^{2+}, Pb^{2+}, Zn^{2+}, and Ba^{2+}. The order of heavy

Table 6.6   Effect of Acid Mist on Pollen Growth in Red Maple Stigmas and Styles[a]

Measure	pH	April 13[b]	April 15[c]
Number of pollen grains germinating	5.6	60 ± 13	43 ± 8
	4.6	61 ± 20	—[d]
	3.6	37 ± 9	34 ± 8
	2.6	19 ± 9	22 ± 7
Number of tubes reaching base of style	5.6	3.8 ± 0.5	4.2 ± 1.0
	4.6	3.0 ± 0.6	—[d]
	3.6	2.4 ± 0.8	3.8 ± 1.0
	2.6	1.0 ± 0.4	2.8 ± 1.4

[a]From Van Ryn *et al.* (1988).
[b]Based on average of five trees (mean ± SE).
[c]Based on average of four trees (mean ± SE).
[d]Data missing owing to technical malfunction.

metal inhibition of tube elongation was $Cd^{2+} > Pb^{2+} > Hg^{2+} = Cu^{2+} > Zn^{2+} > Ba^{2+}$.

A given dosage of a pollutant may or may not influence pollen germination and tube growth similarly. For example, growth of cottonwood pollen tubes in culture was more sensitive than pollen germination to SO_2 (Karnosky and Stairs, 1974). Exposure of eastern white pine pollen in water to O_3 reduced only germination in a culture medium (Benoit *et al.*, 1983). By comparison, Cd^{2+} and Zn^{2+} reduced pollen germination and tube growth in red pine to the same extent, whereas Cu^{2+} and Hg^{2+} reduced pollen tube growth more than pollen germination. The ions Ba^{2+} and Pb^{2+} inhibited germination more than tube growth (Chaney and Strickland, 1984).

Summary

Several environmental factors interact to influence reproductive growth, indirectly by affecting physiological processes (especially carbohydrate and hormone relations) and directly by inducing injury. Among the most important environmental factors that influence reproductive growth are light intensity, water supply, temperature, soil fertility, and pollution.

Only the outer crown of fruit trees receives enough light for maximum yield of fruits of high quality. Light intensity affects formation of flower buds, fruit set, growth of fruits, and abscission of reproductive structures. Much attention has been given to (1) maximizing interception of light by fruit trees because the light that falls in the space between trees does not increase fruit yield, and (2) optimizing light distribution within the canopy. The light environment influences various aspects of fruit quality such as fruit size, color, and concentration of soluble solids. With few exceptions, day length does not appear to directly regulate flowering of many woody plants.

Both drought and flooding of soil can adversely affect each stage of reproductive growth. Adequate soil moisture is necessary for high yields of fruits, cones, and seeds as shown by stimulation of reproductive growth by irrigation. However, a short period of drought during the floral initiation period sometimes stimulates formation of flower buds by inhibiting vegetative growth. Short periods of drought also may break dormancy of flower buds of some species.

Flooding of soil during the growing season decreases flowering, fruit set, crop yield, and fruit quality. Flooding of soil sometimes causes injury to fruits. The adverse effects of flooding vary with species and genotype, age of plants, and time and duration of flooding.

Temperature affects reproductive growth by breaking dormancy of flower buds through exposure to low temperature; by controlling flowering, pollen germination, growth of pollen tubes, fruit set, and fruit growth; and by promoting chilling, freezing, and heat injury. Symptoms of chilling injury to fruits include lesions, discoloration, tissue breakdown, failure to ripen, increased susceptibility to decay organisms, and shortened storage life. Freezing may injure flowers and induce mortality of flower buds and conelets. Freezing injury sometimes is lessened by supercooling of reproductive structures because of the presence of ice-nucleating bacteria and other nucleating sources. Very high temperatures may cause burns and lesions on fruits and adversely affect fruit quality.

Soil fertility often is too low for satisfactory reproductive growth. Because the mineral requirements of reproductive growth are very high, routine applications of fertilizers to fruit orchards, and sometimes seed orchards, are necessary.

By reducing flowering and fruit growth, salinity often inhibits reproductive growth and yield. The adverse effects of salinity on reproductive growth often are preceded or accompanied by necrosis of leaf tips and margins, early leaf shedding, and dieback of twigs. Most fruit trees are sensitive to salinity, but date palm is not. Tolerance of different rootstocks to salinity varies greatly.

Both yield and quality of flowers, fruits, cones, and seeds often are reduced by pollutants in the air and soil. Pollutants adversely influence several stages of reproductive growth by injuring reproductive tissues, affecting mechanisms of flowering and fruiting, inducing abscission of vegetative and reproductive organs, and influencing pollinating insects. Reduction in seed production often is associated with effects of pollutants on germination of pollen and on growth of pollen tubes.

General References

Atkinson, D., Jackson, J. E., Sharples, R. O., and Waller, W. M., eds. (1980). "Mineral Nutrition of Fruit Trees." Butterworth, London.

Bowen, G. O., and Nambiar, E. K. S. (1984). "Nutrition of Plantation Forests." Academic Press, London.

Faust, M. (1989). "Physiology of Temperate Zone Fruit Trees." Wiley, New York.

Jackson, J. E. (1985). Future fruit orchard design: Economics and biology. *In* "Attributes of Trees as Crop Plants" (M. G. R. Cannell and J. E. Jackson, eds.), pp. 441–459. Institute of Terrestrial Ecology, Huntingdon, England.

Jones, H. G. (1992). "Plants and Microclimate," 2nd Ed. Cambridge Univ. Press, Cambridge.

Katterman, F., ed. (1990). "Environmental Injury to Plants." Academic Press, San Diego.

Kozlowski, T. T., ed. (1981). "Water Deficits and Plant Growth," Volume 6: Woody Plant Communities. Academic Press, New York.

Kozlowski, T. T., ed. (1983). "Water Deficits and Plant Growth," Volume 7: Additional Woody Crop Plants." Academic Press, New York.

Kozlowski, T. T., Kramer, P. J., and Pallardy, S. G. (1991). "The Physiological Ecology of Woody Plants." Academic Press, San Diego.

Marshall, C., and Grace, J., eds. (1992). "Fruit and Seed Production: Aspects of Development, Environmental Physiology and Ecology." Cambridge Univ. Press, Cambridge.

Ormrod, D. P. (1986). Gaseous air pollution and horticultural crop production. *Hortic. Rev.* **8**, 1–42.

Rieger, M. (1989). Freeze protection for horticultural crops. *Hortic. Rev.* **11**, 45–109.

Sakai, A., and Larcher, W. (1987). "Frost Survival of Plants." Springer-Verlag, Berlin and New York.

Saure, M. C. (1985). Dormancy release in deciduous fruit trees. *Hortic. Rev.* **7**, 239–300.

Sedgley, M. (1990). Flowering of deciduous perennial fruit crops. *Hortic. Rev.* **12**, 223–264.

Smith, W. H. (1990). "Air Pollution and Forests," 2nd Ed. Springer-Verlag, New York.

Unsworth, M. H., and Ormrod, D. P., eds. (1982). "Effects of Gaseous Air Pollution in Agriculture and Horticulture." Butterworth, London.

7

Cultural Practices and Vegetative Growth

Introduction

As emphasized by Jones (1985) most increases in yield of forest and fruit trees have resulted from improvements in management practices. However, cultural practices are appropriate only if they prevent injury to trees or increase the efficiency of the physiological mechanisms that regulate their growth—that is, if they increase the availability and control the balance of internal growth requirements (e.g., carbohydrates, hormones, water, and mineral nutrients). Hence, foresters and horticulturists need to manage trees to minimize the adverse effects of environmental stresses on tree growth and yield. The most important cultural practices that can influence establishment and growth of trees include site preparation for planting, irrigation, correction of mineral deficiencies, thinning of stands, pruning of branches or roots, application of chemicals such as growth regulators and pesticides, and use of systems of integrated pest management.

Site Preparation

A variety of cultural treatments have been used to improve regeneration of harvested forests and plantations. Site preparation increases establishment and growth of trees by altering soil properties, controlling plant competition, or both. The major site preparation treatments include disposal of logging slash, disking, bedding, subsoiling, windrowing, draining of wet sites, and eliminating competing plants.

Slash disposal decreases fire hazard and stimulates regeneration of forests by eliminating heavy shade, increasing infiltration of water into the soil by reducing interception; it also facilitates planting of seedlings. Seedbeds often can be improved by burning of slash. If an area has not been disturbed by logging, burning generally improves the seedbed. However, if it has been disturbed, burning may not be very beneficial. In an undisturbed seedbed three times more shortleaf pine seedlings emerged from seedbeds that had been burned than from unburned seedbeds (Boggs and Wittwer, 1993). On many sites, root growth of seedlings is restricted by mechanical resistance to penetration of soil by roots and by poor soil aeration. Such sites often can be improved by disking, bedding, or subsoiling. Disking decreases resistance of the surface soil to penetration by roots and increases movement of water and air into the soil. Subsoiling increases infiltration of water and decreases runoff. On subsoiled sites the volume of soil exploited by roots during the first few seasons after planting is increased (Morris and Lowery, 1988). Bedding (mounding up of surface soil, litter, and logging debris into low ridges or beds) improves drainage; increases availability of soil moisture, mineral nutrients, and soil oxygen; reduces plant competition; and accelerates mineralization of organic matter.

There are many examples of beneficial effects of site preparation on establishment and growth of woody plants. After loblolly pine trees in the southern United States are harvested, the sites often are prepared for planting by chopping and burning of vegetation or blading (Stransky *et al.*, 1985). Productivity of loblolly pine plantations in Louisiana was increased by chopping and burning. Harrowing as an additional treatment increased production after 12 years by 394 ft^3 per acre over chopping and burning (Haywood and Burton, 1989). Herbicides controlled several species of broad-leaved trees (red maple, dogwood, red oak, willow oak, and water oak) in preparing sites in Georgia for growth of pines. Mechanical site preparation (shearing, raking, and disking) provided comparable control of competing broad-leaved trees but at a higher cost (Shiver *et al.*, 1990). Site preparation in the western United States suppressed broad-leaved vegetation and increased the amount of soil resources available to ponderosa pine, sugar pine, and white fir (Lanini and Radosevich, 1986). Methods such as brush raking, which reduced shrub development in combination with herbicides, were particularly effective in increasing availability of light and water to the conifers.

The few examples of beneficial effects of site preparation cited here should not obscure some possible harmful effects. The effectiveness of site preparation depends on the skills of forest managers and varies with specific management practices, soils, plant species, and topography. Because potential hazards of site preparation do exist, forest managers should "do as much site preparation as necessary, but as little as possible" (Haines *et al.*, 1975).

In addition to its beneficial effects, disposal of logging slash can be harmful. Thin layers of slash often benefit regeneration of plantations or forests by improving the microclimate and protecting seedlings from desiccation. In contrast, thick and dense layers of slash impede tree establishment. Piling slash off a site in Oregon led to soil compaction and loss of humus. Douglas fir seedlings did not grow as tall as seedlings growing on similar sites but where slash had been burned (Minore and Weatherly, 1990).

Some site preparation treatments can have adverse effects on availability of mineral nutrients. Scraping off the surface soil can greatly impair site productivity if the nutrients (which are concentrated in the litter and upper soil horizons) are moved laterally beyond the future extent of tree roots. Windrowing may be damaging because it moves topsoil and humus (hence mineral nutrients) with plant materials (Morris *et al.*, 1983).

Drainage

Many investigators reported that draining of unproductive wet sites improved site quality and stimulated growth of trees. For example, about half the peatlands of Finland have been drained to increase tree growth. The total volume of peatland forests increased from 6,000 m^3 in the 1920s to 15,000 m^3 in the 1980s and accounted for approximately 22% of the total volume of Finland's forests (Mikola, 1989).

Stimulation of tree growth on drained peatlands is associated with lowering of the groundwater table, increased substrate aeration, and greater rooting depth of some species of plants (Boggie, 1972; Campbell 1980). The lowered water content of drained peat also reduces the heat capacity and thermal conductivity of the substrate. Hence, the surface layers of drained peatlands warm faster than those of undrained peatlands (Pessi, 1958). An initial effect of drainage was improved N relations of black spruce trees (MacDonald and Lieffers, 1990). Trees in drained plots had higher rates of photosynthesis and higher foliar N contents. Both photosynthesis and foliar N contents correlated positively with rates of height growth.

Drainage of peatlands affects both phenology and growth rates of trees. In a drained Alberta peatland, bud opening of trees occurred 2 to 6 days earlier than in an undrained site (Lieffers and Rothwell, 1987). The responses on drained peatlands may be expected to vary with plant species and genotype, stand density, age of plants, and growth parameters measured.

There often is a lag response in tree growth to drainage. For the first 3 to 6 years after a peatland was drained, cambial growth of black spruce trees did not increase (Dang and Lieffers, 1989). Subsequently, cambial growth increased almost linearly to a maximum at 13 to 19 years after drainage. The increases ranged from 76 to 766% of the projected growth of trees on undrained sites.

The site index of balsam fir and red spruce trees increased with improved drainage (Williams *et al.*, 1991). However, site index was not a good indicator of stand biomass production (there was no significant change in biomass production with increase in site index on poorly drained soils). On better drained soils, biomass production decreased with increases in site index, suggesting higher rates of self-thinning with increases in site index.

The response of different species in mixed forests often varies in response to drainage. For example, in a spruce–fir forest in Maine, growth of balsam fir was stimulated by drainage whereas growth of red spruce was not (Meng and Seymour, 1992). Biomass production, however, was much greater on well-drained sites (15.1 Mg ha^{-1}) than on poorly drained sites (9.4 Mg ha^{-1}).

Many peatland sites are mineral deficient. For example, most peatlands in Norway lack adequate amounts of N, P, and frequently B for satisfactory tree growth. Hence, the positive effects of soil drainage can be further enhanced by applications of appropriate fertilizers (Braekke, 1990). Over a 27-year period the cation-exchange capacity of a peatland in Norway increased by 32 to 38% after drainage, and by 37 to 43% after additional fertilization (Braekke, 1987). Drainage of bogs in Norway increased mortality of fine roots. However, growth of both fine roots and large roots was greatly stimulated by additions of NPK fertilizers to the drained sites (Braekke, 1992).

Drainage of wetlands, often together with addition of fertilizers, was followed by increases of 80 to 1,300% in volume growth of pines in the southeastern United States (Terry and Hughes, 1975). With both drainage and fertilization, loblolly pine grew to a height of 80 ft (24 m) in 25 years compared to 45 ft (13.5 m) on unmanaged sites (Campbell and Hughes, 1981). Both height and diameter growth of loblolly and slash pines in Louisiana were related to soil drainage in the winter (Haywood *et al.*, 1990). Slash pine grew better than loblolly pine on poorly drained sites, whereas loblolly pine grew better than slash pine on better drained soils.

Drainage of soils often is necessary for growth of shade and ornamental trees. Hardpans and bedrock beneath shallow soils frequently are encountered in urban soils and restrict root growth and drainage. Harris (1992) and Rolf (1994) include good discussions of methods of improving soil drainage.

It should be remembered that wetlands provide many benefits in addition to timber production. They can produce up to 8 times as much plant growth as the average wheat field (Maltby, 1986). In addition to contributing to production of rice, sago, oil palms, mangroves, crayfish, shrimps, oysters, waterfowl, fish and fur-bearing animals, wetlands support grazing lands, control floods, maintain water quality, absorb toxic chemicals and clean up polluted waters. As Maltby (1986) emphasized, drainage of wetlands may lead to reduction of yield in other parts of the ecosystem and also reduce the capacity of a whole ecosystem for sustained yield of useful products. For these reasons, and because wetlands are the most fragile and threatened ecosystems on earth, the possible benefits of drained wetlands should be very carefully weighed against their potential benefits if left undrained. Richardson (1983) suggested that, in considering whether a wetland should or should not be drained, close attention should be given to sustained yields, irreversibility of drainage, values of undeveloped wetlands to future generations, and mitigation and reclamation procedures.

Herbicides

Because of their effectiveness in control of herbaceous and woody weed plants, together with their low costs and ease of application, herbicides often have been used to decrease competition between crop trees and undesirable herbaceous and woody plants. The elimination of undesirable plants increases the availability of water, light, and mineral nutrients to crop trees. Where herbaceous plants in shortleaf pine stands were controlled by herbicides, soil moisture was greatest and needle moisture stress was lower than in control plots (Yeiser and Barnett, 1991). Stimulation of tree growth following application of herbicides was associated with increased absorption of N, presumably because of decreased plant competition (Ebert and Dumford, 1976). Some herbicides influence mineral nutrition by affecting development of mycorrhizae. Roots of Douglas fir seedlings treated with napromide were infected with a greater diversity of mycorrhizal fungi than seedlings treated with dimethyl-2,3,5,6-tetrachlorophthalate (DCPA) or bifenox (Trappe, 1983).

Some herbicides have been useful in production of nursery stock; others were harmful. Ideally, nursery managers prefer selective preemergence herbicides that can be applied to seedbeds immediately after seeding. Tree seedlings in the cotyledon stage of development are most susceptible to herbicides; hence, some herbicides that have been used in transplant beds or in young plantations should not be used in seedbeds (Kozlowski and Sasaki, 1970).

As emphasized in Chapter 2, some herbicides suppress seed germination and others do not but may injure recently emerged seedlings. When pine seeds had been treated with 2-chloroallyldiethyldithiocarbamate (CDEC) or ethyl *N,N*-di-n-propylthiocarbamate (EPTC), the emerging plants had fused cotyledons; those treated with 2,4-D had swollen stems and chlorotic, shriveled cotyledons (Fig. 7.1). At concentrations of 25 ppm or higher, 2,4,5-trichlophenoxyacetic acid (2,4,5-T) maintained in direct contact with seeds and recently emerged red pine seedlings caused inhibition of root

Figure 7.1 Effects of herbicides on red pine seedlings in the cotyledon stage of development. (Top) Untreated control seedlings are shown at left and herbicide-treated seedlings at right. Note distortion of cotyledons of 20-day-old seedlings by Tordon (4-amino−3,5,6-tri-chloropicolinic acid). (Bottom) Fusion of cotyledons and suppression of their development by CDEC (2-choroallyldiethyldithiocarbamate). From Kramer and Kozlowski (1979).

elongation; proliferation, expansion, and collapse of parenchyma cells in the stem, roots, and cotyledons; formation of callus tissue; inhibition of formation of primary needle primordia and their expansion; as well as distortion of primary needles and fusion of primary needles to cotyledons (Wu *et al.*, 1971).

In nurseries there may be problems with persistence of herbicides and accumulation of toxic residues when beds are used repeatedly to grow a variety of tree species for different lengths of time. The longevity of different herbicides in soil varies greatly and is influenced by leaching, microbial action, adsorption to soil, volatilization, and chemical reaction (Kozlowski, 1979).

There is an extensive literature on beneficial effects of herbicides on establishment and growth of forest trees, and only a few examples will be cited. On adverse heavy clay sites along the southern shore of Lake Superior in Wisconsin, the use of herbicides facilitated establishment of trees (Kuntz and Kozlowski, 1963). In the southeastern United States the growth of young pine trees was greatly reduced by competing vegetation, with growth being inhibited more by competing herbaceous plants than by woody plants (Cain, 1991). In that region the judicious use of herbicides has greatly increased establishment and growth of pine trees. For example, controlling plant competition with herbicides during the first year increased height and diameter growth of loblolly, longleaf, and slash pines. Extending the herbicide treatment to the second year resulted in additional height and diameter increases of 60 and 70%, respectively (Creighton *et al.*, 1987). The effects of four treatments for control of plant competition on growth of loblolly pine trees during the first 5 years are shown in Fig. 7.2. The treatments were (1) no herbicides (site preparation only), (2) control of both hardwood tree species and shrubs with herbicides, (3) control of herbaceous plants with herbicides, and (4) total control of combined treatments 2 and 3 to yield bare ground conditions. After 5 years, volume growth of pine was increased by 400% with control of both herbaceous and woody plants, by 171% with control of herbaceous plants only, and by 67% with control of woody plants only (Miller *et al.*, 1991). Twenty years after young loblolly pine trees were released from competing broad-leaved trees with herbicides, volume growth had increased by 40% over controls, the equivalent of 8 years of growth (Michael, 1980).

Successful use of herbicides requires considerable knowledge of characteristics of the chemical used and of responses of plants to be favored and

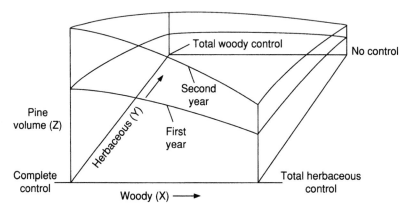

Figure 7.2   Effects of controlling woody or herbaceous vegetation or both with herbicides on volume growth of loblolly pine. For explanation, see text. From Miller *et al.* (1991).

those to be eliminated. The influence of herbicides on plants varies greatly with the compound used; rate, method, time, and number of applications; plant species and genotype; soil type; and weather (Kozlowski, 1960; Kozlowski and Kuntz, 1963; Kozlowski *et al.*, 1967a,b).

When applying herbicides, managers should be aware of the potential hazards of spray drift. Application of herbicides that may injure nontarget plants should be avoided on windy days. Sprayers or spreaders used for applying herbicides, especially hormone-type herbicides, should not be used later for applying other chemicals.

The effectiveness of herbicides on both target and nontarget plants varies with the method of application (Kozlowski *et al.*, 1991). Certain herbicides control weeds effectively when applied to the soil surface. Some remain in the surface soil and hence may not be absorbed by tree roots (Kozlowski and Torrie, 1965). However, if persistent herbicides are applied to the soil surface and later mixed into the soil, the toxic residues may be absorbed by trees and injure them. Nursery managers often prefer herbicides that can be applied to nursery beds after seeding to control weeds. However, tree seedlings in the cotyledon stage of development are extremely sensitive to certain herbicides (Sasaki and Kozlowski, 1968b,c). Whereas DCPA applied at rates up to 4 lb acre^{-1} as a preemergence or postemergence spray controlled several weeds and was not toxic to young red pine seedlings, other herbicides were very toxic (Kuntz *et al.*, 1964; Kozlowski and Torrie, 1965). Direct contact of red pine seeds with various herbicides induced abnormal growth of emerging seedlings (Kozlowski and Sasaki, 1968a,b; Wu *et al.*, 1971; Wu and Kozlowski, 1972; Kozlowski, 1986b).

The actual toxicity of herbicides to plants cannot adequately be characterized by applying them to the soil because they often are lost by evaporation, leaching, microbial or chemical decomposition, and irreversible adsorption on the soil. The inhibitory effects of commercial formulations of herbicides on photosynthesis may be exerted by active or inert ingredients and synergistic effects of both (Sasaki and Kozlowski, 1968b,c).

Irrigation

Humans have depended on irrigation for survival since the beginning of recorded history, and probably even before that time. Evidence of the importance of irrigation is documented in Persian, Turkish, Indian, Chinese, and Roman history. In North America the Hahokam Indians of Arizona used irrigation as early as 1000 B.C. Irrigation on a large scale was introduced into the United States by the Mormons in Utah (Sneed and Patterson, 1983). In modern times irrigation has expanded greatly in Australia, China, India, and Israel. An important change has been the use of supplementary irrigation in the relatively humid climates of England, Western Europe, and the eastern United States. In the decade that ended in 1979, the area irrigated in the eastern United States increased by 77%, and that in the western states increased by 17% (Kramer, 1983).

Forest Trees

On some sites, irrigation can increase several aspects of tree growth, including shoot growth, cambial growth, and root growth. Usually the increases are much greater when irrigation is combined with addition of fertilizers. Growth of Monterey pine trees (as annual increment of wood volume and basal area per unit mass of foliage) increased appreciably following irrigation and almost doubled if irrigation was combined with addition of fertilizer (Cromer, 1980; Cromer *et al.*, 1983, 1984).

Unfortunately, the high costs of irrigating forest trees may be a limiting factor except for nurseries, seed orchards (Chapter 8), short-rotation intensive culture plantations, and certain plantings on infertile soils that can be irrigated with effluent from sewage plants. An economic analysis showed that sprinkler irrigation in Monterey pine stands in Australia was not profitable, and flooding irrigation was only marginally profitable (Stewart and Salmon, 1986). Although irrigation significantly increased biomass production of young hybrid poplars in Wisconsin during the first year after planting, it had little effect on tree survival during the establishment year (Hansen, 1988). In Romania, however, where annual precipitation was only 450 mm and the period of July 15 to September 15 was without rain, irrigation increased poplar yields by 300 to 500% (Papadol, 1982).

Interest has been growing in the use of wastewater and wet or dry sludge from sewage disposal plants to irrigate and fertilize established forests and also to reclaim unfavorable sites such as landfills and strip-mined lands (Wilson *et al.*, 1985). This practice often increases growth of established forests, improves the quality of the runoff water by removing phosphates, nitrates, and other ions from the water, and adds to groundwater supplies. Crohn (1995) suggested that applications of anaerobically digested sludge at 4 Mg ha^{-1} (dry mass) to a 30-year-old northern hardwood forest at 3-year intervals for 50 years may increase production of harvestable stems and branches by approximately 25%.

Ornamental Trees and Shrubs

It often is necessary to water transplanted trees and shrubs for 2 to 3 years while their root systems are becoming established. Most shade trees do not have a fully redeveloped root system until 3 to 6 years after transplanting (Neely and Himelick, 1987). During periods of prolonged drought, even well-established trees benefit greatly from summer irrigation. For example, 6 years into a continuing drought in California, not only was growth impeded in unirrigated valley oaks, but 28% of them had died. In comparison, irrigated trees showed increased growth and none died (Hickman, 1993).

The amount of water to apply can be determined by multiplying the area to be irrigated by the desired depth of irrigation. Harris (1992) recommended that during the growing season a tree in an area without turf should receive enough rain or irrigation water within its dripline to equal twice the amount lost by evapotranspiration. Methods for calculating the amount of water to apply are outlined by Harris (1992).

Young shade trees usually can be irrigated efficiently by drip or mini-sprinkler systems. Where summer irrigation is needed it is useful to prepare a circular mound of soil 7.5 to 10 cm high at the edge of the planting hole. Such a mound serves as the dike of a reservoir that should be filled with water at 7- to 10-day intervals during dry periods in the growing season. The water in the reservoir will adequately soak the backfill and soil ball containing the root system. Frequent (e.g., daily) watering that wets only the surface soil to a depth of an inch or so is of little value because most of the absorbing roots are at a greater soil depth and will be left in dry soil during superficial watering. Wetting the soil thoroughly at longer intervals is more beneficial. However, allowing a hose to run water into the reservoir for prolonged periods can lead to soil waterlogging. Probably more ornamental trees and shrubs are killed by overwatering than by lack of supplementary irrigation (Neely and Himelick, 1987).

Correction of Mineral Deficiencies

Improved ability to recognize mineral deficiencies of plants and identify their causes has led to extensive research on methods of alleviating them. Among the most important of these methods are application of fertilizers, inoculation of plants with mycorrhizal fungi, and use of N-fixing species as interplanted or understory plants in order to increase the supply of N.

Fertilizers

A voluminous literature shows that judicious fertilizer applications increase growth of seedlings as well as older forest, fruit, and shade trees. Increased growth in response to fertilizers typically is associated with higher rates of photosynthesis per unit of leaf area, more and larger leaves, delayed senescence and shedding of leaves, and greater partitioning of carbohydrates to shoots than to roots. One-third to one-fourth of the ^{15}N-labeled nitrate added to a hardwood forest was retained by aboveground plant tissues and surface soil horizons. Nitrogen assimilation into leaves and stem tissues was lower in red spruce than in deciduous species (Nadelhoffer *et al.*, 1995).

Efficient use of fertilizers requires information about which nutrient or nutrients are limiting plant growth and the extent of the deficiency of each nutrient. The nutritional status of woody plants can be diagnosed by plant tissue analysis, soil analysis, or determination of plant responses to fertilizers containing specific elements. Methods of determining the nutritional status of plants are discussed by Mead (1984) and Walker (1991).

Nurseries   Because large amounts of nutrients are leached from nursery soils by irrigation and rain, regular additions of fertilizers are essential to produce sturdy seedlings with high potential for growth and survival. Nursery mineral nutrition may influence growth of outplanted seedlings by affecting storage and retranslocation of nutrients and by influencing seedling response to drought, cold, and disease. Responses of seedlings to fertilizers are influenced by soil and plant nutrient status, length of growing season, amount of soil organic matter, and previous harvesting practices (Fisher and Mexal, 1984). In some areas late-season fertilization is beneficial (Margolis and Waring, 1986). Judicious late-season fertilization in nurseries often greatly improves growth and increases survival of outplanted stock (van den Driessche, 1985, 1988). Late-season fertilizer applications to Douglas fir seedlings increased nutrient concentrations slightly, increased survival, stimulated early bud opening in the year after planting, and increased growth after planting (van den Driessche, 1991a,b). Fertilizing 1-year-old longleaf pine seedlings in late October in North Carolina increased seedling size in the nursery, decreased the percentage of seedlings in the grass stage 2 years after planting, and increased height growth by more than 10% after 8 years (Hinesley and Maki, 1980).

Seedlings of woody plants often are grown in containers. Because the containers usually are filled with media (e.g., sand, peat, vermiculite) that have low capacity to retain ions, frequent applications of fertilizers are needed. Much useful information on application of fertilizers to seedlings in nursery beds and in containers is given by Tinus and McDonald (1979), Duryea and Landis (1984), Wright and Niemiera (1987), and van den Driessche (1988).

Forest Trees   Application of fertilizers at or shortly after planting forest trees often increases growth and decreases tree mortality, providing that other factors such as drought are not limiting. Examples are increases in growth of slash pine, loblolly pine (Bengston, 1979), and Douglas fir in the United States (Carlson and Preissig, 1980) and Monterey pine in Australia (Flinn, 1985). The amount of increase in growth depends on such factors as species and genotype, amount and type of fertilizer, site preparation, water supply, and age of trees (Stone, 1973; Wilhite and McKee, 1985; Kozlowski *et al.*, 1991).

Increases in basal area of Monterey pine in response to N fertilizers were negligible during drought years but up to 300% in wet years (Benson *et al.*, 1992). Growth of a mineral-deficient Engelmann spruce plantation was stimulated by N fertilizer (Brockley, 1992). The effect was small in the first year after treatment, but height growth of trees increased in the next 2 years, with height averaging 41 cm more in fertilized than in unfertilized trees. Most of the growth response was associated with improved N relations. Results of application of N and a complete-mix fertilizer were no better than applying N alone.

Growth of many established plantations is inhibited by mineral deficiencies that can be corrected by application of fertilizers. For example, fertilizers have been effectively used to stimulate growth of conifer plantations in Australia (Attiwill, 1982), New Zealand (Will, 1985), Europe (McIntosh, 1984), and North America (Pritchett, 1979; Miller and Tarrant, 1983; Binkley and Reid, 1984).

Because of the large amounts of mineral nutrients removed per unit of biomass harvested in trees grown on short rotations, fertilizer applications may be necessary to sustain productivity over several rotations. Application of N fertilizers to dense stands of young American sycamore trees increased aboveground biomass production by 2 to 3 times by the end of the growing season (Tschaplinski *et al.*, 1991). Periodic low-level fertilization increased biomass production as much as a single heavy application (450 kg N ha^{-1}), and with a fraction of the total N applied.

Application of municipal sludge stimulated growth of Sitka spruce (Bayes *et al.*, 1987), loblolly pine (McKee *et al.*, 1986), and Corsican pine trees (Moffat *et al.*, 1991). Because of high rates of leaching of nitrates from sludge, some investigators mixed sewage sludge with sawdust or pulp and paper sludge to increase the C/N ratio of the sludge. Such mixtures reduced nitrate losses while stimulating tree growth (McDonald *et al.*, 1994).

Responses of trees following applications of fertilizers sometimes have been spectacular. For example, production of 6-year-old Tasmanian blue gum more than doubled (from 7.2 to 15.4 tons ha^{-1} year^{-1}) in 3.5 years following fertilization with N and P (Cromer and Williams, 1982).

Growth of most trees is increased by a mixture of nitrate N and ammonium N, but the best ratio differs with species and age of plants. Because the availability of nitrate N and ammonium N varies seasonally and with soil type, plants may be seasonally adapted to utilize only one of these two N sources. Some conifers growing on acid soils respond more favorably to fertilization with ammonium N. Both photosynthesis and growth of slash pine seedlings responded better to fertilization with ammonium N than with nitrate N (van den Driessche, 1972). Similarly, growth of jack pine seedlings was increased more by fertilization with ammonium N than with nitrate N (Margolis *et al.*, 1988). However, plants growing on calcareous soils often respond better to nitrate N (Marschner, 1986, 1995).

Responses of trees to N fertilizers are not always positive. Radwan *et al.* (1991) recommended that sites high in soil N and low in P should not be fertilized with N. Application of N fertilizers to such sites may inhibit tree growth. Radwan *et al.* (1991) suggested that application of P fertilizers to conifer stands should be confined to sites with soils low in P when the trees are young (before canopy closure). Applying P fertilizers to such stands probably would accelerate canopy closure and nutrient cycling.

Growth responses to fertilizers vary appreciably during development of forest stands (Kozlowski *et al.*, 1991). Stimulation of growth of plantation trees by fertilizers is most likely before canopy closure occurs and near the end of a rotation (Fig. 7.3). Growth of young trees that are not fully utilizing a site (stage I) depends strongly on soil fertility, and trees may show a deficiency of several mineral elements. At this stage the efficiency of mineral uptake is very low, but the effect of fertilizers on subsequent stand growth, through increased crown development, is very high. By the time canopy closure occurs (stage II), uptake of soil minerals has increased greatly but the absorbed minerals are rapidly recycled. Absorption of atmospheric deposits of nutrients is high. Associated with this increased input and recycling of minerals is a low growth response to fertilizers. When stage III is reached, N deficiency may occur as N is immobilized in the biomass and humus; hence, responses to fertilizers are again likely to increase (Miller, 1981). By the time forest stands that have been fertilized for a long time achieve maximum tree biomass and litter production, soil fertility has increased and responses to fertilizers are low. Therefore, addition of fertilizers should gradually be reduced (Ingestad, 1987).

Shade and Ornamental Trees   Landscape trees are subjected to very severe environmental stresses. Major aboveground stresses include air pollution, high temperatures, and restricted crown space. Belowground stresses include soil of poor structure, too dry or too wet soils, poor soil aeration, adverse effects of agricultural chemicals, and mineral deficiency (Craul, 1985; Kozlowski, 1985b; Kozlowski *et al.*, 1991).

Growth of shade trees is limited more often by N deficiency than by deficiency of any other element. Although complete fertilizers (NPK) often are applied to shade trees, usually P and K are not deficient elements; hence, responses to added P and K are more often the exception rather than the rule (Harris, 1992). When Neely *et al.* (1970) applied various fertilizers (N, PK, NPK, and NPK plus micronutrients) to shade trees, N was the only nutrient that significantly increased stem diameter growth. Neely and Himelick (1987) provided much useful information on fertilizing shade trees.

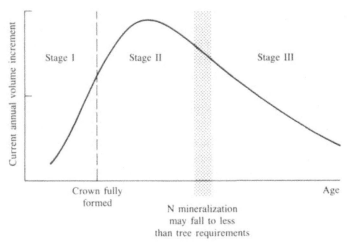

Figure 7.3 Three nutritional stages in the life of a forest stand. After Miller (1981).

Overfertilization Applications of excess fertilizers can be harmful to plants, with the amount of injury varying with species and genotype, type of fertilizer, and time of application. Very high salt concentrations in the soil solution may reduce its osmotic potential sufficiently to decrease water absorption, leading to leaf dehydration, stomatal closure, reduction in photosynthesis, leaf injury, and plasmolysis of root cells (Kozlowski *et al.,* 1991).

Overfertilization in nurseries may result in seedlings of low quality and reduced capacity for growth and survival when outplanted. Conifer seedlings receiving very high dosages of mineral nutrients, especially N, often have low root–shoot ratios and thin tracheid walls. Such seedlings may be very prone to drought and frost injury (Wilde, 1958). High N applications stimulated shoot growth more than root growth of loblolly pine seedlings and reduced their drought tolerance (Pharis and Kramer, 1964). Similarly, heavy applications of N fertilizer to lodgepole pine seedlings reduced their recovery from a 2-week drought by 39% (Etter, 1969).

Heavy applications of N fertilizers can delay bud set and increase susceptibility to autumn frosts. Late-season N fertilization usually increases early bud opening in the following spring and may increase injury by spring frosts (Benzian *et al.,* 1974). However, once seedlings have become cold hardy, application of N fertilizer has little effect on cold hardiness or even may slightly increase it (van den Driessche, 1991a,b).

Inoculation with Mycorrhizal Fungi

The soil in most natural forests and plantations contains abundant mycorrhizae. For example, the top 15 cm of soil under a Monterey pine stand in

Australia contained about 1,421 kg ha^{-1} of mycorrhizae (Marks *et al.*, 1968). In contrast, soils of treeless areas often lack appropriate mycorrhizal fungi. As a result, attempts to establish plantations of exotic species of trees often failed until essential mycorrhizal fungi were introduced (Marx, 1980). Many adverse planting sites also have low mycorrhizal populations.

In most well-managed tree nurseries in the United States, production of tree seedlings is not seriously affected by deficiencies of mycorrhizae. In some nurseries, however, excessive use of fertilizers, soil fumigants, and fungicides can appreciably reduce or eradicate fungal mycelia in the soil. Natural development of mycorrhizae on tree seedlings grown in containers also may be erratic if the seedlings are fertilized and watered very heavily in order to stimulate their growth.

Inoculation with mycorrhizae not only increases mineral uptake and seedling growth (see Chapter 10 of Kozlowski and Pallardy, 1997) but also may prevent Al toxicity. For example, associations of *Pisolithus tinctorius* with roots of pitch pine seedlings prevented Al toxicity from developing when the seedlings were exposed to 200 μM Al in sand culture (Cumming and Weinstein, 1990). The fungal symbiont modulated ionic relations in the rhizosphere so as to reduce Al–P precipitation reactions and Al uptake.

Absorption of mineral nutrients as well as growth and survival of out-planted seedlings sometimes have been dramatically increased by inoculating nursery beds with mycorrhizal fungi (Perry *et al.*, 1987). On poor sites such as coal spoils, pine seedlings that had been inoculated with *Pisolithus tinctorius* in the nursery grew 180 to 480% more than seedlings with only naturally formed mycorrhizae (Marx and Artman, 1979; Berry, 1982).

Inoculation with *Glomus fasciculatus* of slow-growing micropropagated avocado plantlets stimulated formation of a well-developed root system that was converted into a mycorrhizal system. Formation of mycorrhizae was essential for subsequent growth and development of avocado plants (Vidal *et al.*, 1992). For many years mycorrhizal deficiency in forest nurseries was corrected by transfers of soil, humus, or duff from natural forests or plantations (Fig. 7.4). Problems with this method are that the specific introduced fungi cannot be controlled, and very large volumes of soil are needed to inoculate even a single nursery. Furthermore, the soil inoculum may contain noxious weeds and harmful microorganisms as well as mycorrhizal fungi. For these reasons attention has shifted to greater use of pure mycelial cultures of ectomycorrhizae. There are, however, two major problems: a deficiency of adequate amounts of inoculum for large-scale applications of pure mycelial cultures to nurseries, and difficulty in determining the specific fungal species to be introduced to various sites and different host plants. Primary attention has centered on *Pisolithus tinctorius* because this fungus is adapted to adverse environmental conditions, is easily propagated, has a host range of more than 50 species of trees, and occurs naturally in many countries

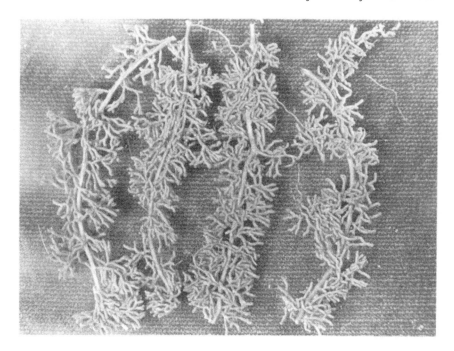

Figure 7.4 Ectomycorrhizae formed by *Pisolithus tinctorius* on loblolly pine after artificial inoculation of a fumigated nursery soil. Note abundant dichotomy of small roots. Photo courtesy of the U.S. Forest Service.

(Marx *et al.,* 1989). Considerable study also has been devoted to inoculation with such other fungi as *Cenococcum geophilum, Hebeloma crustuliniforme, Laccaria bicolor, L. laccata, Suillus granulatus, S. luteus,* and *Thelephora terrestris.*

Responses of trees to mycorrhizal inoculation vary with plant species and genotype, soil type and fertility, and species of mycorrhizal fungi (Fig. 7.5). Mycorrhizal dependency of seedlings varied in the order green ash > red gum > tulip poplar > American sycamore and was influenced by fungal species. Burgess *et al.* (1993) showed large variations in growth responses of Tasmanian blue gum and karri following inoculation with a wide range of ectomycorrhizal fungal species that are common associates of eucalypts in Australia. Growth increases of seedlings to inoculation ranged from 1.56 to 13 times the growth rates of noninoculated seedlings. The largest growth increases were obtained by inoculation with *Pisolithus tinctorius;* moderate increases with *Scleroderma verrucosum, Descolea maculata,* and *Laccaria laccata* (LAC 2); and smaller increases with *Hydnangium carneum, Hymenogaster* spp., *Setchelliogastor* spp., and *Laccaria laccata* (LAC 1). *Hysterangium inflatum* and *Thaxterogaster* spp. formed superficial mycorrhizae but did not stimu-

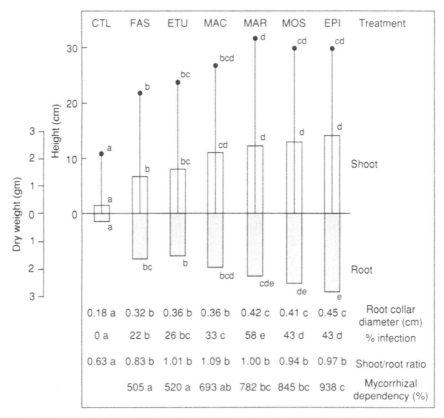

Figure 7.5   Effects of inoculation with mycorrhizal fungi on growth for 32 weeks of *Acacia scleroxyla* seedlings. Treatment means for each growth parameter followed by the same letter are not significantly different. MAR, *Gigaspora margarita*; EPI, *Glomus epigaeum*; MOS, *Glomus mosseae*; MAC, *Glomus macrocarpum*; ETU, *Glomus etunicatum*; FAS, *Glomus fasciculatum*; CTL, uninoculated control. From Borges and Chaney (1988).

late seedling growth. The mycorrhizal fungi that increased seedling growth also increased their P uptake.

Inoculation of tree seedlings grown in containers varied with soil fertility and pH, type of container, growing medium, fungicides, inoculum storage, and frequency of watering (Marx *et al.*, 1982). Thomson *et al.* (1994) screened 47 isolates of ectomycorrhizal fungi for their effects on growth of Tasmanian blue gum seedlings where the supply of P was deficient. The dry weight of inoculated plants ranged from 50 to 350% of the dry weights of uninoculated plants. Early-colonizing fungal species (*Descolea maculata, Hebeloma westraliense, Laccaria laccata,* and *Pisolithus tinctorius*) generally stimulated plant growth more than late-colonizing species (*Cortinarius* spp.

and *Hysterangium* spp.). Plant dry weight increases corresponded with increased P uptake by the plants and were positively correlated with the length of the colonized roots, indicating that the fungi which colonized roots effectively were most effective in increasing plant growth. Marx *et al.* (1982, 1984) presented excellent reviews of problems associated with mycorrhizal inoculation.

Nitrogen-Fixing Plants

Increasing costs of fertilizers and trends toward short rotations have stimulated interest in improving soil fertility through on-site sources. Even though growth of most trees is accelerated by proximity to N-fixing plants, such strategies have not been extensively used in forests, probably because: (1) N-fixing plants are not well adapted to forest conditions, (2) they may become weeds, (3) there is not enough seed or planting stock, and (4) costs and benefits are not always certain (Gordon, 1983). Despite these constraints, a wide range of N-fixing plants have been shown to stimulate growth of trees. Two forms of N-fixing organisms are important: bacteria of the genus *Rhizobium*, which form nodules on roots of legumes, and actinomycetes of the genus *Frankia* which grow on roots of such genera as *Alnus*, *Myrica*, and *Casuarina* (see Chapter 9 of Kozlowski and Pallardy, 1997).

Introduced N-fixing plants may be valuable crop plants (e.g., red alder, black alder, casuarina, autumn olive, and black locust), or they may be plants without commercial value. Nitrogen-fixing plants that are not commercially valuable (e.g., herbaceous legumes, *Ceanothus*, some species of alder, *Myrica*, *Elaeagnus*, *Cercocarpus*, and *Shepherdia*) may be important as short-term green manure plants followed by a plantation of crop trees, as understory plants, or as plants grown above or with other plants for soil stabilization. The use of N-fixing plants in plantations has largely involved interplanting of N-fixing trees with non-N-fixing trees or underplanting trees with N-fixing ground covers.

Interplanting The need for synthetic N fertilizers sometimes can be reduced by interplanting crop trees with N-fixing trees. Interplanting with alders, black locust, or autumn olive stimulated growth of fast-growing poplars, especially on poor sites (Gordon and Dawson, 1979; Hansen and Dawson, 1982). After 10 years cottonwoods interplanted with European black alder were 29% taller and 21% larger in diameter than those without interplanted alder (Plass, 1977). Interplanting of leguminous trees shows considerable potential in tropical forestry. Trees in the genera *Acacia, Acrocarpus, Albizia, Leucaena, Mimosa,* and *Sesbania* show promise for interplanting with important timber species. Valuable leguminous timber trees in the genera *Dalbergia, Intsia, Pericopsis,* and *Pterocarpus* might be advantageously mixed with other species (National Academy of Science, 1979). Other applications of interplanting N-fixing trees are discussed by Mikola (1989) for Europe, Turvey and Smethurst (1983) for Australia, and Domingo (1983) for southwest Asia.

Underplanting Nitrogen-fixing ground covers often stimulate growth of plantation trees by benefits that go beyond those of the added N. Growth of 2-year-old American sycamore trees was appreciably increased when planted with Mississippi subterranean clover (*Trifolium subterraneum*) or common clover (*T. incarnatum*) as ground covers. The benefits accrued from N fixation as well as from control of competing weeds (Haines *et al.*, 1978). Ground covers of leguminous herbaceous plants stimulated growth of poplars by adding N to the soil (Zavitkovski, 1979). Lupine (*Lupinus arboreus*) has an important effect on N nutrition of Monterey pines grown on N-deficient coastal sand dunes in New Zealand. Uptake of N by the pines is increased through the influence of litter and seedling exudates. The decomposing root systems of dead lupine plants increase N availability to Monterey pine litter by N transfer or by stimulating N-fixing activity elsewhere in the ecosystem. Lupine may contribute at least half of the total N accumulated in the biomass and litter (Gadgil, 1971a–c). When herbaceous legumes are planted prior to planting a tree crop, the cost of choosing the site without a commercial crop species on it may be of concern. This problem can be avoided by planting herbaceous noncrop plants at about the same time the tree crop species is planted.

Much more research is needed on reciprocal relations between specific N-fixing plants and crop trees. Information is especially needed on effects of competition of N-fixing plants and crop trees for light, water, and mineral nutrients and on soil properties such as pH and allelopathic effects.

Agroforestry Nitrogen-fixing trees and shrubs have several distinct roles in agroforestry (AF), for example, in improvement of soil fertility and as fodder plants, windbreaks, and plywood and pulpwood trees (Brewbaker, 1987).

In Africa and Asia, herbaceous food crops are grown on sites that previously were forested. At few-year intervals, the land is left to "bush fallow" by natural regeneration or by planting small trees or shrubs, some of which are legumes. Soil fertility also can be improved by interplanting N-fixing trees in rows with low crop plants, a system called alley cropping. In the South Pacific region, the major AF systems include a variety of tree crops such as coffee, cacao, and coconut, together with N-fixing trees (e.g., *Casuarina*, *Gliricidia*, and *Leucaena*) and food plants such as cassava, taro, sweet potato, and yam (Vergara and Nair, 1985). In West Africa, use of *Gliricidia sepium* trees as a live support system for water yam (*Dioscorea alata*) more than doubled the yield of yams. The trees and mulch improved growing conditions for the yam plants, while the trees extracted mineral nutrients from the soil layers that were not penetrated by yam roots (Budelman, 1990). Many additional examples of plant combinations in AF systems in the tropics and temperate zone are described in the book edited by Nair (1989) and an article by Colletti and Schultz (1995).

Much more research is needed on characteristics of N-fixing plants that govern their use in AF systems (Kozlowski and Huxley, 1983). In particular, studies are needed on the physiological ecology of components of AF systems to provide a rationale for use of species mixtures and cultural practices that will maximize the use of radiant energy, optimize water use efficiency, and minimize losses of mineral nutrients. Considerable attention has been given to use of *Leucaena leucocephala* in AF systems. This species is widely adapted to grow from 30°S to 30°N and at elevations up to 1500 m. However, many other promising legumes such as different species of *Acacia, Prosopis, Desmodium, Cassia,* and *Tamarindus* have not been adequately studied in AF combinations (Rachie, 1983). Quantitative studies are needed on the role of a variety of N-fixing plants on growth and yield of the whole agricultural crop, effects of tree–crop interfaces on herbaceous crops and on trees, and growth of trees as a single crop (Huxley, 1983).

Thinning of Forest Stands

During their development forest stands typically progress through three sequential stages (Oliver, 1981) including (1) an "open grown" stage during which all trees grow at essentially similar rates before crown closure of a stand occurs, (2) a "plastic" stage during which crown lengths as well as crown forms are altered, and (3) a "stagnation" stage during which the physiological efficiency of trees declines appreciably (e.g., the ratio of photosynthesis to respiration decreases) because of intense competition among trees for resources, as well as among trees, shrubs, and herbaceous plants. Responses of trees during the stagnation stage include a progressive decrease in the rate of height and volume growth, stratification into crown classes, and changes in susceptibility to insects and diseases. Trees that originally are closely spaced enter each successive developmental stage faster than widely spaced trees do.

The yield of wood in forest stands can be optimized by temporarily reducing stand density by removing some of the trees and thereby stimulating diameter growth of the residual trees (height growth commonly is only slightly affected by thinning of stands). Hence, managed forest stands commonly are thinned and with several objectives in mind (Daniel *et al.,* 1979): (1) to salvage and use wood that otherwise would be lost by tree mortality (such losses typically amount to approximately 25 to 35% of gross production), (2) to increase merchantable yield of wood by redistributing growth to fewer and larger trees, (3) to shorten the rotation age (if rotation age is determined by attainment of a certain tree diameter), and (4) to realize an early return of invested capital and thus increase the rate of return on investment.

In thinned stands the final volume of wood produced is less than it is in unthinned stands. Thinning lengthens the rotation age if it is determined by the age at which the growth rate of individual trees becomes unacceptably low or if it is based on culmination of the mean annual increment.

Thinning and Availability of Resources

Thinning of stands generally results in greater availability of light, water, and mineral nutrients to the residual trees. The amount of light transmitted to the forest floor increases with the intensity of thinning as shown by correlations between light penetration and the number of trees per unit area (Trappe, 1938; Jackson, 1981). Whereas extractable soil water was almost depleted by midsummer under closely spaced loblolly pine trees, it was not used up under widely spaced trees (Fig. 7.6). The soil moisture content also was higher in a thinned Douglas fir stand than in an unthinned stand. Increases in water yield were maintained for 6 years after light thinning of ponderosa pine stands in Arizona and for 10 years after heavy thinning (Baker, 1986). The greater availability of water to trees in thinned stands over those in unthinned stands was attributed largely to high evapotranspiration rates. Eventually, however, growth of the pine trees, as well as an increase in shrubby and herbaceous vegetation, resulted in use of as much water as was used by the original forest overstory, accounting for loss of the stimulatory effect of thinning on tree growth. After Morikawa *et al.* (1986) thinned a hinoki cypress stand, total transpiration of the stand decreased while transpiration of the individual trees increased, further emphasizing increased availability of water to the residual trees.

The effects of thinning on water use are modified by the amount of understory vegetation present (see Chapter 12 in Kozlowski and Pallardy, 1997). Even though the density of a Douglas fir stand was greatly reduced by thinning, the understory vegetation consumed approximately half of the available water. As a result, diameter growth of the trees in the thinned stand was only slightly greater than in the unthinned stand (Black *et al.*, 1980).

Stand thinning increases the supply of mineral nutrients available to the residual trees. This is shown by experiments on spacing of both seedlings and older trees. For example, increasing the spacing among Douglas fir seedlings increased the N content of needles (van den Driessche, 1984). The concentrations of N and P also were lower in needles of 5-year-old Douglas fir trees when they were closely grown than when they were widely spaced (Cole and Newton, 1986).

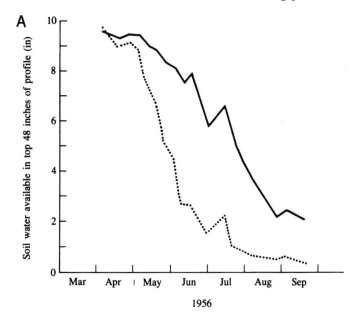

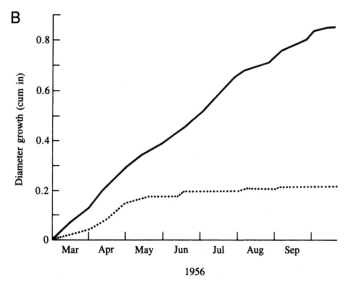

Figure 7.6 Effect of stand thinning on (A) soil water depletion and (B) diameter growth of loblolly pine trees during one growing season. The solid line is for thinned stands, and the dotted line denotes unthinned stands. From Zahner and Whitmore (1960). Reprinted from the *Journal of Forestry* **58**: 628–634, published by the Society of American Foresters, 5400 Grosvenor Lane, Bethesda, MD 20814–2198. Not for further reproduction.

Accelerated rates of photosynthesis (Wang *et al.*, 1995) as well as greater leaf biomass in trees of thinned stands are followed by increased downward transport of carbohydrates and growth hormones in tree stems (see Chapter 5 in Kozlowski and Pallardy, 1997). Computer simulations showed that 21% greater seasonal photosynthesis could occur in thinned over unthinned lodgepole pine stands (Donner and Running, 1986).

Vegetative Growth

There are many examples of increased cambial and volume growth following thinning of forest stands, and only a few examples will be given. In heavily thinned Douglas fir stands, volume growth was 12 times greater than in unthinned stands (Waring *et al.*, 1981). Diameter growth of trees in unthinned loblolly pine stands was only 30% of that in thinned stands (Fig. 7.6)

During plantation establishment, trees typically are planted relatively close together to develop nearly cylindrical stems. After this has been accomplished and competition among trees has become intense, plantations commonly are thinned to concentrate wood production in the lower stems of selected trees. Too-wide original spacing results in trees with long crowns, many knots, and stems with too much taper.

Before competition was severe in young red pine plantations, most wood was produced in the lower stems of trees, which consequently were strongly tapered. When the crowns began to close, more wood was added to the upper stem than the lower stem (Fig. 7.7C). Shortly after the stand was thinned, more carbohydrates were allocated to the lower stem, where wood production accelerated (Fig. 7.7D).

The amount of growth increase and rapidity of response to thinning vary greatly with the physiological efficiency, especially photosynthetic production, of the residual trees. The physiologically efficient, large-crowned dominant trees have the greatest capacity to respond to thinning. The smaller crowned intermediate trees have more limited capacity to respond. The physiologically inefficient suppressed trees with their very small crowns have the lowest capacity to respond and may not show a positive response for many years. Responses to thinning also vary appreciably with site quality. Physiologically inefficient stands on poor sites require a longer time than stands on good sites to respond to thinning.

Thinning Shock Early responses to thinning are not always positive. Sometimes the residual trees of heavily thinned stands show a reduction in height or diameter growth, chlorotic foliage, and injury (e.g., sunscald) or mortality of trees associated with increased exposure. In addition to experiencing a physiological shock, the trees in thinned stands may be subject to damage from increased wind, snow, or ice. The severity and duration of

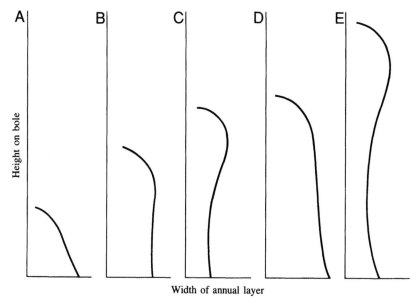

Figure 7.7 Variations in thickness of the annual ring at various stem heights in plantation-grown conifers of varying age: (A) at 8 years, when crowns extend to the base of the tree; (B) crowns closing; (C) lower branches are dead; (D) shortly after thinning when crowns have been exposed to full light; (E) competition is again severe and crowns have closed. The horizontal scale is greatly exaggerated. From Farrar (1961).

such thinning shock vary with species, thinning intensity, and site as well as tree vigor and age.

Some species that grow in very dense stands do not thin naturally but develop a stagnated physiological state, especially on poor sites. Examples are Douglas fir, lodgepole pine, Scotch pine, jack pine, Virginia pine, and sand pine. Stagnated stands of these species often have limited capacity to respond positively to thinning. Immediately after a 27-year-old Douglas fir stand was thinned, the residual trees showed thinning shock characterized by appreciable reduction in their rate of height growth (Harrington and Reukema, 1983). Low capacity of stagnated lodgepole pine trees to respond to thinning was attributed to their high rate of use of photosynthate in root growth and/or respiration (Worrall *et al.*, 1985). The slow response to thinning also may have been associated with photoinhibition of photosynthesis in the suddenly exposed foliage (see Chapter 5 of Kozlowski and Pallardy, 1997).

Wood Quality

Increasing the rate of cambial growth by thinning of forest stands usually improves wood quality. This is because the outer stemwood of trees in thinned stands is stronger, warps less, and is more easily worked than the inner wood. Thinning usually prolongs seasonal diameter growth and increases the proportion of dense latewood, thereby increasing the strength of wood. In some species, the beneficial effects of thinning on wood quality are counteracted by production of epicormic shoots (see Chapter 3 of Kozlowski and Pallardy, 1997), which may produce knots and degrade lumber.

Insects and Diseases

Thinning of stands has variable effects on attacks of trees by insects and fungi that cause disease. Outbreaks of pine beetles often are associated with high stand densities that lead to low tree vigor (Lorio, 1980). By decreasing competition among trees, thinning reduces tree mortality induced by attacks of beetles on southern pines (Nebeker, 1981). Reduction in mortality of loblolly pines in thinned stands was correlated with increased resin flow (Matson *et al.*, 1987; see also Chapter 8 of Kozlowski and Pallardy, 1997).

Thinning a Douglas fir stand increased the abundance of insect vectors of black stain root disease. The increase was attributed to attraction of the insects to resin and the dead inner bark of slash (Harrington and Reukema, 1983; Witkosky *et al.*, 1986). In comparison, thinning of lodgepole pine stands reduced the amount of *Atropellis* canker which causes cambial necrosis and stem deformity (Stanek *et al.*, 1986).

Wind Damage

Trees growing in dense forest stands may become prone to windthrow when the surrounding trees are removed. Open-grown, strongly tapered trees are much more stable (Petty and Worrell, 1981). Because of the dangers of damage to trees by strong winds along the coasts of England and Ireland, conventional thinning is not recommended for many forest stands.

A description of methods and schedules of thinning of forest stands is beyond the scope of this book. As emphasized by Oliver *et al.* (1986), because of different objectives of forest managers and lack of uniformity of forest stands, there is no single correct way to thin a stand. However, an incorrect way can appreciably set back a manager's objectives. For useful discussion of methods of thinning forest stands, the reader is referred to the books by Daniels *et al.* (1979) and Smith (1986).

Pruning

The pruning of branches of trees has been practiced for centuries and was even recorded in the Bible. The effects of pruning of branches on the amount of growth and on stem form may be expected to vary with the severity of pruning. Removing small amounts of foliage generally has little effect on growth, whereas removal of large amounts decreases growth (Rook and Whyte, 1976). When a small amount of leaf surface is removed, photosynthesis of the remaining foliage is increased and growth often is not significantly affected. After some critical amount of foliage is removed, however, compensatory photosynthesis does not occur because the remaining leaves have a maximum photosynthetic capacity. Hence, together with total reduction of photosynthate growth is inhibited (Luxmoore *et al.*, 1995).

Foresters, arborists, and fruit growers prune trees but for somewhat different reasons.

Forest Trees

Foresters prune trees primarily to produce lumber with few knots and to correct the form of strongly tapered tree stems (to produce more cylindrical stems). In agroforestry systems pruning of trees has been used to improve growth of understory plants. For example, pruning of hedgerows of leadtree greatly increased the yield of maize (Field and OeMatan, 1990).

By pruning branches of English oak, Hochbichler *et al.* (1990) obtained clear stem sections 5 to 6 m long. In general, pruning is done selectively on forest trees that are likely to survive because the high costs of pruning may exceed the economic benefits derived, especially as most young forest trees are likely to die because of competition before they reach harvest size. Furthermore, many forest trees shed branches either by physiological abscission (cladoptosis) or through the action of biotic and mechanical agents (see Chapter 3 of Kozlowski and Pallardy, 1997).

The influence of removal of live branches on diameter growth is the opposite of the effect of thinning. Whereas thinning of a stand stimulates diameter growth of the lower stems of residual trees, pruning tends to inhibit diameter growth at the stem base. This is because wood production following removal of branches becomes concentrated in the upper stem. Hence, pruning tends to reduce stem taper, with the amount of reduction varying with pruning severity and timing as well as crown characteristics of the tree before the pruning treatment. The effect of pruning on diameter growth is greater on open-grown trees than on trees in closed stands. The more branches that are removed, the greater is the reduction in wood increment and the upward displacement of growth increment that leads to a decrease in stem taper (Larson, 1963a). Pruning experiments on loblolly pine in which the remaining crowns were 50, 35, or 20% of tree height inhibited diameter growth at breast height much more than at 80% of height (Fig. 7.8).

Correction of stem form by pruning of branches of strongly tapered trees also is modified by tree age, with stems of old trees being less responsive to change in taper than those of young trees. Unfortunately many pruning treatments did not alter stem form either because too few

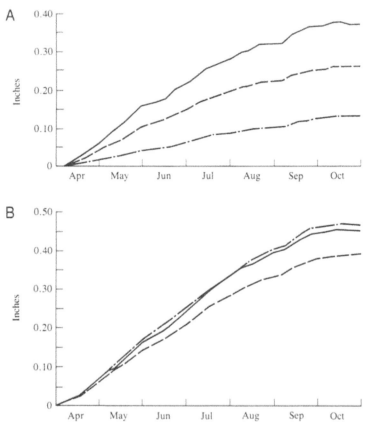

Figure 7.8 Effect on diameter growth of pruning loblolly pine trees to various percentages of their height. (A) Diameter growth at breast height (135 cm). (B) Diameter growth at 80% of tree height. Class 50 trees (solid lines) were pruned to 50% of their height, class 35 trees (dashed lines) to 35% of their height, and class 20 trees (dot–dash lines) to 20% of their height. Reducing the crown size reduced diameter growth much more at breast height than at 80% of their height. Reprinted from the *Journal of Forestry* **50**: 474–479, published by the Society of American Foresters, 5400 Grosvenor Lane, Bethesda, MD 20814-2198. Not for further reproduction.

branches were removed or because pruning was excessively delayed. Larson (1963a,b) emphasized that tree stems become more cylindrical with increasing age and with greater stand density. Hence, a long delay in pruning or removal of only a few branches from trees in closed stands may not be followed by significant changes in stem form.

Pruning of branches usually has little effect on height growth. Increase in tree height depends on carbohydrates and growth hormones produced in the upper crown, and a large portion of the crown can be removed by pruning from below with little or no effect on height growth. For example, removal of 30 to 70% of the live crown of red pine trees had little effect on height growth (Fig. 7.9), and this also was true for loblolly pine.

Many suppressed basal branches with only a few leaves consume in respiration most of the carbohydrates produced in photosynthesis and do not contribute carbohydrates for growth of the main stem (Kozlowski, 1971b). Removal of such lower branches will not affect diameter growth but will improve wood quality by assuring production of knot-free lumber.

Ornamental Trees

Shade trees are routinely pruned with several objectives in mind, such as eliminating dead, injured, diseased, broken, and crowded branches; training for topiary, espalier, or bonsai forms; reducing tree size to prevent crowding, avoid street wires, facilitate spraying, and prevent obstruction of views; influencing flowering and fruiting; invigorating trees; and decreasing water loss from the crowns of transplanted trees (Harris, 1992).

The time to prune depends on the plant species and objectives. Light prunings can be made at any time of the year. Removal of broken, dead, or weak branches has a negligible dwarfing effect on trees irrespective of the time when done. Heavy pruning of shade trees usually is done late in the dormant season. However, for maximum dwarfing, trees should be pruned

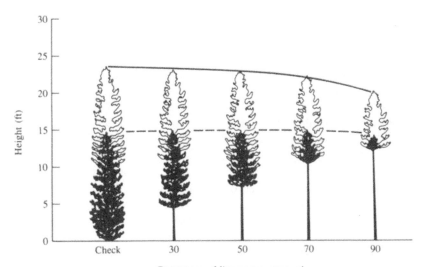

Figure 7.9 Effect of removal of various percentages of the live crown on height growth of red pine. Reprinted from the *Journal of Forestry* **55**: 904–906, published by the Society of American Foresters, 5400 Grosvenor Lane, Bethesda, MD 20814-2198. Not for further reproduction.

when seasonal growth of shoots is almost completed (from late spring to mid summer). Summer pruning should be moderate so as not to stimulate growth of new shoots. Harris (1992) has a good discussion of methods of pruning shade trees.

The form of Christmas trees often is improved by pruning so as to favor loss of apical dominance (see Chapter 3 of Kozlowski and Pallardy, 1997). Young unpruned conifers often have long internodes in the main stem and branches, which give the tree a spindly appearance. Hence growers often "shear" young trees by cutting back the terminal leader and current-year lateral shoots or by debudding them. Shearing inhibits shoot elongation and stimulates expansion of dormant buds as well as bud formation and expansion of newly formed buds into branches. The result is a bushy, high-quality Christmas tree (Fig. 7.10). Species with dormant buds along each internode (e.g., spruces and firs) can be sheared at any season except early summer. Shearing of pines, which generally do not bear dormant buds along each internode, should be done while the shoots are elongating (Smith, 1986).

Fruit Trees

Fruit trees are pruned regularly, largely to ensure harvesting fruit of high quality and for convenience in spraying and picking fruit. To obtain fruit of high quality it is necessary to prune branches to increase exposure of fruits and adjacent leaves in the fruit-bearing part of the canopy to high light intensity (Chapter 8).

Eradicative Pruning

Excision of infected or infested branches has long been practiced to prevent further spread of pathogens or insect pests into trees. Such eradicative pruning often is combined with other treatments in programs of integrated pest management. As emphasized by Svihra (1994), effective eradicative pruning requires considerable skill and knowledge of the biology of the tree as well as that of the attacking insect or pathogen and their vectors. These requirements are important because eradicative pruning (1) may increase the undesirable effects of the insect pest or disease by cutting into healthy tissues, hence permitting spread of the insect or pathogen, (2) necessitates evaluation of visible external symptoms and the extent of internal symptoms, and (3) typically is used with other cultural practices to stimulate production of the host tree of chemical defenses against insects and diseases. Svihra (1994) discussed benefits and problems associated with eradicative pruning.

The wounds caused by pruning are invasion sites for pathogenic fungi. Diseases may be increased by transmission of pathogens on pruning tools. However, the stimulation of growth of succulent shoots (which may be

Figure 7.10 Shaping of Scotch pine trees by cutting back of shoots (shearing). (Left) Spindly, control tree; (right) well-shaped and bushy tree that had been sheared. Photos courtesy of U.S. Forest Service.

more susceptible than normal shoots to invasion by pathogens) may be a more important cause of disease transmission (Mika, 1986).

Pruning of Monterey pine branches greatly increased infection by *Diplodia pinea*, which causes crown wilt and decay of the cambium (Chou and MacKenzie, 1988). Only the upper stem was predisposed to infection. Hence, increasing the amount of pruning provided more wound openings, thereby increasing the chances of infection. Six years after Norway spruce trees were pruned, nearly half the wounds had not healed, and the wood in more than one-third of the pruned trees showed discoloration and decay (Tomiczek, 1992). Cankers associated with decayed branch stubs and poorly healed pruning wounds were good indicators of discoloration and heart-rot of *Acacia mangium* trees (Lee *et al.*, 1988).

The time of year when trees are pruned may influence their susceptibility to infection. In Finland the most susceptible time of pruning for inoculation of Scotch pine tree with the canker fungus, *Phacidium coniferanum*, was between October and December (Glass and McKenzie, 1989).

Root Pruning

During lifting and transport of planting stock many of the small absorbing roots are lost, hence disrupting the previous close contact of the root system with a large volume of water-supplying soil. Such losses of the principal absorbing roots often lead to shoot dehydration, growth inhibition, and frequent mortality of transplanted trees (Kozlowski and Davies, 1975a,b).

Root pruning has been used in nurseries to develop compact root systems of forest and ornamental trees so as to condition them to withstand transplanting shock (Chapter 2). Root pruning eventually stimulates regeneration of roots from each severed root while decreasing shoot growth. Undercutting the roots of Monterey pine seedlings in nursery beds produced plants with a compact mass of fibrous roots and a high root–shoot ratio. When planted in the field the root-pruned plants survived drought better than the control (unpruned) seedlings, to a large extent because the former, with their more extensive roots systems, were more efficient in absorbing water (Rook, 1969, 1971). Pruning of the root system of landscape-size Engelmann spruce trees in the nursery, 5 years before transplanting, was followed by a quadrupling of the root surface area and a doubling of the size of the root system in the root ball (Watson and Sydnor, 1987). A reduction in shoot growth following root pruning has been shown for many species, including peach (Richards and Rowe, 1977a) and holly (Randolph and Wiest, 1981).

Responses of plants to root pruning vary with time after treatment and time of year when the roots are pruned. Root growth of deciduous trees is increased more by root pruning in the spring than in the autumn, in part because the rate of photosynthesis is low in the autumn and carbohydrates are less available for transport to the roots (Geisler and Ferree, 1984). In the short term root pruning decreases the root–shoot ratio and inhibits tree growth. In the longer term root pruning stimulates regeneration of roots from each severed root while decreasing shoot growth (hence increasing the root–shoot ratio). The time required for restoration of plant balance after root pruning varied from 25 days for peach seedlings (Richards and Rowe, 1977a,b) to 80 days for Monterey pine seedlings (Rook, 1971) and several months for 22-year-old eastern white pine trees (Stephens, 1964).

In addition to pruning roots mechanically, some investigators used chemicals, especially copper compounds, to prune roots of plants grown in containers (Pellett *et al.*, 1980; Burdett and Martin, 1982). Plant responses to root pruning with chemicals may be expected to vary with species, container size, growing medium, and concentration of the chemical used (Geisler and Ferree, 1984).

The physiological mechanism regulating growth responses of root-pruned trees is complex and undoubtedly involves changes in availability of water, mineral nutrients, carbohydrates, and hormones at meristematic sites. Removing a portion of the root system rapidly decreases absorption of water and mineral nutrients and induces a plant water stress as shown by a decrease in the xylem Ψ of root-pruned holly trees (Randolph and Wiest, 1981). The reduction in water absorption by root-pruned, transplanted Norway spruce trees lasted for 7 weeks, by which time new roots had regenerated and water absorption increased (Parviainen, 1979). Root regeneration provided more root branches, which increased the absorbing surface.

Root pruning is rapidly followed by a decrease in photosynthesis, with the amount of reduction being correlated with the severity of pruning (Rook, 1971). The rate of photosynthesis of Carib pine decreased during the first 2 weeks after root pruning but began to recover after 4 weeks (Abod *et al.*, 1979). As roots start to regenerate, increased partitioning of photosynthate to roots favors their growth. Within a month after the roots of Monterey pine were pruned, there was a threefold increase in the proportion of photosynthate that was transported to the roots. The strong sink strength of the new roots was associated with reduced shoot growth. A gradual decrease in the proportion of carbohydrates that were transported to the roots (and an increase in carbohydrate retention by the shoots) occurred over the next 2 months (Rook, 1971).

Application of Growth Regulators

Applied plant growth regulators may variously influence growth, development, and survival of woody plants. Actual plant responses vary not only with the growth regulator and dosages applied as well as method and time of application but also with plant species and genotype, stage of plant development, site, and environmental conditions. Enough success in influencing plant growth has been achieved with some growth regulator treatments to warrant continued research, with the objective of extending the use of growth regulators in commercial practice.

Applied growth regulators have been used more by horticulturists than by foresters. However, use of these compounds in both horticulture and forestry is likely to increase as new growth-regulating compounds are developed and more is learned about how they affect the amount and quality of food and fiber crops as well as the aesthetic qualities of woody plants.

A voluminous literature has accumulated on the effects of applied growth regulators on seed germination, vegetative growth, and reproductive growth. Applied growth regulators also influence the flow of latex and oleoresins (see Chapter 8 of Kozlowski and Pallardy, 1997) as well as the resistance of plants to environmental stresses.

Rooting of Cuttings

Root formation at the ends of cuttings necessitates that cells possess the capacity to sequentially dedifferentiate, become meristematic, and differentiate into root initials (see also Chapter 9). Both initiation and acceleration of rooting of cuttings of a wide variety of woody plants have been stimulated by exogenous growth regulators. These have been applied as dilute aqueous solutions in which the bases of cuttings are immersed for many hours, concentrated solutions in which cuttings are immersed for a few seconds, or as concentrates dried into talc (Considine, 1983; Hartmann *et al.*, 1990). The immersion methods often have been more successful than the dry talc methods. Considerable success has been achieved in promoting rooting with the three auxins, indoleacetic acid (IAA), α-naphthyleneacetic acid (NAA), and indolebutyric acid (IBA). However, IBA, which moves slowly in plants and tends to be localized near the site of application, has been most popular (Nickell, 1991). The aryl esters of IAA and IBA often are considered better than their free acids in stimulating root initiation (Haissig, 1979). Mixtures of IBA and NAA have been particularly successful for rooting of cuttings of certain plants.

Control of Tree Form and Size

Both crown form and tree size can be regulated by exogenous growth regulators. Crown form has been modified by altering bud initiation and dormancy, shoot extension, promotion of branching, abscission of leaves and branches, and suppression of root suckers and water sprouts (Looney, 1983; Nickell, 1991).

Crown form has been altered by either increasing or inhibiting shoot growth, with growth suppression with applied growth regulators generally being more successful. Gibberellic acid is the most widely used growth regulator for stimulating shoot growth (Nickell, 1991).

Growth-Suppressing Chemicals Several chemicals that reduce growth have been applied to forest, fruit, and shade trees to control tree height, avoid interference with power lines, better manage fruit trees, reduce the cost of storing and shipping nursery stock, and increase flowering through suppression of vegetative growth (Table 7.1).

There are many examples of growth inhibition of different species of woody plants by growth-suppressing chemicals. These include apple (Greene, 1991), pecan (Marquard, 1985; Wood, 1988a,b), stone fruits (Erez, 1986), grape (Shaltout *et al.*, 1988), citrus (Swietlik, 1986), mango (Kurian and Iyer, 1993a), Douglas fir (Wheeler, 1987), pin oak, red maple, white oak, and black walnut (Sterrett *et al.*, 1989).

Table 7.1 Some Examples of Growth-Inhibiting Chemicals

Common name	Structure	Trade name
Subapical inhibitors		
Daminozide (SADH)	Succinic acid-2,2-dimethylhydrazide	Alar, B-9
Chlormequat (CCC)	(2-Chloroethyl)trimethylammonium chloride	Cycocel
CBBP(Phosphon)	2,4-Dichlorobenzyltributylphosphonium chloride	Phosphon-D
Ancymidol (El-531)	α-Cyclopropyl-2,4-methoxy-propyl-2,5-pyrimidine	A-Rest
Paclobutrazol	1-(4-Chlorophenyl)-4,4-dimethyl-2-(1H-1,2,4-triazol-1-yl)pentan-3-ol	PP333, Cultar
Uniconazole	(*E*)-1-(*p*-Chlorophenyl)-4,4-dimethyl-2-(1,2,4-triazol-1-yl)-1-penten-3-ol	S-3307
Flurprimidol	α-[1-Methylethyl]-α-[4-trifluoro-methoxy)pentyl]-5- pyrimidine-methanol	EL-500
General inhibitors		
Maleic hydrazide (MH)	1,2-Dihydro-3,6-pyridazinedione	Royal Slo-Gro
Chlorflurenols (morphactins)	Methyl-2-chloro-9-hydroxyfluorene-9-carboxylate	Atrinal
Fluoridamide	*N*-4-Methyl-3,3-(1,1,1-trifluoromethyl)sulfonyl aminophenylacetamide	Sustar

Plant growth retardants are chemicals that reduce shoot length, primarily by reducing cell expansion but also by inhibiting cell division. Some growth retardants (e.g., daminozide) affect shoot growth largely by inhibiting internode elongation, with relatively minor effects on the number of leaves or their size (Filipovich and Rowe, 1977). Others, such as paclobutrazol, typically inhibit expansion of both internodes and leaves, with spur leaves being affected more than leaves of long shoots.

In addition to inhibiting growth, the growth retardants also may induce several physiological changes (Grossmann, 1990). These include (1) retardation of senescence together with increases in concentrations of chlorophyll, protein, and minerals, (2) increase in transport of assimilates to seeds, (3) promotion of flowering (Chapter 8), (4) reduction of water absorption, and (5) increases in resistance to cold, heat, drought, and disease.

Growth retardants act, at least in part, by inhibiting gibberellin (GA) synthesis in plants. This is shown by the following lines of evidence (Graebe, 1987): (1) treated plants have lower endogenous GA levels than untreated plants, (2) morphological responses to application of growth retardants are relieved by simultaneous application of GA, and (3) GA biosynthesis in cell-free systems from higher plants is inhibited by application of growth retardants. In addition to inhibiting GA synthesis some growth retardants decrease levels of abscisic acid (ABA) (Kurian *et al.*, 1991), cytokinins (Kurian *et al.*, 1992), and sterols in plants. Both sterols and GAs can relieve growth inhibition caused by these compounds.

The growth retardants include (1) onium compounds such as chlormequat chloride (Cycocel, CCC) and meliquat chloride, (2) pyrimidines such as ancymidol and flurprimidol, and (3) triazoles, including paclobutrazol, uniconazole, BAS III, triadimefon, and triapenthanol.

Another group of compounds, the general growth inhibitors (also called plant growth suppressors) (Rademacher, 1991), arrest activity of apical meristems or kill terminal buds. They also decrease apical dominance and, by stimulating growth of lateral buds, often inhibit terminal growth. In this group are maleic hydrazide, chlorflurenol, mefluidide, and amidochlor (Table 7.1).

In the last decade most research on growth-suppressing chemicals has focused on the triazoles, especially paclobutrazol and uniconazole (Fig. 7.11; Tukey, 1989). When used in appropriate concentrations they inhibit height growth, internode elongation, and leaf area but do not alter the number of internodes. Root growth either is maintained or slightly promoted.

Growth retardants have been used as soil surface applications, soil surface bands, soil injections, trunk and crown drenches, subsoil bands, tree injections, and tree bark paints. The amount of growth control varies with many factors including plant species and genotype, specific compounds,

A   B

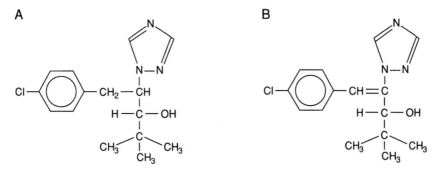

Figure 7.11 Molecular structures of (A) paclobutrazol and (B) uniconazole.

dosage, time of application, soil type, pH, soil organic matter content, tree vigor, rootstock, as well as the amount, frequency, and method of application (Davis and Curry, 1991).

One problem with soil-surface applications of growth retardants is that plant responses may be greatly delayed. Generally an entire season is needed for the compound to be washed into the rhizosphere, absorbed by the roots, and transported to active meristematic tissues. Richardson *et al.* (1986) demonstrated the importance of soil type on tree responses to growth retardants. When apple trees growing in a sandy soil under sprinkler irrigation were treated with 2.0 g/m^2 paclobutrazol, growth was inhibited for 3 years. However, the effects of the same treatment applied to clay or loam soils were evident through the fourth and fifth years.

There is wide variation in the growth-inhibiting power of various triazoles. Triadimefon appears to be least effective and uniconazole the most effective of the triazoles.

Branching The branching habits of trees often present problems for growers. For example, in some strains of apple and pear trees, the first whorl of branches of cut back (headed) trees emerges from a narrow zone below the heading cut. These branches often are too closely spaced and are unsuitable as scaffold branches of orchard trees. By treating dormant buds below the heading cut with appropriate growth regulators such as a mixture of benzyladenine (BA) and GA$_{4/7}$ the headed trees may produce more branches, wider branch angles, and increased shoot growth (Looney, 1983).

Still another problem is a tendency of some apple cultivars to produce too high a proportion of fruiting spurs to nonfruiting branches (hence leading to inefficient use of orchard space). Stimulation of production of more lateral branches and an increase in shoot length have been achieved by early-season applications of appropriate growth regulators such as mixtures of BA and GA$_4$ plus GA$_{4/7}$. Such treatments sometimes have been even more effective when GA$_3$ was added to the mixture (Looney, 1983).

Suppression of Root Suckers and Water Sprouts   Some trees such as apple and pear sometimes produce shoots from adventitious buds located near the soil surface. These root suckers are unsightly, inhibit access to trees, and often harbor pests. Some trees also produce undesirable water sprouts on the main stem and major branches. Unwanted sprouts have been controlled by applications of NAA, chlorflurenol, and maleic hydrazide (Looney, 1983; Harris, 1992).

Integrated Pest Management

When the chlorinated hydrocarbons, organophosphates, carbamates, and many other pesticides were introduced after World War II, pest control shifted from largely biological measures to use of chemicals. Unfortunately, the heavy dependence on chemicals gradually caused serious problems for growers. Many insects and plant pathogens developed resistance to chemical pesticides, and some weeds began to show resistance to herbicides (Muir, 1978). By 1975 approximately three-fourths of the most important insect pests in California had developed resistance to the major insecticides (Harris, 1992). After plants were treated with insecticidal chemicals, insect populations typically decreased abruptly only to increase later to levels higher than those existing before the treatments began. This resurgence reflected both increased resistance of insect pests to insecticides as well as killing by the insecticides of the natural predators of the insect pests.

Following spraying with carbaryl and chlordane to control Japanese beetles near San Diego, California, populations of the wooly whitefly increased by as much as 1200%. Prior to this treatment, introduced parasitoids had kept populations of wooly whitefly under acceptable biological control (Dreistadt *et al.*, 1990). As a result of spray-induced mortality of beneficial arthropods, populations of the citrus red mite and purple scale increased so much that some trees were killed (DeBach and Rose, 1977). Furthermore, public concern with environmental and human health led to considerable restraint in the use of chemical pesticides. Hence, it became obvious that sole reliance on chemicals to control pests was ill-advised.

In the 1980s and 1990s attention has shifted to integrated pest management (IPM) for controlling pests of forest, fruit, and landscape trees. The basic objectives of IPM systems are to decrease plant pests to acceptable levels while protecting people and desirable plants and animals, controlling costs, and preserving plant ecosystems. Integrated pest management relies primarily on natural control of pests and also on a variety of pest-suppressing techniques such as cultural methods, resistant plants, pest specific diseases, sterile insects, insect attractants, parasites of predators, and judicious application of biorational pesticides as needed. Integrated pest management depends heavily on mortality of pests because of natural enemies and weather, and it involves control measures that minimally disrupt these factors. Pesticides are used sparingly and only after monitoring of pest populations and natural control factors indicate a need for them (Flint and van den Bosch, 1981).

Several measures have been incorporated into IPM systems used in forestry. For example, quarantine measures have been used to prevent arrival and establishment of new pests. However, the introduction of many nonindigenous insects such as bark beetles, aphids, and sawflies poses difficulties for use of quarantine measures. Nevertheless, the impacts of introduced insects can be minimized by felling and destroying infected trees. Site selection for planting also is important. For example, birch trees in Europe should not be planted on sandy soils because of the susceptibility of the trees on such sites to *Phyllobius* weevils (Annila, 1977). Soil scarification, use of vigorous plants, and management of clear-cut areas to maximize the distance between breeding areas may reduce weevil populations on replanting sites (Eidmann, 1979). Monitoring of population densities of insects is essential for determining the critical density above which control measures are necessary to avoid serious injury to trees (Wainhouse, 1987).

In Europe the use of insecticides varies greatly in different countries. In several countries DDT has been used but not in Norway (Bakke, 1970). The major use of insecticides in Europe is to protect seedlings from damage by *Hylobius* weevils. Large-scale applications of insecticides are used largely to control Lepidoptera and sawfly defoliators. Conservative use of insecticides is a basic feature of IPM. Accurate timing and placement of insecticides can maximize their effectiveness and minimize their impacts on natural enemies of insect pests and on the environment (Wainhouse, 1987).

Various IPM practices have been successfully used in fruit orchards. Plant breeders have developed varieties of apple trees with resistance to important insect pests and diseases. Resistance to wooly aphid, which originated in the apple variety Northern Spy, has been incorporated in Malling-Merton rootstocks that are readily available to growers. Sources of resistance also exist for powdery mildew and apple scab. In apple orchards IPM programs also incorporate use of selective chemicals for controlling pests while permitting survival of predatory insects (Solomon, 1987). A system for controlling the codling moth involves capture of male moths at pheromone-baited traps. The information obtained on insect numbers is coupled with a time model for insect development and then used to schedule applications of pesticides to control hatching larvae (Croft, 1978).

Various forms of IPM are gradually becoming important in managing landscape trees. Applied compounds that have been substituted for traditional insecticides include dormant and summer oils, azadirachtin, soap, *Bacillus thuringiensis* (*Bt*), and entomogeneous nematodes. The refined horticultural soaps and oils have short periods of toxicity. Summer oils showed little phytotoxicity and controlled the wooly bark adelgid, cottony maple scale, calico scale, golden oak scale, boxwood psyllid, honeylocust plant bug, sycamore lacebug, European pine sawfly, and the euonymus webworm (Baxendale and Johnson, 1988; Davidson *et al.*, 1990). Introduction of a parasitic wasp and a predaceous beetle reduced populations of the ash whitefly (Paine *et al.*, 1993). An IPM program, which controlled plant pests more effectively than application of concentrated pesticides, reduced pesticide use by 94% while maintaining plant quality (Holmes and Davidson, 1984). Principles of IPM are discussed in more detail by Huffaker (1980), Flint and van den Bosch (1981), Sill (1982), Burn *et al.* (1987), Bennett and Owens (1986), and Raupp *et al.* (1992).

Summary

Important cultural practices for increasing yields of forest and fruit trees include site preparations, irrigation, alleviation of mineral deficiencies, thinning of forest stands, pruning of branches, application of growth-regulating chemicals and pesticides, and use of integrated pest management.

Site preparation often improves tree establishment and growth by changing soil properties and controlling plant competition. Useful treatments may include slash disposal, disking, bedding, subsoiling, windrowing, drainage, and use of herbicides. The effectiveness of site preparation varies with management practices, soils, plant species, and topography.

Site quality and tree growth often can be improved by draining unproductive wet sites. Stimulation of tree growth on drained peatlands is associated with lowering of the water table, improved soil aeration, and greater depth of rooting. Because peatlands often are mineral deficient, the effects of drainage can be further improved by addition of fertilizers. However, drainage of certain productive wetlands can be very harmful.

Herbicides often have been used to decrease competition between crop trees and undesirable vegetation. The elimination of competing vegetation increases the availability of light, water, and mineral nutrients to crop trees. However, herbicides sometimes injure crop plants by suppressing seed germination, inducing formation of abnormal seedlings, injuring foliage, decreasing growth, and causing plant mortality.

Irrigation can greatly stimulate growth of woody plants. It often is essential in nurseries and for older ornamental trees, shrubs, and fruit trees.

Unfortunately, the high costs of irrigation of forest trees have restricted its use largely to nurseries, seed orchards, and some short-rotation intensive culture plantations. Overwatering occurs commonly and should be avoided.

Methods of alleviating mineral deficiencies include application of fertilizers, inoculation with mycorrhizal fungi, and use of N-fixing species as interplanted or understory plants. Large amounts of mineral nutrients are leached from nursery soils; hence, regular additions of fertilizers are essential to produce planting stock with high potential for growth. Judicious late-season fertilization in nurseries often stimulates growth and increases survival of outplanted trees. However, overfertilization may result in seedlings of low quality and reduced capacity for growth and survival when outplanted. In some nurseries, excessive use of fertilizers, soil fumigants, and fungicides can reduce mycorrhizal populations.

Addition of fertilizers at or shortly after planting forest trees also may stimulate growth and decrease mortality. Growth of established plantations also can be stimulated by fertilizers. Application of fertilizers before canopy closure and near the end of a rotation is most efficient. Although complete (NPK) fertilizers commonly are applied to shade trees, P and K usually are not deficient elements.

Absorption of mineral nutrients as well as growth and survival of outplanted seedlings have been increased by inoculating nursery beds with mycorrhizal fungi. Responses to inoculation vary with plant species and genotype, soil type and fertility, and species of mycorrhizal fungi.

Interplanting of trees with N-fixing trees (e.g., alder, black locust) can reduce the need for N fertilizers. Planting of N-fixing ground covers also can stimulate growth of trees in plantations and agroforestry systems.

During plantation establishment trees are planted close together to develop nearly cylindrical stems. Too-wide original spacing results in trees with long crowns, many knots, and stems with too much taper. As competition among trees intensifies, plantations generally are thinned to concentrate wood production in the lower stems of trees left standing (to develop more cylindrical stems). Accelerated rates of photosynthesis and greater leaf biomass following stand thinning are associated with increased downward transport of carbohydrates and hormones from the crown; hence, cambial growth in the lower stem is increased. Thinning of stands sometimes induces production of epicormic shoots, alters susceptibility to insects and disease, and increases windthrow of trees.

Branches of forest trees often are pruned to produce lumber with few knots and to correct the form of strongly tapered trees. The effects of pruning are the reverse of thinning (pruning concentrates diameter growth in the upper stem). Pruning of forest trees typically is selective on trees that probably will survive because the costs of pruning may exceed the economic benefits derived. Shade trees are pruned to eliminate dead and injured branches, alter growth form, control tree size, facilitate spraying, prevent obstruction of views, and reduce water loss from transplanted trees. Eradicative pruning to eliminate branch infections or infestations to prevent further spread of a pathogen or insect pest typically is combined with other treatments in programs of integrated pest management.

Roots of seedlings in nurseries sometimes are pruned to develop compact root systems that increase the chances of withstanding transplanting shock. Root pruning stimulates regeneration of roots from the severed roots while decreasing shoot growth. Whereas root pruning decreases plant growth in the short term it stimulates growth in the long term.

Applied plant growth regulators may influence growth and development of woody plants. For example, they may influence seed germination, rooting of cuttings, and tree form and size. Growth regulators also may influence the flow of latex, oleoresin production, and resistance of plants to environmental stresses. A variety of chemicals have been applied to trees to inhibit their growth. By inhibiting gibberellin synthesis, the growth retardants arrest shoot elongation by reducing cell expansion and cell division. Another group of chemicals, the general growth inhibitors, inhibit activity of apical meristems or kill terminal buds.

Since the 1970s, attention has shifted from use of chemicals for control of pests of forest, fruit, and shade trees to reliance on integrated pest management (IPM). Systems of IPM are used to control pests at acceptable levels while protecting people as well as desirable plants and animals and preserving plant ecosystems. Integrated pest management relies on use of several pest-suppressing techniques including cultural methods, resistant plants, pest- specific diseases, sterile insects, insect attractants, parasites of some predators, and judicious use of some pesticides. In addition, IPM depends heavily on natural mortality of pests.

General References

Alexander, A., ed. (1986). "Foliar Fertilization." Martinus Nijhoff, Dordrecht, The Netherlands.

Bacon, P. E., ed. (1995). "Nitrogen Fertilization in the Environment." Dekker, New York.

Bennett, G. W., and Owens, J. M., eds. (1986). "Advances in Urban Pest Management." Van Nostrand-Reinhold, New York.

Binkley, D. (1986). "Forest Nutrition Management." Wiley, New York.

Binkley, D., Carter, R., and Allen, H. L. (1995). Nitrogen fertilization practices in forestry. *In* "Nitrogen Fertilization in the Environment" (P. E. Bacon, ed.), pp. 421–441. Dekker, New York.

Bowen, G. D., and Nambiar, E. K. S., eds. (1984). "Nutrition of Plantation Forests." Academic Press, London.

Burn, A. J., Cooker, T. H., and Jepson, P. C., eds. (1987). "Integrated Pest Management." Academic Press, New York.

Daniel, T. W., Helms, J. A., and Baker, F. S. (1979). "Principles of Silviculture." McGraw-Hill, New York.

Dent, D. (1995). "Integrated Pest Management." Chapman and Hall, London.

Flint, M. L., and van den Bosch, R. (1981). "Introduction to Integrated Pest Management." Plenum, New York and London.

Geisler, D., and Ferree, D. C. (1984). Responses of plants to root pruning. *Hortic. Rev.* **6**, 155–180.

Gordon, J. C., and Wheeler, C. T., eds. (1983). "Biological Nitrogen Fixation in Forest Ecosystems." Martinus Nijhoff/W. Junk, The Hague.

Harris, R. W. (1992). "Arboriculture." 2nd Ed. Prentice-Hall, Englewood Cliffs, New Jersey.

Hartmann, H. T., Kester, D. E., and Davies, F. T., Jr. (1990). "Plant Propagation: Principles and Practices," 5th Ed. Prentice-Hall, Englewood Cliffs, New Jersey.

Huffaker, C. B., ed. (1980). "New Technology of Pest Control." Wiley, New York.

Luckwill, L. C. (1981). "Growth Regulators in Crop Production." Arnold, London.

Mika, A. (1986). Physiological responses of fruit trees to root pruning. *Hortic. Rev.* **8**, 338–378.

Neely, D., and Himelick, E. B. (1987). Fertilizing and watering trees. Illinois Natural History Survey Circular 56, Champaign, Illinois.

Nickell, L. G., ed. (1983). "Plant Growth Regulating Chemicals," Vols. 1 and 2. CRC Press, Boca Raton, Florida.

Pessazaki, M., ed. (1994). "Handbook of Plant Stress." Dekker, New York.

Raupp, M. J., Koehler, C. S., and Davidson, J. A. (1992). Advances in implementing integrated pest management for woody landscape plants. *Annu. Rev. Entomol.* **37**, 561–585.

Rolf, K. (1994). A review of preventative and loosening measures to alleviate soil compaction in tree planting areas. *Arboric. J.* **18**, 431–448.

Sheppherd, K. R. (1986). "Plantation Silviculture." Martinus Nijhoff, Dordrecht, The Netherlands.

Sill, W. H., Jr. (1982). "Plant Protection: An Integrated Interdisciplinary Approach." Iowa State Univ. Press, Ames, Iowa.

Smith, D. M. (1986). "The Practice of Silviculture." Wiley, New York.

Smith, D. M. (1995). Forest stand regeneration: Natural and artificial. *In* "Encyclopedia of Environmental Biology" (W. A. Nierenberg, ed.), Vol. 2, pp. 155–165. Academic Press, San Diego.

Tattar, T. A. (1978). "Diseases of Shade Trees." Academic Press, New York.

Taylor, G. E., Jr., Johnson, D. W., and Andersen, C. P. (1994). Air pollution and forest ecosystems: A regional to global perspective. *Ecol. Appl.* **4**, 662–669.

van den Driessche, R., ed. (1991). "Mineral Nutrition of Conifer Seedlings." CRC Press, Boca Raton, Florida.

8

Cultural Practices and Reproductive Growth

Introduction

A variety of cultural practices, alone and in combination, have been used to regulate reproduction of woody plants. As emphasized by Jackson (1989), the major reasons for manipulating flowering and fruiting are to (1) change the balance between vegetative and reproductive growth in order to favor the latter, (2) stimulate flowering and fruit setting when they are less than optimal, (3) reduce the number of fruits when a heavy fruit load would result in small fruits, low fruit quality, and reduced production of fruit buds (hence lowering fruit yield in the following year), and (4) regulate the season of flowering and fruiting as well as the quality of harvested fruits during storage.

Because fruit quality frequently is more important than fruit yield, management of fruit orchards often is largely directed toward improving fruit quality. In fact some cultural practices, such as thinning of flowers and young fruits, sometimes reduce total crop yield but ensure production of high-quality, high-market-value fruits (Lakso, 1987).

Flowering has been induced in juvenile trees by use of appropriate rootstocks and by manipulating the environment to make young trees grow as rapidly as possible until they achieve the adult stage and critical size necessary for flowering. To promote flowering thereafter, rapid growth should be stimulated because the vigor of adult trees and their flowering capacity are closely correlated. However, this generalization should be viewed with caution because flowering in adult trees also can be triggered by treatments that retard vegetative growth at critical stages of reproductive growth, as discussed later in this chapter.

Growing juvenile plants under long days with high light intensities in a warm greenhouse induced flowering in Japanese larch within 4 years, whereas normally this species does not flower for 10 to 15 years (Wareing and Robinson, 1963). Seedlings of silver birch (Robinson and Wareing, 1969) and apple (Aldwinckle, 1975) flowered in 9 to 18 months when grown continuously under long days. The length of the juvenile phase of crabapple, red pine, and white spruce was greatly reduced by exposure to continuous light (Holst, 1961; Zimmerman, 1971). Under warm greenhouse conditions, citrus seedlings of 13 genetically diverse seedling families flowered in 30 to 48 months, much earlier than they flower in the field. However, the intensity of flowering varied among seedling families. Seedlings of lime, lemon, and mandarin parentage flowered sooner than those of sweet orange, grapefruit, or trifoliate orange parentage (Snowball *et al.*, 1994a). Juvenility of citrus seedlings also was shortened by application of the growth retardant paclobutrazol (Snowball *et al.*, 1994b).

Several treatments have been used to regulate reproductive growth in adult trees. These include judicious arrangement and spacing of trees in orchards and plantations, grafting, application of fertilizers, irrigation, thinning of forest stands, thinning of flowers and young fruits, pruning of branches or roots, scoring or girdling of branches and stems, and application of chemicals such as synthetic growth regulators and inhibitors of vegetative growth. The quality of harvested fruits also can be prolonged by placing them in controlled-atmosphere storages.

Arrangement and Spacing of Trees

Much attention has been given to (1) maximizing interception of light by fruit trees, because the light that falls in the space between trees does not increase fruit yield, and (2) optimizing light distribution within the canopy to increase the efficiency of light in photosynthesis, formation of fruit buds, and fruit coloring (Jackson, 1980).

Interception of light can be increased in young orchards to desirable levels by planting trees that are tall with many lateral branches at the time of planting, and by planting trees at high densities to minimize the amount of space in the alleyways between rows (Jackson, 1980, 1981, 1985). Although it is possible to increase light interception in apple orchards to very high values, this sometimes is inadvisable because it may reduce both fruit yield and quality. For example, an apple orchard with trees 5 m high, spaced at 5.9 by 5.9 m and overlapping the alleyways, intercepted 81% of the available light in the summer, but many of the fruits produced were small and green. Palmer (1989) and Wagenmakers (1991) provide much useful information on methods of regulating light interception in orchards so as to achieve high fruit yield and quality.

High-Density Orchards

Very close spacing of dwarf orchard trees has been used to increase early yield of fruits and nuts, including apples, peaches, apricots, cherries, plums, pecans, and macadamia nuts. Close spacing of trees also facilitates harvesting of fruits and applications of pesticides and growth regulators. Furthermore, high-density orchard systems reduce pruning requirements.

High yields of fruits often are produced in high-density systems. For example, the accumulated yield of fruit over 11 years of a high-density peach orchard was more than twice that of a standard orchard (Phillips and Weaver, 1975). Unfortunately, high-density orchards may become uneconomical sooner than standard orchards because of progressively reduced penetration of light into the canopy (Chalmers *et al.*, 1981; Witt *et al.*, 1989).

Several forms of high-density systems have been used. In hedgerow plantings, the trees are closely spaced in rows with alleys wide enough to handle machinery. In bedding systems, the trees are so closely spaced that they are not identifiable as individual trees either within or between rows. An example is the meadow orchard often used with apple and peach (Luckwill, 1978; Erez, 1978). As many as 70,000 to 100,000 trees ha^{-1} have been used in apple meadow orchards.

Grafting

Precocious flowering can be encouraged by grafting scions on dwarfing rootstocks, grafting on related species, grafting mature scions into seedlings, or grafting or budding in crowns of mature fruiting trees. Vegetative growth of some fruit trees also can be inhibited and flowering stimulated by inserting a dwarfing stem piece (interstock, intermediate stock, or interstem) between a vigorous rootstock and a vigorous scion variety. When scions from adult trees are grafted to seedling rootstocks in the juvenile, nonflowering stage, they usually soon flower.

The Grafting Process

Grafting involves placing the cambium of an excised branch (scion) in contact with the cambium of a rooted plant (stock, rootstock, or understock) in order to bring about a cambial union between the two. This system sometimes is referred to as a "stion" (a combination of stock and scion). Before differentiation of connecting vascular tissues occurs, wound callus tissue, which is formed by both the stock and scion, fills the voids between them. Usually the stock produces most of this callus tissue. As the cells that are damaged during grafting turn brown and die, the callus cells produced by the stock and scion are separated by a boundary line of dead cells, called the isolation layer. Healthy cells soon expand on both sides of the isolation layer and divide rapidly to produce callus tissue. Within a few weeks, tracheary elements differentiate in the callus and across the isolation layer. Shortly thereafter the isolation layer breaks down and is re-sorbed. Eventually vascular cambium is formed through the intermingled callus of the stock and scion. The new cambium usually originates where the existing cambia of the stock and scion are in contact with callus tissue. The newly formed cambium extends until it meets other newly formed, converging cambial cells. When xylem and phloem elements are produced by the new cambium, vascular continuity is established between the stock and scion of compatible combinations. It should be emphasized that heal-ing of a graft union is achieved by cells that develop after the grafting process has been initiated. The protoplasts of the stock and scion do not fuse, and the cells produced by each retain their identity.

Rootstocks

Selected rootstocks often are grafted to scions to increase crop yield, fruit quality, resistance to disease and frost, and adaptability to soil (Wutscher and Dube, 1977; Fallahi *et al.*, 1989). Fruit tree scions have been propa-gated on rootstocks for over 20 centuries. Because fruit yield and quality often are inversely related, the use of rootstocks for increasing yield of fruit sometimes is undesirable. Because of environmental interactions with scion–rootstock combinations, it usually is essential to evaluate different combi-nations under local conditions.

Perhaps the most widely used cultural technique for increasing early flowering in apple and pear is grafting of a scion variety directly to a dwarfing rootstock. Shoot elongation of the compound tree ceases early and abruptly, leading to early and prolific formation of fruit buds. For example, some varieties of apple worked on East Malling IX (M.9) root-stocks may grow to only about one-third of normal size and begin to pro-duce fruit in the second or third year.

On dwarfing rootstocks, a larger proportion of the available carbohy-drates is directed toward reproductive rather than vegetative tissues. For example, on the dwarfing apple stock M.9 about 70% of the assimilates were directed to fruits, largely at the expense of the stem and branches but also of the roots and leaves. In comparison, only 40 to 50% of the assimi-lates were directed to the fruits of later-bearing trees on the vigorous rootstock M.16 (Avery, 1969, 1970). Rootstocks also may influence absorp-tion of mineral nutrients. The pistachio rootstock *Pistachia atlantica* in-creased absorption of P, B, Cu, and Zn by pistachio over the amounts absorbed by plants grafted on other rootstocks (P. H. Brown *et al.*, 1994).

Because trees on dwarfing rootstocks are small and can be closely spaced, the yield of fruit per hectare often is high. Some rootstocks may even cause certain varieties of fruit trees to bear heavily at an early age without materially impeding their vegetative growth. Weak rootstocks also tend to increase the size of fruits, probably by decreasing vegetative growth, and they may hasten fruit ripening. Although the growth rate of the compound tree is intermediate between the growth rates of rootstock and scion, the influence of the rootstock generally is greater than that of the scion (Tubbs, 1973a,b). Victoria plums grown on the French hybrid rootstock Ferlenain formed dwarf trees of smaller stature, but with fruit yields similar to or higher than those on the dwarfing rootstock Pixy. Fruits generally were much larger from trees on Ferlenain rootstocks (Webster and Wertheim, 1993). Clones of *Actinidia helmsleyana, A. eriantha,* and *A. rufa* increased the number of flowers per cane of kiwifruit (cultivar Hayward) by 110, 73, and 30%, respectively. The numbers of floral primordia were similar in the three Hayward–rootstock combinations. Hence, the effects of the rootstocks on flowering were the result of different amounts of floral abortion after flower initiation (Wang *et al.,* 1994a). In another study the roots of flower-promoting rootstocks (*Actinidia helmsleyana* and *A. eriantha*) tended to have more starch grains in parenchyma cells and larger cross-sectional areas of xylem vessels and of phloem idioblasts than roots of the rootstocks *A. rufa, A. deliciosa,* and *A. chinensis,* which had lower capacity to stimulate flowering in kiwi (Wang *et al.,* 1994b).

Specific dwarfing rootstocks may have useful attributes as well as some shortcomings. For example, European plum cultivars on the dwarfing rootstock Pixy have good resistance to silver leaf disease caused by *Chondrostereum purpureum,* but they are not very drought tolerant (Dettori, 1985). In addition, trees on Pixy rootstocks produce smaller fruits than they do on other rootstocks (Webster, 1989; Webster and Wertheim, 1993).

Control of tree size by dwarfing rootstocks appears to be associated with a decrease in gibberellins produced in roots and transported to the tops. The more that a rootstock reduces tree growth, the lower is the amount of gibberellins in the xylem sap. This relation is shown by studies in which strongly dwarfing rootstocks were inhibited more than less dwarfing rootstocks by triazole inhibitors of vegetative growth.

Choice of Rootstocks   In addition to stimulating reproductive growth, the ideal rootstock should be cold hardy, drought tolerant, heat tolerant, and adaptable to a wide variety of climatic conditions. A single rootstock usually does not have all of these attributes. Because of environmental interactions with scion-rootstock combinations, much research has been devoted to evaluating different combinations under local conditions. About 40 rootstocks and interstems are available for apple trees. Most apple rootstocks currently used were developed at the East Malling Research Station in England and are designated M.1, M.2, and so on. Those developed jointly by the East Malling and Merton Stations are designated MM.101, MM.102, etc. The M.9 rootstock is the most popular dwarfing rootstock in the world for one-union trees. However, M.7 has been the preferred rootstock for apple in the United States. It is both disease tolerant and adaptable to a wide range of soils and climates (Ferree and Carlson, 1987).

Pear rootstocks may come from several species of *Pyrus* or a different genus. Seedling rootstocks are widely used with pear. Clonal rootstocks have been largely restricted to selections of quince. In recent years clonal stocks of *Pyrus communis* have become popular in the United States. In cold parts of Europe, pear rootstocks have been domestic or wild seedlings, but in warmer countries, quince rootstocks have been more extensively used (Lombard and Westwood, 1987).

Peach cultivars usually are propagated on peach rootstocks, but sometimes on apricot or almond seedlings (Layne, 1987). The major sources of rootstocks for cherry trees are seedlings or clonal selections of mazzard and mehaleb (Perry, 1987). Rootstocks for cherry in the United States vary regionally. Mazzard rootstocks are favored for cherry growing on marginal soils. In the arid western states, mehaleb rootstocks are popular for cherry on well-drained soils. In the East and Pacific Northwest, sweet cherries are grown mostly on mazzard, and sour cherries on mehaleb.

Rootstocks for almond vary widely (Kester and Grasselly, 1987). Almond seedling rootstocks have been favored in Europe where orchards are grown without irrigation. In California and various countries where almond orchards are irrigated, peach seedlings have been the dominant rootstock. In the last two decades peach–almond hybrids have become very popular for both almond and peach in Europe.

For apricot the most widely used rootstocks have been apricot seedlings. However, rootstocks vary among countries and include apricot, myrobalan plum, and peach (Crossa-Raynaude and Audergon, 1987). For plums the most popular rootstocks in the United States are myrobalan and its hybrids (Okie, 1987). In Europe both myrobalan and Marianna (*P. cerasifera* × *munsoniana*) are the most successful rootstocks.

Almost all citrus trees are now propagated onto rootstock seedlings. Early rootstocks included rough lemon, sour orange, trifoliate orange, sweet orange, and sometimes grapefruit or Cleopatra mandarin (Castle, 1987). In the 1990s there has been much diversification of rootstocks used for citrus throughout the world because of the spread of diseases, especially phytophthora and tristeza, the latter caused by a virus. In California, rough lemon and sour orange rootstocks are being replaced by Carizzo, and almond is used for lemons. In Arizona, rough lemon is the major rootstock, as is sour orange in Texas. In Brazil, most trees are budded to Rangpur; in Italy, to sour orange. In Mexico, where tristeza is not a problem, sour orange is the major stock. In China, a wide range of rootstocks are used: trifoliate orange for satsumas and kumquats, and *Citrus sunki* and other mandarins for satsumas and sweet orange (Castle, 1987).

In forestry, grafting to rootstocks has been used primarily in establishing conifer seed orchards and clone banks because many conifers are not easily propagated by cuttings from mature trees. Rootstocks have been used to increase graft success, reduce incompatibility between stock and scion, modify scion vigor, and increase seed production. However, much less research has been conducted (and much less success obtained) on use of rootstocks for forest trees than for fruit trees (Jayawickrama *et al.*, 1991).

Often, but not always, grafts are more successful when scions are grafted to rootstocks of the same species, as in Pacific silver fir (Karlsson and Carson, 1985). However, some interspecific grafts have been successful, as in *Chamaecyparis* (Hunt and O'Reilly, 1984). Both male and female flowering of loblolly pine were increased by grafting scions on Virginia pine (Schmidtling, 1983). Although plant propagators usually avoid intergeneric grafts in conifers, some have been successful. For example, rootstocks of oriental arborvitae (*Platycladus orientalis*) often have been used for grafting scions of Lawson cypress (Hunt and O'Reilly, 1984).

Graft Incompatibility

Successful grafting depends not only on contact between cambia and other meristematic tissues but also on compatibility between members of the graft union. Incompatibility in graft unions occurs commonly (Fig. 8.1) and has plagued plant propagators for centuries. Grafts within species are more likely to be successful than those between species, genera, or families of plants. Even when members of the graft union are compatible, some woody plants graft more readily than others. Whereas apple and pear trees are easily grafted, apricots are not (Hartmann *et al.*, 1990).

There is considerable difference of opinion about how graft incompatibility should be defined, to a large extent because there are many different causes of incompatibility. As emphasized by Andrews and Serrano-Marquez (1993), many early investigators did not distinguish between graft incompatibility and graft failure, the latter caused by adverse environmental conditions and/or lack of skill of the grafter. Here we focus on graft incompatibility as defined by Andrews and Serrano-Marquez (1993), namely, "failure of a graft combination to form a strong union and to remain healthy due to cellular, physiological intolerance resulting from metabolic, developmental, and/or anatomical differences."

The severity of the symptoms of incompatibility ranges from rapid and complete failure of a stock and scion to graft, or sudden collapse in the first year, to vigorous growth in the first year and a slow growth decline thereafter. Sometimes the symptoms of incompatibility are delayed for many years. Most causes of delayed incompatibility are associated with diseases. An example is the walnut blackline disease caused by cherry leaf roll virus (Mircetich *et al.*, 1980).

Graft incompatibility can be identified by a variety of internal and exter-

Figure 8.1 Incompatibility of graft unions. (A) Six-year-old trees of Oullins gage on common mussel broken smoothly at the union; (B) apricot (var. Croughton) on plum rootstock (Brampton). Photo courtesy of East Malling Research Station.

nal symptoms, with the former appearing first. Internal symptoms may include breakdown of the cortex and phloem, atypical axillary parenchyma cells, lack of axillary parenchyma in the phloem, and necrosis of cortical cells. Common external symptoms may include late bud opening, abnormal leaf structure, premature leaf abscission, inhibition of vegetative growth, shoot dieback, overgrowth of tissues at or about the graft union, and premature plant mortality (Andrews and Serrano-Marquez, 1993).

Causes of Incompatibility Over the years a variety of causes of graft incompatibility have been identified. Herrero (1956) described four classes of incompatibility: (1) graft combinations where the bud failed to grow out, (2) incompatibility due to virus infection, (3) graft combinations with mechanically weak unions, and (4) graft combinations in which growth inhibition was associated with abnormal starch distribution.

Mosse (1962) placed graft incompatibility of fruit trees into two broad groups: translocation incompatibility and localized incompatibility. Neither type was confined to a particular plant species, and both could occur in the same graft combination. Translocation incompatibility was characterized by (1) accumulation of starch above the union and absence below it, (2) phloem degeneration, (3) different responses of reciprocal grafts, (4) normal vascular continuity at the union, although there might be overgrowth of the scion, and (5) early effects on plant growth. Moing and Guadillère (1992) studied carbon and N partitioning in peach/plum graft combinations. For the first 20 days after bud burst (from day 57 to day 78 after grafting), carbon and N reserves were mobilized in the rootstock in both a compatible and incompatible graft. Later, as shoot growth of the incompatible graft declined, carbon export from the scion through the phloem to the rootstock slowed and N assimilation stopped. Localized incompatibility was associated with (1) breaks in cambial vascular continuity, (2) similar behavior of reciprocal combinations, and (3) gradual starvation of the root system and slow development of external symptoms (Mosse, 1962). Hartmann *et al.* (1990) added virus-induced incompatibility to Mosse's translocation and localized incompatibilities.

The major mechanisms of graft incompatibility that have received considerable attention since the 1970s involve cellular recognition, wounding and lignification, hormonal growth regulators, and toxins (Andrews and Serrano-Marquez, 1993). According to the cellular recognition theory, when opposing cells of graft partners come in contact, their cell walls break down and plasmodesmata form. The dissolution of cell walls of contacting cells is accelerated by cell wall-degrading enzymes. After plasmodesmata form, protein molecules which determine incompatibility are released by the plasmalemmas (Andrews and Serrano-Marquez, 1993). Several objections that have been raised to the cellular recognition theory (Moore, 1984) include lack of structural evidence for a recognition system in plant grafts and difficulty of explaining the differential responses of reciprocal grafts. A strong argument against the cellular recognition theory is that cell contact does not appear to be necessary for incompatibility to be expressed.

In the process of compartmentalization, wounded tissues are isolated by release of toxic chemicals that prevent rapid spread of invading microorganisms. Strong compartmentalization is associated with successful grafting. Although compartmentalization is common to both grafts and wound responses, it is not the only process common to both. During formation of both compatible and incompatible grafts, the initial events are similar and include deposition and polymerization of cell walls, increase in tensile strength, disappearance of a large vacuole, cytoplasm vesiculation, loss of cellular integrity, and changes in peroxidase and acid phosphatase patterns (Andrews and Serrano-Marquez, 1993). Nevertheless, prior wounding apparently is not necessary for incompatibility to be expressed, although it can accelerate its symptoms (Moore and Walker, 1983).

Many studies have shown that endogenous growth hormones are involved in incompatibility. Apically produced auxin is transported downward in the phloem. The amount that reaches the roots influences both the kind and amount of cytokinins synthesized and transported to the shoots through the xylem. Auxin in the phloem is oxidized or degraded by compounds in the bark. Variations in amounts of active auxins in different cultivars regulate their growth rates and the dwarfing effects of rootstocks and interstocks. This view is supported by the finding that grafting of a ring of M.26 bark to the scion of a Gravenstein/MM.111 tree produced a tree similar to one with a dwarfing interstock. More dwarfing was induced by grafting of a long piece than a short piece of bark (Lockhard and Schneider, 1981).

Some investigators emphasized that formation of callus and vascular connections across the graft union depend on accumulation of auxin at the base of the scion (Gebhardt and Goldbach, 1988). The essentiality of auxin is emphasized by improved grafting following applications of auxin to stock–scion combinations (Beeson, 1986; Beeson and Proebsting, 1989). Although auxins may not be the only growth hormones that regulate grafting, they usually are very important components of the regulatory hormonal system.

In a number of species, graft incompatibility has been attributed to specific toxins that are produced by plants during the grafting process. In some pear cultivars grafted on quince, a cyanogenic glucoside (prunasin) is catabolized by glycosidase to liberate cyanide at the graft interface. The cyanide destroys both xylem and phloem, events associated with graft incompatibility (Moore, 1986). Studies of almond grafted on plum showed that toxic hydrolytic products of amygdalin and benzaldehyde may account for incompatibility (Heuser, 1987). In some species, isoperoxidase compounds of the stock and scion may account for abnormal lignification and lack of vascular connections at graft unions (Santamour *et al.*, 1986). For example, differences in cambial peroxidase enzymes were associated with graft incompatibility of both Chinese chestnut and northern red oak (Santamour, 1988a,b).

Control of graft incompatibility in many species will be difficult until more is known about specific mechanisms of incompatibility. In species in which the mechanism depends on toxins, graft incompatibility can be partly controlled by selecting rootstocks or scions that produce low amounts of toxic metabolites (Andrews and Serrano-Marquez, 1993). Techniques of interfering with mechanisms of translocated incompatibility or those that are not caused by toxins may be more difficult to develop until the mechanisms involved are better understood. Hence, much more research is needed on mechanisms of graft incompatibility in various species and genotypes.

For excellent discussions of grafting techniques and problems, see Howard (1987) and Hartmann *et al.* (1990).

Fertilizers

It is common practice to apply fertilizers to fruit trees to stimulate floral induction, increase fruit set and growth, develop high-quality fruits, and prolong the storage life of harvested fruits. Most benefits of fertilizers are obtained only if factors other than mineral deficiency are not seriously limiting growth.

It is difficult to sustain yield of fruit trees in the tropics without a fallow period or fertilizer application. Acceptable yields sometimes can be obtained without fertilizers by growing crop trees together with other trees, often N-fixing legumes. In coffee plantations, for example, the roots of selected interplanted trees can absorb mineral nutrients from soil depths beyond the reach of coffee roots. Such nutrients eventually are returned to the soil surface in litter and subsequently become available for absorption by the roots of coffee trees (Jordan, 1985).

Foliar sprays of fertilizers sometimes have been used to supplement soil applications in order to stimulate reproductive growth. Calcium sprays have been applied to correct a wide variety of fruit disorders. Advantages of foliar applications of B, Cu, Mg, Mn, and Zn over soil applications include high effectiveness, rapid plant responses, and reduced toxicity from excessive accumulation of certain elements in the soil. A disadvantage is that the effects of sprays are temporary.

The major element most often deficient for reproductive growth is N. Nearly one-third of the total annual N requirement of 20-year-old apple trees and one-fifth of that of 25-year-old trees were used in the fruit and removed when fruits were harvested (Magness and Regeimbal, 1938; Murneek, 1942). Hence, annual applications of N to fruit trees are standard practices (Sanchez *et al.*, 1995). Potassium and Mg need to be added only when a need for them is demonstrated. Phosphorus is not deficient in many orchards (Westwood, 1993).

Both the rate of fruit growth and size of fruits at harvest are controlled by relations between the numbers of fruits and leaf surface. Application of N fertilizers may decrease the size of harvested fruits if floral initiation and fruit set are increased proportionally more than leaf growth. However, if the added N does not greatly increase floral initiation and fruit set, it usually greatly increases the size of the mature fruits.

Fruit growers must consider the effects of fertilizers on fruit quality, both at harvest and during storage, as well as fruit yield. Excessive applications of N to increase tree growth and fruit yield may result in soft fruits subject to preharvest drop, cork spot, bitter pit, and internal breakdown in storage (Swietlik and Faust, 1984). Both the amount of N and the form in which it is applied are important. Ammonium N is said to reduce Ca uptake and can contribute to Ca deficiency, so $CaNO_3$ fertilization has been recommended for apples. However, Bramlage *et al.* (1980) concluded that fruit quality was affected more by the amount of N than by the form in which it was applied.

Nitrogen fertilizers traditionally have been applied to fruit trees in the spring, but it may be advantageous to apply additional N later in the season. Summer applications of N fertilizers to some fruit trees stimulate more vigorous flowers the following spring, which are more likely to set fruit than those receiving fertilizer only in the spring. Hill-Cottingham and Williams (1967) found essentially no fruit set in the following spring in control apple trees or those given N fertilizer in the preceding spring. In contrast, trees given N fertilizer in the summer showed appreciable fruit set, and those given N in the autumn showed a heavy fruit set in the next spring (Fig. 8.2). Summer and autumn applications of N to apple trees were superior to spring applications because they increased the longevity of ovules and the period during which the stigmas were receptive to pollen (Delap, 1967). Of course, autumn fertilization is most effective on trees with low N reserves (Hennerty *et al.*, 1980). Even though application of N fertilizer usually is beneficial, it must be performed with caution. If too many fruits are set, the demand for carbohydrates by the growing fruits may exceed the photosynthetic capacity of the leaves and result in small fruits.

Forest Trees

Fertilizers have been applied to seed orchards and mature stands of forest trees selected for seed production, often in combination with stand thinning, to increase flowering and seed production. In temperate zones, fertilizers usually are applied in the spring when initiation or differentiation of cones is occurring (Sweet and Hong, 1978). With pines and other trees that require 2 years to mature seed, the effects of fertilizers are seen on flowering in the spring after fertilization and on seed production the second autumn after fertilization. Because the greatest effect is on the next crop, fertilizers need to be applied annually to maintain high yields. Of course, seed production is related to other factors in the environment, especially water supply and unseasonable freezes. For example, a positive flowering response of Monterey pine seed orchard trees to N fertilizers was

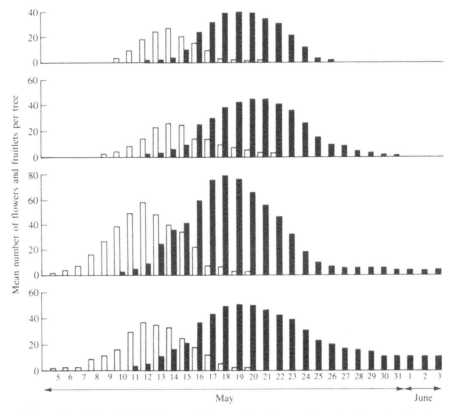

Figure 8.2   Effect of N fertilizers applied at various times on flowering (open bars) and number of fruitlets (solid bars) of Lord Lambourne apple trees. (Top to bottom) Without N; spring application; summer application; autumn application. From Hill-Cottingham and Williams (1967).

demonstrated in years of high precipitation in Australia; in drier years, flowering was stimulated by an interaction of irrigation and N fertilizer. Hence, the response to N fertilizer depended on an adequate water supply (Griffin *et al.*, 1984).

Flowering of clones of trees that tend to flower profusely can be influenced most by fertilizers. Male and female flowering of some species may be affected differently by fertilizers. For example, N fertilizer promoted initiation of female cones in pine, larch, and several angiosperms, but in pine the effect on formation of male cones was negligible or negative (Sweet and Will, 1965; Giertych and Forward, 1966). In comparison, in Douglas fir, N fertilizer increased production of both male and female cones (Smith *et al.*, 1968; Griffith, 1968). In some conifers, the form of N applied is important to flowering. In Douglas fir, nitrate N promoted flowering whereas ammonium N did not (Ebell, 1972a). The effects of the form of N applied may vary with soil conditions as well as species.

The efficacy of N fertilizers on flowering has been attributed to their capacity to increase the proportion of arginine and other guanidines in the soluble N pools of shoots (Ebell and McMullen, 1970). Application of N increased both arginase levels and differentiation of reproductive primordia in trees as different as Douglas fir (Ebell, 1972a,b) and apple (Grasmanis and Edwards, 1974). However, Sweet and Hong (1978) emphasized that neither high N levels nor arginine had a specific role in inducing cone production in Monterey pine. They concluded that the major role of N fertilizer, when it increased cone production, probably was to increase crown size and hence the number of sites in the crown where cones could form. Such a mechanism would be effective only on sites where growth rates and crown size were limited by N deficiency.

Sweet and Hong (1978) concluded that for optimal cone production all mineral nutrients should be maintained at levels that are optimal for vegetative growth. They emphasized that the timing of application of N fertilizers was crucial and that fertilizers should be applied when initiation or differentiation of cones is occurring.

Irrigation

The importance of water supply to reproductive growth is dramatically illustrated by the beneficial effects of irrigation on yields of fruits and seeds (Kozlowski, 1983). Fruit trees are routinely irrigated in California, the Yakima Valley of Washington, and in Florida. Even in humid England, irrigation increased the yield of apples (Table 8.1). However, frequent irrigation in some years decreased yield by the adverse effects of soil anaerobiosis on physiological processes. In three successive years, the yields of irrigated apple trees were 25 to 42% higher than those of unwatered trees (Goode and Ingram, 1971). In 2 of 4 years of irrigation, the water supply (rainfall plus irrigation) in the year before flowering accounted for 71% of the variation in flowering. There also was an interactive effect between irrigation and N application in dry years. Response to N depended on an adequate water supply. Growth and crop yield of Cox's Orange Pippin and Golden Delicious apple trees on each of four rootstocks were increased by irrigation in England over a 4-year period (Fig. 8.3). Pecan orchards in the southeastern United States are irrigated to increase early flowering, nut size, and yield (Worley, 1982; Stein *et al.*, 1989). Large increases in citrus production following irrigation also are well documented (Kriedemann and Barrs, 1981).

Table 8.1   Effect of Irrigation during 1953 to 1960 on Yield of Apples
during Bearing Years[a]

	Yield (kg/tree) after treatment			
Year	Frequent irrigation	Medium irrigation	Infrequent irrigation	Unwatered control
1954	24.5	32.3	30.9	26.5
1956	76.1	76.1	57.0	61.1
1958	113.2	113.4	98.5	78.4
1960	171.1	183.5	164.9	122.0
Total	384.9	405.3	351.3	288.0

[a]From Goode and Hyrycz (1964).

Crop yields are functions of both the amount and timing of irrigation. The best times for irrigating vary somewhat among species. The critical periods are those in which plant water deficits appreciably reduce yields (Table 8.2). Enough water should be applied at each irrigation to wet to field capacity the entire volume of soil that contains roots. Methods of estimating the amounts of water to apply are discussed by Thorne and Thorne (1979).

The cost of irrigation of forest trees may be prohibitive except in nurs-

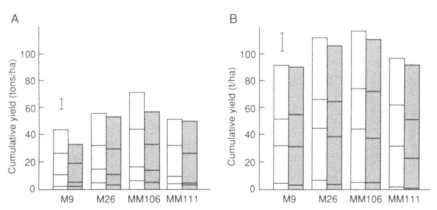

Figure 8.3   Effect of irrigation on crop yield from 1984 to 1987 for (A) Cox's Orange Pippin and (B) Golden Delicious apple trees on four rootstocks. Each portion of the stacked histogram represents the yield each year, with 1984 at the bottom and 1987 at the top. Open bars represent irrigated trees, and shaded bars denote nonirrigated trees. Bars indicate significant differences. From Higgs and Jones (1990).

Table 8.2   Critical Periods of Soil Water Stress[a]

Species	Critical water-stress period
Apricot	Period of flower and bud development
Cherry	Period of rapid growth of fruit prior to maturing
Citrus	Flowering and fruit setting stages; heavy flowering may be induced by withholding irrigation just before flowering
Cotton	Flowering and boll formation > early stages of growth > after bolls form
Peach	Period of rapid fruit growth prior to maturity
Strawberry	Fruit development to ripening

[a]From Doorenbos and Pruitt (1977). Food and Agriculture Organization of the United Nations, FAO Irrigation and Drainage Paper No. 24.

eries, seed orchards, and possibly plantings on infertile soil that can be irrigated with effluent from sewage plants. Some investigators have emphasized the importance of irrigating seed orchards on seed yield, especially in dry regions. In Australia, supplementing rainfall with irrigation stimulated flowering and seed production in Monterey pine seed orchards (Griffin *et al.*, 1984). Harcharik (1981) reported that irrigation plus fertilizer applications increased seed production of pines in the southeastern United States by 30% compared with adding only fertilizers, and it doubled seed production over that by trees that were neither irrigated nor fertilized.

Irrigation Methods

The principal methods of applying water are basin and furrow irrigation, sprinkler irrigation, trickle or drip irrigation, subsurface irrigation, and mist irrigation. Application of water in furrows or basins can only be done on level land, requires much labor, and often results in flooding of low areas in orchards. Sprinkler systems have largely supplanted furrow irrigation because they can be used on rolling land, permit good control of the amount of water applied, protect trees against freezing (Parsons *et al.*, 1991), and require less labor than furrow irrigation does. Fertilizers sometimes are applied in the irrigation water (a process called fertigation), but the possibility of salt injury exists. The wet leaf surfaces also favor spread of leaf pathogens. Sprinkler irrigation sometimes is useful in preventing frost injury in orchards (Parsons *et al.*, 1985, 1991). Trickle or drip irrigation (distributing water through tubing and allowing it to trickle out on the soil through nozzles), allows for release of measured amounts of water near the roots. Subsurface irrigation can be used on sloping land, conserves water, and decreases salt accumulation on plants, but it is expensive to install.

Applying irrigation water as a mist can increase leaf Ψ as much as 0.8 to 1.0 MPa over the Ψ in leaves of soil-irrigated trees. Some of the effect of mist irrigation on leaf Ψ results from increased air humidity (hence decreased evaporation). The major beneficial effects of mist irrigation are that (1) wetting of leaves lowers the leaf temperature (hence decreasing the vapor pressure difference between the leaf and air) and (2) evaporation of water occurs more from the leaf surface than from the leaf interior. Evaporation from the leaf surface decreases both water movement within the plant and the leaf water deficit (Jones *et al.*, 1985). The low water stress of mist-irrigated leaves often persists after the leaves dry, probably because of water storage in the leaves (Goode *et al.*, 1979).

Regulated Deficit Irrigation

Where drainage is adequate, overwatering can result in luxury use of water, creating excessive vegetative growth and thereby intensifying the tendency toward biennial bearing (Cripps, 1981; Klein, 1983). Hence, irrigation techniques have been designed to inhibit excessive vegetative growth by reducing irrigation during rapid vegetative growth.

Crop yields either have not been reduced or have been increased by withholding irrigation during the period of rapid shoot elongation. During the subsequent fruit growth period, moisture stress is minimized. This technique, called regulated deficit irrigation (RDI), increased yield of peach (Chalmers *et al.*, 1984) and pear trees (Mitchell *et al.*, 1984, 1986). By withholding as much as 87.5% of the irrigation requirements of peach trees (as determined from pan evaporation), vegetative growth could be reduced by as much as 75% without reducing fruit yields. Less severe water-withholding treatments increased fruit yield by up to 30% while appreciably reducing vegetative growth (Chalmers *et al.*, 1983). Water use by peach trees under RDI was reduced by one-half during slow fruit growth and by nearly one-third when the trees subsequently were irrigated (Boland *et al.*, 1993). Vegetative growth was reduced by RDI whereas fruit growth was not. Reduction in water use during RDI was attributed to both smaller leaf area and decreased stomatal conductance. Post-RDI reduction in plant water loss was largely explained by reduced leaf area.

The effect of RDI on fruit growth depends on time of treatment. Caspari *et al.* (1994) withheld water from Asian pear trees before or during the period of rapid fruit growth. Water use by trees was reduced by 8% with each treatment over well-watered control trees. Shoot growth was reduced without adverse effects on fruit growth and yield when water was withheld before rapid fruit growth began. In contrast, fruit growth was inhibited by a water deficit during the final weeks of fruit growth without any favorable effect on shoot growth. Peach trees subjected to RDI photosynthesized more and used water more efficiently during the latter part of the stress period than non-water-stressed trees (Girona *et al.*, 1993). However, midday leaf Ψ and stomatal conductance of leaves of RDI trees did not recover to the values of control trees until more than 3 weeks after full irrigation was resumed at the beginning of stage III of fruit growth. Regulated deficit irrigation caused only an 8% reduction in trunk growth but resulted in a 40% saving in irrigation water.

The effects of RDI on reproductive growth may vary appreciably with site. Regulated deficit irrigation of French prune trees growing on a deep soil increased both flowering and fruit yield (Lampinen *et al.*, 1995). In comparison, RDI of trees growing on shallow soil increased flowering but greatly reduced fruit yield. The lowered yield was attributed to increased fruit drop because of the accelerated onset of and more severe water stress.

Developing irrigation regimes requires knowledge of both appropriate timing and amount of water to apply to restore losses by evapotranspiration. Procedures for scheduling irrigation in orchards may involve (1) plant or soil measurements to determine irrigation timing or (2) use of water budgets to determine the soil depth of water application and timing of irrigation. The water budget method requires, in addition to an estimate of evapotranspiration, determination of allowable soil water depletion (also called yield threshold) (Goldhamer and Snyder, 1989).

Several indicators for irrigation timing have been used including the rate of fruit growth, stomatal aperture, stem diameter growth, leaf Ψ, fruit turgor, velocity of sap movement, and canopy to air temperature differences (Shalhevet and Levy, 1990). Gravimetric soil water measurements, tensiometers, and neutron-scattering meters have been useful in determining soil moisture content as an index of irrigation need.

Irrigation Problems

In addition to the difficulties associated with methods of applying water and timing of irrigation, problems with soil anaerobiosis and salt accumulation sometimes accompany irrigation. Prolonged saturation of deep soil horizons often occurs because of too-frequent or heavy irrigation together with poor internal drainage. As mentioned, flooding of soil commonly inhibits formation of flower buds, leads to abscission of flowers and young fruits, arrests fruit development, sometimes induces cracking of fruits, and adversely affects the keeping quality of fruits (Chapter 6).

Irrigation water frequently contains appreciable amounts of salts. For example, water used for irrigation in the southwestern United States may contain 0.1 to 5.0 tons of salt per acre foot (0.25 to 12.5 metric tons ha^{-1}). Hence, 1 to 10 tons of salt acre^{-1} (2.5 to 24.0 metric tons ha^{-1}) may be added to the soil in a year, with most remaining when the water is removed

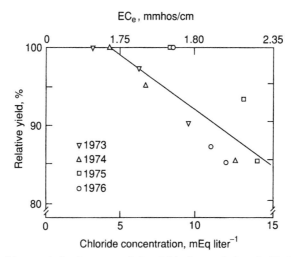

Figure 8.4 Linear relation between relative yield of grapefruit and chloride concentration of the soil saturation extract. From Bielorai *et al.* (1978), with permission of Springer-Verlag.

by evapotranspiration. Salt accumulation can be prevented only by supplying enough water to leach it out. Often this cannot be done because of lack of water or inadequate drainage (Kramer, 1983; Kramer and Boyer, 1995).

When irrigation water is applied to the soil, the fine soil particles (which remain in suspension for a long time) reduce the infiltration rate by clogging soil pores. Evaporative loss of water is increased, causing salts to concentrate in the surface layer. It has been estimated that about one-third of the world's 230×10^6 ha of irrigated land is influenced by salinity (Downton, 1984).

Much research has shown that yield of most fruit trees is reduced by salinity (Chapter 6). For example, yield of grapefruit was linearly related to the chloride concentration in the soil solution (Fig. 8.4). Fruit yield was reduced by 1.45% for each 1 mEq liter^{-1} increase in chloride concentration over a threshold value of 4.5 mEq liter^{-1} (Bielorai *et al.*, 1978).

Thinning of Forest Stands

Foresters have long been concerned about the poor seed crops of suppressed trees growing in dense stands. In many conifer stands, almost all cones are produced by dominant trees and low or negligible amounts by suppressed trees. Thinning of forest stands increases the vigor of the residual trees and increases their capacity for reproductive growth. Wenger

(1954) thinned a loblolly pine stand in 1946–1947. The remaining trees produced an increased number of cones in the year after thinning. In 1949, a very poor seed year, each released tree produced an average of 51 cones, whereas unreleased trees produced only 5 cones each. In 1950 and 1951 the released trees produced 107 and 132 cones per tree; control trees produced only 16 and 48 cones, respectively, during the same 2 years. Crown release of pine trees is followed by an increase in pollen production, which is reflected in more viable seeds per cone (Matthews, 1963).

Pruning

Pruning of Branches

Fruit trees are pruned primarily to ensure production of fruits of high quality. By reducing the number of competing fruits and by increasing penetration of light into the fruit-bearing zone of the crown, pruning generally increases fruit size. The influence of increased light intensity alone within the crown often is modified by changes in temperature and humidity.

The beneficial responses of pruning are associated with greater availability of carbohydrates, hormones, and mineral nutrients to fewer fruits. Pruning of apple trees increased the concentrations of cytokinins, followed by increased amounts of auxins and gibberellins (Grochowska *et al.*, 1984). Nitrogen concentrations of fruits also may increase after pruning (Hansen, 1987).

The effects of pruning vary with species and cultivar, severity and timing of pruning, rootstock, tree vigor, age, and environmental conditions. Dormant pruning is used to increase light penetration into tree crowns. However, the distribution of light in the fruiting zone depends on the extent to which shoot growth is influenced. Normal dormant pruning stimulates shoot growth and decreases yield by reducing the number of flower buds (and hence the number of fruits), but it increases fruit size and improves fruit color. The best color of apples followed exposure of trees to at least 70% of full sun, adequate color to 40–70%, and unacceptable color to less than 40% (Heinicke, 1966). The direct heating of sunlight on apples appears to accelerate anthocyanin production.

The larger fruits of dormant-pruned trees result from better exposure of spur leaves to light, and to higher temperature, as well as greater sink strength of fruits. However, heavy dormant pruning may cause excessive growth of new shoots in the outer crown and hence inhibit penetration of light to the fruit-bearing zone (Mika, 1986). Because carbohydrates from sprouting shoots usually are not translocated to growing fruits, such heavy sprouting may inhibit fruit growth. Heavily pruned trees also may produce very large, soft fruits that are high in N and low in Ca. Such fruits may not store satisfactorily.

Summer pruning often is practiced in high-density orchards to reduce tree size and increase penetration of light into tree crowns (Taylor and Ferree, 1981). Whereas the light intensity within the crown is increased during the year of summer pruning, it often is decreased in the following year because of the vigorous regrowth of shoots.

Summer pruning has been useful in increasing fruit color of poor-coloring apple cultivars in regions where color development of fruits on unpruned trees is unsatisfactory. Because more flower buds are formed on summer-pruned than on dormant-pruned trees, some investigators concluded that flower bud formation is increased by summer pruning. However, trees that are pruned in the summer form fewer buds per tree than unpruned trees do, although summer pruning increases the number of flower buds per unit of branch length (Mika, 1986). Summer pruning should be used with caution to avoid production of small fruits with low soluble solids. For more details on the effects of dormant pruning and summer pruning on reproductive growth, the reader is referred to Mika (1986) and Marini and Barden (1987).

Pruning of Roots

Flowering of trees has been stimulated by root pruning alone or in combination with other treatments (Geisler and Ferree, 1984). Examples are apple (Schumacher *et al.*, 1978), slash pine (Hoekstra and Mergen, 1957), and Japanese larch (Mikami *et al.*, 1980). A combination of girdling of branches and pruning of roots stimulated flowering of eastern white pine more than root pruning alone (Stephens, 1964). The increased number of flowers in root-pruned trees often is associated with reduced fruit set and smaller fruit size (Geisler and Ferree, 1984).

Root pruning of Jonathan apple trees reduced the intensity of biennial bearing and reduced yield, fruit size, and preharvest drop. It also produced firmer fruits with a high soluble solids concentration (Ferree, 1992).

Scoring or Girdling (Ringing) of Branches and Stems

Severing the phloem early in the season by scoring (cutting through the bark on one side of a branch or stem) or girdling (cutting through the bark around the entire branch or stem circumference) to stimulate reproductive growth has been practiced for centuries. When downward phloem translocation of carbohydrates and growth hormones is blocked by scoring or girdling, these compounds tend to diffuse into the xylem and are transported upward in the transpiration stream and concentrate in the leaves and tissues involved in reproduction.

In most cases girdling has been used to increase the number of flowers but also to increase fruit set, increase fruit size, and advance maturity of fruits including grapes, apples, citrus fruits, apricots, nectarines, peaches, and olives (Sedgley and Griffin, 1989). Fruits on girdled branches of apple trees were heavier than fruits on ungirdled branches (Schechter *et al.*, 1994). There was a difference of 20 to 25% in fruit dry weight that ranged from 10 to 25 g, independent of the fruit load. Increases in dry weight were accompanied by increases in fruit size as also reported for nectarines (Day and DeJong, 1990) and citrus fruits (Cohen, 1984). Girdling improved cone yields in Douglas fir seed orchards. Furthermore, it did not adversely affect vegetative growth or seed quality. Flower promotion treatments such as girdling probably improve the genetic quality of orchard seed by creating near-random mating conditions and by preempting unwanted pollen sources (Wheeler *et al.*, 1985).

An increase in flower initiation following phloem blockage by girdling usually is evident in the season following treatment. However, increased yields of fruits and cones also may occur in the year of treatment, the following year, and even several years later (Ebell, 1971). Girdling often has been used as an adjunct treatment with application of gibberellins to conifers. Untreated Douglas fir trees averaged 79 seed cones and 4,500 pollen cones. Girdling increased production of seed cone and pollen cone buds to 325 and 9,300, respectively. Treatment with $GA_{4/7}$ alone was almost as effective as girdling alone. However, combined girdling and application of $GA_{4/7}$ increased production of seed cone and pollen cone buds to 585 and 18,250, respectively (Ross and Bower, 1989). Either $GA_{4/7}$ application or girdling of large Sitka spruce trees stimulated production of both seed cones and pollen cones, but a combination of $GA_{4/7}$ application plus girdling stimulated cone production even more (Philipson, 1985). Stimulation of cone initiation carried over into the second year after treatment but only when $GA_{4/7}$ application and girdling were combined. Overlapping saw-cut girdles and stem injection of $GA_{4/7}$, applied alone and in combination to 10-year-old loblolly pine seed orchard trees, increased pollen cone production the first year after treatment. Girdling, which increased the number of pollen cone buds by 288%, was more important than $GA_{4/7}$ application (Wheeler and Bramlett, 1991).

Determining the optimal time of girdling often is a problem for fruit growers. Girdling of branches of early-season nectarine trees was most effective when done at the beginning of stage II of fruit development (Day and DeJong, 1990). The best results were obtained by girdling 30 days after full bloom, when the seeds were approximately 10 mm long. Girdling at that time increased the eventual weight of fruits by 22.5%, increased the soluble solids concentration by 42%, and more than doubled the percentage of fruits in the largest three size classes. Lavee *et al.* (1983) found that the best time for girdling of olive trees in Israel was in January.

There are dangers in girdling incorrectly, often because of slow healing of the girdled area. If the girdle is too wide or too deep, healing can be slow, and depletion of carbohydrates in roots may be followed by injury to roots and death of trees. In fact, girdling sometimes is used in forestry to kill undesirable trees. In Spain some nectarine cultivars were injured because of poor healing of girdling wounds (Fernandez-Escobar *et al.*, 1987). The rate of formation of callus bridges varied with the cultivar and width of the girdle. When girdles were over 5 mm wide callus formed slowly on some cultivars, and trees of one cultivar died. Lavee *et al.* (1983) reported that both the width of the wound and covering of the girdle influenced the rate of wound healing in olive trees. A 5-mm-wide girdle healed too fast for maximum yield of olives; girdles 20 to 25 mm wide healed too slowly. According to Sedgley and Griffin (1989), thin, overlapping girdles 2 to 2.5 cm wide often are useful in stimulating reproductive growth and heal well if not continued for more than 3 successive years.

Application of Plant Growth Regulators

Applied plant growth regulators may influence some aspects of reproductive growth that cannot be readily controlled by other methods. Exogenous growth regulators may modify sex expression. For example, application of ethephon to papaya plants modified sex expression toward femaleness (Bartholomew and Criley, 1983). Applied growth regulators also may influence the following: transition from a juvenile (nonflowering) stage to an adult (flowering) stage; flowering; fruit set; shedding of flowers and young fruits; fruit ripening; and prevention of chilling injury in cold-stored fruits. The responses of plants to growth regulators vary greatly with species and genotype, the compound applied, dosage and timing of application, mode of application, plant vigor and age, cropping history, and environmental conditions before, during, and after treatment. The most serious limitation to use of growth regulators as standard cultural practices is a lack of consistency in plant response (Bukovac, 1985). Although many different growth regulators have been evaluated, only a few have been adopted for routinely regulating reproductive growth. Use of plant growth regulators has been very valuable in production of apples and pears. Their use with stone fruits has been much less successful. The use of applied growth regulators has been much more limited in forestry than in horticulture. Research on use of growth regulators in forestry has focused largely on (1) stimulating flowering of selected genotypes as early as possible to accelerate testing their progeny and (2) accelerating and regulating production of genetically improved seeds in seed orchards (Bonnet-Masimbert, 1987a,b; Bonnet-Masimbert and Zaerr, 1987).

Flowering

Floral induction in juvenile conifers and floral promotion in adult trees have been achieved by applying gibberellins. In the late 1950s the more polar gibberellins such as GA_3 (with more than one hydroxyl) were effective in inducing flowering of species in nine genera of the Cupressaceae: *Chamaecyparis, Cryptomeria, Cupressus, Juniperus, Metasequoia, Sequoiadendron, Taxodium, Thuja,* and *Thujopsis*) (Ross and Pharis, 1985; Owens, 1991).

In the mid 1970s it was discovered that floral induction was stimulated in the Pinaceae by applications of less polar gibberellins (with only one hydroxyl), primarily by a mixture of GA_4 and GA_7. Induction of both pollen cones and seed cones was demonstrated in 17 species in five genera (*Larix, Picea, Pinus, Pseudotsuga,* and *Tsuga*). The best results were obtained when application of $GA_{4/7}$ was combined with exposure to drought or cultural treatments such as root pruning, girdling of branches and stems, and application of N fertilizers or retardants of vegetative growth, even when the cultural treatments were ineffective by themselves (Ross and Pharis, 1985). Progress in stimulating flowering in the Pinaceae was slower than for the Cupressaceae because (1) the less polar gibberellins were not so readily available and were more expensive than more polar gibberellins, (2) treatments are not so easily applied, (3) timing of application is more important because its effect varies with the stage of bud development, (4) the length of treatment needed often is longer, (5) adjunct cultural treatments often are required, and (6) the sex expression of cones is not readily manipulated (Owens, 1991).

Initiation of seed cones and pollen cones in various species of the Pinaceae may be affected differently by $GA_{4/7}$. For example, $GA_{4/7}$ promoted formation of both pollen cones and seed cones in jack pine (Cecich, 1983; Rottink, 1986), pollen cones alone in southern pines (Hare, 1984), and seed cones alone in white spruce (Marquard and Hanover, 1984). Differences in the effects of $GA_{4/7}$ on seed cones and pollen cones often vary with the time of treatment. Flowering may be enhanced by $GA_{4/7}$ by promoting the number of axillary buds initiated, increasing the number of such buds that develop reproductively, or both. Treatment during early shoot elongation may stimulate initiation of axillary buds and enable development of proportionally more reproductive buds. Early treatments also may operate at the molecular level and commit axillary apices to reproductive development as soon as they are initiated. Treatment during late shoot elongation (near the time of onset of development of reproductive buds) may have a direct morphogenetic effect and induce more buds to become reproductive.

The crucial importance of the time of treatment with $GA_{4/7}$ was shown by Owens and Colangeli (1989). Most pollen cones and seed cones of western hemlock were produced when the trees were sprayed with $GA_{4/7}$ before the vegetative buds opened and shoot elongation began (Fig. 8.5). Two to three weekly applications beginning before and during bud swelling were adequate for stimulating production of pollen cones. At least three weekly applications were necessary for inducing production of seed cones, and such treatments were equally effective whether they were started before the buds swelled, after buds had swollen, or when the buds were opening. The effect of $GA_{4/7}$ treatment during shoot elongation was slight.

Application of $GA_{4/7}$ increased flowering in European and Japanese larches (Table 8.3). May and June applications were best, especially for gibberellin plus girdling treatments. However, some flowering was induced by treatments as late as August and September. Most pollen and seed cones were produced in the upper crown. However, applied $GA_{4/7}$ induced formation of seed cone buds in the lower crown. Girdling in combination with a single injection of $GA_{4/7}$ was a cost-effective treatment for increasing seed production in Douglas fir seed orchards (Ross and Bower, 1989). However, biennial retreatment reduced tree vigor and flowering response. The associated physiological stresses were alleviated by irrigation and fertilization during the off-treatment year. Both $GA_{4/7}$ and root pruning stimulated male and female flowering in mature tamarack trees (Fig. 8.6). The response to $GA_{4/7}$ was similar to that in Douglas fir, western hemlock, and other larches.

The effectiveness of $GA_{4/7}$ often is modified by weather. In cool, wet weather out of doors $GA_{4/7}$ did not stimulate cone initiation in container-grown Sitka spruce even when combined with stem girdling (Philipson, 1992). In contrast, $GA_{4/7}$, in combination with heat and drought, induced production of pollen and seed cones. Girdling of stems further increased cone initiation.

In contrast to their effects on gymnosperms, applied gibberellins inhibit floral induction in many angiosperm trees, including temperate fruit crops, citrus, and mango (Sedgley and Griffin, 1989). Application of GA_3 at any time from early November until bud opening inhibited flowering of sweet orange and satsuma and Clementine mandarins (Guardiola *et al.*, 1982). The extent of flower inhibition in mango by exogenous GA_3 varied among cultivars (Tomer, 1984). When GA_3 was applied to Valencia orange it caused potentially flowering buds to revert to a vegetative state (Lord and Eckard, 1987). However, flowering of some angiosperm trees was stimu-

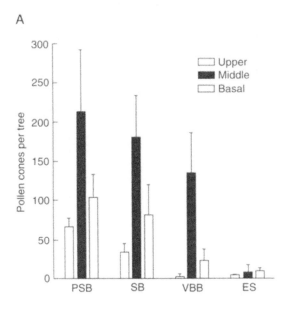

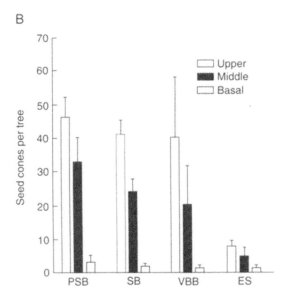

Figure 8.5   Effect of application of $GA_{4/7}$ on formation of pollen cones (A) and seed cones (B) of western hemlock. Data are the average number of cones produced in three crown positions (upper, middle, and basal) following treatments started at four phenological stages: PSB, preswollen buds; SB, swollen buds; VBB, vegetative bud burst; and ES, elongating shoot. Vertical bars represent standard errors. From Owens and Colangeli (1989).

Table 8.3 Effect of Gibberellins and Branch Girdling Applied in 1979 on Male and Female Flowering in 1980 in Upper and Lower Branches of Japanese Larch and European Larch.[a,b]

Species and treatments	Lower branches Male No.[c]	Male %[d]	Lower branches Female No.[c]	Female %[d]	Normal trees Upper branches Male No.[c]	Male %[d]	Upper branches Female No.[c]	Female %[d]
Japanese larch								
Control	19A	60	0A	0	44A	72.7	2.2A	45.4A
Injection of $GA_{4/7}$ (100 mg liter^{-1})	45.6AB	60	2.5A	33.3	125.5B	83.7	11.1AB	48.8AB
Injection + girdling	91.9B	73.3	7.3B	40	138.5B	80.3	19.1B	76.8B
Spray of $GA_{4/7}$ (400 mg liter^{-1})	56.1AB	66.6	0.3A	20	116.3B	73.3	4.1A	45.4A
European larch								
Control	6.5	50	0A	0A	54.5	54.5	7.8A	45.4
Injection of $GA_{4/7}$ (100 mg liter^{-1})	40.3	61.4	7.1B	47.7AB	66.7	74.2	15.4B	74.2
Injection + girdling	53	68.2	6.2B	61.4B	94.4	80.9	29.6C	67.4
Spray of $GA_{4/7}$ (400 mg liter^{-1})	26.9	61.4	4.1AB	27.3AB	111.9	77.5	11.2AB	64.1

[a]Adapted from Bonnet-Masimbert (1982).
[b]Values within each treatment grouping followed by the same or no letter do not differ at $p \leq 0.05$.
[c]Mean number of strobili per treated branch (results averaged over all application dates).
[d]Percentage of flowering branches (results averaged over all application dates).

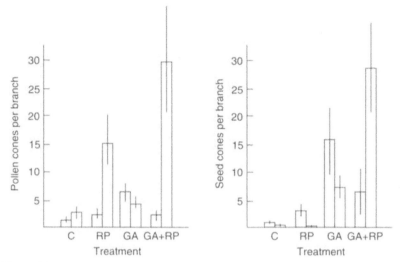

Figure 8.6   Flowering response of greenhouse-grown juvenile (shaded bars) and mature grafts (open bars) of tamarack to root pruning (RP), gibberellin $A_{4/7}$ (GA), and GA plus RP treatments. C, Control. From Eysteinsson and Greenwood (1990).

lated by less polar gibberellins, emphasizing the importance of the type of gibberellin applied (Looney *et al.*, 1985).

Retardants of Vegetative Growth   A number of applied chemicals that inhibit vegetative growth stimulate flowering in a variety of broad-leaved trees. For example, trees treated with daminozide or chlormequat often stop shoot elongation early in the season and exhibit precocious and prolific formation of flower buds. The degree of response varies with species and genotype, the chemical applied, its concentration, and the time of application.

Benzyladenine and daminozide increased flowering on several apple cultivars (Baldwin, Golden Delicious, McIntosh, and Delicious) but not on Early McIntosh (Fig. 8.7). Although daminozide commonly stimulated formation of flower buds on annually bearing cultivars with modest crop loads, it usually did not increase flowering in biennial cultivars with heavy fruit crops (McLaughlin and Greene, 1991b). Some adverse effects of daminozide on fruit quality have been reported. For example, the high rates of daminozide necessary to control vegetative growth of apple trees reduced fruit size (Williams and Edgerton, 1983; Stinchcombe *et al.*, 1984).

With the introduction of growth-inhibiting chemicals that block gibberellin synthesis, such as triazole derivatives like paclobutrazol (PB), some of the problems with other growth retardants were overcome. The mechanisms of action of various triazoles are not identical and are partly mediated by their effects on hormones other than gibberellins. For example, unlike PB, uniconazole appears to lower auxin levels (Lürssen, 1987).

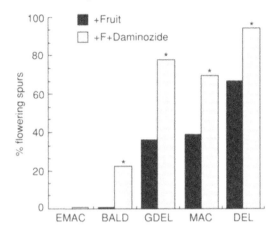

Figure 8.7   Effect of daminozide on flowering of several cultivars of apple: EMAC, Early McIntosh; BALD, Baldwin; GDEL, Golden Delicious; MAC, McIntosh; DEL, Delicious. Asterisk indicates significant difference within cultivar by Duncan's multiple range test, $p = 0.05$. From McLaughlin and Greene (1991a). *J. Am. Soc. Hortic. Sci.* **116,** 446–449.

In addition to indirectly affecting reproductive growth by inhibiting shoot growth, paclobutrazol probably also acts directly because after application it is present in axillary floral buds (Browning *et al.,* 1992).

There are many reports of increased flowering following application of triazole derivatives, for example, in apple, cherry, lemon, pear, avocado, plum, mango, nectarine, and eucalyptus. Dilute foliar applications of PB increased flowering of several apple cultivars in the year after treatment. When PB was applied to the roots it was as effective as foliar application, but the effects were not apparent until a year later (Tukey, 1983, 1986). A single soil application of PB in the spring greatly increased the proportion of flower buds that formed in tart cherry (94% of the buds of treated trees were flower buds versus 23% in controls). The date of full bloom was advanced by 2 to 4 days in the treated trees (Walser and Davis, 1989). Paclobutrazol applied to sweet cherry trees at least doubled the number of floral buds and also increased the number of floral buds per cluster as well as the number of flowers per floral bud (Webster *et al.,* 1986).

Soil applications of PB reduced shoot elongation, suppressed apical dominance, and promoted flowering in young grafts of mango as well as in older, bearing trees. Foliar sprays were ineffective (Kulkarni, 1988). Soil drenches of PB induced early and profuse flowering in 2 successive years and also increased both fruit set and yield in 9-year-old mango trees (Kurian and Iyer, 1993b,c). In sweet cherry, PB increased the number of fruits per spur on 3-year-old spurs but not on 2-year-old spurs (Lauri, 1993). Paclobutrazol also reduced vegetative growth and increased production of flower buds in 2- to 17-year-old Tasmanian blue gum and shining gum trees (Griffin *et al.*, 1993). Responses to stem injections and root collar drenches persisted for up to 6 years. Vegetative growth was reduced in the year of application, but flowering responses were not evident until the next year. There were no obvious effects on seed quality.

Sensitivity to PB varies appreciably among species and cultivars. It was greater in pecan than in apple, peach, or citrus, in which the effect on flowering was not evident as early or is dissipated by the second year. In pecan, the effect of PB may persist for 3 or 4 years (Wood, 1988a). The effects of PB were greater on the plum cultivars Opal and Cambridge Gage than on Rivers Early Prolific (Webster, 1990).

Delay of Flowering A wide variety of growth regulators have been applied to delay flowering until the danger of frost is over. Application of ethephon in the autumn variously delayed flowering in several stone fruits (Dennis *et al.*, 1977), almond (Browne *et al.*, 1978), grape (Iwaseki, 1980), apricot (Buban and Turi, 1985), and peach (Durner and Gianfagna, 1988; Gianfagna, 1989) but not apple (Anderson and Seeley, 1993). The delay in flowering after ethephon treatment is correlated with ethylene levels in dormant buds. Ethephon-treated buds of peach had 65% more ethylene than buds from untreated trees (Crisosto *et al.*, 1987).

Success in delaying flowering with ethephon depends largely on the dosage and time of application. Best results generally were obtained when ethephon was applied between the time of vegetative maturity and early leaf abscission in the autumn preceding flowering (Anderson and Seeley, 1993). In many experiments, ethephon caused injury or induced other undesirable responses, including gummosis, abscission of buds, failure of flowers to open, and reduced fruit set. Addition of gibberellic acid (GA) to ethephon sprays may reduce phytotoxicity while delaying flowering (Murdock and Ferguson, 1990).

While stimulating vegetative growth exogenous gibberellins may inhibit flowering, but the plant responses often are inconsistent and accompanied by undesirable plant responses (including abscission of flower buds and reduced fruit set). Treatments with GA in late summer and early autumn delayed flowering of stone fruits. Best results were obtained with treatments several weeks before the leaves were shed (before the best time for applying ethephon to delay flowering) (Painter and Stembridge, 1972).

Auxins such as naphthaleneacetic acid (NAA) applied late in the summer delayed flowering of apricot, cherry, and peach by 1 to 5 days (Gertych, 1953). The delay in flowering was accompanied by injury to leaves and leaf buds (Marth *et al.*, 1947). Anderson and Seeley (1993) concluded that exogenous auxins, like gibberellins, delay flowering for too short a time and cause too much injury to be practical.

Several growth-retarding chemicals have been tested for delaying flowering, with variable results. If maleic hydrazide (MH) was applied to red raspberry plants when the leaflets were very small (1 cm), flowering was delayed without injury (Kennard *et al.*, 1951); however, if MH was applied at other developmental stages, flowering was not delayed and the plants were injured (Ferres, 1952; Modlibowska and Ruxton, 1953).

Autumn application of daminozide delayed bloom in apple for 4 to 5 days (Sullivan and Widmoyer, 1968, 1970), in apricot for 4 to 5 days (Ogasanovic *et al.*, 1983), and in peach for 3 to 5 days (Morini *et al.*, 1976). Treating flower buds of apple, sour cherry, peach, and plum with aminoethoxyvinylglycine (AVG), an inhibitor of ethylene synthesis, delayed flowering (Dennis *et al.*, 1977). Applications of AVG in the field postponed flowering of rabbiteye blueberry (Dekazos, 1981) and apple without a reduction in fruit quality (Dekazos, 1982).

Fruit Set

The failure of many fruit trees to set fruits adequately is of much concern to growers. Hence, much attention has been given to stimulation of fruit set with plant growth regulators. A variety of exogenous plant growth regulators have been used, including auxins (e.g., NAA), 2,3,5-triiodobenzoic acid (TIBA), gibberellins, cytokinins [e.g., benzyladenine (BA), kinetin, zeatin], and various mixtures [e.g., GA_3 + NAA; GA_3 + DPU (diphenylurea) + NOXA (α-naphthoxyacetic acid)] (Faust, 1989).

Fruits of fig, grape, avocado, apple, and pear can be set by applying auxins. However, most species of woody plants do not set fruit following application of auxins. Exogenous auxins induce fruit set in apricot but not in other stone fruits such as cherry, peach, and plum.

Fruits of some species of woody plants that can be set with auxins can be set as well or better with exogenous gibberellins. The gibberellins are among the most active chemicals available to increase fruit set in apple (Dennis, 1986). On apples, $GA_{4/7}$ usually is more effective than GA_3 (Goldwin, 1978). Treatment with gibberellins often increases fruit set when pollination is prevented or the flowers are injured by frost. The effects of gibberellins on trees with uninjured flowers vary appreciably, with fruit set being increased, decreased, or negligibly influenced. Such differences often are traceable to the types of gibberellins used, dosage, time of application, and flower quality.

When $GA_{4/7}$ was applied to open-pollinated apple flowers from 6 days before to 34 days after bloom, fruit set was increased (Greene, 1989). Applications near the time of bloom influenced initial fruit set; applications 10 days after full bloom increased fruit set by reducing June drop of apples. Flower quality is important in influencing fruit set by applied gibberellins. For example, ovule vigor was necessary for fruit development in apple after application of hormone sprays (Goldwin, 1985). Application of gibberellin inhibitors, such as paclobutrazol, often reduces fruit set (Dheim and Browning, 1987).

Often mixtures of gibberellins and auxins have been applied to promote fruit set. For example, mixtures of GA_3 plus auxin induced fruit set of apples, cherries, and plums (Jackson, 1989). There are many reports of increased fruit set following application of some growth retardants. For example, application of 2.5 g of paclobutrazol per mango tree was followed by increases over control trees of 95 and 96% in crop yield in the first and second years, respectively. This increase in yield was attributed largely to higher fruit set from enhanced flowering and hermaphrodite flowers. A higher dosage (10 g per tree) reduced fruit set (Kurian and Iyer, 1993b,c).

Fruit Growth

Increase in size of fruits may be influenced by application of several individual growth regulators and mixtures. Fruit enlargement depends on availability of gibberellins (Chapter 4). Growth of apples is increased by exogenous gibberellins, and treatment with inhibitors of gibberellin synthesis results in small fruits (Stinchcombe *et al.*, 1983).

Auxin applications to apricots, blackberries, grape, and orange stimulate fruit enlargement. Applied auxins may increase fruit size by different mechanisms, depending on the time of treatment. For example, NAA applied to satsuma mandarin trees before fruit drop stimulated fruit abscission and increased fruit size by reducing competition among fruits for resources. When applied after June drop, NAA directly stimulated growth of fruits (Ortola *et al.*, 1991).

Spraying of polyamines (spermine, spermidine, or putrescine) 9 days after full bloom on flowers of apple trees increased both fruit set and crop yield per tree. These compounds increased the rate of fruit growth during cell division (stage I) but not during subsequent cell enlargement (Costa and Bagni, 1983).

Inhibitors of vegetative growth ideally should not reduce the size or shape of fruits (Quinlan, 1981). However, depending on the concentration and time of application, paclobutrazol (PB) may cause some undesirable effects such as reduced size of fruits, decreased length–diameter ratio, reduced fruit soluble solids, and russeting. The undesirable effects may be reduced or eliminated if application of PB is delayed until late in the season. Some of the harmful effects of PB also can be counteracted by applications of promalin (Greene, 1991). In addition, undesirable effects may be reduced by using moderate rates of PB and treatment with GA_3 or a combination of $GA_{4/7}$ plus BA at or near anthesis (Tromp, 1987; Curry,

1988). Much more research is needed to explore the full potential of growth retardants in promoting reproductive growth.

Fruit Ripening

Ripening of many fruits can be stimulated by exposure to ethylene or ethephon. For example, exposure of bananas to ethephon increased ripening without a decrease in quality (Bondad, 1976). Ethephon applied to coffee trees 6 to 8 weeks before the main flush of berry ripening greatly accelerated ripening (Bondad, 1976). Applying ethephon to mango fruits when they are small may advance ripening by as much as 10 days (Bartholomew and Criley, 1983).

The tissues of all fruits are exposed to ethylene throughout their development. The response to additional ethylene varies appreciably among plant species, genotypes, and developmental stages of fruits. In unripe (preclimacteric) fruits of the climacteric type (e.g., apples, bananas), ethylene treatments accelerate the beginning of ripening and also increase respiration and ethylene synthesis without changing the climacteric pattern and amount of respiration and ethylene evolution. In nonclimacteric fruits, the extent of the respiratory response increases with higher concentrations of applied ethylene, but the increase in respiration is not accompanied by greater ethylene production or other ripening changes (Yang, 1987).

Shedding of Flowers and Young Fruits

Woody plants initiate many more fruits and seeds than can develop to maturity with the resources available to them. Hence, competition within trees for resources is inevitable, and subsequent growth of fruits, or abortion, will vary with their sink strength. Overproduction of fruits and seeds has been considered by some investigators to be an adaptation to unpredictability of the environment, the amount of pollen and its quality, and predation of fruits and seeds. Another prevailing view is that plants initiate too many fruits only because selection favors overinitiation of flowers.

Fruit growers often thin out some of the flowers and young fruits on trees, thereby reducing competition among reproductive structures for resources. The trees respond by partitioning more carbohydrates and mineral nutrients into the remaining fruits, hence increasing their growth and that of the seeds they contain. Thinning of flowers also tends to maintain annual rather than biennial bearing and decreases injury that might be caused by breaking of branches that bear heavy fruit loads. The effects of thinning of reproductive structures are greatest in trees that bear large fruits, including apples, pears, peaches, and plums. In trees with small fruits the effects of fruit number on fruit development are less pronounced.

For a long time, thinning of reproductive structures was done by hand. Since the 1970s a variety of chemicals have been used to induce abortion and abscission of flowers and young fruits. These include dinitrophenols, synthetic auxins such as NAA and naphthalene acetamide (NAAm), 3-chlorophenoxypropionic acid and its amide (3-CPA), 1-naphthyl *N*-methyl carbamate (carbaryl), dinitro-*ortho*-cresol (DNOC), and ethephon (Edgerton, 1973; Nickell, 1983a,b).

Ideally, thinning sprays should remove only enough fruits to ensure adequate flowering during the next season. Chemical thinning usually is done at blossom time and during the early postbloom period. In the state of Washington, chemicals for thinning of apples and pears generally are applied between 10 and 25 days after full bloom (Childers, 1983). Strongly biennially blooming apple cultivars may require both bloom and postbloom treatments. Some postbloom chemicals are hormonal growth regulators that inhibit endogenous hormones which control the flow of nutrients into growing fruits. Other postbloom chemicals are compounds that cause stress and embryo abortion (Williams, 1979).

Chemical thinning sprays commonly increase fruit yields for 2 years or more by altering the habit of biennial bearing. The sprays reduce fruit set early in the growing season, and more fruit buds form for the next year's crop. Chemical thinning compounds sometimes have been used together with growth retardants to counteract the biennial bearing habit (Williams, 1979).

The efficiency of chemical thinning treatments varies greatly with plant species, the chemical and concentration used, method and time of application, age and vigor of trees, and weather, as discussed by Edgerton (1973) and Williams (1979). Whereas NAA or carbaryl are effective for postblossom thinning of apple, they generally are not effective on stone fruits. However, stone fruits can be effectively treated with DNOC on blossoms or 3-CPA on fruitlets (Webster, 1980). In general, chemical thinning is less reliable with stone fruits than with apple (Jackson, 1989).

The most effective time of thinning of Fuji apple flowers with benzyladenine (BA) was 20 days after full bloom, when increasing the BA concentration also stimulated thinning. Such treatment also increased fruit size and weight. When BA at high concentrations was applied 30 days after full bloom, the weight of individual fruits was reduced (Bound *et al.*, 1991). Similarly, late thinning of peach fruits may not be followed by increased fruit size.

The mechanisms of action of fruit-thinning chemicals are not identical. Dinitro-*ortho*-cresol causes necrosis of flower parts and stimulates production of ethylene. At high concentrations DNOC acts as a pollenicide; at low concentrations it is an inhibitor of the growth of pollen tubes. Naphthaleneacetic acid at first temporarily inhibits abscission of young fruits by augmenting the amount of endogenous auxin. It subsequently causes injury that impedes transport of auxins and other hormones from young fruits to their abscission zones. Naphthaleneacetic acid and other auxinlike substances also operate by increasing production of ethylene. Applications of inhibitors of ethylene production negate the thinning effect. Most fruit-thinning chemicals have not been completely reliable because they sometimes induced excessive abscission of young fruits (Addicott, 1982). For more detailed information on use of fruit-thinning chemicals, readers are referred to Williams (1979), Nickell (1983a,b, 1991), and Ryugo (1988).

Prevention of Preharvest Fruit Drop   In addition to their use in thinning out excessive flowers and young fruits, exogenous growth regulators have been used to prevent preharvest fruit drop. Daminozide, NAA, and fenoprop have been used on apples; daminozide, NAA, and 2,4-D, on maturing pears. Prevention of fruit drop has been quite successful following one or two properly timed high-volume sprays (Bukovac, 1985).

Storage of Harvested Fruits

Successful storage of harvested fruits is related to slowing down their rate of metabolism and the onset of senescence. Hence, much research has been done on methods of controlling the environment of fruit storages to reduce respiration and undesirable biochemical changes so as to prolong the storage life and quality of fruits. Storage in controlled atmospheres has been used primarily for apples but is gradually being used for other fruits as well. Attempts to store soft fruits that ripen and senesce rapidly, such as peaches and nectarines, for long periods have not been very successful because of chilling injury and disease problems (Lill *et al.*, 1989). The beneficial effect of low temperature on fruit storage often has been supplemented by an atmosphere high in CO_2 and low in O_2. Such atmospheres are effective because low O_2 levels decrease ethylene production and fruit metabolism and CO_2 counteracts the action of ethylene. The life of fruits during shipping or storage also has been increased by removing ethylene from the fruit environment by chemicals such as potassium permanganate or ozone. Low-pressure (hypobaric) storage also has been useful in extending the storage life of fruits. In the process of exposing fruits to a slow stream of air maintained at low pressure, air is introduced into the low-pressure chamber from the exterior and then expelled. Thus, respiratory gases are prevented from accumulating. The storage life of fruit is extended because ethylene is removed and its production and activity reduced.

A number of physiological effects of controlled atmospheres (CA) on fruits were described by Smock (1979), including (1) reduction in loss of normal acidity, (2) decrease in loss of chlorophyll, (3) decrease in production of volatile compounds, not only during storage but also after removal of fruit from CA, (4) slowing of decomposition of carbohydrates, (5) reduction in protein synthesis, and (6) conversion of nitrate to nitrite and eventually to other N compounds.

Controlled-atmosphere storage helps to maintain the texture of fruits. For example, firmness of apples was retained better by storing them in 3% O_2 rather than in air, and even more so if the CO_2 level was increased (Seipp, 1974; Lange and Fica, 1982). The faster the O_2 content was reduced, the better was retention of fruit texture as shown for apples, pears, strawberries, peaches, avocados, and mangos (Lau and Looney, 1982; Weichmann, 1986).

Most fruits are highly resistant to postharvest pathogens during most of their lives. As they ripen and senesce, however, they become increasingly susceptible to pathogens. Decay of harvested fruits can be reduced by the combined effects of low temperature, low O_2, and high CO_2 of CA storage (El-Goorani and Sommer, 1981). To prevent decay of fruits in storage, it is necessary to delay the onset of fruit senescence as long as possible. In CA storage the metabolic activity of both the fruit and pathogens is decreased, thus extending the period of resistance of the fruit to pathogens. Lowering the O_2 level to the 2 or 3% often used in modified atmospheres has only limited effects on pathogen activity. Oxygen concentrations near 1% or less generally are necessary for significant reduction in growth of pathogens as well as in formation and germination of spores of many postharvest fungi.

During prolonged storage of fruits in CA, the CO_2 content usually is not maintained above 5% because of possible injury at higher CO_2 concentrations. Nevertheless, fruits can tolerate high CO_2 concentrations (20 to 30%) for short periods, and such treatments may suppress certain diseases. Whereas rots of oranges caused by *Penicillium digitatum* were suppressed by high CO_2 concentrations, those caused by *Diplodia natatensis* were not (Brooks *et al.*, 1932; Aharoni and Lattar, 1972). For specific recommendations on the temperatures and gas concentrations for storage of fruits, see Dewey (1977) and Smock (1979).

Prevention of Freezing and Chilling Injury

Cultural practices that protect fruit crops from freezing have been used for centuries. A distinction often is made between passive (indirect) and active (direct) methods (Rieger, 1989). Passive protection is achieved by decreasing the probability or severity of a freeze or rendering plants less susceptible to freezing injury. Examples of passive treatments, which generally are applied well in advance of a freeze, include use of hardy cultivars and selection of planting sites that are free of freezing temperatures during the growing season. Active treatments are applied just before or during a freeze to prevent injurious temperatures or reduce ice formation in plant tissues. Detailed discussion of the many methods used to prevent freezing injury to reproductive tissues is beyond the scope of this volume. Such methods, which are discussed in detail by Rieger (1989), include pruning, use of appropriate rootstocks, judicious use of fertilizers, application of various chemicals (e.g., ethephon, gibberellic acid, retardants of vegetative growth, antitranspirants), sprinkling to delay budbreak, control of bacterial ice nucleation, and active treatments such as heating, irrigation flooding, and use of sprinklers, wind machines, and insulation (mulching, use of wraps, foam, plastics, and tree and orchard covers).

Several methods have been used to prevent chilling injury of fruits in cold storage. These include cool temperature conditioning, intermittent warming, and treatment with growth regulators and other chemicals.

The temperatures to which fruits are exposed just before they are placed in cold storage influences their tolerance of chilling. Conditioning at temperatures slightly higher than the critical chilling range reduced the sensitivity to chilling of papayas (Chen and Paull, 1986), grapefruit (Chalutz *et al.*, 1985), lemons (Houck *et al.*, 1990), and limes (Spalding and Reeder, 1983). A stepwise lowering of temperature usually was more effective in reducing chilling injury than was a single reduction in temperature (Wang and Zhu, 1981). Biochemical changes associated with low-temperature preconditioning include increases in proline, squalene, long-chain aldehydes, sugar, and starch, as well as decreases in RNA, proteins, and lipids.

Exposing fruits to high temperatures also may reduce subsequent chilling injury. For example, exposure to heat increased antifungal compounds and reduced decay by *Penicillium* rot in cold-stored citrus fruits (Ben-Yehoshua *et al.*, 1987). Maintaining high atmospheric humidity is important during heat treatment (Wang, 1993).

The storage life of some chilling-sensitive fruits also has been increased by interrupting the exposure to low temperature with periods of warm temperatures. It is essential, however, that the warm temperature treatments are imposed before chilling injury becomes irreversible. Warming of chilled tissues for brief periods tends to repair injury to membranes, organelles, and metabolic pathways (Lyons and Breidenbach, 1987). Presumably, the toxic compounds that accumulate during chilling are removed during the short periods of warming.

Growth Regulators

Prevention of chilling injury by abscisic acid (ABA) was reported for grapefruit (Kawada *et al.*, 1979). Such a protective effect may involve ABA action in stabilizing the microtubular network, inhibiting loss of glutathione, and stabilizing membranes (Wang, 1993). Triazole growth regulators (e.g., paclobutrazol) increased the tolerance of plants to chilling by protecting membranes from oxidative damage and lipid peroxidation as well as by decreasing gibberellin synthesis (Waldman *et al.*, 1975).

Fruits such as papaya that develop resistance to chilling injury during ripening may benefit by exposure to ethylene before storage. In contrast, fruits that show decreased resistance to chilling injury during ripening (e.g., avocado) do not increase tolerance to injury during cold storage (Chaplin *et al.*, 1983).

Treatment of some fruits with polyamines before they were placed in cold storage not only increased their polyamine levels but also reduced chilling injury. The protective effects of polyamines were associated with their antioxidant activity and stabilizing influence on membranes. Several other compounds (e.g., Ca, mineral or vegetable oils, fungicides, and radical scavengers) also delayed development of chilling injury. These compounds may act by suppressing oxidative processes, increasing the ratio of unsaturated to saturated fatty acids, and reducing water loss (Wang, 1993).

Summary

Reproductive growth of woody plants has been improved by a variety of cultural practices. Flowering of juvenile trees has been induced by use of appropriate rootstocks and by growing plants under high light intensities and long days.

Interception of light can be increased in fruit and nut orchards by (1) planting trees that are tall with many lateral branches and (2) planting trees at high densities to minimize the amount of space in the alleyways between trees. However, practices that increase light interception to unusually high levels may reduce fruit yield and quality. Planting of dwarf trees at close spacing has greatly increased yields of fruits and nuts.

Selected rootstocks are used to increase not only yield and quality of fruits but also resistance to disease and adaptability to soil. Early flowering can be induced by grafting scions on dwarfing rootstocks, grafting on related species, grafting mature scions onto seedlings, and grafting or budding in crowns of mature trees. Flowering also can be stimulated by inserting a dwarfing stem piece between a vigorous rootstock and a vigorous scion variety. Successful grafting depends on compatibility between members of the graft union.

Fertilizers commonly are applied to fruit trees to stimulate floral induction, increase fruit set and growth, develop high-quality fruits, and prolong the storage life of harvested fruits. Nitrogen is the element most often deficient and often is applied in the spring, but additional N applied later in the season may be useful. However, excessive applications of N may result in soft fruits of low quality. Fertilizers frequently are used in forest seed orchards and mature stands selected for seed production, often in combination with thinning of stands.

Beneficial effects of irrigation on both yield and quality of fruits are well documented, even in humid regions. Water may be applied by basin and furrow irrigation, sprinkler irrigation, trickle or drip irrigation, subsurface irrigation, and mist irrigation. Fertilizers sometimes are applied in irrigation water. Crop yields are functions of both the amount and timing of irrigation. By withholding irrigation water during the period of shoot elongation (regulated deficit irrigation or RDI), fruit yield may be increased while vegetative growth is inhibited. Scheduling of irrigation in orchards may involve (1) use of plant or soil measurements to determine when to irrigate or (2) use of a water budget to estimate the depth of water application and timing. Soil anaerobiosis and accumulated soil salts may adversely influence growth of irrigated trees.

In many conifer stands almost all the cones are produced by dominant trees. Thinning of forest stands increases the amount of light, water, and mineral nutrients available to the residual trees and increases their capacity for reproductive growth. Crown release by thinning of stands also is followed by an increase in pollen production, which is associated with more viable seeds per cone.

Pruning of branches increases fruit size and quality by increasing penetration of light into tree crowns and reducing the number of competing fruits. The beneficial effects of pruning are mediated by the greater availability of carbohydrates, hormones, and mineral nutrients to fewer fruits. Both dormant-season and summer pruning have been used. Dormant pruning usually stimulates shoot growth and decreases total fruit yield by reducing the number of floral buds, but it increases fruit size and improves fruit color. However, very heavy dormant pruning may stimulate excessive growth of new shoots in the outer crown (hence inhibiting penetration of light to the fruit-bearing zone). Such heavy sprouting often creates strong vegetative sinks for carbohydrates and inhibits fruit growth. Summer pruning in high-density orchards reduces tree size and increases penetration of light into tree crowns during the year of pruning. Because of vigorous regrowth of shoots, however, the light intensity within the crown generally is reduced in the following year. Summer pruning should be used with caution to avoid production of small fruits with low soluble solids content. Flowering can be stimulated by root pruning alone or in combination with other treatments (e.g., girdling of branches).

Blocking of downward transport of carbohydrates and growth hormones in the phloem by girdling of branches or stems often stimulates flowering and fruit set, and it may increase growth of fruits and cones. Increased yields may occur in the year of phloem blockage, the following year, or even up to several years later. Phloem blockage together with application of gibberellins often is more effective than either treatment alone.

Flowering in conifers has been induced by applied gibberellins. The best results were obtained when trees were treated with $GA_{4/7}$ together with other treatments, especially water stress, root pruning, girdling of branches and stems, and applications of fertilizers or retardants of vegetative growth. Flowering is enhanced by $GA_{4/7}$ by promoting initiation of axillary buds, increasing the proportion of buds that develop reproductively, or both.

Several applied chemicals inhibit vegetative growth and stimulate flowering in many broad-leaved trees. Primary attention has been given to the triazoles (e.g., paclobutrazol, uniconazole), which act by blocking gibberellin synthesis. The mechanisms of action of triazoles are partly mediated by their effects on hormones other than gibberellins. Sensitivity to triazoles varies among species and genotypes, with concentration and specific triazole used, and with time of application. Sometimes triazoles cause undesirable effects such as reduced fruit size, decreased soluble solids of fruits, and russeting. These effects may be reduced by lowered rates of application, application of promalin, treatment with gibberellins, or use of a combination of $GA_{4/7}$ plus benzyladenine at or near anthesis.

A wide variety of growth regulators have been applied to trees to delay flowering until the danger of frost is over. For example, ethephon application delayed flowering in stone fruits but sometimes caused injury (gummosis, bud abscission, failure of flowers to open, and reduction in fruit set). Addition of gibberellic acid to ethephon sprays may reduce phytotoxicity. Flowering also has been delayed by applications of auxins, gibberellins, and growth retardants; however, the results have been inconsistent, and plants sometimes were injured.

Fruit growth and yield can be increased by thinning out some of the flowers and young fruits. Thinning of reproductive structures also tends to maintain annual rather than biennial bearing. Hand thinning of reproductive structures has been largely replaced by use of chemicals such as dinitrophenols, synthetic auxins, carbaryl, and ethephon. Use of thinning sprays often increases fruit yields for 2 years or longer.

A variety of exogenous plant growth regulators have been used to stimulate fruit set. These include auxins, gibberellins, cytokinins, and various mixtures. Fruits of fig, grape, avocado, apple, and pear can be set by applying auxins. However, most species of woody plants do not set fruit following application of auxins. Fruits of some species that can be set with auxins can be set better with applied gibberellins. Fruits of some species can be set by applications of growth retardants (e.g., paclobutrazol) and polyamines.

Increases in the size of fruits can be influenced by several exogenous growth regulators including gibberellins, auxins, and polyamines.

To prolong their quality, certain harvested fruits can be placed in storages maintained at low temperature, low O_2, and high CO_2. The life of fruits during shipping or storage also can be prolonged by removing ethylene with potassium permanganate or ozone. The beneficial effects of controlled-atmosphere storage are mediated by reducing the rate of respiration and undesirable biochemical changes as well as by arresting the activity of pathogens that induce decay. Controlled-atmosphere storage has been used primarily for apples but is gradually being adopted for other fruits. Controlled-atmosphere storage for soft fruits that ripen and senesce rapidly (e.g., peaches and nectarines) has not been very successful.

Both passive (indirect) and active (direct) methods have been used to protect fruit crops from freezing. Passive protection involves decreasing the severity of a freeze or making plants less susceptible to freezing injury. Passive treatments include use of hardy cultivars and selecting sites free of freezing temperatures. Active treatments are applied just before or during a freeze to prevent injurious temperatures during the growing season. Treatments may include use of appropriate rootstocks, judicious use of fertilizers, use of chemicals or sprinkling to delay budbreak, control of bacterial ice nucleation, heating, irrigation, flooding, use of sprinklers, wind machines, and insulation (e.g., mulches, wraps, foam, plastics, and tree covers). Methods of preventing chilling injury of fruits in storage include cool temperature conditioning, intermittent warming, and application of growth regulators or other chemicals.

General References

Andrews, P. K., and Serrano-Marquez, C. (1993). Graft incompatibility. *Hortic. Rev.* 15, 183–232.

Bacon, P. E., ed. (1995). "Nitrogen Fertilization in the Environment." Dekker, New York.

Childers, N. F. (1983). "Modern Fruit Science." Horticultural Publications, Gainesville, Florida.

Dick, J. M. (1995). Flower induction in tropical and subtropical forest trees. *Comm. For. Rev.* 74, 115–120.

Edgerton, L. J. (1973). Chemical thinning of flowers and fruits. *In* "Shedding of Plant Parts" (T. T. Kozlowski, ed.), pp. 435–474. Academic Press, New York.

El-Goorani, M. A., and Sommer, N. F. (1981). Effects of modified atmospheres on postharvest pathogens of fruits and vegetables. *Hortic. Rev.* 3, 412–461.

Faust, M. (1989). "Physiology of Temperate Zone Fruits." Wiley, New York.

Galletta, G. J., and Himelrick, D. G., eds. (1990). "Small Fruit Crop Management." Prentice-Hall, Englewood Cliffs, New Jersey.

Hartmann, H. T., Kester, D. E., and Davies, F. T., Jr. (1990). "Plant Propagation: Principles and Practices," 5th Ed. Prentice-Hall, Englewood Cliffs, New Jersey.

Jackson, J. E. (1985). Future fruit orchard design: Economics and biology. *In* "Attributes of Trees as Crop Plants" (M. G. R. Cannell and J. F. Jackson, eds.), pp. 441–459. Institute of Terrestrial Ecology, Huntingdon, England.

Jayawickrama, K. J. S., Jett, J. B., and McKeand, S. E. (1991). Rootstock effects in grafted conifers, a review. *New Forests* 5, 157–173.

Kay, S. J. (1991). "Postharvest Physiology of Perishable Plant Products." Van Nostrand-Reinhold, New York.

Kozlowski, T. T., Kramer, P. J., and Pallardy, S. G. (1991). "The Physiological Ecology of Woody Plants." Academic Press, San Diego.

Lockhard, R. G., and Schneider, G. W. (1981). Stock and scion growth relationship and the dwarfing mechanism in apple. *Hortic. Rev.* 3, 315–375.

Luckwill, L. C. (1981). "Growth Regulators in Crop Production." Edward Arnold, London.

Marini, R. P., and Barden, J. A. (1987). Summer pruning of apple and peach trees. *Hortic. Rev.* 9, 351–375.

Mika, A. (1986). Physiological responses of fruit trees to pruning. *Hortic. Rev.* 8, 337–378.

Nickell, L. G., ed. (1983). "Plant Growth Regulating Chemicals," Vols. 1 and 2. CRC Press, Boca Raton, Florida.

Nickell, L. G. (1991). Use of growth regulating chemicals. *In* "Physiology of Trees" (A. S. Raghavendra, ed.), pp. 467–487. Wiley, New York.

Rieger, M. (1989). Freeze protection for horticultural crops. *Hortic. Rev.* 11, 45–109.

Rom, R. C., and Carlson, R. F., eds. (1987). "Rootstocks for Fruit Crops." Wiley, New York and Chichester.

Ryugo, K. (1988). "Fruit Culture: Its Science and Art." Wiley, New York.

Sedgley, M., and Griffin, A. R. (1989). "Sexual Reproduction of Tree Crops." Academic Press, London and San Diego.

Smock, R. M. (1979). Controlled atmosphere storage of fruits. *Hortic. Rev.* 1, 301–336.

Stewart, B. A., and Nielsen, D. R., eds. (1990). "Irrigation of Agricultural Crops." Am. Soc. Agronomy, Madison, Wisconsin.

Swietlik, D., and Faust, M. (1984). Foliar nutrition of fruit crops. *Hortic. Rev.* 6, 287–355.

Wagenmakers, P. S. (1991). Planting systems for fruit trees in temperate climates. *Crit. Rev. Plant Sci.* 10, 369–385.

Wang, C. Y. (1993). Approaches to reducing chilling injury of fruits and vegetables. *Hortic. Rev.* 15, 63–95.

Weichmann, J. (1986). The effect of controlled-atmosphere storage on the sensory and nutritional quality of fruits and vegetables. *Hortic. Rev.* 8, 101–127.

Wright, C. J. (1989). "Manipulation of Fruiting." Butterworth, London.

9

Biotechnology

Introduction: The Scope and Potential of Biotechnology

The discovery of the double helical structure of DNA, the genetic codes contained in its base sequences, and the molecular basis of inheritance constitutes three of the most significant scientific discoveries of the twentieth century. As technical capability has developed to the point where researchers can routinely sequence DNA and manipulate these molecules in a variety of ways, there has been rapid progress in the understanding of gene expression and its regulation by environmental and developmental factors. New techniques have permitted the isolation of sequences of DNA (genes) that are associated with specific plant responses. These fragments of DNA can be sequenced and expressed so that the proteins which are gene products can be analyzed. This approach has great potential to link gene expression to plant function.

Genes are now routinely transferred from one species to another, generally without the need for sexual reproduction (thus the term genetic engineering). Hence, the breeding barriers that formerly restricted gene transfer to closely related individuals theoretically no longer limit plant improvement. For example, bacterial genes coding for insecticidal compounds have been incorporated into plant genomes, increasing the insect resistance of plants without the need for intervention with pesticides (McCown et al., 1991). A gene that increases plant tolerance to the herbicide glyphosate has been introduced and successfully expressed in poplars (Fillatti et al., 1987, 1988). Improvements in seed protein quality and preservation of flower and fruit freshness also are possible. Transgenic tomato plant lines have been developed that exhibit superior shelf life through genetic alterations that reduce either the accumulation of a precursor of ethylene or the amount of an enzyme that promotes cell wall softening in ripening fruits (Giovannoni et al., 1992; Greenberg and Glick, 1993). Nevertheless, much remains to be learned about gene function, particularly with regard to plant traits that depend on the expression of many genes, before researchers can intelligently "engineer" complex but highly desirable traits.

The possibilities for propagation of desirable plants also have expanded greatly beyond controlled breeding and simple techniques for asexual propagation such as rooting of cuttings and grafting. For example, micropropagation of valuable genotypes has become feasible in some woody plants through *in vitro* callus and organ culture and somatic embryogenesis, but it remains unobtainable in other species despite many years of research. *In vitro* microculture techniques also are necessary stages in most genetic engineering strategies. Hence, genetic engineering of species recalcitrant to culture will be hindered until suitable techniques are devised.

With respect to woody plants, "biotechnology" can be conveniently divided into three primary areas: (1) *in vitro* culture of plants; (2) immunology, that is, the use of antibodies in plant disease detection and quantitative analysis of metabolites; and (3) genetic engineering, that is, the application of novel techniques to transfer genes from one organism to another (Schaff, 1991). Some procedures associated with transfer of DNA between organisms also have been successfully used in analysis of genetic variation at the level of DNA. In this chapter we provide necessary background and an overview of techniques associated with biotechnology and its potential and current limitations with regard to growth of woody plants.

Basic Aspects of Molecular Biology

An appreciation of the basic events involved in nucleic acid metabolism in cells is required to understand biotechnology and the techniques associated with it. In the following sections we provide an outline of the processes of DNA replication, DNA transcription, and translation of mRNA on ribosomes that result in the synthesis of proteins:

DNA
↓ transcription
RNA
↓ translation
Protein

DNA Replication

Daughter cells receive one set of the original strands of DNA of a parent cell and one set newly synthesized during the events of cell division. Replication of DNA occurs simultaneously at numerous locations along the genome, as double-stranded DNA molecules are "nicked" and unwound from a tightly coiled double helix by a special enzyme (a helicase) (Anderson and Beardall, 1991). Then the enzyme DNA polymerase directs synthesis of new strands using deoxynucleotide molecules complementary to the existing DNA molecule. Three DNA polymerases have been identified in higher plants, two of which apparently are active in DNA replication and one of which may function in DNA repair. Short RNA primer molecules that complement specific regions on both single strands of DNA provide a starting point for action by DNA polymerase. Ultimately, RNA primers are removed and replaced with DNA.

Transcription

Gene expression first involves synthesis of messenger RNA (mRNA). The essential part of the message consists of numerous three-base sequences that code for amino acids in a protein. However, the gene from which an mRNA molecule is synthesized is structurally more complex than a simple strand of DNA that is complementary to the amino acid-coding bases of mRNA (Robinson *et al.*, 1993) (Fig. 9.1). Genes possess regions of base sequences "upstream" (i.e., in the 5′ direction) that respond to various kinds of regulatory molecules, and they contain special DNA sequences that indicate start and stop positions for RNA synthesis. Additionally, in eukaryotic cells there are regions of base sequences (introns) within the coding region that are not retained in a mature mRNA molecule. Conversely, the nucleotide regions in DNA that do code for amino acids in a mature mRNA molecule are called exons.

Three RNA polymerase enzymes are involved in transcription of DNA. In the nucleus, DNA is transcribed by all three enzymes, but each transcribes either ribosomal RNA (rRNA), transfer RNA (tRNA), or structural genes encoding proteins (mRNA). Normally, chromosomal DNA exists in a complex structural arrangement (Fig. 9.2) that is not transcribable. It must be unwound sufficiently to interact with regulatory molecules and RNA polymerase before transcription can occur.

Even if a gene is accessible, transcription will be low unless special proteins are present that are coded elsewhere on the chromosomes (sometimes called *trans*-acting factors). These proteins bind to nearby regulatory sequences of the gene (*cis*-acting DNA sequences). Once induced, transcription by RNA polymerase is directed to a specific start point near a sequence of four bases (the TATA box, where T and A represent the nucleotide bases thymine and adenine, respectively) (Fig. 9.1). The RNA polymerase molecule, along with several proteins known as transcription factors, are positioned at the transcription start site. Proper RNA polymerase positioning may be promoted or inhibited by *trans*-acting factors

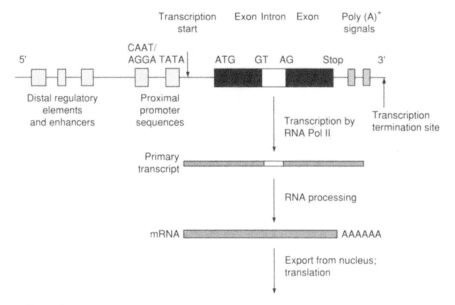

Figure 9.1 Conceptual representation of a plant gene and its expression. From Regulation of gene expression. Robinson, N. J., Shirsat, A. H., and Gatehouse, J. A., *In* "Plant Biochemistry and Molecular Biology" (P. J. Lea and R. C. Leegood, eds.). Copyright 1993 John Wiley & Sons. Reprinted by permission of John Wiley & Sons, Ltd.

(DNA-binding proteins) that bind to regulatory sequences of DNA. Other short base sequences upstream from the start site, called enhancers, also may promote transcription of a gene.

The initial product of transcription is heteronuclear RNA (hnRNA) which differs from mRNA in that it contains (1) upstream and downstream RNA base sequences and (2) introns that are not translated on the ribosomes during protein synthesis (Watson and Murphy, 1993). This hnRNA molecule has a 7-methylguanosine "cap" added to the 5′ end that protects it from degradation by nucleases and also functions in binding of the mature mRNA to ribosomes. Additionally, a 13- to 23-base pair polyadenosine "tail" is added to the 3′ end of the RNA. The poly(A) tail also promotes stability, again presumably by protecting the mRNA from nuclease attack. The final stage of processing involves removal of intron regions of the RNA and resplicing of the molecule at the exon–exon junction (Fig. 9.1).

Translation

Three steps are involved in translation of mRNA on ribosomes: initiation, elongation, and release. The stage for translation is set when a 40 S ribosome in the cytoplasm associates with a special molecule of transfer RNA

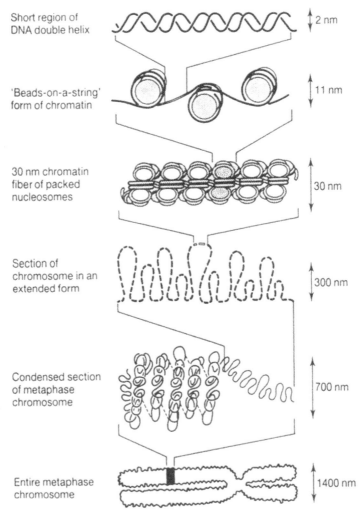

Short region of
DNA double helix

2 nm

'Beads-on-a-string'
form of chromatin

11 nm

30 nm chromatin
fiber of packed
nucleosomes

30 nm

Section of
chromosome in an
extended form

300 nm

Condensed section
of metaphase
chromosome

700 nm

Entire metaphase
chromosome

1400 nm

Figure 9.2   Hierarchy of chromatin packing believed to give rise to the condensed metaphase chromosome in plants. From Watson and Murphy (1993). Genome organization, protein synthesis and processing in plants. *In* "Plant Biochemistry and Molecular Biology" (P. J. Lea and R. C. Leegood, eds.), pp. 197–219. Copyright 1993 John Wiley & Sons. Reprinted by permission of John Wiley & Sons, Ltd.

(tRNA) linked to the amino acid methionine. The 60 S ribosome binds to this complex, forming an 80 S–Met–tRNA–mRNA complex that begins the elongation phase of translation. Amino acid molecules that correspond to the three-base codons on the mRNA molecule are added in sequence to the growing polypeptide chain as the ribosome travels along the mRNA.

Chain termination codons and specific polypeptide molecules end the elongation step and release the newly synthesized polypeptide. The polypeptide often contains more amino acids than will constitute the mature protein. The extra amino acids facilitate movement of the protein out of the nucleus and direct it to specific locations within the cell, a process that frequently involves transfer across membranes into organelles such as chloroplasts and mitochondria.

Insertion of Foreign Genes: Transformation

Developments in biotechnology now allow routine transfer of DNA from one organism to another without the need for sexual reproduction. Successful genetic transformation of an organism depends on the stable incorporation of a novel gene into the genome of the target plant and its transmission to subsequent generations of progeny. Through transformation, genetically based traits can be transferred between quite different plants, and, theoretically, between any two organisms.

Other approaches to genetic engineering through modification of gene expression also have been discovered. For example, it is possible to transfer DNA that is an inverted version of a sequence or segment of a sequence to a gene of interest in a particular organism. This "gene," when transcribed, produces an antisense RNA molecule that prevents production of a protein gene product by interfering with transcription or normal mRNA processing, transport, or translation (Woodson, 1991; Bird and Ray, 1991). Interestingly, reduction in gene expression of an endogenous gene also may occur when additional copies of the same gene are incorporated into the genome in a phenomenon called sense suppression or cosuppression (van der Krol *et al.*, 1988; Flavell, 1994; Finnegan and McElroy, 1994).

Although transformation of plants is becoming more common, from a practical standpoint there often are technical barriers that must be overcome. The following sections provide a brief introduction to the methods employed in transferring genes in woody plants.

Vectors

Vectors are biological agents that promote transfer of foreign DNA into cells. A variety of cloning vectors are available, including plasmids, viruses, and bacteriophages, all of which have been modified genetically to carry foreign DNA. They generally are introduced into plant cells where the desire is to incorporate a specific gene into the plant's chromosomes. Vectors usually contain additional genetic information that (1) stimulates transcription of transferred DNA in the target cell, (2) permits positive identification of target cells that have taken up the genetic material of the vector, and (3) greatly increases the relative abundance of transformed cells as cell division proceeds. Identification of transformed cells usually is accomplished by action of a reporter gene. For example, one common reporter gene codes for an enzyme (β-glucuronidase) that catalyzes a color-producing reaction when cells are exposed to appropriate chemical reagents. Assurance of the exclusive presence of transformed cells is promoted by inclusion of a selectable marker gene in the vector. Frequently, this gene confers resistance to an antibiotic to which wild-type cells are susceptible. Hence, by culturing colonies in the presence of this antibiotic only transformed cells survive.

Genomic and cDNA Libraries

For genes to be effectively studied and manipulated by molecular techniques, many copies need to be obtained. These objectives are accomplished through the process of cloning individual genes, generally in bacterial cells. The collections of these clones created from plant DNA are called libraries. Genomic libraries of plants are collections of bacterial populations containing inserts of plant DNA derived from the entire plant genome. To produce a genomic library, total plant DNA is isolated and then treated with an enzyme that cuts all DNA molecules at particular sequences of nucleotides, producing a large number of DNA fragments with identical, distinctive nucleotide sequences "dangling" at each end. A selected vector is treated with the same enzyme, producing vector DNA with "sticky" ends to the plant DNA fragments. Plant DNA fragments are then taken up by the vector, spontaneously matched, and terminal nucleotide sequences are joined covalently by the enzyme DNA ligase to produce recombinant vector DNA.

The vector is then introduced into *Escherichia coli*, and the bacterium is transformed by incorporation of vector DNA into the bacterial chromosome. *Escherichia coli* can then be grown in the presence of an antibiotic for which only transformed cells have an antibiotic resistance gene introduced via the vector. As the vector molecules introduce many different fragments of the plant genome into different *E. coli* populations, the resulting collection of clones forms the genomic library.

A cDNA library differs from a genomic library in the source of the plant DNA. In fact, cDNA libraries are not derived from plant DNA at all; rather, they are obtained from mRNA actively being transcribed in plant tissues that are expressing genes of interest. In the presence of the enzyme reverse transcriptase, strands of DNA complementary to the mRNA molecules (cDNA) are produced, and these molecules are then cloned into *E. coli* as noted above. Hence, there are several differences between genomic and cDNA libraries because the latter lack the regulatory sequences, noncoding regions, and introns of genomic DNA.

Requirements for Expression of Transferred Genes in Plants

Once a new gene has been introduced into a plant, it will not necessarily be expressed. As mentioned previously, to assure its expression, genes are transferred with other segments of DNA (promoters) known to stimulate transcription of adjacent DNA. One such promoter sequence very commonly employed is a segment of DNA derived from the cauliflower mosaic virus. Inhibition of gene expression in transformed plants also may arise from variation among organisms in base sequences that code for the same amino acids (codon bias). For example, expression in plants of the gene from *Bacillus thuringiensis* that codes for a protein with insecticidal properties initially was poor. Apparently, bacterial codons for amino acids do not function well in the protein synthesis apparatus of plants. When the wild type gene was modified to convert codons to plant-preferred sequences for the same amino acids, expression of the protein in tomato and tobacco was increased up to 100-fold (Perlak *et al.*, 1991). Additional situations in which expression of transferred genes is inhibited, including sense suppression or cosuppression and gene inactivation, are discussed elsewhere in this chapter.

Techniques of DNA Transfer

The desired transforming agent generally is a specifically "engineered" set of DNA sequences in a form suitable for transfer to target plant cells. Several methods of introduction of foreign DNA into target plant cells have been developed. The potential utility and hazards of such DNA transfer are substantial, as the natural barriers to gene flow between taxonomically diverse plants are circumvented. The most frequently used transformation techniques are described in the following sections.

Agrobacterium-Mediated Transfer *Agrobacterium tumefaciens* and *A. rhizogenes* are pathogenic bacteria that cause the tumors on dicotyledonous plants known as crown gall and hairy root disease, respectively. These organisms possess loops of DNA called plasmids that are required for tumor induction (Walden, 1993). The plasmids contain many genes that participate in the biochemical events of infection and tumor development, and some of the DNA in a particular portion of the plasmids (T-DNA) is transferred to the plant genome where the genes it contains are expressed by the plant. Hence, *Agrobacterium* can serve as a vector to transform plant cells. The bacteria can be manipulated in such a way that specific DNA sequences, along with antibiotic resistance genes, promoter sequences, and reporter genes, can be spliced into the T-DNA region of the plasmid and transferred to the plant genome (Fig. 9.3).

Successful transfer of genes via *Agrobacterium*-mediated transformation has been obtained for numerous woody species including apple, aspen, grape,

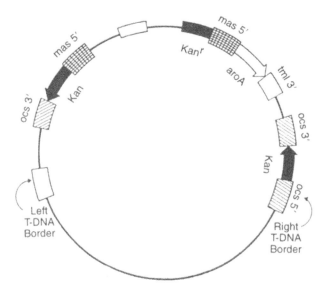

Figure 9.3   Genetic map of the pPMG85/587 plasmid of *Agrobacterium tumefaciens* used to transform hybrid poplar. Between the left and right T-DNA borders are genes that were incorporated into the plant genome. The *aroA* region contains a gene that codes for an enzyme (5-enolpyruvylshikimate-3-phosphate synthase) which is less sensitive to the herbicide glyphosate than the native enzyme of hybrid poplar. *Kan* and *Kan*ʳ regions code for neomycin phosphotransferase that confers resistance to the antibiotic kanamycin allowing selection for transformed cells. The *ocs* and *mas* regions represent DNA sequences of the promoter regions of octopine synthase and mannopine synthetase genes that boost transcription of inserted genes in transformed cells. From *Mol. Gen. Genet., Agrobacterium*-mediated transformation and regeneration of *Populus*. Fillatti, J. J., Sellmer, J., McCown, B., Haissig, B., and Comai, L., **206**, 192–199, Figure 1, 1987. Copyright Springer-Verlag.

European larch, mango, plum, almond, poplar, rose, and walnut (Huang *et al.*, 1991; Mante *et al.*, 1991; Mathews *et al.*, 1992; Miki *et al.*, 1993; Firoozabady *et al.*, 1994; Archilletti *et al.*, 1995). *Agrobacterium*-mediated transformation of important gymnosperms also has been accomplished (e.g., white spruce, Ellis *et al.*, 1989). Significantly, nearly all monocotyledonous plants are not infected by species of *Agrobacterium*. Hence, other methods must be applied to obtain DNA transfer.

Direct DNA Transfer   Several techniques have been developed for introduction of foreign genes into cells without the use of biological vectors. This class of methods often is referred to as DNA-mediated or direct gene transfer.

In some cases DNA is taken up directly into protoplasts with a high rate of foreign gene incorporation into the host cell genome (Lurquin, 1989). Chemical treatment of protoplasts with polyethylene glycol and divalent cations such as calcium promote passage across the plasmalemma of free DNA or of DNA sequestered in membrane-bound structures known as liposomes. Transfer across plasma membranes also is facilitated by high-temperature treatments and exposure of the protoplasts to electric fields (electroporation). These treatments likely influence DNA–membrane interactions and/or protect the free DNA from destruction by internal nucleases (Potrykus *et al.*, 1987). The percentage of transformed protoplasts is relatively small, but where transformation does occur the transferred DNA appears to be randomly located on chromosomes, is stable, and remains in the genome through meiosis.

These methods usually require protoplasts for transfer, and hence protoplast culture procedures and the capability for regenerating plants from protoplasts must be available for successful creation of transgenic plants. Other methods allow DNA transfer in some species (e.g., maize and soybean) for which protoplast culture techniques are difficult or impossible. One technique employs bombardment of target tissues with DNA-coated gold or tungsten particles. This procedure can transform a wide variety of target plant tissues. Such tissues need only be (1) meristematic or possess actively dividing cells and (2) be capable of regenerating whole plants. Operationally, particles are "fired" at explant tissue in an apparatus that uses blank gunpowder cartridges or compressed helium (Fig. 9.4). Rochange *et al.* (1995) compared the effectiveness of these two particle bombardment procedures in obtaining transient expression of foreign genes in Tasmanian blue gum. Alternatively, the particles may be accelerated by high-voltage fields (McCown *et al.*, 1991).

The capacity for expression of newly introduced genes in transformed plants may be transient or long-term. Transient gene expression, usually assumed to indicate placement of DNA into the nuclei of target plant cells, is more easily obtained than stable transformation and may be useful in some studies of gene expression. For example, Ellis *et al.* (1991) were able to demonstrate the capacity of angiosperm promoter sequences to stimulate gene expression in a gymnosperm species (white spruce) even though promoter expression was transient. Long-term capacity for gene expression is an obvious requirement for many introduced traits such as insect resistance and growth characteristics. Yellow poplar, loblolly pine, Norway spruce, white spruce, cranberry, and hybrid poplar have been stably transformed using particle bombardment-based introduction of foreign DNA (Stomp *et al.*, 1991; Wilde *et al.*, 1992; Serres *et al.*, 1992; Ellis *et al.*, 1993; Miki *et al.*, 1993).

The transformed tissues and regenerated transformed plants produced by particle bombardment may be chimeric (i.e., contain both transformed and untransformed cells), but as mentioned previously this possibility is minimized by incorporating a selectable marker gene into the transferred DNA that confers resistance to a normally lethal compound (e.g., kanamy-

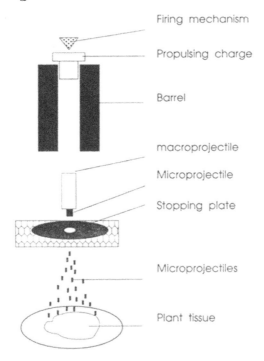

Firing mechanism

Propulsing charge

Barrel

macroprojectile

Microprojectile

Stopping plate

Microprojectiles

Plant tissue

Figure 9.4   Representation of the Biolistic microprojectile gun. The apparatus is enclosed in a vacuum chamber. Microprojectiles are DNA-coated particles of gold or tungsten. The force for particle acceleration can be derived from exploding gunpowder (as shown here), compressed helium, or electrical fields. Reprinted with permission from Miki, B. L., Fobert, P. F., Charest, P. J., and Iyer, V. N. (1993). Procedures for introducing foreign DNA into plants. *In* "Methods in Plant Molecular Biology and Biotechnology" (B. R. Glick and J. E. Thompson, eds.), pp. 67–88. Copyright CRC Press, Boca Raton, Florida.

cin). After transformation, tissues are cultured on medium containing this compound, and untransformed cells do not survive (McCown *et al.*, 1991).

Unlike vector-mediated transformation, direct gene transfer frequently results in insertion of multiple copies of a gene into transformed cells. This attribute of direct gene transfer may be advantageous, as when it is desired to suppress gene expression by adding copies of an existing gene to the genome by sense suppression (see above). Insertion of multiple gene copies may also be disadvantageous if expression of the introduced gene is desired.

Confirmation of Gene Presence and Expression

Southern Blots (DNA Hybridization)   There often exists a need to verify the presence of a specific gene in a plant, and this is a routine need in confirming transformation. One widely used technique, first described by Southern (1975), involves hybridization of known DNA sequences with a population of DNA fragments obtained from sample tissues. Genomic DNA is first digested with restriction enzymes and then separated by size using gel electrophoresis. Denatured DNA fragments on the gel are subsequently transferred to a porous filter by capillary movement. Filters, which may be made of nitrocellulose, nylon, or other materials, are charged and tend to bind DNA. After transfer, DNA is fixed further by baking in an oven.

The presence of the gene in the DNA on the filter is detected by incubating the filter with radioactively labeled DNA or RNA fragments containing the nucleotide sequence of the gene. If the gene is present in the digested DNA, the radioactively labeled matching sequences will hybridize tightly. After washing to remove nonmatching DNAs, the filters are exposed to X-ray film to produce autoradiographs that localize radioactivity so that bands on the original gel can be confirmed as the DNA segment containing the gene (Fig. 9.5).

Northern Blots (RNA Hybridization) Researchers also often need to know whether certain genes are being transcribed in plant tissues. This type of

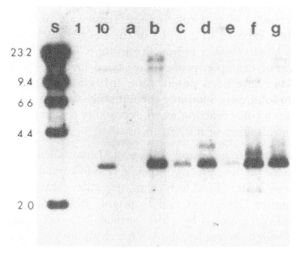

Figure 9.5 Autoradiographs of a Southern blot of DNA from yellow poplar callus lines, some of which were transformed and contained the β-glucuronidase (GUS) gene. Genomic yellow poplar DNA was digested with restriction enzymes. Lanes 1 and 10 represent two different amounts (lane 10 = 10× lane 1) of GUS gene DNA digested with the same enzymes. Transformed callus lines (lanes b–g) show a varying number (3–30) of copies of the GUS gene present. Lane a represents results with an untransformed callus line. Molecular weight markers and their corresponding kilobase values are shown at far left (lane S). Transformation was accomplished using microparticle bombardment. From Wilde *et al.* (1992).

knowledge is especially useful in comparative studies of plant growth and development where, for example, an investigator would like to know which genes participate in flower or fruit development as opposed to those associated with root growth or xylem element differentiation. In Northern blot hybridizations, specific mRNAs can be isolated and identified, allowing inferences as to which genes are "switched on" during a particular developmental stage.

The method is similar to that employed in Southern hybridizations, but RNA is isolated instead of DNA. This is a somewhat more difficult task than DNA extraction, as the environment (glassware surfaces, fingertips, etc.) tends to be filled with ribonucleases that degrade sample RNA. Routinely, Northern hybridizations are performed on total RNA isolates, but mRNA may be separated from other RNA species (rRNA, tRNA) if desired. In a procedure that is similar to Southern hybridizations, Northern blots involve transfer of RNA to a filter membrane by capillary action. After mRNA binding, filters are incubated with radioactively labeled probes that contain nucleotide sequences that correspond to the mRNA of interest. Subsequent autoradiography confirms the presence of the mRNA of interest on the filter.

In situ hybridization of mRNA within tissues is another method that can be employed in developmental studies to pinpoint the physical location of intense transcription of particular genes. Tissues are first fixed and sectioned and then exposed to a probe of the desired nucleotide sequence that has been labeled with a radioactive isotope. Other types of tags also may be used. For example, probes have been linked with molecules to which antibodies have been raised. After hybridization, sectioned tissues are exposed to solutions containing the antibody. An enzyme linked to the antibody causes a color reaction when exposed to a substrate (see below), allowing the localization of mRNAs in tissues.

Immunological Methods to Identify Molecules   Immunological methods of analysis are based on the capacity of special cells in animals, B lymphocytes, to produce proteins that bind to foreign molecules (antigens) introduced into the organism (Dumbroff and Gepstein, 1993). This "immune response" is a primary means by which animals fight off infections. Once triggered by foreign molecules, the DNA in lymphocytes undergoes rapid mutation and recombination, and when individual lymphocytes differentiate they produce a family of glycoprotein antibodies of varying composition (polyclonal antibodies). Each antibody type has an affinity for the antigen that elicited the immune response. The immune system generally responds to relatively large molecules, although each type of antibody may bind to a limited region of the antigen.

Western Blots (Protein–Antibody Blot)   The polypeptide translation products of particular genes can be identified if antibodies to the polypeptide have been prepared. Protein preparations are first subjected to gel electrophoresis to physically separate proteins by molecular weight. Proteins are then transferred to membranes, often made of nitrocellulose paper, by blotting, and the membranes are subsequently exposed to antibodies to the protein of interest. Detection of the protein–antibody complex on membranes often is accomplished by prior linking to the antibody of an enzyme that catalyzes a color-producing reaction. After binding of the antibody to the target protein, addition of enzyme substrates localizes the target protein by color development.

ELISA (Enzyme-Linked Immunosorbent Assay)   Antibody-based analytical assays have made it possible to do quantitative analysis of many important plant metabolites without the expenses associated with chemical analyses. The range of molecules capable of detection is widened by linking molecules that would ordinarily be too small to elicit the immune response (e.g., plant hormones) to larger proteins. Specific subpopulations of antibodies can be identified for which the plant metabolite provides an epitope (the key binding site of the antibody), and these so-called monoclonal antibodies thus are quite specific to the molecule of interest. Once identified, large amounts of pure monoclonal antibody can be produced by hybridoma cells (fusion products of cancerous B lymphocytes and spleen cells of immunized mice) that are kept in culture. Polyclonal antibodies, which are easier to produce, also may be used in assays, but at the cost of reduced specificity.

The binding of antibodies per se does not yield an easily measured indicator, and hence substrates must be linked with other reporter molecules. These reporter molecules frequently are enzymes that catalyze reactions resulting in development of color, hence the name enzyme-linked immunosorbent assay (ELISA). For example, alkaline phosphatase may be conjugated to a plant hormone to which antibodies have been raised, and when *p*-nitrophenyl phosphate is added to the solution, a yellow color develops. Alternatively, antigen molecules may be radioactively labeled (radioimmunoassay) and measured by scintillation counting.

Operationally, ELISA-based techniques fall into three basic categories (Dumbroff and Gepstein, 1993).

Competitive Antigen Capture   For competitive antigen capture, antibodies are coated on the walls of small, cuplike reaction wells. Molecules of the plant metabolite (e.g., abscisic acid, ABA) to which a reporter enzyme has been linked (conjugated) are added and are bound by the antibodies in the well. The reporter enzyme (e.g., alkaline phosphatase) commonly catalyzes a color-producing reaction. After removing the solution containing the conjugated ABA, the sample of interest then is added and any unconjugated ABA in the sample enters into competition for antibody binding sites with conjugated ABA. Displacement of conjugated ABA will be proportional to the amount of free ABA in the sample. After washing, color-producing substrate is added, and the intensity of color development is measured in a spectrophotometer. Intensity of color development is inversely proportional to the sample ABA concentration.

Double Antibody Sandwich, Antigen Capture   As in competitive antigen capture, antibodies are coated on reaction well walls. However, a sample to be assayed is added next instead of plant metabolite conjugated to a reporter enzyme. Molecules of metabolite in the sample bind to antibodies attached to the wall in proportion to their concentration. Subsequently, another antibody to which a reporter enzyme has been linked is added and binds to the immobilized antigenic molecules at another site. Hence, the plant metabolite molecule is "sandwiched" between two antibodies. Finally color-developing reagents are added. In this assay, the concentration of the antigenic molecule is directly proportional to color intensity.

Antibody Capture, Indirect Detection   The antigenic plant metabolite molecules of sample unknowns are coated onto well walls and exposed to a primary antibody. The primary antibody then is bound to a second polyclonal antibody to which the reporter enzyme has been conjugated. Again, color intensity is directly proportional to the concentration of antigen. This method requires large amounts of plant metabolite in the sample unless the antibodies are very specific.

Molecular Analysis of Genetic Structure and Variation

The analysis of genetic variation among and within populations of woody plants is greatly facilitated by molecular techniques. Isozyme analysis, wherein variation in genes is detected by starch gel electrophoresis of enzyme gene products, allows characterization of genetic variation based on the enzymes studied (Adams, 1983). This approach has been widely applied in studies of population genetics in forest trees. For example, Young *et al.* (1993) studied isozyme variation of nine enzymes in range-wide, regional, and local sugar maple populations from southeastern Canada. Although genetic variation was observed at all three spatial scales examined, there was slightly less variation detected than in other North American woody angiosperms. These data were considered valuable in prescribing management unit sizes for conservation of genetic diversity within sugar maple.

A substantial disadvantage of isozyme analysis, however, is that only a limited number of enzymes can be studied. Further, variation in portions of the plant genome not coding for enzymes remains inaccessible by this method. In contrast, molecular methods are available that open the full range of the plant genome, of both the nucleus and organellar DNA in mitochondria and chloroplasts, to studies of genetic variation (Cheliak and Rogers, 1990; Neale and Williams, 1991). These techniques and some applications are discussed below.

RFLP (Restriction Fragment Length Polymorphism) Analysis

In the RFLP approach, genetic variation at the DNA sequence level can be assessed by use of restriction endonuclease enzymes that "cut" DNA strands at locations where unique sequences of nucleotides occur. For example, the restriction endonuclease *Hin*dIII, obtained from the bacterium *Haemophilus influenzae*, cleaves DNA sequences containing GGATCC in the 5' to 3' direction, leaving two fragments of DNA with four-nucleotide single-stranded regions at the ends (underlined, above, for one fragment). Thus, cleavage of DNA by the *Hin*dIII enzyme is restricted to certain sites on a target DNA molecule (hence the terms restriction sites and restriction enzymes).

If the nucleotide sequences between two individuals vary, the fragments produced by restriction enzymes may be of different lengths because of loss or gain of matching restriction sites or because the intervening fragment segments are of different nucleotide sequence length (Fig. 9.6). These changes can occur by point mutations within restriction sites or in other segments of DNA which become matching restriction sites as a result of the mutation. Insertions, rearrangements, and deletions of DNA between restriction sites also may occur (Neale and Williams, 1991). In any case, separation by electrophoresis will show fragments of different lengths.

The utility of RFLP analysis in phylogenetic, taxonomic, population genetics, and breeding research has been shown in an increasing number of studies, and only a few are given here as examples. Steane *et al.* (1991) employed RFLP analysis of chloroplast DNA in *Eucalyptus* to investigate genetic variation among different subgeneric taxa. Confirming previous classification schemes, subgenera *Monocalyptus* and *Symphomyrtus* were clearly separated by RFLP analysis, where, with the exception of a single individual of one species, 45% of RFLP fragments were specific to one subgenus or another. Fragment differences progressively decreased at lower taxonomic levels (series and species). Restriction fragment length polymorphism analysis also was employed by Waters and Schaal (1991), who could not detect any within- or between-population variation in more than 150 restriction enzyme sites in the chloroplast genome of two populations of Torrey pine separated by 280 km for at least 8000 years. The authors concluded that the two populations were relics of a larger and very genetically uniform population. Finally, Hubbard *et al.* (1992) used RFLP analysis and Southern hybridization with cultivar-specific probes to unequivocally identify rose cultivars in developing a potentially useful tool to protect patent rights to new cultivars.

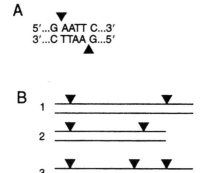

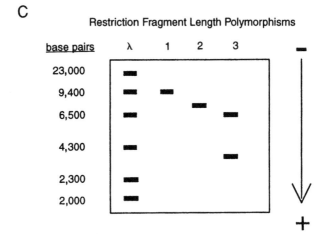

Figure 9.6   Schematic representation of restriction fragment length polymorphism (RFLP) analysis. (A) Recognition sequence for the *Eco*RI restriction endonuclease, showing the location of separation of nucleotide chains of double-stranded DNA. (B) DNA of three trees (1, 2, 3) digested with *Eco*RI. Tree 2 exhibits a deletion of a small amount of DNA between the two sites of cleavage by *Eco*RI, whereas tree 3 has gained an additional site of cleavage relative to trees 1 and 2. (C) Representation of the result of gel electrophoresis of DNA fragments from all three trees after digestion. Only fragments between cleavage sites are shown. Molecular weight standards appear at left (lane λ). From Neale and Williams (1991).

Exposure of genomic DNA to several different restriction enzymes, singly and in combination, allows an overlapping series of DNA sequences to be identified, and ultimately the entire DNA sequence of a chromosome or genome can be deduced. If sequences for specific genes have been cloned, probes can be made that allow for locating by Southern hybridizations (see above) genes on restriction fragments (Fig. 9.7). Thus, genes can be

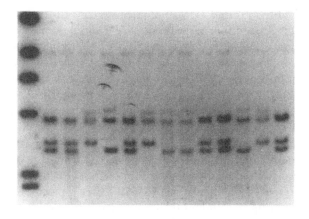

1   2   3   4   5   6   7   8   9   10   11   12   13   14

Figure 9.7   Autoradiograph of a Southern hybridization of DNA from 13 progeny of a full-sib cross of loblolly pine trees. Genomic DNA was digested with the *Hin*dIII restriction endonuclease enzyme, and fragments were subjected to electrophoresis. After blotting, transferred DNA fragments were hybridized with a cDNA probe from loblolly pine. The results illustrate the presence of polymorphisms in DNA fragment length following treatment with restriction enzymes (RFLPs). The far left-hand lane represents a series of DNA standards of known molecular weight (23, 9.4, 6.5, 4.3, 2.3, and 2.0 kilobases from top to bottom). From Neale and Williams (1991).

mapped to specific regions of chromosomes, and detailed genetic linkage maps can be constructed.

Despite its utility, genetic analysis by RFLP techniques has some deficiencies, including the large amounts of time required, the necessity for a large laboratory and for personnel equipped and trained for DNA manipulation, and the need for radioactive isotopes (Landry, 1993). Additionally, relatively large amounts of tissue are needed for RFLP analysis, often making genetic analysis of seedlings impractical. Two additional techniques, one the outgrowth of the other, have overcome many of these drawbacks.

PCR (Polymerase Chain Reaction) and RAPD (Random Amplified Polymorphic DNA) Analysis

The first analysis method, the polymerase chain reaction (PCR), has greatly reduced the amount of tissue needed for molecular-based genetic analysis (Saiki *et al.*, 1985). PCR involves repeated *in vitro* synthesis of a known polymorphic restriction fragment. Synthetic primers corresponding to known sequences of the fragment are added, and the double-stranded DNA of the fragment is separated by heating; subsequent cooling allows the primer to anneal to the complementary DNA sequences on the fragment. The reaction mixture is then heated to the optimal temperature for synthesis of double-stranded DNA in the presence of a heat-stable DNA polymerase. This procedure requires only a few minutes and is repeated in a cyclical fashion many times, resulting in a doubling in number of the DNA fragments with each cycle. Theoretically, 30 cycles would yield over a billionfold amplification of DNA, allowing analysis of a single sequence from a single cell (Arnheim *et al.*, 1991). After amplification, fragments can be separated by electrophoresis and visualized by staining with ethidium bromide or silver, avoiding the necessity for Southern hybridizations and use of radioactive isotopes.

In the randomly amplified polymorphic DNA (RAPD) procedure, PCR is modified by using oligonucleotide primers of arbitrary rather than preselected sequences (Williams *et al.*, 1990). After separation of double-stranded DNA during the initial heating phase of the amplification procedure, the primer will anneal to complementary sequences of single-stranded DNA. The PCR procedure is otherwise conducted as usual, and any DNA segment of length less than about 3000 bases (the length limit of DNA synthesis during the temperature cycles) bounded or led by primer sequences will be amplified. If DNA of two plants differs by loss or gain of primer sites, or if there are deletions, inversions, or insertions of nucleotides within amplified regions, subsequent electrophoresis of amplification products will reveal fragments of varying nucleotide length (polymorphisms).

RAPD-based analysis has much potential in establishing genetic relationships among important woody crop plants (e.g., grapes, Striem *et al.*, 1994; Büscher *et al.*, 1994; poplars, Rani *et al.*, 1995; cranberry, Novy and Vorsa, 1995). In the case of cranberry, RAPD analysis was especially effective in demonstrating that there was much genetic variation within presumed cultivars of this woody fruit species (Novy and Vorsa, 1995). For decades, propagation of cranberry has largely been based on asexual propagation by transplanting of stolons from existing bogs assumed to be a single cultivar. Morphological differences among cultivars are quite small and often masked by environmental influences. Conventional isozyme analysis in cranberry is limited by low genetic variation among available enzymes. However, RAPD analysis of samples of 22 cranberry cultivars revealed that each cultivar was represented by multiple genotypes, many of which did not appear to be closely related. In fact, some clones of a presumed cultivar had the same DNA fragment profile as that of other cultivars, indicating cultivar misclassification.

The RAPD procedure also may be used in developing genetic markers for breeding. DNA sequences that are clearly segregated by RAPD analysis may be quite close to genes controlling desirable traits. Some traits, such as disease resistance or subtle morphological differences, are difficult or expensive to select under field conditions (Kelly, 1995). The RAPD procedure may thus be used to complement conventional selection programs, providing clear indicators of gene presence in progeny of selected matings.

Another objective of breeding research is to sequence important genes so that gene products can be analyzed to establish the reasons for variation in the trait controlled by the gene. RAPD-generated genetic markers can be used with plants of nearly identical genotypes but that differ in a single phenotypic trait (near isogenic lines, NILs). Although differing only in a single phenotypic trait, NILs also usually have additional flanking DNA to the gene of interest that may be variable in nucleotide sequence. RAPD markers produced on NILs differing in that single trait can identify DNA fragments that are located very close to the gene. Once located relative to the marker, overlapping DNA fragments from genomic libraries can be hybridized to the target DNA to "walk" down the chromosome to the gene, which can then be analyzed to obtain its nucleotide sequence. Once the nucleic acid sequence has been characterized, the amino acid sequence of the protein product can be determined. Identification of this protein may allow the biochemical basis of the differential trait expression to be revealed.

Propagation

A wide variety of methods are available for obtaining genetically uniform populations of woody plants. There are several reasons for producing such populations. As discussed above, plant transformation techniques require protoplasts, cells, or meristematic tissues for transfer of novel genes. Hence, regeneration and mass propagation of transgenic plants are integral parts of genetic engineering. Propagation of superior individuals in a conventional breeding program hastens delivery of planting materials to the field (Ahuja, 1993). It also captures desirable nonadditive genetic variation (i.e., superior traits arising from particular gene interactions that would be lost in further breeding). Many woody species, such as poplars and willows, may simply be propagated by rooting of cuttings. Many valuable fruit and nut cultivars are propagated by grafting of scions onto seedling rootstocks (Chapter 8). However, in other cases *in vitro* culture (i.e., culture of plant cells or parts using artificial media) is employed to propagate desirable genotypes. These methods, and their utility and limitations, are discussed below.

Macropropagation

Vegetative propagation by rooting of stem cuttings has been practiced for centuries. For example, Toda (1974) reported that propagation of sugi and other species of conifers by rooting of cuttings has been practiced in Japan for at least 1200 years, and that sugi plantations arising from planting of rooted cuttings appeared in the early 1400s. Rooting involves placing stem pieces possessing at least one bud under conditions that favor formation of adventitious roots (i.e., roots that arise at points other than the root pole) (Kramer and Kozlowski, 1979). Adventitious roots may arise from newly "induced" primordia or from preformed primordia laid down during shoot development. Millions of rooted cuttings are now produced each year for outplanting, mostly from easily rooted angiosperms, such as members of the Salicaceae, and economically important conifers such as Monterey pine, sitka, Norway, and black spruces, and Douglas fir (Zsuffa *et al.*, 1993; Talbert *et al.*, 1993).

Capacity for rooting is complex from a physiological standpoint (Chapter 3) and depends on number of factors. There is wide variation in capacity for rooting among species, with willows and many poplars being easily rooted, whereas oaks, ashes, and walnuts are quite difficult to root (Table 9.1). Considerable variation in rooting capacity also exists within taxa considered as easy to root. For example, within the genus *Populus* most species of sections Aigeiros and Tacamahaca root easily, whereas only European white poplar in section Leuce can be propagated by dormant stem cuttings (Zsuffa *et al.*, 1993). There also is intraspecific variation in rooting capacities of *Populus*, with some clones of eastern cottonwood being much easier to root than others.

The phenomenon of maturation in plants is important in both macro- and micropropagation. Juvenile and mature plants exhibit substantial differences in several morphological and physiological attributes, including leaf morphology (see Chapter 2 of Kozlowski and Pallardy, 1997) and capacity to flower (Greenwood and Hutchison, 1993). In micropropagation (see below), it is more difficult to induce organogenesis and embry-

Table 9.1   Variations in Rooting Capacity of Forest Trees[a]

Very easy to root	Easy to root	Moderately difficult to root	Difficult to root	Very difficult to root
Willows	Monterey pine	Aspens	Most eucalyptus	Most oaks
Non-aspen poplars	Cryptomeria	Red maple	Pines	Beeches
	Yews	Some birches	Spruces	Chestnuts
	Junipers	Some hemlocks	Larches	Ashes
			Birches	Walnuts
			Maples	
			Black locust	

[a]Adapted from Wright (1976); from Kramer and Kozlowski (1979).

ogenesis in explants from mature than from juvenile plant material (Bonga and von Aderkas, 1993). Additionally, juvenile plants are more easily propagated by rooted cuttings than are mature plants. Stem cuttings from the latter often are impossible to root. Even if they are successfully rooted, cuttings from mature plant material often display undesirable characteristics such as slow growth rates and plagiotropy (the tendency to grow at oblique or horizontal angles).

The loss of rooting capacity in mature plants is an unfortunate circumstance, as propagation of mature plant material has some inherent practical advantages. In dioecious species, it is of obvious advantage to propagate female trees if high fruit yields are desired, but the sex of an individual tree usually is not known during the juvenile phase (Bonga, 1987). Additionally, plant propagators, geneticists, and others wish to improve other yield characteristics, such as volume and quality of wood produced, that are only assessable for mature plants. In large-scale operations, juvenility of superior source plants is sometimes preserved by repeated propagation of cuttings from cuttings (serial propagation) or by severe pruning ("hedging") treatments. The latter procedure has been used with some success in Monterey and loblolly pine and sitka spruce (Talbert *et al.*, 1993). Monterey pine hedges have been reported to maintain juvenility for more than 27 years. Similarly, Norway spruce has been maintained in a juvenile state through five cycles of rooting of cuttings at 3-year intervals (St. Clair *et al.*, 1985).

In Vitro Culture and Micropropagation

Prolonged *in vitro* culture of plants and their parts, often called microculture, has been possible since the need was recognized for several plant growth regulators (PGRs) to sustain cultured explants. Since the 1960s, techniques have been developed to liberate protoplasts from their enclosing cell walls and subsequently to sustain them in culture. Regeneration of whole plants from *in vitro* cultures may be obtained by induction of shoots and roots (organogenesis) (McCown, 1985; Han and Keathley, 1989) or by induction of embryolike structures (somatic embryogenesis) (Durzan, 1988). These methods either have or possess the potential for micropropagation (the production of large numbers of plants for distribution) and for providing suitable plant materials for transformation. Over 50 tree species have been propagated by these techniques; however, in only a few cases have these methods reached an operational stage (Ahuja, 1993; Talbert *et al.*, 1993). These approaches are discussed below.

Callus Culture   Undifferentiated callus may be induced to differentiate by shifting the balance among concentrations of PGRs (Fig. 9.8). Skoog and co-workers (e.g., Skoog and Miller, 1957) elegantly demonstrated that tobacco stem callus that was exposed to high cytokinin–auxin ratios in

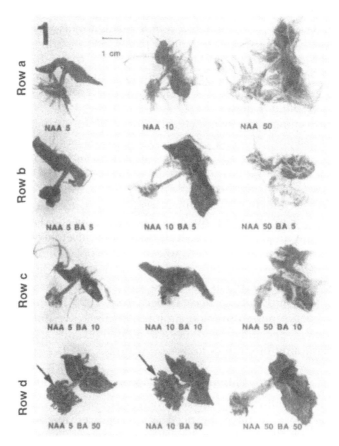

Figure 9.8   Adventitious shoot bud differentiation in excised leaves of paulownia (*Paulownia fortunei*) cultured on media with different concentrations of auxin (naphthaleneacetic acid, NAA) and cytokinin (benzyladenine, BA). Concentrations (micromolar) of growth regulators to which cultured buds were exposed are given below the photos. Arrows indicate location of cut petiolar end. From Rao *et al.*, "High Frequency Plant Regeneration from Excised Leaves of *Paulownia fortunei*," *In Vitro Cellular & Developmental Biology—Plant*, Vol. 29P, No. 2, pp. 72–76. Copyright © 1993 by the Tissue Culture Association, Inc. Reprinted with permission of the copyright owner.

culture produced adventitious shoots, whereas that exposed to low ratios produced roots. Hicks (1994) reviewed the developmental aspects of organogenesis.

Since the delineation of requisite culture materials and protocols, there has been a proliferation of species that have been successfully cultured. The basic culture medium includes five types of materials: inorganic nutrients (Table 9.2), organic supplements (e.g., amino acids, polyamines, protein hydrolysates, and coconut milk), vitamins (e.g., pantothenic acid, thiamine), a carbon source (most often sucrose), and phytohormones (e.g., auxins such as 2,4-D or naphthaleneacetic acid, cytokinins such as kinetin, zeatin, or benzylaminopurine, and sometimes ABA) (Gamborg *et al.*, 1976; Bonga and Durzan, 1987; Bonga and von Aderkas, 1992).

The undifferentiated cells in callus cultures are well-suited to transformation. For example, cocultivation of callus tissues with genetically modified *Agrobacterium tumefaciens* is a convenient method to obtain transformed cells that will eventually give rise to nonchimeric plants. Callus also may be transformed by chemical and ballistic methods. For species for which plants can be regenerated from callus culture, genetic engineering via transformation offers much potential for plant improvement.

Callus culture techniques may also be employed to create haploid plants. Anthers, microspores, pollen grains, and megagametophytes have been used as sources from which haploid callus is cultured. Skirvin *et al.* (1993) listed several advantages to woody plant geneticists of creating haploid plants: (1) homozygous plants can be produced in a single generation, even if a species is self-sterile; (2) large numbers of haploid individuals can be obtained; (3) it is much easier to induce and identify mutations in haploid plants; and (4) the possibility of chimeric origin in cultures of single pollen grains is reduced. Haploid plants of woody angiosperm (poplar) and gymnosperm (European larch) species have been regenerated. Chromosome counts of regenerated plants often indicate spontaneous chromosome doubling so that the putative haploid plant actually is a combination of diploid and haploid cells (von Aderkas and Bonga, 1993). Haploid plants also may exhibit low vigor or lack of viability because of the frequent presence of deleterious or lethal recessive genes (Skirvin *et al.*, 1993).

There are some drawbacks that have limited the utility of undifferentiated callus cultures in research and to commercial plant propagators. Each species (even each genotype) requires a specific culture medium and protocol to successfully initiate and sustain callus in culture. Hence, for each new material there is an arduous search for the right combination of basal medium constituents and concentrations, PGR concentrations, and temporal sequencing of all the components. Regeneration of plants from culture requires similar empirical approaches, and even with much research there are many woody species that have not been successfully regenerated to whole plants from callus.

Undifferentiated cells, such as those of callus and protoplasts, also tend to be genetically unstable, producing genetic variation (called somaclonal variation) in cultures of vegetative plant cells or tissues (Dodds, 1983).

Table 9.2 Comparison of Inorganic Nutrient Ion Concentration of 11 Media Commonly Used for Woody Plant Microculture[a]

Nutrient[b]	1981 Litvay conifer cell	1962 MS general	1975 Anderson Rhododendron shoot	1977 Quoirin–Lepoivre Malus shoot	1973 Durzan conifer cell/shoot	1968 Gamborg-B5 general cell shoot	1981 WPM general cell/shoot	1972 Gresshoff–Doy general shoot	1980 Zimmerman blueberry shoot	1985 DCR conifer shoot	1943 White general shoot/rooting
NH_4^+	20.60	20.61	24.98	5.00	20.00	2.02	4.94	3.17	5.00	5.00	—
K^+	22.52	20.04	10.65	19.80	2.90	24.70	12.61	12.58	5.00	4.62	1.67
Ca^{2+}	0.15	2.99	2.99	5.10	4.15	1.02	3.00	0.680	3.00	2.69	1.27
Mg^{2+}	7.50	1.50	1.50	1.50	1.50	1.01	1.50	1.01	1.50	1.50	2.92
Mn^{2+}	0.124	0.132	0.100	0.005	0.132	0.059	0.132	0.059	0.075	0.132	0.031
Zn^{2+}	0.150	0.029	0.030	0.034	0.030	0.007	0.030	0.010	0.030	0.030	0.019
Na^+	0.010	0.224	1.34	0.202	—	1.10	0.224	1.11	0.402	0.102	2.94
Fe^{2+}	0.050	0.100	0.100	0.100	0.100	0.050	0.100	0.100	0.200	0.100	0.012
NO_3^-	39.40	39.40	34.38	33.00	30.00	24.70	9.64	9.89	10.00	10.48	3.33

SO$_4^{2-}$	7.50	1.73	1.64	1.50	2.04	7.44	2.67	2.67	1.76	5.81
PO$_4^{3-}$	2.50	2.48	2.00	1.30	1.10	1.25	0.866	3.00	1.25	0.144
BO$_3^{2-}$	0.501	0.100	0.10	0.100	0.049	0.100	0.049	0.100	0.100	0.024
Cl$^-$	0.300	5.98	0.0002	—	2.04	1.31	4.04	0.0002	1.16	0.872
Fe EDTA^{3-}	0.110	0.110	0.100	0.110	0.083	0.110	0.110	0.200	0.110	—
Co^{2+}	0.60	0.100	0.10	0.105	0.110	—	1.00	0.100	0.105	—
Cu^{2+}	2.00	0.100	0.10	0.100	0.100	0.100	1.0	0.1	0.100	—
MoO$_4^{2-}$	5.17	1.00	1.00	1.03	1.03	1.03	1.0	1.00	1.61	5.00
I$^-$	25.00	5.00	5.00	5.00	4.52	—	4.51	5.00	5.00	—
Ni^{2+}	—	—	—	—	—	—	—	—	0.193	—
TOTAL N (mM)	60.00	59.36	38.00	50.00	26.72	14.58	13.06	15.00	15.48	3.33
NH$_4^+$/NO$_3^-$ (mM)	0.52	0.727	0.15	0.67	0.08	0.51	0.32	0.50	0.50	0
Total (mM)	101.45	86.48	68.58	61.83	59.98	42.39	36.34	31.18	29.05	19.04

[a]From McCown and Sellmer (1987).
[b]Macroelements are listed in millimolar (mM) and trace elements in micromolar (μM).

Such mutations can occur unusually frequently when cells are in an undifferentiated state, apparently through point mutations, transpositions, and by induction of chromosome abnormalities (Libby and Ahuja, 1993). This is an undesirable attribute if propagation of a single genotype is desired. However, callus (and protoplast) culture may intentionally be utilized to produce new somaclonal variants for propagation (Ahuja, 1987; Serres *et al.*, 1991; Saieed *et al.*, 1994a,b). Additionally, mutagenic agents such as gamma rays, X rays, ethyl methanesulfonate, and colchicine may be employed to increase the induction of genetic alterations (Michler, 1991). Some traits of mutants may be transmitted to progeny in normal sexual reproduction; in other instances vegetative propagation is necessary. Exposure of cultures to selective agents such as high salt and herbicide concentrations (e.g., in citrus and poplar) has resulted in somaclonal selection of new genotypes that are more resistant to these agents (Spiegel-Roy and Kochba, 1980; Michler and Haissig, 1988).

Finally, callus tends to lose its capacity to regenerate to whole plants the longer it is maintained in culture. Partly because of these deficiencies, additional methods have been developed for propagating desirable genotypes *in vitro*.

Shoot and Meristem Culture Shoot tips that include the apical meristem and other apical tissues may be easily cultured in many species. True meristems are obtained by dissection of the shoot tip so that only tissues apical to the first leaf primordium are cultured (Dodds, 1983). If other tissues are cultured as well, the culture is more appropriately called shoot tip culture. Shoot tip and meristem cultures have been widely employed to eradicate virus infections of plants (Morel, 1960). Viruses often are introduced into important plant selections because infection occurs by physical contact with contaminated instruments used to excise plant material for vegetative propagation. Virus infections usually do not extend to apical meristems, however, and meristem culture provides an important means of "rescuing" a desirable genotype.

Shoot cultures also may be used to micropropagate desirable genotypes (Murashige, 1974; McCown, 1985). Shoot tips, stem nodes, leaves, and apical meristems can be induced to proliferate numerous shoots in culture by careful adjustment of PGRs (Fig. 9.9). Shoots may be traced to existing axillary buds of the cultured explant or to shoots derived from newly differentiated adventitious buds. High cytokinin concentrations tend to promote more adventitious shoots, which may be more unstable genetically because they arise from undifferentiated tissues (McCown and Lloyd, 1983).

Once shoot proliferation is obtained, repeated subculturing is performed in a procedure known as "stabilization" (Hartmann *et al.*, 1990). During

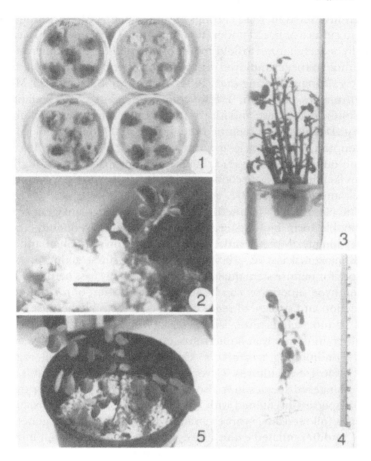

Figure 9.9   Stages in organogenesis and plantlet production in black locust. (1) Culture of callus derived from xylem tissues. (2) Shoot induction *in vitro* (bar: 3 mm). (3) Shoot multiplication on agar medium with 0.5 μ*M* benzylaminopurine (tube diameter: 2.5 cm). (4) Rooted shoot from tube culture in 3 (scale same as in 3). (5) Regenerated black locust plantlet in pot (pot diameter: 25 cm). From *Plant Cell Rep.,* Cambial tissue culture and subsequent shoot regeneration from mature black locust (*Robinia pseudoacacia* L.). Han, K. H., Keathley, D. E., and Gordon, M. P., **12**, 185–188, Figs. 1–5 (1993). Copyright Springer-Verlag.

subculturing, shoot growth rates progressively increase and shoots become more uniform in appearance, with smaller leaves and more narrow stems (McCown, 1985). Shoot cultures derived from juvenile plants stabilize more quickly than those from plants that are more mature, suggesting that some degree of rejuvenation occurs during stabilization. This assertion is supported by the greater capacity of shoot "microcuttings" to root adventitiously, a distinctly juvenile trait, than cuttings of source plants. Microcuttings can be rooted in culture or in soil under high humidity. The tender plants can then gradually be acclimatized to field conditions for planting. This micropropagation method has been employed in production of woody ornamental and fruit crops (e.g., banana, Hwang *et al.*, 1984; birches, Smith *et al.*, 1986; McCown, 1989a,b; maples, McCown, 1989b; sourwood, Banko and Stefani, 1991; rhododendron, McCown and Lloyd, 1983; apple, Liu *et al.*, 1983) and forest trees (e.g., Douglas fir, Mohammed and Vidaver, 1990; loblolly pine, McKeand and Frampton, 1984; black locust, Han *et al.*, 1993; black cherry, Tricoli *et al.*, 1985; mesquite, Shekhawat *et al.*, 1993). Micropropagation through organogenesis is easier and more economical for angiosperms than gymnosperms (Cheliak and Rogers, 1990).

In vitro culture may be used to conserve specific plant genotypes in a propagatable form by long-term cryopreservation. Frequently, there is a need for long-term preservation of specific genotypes (Chen and Kartha, 1987; Klimaszewska *et al.*, 1992). For example, in selecting among tree genotypes for mature stem qualities, many years may pass before judgments as to genotype superiority can be made. Preservation of a given genotype by collection and storage of seeds usually is not feasible because most tree species cannot self-fertilize. Viable shoot or meristem cultures may be maintained indefinitely at liquid nitrogen temperature ($-196°C$) after certain preconditioning treatments. On thawing, cultures will regenerate shoots as do fresh cultures. Cryopreservation of callus cultures derived from endangered species may also offer a means of preserving the germplasm of species threatened with extinction (Fay, 1992). Callus culture of genotypes followed by cryopreservation is feasible, but the genetic instability of undifferentiated callus during *in vitro* culture makes it more likely that the original genotype will not be preserved.

Shoot and callus cultures are subject to a physiological disorder known as vitrification. Callus cultures assume the appearance of a watery mass. Vitrified plantlets from shoot cultures have broad translucent stems and fragile, translucent, often elongated leaves that are thick and wrinkled or curled (Gaspar *et al.*, 1987; Jones, 1993). Vascular tissues are not properly lignified, and plantlets exhibit slow growth rates. When it appears, vitrification may be remedied in a variety of ways, including changes in agar concentration and supplier, carbon source, and concentration of NH_4^+. Cold treatments and addition of phloridzin may also cause cultures to revert to a normal state (Gaspar *et al.*, 1987; Pâques, 1991).

Embryogenesis   Under certain culture conditions, tissues undergo developmental differentiation, in a process called somatic embryogenesis, into embryolike structures possessing root and shoot poles connected by vascular tissue (Cheliak and Rogers, 1990) (Fig. 9.10). In contrast to the sequen-

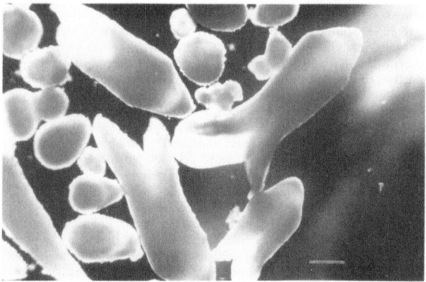

Figure 9.10 Somatic embryos derived from embryogenic callus of spruce (top) and yellow poplar (bottom). Bar in bottom photo: 500 μm. From Cheliak and Rogers (1990), with permission of National Research Council Canada, and Merkle (1991), with permission of Plenum Press.

tial differentiation of shoots and roots associated with organogenesis, somatic embryos exhibit simultaneous differentiation of roots and shoots similar to that seen in development of zygotic embryos of seeds (Litz and Gray, 1992; Ahuja, 1993).

Several types of plant tissues can be used to generate somatic embryos, including leaves, microspores, anthers, endosperm, zygotic embryos, cotyledons, and megagametophytes, but most somatic embryogenesis research has focused on use of zygotic embryos (Tulecke, 1987; Druart, 1990; Piagnani and Eccher, 1990; Von Aderkas and Bonga, 1993). In some species, immature zygotic embryos are more suitable than those that are mature as tissue sources for producing somatic embryos. For example, maximum embryogenic potential of yellow poplar zygotic embryos in culture was reached during the eighth week after fertilization but decreased to near zero in mature embryos (Sotak *et al.*, 1991). However, there appears to be a lower age limit below which embryogenic potential of cultured zygotic embryos is reduced. In loblolly pine, best initiation of embryogenic tissues occurred at the precotyledonary stage of embryo development, whereas in Norway spruce optimal initiation was observed at the time when cotyledon primordia formed (Becwar *et al.*, 1988, 1990).

There are several phases in embryogenesis *in vitro*. First, initiation of embryo production in cultured tissues must be induced, typically in a medium containing high concentrations of auxin and with a high auxin–cytokinin ratio (Talbert *et al.*, 1993). This treatment can induce the development of differentiated proembryonic structures (Becwar, 1993; Gupta and Durzan, 1987). Additionally, there may be other modes of initiation (Hakman *et al.*, 1987; Litz and Gray, 1992). Once initiated, proembryo production will continue indefinitely, even at lower hormone concentrations, if media are changed periodically. The multiplication phase can be sustained for several years (Talbert *et al.*, 1993). Embryogenesis may be obtained directly from an explant, with no intervening necessity for development of callus, as for citrus. Alternatively, embryoids may arise from callus after establishment in culture, as in species of conifers (Cheliak and Rogers, 1990). Somatic embryos may subsequently become embryogenic themselves, allowing many plantlets to be produced from a single genetic stock (Tulecke and McGranahan, 1985; Merkle *et al.*, 1987).

Maturation is the final stage in somatic embryogenesis. Whereas in some cases somatic embryos will mature without special treatments, in many species, exposure to ABA and low osmotic potentials are required for maturation (Litz and Gray, 1992; Livingston *et al.*, 1992). These requirements for complete development of somatic embryos are roughly similar to those for zygotic embryos, as in the development and maturation of the latter ABA concentrations increase until maturation and dehydration of the embryo occurs. Somatic embryos may enter a state of dormancy on maturation, as cold treatments similar to stratification regimes result in declines in ABA concentration within embryos and enhance germination rates (Rajasekaran *et al.*, 1982; Litz and Gray, 1992). Mature embryos will "germinate" on specific culture media (Fig. 9.11), giving rise to seedling plants that may be acclimatized to planting environments (a process called conversion). The mechanisms controlling the switch between embryogenic and maturation modes and the development of somatic embryos remain largely unknown.

Somatic embryogenesis offers a potential means of mass micropropagation of desirable genotypes. It also provides the opportunity to create artificial seeds, where embryos are enclosed in an easily planted capsule, usually consisting of a hygroscopic gel that may contain mineral and organic nutrients (Murashige, 1977; Redenbaugh *et al.*, 1986, 1987; Talbert *et al.*, 1993; Dupuis *et al.*, 1994). Somatic embryogenesis may be of particular value in micropropagation of conifers, which are difficult to propagate by conventional *in vitro* culture (Fowke *et al.*, 1995). Embryogenic somatic embryos or callus of English walnut, pecan, and cherry have been transformed by inoculation with *Agrobacterium tumefaciens* (McGranahan *et al.*, 1993; da Câmara Machado *et al.*, 1995). Subsequently, transformed cells in these tissues gave rise to embryos that produced transgenic regenerated plants. Production of large numbers of somatic embryos in liquid culture systems

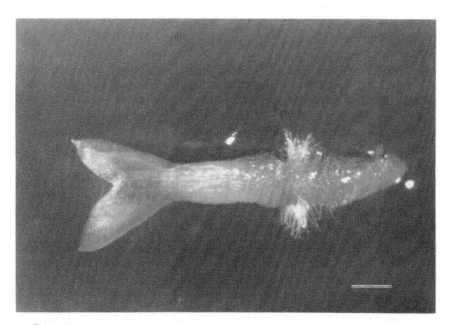

Figure 9.11   Germinating somatic embryo of yellow poplar 5 days after transfer to germination medium. Bar: 500 μm. From Merkle (1991), with permission of Plenum Press.

has been attained. For example, Becwar *et al.* (1988) produced liquid cultures of conifers that contained 10^6 embryos per liter of solution. Liquid culture appears to offer the greatest economic potential for micropropagation of woody plants, primarily because it can be automated and therefore reduces labor costs (Zandvoort and Holdgate, 1991; Talbert *et al.*, 1993; Tautorus *et al.*, 1994).

Protoplast Isolation and Culture   Plant protoplasts are living cells that lack walls. They are produced by treating plant tissues with mixtures of enzymes that digest cell walls. As in other forms of *in vitro* culture, a variety of source tissues have been used to produce protoplasts, including nucellar tissue, pollen grains, roots, petioles, leaf laminae, cotyledons, shoots, flower petals, and xylem (Eriksson, 1985; Vardi and Galun, 1988; Hidano and Niizeki, 1988; McCown, 1988; Kirby, 1988; Leinhos and Savidge, 1993). Seedling cotyledon tissues are most often used in conifer protoplast research (Kirby, 1988). Often callus or cell suspension cultures are developed from these tissues, and protoplasts are derived from these sources (Russell, 1993). In citrus, for example, cultured callus derived from nucellar tissues of the ovule is used to produce protoplasts (Vardi *et al.*, 1982). An unusual requirement for low ammonium concentrations in the media appears to be a feature of successful protoplast culture in some (McCown, 1988) but not all cases (Ochatt *et al.*, 1987).

The exact protoplast isolation procedure is quite genotype specific, but there are common features. Tissues often are finely dissected or shredded to enhance penetration of enzyme solutions. Plasmolysis also is induced by exposing the source tissue to solutions with low osmotic potential. The enzymes used to digest the cell wall matrix usually consist of various mixtures of cellulase, hemicellulase, and pectinase. They are obtained mostly from fungi or bacteria grown on cell wall substrates (Eriksson, 1985). Addition of a protein, bovine serum albumin (BSA), often improves protoplast yields, and it is thought that BSA serves as an alternate substrate for protease molecules that would otherwise attack proteins on the plasmalemma surface.

Liberated protoplasts have a distinctive spherical appearance (Fig. 9.12), with chloroplasts, a thin layer of peripheral cytoplasm, and a large vacuole (David, 1987). In this state, numerous genetic manipulations can be accomplished with protoplasts. For example, protoplasts can be induced to take up DNA either spontaneously or through treatments such as electroporation and microinjection.

Somaclonal selection can be conducted by exposing the millions of protoplasts in a small volume of culture medium to selective pressures such as high NaCl or heavy metal concentrations (Dodds, 1983; Tal, 1994). Fusion may occur in culture when two protoplasts come into contact in the pres-

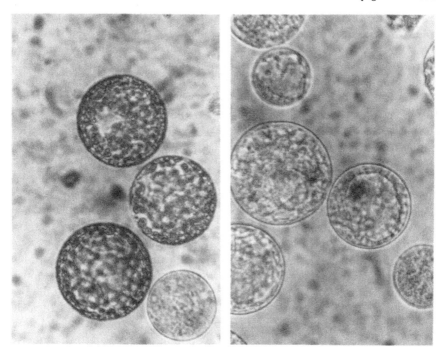

Figure 9.12 Protoplasts of cluster pine (left) and giant sequoia (right). From David (1987). Conifer protoplasts. *In* "Cell and Tissue Culture in Forestry" (J. M. Bonga and D. J. Durzan, eds.), Vol. 2, pp. 2–15. Reprinted by permission of Kluwer Academic Publishers.

ence of polyethylene glycol. Application of an electric field also tends to promote protoplast fusion (Zimmermann and Scheurich, 1981; Zachrisson and Bornman, 1984). Although the fusion product has a common cytoplasm, nuclei may or may not fuse. One or the other nucleus may be lost or destroyed by irradiation; in this case the product is known as a cytoplasmic hybrid or cybrid. If there is fusion of nuclei, a true somatic hybrid is created. The development of somatic hybrids is most advanced in the Rutaceae, and regenerated plants of true somatic hybrids between *Citrus sinensis* and *Poncirus trifoliata* cultivars have been produced (Ohgawara *et al.*, 1985). Intergeneric somatic hybrids between wild pear and Colt cherry also have been obtained (Ochatt *et al.*, 1987).

Once protoplasts are liberated and removed from the enzyme solutions, they are purified to remove contaminating materials such as cell wall debris (Eriksson, 1985; Russell, 1993). Subsequently, protoplasts may be cultured in liquid or on semisolid media, again with genotype-specific requirements for media composition and protocols. Culture of protoplasts on a more solid matrix provides the physical isolation necessary to produce genetically identical colonies, and there is an optimal plating density above and below which cultures do not grow well. Low light levels are preferred for protoplast culture, perhaps because high light bleaches chloroplasts.

Protoplasts begin to synthesize a new cell wall within hours, and some regenerated cells may begin to divide within 1 to 2 weeks (David *et al.*, 1984). Only a small proportion of cells may divide. For example, only 10 to 15% of cells regenerated from protoplasts of Douglas fir and lodgepole and cluster pines divided during the first week of culture (David, 1987). Applied electrical fields can stimulate protoplasts to divide (Rech *et al.*, 1987), and higher osmotic potential in the culture medium, alteration of nitrogen nutrition, and other modifications are required for continued cell division and development of microcalli.

The ability of researchers to cycle woody plants through protoplast isolation to regeneration of plantlets remains limited, but the number of successful examples is increasing. For instance, whereas McCown and Russell (1987) reported only four such examples through Spring 1986, whole plants had been regenerated from protoplasts isolated from 45 woody plants by the time of a review by Russell (1993). These systems were developed for a wide variety of plants, including gymnosperms (e.g., hybrid larch and white spruce), temperate deciduous angiosperms that are important timber trees (e.g., yellow poplar, black poplar, and aspen), and numerous fruit and ornamental trees (e.g., hybrid elm, citrus, apple, cherry, pear, and coffee).

Accomplishments and Prospects

Genetic Engineering of Plants

Transformation of plants is now widely reported, if not routine. As noted previously, several taxa of woody plants are included among the successful attempts. Barriers to transformation provided by host range restrictions of *Agrobacterium* spp. have partially been overcome by alternative means that introduce DNA directly. In other situations, it has been found that coculture of *A. tumefaciens* with plant tissues is all that is necessary for transformation; there need not be formation of tumors. For example, only a 48-hr cocultivation period of *A. tumefaciens* with leaf pieces was needed to obtain transformation in almond, apple, and sweet orange (Archilletti *et al.*, 1995; Pena *et al.*, 1995; Yao *et al.*, 1995).

New approaches to achieving transformation also are appearing. Songstad *et al.* (1995) reviewed four promising advances in DNA delivery techniques. First is a technique whereby a mixture of plant cells, plasmid DNA containing genes to be transferred, and small silicon carbide fibers ("whiskers" about 0.6 μm in mean diameter and 10–80 μm long) are vortexed together. The fibers apparently facilitate delivery of foreign DNA through cell walls during high-speed mixing. Silicon carbide is extremely hard, and fractured fragments have very sharp edges. Whiskers likely perforate and abrade cells like miniature needles during mixing, facilitating entry of extracellular DNA (Wang *et al.*, 1995). The fibers are negatively charged, and thus it is not likely that the negatively charged DNA adheres to the whiskers as they penetrate the cell. Instead, the cell is probably made more receptive to passage of nonadherent DNA. A possible disadvantage of this technique, noted by Songstad *et al.* (1995), was the similarity of the silicon carbide fibers to those of asbestos, suggesting a potential risk of lung damage and carcinogenesis. Second, there are efforts to achieve stable transformation by subjecting plant tissues to electroporation treatments (thereby obviating the need for protoplasts). Third, transformation is being attempted using an electrophoresis apparatus that draws DNA through embryos (Fig. 9.13). Finally, stable transformation through microinjection of DNA into plant cells may be feasible.

There are several areas in which genetic engineering appears poised to improve the growth performance and products of woody plants. For example, transgenic poplars already have been engineered for high tolerance of the herbicides glyphosate and sulfometuron methyl (Michler and Haissig, 1988). Such genotypes could be used in short-rotation intensive culture systems where herbicides used in weed management may injure young poplar plants. Woody plants may also be engineered for increased resistance to virus infections by incorporating genes coding for viral coat proteins into the plant genome. When expressed, viral coat proteins accumulate in plant cells, and their presence seems to inhibit the capacity of viruses to replicate themselves (Nelson *et al.*, 1988; Kunik *et al.*, 1994; Krastanova *et al.*, 1995).

Increasing insect resistance in plants may be associated with manipula-

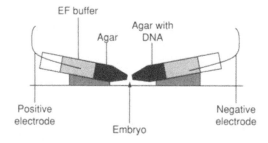

Figure 9.13 Apparatus for introducing DNA into the meristem region of plant embryos by electrophoresis. EF Buffer is electrophoresis buffer. From Songstad, D. D., Somers, D. A., and Griesbach, R. J. (1995). Advances in alternative DNA delivery techniques. *Plant Cell, Tissue Organ Cult.* **40**, 1–15. Reprinted by permission of Kluwer Academic Publishers.

tion of many genes from both plant genomes and foreign sources (Fig. 9.14). Most emphasis has been placed on transfer and expression in plants of a gene from the bacterium *Bacillus thuringiensis* (*Bt*) that codes for the production of a protein that is toxic to larvae in specific insect taxa. Suspensions of *Bt* have been used for over 30 years as a biological insecticide in horticultural crops and forest trees (Beckwith *et al.*, 1988), but internal production of the toxin has a number of technical and environmental advantages (Strauss *et al.*, 1991). Transgenic tomato, tobacco, and poplar plants have already been developed that express the *Bt* toxin gene sufficiently to provide effective protection (Perlak *et al.*, 1991; McCown *et al.*, 1991; McBride *et al.*, 1995; Kleiner *et al.*, 1995).

Manipulation of a group of proteins that interfere with insect digestive enzymes also is considered a potentially valuable strategy to increase plant resistance to insects. This class of proteins, called proteinase inhibitors, if present in high concentration in food sources, decreases weight and fitness and increases mortality of insects. For example, the trypsin inhibitor pro-

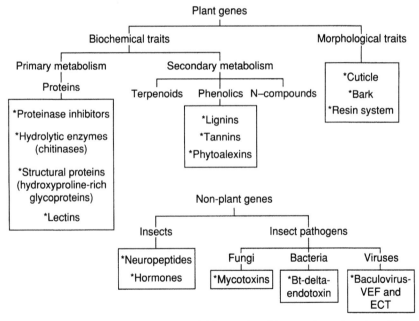

Figure 9.14 Sources of genes that could be employed in genetic engineering of insect resistance in trees. VEF, Virulence enhancing factor; EGT, ecdysteroid glucosyltransferase; Bt, *Bacillus thuringiensis*. Reprinted from *For. Ecol. Manage.*, **43**. Strauss, S. H., Howe, G. T., and Goldfarb, B., Prospects for genetic engineering of insect resistance in forest trees, 181–209. Copyright 1991, with kind permission of Elsevier Science-NL, Sara Burgerhartstraat 25, 1055 KV Amsterdam, The Netherlands.

tein from cowpea has been transferred to tobacco, with a resultant increase in resistance of transgenic plants to insect herbivory (Hilder *et al.*, 1987). However, some insects apparently respond to these inhibitors with induction of different forms of proteinase that show little or no inhibition (Jongsma *et al.*, 1995).

Genetic engineering also shows promise in producing new types of floricultural crops. With regard to flower color, the pigment synthesis pathway for flavonoid compounds has been extensively studied (Woodson, 1991), and flower color alterations have been obtained from manipulation of genes coding for key enzymes in the pigment pathways. For example, transformed Moneymaker pink-flowering chrysanthemum plants were developed with white instead of pigmented flowers. This color change was the result of sense suppression obtained by adding additional copies of the gene that codes for an enzyme which catalyzes a reaction early in the flavonoid pigment synthesis pathway (Courtney-Gutterson *et al.*, 1994). Similarly, transformed rose plants have also been produced by sense suppression techniques (Gutterson, 1995). Other possible benefits of genetic engineering in floriculture include (1) alterations in plant architecture (e.g., changes in flower petal number and size), (2) better control of flowering, and (3) increased flower longevity by alteration of genes involved in senescence (Woodson, 1991; Savin *et al.*, 1995).

Genetic engineering techniques also may greatly reduce the long generation times that commonly hinder conventional tree breeding. For example, precocious flowering in *Arabidopsis* and aspen plants has been obtained by insertion of a gene that controls floral induction (Weigel and Nilsson, 1995).

Alterations in fruit ripening in tree fruit crops may be possible through transformation. As previously noted in work with nonwoody plants, the commercial shelf life of tomatoes has been lengthened by genetic engineering of varieties with an antisense copy of a fragment or the entire gene that codes for polygalacturonase (PG). This enzyme increases in activity during tomato ripening and catalyzes the breakdown of pectins within cell walls, which causes fruit softening (Giovannoni *et al.*, 1992). Ripening fruits of transformed plants showed only 5–10% the level of PG activity of untransformed plants, with no detectable difference in accumulation of the primary red pigment, lycopene, in tomato fruits (Smith *et al.*, 1988). Reductions in the levels of enzymes responsible for tissue browning also have been obtained via similar approaches (e.g., in potato tubers, Bachem *et al.*, 1994). If similar successes can be repeated in tree fruit crops, the marketability and appearance of tree fruits under normal handling might be improved.

Improvement in stress tolerance of plants also may be obtained via plant transformation. In an example using tobacco, transgenic lines were developed that contained the gene for superoxide dismutase (SOD) along with the cauliflower mosaic virus promoter (van Camp *et al.*, 1994). Superoxide dismutase converts harmful superoxide radicals into less toxic H_2O_2. The SOD levels in some transformed plants increased substantially, presumably in response to higher levels of expression associated with presence of the promoter. These higher levels of SOD activity were correlated with reduced visible damage to leaves when plants were exposed to moderately high levels of O_3. Increases in resistance of transgenic tobacco to fungal attack also were enhanced by expression of two inducible enzymes (a chitinase from rice and a glucanase from alfalfa), the genes for which were incorporated into the tobacco genome (Zhu *et al.*, 1994).

Although genetic engineering is a reality, technological and other problems remain. Following transformation, newly acquired traits in transgenic plants may not be retained following propagation, and loss of these traits is not associated with loss of the transferred genes but rather with gene inactivation (Finnegan and McElroy, 1994). The latter phenomenon may be associated with targeted methylation (and thus prevention of transcription) of the transgene DNA or its promoter sequence, as the plant somehow recognizes the transferred gene as "foreign." Alternatively, when additional copies of an already present gene are incorporated into the genome, expression of all these genes may be suppressed in a previously mentioned phenomenon called sense suppression or cosuppression.

The difficulty in *in vitro* culture of woody plants, especially with regard to poor regeneration, presents a significant barrier to routine application of transformation techniques such as are attained with agronomic crop species. Researchers and others also must be concerned with how, once transformed plants are taken to the field, transgenes might move into natural populations of plants that are interfertile with the transformed populations. Such gene flow would occur most likely through hybridization between cultivated plants and closely related weed species (Ellstrand and Hoffman, 1990). Although traits thus transferred might be maladaptive, others (e.g, reproductive output, tolerance of insects and diseases) could increase weed problems (Klinger and Ellstrand, 1994). Hence, ideally, genetically engineered plants should be sterile to prevent such events (Strauss *et al.*, 1991, 1995). Tiedje *et al.* (1989) listed additional possible undesirable outcomes of the introduction of transformed organisms: (1) broadened host ranges of pathogens, (2) disruptive effects on communities as a result of alteration in competitive abilities of species, (3) adverse effects on ecosystem processes, and (4) inadvertent promotion of the evolution of pest resistance.

Finally, public acceptance of genetically engineered plants is not assured. Boulter (1995) noted that the public exhibited both support for and concerns about plant biotechnology. Concerns centered mainly on the risks and morality of genetic engineering. The perceived risk of a technology or event is the actual risk as modified by an "outrage" reaction. The outrage reaction increases the perception of risk if a technology or event is unfamiliar, poorly understood, easily envisioned to go wrong, perceived to hold the possibility of dreaded results, likely to influence everyone without an element of personal control, perceived as artificial, concentrated in its effects in time or space, or uncontrolled. The morality of genetic engineering has been questioned on religious grounds as a blasphemous usurping of the right of God as Creator. Others have asserted that genetic engineering is morally objectionable because it is unnatural, disrespectful of life, and unfairly tilted toward conferring benefits on those who are at lower risk of suffering the negative effects of the technology. Boulter (1995) suggested that scientists could help resolve some of these contentious issues both by entering into a respectful, candid dialog with the public on a personal level and by aiding in efforts to increase the understanding of science and biotechnology in schools and, through the mass media, among the general public.

Plant Propagation

Vasil (1994) reported that nearly 600 companies around the world produce 500 million plants, representing 50,000 different varieties, mostly through *in vitro* shoot culture methods. Most plants produced were ornamental and vegetable species, although considerable numbers of micropropagated plantlings of Monterey pine are now being produced in New Zealand (Gleed, 1993). Production of stecklings (rooted cuttings) from conifers such as Monterey pine, sugi, and Norway spruce and from woody angiosperms such as poplars and eucalypts has made clonal forestry an increasingly feasible forest production system in many parts of the world (Ahuja and Libby, 1993a,b).

Micropropagation also has been employed commercially to produce trees in the genera *Acer, Betula, Syringa, Malus, Prunus, Pyrus, Kalmia, Populus, Ulmus,* and *Amelanchier* (McCown, 1989b). In 1994, unit costs for micropropagated plants were $0.10–0.15 per plant, a value that is substantially higher than would be required for widespread penetration of micropropagated stock into forest planting programs ($0.01–0.05/plant) (Vasil, 1994). The labor-intensive nature of *in vitro* production of plantlings remains an economic barrier to the use of micropropagated stock in forest planting. Decreases in production costs will largely depend on production of propagules (e.g., somatic embryos, shoot meristems) in large volumes of growth media and on automation of handling. McCown (1989b) noted that micropropagated plantlings can also be effectively used as stock plants, providing genetically uniform, rejuvenated, and disease-free material for macropropagation by cuttings.

Although previous work has shown that micropropagation of desirable plants by various means is possible, much research remains to be done before micropropagation for certain types of materials can be undertaken with confidence of success. For example, propagation of mature plant materials remains difficult, and progress in this area awaits a better fundamental understanding of the process of ontogenetic aging in plants. The large-scale production of somatic embryos of superior genotypes in large culture vessels followed by encapsulation into artificial seeds appears to be an attractive goal in efforts to make commercial clonal production of forest species a reality. However, although obtained for more than 200 species, somatic embryogenesis is not well understood, and yields of viable plants from somatic embryos often are low (Vasil, 1994). Embryos vary substantially in size, shape, and morphology and with regard to timing and efficiency with which they mature and germinate. Production of plants from somatic embryos in artificial seeds also has not progressed sufficiently for commercial adaptation, but improvements in techniques (e.g., adaptation of modified pharmaceutical capsules, Dupuis *et al.*, 1994) may speed this process.

Summary

Biotechnology encompasses three aspects of woody plant biology: (1) *in vitro* culture, (2) use of immunological methods to quantify plant metabolites, and (3) genetic engineering, that is, the transfer of genes between organisms without normal sexual reproduction. Rapid developments in these areas have made feasible the production and mass propagation of new genotypes of important forestry and horticultural crop plants.

The procedures of biotechnology largely depend on an understanding of the molecular biology of DNA replication, transcription, and translation. In association with cell division, replication of DNA occurs via DNA polymerases simultaneously at many locations on the genome where chromosomal DNA has been unwound. Gene expression begins with transcription, the synthesis of messenger RNA (mRNA) in a series of reactions involving RNA polymerases and various regulatory molecules. This process ensures transcription of specific genes, thus assuring proper developmental coordination of gene expression. Protein products arise from translation of mRNA on ribosomes in the cytoplasm.

In the process of genetic transformation, new genes are stably introduced into the genome of target cells. There are a variety of methods to achieve transformation, including those associated with biological vectors and physical methods of introduction through direct exposure of cells or protoplasts to DNA and the acceleration of small DNA-coated projectiles into plant tissues. The most frequent biological method of transformation in plants employs species of *Agrobacterium* that carry a plasmid containing a region of DNA that is readily integrated into the plant genome. Cocultivation of the bacterium with plant cells or protoplasts results in incorporation of the gene of interest, along with (1) DNA sequences that direct its transcription, (2) a reporter gene that confirms the presence of the desired gene after elimination of the bacterium from culture, and (3) a gene conferring resistance to subsequently applied chemicals so that nontransformed cells can be eradicated. Direct transfer of similar DNA constituents is promoted by treatments such as exposure of protoplasts to polyethylene glycol and pulsed electric fields (electroporation). Ballistic methods of DNA introduction can be achieved with tissues if cells in meristematic regions can be transformed.

Several methods of confirmation of gene transfer and expression are available. The presence of a specific segment of DNA in the genome of a plant cell can be confirmed by Southern hybridization, in which plant DNA is cleaved into fragments by specific restriction enzymes followed by hybridization of the fragments with radioactively labeled sequences of DNA of the gene of interest (probes). Transcription of genes can be confirmed by Northern hybridizations, a procedure similar to Southern hybridization analysis, in which RNA is cleaved at certain sequences by enzymes and probed with known radioactively labeled nucleic acid sequences that correspond the mRNA of interest.

Protein products are frequently detected and quantified by antibody-based assays. Antibodies are produced in animals in response to the introduction of foreign macromolecules into the body. These antibodies can be isolated and purified, and they will bind to plant metabolites to which they were raised. Linkage of the antibodies with enzymes that catalyze color-producing reactions allows simple colorimetric measurement of concentration.

Molecular biology techniques have greatly facilitated genetic analysis. Genetic variation at the DNA level can be detected by restriction fragment length polymorphism (RFLP) analysis wherein plant DNA is cleaved by restriction enzymes at specific base sequences and the fragments are separated by electrophoresis. Base sequence variations between different genotypes will cause fragments to be of different length and thus travel different distances in gels. The amount of plant DNA needed for analysis has been greatly reduced by development of the polymerase chain reaction (PCR), which employs *in vitro* cyclical replication of specific restriction fragment DNA sequences to greatly increase the number of fragments of isolated DNA. Another variation of this technique, the randomly amplified polymorphic DNA procedure (RAPD), employs random-sequence primers to direct replication of short fragments of DNA.

Genetic engineering often is conducted on cells and protoplasts cultured *in vitro*, which must be regenerated to whole plants. In addition, it often is desirable to propagate en mass new genotypes generated by traditional breeding and selection methods. Hence, there has been substantial effort directed toward developing means of multiplying plant materials vegetatively for distribution. Such methods may be associated with traditional techniques such as rooting of cuttings or *in vitro* culture techniques. Many woody plants, such as willows and poplars, are easily propagated by rooted cuttings. In other cases rooting is not so readily obtained; however, there has been sufficient progress in rooting important trees, such as Monterey pine and Norway spruce, that commercial availability of superior genotypes of these species from cuttings is a reality.

Micropropagation may be accomplished through culture of undifferentiated callus tissues and subsequent induction of organs by manipulation of culture media and plant growth regulator concentrations. Alternatively, propagules can be obtained directly from rooting of microcuttings derived from shoot culture of differentiated tissues such as shoots, buds, or leaves. Somatic embryos also have been produced through appropriate manipulation of culture media to resemble the developmental regime *in situ*. Indefinite proembryoid production can be sustained, and proembryos produced may be matured through exposure to ABA and low osmotic potential.

Micropropagation also may be achieved through protoplast isolation, culture, and regeneration. Protoplasts are isolated by exposure of cells or tissues to cell wall-digesting enzymes. Thereafter, free protoplasts may be manipulated to fuse together (sometimes giving true somatic hybrids) or genetically transformed. Cell walls of protoplasts soon form, and genetically uniform colonies can be obtained by subsequent culture of protoplasts.

Potential plant improvements that may be obtained from genetic engineering are numerous, including increased tolerance of herbicides, greater resistance to insect, fungus, and virus pests, precocious flowering, delay of fruit degradation and flower senescence, color alteration in floricultural crops, and enhanced resistance to pollutants. Expression of traits in transformed plants is not always stable because transcription of introduced genes may be suppressed. Additionally, as methods to achieve plant transformation usually require *in vitro* culture, these techniques may not be available for many recalcitrant woody plants. Appropriate controls must be developed to prevent movement of foreign genes into natural plant populations.

Many woody species, including important angiosperm and gymnosperm species, have been micropropagated. However, although quite useful, micropropagation of many woody plants is a formidable undertaking. Culture of plants *in vitro* is a largely empirical endeavor, and a tedious search for appropriate media and protocols attends the culture of nearly every new plant material. Regeneration to plants also frequently is difficult. Although it is desirable, from the standpoint of selection of superior genotypes, to obtain cultures from mature plants, it often has proved quite difficult to micropropagate mature plants. Moreover, micropropagation costs are still high compared with conventional propagation by seed.

General References

Ahuja, M. R., ed. (1988). "Somatic Cell Genetics of Woody Plants." Kluwer, Dordrecht, The Netherlands.

Ahuja, M. R., ed. (1991). "Woody Plant Biotechnology." Plenum, New York.

Ahuja, M. R., ed. (1993). "Micropropagation of Woody Plants." Kluwer, Dordrecht and Boston.

Ahuja, M. R., and Libby, W. J., eds. (1993). "Clonal Forestry," Vols. 1 and 2. Springer-Verlag, Berlin.

Bennett, A. B., and O'Neill, S. D., eds. (1990). "Horticultural Biotechnology." Wiley–Liss, New York.

Bonga, J. M., and Durzan, D. J., eds. (1987). "Cell and Tissue Culture in Forestry," Vols. 1–3. Martinus Nijhoff, Dordrecht, The Netherlands.

Bonga, J. M., and von Aderkas, P., eds. (1992). "*In Vitro* Culture of Trees." Kluwer, Dordrecht, The Netherlands.

Debergh, P. C., and Zimmerman, R. H., eds. (1991). "Micropropagation: Technology and Application." Kluwer, Dordrecht, The Netherlands.

Dodds, J. H., ed. (1983). "Tissue Culture of Trees." AVI Publ., Westport, Connecticut.

Dunstan, D. I. (1988). Prospects and progress in conifer biotechnology. *Can. J. For. Res.* **18**, 1497–1506.

Fowke, L. C., and Constabel, F., eds. (1985). "Plant Protoplasts." CRC Press, Boca Raton, Florida.

Glick, B. R., and Thompson, J. E., eds. (1993). "Methods in Plant Molecular Biology and Biotechnology." CRC Press, Boca Raton, Florida.

Hammerschlag, F. A., and Litz, R. E., eds. (1992). "Biotechnology of Perennial Fruit Crops." Commonwealth Agricultural Bureaux International, Oxon, England.

Hanover, J. W., and Keithley, D. E., eds. (1987). "Genetic Manipulation of Woody Plants." Plenum, New York.

Hartmann, H. T., Kester, D. E., Davies, F. J., Jr. (1990). "Plant Propagation: Principles and Practices," 5th Ed. Regents/Prentice-Hall, Englewood Cliffs, New Jersey.

Lea, P. J., and Leegood, R. C., eds. (1993). "Plant Biochemistry and Molecular Biology." Wiley, Chichester and New York.

National Academy of Sciences. (1987). "Introduction of Recombinant DNA-Engineered Organisms into the Environment: Key Issues." National Academy Press, Washington, D.C.

Pollard, J. W., and Walker, J. M., eds. (1990). "Plant Cell and Tissue Culture." Humana, Clifton, New Jersey.

Sambrook, J., Fritsch, E. F., and Maniatis, T. (1989). "Molecular Cloning: A Laboratory Manual," 2nd Ed. Cold Spring Harbor Press, Cold Spring Harbor, New York.

Schuler, M. A., and Zielinski, R. E. (1989). "Methods in Plant Molecular Biology." Academic Press, San Diego.

Shaw, C. H., ed. (1988). "Plant Molecular Biology: A Practical Approach." IRL Press, Eynsham, Oxford, England.

Weissbach, A., and Weissbach, H., eds. (1988). "Methods for Plant Molecular Biology." Academic Press, San Diego.

Scientific and Common Names of Plants

Scientific Names of Plants

Abies Mill. spp.	fir
Abies alba Mill.	European silver fir
Abies amabilis Dougl. ex J. Forbes	Pacific silver fir, silver fir
Abies balsamea (L.) Mill.	balsam fir
Abies concolor (Gord.) Lindl. ex Hildebr.	white fir
Abies fraseri (Pursh) Poir.	Fraser fir
Abies grandis (D. Don ex Lamb.) Lindl.	grand fir
Abies lasiocarpa (Hook.) Nutt.	alpine fir
Abies magnifica A. Murr.	California red fir, red fir
Acacia Mill. spp.	acacia
Acacia albida Delile	apple-ring acacia
Acacia farnesiana (L.) Willd.	sweet acacia
Acacia karroo Hayne	karroo thorn
Acacia macracantha Humb. & Bonpl. ex Willd.	steel acacia
Acacia mangium Willd.	
Acacia pulchella R. Br.	prickly moses
Acacia scleroxyla Tuss.	
Acacia tortilis (Forssk.) Hayne	umbrella-thorn
Acer L. spp.	maple
Acer cappadocicum Gleditsch	Kolchischer ahorn
Acer davidii Franchet	Father David's maple
Acer ginnala Maxim.	Amur maple
Acer griseum (Franco) Pax	paperbark maple
Acer hersii Rehder [*A. davidii* ssp. *grosseri* (Pax) DeJong.]	
Acer macrophyllum Pursh.	bigleaf maple
Acer negundo L.	box elder
Acer pensylvanicum L.	striped maple
Acer platanoides L.	Norway maple
Acer pseudoplatanus L.	sycamore maple
Acer rubrum L.	red maple
Acer saccharinum L.	silver maple
Acer saccharum Marsh.	sugar maple

Acer tataricum L.	Tatarian maple
Acer velutinum Boissier	velvet maple
Acrocarpus Wight ex Arn. spp.	
Acrocarpus fraxinifolius Wight. & Arn.	shingle tree, Kenya coffee shade tree, pink cedar
Actinidia Lindl. spp.	kiwifruit
Actinidia chinensis Planch.	kiwifruit
Actinidia deliciosa (A. Chev.) C. F. Lang ex A. R. Ferguson	kiwifruit
Actinidia eriantha Benth.	
Actinidia helmsleyana Dunn	
Actinidia rufa Franch. & Sav.	
Adenostoma fasciculatum Hook. and Arn.	greasewood
Aesculus L. spp.	horse chestnut, buckeye
Aesculus californica (Spach) Nutt.	California buckeye
Aesculus hippocastanum L.	horse chestnut
Agathis australis Hort. ex Lindl.	New Zealand kauri
Ailanthus altissima (Mill.) Swingle	ailanthus, tree of heaven
Albizia Durrazz. spp.	albizia
Aleurites fordii Hemsl.	tung
Alnus B. Ehrh. spp.	alder
Alnus crispa (Ait.) Pursh	green alder
Alnus glutinosa (L.) Gaertn.	European black alder, black alder, European alder
Alnus incana (L.) Moench	gray alder
Alnus japonica Steud.	Japanese alder
Alnus rubra Bong. (*A. oregona* Nutt.)	red alder
Amelanchier Medic. spp.	shadbush
Antiaris africana Engl.	African upas tree
Arabidopsis Heynh. spp.	mouse-ear cress
Aralia L. spp.	
Aralia spinosa L.	Devil's-walking stick, Hercules'-club
Araucaria Juss. spp.	araucaria
Araucaria araucana (Mol.) C. Koch	monkey-puzzle tree
Araucaria cunninghamia Sweet.	hoop pine
Artocarpus J.R. & G. Forst. spp.	breadfruit
Aster L. spp.	aster
Atriplex patula L.	saltbush
Avicennia L. spp.	black mangrove
Avicennia marina (Forsk.) Vierh.	black mangrove
Bambusa Schreb. spp.	bamboo
Bernardia Endl. (*Bernardia* Vill.)	
Bertholletia excelsa Humb. & Bonpl.	Brazil nut
Betula L. spp.	birch
Betula alleghaniensis Britt.	yellow birch
Betula costata Trautv.	
Betula davurica Pall.	

Betula mandschurica (Regal) Nakai (*B. platyphylla* Sukachev)	Japanese white birch
Betula nigra L.	river birch
Betula papyrifera Marsh.	paper birch, white birch, canoe birch
Betula pendula Roth. (*B. verrucosa* J.F. Ehrh.)	silver birch
Betula populifolia Marsh.	gray birch
Betula pubescens J.F. Ehrh.	European white birch, European birch, hairy birch, white birch
Bougainvillea Comm. Ex Juss. spp.	bougainvillea
Bruguiera gymnorhiza Lam.	Burma mangrove
Buxus L. spp.	box
Calluna Salisb. spp.	heather
Calocedrus decurrens (Torr.) Florin (*Libocedrus decurrens* Torr.)	incense cedar
Camellia L. spp.	
Camellia sinensis (L.) O. Kuntze	tea
Carica papaya L.	papaya
Carya Nutt. spp.	hickory
Carya illinoensis (Wangenh.) K. Koch	pecan, pecan hickory
Casearia Jacq. spp.	
Cassia L. spp.	senna
Castanea Mill. spp.	chestnut
Castanea dentata (Marsh.) Borkh.	American chestnut
Castanea mollissima Blume	Chinese chestnut
Casuarina L. ex Adans. spp.	beefwood, Australian pine, she-oak
Casuarina cristata Miq.	Belah beefwood
Casuarina cunninghamiana Miq.	river she-oak, creek-oak
Casuarina decaisneana F. Muell.	desert-oak
Casuarina distyla Vent.	stunted beefwood
Casuarina equisetifolia L. ex J.R. & G. Forst.	coast she-oak
Casuarina glauca Sieb. ex K. Spreng	swamp she-oak
Casuarina inophloia F. Muell & Bailey	
Casuarina littoralis Salisb.	black-oak
Casuarina luehmannii R.T. Bak.	bull-oak
Casuarina stricta Ait.	she-oak
Casuarina torulosa Ait.	forest-oak
Catalpa Scop. spp.	catalpa
Catalpa bignonioides Walt.	southern catalpa
Catalpa speciosa Warder	northern catalpa
Ceanothus L. spp.	redroot
Cecropia Loefl. spp.	
Cecropia obtusifolia Bertolini	cecropia
Cecropia peltata L.	trumpet tree
Cedrus L. spp.	cedar
Celtis L. spp.	hackberry

Celtis laevigata Willd.	sugarberry, sugar hackberry
Celtis occidentalis L.	common hackberry
Cercis L. spp.	redbud
Cercis canadensis L.	eastern redbud
Cercis siliquastrum L.	Judastree redbud
Cercocarpus H.B.K. spp.	mountain mahogany
Chamaecrista chamaecristoides (Collad)	partridge pea
Chamaecyparis Spach	white cedar, false cypress
Chamaecyparis lawsoniana (A. Murr.) Parl.	Lawson cypress, Port Orford cedar
Chamaecyparis nootkatensis (D. Don) Spach.	Alaska cedar, yellow cedar
Chamaecyparis obtusa (Siebold & Zucc.) Endl.	hinoki cypress
Chrysanthemum [Tourn.] L. spp.	chrysanthemum
Cinchona L. spp.	
Cinchona ledgeriana Moens. ex Trimen	cinchona bark tree
Citrus L. spp.	citrus, lemon, lime, grapefruit, orange
Citrus aurantifolia (Christm.) Swingle	sweet lime, Tahiti lime
Citrus aurantium L.	sour orange
Citrus limon (L.) Burm.f.	rough lemon
Citrus paradisi Macfad.	grapefruit
Citrus reticulata Blanco 'Cleopatra'	tangarine, Cleopatra mandarin orange
Citrus sinensis L. 'Valencia'	Valencia orange, sweet orange, spice orange
Citrus sunki Hort. ex Tan.	Sunki mandarin
Cladrastis lutea (Mich.f.) K. Koch [*C. kentukea* (Dum.-Cours.) Rudd]	yellowwood
Clematis L. spp.	clematis
Cocos L. spp.	coconut
Cocos nucifera L.	coconut
Coffea L. spp.	coffee
Coffea arabica L.	Arabian coffee
Colophospermum mopane Benth.	butterfly tree
Combretum apiculatum Sond.	red bushwillow
Cornus L. spp.	dogwood
Cornus florida L.	flowering dogwood
Cornus stolonifera Michx. (*C. sericea* L.)	red osier dogwood, red stemmed dogwood
Corylus L. spp.	hazel, filbert
Corylus avellana L.	European hazel
Cotinus Mill spp.	smoke tree
Cotinus coggyria Scop.	smoketree
Crataegus L. spp.	hawthorn
Crotolaria incana L.	shackshack crotalaria
Croton L. spp.	
Cryptomeria japonica (L.f.) D. Don	Japanese cryptomeria, sugi

Cupressus arizonica Greene	Arizona cypress
Cydonia Mill. spp.	quince
Dacrydium cupressinum Soland. ex Forst.	rimu, Imou pine
Dalbergia L.f. spp.	rosewood
Dennstaedtia punctilobula (Michx.)	hay-scented fern
Deschampsia flexuosa (L.) Trin.	wavy hairgrass
Desmodium Desv. spp.	tick clover
Dialinum maingayi Baker	keranji
Dioscorea alata L.	water yam
Diospyros virginiana L.	common persimmon
Dipterocarpus alatus Roxb.	Gurjuniol tree
Elaeagnus L.	elaeagnus
Elaeagnus latifolia L.	oleaster
Elaeagnus umbellata Thunb.	autumn olive
Elaeis guineensis Jacq.	oil palm
Empetrum hermaphroditum Hagerup	crowberry
Erythrina L. spp.	erythrina
Erythroxylum P. Br. spp.	
Erythroxylum coca Lam.	cocaine plant
Eucalyptus L'Hér. spp.	eucalyptus
Eucalyptus argophloia Blakely	Queensland white gum
Eucalyptus camaldulensis Dehn. (see *E. rostrata* Schlecht.)	
Eucalyptus citriodora Hook.	lemon-scented gum
Eucalyptus cloeziana F. Muell.	gympie messmate
Eucalyptus delegatensis R.T. Bak. (*E. gigantea* Hook.f.)	alpine ash, mountain ash, woollybutt
Eucalyptus diversicolor F. Muell.	karri
Eucalyptus drepanophylla F. Muell. ex Benth.	Queensland grey iron bark
Eucalyptus erythronema Turcz.	crimson mallee
Eucalyptus fastigiata Deane & Maiden	brown barrel
Eucalyptus gigantea L. (*E. delegatensis* R.T. Bak.)	alpine ash
Eucalyptus globulus Labill.	Tasmanian blue gum, blue gum
Eucalyptus grandis W. Hill. ex Maiden	rose gum
Eucalyptus intermedia R. T. Baker	pink bloodwood
Eucalyptus landsdowneana F. Muell. & J.E. Brown var. *landsdowneana*	crimson mallee box
Eucalyptus maculata Hook.	spotted gum
Eucalyptus miniata A. Cunn. ex Schau.	Darwin woollybutt
Eucalyptus moluccana Roxb.	grey box, gum-topped box
Eucalyptus nitens Maiden	shining gum
Eucalyptus paniculata Smith	grey iron bark
Eucalyptus pellita F. Muell.	large-fruited red mahogany
Eucalyptus peltata Benth. ssp. *peltata*	rusty jacket, yellow jacket
Eucalyptus pilularis Sm.	blackbutt

Eucalyptus polycarpa F. Muell.	long-fruited bloodwood
Eucalyptus regnans F. Muell.	mountain ash
Eucalyptus robusta Sm.	swamp mahogany
Eucalyptus rostrata Schlecht. (*E. camaldulensis* Dehnh.)	Murray red gum, red gum, river red gum
Eucalyptus saligna Sm.	Sydney blue gum
Eucalyptus tetrodonta F. Muell.	Darwin stringy bark
Eucalyptus urophylla S. T. Blake	Timor or white gum
Eucalyptus viminalis Labill.	manna gum
Euphorbia L. spp.	spurge
Fagus L. spp.	beech
Fagus grandifolia J.F. Ehrh.	American beech
Fagus sylvatica L.	European beech
Forsythia viridissima Lindl.	greenstem forsythia
Fortunella Swingle spp.	kumquat
Fragaria L. spp.	strawberry
Fraxinus L. spp.	ash
Fraxinus americana L.	white ash, American ash
Fraxinus excelsior L.	European ash
Fraxinus mandshurica Rupr.	Manchurian ash
Fraxinus pennsylvanica Marsh.	green ash, red ash
Ginkgo L. spp.	ginkgo
Ginkgo biluba L.	ginkgo
Gleditsia triacanthos L.	honey locust
Gliricidia H.B. & K. Spp.	
Gliricidia sepium (Jacq.) Kunth ex Griseb.	madre
Glycine max (L.) Merr.	soybean
Gossypium L. spp.	cotton
Gossypium hirsutum L.	cotton
Grevillea robusta A. Cunn. ex R. Br.	silk-oak
Hamamelis virginiana L.	witch hazel
Hevea Aubl. spp.	
Hevea brasiliensis (Willd. ex A. Juss) Mull-Arg.	Brazilian rubber tree
Hibiscus L. spp.	hibiscus, mallow
Hiraea Jacq. spp.	
Hopea odorata Roxb.	sweet leaf
Hovea heterophylla A. Cunn. ex Hook.	hovea
Ilex L. spp.	holly
Ilex decidua Walt.	deciduous holly, possum haw
Ilex opaca Ait.	American holly
Ilex vomitoria Ait.	yaupon holly
Indigo suffruticosa Mill.	Anil indigo
Intsia Thou. spp.	
Intsia palembanica Miq.	merbau, kwila
Ipomoea batatas (L.) Lam.	sweet potato
Juglans L. spp.	walnut

Juglans cinerea L.	butternut
Juglans nigra L.	black walnut
Juglans regia L.	English walnut
Juniperus L. spp.	juniper, red cedar
Juniperus communis L.	common juniper
Juniperus virginiana L.	eastern red cedar
Kalmia L. spp.	
Kalmia latifolia L.	mountain laurel
Khaya ivorensis A. Chev.	Lagos mahogany
Khaya senegalensis (Desr.) A. Juss	kuka, Senegal mahogany
Lactuca sativa L.	head lettuce
Larix Mill. spp.	larch, tamarack
Larix dahurica Turcz.	Dahurian larch
Larix decidua Mill.	European larch
Larix kaempferi (Lamb.) Carr. (*L. leptolepis* (Siebold & Zucc.) Gord.]	Japanese larch
Larix laricina (DuRoi) K. Koch	eastern larch, tamarack
Larix lyallii Parl.	subalpine larch
Larix occidentalis Nutt.	western larch
Larrea Cav. spp.	creosote bush
Leucaena Benth. spp.	
Leucaena leucocephala (Lam.) de Wit	leucaena, leadtree
Lindera benzoin Meissn.	spicebush
Liquidambar styraciflua L.	sweet gum, red gum
Liriodendron tulipifera L.	yellow poplar, tulip poplar, tulip tree
Litchi Sonn. spp.	
Litchi chinensis Sonn.	litchi, lychee
Lithocarpus densiflorus (Hook. & Arn.) Rehd.	tan oak
Lonchocarpus Benth. spp.	spear fruit
Lonicera L. spp.	honeysuckle
Lophostemon confertus (*Tristania conferta* R. Br.)	brush box, pink box, Queensland box
Lupinus arboreus Sims	lupine
Lysiloma Benth. spp.	
Macadamia F. J. Muell. spp.	
Macadamia integrifolia Maiden & Betche	Macadamia nut
Macadamia ternifolia F. Muell.	Queensland nut
Macfadyena A. DC. spp.	
Maclura pomifera (Raf.) Schneid.	osage orange
Macroptilium atropurpureum (DC.) Urban	
Magnolia L. spp.	magnolia
Magnolia grandiflora L.	southern magnolia
Malus Mill. spp.	apple, crabapple
Malus domestica Borkh. [*M. pumila* Mill.; *M. sylvestris* (L.) Mill.]	apple

Mangifera indica L.	mango
Manihot esculenta Crantz	cassava
Margaritaria Opiz. spp.	
Melaleuca quinquenervia (Cav.) S.T. Blake	paperbark tree
Metasequoia glyptostroboides Hu & Cheng	dawn redwood, metasequoia
Mimosa L. spp.	mimosa
Mimosa chaetocarpa Brandegee	
Montezuma Sessé & Moc. spp.	montezuma
Morus L. spp.	mulberry
Morus rubra L.	red mulberry, American mulberry
Musa L. spp.	banana
Myrica L. spp.	wax myrtle
Nicotiana tabacum L.	tobacco
Nothofagus cunninghamia Oerst.	Tasmanian false-beech
Nyssa L. spp.	gum, tupelo
Nyssa aquatica L.	water tupelo, swamp tupelo
Nyssa sylvatica Marsh.	black gum, tupelo gum
Nyssa sylvatica var. *biflora* (Walt.) Sarg.	swamp black tupelo, swamp tupelo
Olea L. spp.	olive
Olea europaea L.	olive
Ostrya virginiana (Mill.) K. Koch	eastern hop hornbeam, ironwood
Oxydendrum arboreum (L.) DC.	sourwood
Parkia R. Br. spp.	nittatree
Parkia discolor Spruce ex Benth.	
Parkia javanica (Lam.) Merrill	Java locust
Parkia pendula (Willd.) Benth. ex Walp	
Parthenium argentatum A. Gray	guayule
Paulownia Siebold & Zucc. spp.	paulownia
Paulownia fortunei (Seem.) Hance	paulownia
Paulownia tomentosa (Thunb.) Siebold & Zucc. ex Steud.	royal paulownia, princess tree
Pavetta gardeniifolia A. Rich (*P. assimilis*)	brides-bush
Pelargonium L'Hér, ex. Ait. spp.	geranium, pelargonium
Pericopsis Thw. spp.	
Persea americana Mill.	avocado
Phellodendron Rupr. spp.	cork tree
Phellodendron amurense Rupr. var. *wilsonii* (Hayata & Kanchira) C.E. Chang	cork tree
Phellodendron wilsonii (*Phellodendron amurense* var. *wilsonii*)	
Phoenix dactylifera L.	date palm
Picea A. Dietr. spp.	spruce
Picea abies (L.) Karst.	Norway spruce

Picea engelmannii Parry ex Engelm.	Engelmann spruce
Picea glauca (Moench) Voss	white spruce
Picea glauca (Moench) Voss var. *Engelmannii* (Parry) Boivin	interior spruce
Picea jezoensis (Siebold & Zucc.) Carriere	Yeddo spruce
Picea mariana (Mill.) B.S.P.	black spruce
Picea pungens Engelm.	blue spruce, Colorado blue spruce, Colorado spruce
Picea rubens Sarg.	red spruce
Picea sitchensis (Bong.) Carr.	Sitka spruce
Pinus L. spp.	pine
Pinus albicaulis Engelm.	whitebark pine
Pinus banksiana Lamb.	jack pine
Pinus caribaea Morelet	Caribbean pine
Pinus caribaea var. *hondurensis* Morelet	Carib pine
Pinus cembra L.	Swiss stone pine
Pinus cembroides Zucc.	Mexican pinyon pine
Pinus clausa (Chapm.) Vasey	sand pine
Pinus contorta Dougl. ex Loud.	lodgepole pine
Pinus densiflora Sieb. and Zucc.	Japanese red pine
Pinus echinata Mill.	shortleaf pine
Pinus edulis Engelm.	pinyon pine
Pinus eldarica L.	eldarica pine
Pinus elliottii Engelm.	slash pine
Pinus halepensis Mill.	Aleppo pine
Pinus jeffreyi Grev. & Balf.	Jeffrey pine
Pinus lambertiana Dougl.	sugar pine
Pinus montana Mill. (*P. mugo* Turra)	mugo pine, Swiss mountain pine
Pinus monticola Dougl. ex D. Don	western white pine
Pinus mugo (see *P. montana*)	
Pinus nigra Arnold	Austrian pine
Pinus nigra var. *calabrica* (Loud.) Schneider	Corsican pine
Pinus palustris Mill.	longleaf pine
Pinus pinaster Ait.	cluster pine
Pinus pinea L.	Italian stone pine
Pinus ponderosa Dougl. ex P. Laws. & C. Laws.	ponderosa pine, western yellow pine
Pinus pungens Lamb.	Table Mountain pine
Pinus radiata D. Don	Monterey pine
Pinus resinosa Ait.	red pine
Pinus rigida Mill.	pitch pine
Pinus sabiniana Dougl.	digger pine
Pinus serotina Michx.	pond pine
Pinus strobus L.	eastern white pine, white pine
Pinus sylvestris L. (*P. silvestris*)	Scots pine, Scotch pine

Pinus taeda L.	loblolly pine
Pinus thunbergii Parl.	Japanese black pine
Pinus virginiana Mill.	Virginia pine
Piper auritum H.B. & K.	momo, acayo
Pistachia atlantica Desf.	Mount Atlas mastic tree
Platanus L. spp.	sycamore, planetree
Platanus occidentalis L.	American sycamore
Platycladus orientalis (L.) Franco	oriental arborvitae
Podocarpus L'Hér. ex Pers. spp.	podocarpus
Poncirus trifoliata L.	trifoliate orange
Populus L. spp.	aspen, cottonwood, poplar
Populus balsamifera L.	balsam poplar
Populus davidiana Dode	
Populus deltoides Bartr. ex Marsh.	eastern cottonwood
Populus grandidentata Michx.	bigtooth aspen
Populus nigra var. *italica* Moench.	Lombardy poplar
Populus simonii Carr.	Simon aspen
Populus tremula L.	European aspen
Populus tremuloides Michx.	trembling aspen, quaking aspen
Populus trichocarpa Torr. & A. Gray	black cottonwood
Prockia P. Br. ex. Linn.	
Prosopis L. spp.	mesquite
Prosopis chilensis (Mol.) Stuntz	mesquite
Prosopis cineraria (L.) McBride	jhand
Prunus L. spp.	plum, prune, apricot, cherry, almond, peach
Prunus armeniaca L.	apricot
Prunus avium (L.) L.	sweet cherry, mazzard
Prunus cerasifera J.F. Ehrh.	cherry plum
Prunus cerasus L.	sour cherry, Montmorency or tart cherry
Prunus davidiana (Carriere) Franch.	peach
Prunus domestica L.	garden plum
Prunus dulcis var. *dulcis* (Mill.) D.A. Webb	almond
Prunus laurocerasus L.	cherry laurel
Prunus mahaleb L.	mahaleb
Prunus padus L.	garden plum, bird cherry
Prunus pauciflora Bunge. (*P. pseudocerasus* Lindl.)	cherry
Prunus persica (L.) Batsch.	peach, nectarine
Prunus salicina Lindl.	Japanese plum
Prunus serotina J.F. Ehrh.	black cherry
Prunus serrulata Lindl.	flowering cherry
Pseudotsuga Carrière spp.	Douglas fir
Pseudotsuga macrocarpa (Vasey) Mayr.	bigcone Douglas fir
Pseudotsuga menziesii (Mirb.) Franco	Douglas fir

Psychotria L. spp.	wild coffee
Psychotria furcata DC.	
Psychotria limonensis Krause	wild coffee
Psychotria marginata Swartz	
Pterocarpus Jacq. spp.	paduk
Pyrus L. spp.	pear
Pyrus betulaefolia Bunge	
Pyrus calleryana Decne.	Bradford pear, Callery pear
Pyrus communis L.	common pear
Pyrus pyrifolia (Burm.f.) Nakai (*P. serotina* Rehd.)	Japanese pear, Asian pear
Quercus L. spp.	oak
Quercus agrifolia Nee	California live oak
Quercus alba L.	white oak
Quercus coccinea Muench.	scarlet oak
Quercus ellipsoidalis E.J. Hill	northern pin oak
Quercus lobata Nee	valley oak
Quercus lyrata Walt.	overcup oak
Quercus macrocarpa Michx.	bur oak
Quercus nigra L.	water oak
Quercus pedunculata Ehrh.	pedunculate oak (European)
Quercus phellos L.	willow oak
Quercus prinus L.	chestnut oak
Quercus robur L.	English oak
Quercus rubra L. (*Q. borealis* Michx. f.)	northern red oak, red oak
Quercus velutina Lam.	black oak
Quercus virginiana Mill.	live oak
Randia Houst. ex. Linn. spp.	
Randia spinosa Poir.	
Randia submontana K. Krause	
Raphia pedunculata Beauv.	raffia or Madagascar palm
Rhizophora mangle L.	American mangrove
Rhododendron L. spp.	rhododendron, azalea
Rhododendron ponticum L.	mountain rose bay
Rhus L. spp.	sumac
Rhus glabra L.	smooth sumac
Rhus typhina L.	staghorn sumac
Ribes L. spp.	currant, gooseberry
Robinia L. spp.	locust
Robinia pseudoacacia L.	black locust
Rosa L. spp.	rose
Rubus L. spp.	blackberry, raspberry
Rubus chamaemorus L.	cloudberry
Salix L. spp.	willow
Salix fragilis L.	brittle willow
Salix nigra Marsh.	black willow

Sebastiana Spreng. spp.	sebastian bush
Sebastiana fruticosa (Bartr.) Fern.	sebastian bush
Sequoia sempervirens (D. Don) Endl.	redwood, coast redwood
Sequoiadendron giganteum (Lindl.) Buchh.	giant sequoia, big tree
Sesbania grandiflora (L.) Poir.	agafi
Shepherdia Nutt. spp.	buffaloberry
Simmondsia chinensis (Link) C.K. Schneid.	jojoba
Sindora coriacea Prain	saputi, minyak, sepetir
Solanum [Tourn.] L. spp.	
Solanum tuberosum L.	potato
Solidago L. spp.	goldenrod
Sonneratia Linn. f. spp.	mangrove
Sonneratia alba Griff.	
Sonneratia lanceolata Blume	
Sorbus L. spp.	mountain ash
Sorbus americana Marsh.	American mountain ash
Sorbus aucuparia L.	European mountain ash
Syringa L. spp.	
Syringa vulgaris L.	lilac
Tabebuia serratifolia (Vahl.) Nichols	yellow poui
Tamarindus L. spp.	
Tamarindus indica L.	tamarindo
Taxodium distichum (L.) L. Rich	bald cypress
Taxus L. spp.	yew
Taxus brevifolia Nutt.	Pacific yew
Terminalia L. spp.	myrobalan
Terminalia ivorensis A. Chevalier	idigbo
Terminalia spinosa Northrop	
Terminalia superba Engl. and Diels	afara
Thea L. spp. [see *Camellia* L. spp.]	
Theobroma L. spp.	
Theobroma cacao L.	cacao
Thuja L. spp.	thuja
Thuja occidentalis L.	northern white cedar
Thuja orientalis L. [see *Platycladus orientalis* (L.) Franco]	
Thuja plicata J. Donn ex D. Don	western red cedar
Thujopsis Siebold & Zucc. (*Thuja* spp.)	arborvitae
Tilia L. spp.	basswood, linden
Tilia americana L.	American basswood
Tilia amurensis Rupr.	Amur linden
Tilia cordata Mill.	littleleaf linden
Tilia heterophylla Vent.	white basswood
Tovomita Aubl. spp.	

Trema Lour. spp.

   Trema micrantha (L.) Blume   Florida trema

   Trema orientalis (L.) Blume   Oriental trema, anabiong trema

Trifolium L. spp.   clover

   Trifolium incarnatum L.   common clover

   Trifolium subterraneum L.   subterranean clover

Triplochiton scleroxylon K. Schum.   obeche

Tristania conferta R. Br.   brush box, pink box, Queensland box

Tsuga (Endl.) Carr. spp.   hemlock

   Tsuga canadensis (L.) Carr.   eastern hemlock, Canadian hemlock

   Tsuga heterophylla (Raf.) Sarg.   western hemlock

Ulmus L. spp.   elm

   Ulmus americana L.   American elm

Vaccinium L. spp.   blueberry

   Vaccinium macrocarpon Ait.   cranberry

   Vaccinium myrtillus L.   myrtle whortleberry

   Vaccinium vitis-idaea L.   lingonberry

Viburnum carlesii Hemsl.   Koreanspice viburnum

Vitis L. spp.   grape

Xylophragma Sprague spp.

Common Names of Plants

acacia   *Acacia* Mill. spp.

   apple-ring   *Acacia albida* Delile

   Karroo thorn   *Acacia karroo* Hayne

   prickly moses   *Acacia pulchella* R. Br.

   steel   *Acacia macracantha* Humb. & Bonpl. ex Willd.

   sweet   *Acacia farnesiana* (L.) Willd.

   umbrella-thorn   *Acacia tortilis* (Forssk.) Hayne

afara   *Terminalia superba* Engl. and Diels

African upas tree   *Antiaris africana* Engl.

agafi   *Sesbania grandiflora* (L.) Poir.

ailanthus, tree of heaven   *Ailanthus altissima* (Mill.) Swingle

albizia   *Albizia* Durrazz. spp.

alder   *Alnus* B. Ehrh. spp.

   European   *Alnus incana* (L.) Moench

   European black or black   *Alnus glutinosa* (L.) Gaertn.

   green   *Alnus crispa* (Ait.) Pursh

   red   *Alnus rubra* Bong. (*A. oregona* Nutt.)

almond   *Prunus dulcis* var. *dulcis* (Mill.) D.A. Webb

apple, crabapple   *Malus* Mill. spp.; *Malus domestica*

	Borkh. [*Malus pumila* Mill.; *M. sylvestris* (L.) Mill.]
apricot	*Prunus armeniaca* L.
araucaria	*Araucaria* Juss. spp.
arborvitae, eastern	*Thuja occidentalis* L.
arborvitae, Oriental	*Platycladus orientalis* (L.) Franco
ash	*Fraxinus* L. spp.; *Eucalyptus* L'Her. spp.
alpine	*Eucalyptus gigantea* Hook. f.
European	*Fraxinus excelsior* L.
green or red	*Fraxinus pennsylvanica* Marsh.
Manchurian	*Fraxinus mandshurica* Rupr.
white or American	*Fraxinus americana* L.
aspen (see also cottonwood, poplar)	*Populus* L. spp.
bigtooth	*Populus grandidentata* Michx.
European	*Populus tremula* L.
Simon	*Populus simonii* Carr.
trembling or quaking	*Populus tremuloides* Michx.
aster	*Aster* L. spp.
avocado	*Persea americana* Mill.
azalea (see rhododendron)	
bamboo	*Bambusa* Schreb. spp.
banana	*Musa* L. spp.
basswood (see also linden)	*Tilia* L. spp
American	*Tilia americana* L.
white	*Tilia heterophylla* Vent.
beech	*Fagus* L. spp.
American	*Fagus grandifolia* J.F. Ehrh.
European	*Fagus sylvatica* L.
beefwood (Australian pine, she-oak)	*Casuarina* L. ex Adans. spp.
Belah	*Casuarina cristata* Miq.
stunted	*Casuarina distyla* Vent.
birch	*Betula* L. spp.
European white, European, hairy, or white	*Betula pubescens* J.F. Ehrh.
gray	*Betula populifolia* Marsh.
hairy (see birch, European, hairy, or white)	
Japanese white	*Betula mandshurica* (Regal) Nakai (*B. platyphylla* Sukachev)
paper, white, or canoe	*Betula papyrifera* Marsh.
river	*Betula nigra* L.
silver	*Betula pendula* Roth. (*Betula verrucosa* J.F. Ehrh.)
yellow	*Betula alleghaniensis* Britt.
black mangrove	*Avicennia* L. spp.; *Avicennia marina* (Forsk.) Vierh.
black-oak	*Casuarina littoralis* Salisb.

blackberry, raspberry	*Rubus* L. spp.
bloodwood, see eucalyptus	
blue beech	*Carpinus caroliniana* Walt.
blueberry	*Vaccinium* L. spp.
bougainvillea	*Bougainvillea* Comm. ex Juss. spp.
box	*Buxus* L. spp.
box elder	*Acer negundo* L.
Brazil nut	*Bertholletia excelsa* Humb. & Bonpl.
breadfruit	*Artocarpus* J.R. & G. Forst. spp.
brides-bush	*Pavetta gardeniifolia* A. Rich (*P. assimilis*)
Brush box	*Tristania conferta* R. Br.
buckeye	*Aesculus* L. spp.
California	*Aesculus californica* (Spach) Nutt.
buffaloberry	*Shepherdia* Nutt. spp.
bull-oak	*Casuarina luehmannii* R.T. Bak.
bushwillow, red	*Combretum apiculatum* Sond.
butterfly tree	*Colophospermum mopane* Benth.
butternut or white walnut	*Juglans cinerea* L.
cacao	*Theobroma cacao* L.
cassava	*Manihot esculenta* Crantz
catalpa	*Catalpa* Scop. spp.
northern	*Catalpa speciosa* Warder
southern	*Catalpa bignonioides* Walt.
ceanothus	*Ceanothus* L. spp.
cecropia	*Cecropia obtusifolia* Bertolini
cedar	*Cedrus* L. spp.; *Thuja* L. spp.; *Juniperus* L. spp.; *Chamaecyparis* Spach
Alaska or yellow	*Chamaecyparis nootkatensis* (D. Don) Spach
eastern red	*Juniperus virginiana* L.
incense	*Calocedrus decurrens* (Torr.) Florin (*Libocedrus decurrens* Torr.)
northern white	*Thuja occidentalis* L.
Port Orford	*Chamaecyparis lawsoniana* (A. Murr.) Parl.
western red	*Thuja plicata* J. Donn ex D. Don
cherry	*Prunus* L. spp.
black	*Prunus serotina* J.F. Ehrh.
flowering	*Prunus serrulata* Lindl.
sour, tart, or Montmorency	*Prunus cerasus* L.
sweet	*Prunus avium* (L.) L.
cherry laurel	*Prunus laurocerasus* L.
chestnut	*Castanea* Mill. spp.
American	*Castanea dentata* (Marsh.) Borkh.
Chinese	*Castanea mollissima* Blume
chrysanthemum	*Chrysanthemum* [Tourn.] L. spp.

cinchona bark tree	*Cinchona ledgeriana* Moens. ex Trimen
clematis	*Clematis* L. spp.
cloudberry	*Rubus chamaemorus* L.
clover	
common	*Trifolium incarnatum* L.
subterranean	*Trifolium subterraneum* L.
tick	*Desmodium* Desv. spp.
coca	*Erythroxylum* P. Br. spp.
cocaine plant	*Erythroxylum coca* Lam.
coconut	*Cocos nucifera* L.
coffee	*Coffea* L. spp.
Arabian	*Coffea arabica* L.
wild	*Psychotria limonensis* Krause
cork tree	*Phellodendron amurense* Rupr. var. *wilsonii* (Hayata & Kanchira) C.E. Chang
cotton	*Gossypium* L. spp.; *Gossypium hirsutum* L.
cottonwood (see also aspen, poplar)	*Populus* L. spp.
eastern	*Populus deltoides* Bartr. ex Marsh.
cranberry	*Vaccinium macrocarpon* Ait.
creosote bush	*Larrea* Cav. spp.
crotalaria, shackshack	*Crotalaria incana* L.
crowberry	*Empetrum hermaphroditum* Hagerup
cryptomeria, Japanese	*Cryptomeria japonica* (L.f.) D. Don
currant, gooseberry	*Ribes* L. spp.
cypress (see also cedar)	*Chamaecyparis* Spach; *Cupressus* L. spp; *Taxodium* Rich. spp.
Arizona	*Cupressus arizonica* Greene
bald	*Taxodium distichum* (L.) L. Rich
hinoki	*Chamaecyparis obtusa* (Siebold & Zucc.) Endl.
Lawson (see Port Orford cedar)	
desert-oak	*Casuarina decaisneana* F. Muell.
Devil's-walking stick, Hercules-club	*Aralia spinosa* L.
dogwood	*Cornus* L. spp.
flowering	*Cornus florida* L.
red-osier or red-stemmed	*Cornus stolonifera* Michx. (*C. sericea* L.)
Douglas fir	*Pseudotsuga menziesii* (Mirb.) Franco
bigcone	*Pseudotsuga macrocarpa* (Vasey) Mayr.
elaeagnus	*Elaeagnus* L.
elm	*Ulmus* L. spp.
American	*Ulmus americana* L.
erythrina	*Erythrina* L. spp.
eucalyptus	*Eucalyptus* L'Hér. spp.
alpine ash, mountain ash, woollybutt	*Eucalyptus delegatensis* R.T. Bak. (*E. gigantea* Hook. f.)

blackbutt	*Eucalyptus pilularis* Sm.
brown barrel	*Eucalyptus fastigiata* Deane & Maiden
crimson mallee	*Eucalyptus erythronema* Turcz.
crimson mallee box	*Eucalyptus landsdowneana* F. Muell. & J.E. Brown var. *landsdowneana*
Darwin stringy bark	*Eucalyptus tetrodenta* F. Muell.
Darwin woollybutt	*Eucalyptus miniata* A. Cunn. ex Schau.
grey box	*Eucalyptus moluccana* Roxb.
grey iron box	*Eucalyptus paniculata* Smith
gympie messmate	*Eucalyptus cloeziana* F. Muell.
karri	*Eucalyptus diversicolor* F. Muell.
large-fruited red mahogany	*Eucalyptus pellita* F. Muell.
lemon-scented gum	*Eucalyptus citriodora* Hook.
long-fruited bloodwood	*Eucalyptus polycarpa* F. Muell.
manna gum	*Eucalyptus viminalis* Labill.
mountain ash	*Eucalyptus regnans* F. Muell.
Murray red gum, red gum, river red gum	*Eucalyptus rostrata* Schlechtend. (*E. camaldulensis* Dehnh.)
pink bloodwood	*Eucalyptus intermedia* R. T. Baker
Queensland grey iron bark	*Eucalyptus drepanophylla* F. Muell. ex Benth.
Queensland white gum	*Eucalyptus argophloia* Blakely
rose gum	*Eucalyptus grandis* W. Hill. ex Maiden
shining gum	*Eucalyptus nitens* Maiden
spotted gum	*Eucalyptus maculata* Hook.
swamp mahogany	*Eucalyptus robusta* Sm.
Sydney blue gum	*Eucalyptus saligna* Sm.
Tasmanian blue gum, blue gum	*Eucalyptus globulus* Labill.
false-beech, Tasmanian	*Nothofagus cunninghamia* Oerst.
false cypress (see cedar)	
fir	*Abies* Mill. spp.
alpine	*Abies lasiocarpa* (Hook.) Nutt.
balsam	*Abies balsamea* (L.) Mill.
California red or red	*Abies magnifica* A. Murr.
European silver	*Abies alba* Mill.
Fraser	*Abies fraseri* (Pursh) Poir.
grand	*Abies grandis* (D. Don ex Lamb.) Lindl.
Pacific silver or silver	*Abies amabilis* Dougl. ex J. Forbes
white	*Abies concolor* (Gord.) Lindl. ex Hildebr.
forest-oak	*Casuarina torulosa* Ait.
forsythia	*Forsythia* Vahl. spp.
greenstem	*Forsythia viridissima* Lindl.
ginkgo	*Ginkgo* L. spp.
goldenrod	*Solidago* L. spp.
gooseberry (see currant)	
grape	*Vitis* L. spp.

grapefruit	*Citrus paradisi* Macfad.
greasewood	*Adenostoma fasciculatum* Hook. & Arn.
guayule	*Parthenium argentatum* A. Gray
gum (see also tupelo and eucalyptus)	*Nyssa* L. spp.; *Liquidambar* L. spp.
black (see tupelo, black)	
sweet or red	*Liquidambar styraciflua* L.
gurjuniol tree	*Dipterocarpus alatus* Roxb.
hackberry	*Celtis* L. spp.
common	*Celtis occidentalis* L.
hairgrass, wavy	*Deschampsia flexuosa* (L.) Trin.
hawthorn	*Crataegus* L. spp.
hay-scented fern	*Dennstaedtia punctilobula* Michx.
hazel, filbert	*Corylus* L. spp.
European	*Corylus avellana* L.
witch	*Hamamelis virginiana* L.
head lettuce	*Lactuca sativa* L.
heather	*Calluna* Salisb. spp.
hemlock	*Tsuga* (Endl.) Carr. spp.
eastern or Canadian	*Tsuga canadensis* (L.) Carr.
western	*Tsuga heterophylla* (Raf.) Sarg.
Hercules'-club	*Aralia spinosa* L.
hibiscus	*Hibiscus* L. spp.
hickory	*Carya* Nutt. spp.
holly	*Ilex* L. spp.
American	*Ilex opaca* Ait.
deciduous, possum haw	*Ilex decidua* Walt.
yaupon	*Ilex vomitoria* Ait.
honeysuckle	*Lonicera* L. spp.
hop hornbeam, eastern or ironwood	*Ostrya virginiana* (Mill.) K. Koch
horse chestnut (see also buckeye)	*Aesculus* L. spp.; *Aesculus hippocastanum* L.
hovea	*Hovea heterophylla* A. Cunn. ex Hook.
idigbo	*Terminalia ivorensis* A. Chevalier
indigo, Anil	*Indigo suffruticosa* Mill.
ironbark, see eucalyptus	
Java locust	*Parkia javanica* (Lam.) Merrill
jhand	*Prosopis cineraria* (L.) McBride
jojoba	*Simmondsia chinensis* (Link) C.K. Schneid.
juniper	*Juniperus* L. spp.
common	*Juniperus communis* L.
karri	*Eucalyptus diversicolor* F. Muell.
kauri, New Zealand	*Agathis australis* Hort. ex Lindl.
keranji	*Dialinum maingayi* Baker
kiwifruit	*Actinidia* Lindl. spp.
Kolchischer ahorn	*Acer cappadocicum* Gleditsch
kuka (see mahogany, Senegal)	

kumquat   *Fortunella* Swingle spp.
kwila (see merbau)
larch, tamarack   *Larix* Mill. spp.
  Dahurian   *Larix dehurica* Turcz.
  eastern, tamarack   *Larix laricina* (DuRoi) K. Koch
  European   *Larix decidua* Mill.
  Japanese   *Larix kaempferi* (Lamb.) Carr. [*L. leptolepis* (Siebold & Zucc.) Gord.]
  subalpine   *Larix lyallii* Parl.
  western   *Larix occidentalis* Nutt.
laurel, cherry   *Prunus laurocerasus* L.
leadtree (see leucaena)
lemon, rough   *Citrus limon* (L.) Burm. f.
leucaena, leadtree   *Leucaena leucocephala* (Lam.) de Wit
lilac   *Syringa vulgaris* L.
lime, Tahiti sweet   *Citrus aurantiifolia* (Christm.) Swingle
linden (see also basswood)   *Tilia* L. spp
  Amur   *Tilia amurensis* Rupr.
  littleleaf   *Tilia cordata* Mill.
lingonberry   *Vaccinium vitis-idaea* L.
litchi, lychee   *Litchi chinensis* Sonn.
locust (see also Java-locust)   *Robinia* L. spp.; *Gleditsia* L. spp.
  black   *Robinia pseudoacacia* L.
  honey   *Gleditsia triacanthos* L.
lupine   *Lupinus arboreus* Sims
macadamia nut   *Macadamia integrifolia* Maiden & Betche
madre   *Gliricidia sepium* (Jacq.) Kunth ex Griseb.
magnolia   *Magnolia* L. spp.
  southerm   *Magnolia grandiflora* L.
mahaleb   *Prunus mahaleb* L.
mahogany
  Lagos   *Khaya ivorensis* A. Chev.
  mountain   *Cercocarpus* H.B. & K. spp.
  Senegal or kuka   *Khaya senegalensis* (Desr.) A. Juss
mandarin (see orange)
mango   *Mangifera indica* L.
mangrove, American   *Rhizophora mangle* L.
mangrove, Burma   *Bruguiera gymnorhiza* Lam.
maple   *Acer* L. spp.
  Amur   *Acer ginnala* Maxim.
  bigleaf   *Acer macrophyllum* Pursh.
  Father David's   *Acer davidii* Franchet
  Norway   *Acer platanoides* L.
  paperbark   *Acer griseum* (Franco) Pax.
  red   *Acer rubrum* L.

silver	*Acer saccharinum* L.
striped	*Acer pensylvanicum* L.
sugar	*Acer saccharum* Marsh.
sycamore	*Acer pseudoplatanus* L.
Tatarian	*Acer tataricum* L.
velvet	*Acer velutinum* Boissier
mazzard (see cherry, sweet)	
melaleuca (see also paperbark tree)	*Melaleuca* L. spp.
merbau, kwila	*Intsia palembanica* Miq.
mesquite	*Prosopis* L. spp.
metasequoia (see redwood, dawn)	
mimosa	*Mimosa* L. spp,
momo, acayo	*Piper auritum* H.B. & K.
Montezuma	*Montezuma* Sessé & Moc. spp.
Mount Atlas mastic tree	*Pistachia atlantica* Desf.
mountain ash (see also eucalyptus)	*Sorbus* L. spp.
American	*Sorbus americana* Marsh
European	*Sorbus aucuparia* L
mountain laurel	*Kalmia latifolia* L.
mouse-ear cress	*Arabidopsis* Heynd. spp.
mulberry	*Morus* L. spp.; *Morus rubra* L.
myrobalan	*Terminalia* L. spp.
myrtle whortleberry	*Vaccinium myrtillus* L.
nectarine (see peach)	
nittatree	*Parkia* R. Br. spp.
oak	*Quercus* L. spp.
black	*Quercus velutina* Lam.
bur	*Quercus macrocarpa* Michx.
California live	*Quercus agrifolia* Neé
chestnut	*Quercus prinus* L.
English	*Quercus robur* L.
live	*Quercus virginiana* Mill.
northern pin	*Quercus ellipsoidalis* E.J. Hill
northern red or red	*Quercus rubra* L. (*Q. borealis* Michx. f.)
overcup	*Quercus lyrata* Walt.
pedunculate or European	*Quercus pedunculata* Ehrh.
scarlet	*Quercus coccinea* Muench.
tan	*Lithocarpus densiflorus* (Hook. & Arn.) Rehd.
valley	*Quercus lobata* Nee
water	*Quercus nigra* L.
white	*Quercus alba* L.
willow	*Quercus phellos* L.
obeche	*Triplochiton scleroxylon* K. Schum.
oleaster	*Elaeagnus latifolia* L.
olive	*Olea* L. spp.; *Olea europaea* L.
autumn	*Elaeagnus umbellata* Thunb.

orange (see also citrus, lemon, lime, grapefruit)	*Citrus* L. spp.
Cleopatra mandarin, tangerine	*Citrus reticulata* Blanco 'Cleopatra'
osage	*Maclura pomifera* (Raf.) Schneid.
sour	*Citrus aurantium* L.
Valencia, sweet, or spice	*Citrus sinensis* L. 'Valencia'
paduk	*Pterocarpus* Jacq. spp.
palm	
attalea	*Attalea excelsa* Mart.
date	*Phoenix dactylifera* L.
Madagascar or raffia	*Raphia pedunculata* Beauv.
oil	*Elaeis guineensis* Jacq.
papaya	*Carica papaya* L.
paperbark tree	*Melaleuca quinquenervia* (Cav.) S.T. Blake
paulownia, royal, princess tree	*Paulownia tomentosa* (Thunb.) Siebold & Zucc. ex Steud.
pea, partridge	*Chamaecrista chamaecristoides* (Collad)
peach or nectarine	*Prunus persica* (L.) Batsch.
pear	*Pyrus* L. spp.
Bradford or Callery	*Pyrus calleryana* Decne.
common	*Pyrus communis* L.
Japanese	*Pyrus pyrifolia* (Burm. f.) Nakai
pecan, pecan hickory	*Carya illinoensis* (Wangenh.) K. Koch
persimmon, common	*Diospyros virginiana* L.
pine	*Pinus* L. spp.
Aleppo	*Pinus halepensis* Mill.
Australian (see beefwood)	
Austrian	*Pinus nigra* Arnold
Carib	*Pinus caribaea* var. *hondurensis* Morelet
Caribbean	*Pinus caribaea* Morelet
cluster	*Pinus pinaster* Ait.
Corsican	*Pinus nigra* var. *calabrica* (Loud.) Schneider
digger	*Pinus sabiniana* Dougl.
eastern white or white	*Pinus strobus* L.
eldarica	*Pinus eldarica* L.
hoop	*Araucaria cunninghamia* Sweet.
Imou (see rimu)	
Italian stone	*Pinus pinea* L.
jack	*Pinus banksiana* Lamb.
Japanese black	*Pinus thunbergii* Parl.
Japanese red	*Pinus densiflora* Siebold & Zucc.
Jeffrey	*Pinus jeffreyi* Grev. & Balf.
loblolly	*Pinus taeda* L.
lodgepole	*Pinus contorta* Dougl. ex Loud.
longleaf	*Pinus palustris* Mill.

Mexican pinyon	*Pinus cembroides* Zucc.
Monterey	*Pinus radiata* D. Don
mugo or Swiss mountain	*Pinus montana* Mill. (*P. mugo* Turra)
pinyon	*Pinus edulis* Engelm.
pitch	*Pinus rigida* Mill.
pond	*Pinus serotina* Michx.
ponderosa or western yellow	*Pinus ponderosa* Dougl. ex P. Laws. & C. Laws.
red	*Pinus resinosa* Ait.
sand	*Pinus clausa* (Chapm.) Vasey
Scots or Scotch	*Pinus sylvestris* L. (*P. silvestris*)
shortleaf	*Pinus echinata* Mill.
slash	*Pinus elliottii* Engelm.
sugar	*Pinus lambertiana* Dougl.
Swiss stone	*Pinus cembra* L.
Table Mountain	*Pinus pungens* Lamb.
Virginia	*Pinus virginiana* Mill.
western white	*Pinus monticola* Dougl. ex D. Don.
whitebark	*Pinus albicaulis* Engelm.
planetree (see sycamore)	
plum (also prune, apricot, cherry, almond, peach)	*Prunus* L. spp.
cherry	*Prunus cerasifera* J.F. Ehrh.
garden, bird cherry	*Prunus domestica* L.; *P. padus* L.
Japanese	*Prunus salicina* Lindl.
podocarpus	*Podocarpus* L'Hér. ex Pers. spp.
poplar (see also Aspen, Cottonwood)	*Populus* L. spp.
balsam	*Populus balsamifera* L.
Lombardy	*Populus nigra* var. *italica* Moench.
yellow or tulip, tulip tree	*Liriodendron tulipifera* L.
potato	*Solanum tuberosum* L.
prune	*Prunus* L. spp.
Queensland nut	*Macadamia ternifolia* F. Muell.
quince	*Cydonia* Mill. spp.
redbud	*Cercis* L. spp.
eastern	*Cercis canadensis* L.
Judastree	*Cercis siliquastrum* L.
red cedar (see cedar)	
redroot (see Ceanothus)	
redwood	
coast	*Sequoia sempervirens* (D. Don) Endl.
dawn	*Metasequoia glyptostroboides* Hu & Cheng
rhododendron, azalea	*Rhododendron* L. spp.
rimu	*Dacrydium cupressinum* Soland. ex Forst.
rose	*Rosa* L. spp.
rose bay, mountain	*Rhododendron ponticum* L.
rosewood	*Dalbergia* L. f. spp.

rubber	*Hevea* Aubl. spp.
Brazilian rubber tree	*Hevea brasiliensis* (Willd. ex A. Juss) Mull-Arg.
saltbush	*Atriplex patula* L.
saputi, minyak, sepetir	*Sindora coriacea* Prain
sebastian bush	*Sebastiana fruticosa* (Bartr.) Fern.
senna	*Cassia* L. spp.
sequoia, giant, big tree	*Sequoiadendron giganteum* (Lindl.) Buchh.
shadbush	*Amelanchier* Medic. spp.
she-oak (see also beefwood)	*Casuarina* L. ex Adans. spp.; *Casuarina stricta* Ait.
coast	*Casuarina equisetifolia* L. ex J.R. & G. Forst.
river, creek-oak	*Casuarina cunninghamiana* Miq.
swamp	*Casuarina glauca* Sieb. ex K. Spreng
shingle tree	*Acrocarpus fraxinifolius* Wight. & Arn.
silk-oak	*Grevillea robusta* A. Cunn. ex R. Br.
smoke tree	*Cotinus* Mill spp.; *Cotinus coggyria* Scop.
sourwood	*Oxydendrum arboreum* (L.) DC.
soybean	*Glycine max* (L.) Merr.
spear fruit	*Lonchocarpus* Benth. spp.
spicebush	*Lindera benzoin* Meissn.
spruce	*Picea* A. Dietr. spp.
black	*Picea mariana* (Mill.) B.S.P.
blue, Colorado blue, or Colorado	*Picea pungens* Engelm.
Engelmann	*Picea engelmannii* Parry ex Engelm.
Norway	*Picea abies* (L.) Karst.
red	*Picea rubens* Sarg.
sitka	*Picea sitchensis* (Bong.) Carr.
white	*Picea glauca* (Moench) Voss
Yeddo	*Picea jezoensis* (Siebold & Zucc.) Carriere
strawberry	*Fragaria* L. spp.
sugarberry, sugar hackberry	*Celtis laevigata* Willd.
sugi, Japanese cedar	*Cryptomeria japonica* (L.f.) D. Don
sumac	*Rhus* L. spp.
smooth	*Rhus glabra* L.
staghorn	*Rhus typhina* L.
sweet leaf	*Hopea odorata* Roxb.
sweet potato	*Ipomoea batatas* (L.) Lam.
sycamore	*Platanus* L. spp.
American	*Platanus occidentalis* L.
tamarack (see larch)	
tamarindo	*Tamarindus indica* L.
tangerine (see Cleopatra mandarin orange)	

tea	*Camellia thea* (L.) Ktze.
tobacco	*Nicotiana tabacum* L.
trema	
Florida	*Trema micrantha* (L.) Blume
Oriental, anabiong	*Trema orientalis* (L.) Blume
trumpet tree	*Cecropia peltata* L.
tulip tree (see poplar, yellow)	
tung	*Aleurites fordii* Hemsl.
tupelo (or gum)	*Nyssa* L. spp.
black	*Nyssa sylvatica* Marsh.
swamp black or swamp	*Nyssa sylvatica* var. *biflora* (Walt.) Sarg.
water or swamp	*Nyssa aquatica* L.
viburnum	*Viburnum* L. spp.
Koreanspice	*Viburnum carlesii* Hemsl.
walnut (see also butternut)	*Juglans* L. spp.
black	*Juglans nigra* L.
English	*Juglans regia* L.
wax myrtle	*Myrica* L. spp.
white cedar (see cedar)	
willow	*Salix* L. spp.
bay	*Salix pentandra* L.
black	*Salix nigra* Marsh.
brittle	*Salix fragilis* L.
yam, water	*Dioscorea alata* L.
yellow poui	*Tabebuia serratifolia* (Vahl.) Nichols
yellowwood	*Cladrastis lutea* (Mich. f.) K. Koch [*C. kentukea* (Dum.-Cours.) Rudd]
yew	*Taxus* L. spp.
Pacific	*Taxus brevifolia* Nutt.

Bibliography

Abbott, D. L. (1984). "The Apple Tree: Physiology and Management." Grower Books, London.

Abbott, J. D., and Gough, R. E. (1987a). Reproductive response of the highbush blueberry to root-zone flooding. *HortScience* **22**, 40–42.

Abbott, J. D., and Gough, R. E. (1987b). Prolonged flooding effects on anatomy of highbush blueberry. *HortScience* **22**, 622–625.

Abeles, F. B. (1973). "Ethylene in Plant Biology." Academic Press, New York.

Aber, J. D., Nadelhoffer, J., Steudler, P., and Melillo, J. (1989). Nitrogen saturation in northern forest ecosystems. *BioScience* **39**, 378–386.

Abod, S. T., Shepherd, V. R., and Bachelard, E. P. (1979). Effects of light intensity, air and soil temperatures on root regenerating potential of *Pinus caribaea* var. *hondurensis* and *Pinus kesiya* seedlings. *Aust. For. Res.* **9**, 173–184.

Abrams, M. D., and Kubiske, M. E. (1990). Leaf structural characteristics of 31 hardwood and conifer tree species in central Wisconsin: Influence of light regime and shade-tolerance rank. *For. Ecol. Manage.* **31**, 245–253.

Ackerman, R. F., and Farrar, J. L. (1965). The effect of light and temperature on the germination of jack pine and lodgepole pine seeds. *Univ. Toronto, Fac. For., Tech. Rep. 5.*

Adams, F. (1980). Interactions of phosphorus with other elements in soil and plants. *In* "The Role of Phosphorus in Agriculture" (F. E. Khasawneh, *et al.*, eds.), pp. 655–680. Am. Soc. Agron., Madison, Wisconsin.

Adams, G. T., Perkins, T. D., and Klein, R. M. (1991). Anatomical studies on first-year winter injured red spruce foliage. *Am. J. Bot.* **78**, 1199–1206.

Adams, W. T. (1983). Application of isozymes in tree breeding. *In* "Isozymes in Plant Genetics and Breeding" (S. D. Tanksley and T. S. Orton, eds.), pp. 381–400. Elsevier, Amsterdam, The Netherlands.

Adams, W. T., Strauss, S. H., Copes, D. L., and Griffin, A. R., eds. (1992). "Population Genetics of Forest Trees." Kluwer, Dordrecht, The Netherlands.

Addicott, F. T. (1982). "Abscission." Univ. of California Press, Berkeley.

Addicott, F. T., ed. (1983). "Abscisic Acid." Praeger, New York.

Addicott, F. T. (1991). Abscission: Shedding of parts. *In* "Physiology of Trees" (A. S. Raghavendra, ed.), pp. 273–300. Wiley, New York.

Ågren, G. I. (1985a). Limits to plant production. *J. Theor. Biol.* **113**, 89–92.

Ågren, G. I. (1985b). Theory for growth of plants derived from the nitrogen productivity concept. *Physiol. Plant.* **64**, 17–28.

Aharoni, Y., and Lattar, F. S. (1972). The effect of various storage atmospheres on the occurrence of rots and blemishes on Shamouti oranges. *Phytopath. Z.* **73**, 371–374.

505

Ahlgren, C. E. (1972). Some effects of inter and intraspecific grafting on growth and flowering of some five-needle pines. *Silvae Genet.* **21**, 122–126.

Ahlgren, I. F. (1974). The effects of fire on soil organisms. *In* "Fire and Ecosystems" (T. T. Kozlowski and C. E. Ahlgren, eds.), pp. 47–72. Academic Press, New York.

Ahuja, M. R. (1987). Somaclonal variation. *In* "Cell and Tissue Culture in Forestry" (J. M. Bonga and D. J. Durzan, eds.), Vol. 1, pp. 272–285. Martinus Nijhoff, Dordrecht, The Netherlands.

Ahuja, M. R. (1993). Biotechnology and clonal forestry. *In* "Clonal Forestry" (M. R. Ahuja and W. J. Libby, eds.), Vol. 1, pp. 135–144. Springer-Verlag, Berlin.

Ahuja, M. R., and Libby, W. J., eds. (1993a). "Clonal Forestry, Volume 1: Genetics and Biotechnology." Springer-Verlag, Berlin.

Ahuja, M. R., and Libby, W. J., eds. (1993b). "Clonal Forestry, Volume 2: Conservation and Application." Springer-Verlag, Berlin.

Aldwinckle, H. S. (1975). Flowering of apple seedlings 16–20 months after germination. *HortScience* **10**, 124–126.

Alexandre, D. Y. (1980). Caractere saisonnier de la fructification dans une foret hygrophile de Côte d'Ivoire. *Rev. Ecol.* (*Terre et Vie*) **34**, 335–350.

All, J. N., and Benjamin, D. J. (1975). Influence of needle maturity on larval feeding preference and survival of *Neodiprion swainei* and *N. rugifrons* on jack pine, *Pinus banksiana. Ann. Entomol. Soc. Am.* **68**, 579–584.

Allen, G. S. (1960). Factors affecting the viability and germination behavior of coniferous seed. IV. Stratification period and incubation temperature, *Pseudotsuga menzeisii* (Mirb.) Franco. *For. Chron.* **36**, 18–29.

Allen, J. A., Chambers, J. L., and Stine, M. (1994). Prospects for increasing the salt tolerance of forest trees: A review. *Tree Physiol.* **14**, 843–853.

Allen, R., and Wardrop, A. B. (1964). The opening and shedding mechanism of the female cones of *Pinus radiata. Aust. J. Bot.* **12**, 125–134.

Aloni, R. (1987). Differentiation of vascular tissues. *Annu. Rev. Plant Physiol.* **38**, 179–204.

Aloni, R., and Zimmermann, M. H. (1983). The control of vessel size and density along the plant axis—A new hypothesis. *Differentiation* **24**, 203–208.

Alston, F. H., and Spiegel-Roy, P. (1985). Fruit tree breeding: Strategies, achievements and constraints. *In* "Attributes of Trees as Crop Plants" (M. G. R. Cannell and J. E. Jackson, eds.), pp. 49–67. Institute of Terrestrial Ecology, Huntingdon, England.

Alvim, P. de T. (1977). Cacao. *In* "Ecophysiology of Tropical Crops" (P. de T. Alvim and T. T. Kozlowski, eds.), pp. 279–313. Academic Press, New York.

Alvim, P. de T., and Alvim, R. (1978). Relation of climate to growth periodicity in tropical trees. *In* "Tropical Trees as Living Systems" (P. B. Tomlinson and M. H. Zimmermann, eds.), pp. 445–464. Cambridge Univ. Press, Cambridge.

Andersen, P. C., Buchanan, D. W., and Albrigo, L. G. (1979). Water relations and yields of three rabbiteye blueberry cultivars with and without drip irrigation. *J. Am. Soc. Hortic. Sci.* **104**, 731–736.

Andersen, P. C., Lombard, P. B., and Westwood, M. N. (1984). Effect of root anaerobiosis on the water relations of several *Pyrus* species. *Physiol. Plant.* **62**, 245–252.

Anderson, J. L., and Seeley, S. D. (1993). Bloom delay in deciduous fruits. *Hortic. Rev.* **15**, 97–144.

Anderson, J. L., Richardson, E. A., and Kesner, C. D. (1986). Validation of chill unit and flower bud phenology models for 'Montmorency' sour cherry. *Acta Hortic.* **184**, 71–78.

Anderson, J. W., and Beardall, J. (1991). "Molecular Activities of Plant Cells." Blackwell, Oxford.

Andrews, P. K., and Serrano-Marquez, C. (1993). Graft incompatibility. *Hortic. Rev.* **15**, 183–232.

Anekonda, T. S., Criddle, R. S., Libby, W. J., and Hansen, L. D. (1993). Spatial and temporal relationships between growth traits and metabolic heat rates in coast redwood. *Can. J. For. Res.* **23**, 1793–1798.

Anekonda, T. S., Criddle, R. S., Libby, W. J., Breidenbach, R. W., and Hansen, L. D. (1994). Respiration rates predict differences in growth of coast redwood. *Plant Cell Environ.* **17**, 197–203.

Angeles, G., Evert, R. F., and Kozlowski, T. T. (1986). Development of lenticels and adventitious roots in flooded *Ulmus americana* seedlings. *Can. J. For. Res.* **16**, 585–590.

Annila, E. (1977). Damage by *Phyllobius* weevils (Coleoptera: Curculionidae) in birch plantations. (In Finnish.) *Metsantutkimuslaitoksen Julk.* **97**(3), 1–20.

Anonymous (1948). Woody plant seed manual. *U.S. Dep. Agric. Misc. Publ.* **654**.

Archbold, D. D., and Dennis, F. G., Jr. (1985). Strawberry receptacle growth and endogenous IAA content as affected by growth regulator application and achene removal. *J. Am. Soc. Hortic. Sci.* **110**, 816–820.

Archilletti, T., Lauri, P., and Damiano, C. (1995). *Agrobacterium*-mediated transformation of almond leaf pieces. *Plant Cell Rep.* **14**, 267–272.

Armstrong, W. (1968). Oxygen diffusion from the roots of woody species. *Physiol. Plant.* **21**, 539.

Arnheim, N., Li, H., and Cui, X. (1991). Genetic mapping by single sperm typing. *Anim. Genet.* **22**, 105–115.

Arora, R. K., and Yamadigni, R. (1986). Effects of different doses of nitrogen and zinc sprays on flowering, fruit set and final fruit retention in sweet lime (*C. limettioides* Tanaka). *Haryana Agric. Univ. J. Res.* **16**, 233–239.

Ashton, P. M. S., and Berlyn, G. P. (1992). Leaf adaptations of some *Shorea* species to sun and shade. *New. Phytol.* **121**, 587–596.

Ashton, P. M. S., and Berlyn, G. P. (1994). A comparison of leaf physiology and anatomy of *Quercus* (section *Erythrobalanus*—Fagaceae) species in different light environments. *Am. J. Bot.* **81**, 589–597.

Ashworth, E. N. (1990). The formation and distribution of ice within forsythia flower buds. *Plant Physiol.* **92**, 718–725.

Attiwill, P. M. (1995). Nutrient cycling in forests. *In* "Encyclopedia of Environmental Biology" (W. A. Nierenberg, ed.), Vol. 2, pp. 625–639. Academic Press, San Diego.

Attiwill, P. M. (1982). Phosphorus in Australian forests. *In* "Phosphorus in Australia" (A. B. Costin and C. H. Williams, eds.), pp. 113–134. Univ. of Melbourne, Australia.

Aubertin, G. M., and Patric, J. H. (1974). Water quality after clearcutting a small watershed in West Virginia. *J. Environ. Qual.* **3**, 243–245.

Aussenac, G., and Granier, A. (1988). Effects of thinning on water stress and growth in Douglas-fir. *Can. J. For. Res.* **18**, 100–105.

Austin, M. E., Mullinix, B. G., and Mason, J. S. (1982). Influence of chilling on growth and flowering of rabbiteye blueberries. *HortScience* **17**, 768–769.

Avery, D. J. (1966). The supply of air to leaves in assimilation chambers. *J. Exp. Bot.* **17**, 655–677.

Avery, D. J. (1969). Comparisons of fruiting and deblossomed maiden apple trees and of nonfruiting trees on a dwarfing and an invigorating root stock. *New Phytol.* **68**, 323–336.

Avery, D. J. (1970). Effects of fruiting on the growth of apple trees on four rootstock varieties. *New Phytol.* **69**, 19–30.

Awad, M., and Young, R. E. (1979). Postharvest variation in cellulase, polygalacturonase, and pectinmethylesterase in avocado (*Persea americana* Mill. cv. Fuerte) fruits in relation to respiration and ethylene production. *Plant Physiol.* **64**, 306–308.

Axelrod, M. C., Coyne, P. I., Bingham, G. E., Kircher, J. R., Miller, P. R., and Hung, R. C. (1980). Canopy analysis of pollutant injured ponderosa pine in the San Bernardino National Forest. *U.S.D.A. For. Serv. Gen. Tech. Rep. PSW* **PSW-43**, 227.

Bachelard, E. P., and Stowe, B. B. (1963). Rooting of cuttings of *Acer rubrum* L. and *Eucalyptus camaldulensis* Dehn. *Aust. J. Biol. Sci.* **16**, 751–767.

Bachem, C. W. B., Speckmann, G. J., van der Linde, P. C. G., Verheggen, F. T. M., Hunt, M. D., Steffens, J. C., and Zabeau, M. (1994). Antisense expression of polyphenol oxidase genes inhibits enzymatic browning in potato tubers. *Bio/Technology* **12**, 1101–1105.

Bäck, J., Neuvonen, S., and Huttunen, S. (1994). Pine needle growth and fine structure after prolonged acid rain treatment in the subarctic. *Plant Cell Environ.* **17**, 1009–1021.

Baer, N. W. (1989). Shelterbelts and windbreaks in the Great Plains. *J. For.* **87**, 32–36.

Baig, M. N., and Tranquillini, W. (1980). The effects of wind and temperature on cuticular transpiration of *Picea abies* and *Pinus cembra* and their significance in desiccation damage at the alpine treeline. *Oecologia* **47**, 252–256.

Baker, F. S. (1950). "Principles of Silviculture." McGraw-Hill, New York.

Baker, J. B., and Blackman, B. G. (1977). Biomass and nutrient accumulation in a cottonwood plantation—The first growing season. *Soil Sci. Am. J.* **41**, 633–636.

Baker, M. B. (1986). Effects of ponderosa pine treatments on water yield in Arizona. *Water Resour. Res.* **22**, 67–73.

Bakke, A. (1970). The use of DDT in Norwegian forestry. Review and present situation. (In Norwegian.) *Tidsskr. Skogbruk* **78**, 304–309.

Balboa-Zavala, O., and Dennis, F. G. (1977). Abscisic acid and apple seed dormancy. *J. Am. Soc. Hortic. Sci.* **102**, 633–637.

Baldocchi, D., and Collineau, S. (1994). The physical nature of solar radiation in heterogeneous canopies: Spatial and temporal attributes. *In* "Exploitation of Environmental Heterogeneity in Plants. Ecophysiological Processes Above and Below Ground" (M. M. Caldwell and R. W. Pearcy, eds.), pp. 21–71. Academic Press, San Diego.

Baldwin, T. T., and Schultz, J. C. (1983). Rapid changes in leaf chemistry induced by damage: Evidence for communication between plants. *Science* **221**, 277–279.

Ball, M. C., and Pidsley, S. M. (1995). Growth responses to salinity in relation to distribution of two mangrove species, *Sonneratia alba* and *S. lanceolata*, in northern Australia. *Funct. Ecol.* **9**, 77–85.

Ballardi, R., Cristoferi, G., Facini, O., and Lercari, B. (1992). The effect of light quality in *Prunus cerasus*. I. Photoreceptors involved in internode elongation and leaf expansion in juvenile plants. *Photochem. Photobiol.* **56**, 541–544.

Banko, T. J., and Stefani, M. A. (1991). *In vitro* flowering of *Oxydendrum arboreum*. *HortScience* **26**, 1425.

Bannan, M. W., and Bindra, M. (1970). The influence of wind on ring width and cell length in conifer stems. *Can. J. Bot.* **48**, 255–259.

Bao, W., O'Malley, D. M., Whetten, R., and Sederoff, R. R. (1993). A laccase associated with lignification in loblolly pine xylem. *Science* **260**, 672–674.

Barber, J. E. (1979). Growth and wood properties of *Pinus radiata* in relation to applied ethylene. *N. Z. J. For. Sci.* **9**, 15–19.

Barbera, G., Fatta del Bosco, G., and LoCascio, B. (1985). Effects of water stress on lemon summer bloom: The "Forzatura" technique in the Sicilian citrus industry. *Acta Hortic.* **171**, 391–397.

Barclay, A. M. M., and Crawford, R. M. M. (1982). Winter desiccation stress and resting bud viability in relation to high altitude survival in *Sorbus aucuparia* L. *Flora* **172**, 21–34.

Barker, J. E. (1979). Growth and wood properties of *Pinus radiata* in relation to applied ethylene. *N. Z. J. For. Sci.* **9**, 15–17.

Barker, J. R., and Tingey, D. T. (1992). The effects of air pollution on biodiversity: A synopsis. *In* "Air Pollution Effects on Biodiversity" (J. R. Barker and D. T. Tingey, eds.), pp. 3–9. Van Nostrand-Reinhold, New York.

Barker, J. R., Henderson, S., Noss, R. F., and Tingey, D. T. (1991). Biodiversity and human impacts. *In* "Encyclopedia of Earth System Science" (W. A. Nierenberg, ed.), Vol. 1, pp. 353–362. Academic Press, San Diego.

Barnard, E. L., and Jorgensen, J. R. (1977). Respiration of field-grown loblolly pine roots as influenced by temperature and root type. *Can. J. Bot.* **55**, 740–743.

Barnes, J. D., Davison, A. W., and Booth, T. A. (1988). Ozone accelerates structural degradation of epicuticular wax on Norway spruce needles. *New Phytol.* **110**, 309–318.

Barnett, J. P., and Vozzo, J. A. (1985). Viability and vigor of slash and shortleaf pine seeds after 50 years of storage. *For. Sci.* **31**, 316–320.

Barritt, B. H., Rom, C. R., Konishi, B. J., and Dilley, M. A. (1991). Light level influences spur quality and canopy development and light interception influence fruit production in apple. *HortScience* **26**, 993–999.

Barros, R. S., and Neill, S. J. (1986). Periodicity of response to abscisic acid in lateral buds of willow (*Salix viminalis* L.). *Planta* **168**, 530–535.

Barros, R. S., and Neill, S. J. (1988). Effect of chilling on the opening and abscisic acid content of dormant lateral buds of willow. *Biol. Plant.* **30**, 264–267.

Bartholomew, D. P., and Criley, R. A. (1983). Tropical fruit and beverage crops. *In* "Plant Growth Regulating Chemicals" (L. G. Nickell, ed.), Vol. 2, pp. 1–34. CRC Press, Boca Raton, Florida.

Bassman, J., Myers, W., Dickmann, D., and Wilson, L. (1982). Effects of simulated insect damage on early growth of nursery-grown hybrid poplars in northern Wisconsin. *Can. J. For. Res.* **12**, 1–9.

Batjer, L. P., and Westwood, M. N. (1963). Effects of pruning, nitrogen, and scoring on growth and bearing characteristics of young Delicious apple trees. *Proc. Am. Soc. Hortic. Sci.* **82**, 5–10.

Battaglia, M. (1993). Seed germination physiology of *Eucalyptus delegatensis* R. T. Baker in Tasmania. *Aust. J. Bot.* **41**, 119–136.

Bauer, T., Blechschmidt-Schneider, S., and Eschrich, W. (1991). Regulation of the photoassimilate allocation in *Pinus sylvestris* seedlings by the nutritional status of the mycorrhizal fungus *Suillus variegatus. Trees* **5**, 36–43.

Bavaresco, L., Fraschini, P., and Perino, A. (1993). Effect of the rootstock on the occurrence of lime-induced chlorosis of potted *Vitis vinifera* L. cv. 'Pinot Blanc.' *Plant Soil* **157**, 305–311.

Baxendale, R. W., and Johnson, W. T. (1988). Evaluation of summer oil spray on amenity plants. *J. Arboric.* **14**, 220–225.

Bayes, C. D., Davis, J. M., and Taylor, C. M. A. (1987). Sewage sludge as a forest fertilizer: Experiences to date. *J. Inst. Water Pollut. Control* **86**, 158–171.

Bazzaz, F. A., Chiariello, N. R., Coley, P. D., and Pitelka, L. F. (1987). Allocating resources to reproduction and defense. *BioScience* **37**, 58–67.

Beckwith, R., Daterman, G., and Strauss, S. (1988). *Bacillus thuringiensis* research in the Pacific Northwest. *Northwest Environ. J.* **4**, 341–342.

Becwar, M. R. (1993). Conifer somatic embryogenesis and clonal forestry. *In* "Clonal Forestry" (M. R. Ahuja and W. J. Libby, eds.), Vol. 1, pp. 200–223. Springer-Verlag, Berlin.

Becwar, M. R., Wann, S. R., Johnson, M. A., Verhagen, V. A., Feirer, R. P., and Nagmani, R. (1988). Development and characterization of *in vitro* embryogenic systems in conifers. *In* "Somatic Cell Genetics of Woody Plants" (M. R. Ahuja, ed.), pp. 1–18. Kluwer, Dordrecht, The Netherlands.

Becwar, M. R., Nagmani, R., and Wann, S. R. (1990). Initiation of embryogenic cultures and somatic embryo development in loblolly pine (*Pinus taeda*). *Can. J. For. Res.* **20**, 810–817.

Beeson, R. C. (1986). The physiology of *Picea* grafts. Ph.D. Dissertation, Oregon State University, Corvallis.

Beeson, R. C., and Proebsting, W. M. (1989). *Picea* graft success: Effects of environment, rootstock disbudding, growth regulators, and antitranspirants. *HortScience* **24**, 253–254.

Beets, P. N., and Pollock, D. S. (1987a). Accumulation and partitioning of dry matter in *Pinus radiata* as related to stand age and thinning. *N. Z. J. For. Sci.* **17**, 246–271.

Beets, P. N., and Pollock, D. S. (1987b). Uptake and accumulation of N in *Pinus radiata* stands as related to age and thinning. *N. Z. J. For. Sci.* **17**, 353–371.

Behboudian, M. H., Törökfalvy, E., and Walker, R. R. (1986). Effect of salinity on ionic content, water relations and gas exchange parameters in some citrus scion–rootstock combinations. *Sci. Hortic.* **28**, 105–116.

Behrens, V. (1987). Kühllagerung von unbewurzelten Koniferenstecklingen. III. Zusammenhang Reservestoffgehate und Bewurzelung. *Gartenbauwissenschaft* **52**, 161–165.

Bell, A. A. (1981). Biochemical mechanisms of disease resistance. *Annu. Rev. Plant Physiol.* **32**, 21–81.

Bell, A. A. (1982). Plant–pest interaction with environmental stress and breeding for pest resistance: Plant diseases. *In* "Breeding Plants for Less Favorable Environments" (M. N. Christiansen and C. F. Lewis, eds.), pp. 335–363. Wiley, New York.

Bella, I. E., and Navratil, S. (1987). Growth losses from winter drying (red belt damage) in lodgepole pine stands on the east slopes of the Rockies in Alberta. *Can. J. For. Res.* **17**, 1289–1292.

Ben-Yehoshua, A., Shapiro, B., and Moran, R. (1987). Individual seed-packaging enables the use of curing at high temperatures to reduce decay and heal injury of citrus fruits. *HortScience* **22**, 777–783.

Bengston, G. W. (1979). Forest fertilization in the United States: Progress and outlook. *J. For.* **78**, 222–229.

Benjamin, D. M. (1955). The biology and ecology of the red-headed pine sawfly. *U.S. Dep. Agric. Tech. Bull.* **1118**.

Bennett, G. W., and Owens, J. M., eds. (1986). "Advances in Urban Pest Management." Van Nostrand-Reinhold, New York.

Bennett, J. H., Lee, E. H., and Heggestad, H. E. (1984). Biochemical aspects of plant tolerance to ozone and oxyradicals: Superoxide dismutase. *In* "Gaseous Air Pollutants and Plant Metabolism" (M. J. Koziol and F. R. Whatley, eds.), pp. 413–424. Butterworth, London.

Benoit, L. F., Skelly, J. M., Moore, L. D., and Dochinger, L. S. (1983). The influence of ozone on *Pinus strobus* L. pollen germination. *Can. J. For. Res.* **13**, 184–187.

Benson, M. L., Meyers, B. J., and Raison, R. J. (1992). Dynamics of stem growth of *Pinus radiata* as affected by water and nitrogen supply. *For. Ecol. Manage.* **52**, 117–137.

Benzian, B., Brown, R. M., and Freeman, S. C. R. (1974). Effect of late-season top-dressings of N (and K) applied to conifer transplants in the nursery on their survival and growth on British forest sites. *Forestry* **47**, 153–184.

Berlyn, G. P. (1961). Recent advances in wood anatomy. The cell wall in secondary xylem. *For. Prod. J.* **14**, 467–476.

Berry, C. R. (1982). Survival and growth of pine hybrid seedlings with *Pisolithus* ectomycorrhizae on coal spoils in Alabama and Tennessee. *J. Environ. Qual.* **11**, 709–715.

Berry F. H. (1969). Decay in the upland oak stands of Kentucky. *U.S.D.A. For. Serv. Res. Pap. NE* **NE-126**.

Berryman, A. A. (1986). "Forest Insects, Principles and Practice of Population Management." Plenum, New York.

Berüter, J., and Droz, P. L. (1991). Studies on locating the signal for fruit abscission in the apple. *Sci. Hortic.* (*Amsterdam*) **46**, 201–214.

Bewley, J. D., and Black, M. (1978). "Physiology and Biochemistry of Seeds, Volume 1: Development, Germination, and Growth." Springer-Verlag, New York and Berlin.

Bewley, J. D., and Black, M. (1982). "Physiology and Biochemistry of Seeds, Volume 2: Viability, Dormancy, and Environmental Control." Springer-Verlag, New York and Berlin.

Bewley, J. D., and Black, M. (1985). "Seeds: Physiology of Development and Germination." Plenum, New York.

Bielorai, H., Shalhevet, J., and Fevy, Y. (1978). Grapefruit response to variable salinity in irrigation water and soil. *Irrig. Sci.* **1**, 61–70.

Bigras, F. J., and D'Aoust, A. L. (1993). Influence of photoperiod on shoot and root frost tolerance and bud phenology of white spruce seedlings (*Picea glauca*). *Can. J. For. Res.* **23**, 219–228.

Billings, W. D. (1978). "Plants and the Ecosystem." Wadsworth, Belmont, California.

Billow, C., Matson, P., and Yoder, B. (1994). Seasonal biochemical changes in coniferous forest canopies and their response to fertilization. *Tree Physiol.* **14**, 563–574.

Binder, W. D., and Fielder, P. (1995). Heat damage in boxed white spruce [*Picea glauca* (Moench.) Voss] seedlings: Its pre-planting detection and effect on field performance. *New Forests* **9**, 237–259.

Bingham, F. T., and Garber, M. J. (1960). Solubility and availability of micronutrients in relation to phosphorus fertilzation. *Soil Sci. Soc. Am. Proc.* **24**, 209–213.

Bingham, R. T., and Squillace, A. E. (1957). Phenology and other features of flowering of pines with special reference to *Pinus monticola* Dougl. *U.S.D.A. For. Serv. Res. Pap. INT* **INT-531**.

Binkley, D., and Reid, P. (1984). Long-term responses of stem growth and leaf area to thinning and fertilization in a Douglas-fir plantation. *Can. J. For. Res.* **14**, 656–660.

Biocca, M., Tainter, F. H., Starkey, D. A., Oak, S. W., and Williams, J. G. (1993). The persistence of oak decline in the western North Carolina Nantahala mountains. *Castanea* **58**, 178–184.

Bird, C. R., and Ray, J. A. (1991). Manipulation of gene expression by antisense RNA. *In* "Biotechnology and Genetic Engineering Reviews" (M. P. Tombs, ed.), Vol. 9, pp. 207–227. Intercept, Southampton, England.

Birk, E. M., and Vitousek, P. M. (1986). Nitrogen availability and nitrogen use efficiency in loblolly pine stands. *Ecology* **67**, 69–79.

Birkholz-Lambrecht, A. F., Lester, D. T., and Smalley, E. B. (1977). Temperature, host genotype, and fungus genotype in early testing for Dutch elm disease resistance. *Plant Dis. Rep.* **51**, 238–242.

Biswell, H. H. (1974). Effects of fire on chaparral. *In* "Fire and Ecosystems" (T. T. Kozlowski and C. E. Ahlgren, eds.), pp. 321–364. Academic Press, New York.

Biswell, H. H. (1989). "Prescribed Burning in California Wildlands Vegetation Management." Univ. of California Press, Berkeley.

Björkman, E. (1970). Forest tree mycorrhiza: The conditions for its formation and the significance for tree growth and afforestation. *Plant Soil* **32**, 589–610.

Björkman, O. (1981). Responses to different quantum flux densities. *Encycl. Plant Physiol. New Ser.* **12A**, 57–107.

Black, M. (1980/1981). The role of endogenous hormones in germination and dormancy. *Isr. J. Bot.* **29**, 181–192.

Black, M., and Wareing, P. F. (1959). The role of germination inhibitors and oxygen in the dormancy of the light-sensitive seed of *Betula* spp. *J. Exp. Bot.* **10**, 134–145.

Black, T. A., Tan, C. S., and Nnyamah, J. U. (1980). Transpiration rate of Douglas-fir trees in thinned and unthinned stands. *Can. J. Soil Sci.* **60**, 625–631.

Blakesley, D., Weston, G., Alderson, P., Hall, J., and Elliott, M. C. (1985). The effect of time of year and the induction of etiolation on rooting and the concentration of IAA and cytokinin in cuttings of *Cotinus coggyria* 'Royal purple.' *News Bulletin, British Plant Growth Regulator Group* **7**, 51–52.

Blechschmidt-Schneider, S. (1990). Phloem transport in *Picea abies* (L.) Karst. in mid-winter. I. Microautoradiographic studies on ^{14}C-assimilate translocation in shoots. *Trees* **4**, 179–186.

Boardman, K. (1977). Comparative photosynthesis of sun and shade plants. *Annu. Rev. Plant Physiol.* **28**, 355–377.

Bogatek, R., and Rychter, A. (1984). Respiratory activity of apple seeds during dormancy removal and germination. *Physiol. Veg.* **22**, 181–191.

Boggie, R. (1972). Effect of water table height on root development of *Pinus contorta* on deep peat in Scotland. *Oikos* **23**, 304–312.

Boggs, J. A., and Wittwer, R. F. (1993). Emergence and establishment of shortleaf pine seeds under various seedbed conditions. *South. J. Appl. For.* **17**, 44–48.

Boland, A.-M., Mitchell, P. D., Jerie, J. H., and Goodwin, I. (1993). The effect of regulated deficit irrigation on tree water use and growth of peach. *J. Hortic. Sci.* **68**, 261–274.

Bond, W. J. (1989). The tortoise and the hare: Ecology of angiosperm dominance and gymnosperm persistence. *Biol. J. Linn. Soc.* **36**, 227–249.

Bondad, N. D. (1976). Response of some tropical and subtropical fruits to pre- and post-harvest applications of ethephon. *Econ. Bot.* **30**, 67–80.

Bonga, J. M. (1987). Clonal propagation of mature trees: Problems and possible solutions. *In* "Cell and Tissue Culture in Forestry" (J. M. Bonga and D. J. Durzan, eds.), Vol. 1, pp. 249–271. Martinus Nijhoff, Dordrecht, The Netherlands.

Bonga, J. M., and von Aderkas, P. (1992). "*In Vitro* Culture of Trees." Kluwer, Dordrecht, The Netherlands.

Bonga, J. M., and von Aderkas, P. (1993). Rejuvenation of tissues from mature conifers and its implications for propagation *in vitro*. *In* "Clonal Forestry" (M. R. Ahuja and W. J. Libby, eds.), Vol. 1, pp. 182–199. Springer-Verlag, Berlin.

Bonga, J. M., and Durzan, D. J., eds. (1987). "Cell and Tissue Culture in Forestry." Vol. 1. Martinus Nijhoff, Dordrecht, The Netherlands.

Bongarten, B. C., and Teskey, R. O. (1987). Dry weight partitioning and its relationship to productivity in loblolly pine seedlings from seven sources. *For. Sci.* **33**, 255–267.

Bonner, F. T. (1986). Measurement of seed vigor for loblolly pine and slash pine. *For. Sci.* **32**, 170–178.

Bonnet-Masimbert, M. (1982). Effect of growth regulators, girdling, and mulching on flowering of young European and Japanese larches under field conditions. *Can. J. For. Res.* **12**, 270–279.

Bonnet-Masimbert, M. (1987a). Floral induction in conifers: A review of available techniques. *For. Ecol. Manage.* **19**, 135–146.

Bonnet-Masimbert, M. (1987b). Preliminary results on gibberellin induction of flowering of seedlings and cuttings of Norway spruce indicate some carry-over effects. *For. Ecol. Manage.* **19**, 163–171.

Bonnet-Masimbert, M., and Zaerr, J. B. (1987). The role of plant growth regulators in promotion of flowering. *Plant Growth Regul.* **6**, 13–35.

Bonomy, P. A., and Dennis, F. G. (1977). Abscisic acid levels in seeds of peach. I. Changes during maturation and storage. *J. Am. Soc. Hortic. Sci.* **102**, 23–26.

Bonser, S. P., and Aarssen, L. W. (1994). Plastic allometry in young sugar maple (*Acer saccharum*): Adaptive responses to light availability. *Am. J. Bot.* **81**, 400–406.

Bonte, J. (1982). Effects of air pollutants on flowering and fruiting. *In* "Effects of Gaseous Air Pollution in Agriculture and Horticulture" (M. H. Unsworth and D. P. Ormrod, eds.), pp. 207–223. Butterworth, London.

Bonte, J., Bonte, C., and Cormis, L. de. (1980). Les composés fluorés atmosphèriques et la fructification. Approche des mecanismes d'action sur le développement des akènes et du réceptacle de la fraise. *C. R. Seances Acad. Agric. Fr.* **66**, 80–89.

Borchert, R. (1983). Phenology and control of flowering in tropical trees. *Biotropica* **15**, 81–89.

Borchert, R. (1991). Growth periodicity and dormancy. *In* "Physiology of Trees" (A. S. Rhagavendra, ed.), pp. 221–245. Wiley, New York.

Borchert, M. I., Davies, F. W., and Michaelsen, J. (1989). Interactions of factors affecting seedling recruitment of blue oak (*Quercus douglasii*) in California. *Ecology* **70**, 389–404.

Borchert, R. (1994). Water status and development of tropical trees during seasonal drought. *Trees* **8**, 115–125.

Borger, G. A., and Kozlowski, T. T. (1972a). Effects of water deficits on first periderm and xylem development in *Fraxinus pennsylvanica. Can. J. For. Res.* **2**, 144–151.

Borger, G. A., and Kozlowski, T. T. (1972b). Effect of light intensity on early periderm and xylem development in *Pinus resinosa, Fraxinus pennsylvanica,* and *Robinia pseudoacacia. Can. J. For. Res.* **2**, 190–197.

Borger, G. A., and Kozlowski, T. T. (1972c). Effects of temperature on first periderm and xylem development in *Fraxinus pennsylvanica, Robinia pseudoacacia,* and *Ailanthus altissima. Can. J. For. Res.* **2**, 198–205.

Borger, G. A., and Kozlowski, T. T. (1972d). Effects of photoperiod on early periderm and xylem development in *Fraxinus pennsylvanica, Robinia pseudoacacia* and *Ailanthus altissima* seedlings. *New Phytol.* **71**, 703–708.

Borges, R. G., and Chaney, W. R. (1988). The response of *Acacia scleroxyla* Tuss. to mycorrhizal inoculation. *Int. Tree Crops J.* **5**, 191–201.

Borges, R. G., and Chaney, W. R. (1989). Root temperature affects mycorrhizal efficacy in *Fraxinus pennsylvanica* Marsh. *New Phytol.* **112**, 411–417.

Bormann, F. H. (1990). Air pollution and temperate forests: Creeping degradation. *In* "The Earth in Transition: Patterns and Processes of Biotic Impoverishment" (G. M. Woodwell, ed.), pp. 25–44. Cambridge Univ. Press, Cambridge.

Bormann, F. H., and Kozlowski, T. T. (1962). Measurement of tree ring growth with dial-gauge dendrometers and vernier tree ring bands. *Ecology* **43**, 289–294.

Bormann, F. H., Likens, G. E., and Mellillo, J. M. (1977). Nitrogen budget for an aggrading northern hardwood forest ecosystem. *Science* **196**, 981–982.

Borochev, A., and Woodson, W. R. (1989). Physiology and biochemistry of flower petal senescence. *Hortic. Rev.* **11**, 15–43.

Botkin, D. B. (1993). "Forest Dynamics: An Ecological Model." Oxford Univ. Press, Oxford and New York.

Boulter, D. (1995). Plant biotechnology: Facts and public perception. *Phytochemistry* **40**, 1–9.

Bound, S. A., Jones, K. M., Koen, T. B., and Oakford, M. J. (1991). The thinning effect of benzyladenine on red 'Fuji' apple trees. *J. Hortic. Sci.* **66**, 789–794.

Bouyon, C., and Bulard, C. (1986). Heterogeneity of dormancy in apple embryos: A link with chlorophyll formation and content of abscisic acid. *J. Exp. Bot.* **37**, 1643–1651.

Bowen, G. D., and Theodorou, C. (1973). Growth of ectomycorrhizal fungi around seeds and roots. *In* "Ectomycorrhizae" (G. C. Marks and T. T. Kozlowski, eds.), pp. 107–150. Academic Press, New York.

Boyer, J. S. (1976a). Photosynthesis at low water potentials. *Philos. Trans. R. Soc. London* **B273**, 501–522.

Boyer, J. S. (1976b). Water deficits and photosynthesis. *In* "Water Deficits and Plant Growth" (T. T. Kozlowski, ed.), Vol. 4, pp. 154–190. Academic Press, New York.

Boyer, J. S. (1985). Water transport. *Annu. Rev. Plant Physiol.* **36**, 473–516.

Brady, C. J., and Spiers, J. (1991). Ethylene in fruit ontogeny and abscission. *In* "The Plant Hormone Ethylene" (A. K. Mattoo and J. C. Suttle, eds.), pp. 235–258. CRC Press, Boca Raton, Florida.

Braekke, F. H. (1987). Nutrient relationships in forest stands—Effects of drainage and fertilization on surface peat layers. *For. Ecol. Manage.* **21**, 269–284.

Braekke, F. H. (1990). Nutrient accumulation and role of atmospheric deposition in coniferous stands. *For. Ecol. Manage.* **30**, 351–359.

Braekke, F. H. (1992). Root biomass changes after drainage and fertilization of a low-shrub pine bog. *Plant Soil* **143**, 33–43.

Braekke, F. H., and Kozlowski, T. T. (1975). Shrinking and swelling of stems of *Pinus resinosa* and *Betula papyrifera* in northern Wisconsin. *Plant Soil* **43**, 387–410.

Bramlage, W. J., and Thompson, A. H. (1962). The effects of early-season sprays of boron on fruit set, color, finish, and storage life of apples. *Proc. Am. Soc. Hortic. Sci.* **80**, 64–72.

Bramlage, W. J., Drake, M., and Lord, W. J. (1980). The influence of mineral nutrition on the quality and storage performance of pome fruits grown in North America. *In* "Mineral Nutrition of Fruit Trees" (D. Atkinson, J. E. Jackson, R. O. Sharples, and W. M. Waller, eds.), pp. 29–39. Butterworth, London.

Braun, G. (1977a). Über die Ursachen und Kriterien der Immissionsresistenz bei Fichte, *Picea abies* (L.) Karst. I. Morphologischanatomische Immissionsresistenz. *Eur. J. For. Pathol.* **7**, 23–43.

Braun, G. (1977b). Über die Ursachen und Kriterien der Immissionsresistenz bei Fichte, *Picea abies* (L.) Karst. II. Reflectorische Immissionsresistenz. *Eur. J. For. Pathol.* **7**, 129–152.

Brewbaker, J. L. (1987). Significant nitrogen fixing trees in agroforestry systems. *In* "Agroforestry: Realities Possibilities and Potentials" (H. L. Gholz, ed.), pp. 31–45. Martinus Nijhoff, Dordrecht, The Netherlands.

Brian, P. W., Petty, J. H. P., and Richmond, P. T. (1959). Effects of gibberellic acid on development of autumn color and leaf-fall of deciduous woody plants. *Nature* (*London*) **183**, 58–59.

Bridgham, S. D., Johnston, C. A., Pastor, J., and Updegraff, K. (1995). Potential feedbacks of northern wetlands on climate change. *BioScience* **45**, 262–274.

Brix, H. (1981). Effects of thinning and nitrogen fertilization on branch and foliage production of Douglas-fir. *Can. J. For. Res.* **11**, 502–511.

Brix, H. (1983). Effects of thinning and nitrogen fertilization on growth of Douglas-fir: Relative contribution of foliage quality and efficiency. *Can. J. For. Res.* **13**, 167–175.

Brix, H., and Ebell, L. F. (1969). Effects of nitrogen fertilization on growth, leaf area, and photosynthesis rate in Douglas-fir. *For. Sci.* **15**, 189–196.

Brix, H., and Mitchell, A. K. (1986). Thinning and nitrogen fertilization effects on soil and tree water stress in a Douglas-fir stand. *Can. J. For. Res.* **16**, 1334–1338.

Brockley, R. P. (1992). Effects of fertilization on the nutrition and growth of a slow-growing Engelmann spruce plantation in south central British Columbia. *Can. J. For. Res.* **22**, 1617–1622.

Brodl, M. R. (1990). Biochemistry of heat shock responses in plants. *In* "Environmental Injury to Plants" (P. Katterman, ed.), pp. 113–135. Academic Press, San Diego.

Brooks, C., Miller, E. V., Bratley, C. O., Cooley, J. S., Mook, P. V., and Johnson, H. B. (1932). Effect of solid and gaseous carbon dioxide upon transit diseases of certain fruits and vegetables. *U.S. Dep. Agric. Tech. Bull.* **318**.

Brough, D. H., Jones, H. G., and Grace, J. (1986). Diurnal changes in water content of the stems of apple trees as influenced by irrigation. *Plant Cell Environ.* **9**, 1–7.

Brown, D. A., Windham, M. T., Anderson, R. L., and Trigiano, R. N. (1994). Influence of simulated acid rain on the flowering dogwood (*Cornus florida*) leaf surface. *Can. J. For. Res.* **24**, 1058–1062.

Brown, D. S. (1952). Climate in relation to deciduous fruit production in California. V. The use of temperature records to predict the time of harvest of apricots. *Proc. Am. Soc. Hortic. Sci.* **60**, 197–203.

Brown, D. S. (1953). The effects of irrigation on flower bud development and fruiting in the apricot. *Proc. Am. Soc. Hortic. Sci.* **61**, 119–124.

Brown, G. N., and Bixby, J. A. (1975). Soluble and insoluble protein patterns during induction of freezing tolerance in black locust seedlings. *Physiol. Plant.* **34**, 187–191.

Brown, J. C., and Jolley, V. D. (1989). Plant metabolic responses to iron deficiency stress. *BioScience* **39**, 546–551.

Brown, J. H., Paliyath, G., and Thompson, J. E. (1991). Physiological mechanisms of plant senescence. *In* "Plant Physiology" (F. C. Steward, ed.), Vol. 10, pp. 227–275. Academic Press, San Diego.

Brown, K. M., and Leopold, A. C. (1973). Ethylene and the regulation of growth in pine. *Can. J. For. Res.* **3**, 143–145.

Brown, P. H., Zhang, Q., and Ferguson, L. (1994). Influence of rootstock on nutrient acquisition by pistachio. *J. Plant Nutr.* **17**, 1137–1148.

Brown, R. T. (1967). Influence of naturally occurring compounds on germination and growth of jack pine. *Ecology* **48**, 542–546.

Browne, L. T., Leavitt, G., and Gerdts, M. (1978). Delayed almond bloom with ethephon. *Calif. Agric.* **32**(3), 6–7.

Browning, G., Kuden, A., and Blake, P. (1992). Site of (2*RS*,3*RS*)-paclobutrazol promotion of axillary flower initiation in pear cv. Doyenne du Comice. *J. Hortic. Sci.* **67**, 121–128.

Brubaker, L. B. (1986). Responses of tree populations to climatic change. *Vegetatio* **67**, 119–130.

Bryson, R. A., and Murray, J. T. (1977). "Climates of Hunger." Univ. of Wisconsin Press, Madison.

Buban, T., and Turi, I. (1985). Delayed bloom in apricot and peach. *Acta Hortic.* **192**, 57–63.

Buckley, R. C. (1982). Seed size and seedling establishment in tropical arid dunecrest plants. *Biotropica* **14**, 314–315.

Budelman, A. (1990). Woody legumes as live support systems in yam cultivation. II. The yam–*Gliricidia-sepium* association. *Agroforestry Systems* **10**(No. 1), 61–69.

Buell, M. F., Buell, H. F., Small, J. A., and Monk, C. D. (1961). Drought effect on radial growth of trees in the William L. Hutcheson Memorial Forest. *Bull. Torrey Bot. Club* **88**, 176–180.

Bukovac, M. J. (1985). Plant growth regulators in deciduous tree fruit production. *In* "Agricultural Chemicals of the Future" (J. L. Hilton, ed.), pp. 75–90. Eighth Beltsville Symp., Agric. Res. Rowman and Allenheld, Totowa, New Jersey.

Bukovac, M. J., and Wittwer, S. H. (1957). Absorption and mobility of foliar applied nutrients. *Plant Physiol.* **32**, 428–435.

Bukovac, M. J., and Wittwer, S. H. (1959). Absorption and distribution of foliar applied nutrients as determined with radioisotopes. *Proc. 3rd Colloq. Plant Analysis Fertilizer Problems (Montreal)*, 215–230.

Burdett, A. N., and Martin, P. A. F. (1982). Chemical root pruning of coniferous seedlings. *HortScience* **17**, 622–624.

Burdon, J. N., and Sexton, R. (1989). The role of petal abscission in red raspberry. *In* "Cell Separation in Plants: Physiology, Biochemistry and Molecular Biology" (D. J. Osborne and M. B. Jackson, eds.), pp. 371–375. Springer-Verlag, Berlin, Heidelberg, and New York.

Burdon, J. N., and Sexton, R. (1990). The role of ethylene in the shedding of red raspberry fruit. *Ann. Bot.* **66**, 111–120.

Burgess, T. I., Malajczuk, N., and Grove, T. S. (1993). The ability of 16 ectomycorrhizal fungi to increase growth and phosphorus uptake of *Eucalyptus globulus* Labill. and *E. diversicolor* F. Muell. *Plant Soil* **153**, 155–164.

Burke, J. J., and Orzech, K. A. (1988). The heat-shock response in higher plants: A biochemical model. *Plant Cell Environ.* **11**, 441–444.

Burley, J. (1976). Genetic systems and genetic conservation of tropical pines. *In* "Tropical Trees: Variation, Breeding and Conservation" (J. Burley and B. T. Styles, eds.), pp. 85–100. Academic Press, London.

Burley, J., and Styles, B. T., eds. (1976). "Tropical Trees: Variation, Breeding and Conservation." Academic Press, London.

Burn, A. J., Coaker, T. H., and Jepson, P. C., eds. (1987). "Integrated Pest Management." Academic Press, London.

Burton, J. D., and Smith, D. M. (1972). Guying to prevent windsway influences loblolly pine growth and wood properties. *U.S. For. Serv. Res. Pap. SO* **SO-80**.

Burton, P. J., and Bazzaz, F. A. (1991). Tree seedling emergence on interactive temperature and moisture gradients and in patches of old-field vegetation. *Am. J. Bot.* **78**, 131–149.

Büscher, N., Zyprian, E., Bachmann, O., and Blaich, R. (1994). On the origin of the grapevine variety Muller–Thurgau as investigated by the inheritance of random amplified polymorphic DNA (RAPD). *Vitis* **33**, 15–17.

Byram, G. M., and Doolittle, W. T. (1950). A year of growth for a shortleaf pine. *Ecology* **31**, 27–35.

Byres, D. P., Dean, T. J., and Johnson, J. D. (1992). Long term effects of ozone and simulated acid rain on the foliage dynamics of *Pinus elliottii* var. *elliottii* Engelm. *New Phytol.* **120**, 61–67.

Caborn, J. M. (1965). "Shelterbelts and Windbreaks." Faber and Faber, London.

Cain, M. D. (1991). The influence of woody and herbaceous competition on early growth of naturally regenerated loblolly and shortleaf pines. *South. J. Appl. For.* **15**, 179–184.

Caldwell, J. M., Sucoff, E. I., and Dixon, R. K. (1995). Grass interference limits resource availability and reduces growth of juvenile red pine in the field. *New Forests* **10**, 1–15.

Callaham, R. Z. (1964). *Pinus ponderosa*: Geographic variation in germination response to temperature. *Proc. Int. Bot. Congr., 9th, 1959* **2**, 57–58.

Camm, E. L., Goetze, D. C., Silim, S. N., and Lavender, D. P. (1994). Cold storage of conifer seedlings: An update from the British Columbia perspective. *For. Chron.* **70**, 311–316.

Campbell, J. A. (1980). Oxygen flux measurements in organic soils. *Can. J. Soil Sci.* **60**, 641–650.

Campbell, R. G., and Hughes, J. H. (1981). Forest management systems in North Carolina pocosins: Weyerhaeuser. *In* "Pocosin Wetlands" (C. Richardson, ed.), pp. 199–213. Hutchinson Ross, Stroudsburg, Pennsylvania.

Campbell, R. J., and Marini, R. P. (1992). Light environment and time of harvest affect 'Delicious' apple fruit quality characteristics. *J. Am. Soc. Hortic. Sci.* **117**, 551–557.

Candolfi-Vasconcelos, M. C., Candolfi, M. P., and Koblet, W. (1994). Retranslocation of carbon reserves from the woody storage tissues into the fruit as a response to defoliation stress during the ripening period in *Vitis vinifera* L. *Planta* **192**, 567–573.

Cannell, M. G. R. (1971a). Production and distribution of dry matter in trees of *Coffea arabica* L. in Kenya as affected by seasonal climatic differences and presence of fruits. *Ann. Appl. Biol.* **67**, 99–120.

Cannell, M. G. R. (1971b). Changes in the respiration and growth rates of developing fruits of *Coffea arabica* L. *J. Hortic. Sci.* **46**, 263–272.

Cannell, M. G. R. (1972). Photoperiodic response of mature trees of Arabica coffee. *Turrialba* **22**, 198–206.

Cannell, M. G. R. (1986). Physiology of southern pine seedlings. *In* "Nursery Management Practices for the Southern Pines" (D. B. South, ed.), pp. 251–274. Auburn University, Auburn, Alabama.

Cannell, M. G. R. (1989). Physiological basis of wood production: A review. *Scand. J. For. Res.* **4**, 459–490.

Cannell, M. G. R., and Jackson, J. E., eds. (1985). "Attributes of Trees as Crop Plants." Institute of Terrestrial Ecology, Huntingdon, England.

Cannell, M. G. R., and Sheppard, L. J. (1982). Seasonal changes in the frost hardiness of provenances of *Picea sitchensis* in Scotland. *Forestry* **55**, 137–153.

Cannell, M. G. R., Tabbush, P. M., Deans, J. D., Hollingsworth, M. K., Sheppard, L. J., Philipson, J. J., and Murray, M. B. (1990). Sitka spruce and Douglas-fir seedlings in the nursery and in cold storage: Root growth potential, carbohydrate content, dormancy, frost hardiness, and mitotic index. *Forestry* **63**, 9–27.

Cardemil, L., and Varner, J. E. (1984). Starch degradation metabolism towards sucrose synthesis in germinating *Araucaria araucana* seeds. *Plant Physiol.* **76**, 1047–1054.

Carlson, C. (1980). Kraft mill gases damage Douglas-fir in western Montana. *Eur. J. For. Pathol.* **10**, 145–151.

Carlson, R. W., and Bazzaz, F. A. (1977). Growth reduction in American sycamore (*Platanus occidentalis* L.) caused by Pb–Cd interaction. *Environ. Pollut.* **12**, 243–253.

Carlson, W. C. (1985). Effects of natural chilling and cold storage on bud break and root growth potential of loblolly pine (*Pinus taeda* L.). *Can. J. For. Res.* **15**, 651–656.

Carlson, W. C., and Preissig, C. L. (1980). Effects of controlled-release fertilizers on the shoot and root development of Douglas-fir seedlings. *Can. J. For. Res.* **11**, 230–242.

Caspari, H. W., Behboudian, M. H., and Chalmers, D. J. (1994). Water use, growth, and fruit yield of 'Hosui' Asian pears under deficit irrigation. *J. Am. Soc. Hortic. Sci.* **119**, 383–388.

Casperson, G. (1968). Wirkung von Wuchs und Hemmstoffen auf die Kambium-tätigkeit und Reaktionsholzbildung. *Physiol. Plant.* **21**, 1312–1321.

Castello, J. D., Silverborg, S. B., and Manion, P. D. (1985). Intensification of ash decline in New York State from 1962 through 1980. *Plant Dis.* **69**, 243–246.

Castello, J. D., Leopold, D. J., and Smallidge, P. J. (1995). Pathogens, patterns, and processes in forest ecosystems. *BioScience* **45**, 16–24.

Castle, W. S. (1987). Citrus rootstocks. *In* "Rootstocks for Fruit Crops" (R. C. Rom and R. F. Carlson, eds.), pp. 361–399. Wiley, New York and Chichester.

Cathey, H. M., and Campbell, L. E. (1975). Effectiveness of five fission-lighting sources on photoregulation of 22 species of ornamental plants. *J. Am. Soc. Hortic. Sci.* **100**, 65–71.

Cecich, R. A. (1983). Flowering in a jack pine seedling seed orchard increased by spraying with gibberellin $A_{4/7}$. *Can. J. For. Res.* **13**, 1056–1062.

Ceulemans, R., and Mousseau, M. (1994). Effects of elevated atmospheric CO_2 on woody plants. *New Phytol.* **127**, 425–446.

Ceulemans, R., Impens, I., and Steenackers, V. (1984). Stomatal and anatomical leaf characteristics of 10 *Populus* clones. *Can. J. Bot.* **62**, 513–518.

Ceulemans, R., Impens, I., and Steenackers, V. (1987). Variations in photosynthetic, anatomical, and enzymatic leaf traits and correlations with growth in recently selected *Populus* hybrids. *Can. J. For. Res.* **17**, 273–283.

Ceulemans, R., Scarascia-Mugnozza, G., Wiard, B. M., Braatne, J. H., Hinckley, T. M., Stettler, R. F., Isebrands, J. G., and Heilman, P. E. (1992). Production physiology and morphology of *Populus* species and their hybrids grown under short rotation. I. Clonal comparisons of 4-year growth and phenology. *Can. J. For. Res.* **22**, 1937–1948.

Chailakhyan, M. K. (1975). Forty years of research on the hormonal basis of flowering—Some personal reflections. *Bot. Rev.* **41**, 1–29.

Chalmers, D. J., Mitchell, P. D., and van Heek, L. (1981). Control of peach tree growth and productivity by regulated water supply, tree density, and summer pruning. *J. Am. Soc. Hortic. Sci.* **106**, 307–312.

Chalmers, D. J., Olsson, K. A., and Jones, T. R. (1983). Water relations of peach trees and orchards. *In* "Water Deficits and Plant Growth" (T. T. Kozlowski, ed.), Vol. 7, pp. 197–232. Academic Press, New York.

Chalmers, D. J., Mitchell, P. D., and Jerie, P. H. (1984). The physiology of growth of peach and pear trees using reduced irrigation. *Acta Hortic.* **146**, 143–149.

Chalutz, E., Waks, J., and Schiffmann-Nadel, M. (1985). Reducing susceptibility of grapefruit to chilling injury during cold treatment. *HortScience* **20**, 226–228.

Chaney, W. R., and Kozlowski, T. T. (1969a). Seasonal and diurnal changes in water balance of fruits, cones, and leaves of forest trees. *Can. J. Bot.* **47**, 1407–1417.

Chaney, W. R., and Kozlowski, T. T. (1969b). Seasonal and diurnal expansion and contraction of *Pinus banksiana* and *Picea glauca* cones. *New Phytol.* **68**, 873–882.

Chaney, W. R., and Kozlowski, T. T. (1971). Water transport in relation to expansion and contraction of leaves and fruits of calamondin orange. *J. Hortic. Sci.* **46**, 71–81.

Chaney, W. R., and Strickland, R. C. (1974). Effect of cadmium and sulfur dioxide on pollen germination. *Proc. Third North Am. For. Biol. Workshop*, 372–373.

Chaney, W. R., and Strickland, R. C. (1984). Relative toxicity of heavy metals to red pine pollen germination and germ tube elongation. *J. Environ. Qual.* **13**, 391–393.

Chapin, F. S. (1991). Integrated responses of plants to stress. *BioScience* **41**, 29–36.

Chaplin, G. R., Wills, R. B. H., and Graham, D. (1983). Induction of chilling injury in stored avocados with exogenous ethylene. *HortScience* **18**, 952–953.

Chaplin, M. H., Stebbins, R. L., and Westwood, M. N. (1977). Effect of fall applied boron sprays on fruit set and yield of 'Halian' prune (*Prunus domestica* L.). *HortScience* **12**, 500–501.

Chapman, K. R. (1973). Effect of four water regimes in performance of glasshouse-grown nursery apple trees. *Queensl. J. Agric. Anim. Sci.* **30**, 125–135.

Chazen, O., and Neumann, P. M. (1994). Hydraulic signals from the roots and rapid cell-wall hardening in growing maize (*Zea mays* L.) leaves are primary responses to polyethylene glycol-induced water deficits. *Plant Physiol.* **104**, 1385–1392.

Cheliak, W. M., and Rogers, D. L. (1990). Integrating biotechnology into tree improvement programs. *Can. J. For. Res.* **20**, 452–463.

Chen, N. M., and Paull, R. E. (1986). Development and prevention of chilling injury in papaya fruit. *J. Am. Soc. Hortic. Sci.* **111**, 639–643.

Chen, P., Li, P. H., and Weiser, C. J. (1975). Induction of frost hardiness in red-osier dogwood stems by water stress. *HortScience* **10**, 372–374.

Chen, P. M., Li, P. H., and Burke, M. J. (1977). Induction of frost hardiness in stem cortical tissues of *Cornus stolonifera* by water stress. I. Unfrozen water in cortical tissues and water status in plants and soil. *Plant Physiol.* **59**, 236–239.

Chen, T. T. H., and Kartha, K. K. (1987). Cryopreservation of woody species. *In* "Cell and Tissue Culture in Forestry" (J. M. Bonga and D. J. Durzan, eds.), Vol. 2, pp. 305–319. Martinus Nijhoff, Dordrecht, The Netherlands.

Chiba, O., and Tanaka, T. (1968). The effect of sulphur dioxide on the development of pine needle blight caused by *Rhizosphaera kalkhoffii* Bubak (I.). *J. Jpn. For. Soc.* **50**, 135–139.

Childers, N. F. (1961). "Modern Fruit Science." Horticultural Publ. Rutgers Univ., New Brunswick, New Jersey.

Childers, N. F. (1983). "Modern Fruit Science." Horticultural Publications, Gainesville, Florida.

Ching, T. M. (1966). Compositional changes of Douglas-fir seeds during germination. *Plant Physiol.* **41**, 1313–1319.

Ching, T. M. (1972). Metabolism of germinating seeds. *In* "Seed Biology" (T. T. Kozlowski, ed.), Vol. 2, pp. 103–218. Academic Press, New York.

Ching, T. M. (1982). Adenosine diphosphate and seed vigor. *In* "The Physiology and Biochemistry of Seed Development, Dormancy and Germination" (A. A. Khan, ed.), pp. 487–506. Elsevier, Amsterdam.

Ching, T. M., and Ching, K. K. (1972). Content of adenosine phosphates and adenylate energy charge in germinating ponderosa pine seeds. *Plant Physiol.* **50**, 536–540.

Choinski, J. S., Jr., and Tuohy, J. M. (1991). Effect of water potential and temperature on the germination of four species of African savanna trees. *Ann. Bot.* **68**, 227–233.

Chou, C. K. S., and MacKenzie, M. (1988). Effect of pruning intensity and season on *Diplodia pinea* infection of *Pinus radiata* stem through pruning wounds. *Eur. J. For. Pathol.* **18**, 437–444.

Christensen, N. L. (1985). Shrubland fire regimes and their evolutionary consequences. *In* "The Ecology of Natural Disturbance and Patch Dynamics" (S. T. A. Pickett and P. S. White, eds.), pp. 85–100. Academic Press, Orlando, Florida.

Christersson, L., and von Fircks, H. A. (1988). Injuries to conifer seedlings caused by simulated summer frost and winter desiccation. *Silva Fenn.* **22**, 195–201.

Christiansen, E., Waring, R. H., and Berryman, A. A. (1987). Resistance of conifers to bark beetle attacks: Searching for general relationships. *For. Ecol. Manage.* **22**, 89–106.

Christiansen, M. N. (1964). Influence of chilling upon seedling development of cotton. *Plant Physiol.* **38**, 520–522.

Chung, H. H., and Kramer, P. J. (1975). Absorption of water and ^{32}P through suberized and unsuberized roots of loblolly pine. *Can. J. For. Res.* **5**, 229–235.

Clark, D. A., and Clark, D. B. (1992). Life history diversity of canopy and emergent trees in a neotropical rain forest. *Ecol. Monogr.* **62**, 315–344.

Clausen, J. J., and Kozlowski, T. T. (1965). Seasonal changes in moisture contents of gymnosperm cones. *Nature (London)* **206**, 112–113.

Clausen, J. J., and Kozlowski, T. T. (1967a). Food sources for growth of *Pinus resinosa* shoots. *Adv. Front. Plant Sci.* **18**, 23–32.

Clausen, J. J., and Kozlowski, T. T. (1967b). Seasonal growth characteristics of long and short shoots of tamarack. *Can. J. Bot.* **45**, 1643–1651.

Clausen, J. J., and Kozlowski, T. T. (1970). Observations on growth of long shoots of *Larix laricina. Can. J. Bot.* **48**, 1045–1048.

Cleary, B. D., and Waring, R. H. (1969). Temperature: Collection of data and its analysis for the interpretation of plant growth and distribution. *Can. J. Bot.* **47**, 167–173.

Clemens, J., and Jones, P. G. (1978). Modification of drought resistance by water stress conditioning in *Acacia* and *Eucalyptus. J. Exp. Bot.* **29**, 895–904.

Clemens, J., Campbell, L. C., and Nurisjah, S. (1983). Germination, growth and mineral ion concentration of *Casuarina* species under saline conditions. *Aust. J. Bot.* **31**, 1–9.

Clements, J. R. (1970). Shoot responses of young red pine to watering applied over two seasons. *Can. J. Bot.* **48**, 75–80.

Cobb, F. W., Jr., Wood, D. L., Stark, R. W., and Parmeter, J. R., Jr. (1968). Theory on the relationship between oxidant injury and bark beetle infestation. *Hilgardia* **39**, 141–152.

Cohen, A. (1984). Effect of girdling date on fruit size of Marsh seedless grapefruit. *J. Hortic. Sci.* **59**, 567–573.

Cohen, E., Shapiro, B., Shalom, Y., and Klein, J. D. (1994). Water loss: A non-destructive indicator of enhanced cell membrane permeability of chilling-injured *Citrus* fruit. *J. Am. Soc. Hortic. Sci.* **119**, 983–986.

Cohen, J. D., and Bandurski, R. S. (1982). Chemistry and physiology of the bound auxins. *Annu. Rev. Plant Physiol.* **33**, 403–430.

Cohn, J. P. (1989). Gauging the biological impacts of the greenhouse effect. *BioScience* **39**, 142–146.

Cole, E. C., and Newton, M. (1986). Nutrient, moisture, and light relations in 5-year-old Douglas-fir plantations under variable competition. *Can. J. For. Res.* **16**, 727–732.

Coley, P. D. (1983). Herbivory and defensive characteristics of tree species in a lowland tropical forest. *Ecol. Monogr.* **53**, 209–233.

Coley, P. D., Bryant, J. P., and Chapin, F. S. (1985). Resource availability and plant antiherbivore defense. *Science* **230**, 895–899.

Collada, C., Caballero, R. G., Casado, R., and Aragoncillo, C. (1988). Different types of major storage proteins in Fagaceae species. *J. Exp. Bot.* **39**, 1751–1758.

Collada, C., Allona, I., Aragoncillo, P., and Aragoncillo, C. (1993). Development of protein bodies in cotyledons of *Fagus sylvatica*. *Physiol. Plant.* **89**, 354–359.

Collet, G. F., and Le, C. L. (1987). Role of auxin during *in vitro* rhizogenesis of rose and apple trees. *Acta Hortic.* **212**, 273–280.

Colletti, J. P., and Schultz, R. C., eds. (1995). Opportunities for agroforestry in the temperate zone. *Agroforestry Systems* **29**, 181–340.

Colombo, S. J. (1990). Bud dormancy status, frost hardiness, shoot moisture content and readiness of black spruce container seedlings for frozen storage. *J. Am. Soc. Hortic. Sci.* **115**, 302–307.

Colombo, S. J., and Timmer, V. R. (1992). Limits of tolerance to high temperatures causing direct and indirect damage to black spruce. *Tree Physiol.* **11**, 95–104.

Colombo, S. J., Glerum, C., and Webb, D. P. (1989). Winter hardening in first-year black spruce (*Picea mariana*) seedlings. *Physiol. Plant.* **76**, 1–9.

Colombo, S. J., Colclough, M. L., Timmer, V. R., and Blumwald, E. (1992). Clonal variation in heat tolerance and heat shock protein expression in black spruce. *Silvae Genet.* 41, 234–238.

Conard, S. G., and Radosevich, S. R. (1982). Growth responses of white fir to decreased shading and root competition by montane chaparral shrubs. *For. Sci.* **28**, 309–320.

Conroy, J. P., Milham, P. J., Mazur, M., and Barlow, F. W. R. (1990). Growth, dry weight partitioning and wood properties of *Pinus radiata* D. Don after 2 years of CO_2 enrichment. *Plant Cell Environ.* **13**, 329–337.

Considine, J. A. (1983). Concepts and practice of use of plant growth regulating chemicals in viticulture. *In* "Plant Growth Regulating Chemicals" (L. G. Nickell, ed.), Vol. 1, pp. 89–183. CRC Press, Boca Raton, Florida.

Constantinidou, H. A. (1992). Evaluation of biological ice nucleation and of nucleation control methods in citrus tissues. *Agricoltura Mediterranea* **122**, 157–163.

Constantinidou, H. A., and Kozlowski, T. T. (1979a). Effects of sulfur dioxide and ozone on *Ulmus americana* seedlings. I. Visible injury and growth. *Can. J. Bot.* **57**, 170–175.

Constantinidou, H. A., and Kozlowski, T. T. (1979b). Effects of sulfur dioxide and ozone on *Ulmus americana* seedlings. II. Carbohydrates, proteins, and lipids. *Can. J. Bot.* **57**, 176–184.

Constantinidou, H. A., and Menkissoglu, O. (1992). Characteristics and importance of heterogenous ice nuclei associated with citrus fruits. *J. Exp. Bot.* **43**, 585–591.

Constantinidou, H. A., Kozlowski, T. T., and Jensen, K. (1976). Effects of sulfur dioxide on *Pinus resinosa* seedlings in the cotyledon stage. *J. Environ. Qual.* **5**, 141–144.

Constantinidou, H. A., Menkissoglu, O., and Stergiadou, H. C. (1991). The role of ice nucleation active bacteria in supercooling of citrus tissues. *Physiol. Plant.* **81**, 548–554.

Coombe, B. G. (1960). Relationship of growth and development to changes in sugars, auxins, and gibberellins in fruit of seeded and seedless varieties of *Vitis vinifera*. *Plant Physiol.* **35**, 241–250.

Coombe, B. G. (1976). The development of fleshy fruits. *Annu. Rev. Plant Physiol.* **27**, 507–528.

Coombe, B. G., and Hales, C. R. (1973). Effects of ethylene and 2-chloroethylphosphonic acid on the ripening of grapes. *Plant Physiol.* **45**, 620–623.

Cortes, P. M., and Sinclair, T. R. (1985). The role of osmotic potential in spring sap flow of mature sugar maple trees (*Acer saccharum* Marsh). *J. Exp. Bot.* **36**, 12–24.

Cosgrove, D. J. (1993a). Water uptake by growing cells: An assessment of the controlling roles of wall relaxation, solute uptake, and hydraulic conductance. *Int. J. Plant Sci.* **154**, 10–21.

Cosgrove, D. J. (1993b). How do plant cell walls extend? *Plant Physiol* **102**, 1–6.

Costa, G., and Bagni, N. (1983). Effects of polyamines on fruit set of apples. *HortScience* **18**, 59–61.

Côté, B., and Dawson, J. O. (1986). Autumnal changes in total nitrogen, salt-extractable proteins and amino acids in leaves and adjacent bark of black alder, eastern cottonwood, and white basswood. *Physiol. Plant.* **67**, 102–108.

Côté, W. A., and Day, A. C. (1965). Anatomy and ultrastructure of reaction wood. *In* "Cellular Ultrastructure of Woody Plants" (W. A. Côté, ed.), pp. 391–418. Syracuse Univ. Press, Syracuse, New York.

Cottignies, A. (1986). The hydrolysis of starch as related to the interruption of dormancy in the ash bud. *J. Plant Physiol.* **123**, 381–388.

Coulson, R. N., and Witter, J. A. (1984). "Forest Entomology: Ecology and Management." Wiley (Interscience), New York and Chichester.

Courtney-Gutterson, N., Napoli, C., Lemiuex, C., Morgan, A., Firoozabady, E., and Robinson, K. E. P. (1994). Modification of flower color in florist's chrysanthemum: Production of a white-flowering variety through molecular genetics. *Bio/Technology* **12**, 268–271.

Coutts, M. P. (1982). The tolerance of tree roots to waterlogging. V. Growth of woody roots of Sitka spruce and lodgepole pine in waterlogged soil. *New Phytol.* **90**, 467–476.

Coutts, M. P., and Grace, J., eds. (1995). "Wind and Trees." Cambridge Univ. Press, Cambridge.

Coutts, M. P., and Philipson, J. J. (1976). The influence of mineral nutrition on the root development of trees. 1. The growth of Sitka spruce with divided root systems. *J. Exp. Bot.* **27**, 1102–1111.

Coutts, M. P., and Philipson, J. J. (1978a). Tolerance of tree roots to waterlogging. I. Survival of Sitka spruce and lodgepole pine. *New Phytol.* **80**, 63–69.

Coutts, M. P., and Philipson, J. J. (1978b). Tolerance of tree roots to waterlogging. II. Adaptation of Sitka spruce and lodgepole pine to waterlogged soil. *New Phytol.* **80**, 71–77.

Couvillon, G. A., and Erez, A. (1985a). Influence of prolonged exposure to chilling temperatures on bud break and heat requirement for bloom of several fruit tree species. *J. Am. Soc. Hortic. Sci.* **110**, 47–50.

Couvillon, G. A., and Erez, A. (1985b). Effect of level and duration of high temperatures on rest in the peach. *J. Am. Soc. Hortic. Sci.* **110**, 579–581.

Cox, G., Sanders, F. E., Tinker, P. B., and Wild, J. A. (1975). Ultrastructural evidence relating to host endophyte transfer in a vesicular–arbuscular mycorrhiza. *In* "Endomycorrhizas" (F. E. Sanders, B. Mosse, and P. B. Tinker, eds.), pp. 297–312. Academic Press, London.

Cox, R. M. (1983). Sensitivity of forest plant reproduction to long range transported air pollutants. *In vitro* sensitivity of pollen to simulated acid rain. *New Phytol.* **95**, 269–276.

Coyne, P. L., and Bingham, G. E. (1982). Variation in photosynthesis and stomatal conductance in an ozone-stressed ponderosa pine stand: Light response. *For. Sci.* **28**, 257–273.

Craig, G. F., Bell, D. T., and Atkins, C. A. (1990). Response to salt and waterlogging stress of ten taxa of *Acacia* selected from naturally saline areas of western Australia. *Aust. J. Bot.* **38**, 619–630.

Craul, P. J. (1985). A description of urban soils and their desired characteristics. *J. Arboric.* **11**, 330–339.

Crawford, R. M. M. (1967). Alcohol dehydrogenase activity in relation to flooding tolerance in roots. *J. Exp. Bot.* **18**, 458–464.

Crawford, R. M. M. (1978). Metabolic adaptations to anoxia. *In* "Plant Life in Anaerobic Environments" (D. D. Hook and R. M. M. Crawford, eds.), pp. 119–136. Ann Arbor Sci. Publ., Ann Arbor, Michigan.

Creasy, I. I. (1985). Biochemical responses of plants to fungal attack. *Rec. Adv. Phytochem.* **19**, 47–79.

Creber, G. T., and Chaloner, W. G. (1984). Influence of environmental factors on the wood structure of living and fossil trees. *Bot. Rev.* **50**, 357–448.

Creelman, R. A., Mason, S., Bensen, R. J., Boyer, J. S., and Mullet, J. E. (1990). Water deficit and abscisic acid cause differential inhibition of shoot versus root growth in soybean seedlings. Analysis of growth, sugar accumulation, and gene expression. *Plant Physiol.* **92**, 205–214.

Creighton, J. L., Zutter, B. R., Glover, G. R., and Gjerstad, D. H. (1987). Planted pine growth and survival responses to herbaceous vegetation control, treatment duration, and herbicide application technique. *S. J. Appl. For.* **11**, 223–227.

Criley, R. A. (1969). Effect of short photoperiods, cycocel, and gibberellic acid upon flower bud initiation and development in azalea 'Hexe.' *J. Am. Soc. Hortic. Sci.* **94**, 392–396.

Cripps, J. E. L. (1981). Biennial patterns in apple tree growth and cropping as related to irrigation and thinning. *J. Hortic. Sci.* **56**, 161–168.

Crisosto, C. H., Lombard, P. B., and Fuchigami, L. H. (1987). Effect of fall ethephon and hand defoliation on dormant bud ethylene levels, bloom delay, and yield components of 'Redhaven' peach. *Acta Hortic.* **201**, 203–211.

Crisosto, C. H., Grantz, D. A., and Meinzer, F. C. (1992). Effects of water deficit on flower opening in coffee (*Coffea arabica* L). *Tree Physiol.* **10**, 127–139.

Crist, C. R., and Schoeneweiss, D. F. (1975). The influence of controlled stresses on susceptibility of European white birch stems to attack by *Botryosphaeria dothidea*. *Phytopathology* **65**, 369–373.

Critchfield, W. B. (1957). Geographic variation in *Pinus contorta*. *Maria Moors Cabot Foundation* **Publ. 3**. Harvard Univ., Cambridge, Massachusetts.

Crocker, W., and Barton, L. V. (1953). "Physiology of Seeds." Chronica Botanica, Waltham, Massachusetts.

Croft, B. A. (1978). Potentials for research and implementation of integrated pest management in deciduous tree fruits. *In* "Plant Control Strategies" (E. H. Smith and D. Pimentel, eds.), pp. 101–115. Academic Press, New York.

Crohn, D. M. (1995). Sustainability of sewage sludge land application to northern hardwood forests. *Ecol. Appl.* **5**, 53–62.

Cromer, R. N. (1980). Irrigation of radiata pine with waste water: A review of the potential for tree growth and water renovation. *Aust. For.* **43**, 87–100.

Cromer, R. N., and Williams, E. R. (1982). Biomass and nutrient accumulation in a planted *E. globulus* (Labill.) fertilizer trial. *Aust. J. Bot.* **30**, 265–278.

Cromer, R. N., Tompkins, D., and Barr, N. J. (1983). Irrigation of *Pinus radiata* with waste water: Tree growth in a response to treatment. *Aust. For. Res.* **13**, 57–65.

Cromer, R. N., Tompkins, D., Barr, N. J., Williams, E. R., and Stewart, H. T. L. (1984). Litter-fall in a *Pinus radiata* forest: The effect of irrigation and fertilizer treatments. *J. Appl. Ecol.* **21**, 313–326.

Cronshaw, J., and Morey, P. R. (1965). Induction of tension wood by 2,3,5-triiodobenzoic acid. *Nature (London)* **205**, 816–818.

Crossa-Raynaud, P., and Audergon, J. M. (1987). Apricot rootstocks. *In* "Rootstocks for Fruit Crops" (R. C. Rom and R. F. Carlson, eds.), pp. 295–320. Wiley, New York and Chichester.

Crossley, A., and Fowler, D. (1986). The weathering of Scots pine epicuticular wax in polluted and clean air. *New Phytol.* **103**, 207–218.

Cuevas, J., Rallo, L., and Rappaport, H. F. (1994). Initial fruit set at high temperature in olive, *Olea europaea* L. *J. Hortic. Sci.* **69**, 665–672.

Cumming, J. R., and Weinstein, L. H. (1990). Aluminum–mycorrhizal interactions in the physiology of pitch pine seedlings. *Plant Soil* **125**, 7–18.

Curran, M., Cole, M., and Allaway, W. G. (1986). Root aeration and respiration in young mangrove plants [*Avicennia marina* (Forsk.) Vierh.]. *J. Exp. Bot.* **37**, 1225–1233.

Currie, D. J., and Paquin, U. (1987). Large-scale geographical patterns of species richness of trees. *Nature (London)* **329**, 326–327.

Curry, E. A. (1988). Chemical control of vegetative growth of deciduous fruit trees with paclobutrazol and RSW0411. *HortScience* **23**, 470–473.

Curry, J. R., and Church, T. W. (1952). Observations on winter drying of conifers in the Adirondacks. *J. For.* **50**, 114–116.

da Câmara Machado, A., Puschman, M., Pühringer, H., Kremen, R., Katinger, H., and da Câmara Machado, M. L. (1995). Somatic embryogenesis of *Prunus subhirtella autumno rosa* and regeneration of transgenic plants after *Agrobacterium*-mediated transformation. *Plant Cell Rep.* **14**, 335–340.

Dang, Q. L., and Lieffers, V. J. (1989). Assessment of patterns of response of tree ring growth of black spruce following peatland drainage. *Can. J. For. Res.* **19**, 924–929.

Daniel, T. W., Helms, J. A., and Baker, F. S. (1979). "Principles of Silviculture." McGraw-Hill, New York.

Darley, E. F. (1966). Studies on the effect of cement-kiln dust on vegetation. *J. Air Pollut. Control Assoc.* **16**, 145–150.

Dathe, W., Sembdner, G., Yamaguchi, I., and Takahashi, N. (1982). Gibberellins and growth inhibitors in spring bleeding sap, roots and branches of *Juglans regia* L. *Plant Cell Physiol.* **23**, 115–123.

Daubenmire, R. (1978). "Plant Geography." Academic Press, New York.

Davenport, T. L., and Manners, M. M. (1982). Nucellar senescence and ethylene production as they relate to avocado fruitlet abscission. *J. Exp. Bot.* **33**, 815–825.

David, A. (1987). Conifer protoplasts. *In* "Cell and Tissue Culture in Forestry" (J. M. Bonga and D. J. Durzan, eds.), Volume 2, pp. 2–15. Martinus Nijhoff, Dordrecht, The Netherlands.

David, H., Jarlet, E., and David, A. (1984). Effects of nitrogen source, calcium concentration and osmotic stress on protoplasts and protoplast-derived cell cultures of *Pinus pinaster* cotyledons. *Physiol. Plant.* **61**, 477–482.

Davidson, J. A., Gill, S. A., and Raupp, M. J. (1990). Foliar and growth effects of repetitive summer horticultural oil sprays on trees and shrubs under drought stress. *J. Arboric.* **16**, 77–81.

Davies, F. S., and Buchanan, D. W. (1979). Fruit set and bee activity in four rabbiteye blueberry cultivars. *Proc. Fla. State Hortic. Soc.* **92**, 246–247.

Davies, F. T., Jr., and Hartmann, H. T. (1988). The physiological basis of adventitious root formation. *Acta Hortic.* **227**, 113–120.

Davies, H. V., and Pinfield, N. J. (1980). Utilisation of food reserves by embryos of *Acer platanoides* L. during after-ripening. *Z. Pflanzenphysiol.* **96**, 59–65.

Davies, W. J., and Kozlowski, T. T. (1974). Stomatal responses of five woody angiosperms to light intensity and humidity. *Can. J. Bot.* **52**, 1525–1534.

Davies, W. J., and Kozlowski, T. T. (1977). Variations among woody plants in stomatal conductance and photosynthesis during and after drought. *Plant Soil* **46**, 435–444.

Davies, W. J., Kozlowski, T. T., and Pereira, J. (1974). Effect of wind on transpiration and stomatal aperture of woody plants. *Bull. R. Soc. N. Z.* **12**, 433–438.

Davis, J. T., and Sparks, D. (1974). Assimilation and translocation patterns of carbon-14 in the shoots of fruiting pecan trees *Carya illinoensis* Koch. *J. Am. Soc. Hortic. Sci.* **99**, 468–480.

Davis, L. D. (1957). Flowering and alternate bearing. *Proc. Am. Soc. Hortic. Sci.* **70**, 545–556.

Davis, M. B. (1983). Holocene vegetational history of the eastern United States. *In* "Late-Quarternary Environments of the United States" (H. E. Wright, Jr., ed.), Vol. 2, pp. 166–181. Univ. of Minnesota Press, Minneapolis.

Davis, T. D., and Curry, E. A. (1991). Chemical regulation of vegetative growth. *Crit. Rev. Plant Sci.* **10**, 151–188.

Davison, R. M. (1971). Effect of early season sprays of trace elements on fruit setting of apples. *N. Z. J. Agric. Res.* **14**, 931–935.

Davison, R. M., Rudnicki, R., and Bukovac, M. J. (1976). Endogenous plant growth substances in developing fruit of *Prunus cerasus* L. V. Changes in inhibitor (ABA) levels in the seed and pericarp. *J. Am. Soc. Hortic. Sci.* **101**, 519–523.

Dawkins, H. C. (1956). Rapid detection of aberrant growth increment of rain-forest trees. *Emp. For. Rev.* **55**, 448–454.

Day, K. R., and DeJong, T. M. (1990). Girdling of early season 'Mayfire' nectarine trees. *J. Hortic. Sci.* **65**, 529–534.

Dean, T. J., Pallardy, S. G., and Cox, G. S. (1982). Photosynthetic responses of black walnut (*Juglans nigra*) to shading. *Can. J. For. Res.* **12**, 725–730.

Deans, J. D., Lundberg, C., Tabbush, P. M., Cannell, M. G. R., Sheppard, L. J., and Murray, M. B. (1990). The influence of desiccation, rough handling and cold storage on the quality and establishment of Sitka spruce planting stock. *Forestry* **63**, 129–141.

DeBach, P., and Rose, M. (1977). Environmental upsets caused by chemical eradication. *Calif. Agric.* **31**(7), 8–10.

DeCarli, M. E., Baldan, P., Mariano, P., and Rascoi, N. (1987). Subcellular and physiological changes in *Picea excelsa* seeds during germination. *Cytobios* **50**, 29–39.

DeCastro, M. F.-G., and Martinez-Honduvilla, C. J. (1982). Biochemical changes in *Pinus pinea* seeds during storing. *Rev. Esp. Fisiol.* **38**, 13–20.

DeCastro, M. F.-G., and Martinez-Honduvilla, C. J. (1984). Ultrastructural changes in naturally aged *Pinus pinea* seeds. *Physiol. Plant.* **62**, 581–588.

DeJong, T. M., and Doyle, J. F. (1985). Seasonal relationships between leaf nitrogen content (photosynthetic capacity) and leaf canopy light exposure in peach (*Prunus persica*). *Plant Cell Environ.* **8**, 701–706.

Dekazos, E. D. (1981). Effects of aminoethoxyvinylglycine (AVG) on bloom delay, fruit maturity and quality of 'Loring' and 'Rio Oso Gem' peaches. *HortScience* **16**, 520–522.

Dekazos, E. D. (1982). Effect of aminoethoxyvinylglycine on bloom delay, fruit quality, and chemical "pruning" of apple trees. *Proc. Fla. State Hortic. Soc.* **95**, 136–138.

Del Arco, J. M., Escudero, A., and Vega Garrido, M. V. (1991). Effects of site characteristics on nitrogen retranslocation from senescing leaves. *Ecology* **72**, 701–708.

Delap, A. V. (1967). The effect of supplying nitrate at different seasons on the growth, blossoming and nitrogen content of young apple trees in sand culture. *J. Hortic. Sci.* **42**, 149–167.

Delmer, D. P., and Stone, B. A. (1988). Biosynthesis of plant cell walls. *In* "Biochemistry of Plants" (J. Preiss, ed.), Vol. 14, pp. 373–420. Academic Press, San Diego.

del Rio, E., Rallo, L., and Caballero, J. M. (1991). Effects of carbohydrate content on the seasonal rooting of vegetative and reproductive cuttings of olive. *J. Hortic. Sci.* **66**, 301–309.

DeLucia, E. H., and Berlyn, G. P. (1984). The effect of increasing elevation on leaf cuticle thickness and cuticular transpiration in balsam fir. *Can. J. Bot.* **62**, 2423–2431.

DeLucia, E. H., Callaway, R. M., and Schlesinger, W. H. (1994). Offsetting changes in biomass allocation and photosynthesis in ponderosa pine (*Pinus ponderosa*) in response to climate change. *Tree Physiol.* **14**, 669–677.

Demaree, J. B., and Wilcox, M. S. (1947). Fungi pathogenic to blueberries in the eastern United States. *Phytopathology* **37**, 487–506.

De Mason, D. A., and Thomson, W. W. (1981). Structure and ultrastructure of the cotyledon of date palm (*Phoenix dactylifera* L.). *Bot. Gaz.* **142**, 320–328.

DeMoranville, C. J., and Deubert, K. H. (1987). Effect of commercial calcium-boron and manganese–zinc formulations on fruit set of cranberries. *J. Hortic. Sci.* **62**, 163–169.

Denmead, O. T. (1991). Sources and sinks of greenhouse gases in the soil–plant environment. *Vegetatio* **91**, 73–86.

Denne, M. P., and Wilson, J. E. (1977). Some quantitative effects of indoleacetic acid on the wood production and tracheid dimensions of *Picea*. *Planta* **134**, 223–228.

Dennis, F. G., Jr. (1976). Trials of ethephon and other growth regulators for delaying bloom in fruit trees. *J. Am. Soc. Hortic. Sci.* **101**, 241–245.

Dennis, F. G., Jr. (1979). Factors affecting yield in apple with emphasis on 'Delicious.' *Hortic. Rev.* **1**, 395–422.

Dennis, F. G., Jr. (1986). Apple. *In* "CRC Handbook of Fruit Set and Development" (S. P. Monselise, ed.), pp. 1–44. CRC Press, Boca Raton, Florida.

Dennis, F. G., Jr., Crews, C. E., and Buchanan, D. W. (1977). Bloom delay in stone fruits and apple with rhizobitoxine analogue. *HortScience* **12**, 386.

De Souza, S. M., and Felker, P. (1986). The influence of stockplant fertilization on tissue concentrations of N, P and carbohydrates and the rooting of *Prosopis alba* cuttings. *For. Ecol. Manage.* **16**, 181–190.

Dettori, S. (1985). Leaf water potential, stomatal resistance and transpiration response to different watering in almond, peach and 'Pixy' plum. *Acta Hortic.* **171**, 181–186.

Dewey, D. H., ed. (1977). "Controlled Atmospheres for the Storage and Transport of Perishable Agricultural Commodities," Hortic. Rep. No. 28. Michigan State University, East Lansing.

de Yoe, D. R., and Brown, G. N. (1979). Glycerolipid and fatty acid changes in eastern white pine chloroplast lamellae during the onset of winter. *Plant Physiol.* **64**, 924–929.

Dheim, M. A., and Browning, G. (1987). The mode of action of (2*RS*,3*RS*)-paclobutrazol on the fruit set of Doyenne du Comice pear. *J. Hortic. Sci.* **62**, 313–327.

Dickinson, R. E., and Cicerone, R. J. (1986). Future global warming from atmospheric trace gases. *Nature* (*London*) **319**, 109–115.

Dickmann, D. I. (1971). Chlorophyll, ribulose-1,5-diphosphate carboxylase, and Hill reaction activity in developing leaves of *Populus deltoides*. *Plant Physiol.* **48**, 143–145.

Dickmann, D. I. (1985). The ideotype concept applied to forest trees. *In* "Attributes of Trees as Crop Plants" (M. G. R. Cannell and J. E. Jackson, eds.), pp. 89–101. Institute of Terrestrial Ecology, Huntingdon, England.

Dickmann, D. I., and Kozlowski, T. T. (1968). Mobilization by *Pinus resinosa* cones and shoots of ^{14}C-photosynthate from needles of different ages. *Am. J. Bot.* **55**, 900–906.

Dickmann, D. I., and Kozlowski, T. T. (1969a). Seasonal growth patterns of ovulate strobili of *Pinus resinosa* in central Wisconsin. *Can. J. Bot.* **47**, 839–848.

Dickmann, D. I., and Kozlowski, T. T. (1969b). Seasonal changes in the macro- and micronutrient composition of ovulate strobili and seeds of *Pinus resinosa* Ait. *Can. J. Bot.* **47**, 1547–1554.

Dickmann, D. I., and Kozlowski, T. T. (1970a). Mobilization and incorporation of photoassimilated ^{14}C by growing vegetative and reproductive tissues of adult *Pinus resinosa* Ait. trees. *Plant Physiol.* **45**, 284–288.

Dickmann, D. I., and Kozlowski, T. T. (1970b). Photosynthesis by rapidly expanding green strobili of *Pinus resinosa*. *Life Sci.* **9** (Part 2, No. 10), 549–552.

Dickmann, D. I., and Kozlowski, T. T. (1973). "Water, Nutrient and Carbohydrate Relations in Growth of *Pinus resinosa* Ovulate Strobili," Proc. First All Union Symposium on Sexual Growth in Conifers. pp. 195–209. Novosibirsk, U.S.S.R.

Dickmann, D. I., and Stuart, K. W. (1983). "The Culture of Poplars." Dept. of Forestry, Michigan State Univ., East Lansing.

Dickson, R. E. (1989). Carbon and nitrogen allocation in trees. *Ann. Sci. For.* **46** (Suppl.), 631s–647s.

Dickson, R. E. (1991). Assimilate distribution and storage. *In* "Physiology of Trees" (A. S. Raghavendra, ed.), pp. 51–85. Wiley, New York.

Dickson, R. E., and Isebrands, J. G. (1991). Leaves as regulators of stress responses. *In* "Response of Plants to Multiple Stresses" (H. A. Mooney, W. E. Winner, and E. J. Pell, eds.), pp. 3–34. Academic Press, San Diego.

Dickson, R. E., and Shive, J. B. (1982). ^{14}CO$_2$ fixation, translocation, and carbon metabolism in rapidly expanding leaves of *Populus tremuloides*. *Ann. Bot.* **50**, 37–42.

Dickson, R. E., Isebrands, J. G., and Tomlinson, P. T. (1990). Distribution and metabolism of current photosynthate by single-flush northern red oak seedlings. *Tree Physiol.* **7**, 65–77.

Dierberg, F. E., and Brezonik, P. L. (1983). Nitrogen and phosphorus mass balances in natural and sewage-enriched cypress domes. *J. Appl. Ecol.* **20**, 323–337.

Dierberg, F. E., Straub, P. A., and Hendry, C. D. (1986). Leaf- to-twig transfer conserves nitrogen and phosphorus in nutrient poor and enriched cypress swamps. *For. Sci.* **32**, 900–913.

Digby, J., and Wareing, P. F. (1966). The effect of applied growth hormones on cambial division and the differentiation of the cambial derivatives. *Ann. Bot.* **30**, 539–548.

Dirr, M. A. (1976). Selection of trees for tolerance to salt injury. *J. Arboric.* **2**, 209–216.

Dirr, M. A. (1978). Tolerance of seven woody ornamentals to soil-applied sodium chloride. *J. Arboric.* **4**, 162–165.

Dixon, M., and Grace, J. (1984). Effect of wind on the transpiration of young trees. *Ann. Bot. (London) New Ser.* **53**, 811–819.

Dixon, M., Webb, E. C., Thorne, C. J., and Tipton, K. F. (1979). "Enzymes." Longmans, London.

Dodds, J. H., ed. (1983). "Tissue Culture of Trees." AVI Publ., Westport, Connecticut.

Doley, D. (1981). Tropical and subtropical forests and woodlands. *In* "Water Deficits and Plant Growth" (T. T. Kozlowski, ed.), Vol. 6, pp. 209–323. Academic Press, New York.

Doley, D., and Leyton, L. (1968). Effects of growth regulating substances and water potential on the development of secondary xylem in *Fraxinus. New Phytol.* **67**, 579–594.

Domanski, R., and Kozlowski, T. T. (1968). Variations in kinetin-like activity in buds of *Betula* and *Populus* during release from dormancy. *Can. J. Bot.* **46**, 397–403.

Domingo, I. L. (1983). Nitrogen fixation in Southeast Asian forestry: Research and practice. *In* "Biological Nitrogen Fixation in Forest Ecosystems" (J. C. Gordon and C. T. Wheeler, eds.), pp. 295–315. Martinus Nijhoff and W. Junk, The Hague.

Domir, S. C., and Roberts, B. R. (1983). Tree growth retardation by injection of chemicals. *J. Arboric.* **9**, 217–224.

Donnelly, J. R. (1974). Seasonal changes in photosynthate transport within elongating shoots of *Populus grandidentata. Can. J. Bot.* **52**, 2547–2559.

Donnelly, J. R., and Shane, J. B. (1986). Forest ecosystem responses to artificially induced soil compaction. I. Soil physical properties and tree diameter growth. *Can. J. For. Res.* **16**, 750–754.

Donner, S. L., and Running, S. W. (1986). Water stress response after thinning *Pinus contorta* stands in Montana. *For. Sci.* **32**, 614–625.

Dormling, I. (1982). Frost resistance during bud flushing and shoot elongation in *Picea abies. Silva Fenn.* **16**, 167–177.

Doorenbos, J., and Pruitt, W. O. (1977). Guidelines for predicting crop water requirements. Irrigation and Drainage Paper 24. FAO, Rome, Italy.

Douds, D. D., Jr., and Chaney, W. R. (1986). The effect of high nutrient addition upon seasonal patterns of mycorrhizal development, host growth, and root phosphorus and carbohydrate content in *Fraxinus pennsylvanica* Marsh. *New Phytol.* **103**, 91–106.

Doumas, P., Morris, J. W., Chien, C., Bonnet-Masimbert, M., and Zaerr, J. B. (1986). A possible relationship between cytokinin conjugate and flowering in Douglas-fir. *Proceedings of the 9th North American Forest Biology Workshop, Stillwater, Oklahoma,* pp. 285–296.

Downs, R. J. (1962). Photocontrol of growth and dormancy in woody plants. *In* "Tree Growth" (T. T. Kozlowski, ed.), pp. 133–148. Ronald, New York.

Downton, W. J. S. (1984). Salt tolerance of food crops: Prospectives for improvement. *Crit. Rev. Plant Sci.* **1**, 183–201.

Dreistadt, S. H., Dahlsten, D. L., and Frankie, G. W. (1990). Urban forests and insect ecology. *BioScience* **40**, 192–198.

Druart, P. (1990). Effect of culture conditions and leaf selection on organogenesis of *Malus domestica* cv. McIntosh "WIJCIK" and *Prunus canescens* Bois. GM79. *Acta Hortic.* **280**, 117–124.

Dua, R. S., Mitra, S. K., Sen, S. K., and Bose, T. K. (1983). Changes in endogenous growth substances, cofactors and metabolites in rooting of mango cuttings. *Acta Hortic.* **134**, 147–161.

Duchelle, S. F., Skelly, J. M., and Chevone, B. I. (1982). Oxidant effects on forest tree seedling growth in the Appalachian Mountains. *Water Air Soil Pollut.* **18**, 363–373.

Duczmal, K. (1963). Badania nad przemianami fizjologicznymi w stratyfikowanych nasionach jabloni. *Szczecin. Tow. Nauk., Wydz. Nauk Lek., [Pr.]* **18**(2), 1–48.

Duff, G. A., Berryman, C. A., and Eamus, D. (1994). Growth, biomass allocation and foliar nutrient contents of two *Eucalyptus* species of the wet–dry tropics of Australia grown under CO_2 enrichment. *Funct. Ecol.* **8**, 502–508.

Dumbroff, E. B., and Gepstein, S. (1993). Immunological methods for assessing protein expression in plants. *In* "Methods in Plant Molecular Biology and Biotechnology" (B. R. Glick and J. E. Thompson, eds.), pp. 207–223. CRC Press, Boca Raton, Florida.

Dunberg, A. (1976). Changes in gibberellin-like substances and indole-3-acetic acid in *Picea abies* during the period of shoot elongation. *Physiol. Plant.* **38**, 186–190.

Dunberg, A., Hsihan, S., and Sandberg, G. (1981). Auxin dynamics and the rooting of cuttings of *Pinus sylvestris*. *Plant Physiol.* **67** (Suppl.), 5.

Dunlap, J. M., Heilman, P. E., and Stettler, R. F. (1992). Genetic variation and productivity of *Populus trichocarpa* and its hybrids. 5. The influence of ramet position on 3-year growth variables. *Can. J. For. Res.* **22**, 849–857.

Dupuis, J. M., Roffat, C., DeRose, R. T., and Molle, F. (1994). Pharmaceutical capsules as a coating system for artificial seeds. *Bio/Technology* **12**, 385–389.

Dure, L. S. (1975). Seed formation. *Annu. Rev. Plant Physiol.* **26**, 259–278.

Durner, E. F., and Gianfagna, T. J. (1988). Fall ethephon application increases peach flower bud resistance to low-temperature stress. *J. Am. Soc. Hortic. Sci.* **113**, 404–406.

Duryea, M. L., and Brown, G. N., eds. (1984). "Seedling Physiology and Reforestation Success." Martinus Nijhoff and W. Junk, The Hague.

Duryea, M. L., and Dougherty, P. M., eds. (1991). "Forest Regeneration Manual." Kluwer, Dordrecht, Boston, and London.

Duryea, M. L., and Landis, T. D., eds. (1984). "Forest Nursery Manual." Martinus Nijhoff and W. Junk, The Hague, Boston, and Lancaster.

Durzan, D. J. (1988). Somatic polyembryogenesis for the multiplication of tree crops. *Biotechnol. Genet. Eng. Rev.* **6**, 341–378.

Durzan, D. J., and Chalupa, V. (1968). Free sugars, amino acids, and soluble proteins in the embryo and female gametophyte of jack pine as related to climate at the seed source. *Can. J. Bot.* **46**, 417–428.

Eagles, C. F., and Wareing, P. F. (1964). The role of growth substances in the regulation of bud dormancy. *Physiol. Plant.* **17**, 697–709.

Ebell, L. F. (1971). Girdling: Its effect on carbohydrate status and on reproductive bud and cone development of Douglas-fir. *Can. J. Bot.* **49**, 453–466.

Ebell, L. F. (1972a). Cone induction response of Douglas-fir to form of nitrogen fertilizer and time of treatment. *Can. J. For. Res.* **2**, 317–326.

Ebell, L. F. (1972b). Cone production and stem growth response of Douglas-fir to rate and frequency of nitrogen fertilization. *Can. J. For. Res.* **2**, 327–338.

Ebell, L. F., and McMullen, E. E. (1970). Nitrogenous substances associated with differential cone production responses of Douglas-fir to ammonium and nitrate fertilization. *Can. J. Bot.* **48**, 2169–2177.

Ebert, E., and Dumford, S. W. (1976). Effects of triazine herbicides on the physiology of plants. *Residue Rev.* **65**, 1–103.

Edgerton, L. J. (1973). Chemical thinning of flowers and fruits. *In* "Shedding of Plant Parts" (T. T. Kozlowski, ed.), pp. 435–474. Academic Press, New York.

Edwards, C. A., and Mumford, P. M. (1985). Factors affecting the oxygen consumption of sour orange (*Citrus aurantium* L.) seeds during imbibed storage and germination. *Seed Sci. Technol.* **13**, 201–212.

Edwards, G. R. (1986). Ammonia, arginine, polyamines and flower initiation in apple. *Acta Hortic.* **179**, 363.

Edwards, G. S., Friend, A. L., O'Neill, E. G., and Tomlinson, P. T. (1992). Seasonal patterns of biomass accumulation and carbon allocation in *Pinus taeda* seedlings exposed to ozone, acidic precipitation, and reduced soil Mg. *Can. J. For. Res.* **22**, 640–646.

Ehlig, C. F. (1960). Effect of salinity on four varieties of table grapes grown in sand culture. *Proc. Am. Soc. Hortic. Sci.* **76**, 323–335.

Eidmann, H. H. (1979). Integrated management of pine weevil (*Hylobius abietis* L.) populations in Sweden. *U.S.D.A. Gen. Tech. Rep. WO* **WO-8**, 103–109.

Eklund, L. (1991). Relations between indoleacetic acid, calcium ions and ethylene in the regulation of growth and cell-wall composition in *Picea abies. J. Exp. Bot.* **42**, 785–789.

Eklund, L., and Little, C. H. A. (1995). Interaction between indole-3-acetic acid and ethylene in the control of tracheid production in detached shoots of *Abies balsamea. Tree Physiol.* **15**, 27–34.

El-Goorani, M. A., and Sommer, N. F. (1981). Effects of modified atmospheres on postharvest pathogens of fruits and vegetables. *Hortic. Rev.* **3**, 412–461.

Eldridge, K. (1976). Breeding systems, variation and genetic improvement of tropical eucalypts. *In* "Tropical Trees: Variation, Breeding and Conservation" (J. Burley and B. T. Styles, eds.), pp. 101–108. Academic Press, London.

Elk, B. C. M. van (1968). Grafting maple, birch, junipers, *Liriodendron*, spruce, and *Taxodium ascendens* 'Nutans.' (Extracted from *Jaarb., Proefsta. Boomwek., Boskoop*, 1966.) *For. Abstr.* **29** (No. 78).

El Kohen, A., Venet, L., and Mousseau, M. (1993). Growth and photosynthesis of two deciduous forest tree species exposed to elevated carbon dioxide. *Funct. Ecol.* **7**, 480–486.

Ellis, D., Roberts, D., Sutton, B., Lazaroff, W., Webb, D., and Flinn, B. (1989). Transformation of white spruce and other conifer species by *Agrobacterium tumefaciens. Plant Cell Rep.* **8**, 16–20.

Ellis, D. D., McCabe, D. E., Russell, D., Martinell, B., and McCown, B. H. (1991).

Expression of inducible angiosperm promoters in a gymnosperm, *Picea glauca* (white spruce). *Plant Mol. Biol.* **17**, 19–27.

Ellis, D. D., McCabe, D. E., McInnis, S., Ramachandran, R., Russell, D. R., Wallace, K. M., Martinell, B. J., Roberts, D. R., Raffa, K. F., and McCown, B. H. (1993). Stable transformation of *Picea glauca* by particle transformation. *Bio/Technology* **11**, 84–89.

Ellstrand, N. C., and Hoffman, C. (1990). Hybridization as an avenue of escape for engineered genes. *BioScience* **40**, 438–442.

Endo, M. (1973). Studies on the daily change in fruit size of the Japanese pear. I. Diurnal fluctuation of fruit diameter as affected by climatic factors. *J. Jpn. Soc. Hortic. Sci.* **42**, 91–103.

Endo, M., and Ogasawara, S. (1975). Studies on the daily change in fruit size of the Japanese pear. V. Diurnal fluctuation of fruit size as affected by rainfall or water sprinkling. *J. Jpn. Soc. Hortic. Sci.* **43**, 359–367.

Epstein, E. (1980). Response of plants to saline environments. *In* "Genetic Engineering of Osmoregulation" (D. W. Rains, R. C. Valentine, and A. Hollaender, eds.), pp. 7–21. Plenum, New York.

Erez, A. (1978). Adaptation of the peach to the meadow orchard system. *Acta Hortic.* **65**, 245–250.

Erez, A. (1986). Effect of soil-applied paclobutrazol in drip irrigated peach orchards. *Acta Hortic.* **179**, 513–520.

Erez, A., Couvillon, G. A., and Hendershott, C. H. (1979). The effect of cycle length on chilling negation by high temperatures in dormant peach buds. *J. Am. Soc. Hortic. Sci.* **104**, 573–576.

Erez, A., Couvillon, G. A., and Kays, S. J. (1980). The effect of oxygen concentration on the release of peach leaf buds from rest. *HortScience* **15**, 39–41.

Ericsson, A. (1984). Effects of low temperature and light treatment, following winter cold storage, on starch accumulation in Scots pine seedlings. *Can. J. For. Res.* **14**, 114–118.

Eriksson, T. R. (1985). Protoplast isolation and culture. *In* "Plant Protoplasts" (L. C. Fowke and F. Constabel, eds.), pp. 1–20. CRC Press, Boca Raton, Florida.

Erstad, J. L. F., and Gislerod, H. R. (1994). Water uptake of cuttings and stem pieces as affected by different anaerobic conditions in the rooting medium. *Sci. Hortic.* **58**, 151–160.

Esau, K. (1965). "Plant Anatomy." Wiley, New York.

Esau, K. (1968). "Viruses in Plant Hosts." Univ. of Wisconsin Press, Madison.

Etter, H. M. (1969). Growth, metabolic components and drought survival of loblolly pine seedlings at three nitrate levels. *Can. J. Plant Sci.* **49**, 393–402.

Evans, G. C. (1972). "The Quantitative Analysis of Plant Growth." Univ. of California Press, Berkeley.

Evans, M. L. (1985). The action of auxin on plant cell elongation. *Crit. Rev. Plant Sci.* **2**, 317–366.

Evelyn, J. (1670). "Sylva." Allestry, London.

Evenari, M. (1960). Plant physiology and arid zone research. *Arid Zone Res.* **18**, 175–195.

Evert, R. F. (1977). Phloem structure and histochemistry. *Annu. Rev. Plant Physiol.* **28**, 199–222.

Evert, R. F., and Kozlowski, T. T. (1967). Effect of isolation of bark on cambial activity and development of xylem and phloem in trembling aspen. *Am. J. Bot.* **54**, 1045–1055.

Evert, R. F., Kozlowski, T. T., and Davis, J. D. (1972). Influence of phloem blockage on cambial growth of sugar maple. *Am. J. Bot.* **59**, 632–641.

Eysteinsson, T., and Greenwood, M. S. (1990). Promotion of flowering in young *Larix laricina* grafts by gibberellin $A_{4/7}$ and root pruning. *Can. J. For. Res.* **20**, 1448–1452.

Facelli, J. M., and Pickett, S. T. A. (1991a). Indirect effects of litter on woody seedlings subject to herb competition. *Oikos* **62**, 129–138.

Facelli, J. M., and Pickett, S. T. A. (1991b). Plant litter: Its dynamics and effects on plant community structure. *Bot. Rev.* **57**, 1–32.

Facteau, T. J., and Rowe, K. E. (1977). Effects of hydrogen fluoride and hydrogen chloride on pollen tube growth and sodium fluoride on pollen germination in 'Tilton' apricot. *J. Am. Soc. Hortic. Sci.* **102**, 95–96.

Facteau, T. J., and Rowe, K. E. (1981). Response of sweet cherry and apricot pollen tube growth to high levels of sulfur dioxide. *J. Am. Soc. Hortic. Sci.* **106**, 77–79.

Facteau, T. J., Wang, S. Y., and Rowe, K. E. (1973). The effect of hydrogen fluoride on pollen germination and pollen tube growth in *Prunus avium* L. cv. 'Royal Ann.' *J. Am. Soc. Hortic. Sci.* **98**, 234–236.

Fahn, A. (1988). Secretory tissues and factors influencing their development. *Phyton* **28**, 13–26.

Fahn, A., and Zamski, E. (1970). The influence of pressure, wind, wounding and growth substances on the rate of resin duct formation in *Pinus halepensis* wood. *Is. J. Bot.* **19**, 429–446.

Fahn, A., Burley, J., Longman, K. A., Mariaux, A., and Tomlinson, P. B. (1981). Possible contributions of wood anatomy to the determination of the age of tropical trees. *In* "Age and Growth of Tropical Trees" (F. H. Bormann and G. Berlyn, eds.), pp. 31–54. Yale Univ. School of Forestry Bull. 94. New Haven, Connecticut.

Fallahi, E., Moon, J. W., Jr., and Rodney, D. R. (1989). Yield and qualtity of 'Redblush' grapefruit on twelve rootstocks. *J. Am. Soc. Hortic. Sci.* **114**, 187–190.

Farrar, J. L. (1961). Longitudinal variations in the thickness of the annual ring. *For. Chron.* **37**, 323–331.

Faust, M. (1989). "Physiology of Temperate Zone Fruit Trees." Wiley, New York.

Faust, M., and Shear, C. B. (1973). Calcium transport patterns in apples. *Pr. Inst. Sadow., Skierniewice*, Ser. E **No. 3**, 423–436.

Fay, M. (1992). Conservation of rare and endangered plants using *in vitro* methods. *In Vitro Cell Dev. Biol.* **28P**, 1–4.

Federer, C. A., and Tanner, C. B. (1966). Spectral distribution of light in the forest. *Ecology* **47**, 555–560.

Feeny, P. (1976). Plant apparency and chemical defense. *Rec. Adv. Phytochem.* **10**, 1–40.

Fenner, M. (1987). Seedlings. *New Phytol.* **106**(Suppl.), 35–47.

Ferlin, P., Flühler, H., and Palomski, J. (1982). Immissions bedingte Fluorbelastung eines Föhrenstandortes im unteren Pfynwald. *Schweiz. Z. Forstwesen* **133**, 139–157.

Fernandez-Escobar, R., Martin, R., Lopez-Rivares, P., and Paz Suarez, M. (1987).

Girdling as a means of increasing fruit size and earliness in peach and nectarine cultivars. *J. Hortic. Sci.* **62**, 463–468.

Ferree, D. C. (1992). Time of root pruning influences vegetative growth, fruit size, biennial bearing, and yield of 'Jonathan' apple. *J. Am. Soc. Hortic. Sci.* **117**, 198–202.

Ferree, D. C., and Carlson, R. F. (1987). Apple rootstocks. *In* "Rootstocks for Fruit Crops" (R. C. Rom and R. F. Carlson, eds.), pp. 107–143. Wiley, New York and Chichester.

Ferrell, G. T. (1978). Moisture stress threshold of susceptibility to fir engraver beetles in pole-size white fir. *For. Sci.* **24**, 85–94.

Ferres, H. M. (1952). The effect of maleic hydrazide in delaying flowering and fruiting. *Annu. Rep. Long Ashton Res. Sta. 1951*, 40–42.

Field, S. P., and OeMatan, S. S. (1990). The effect of cutting height and pruning frequency of *Leucaena leucocephala* hedgerows on maize production. *Leucaena Res. Rep.* **11**, 68–69.

Fillatti, J. J., Sellmer, J., McCown, B., Haissig, B., and Comai, L. (1987). *Agrobacterium*-mediated transformation and regeneration of *Populus. Mol. Gen. Genet.* **206**, 192–199.

Fillatti, J. J., Haissig, B., McCown, B., Comai, L., and Riemenschneider, D. (1988). Development of glyphosate-tolerant *Populus* plants through expression of mutant *aroA* gene from *Salmonella typhimurium. In* "Genetic Manipulation of Woody Plants" (J. W. Hanover and D. E. Keathley, eds.), pp. 243–250. Plenum, New York.

Filipovich, S. D., and Rowe, R. N. (1977). Effect of succinic acid 2,2-dimethyl hydrazide (SADH) on starch accumulation in young apple trees. *J. Hortic. Sci.* **52**, 367–370.

Finazzo, S. F., Davenport, T. L., and Schaffer, B. (1994). Partitioning of photoassimilates in avocado (*Persea americana* Mill.) during flowering and fruit set. *Tree Physiol.* **14**, 153–164.

Finnegan, J., and McElroy, D. (1994). Transgene inactivation: Plants fight back! *Bio/Technology* **12**, 883–888.

Firoozabady, E., Moy, Y., Courtney-Gutterson, N., and Robinson, K. (1994). Regeneration of transgenic rose (*Rosa hybrida*) plants from embryogenic tissue. *Bio/Technology* **12**, 609–613.

Fischer, R. A., and Turner, N. C. (1978). Plant production in the arid and semiarid zones. *Annu. Rev. Plant Physiol.* **29**, 277–317.

Fisher, J. B. (1986). Branching patterns and angles in trees. *In* "On the Economy of Plant Form and Function" (T. J. Givnish, ed.), pp. 493–523. Cambridge Univ. Press, Cambridge.

Fisher, J. B., and Hibbs, D. E. (1982). Plasticity of tree architecture: Specific and ecological variations found in Aubréville's model. *Am. J. Bot.* **69**, 690–702.

Fisher, J. T., and Mexal, J. G. (1984). Nutrition management: A physiological basis for yield improvement. *In* "Seedling Physiology and Reforestation Success" (M. L. Duryea and G. N. Brown, eds.), pp. 271–299. Martinus Nijhoff/Junk, Dordrecht, Boston, and London.

Fisher, R. F., Woods, R. A., and Glavicic, M. R. (1979). Allelopathic effects of goldenrod and aster on young sugar maple. *Can. J. For. Res.* **8**, 1–9.

Fivaz, A. E. (1931). Longevity and germination of seeds of *Ribes*, particularly *R. rotundifolium*, under laboratory and natural conditions. *U.S. Dep. Agric. Tech. Bull.* **261**, 1–40.

Flavell, R. B. (1994). Inactivation of gene expression in plants as a consequence of specific sequence duplication. *Proc. Natl. Acad. Sci. U.S.A.* **91**, 3490–3496.

Flinn, D. W. (1985). Practical aspects of the nutrition of exotic conifer plantations and native eucalypt forests in Australia. *In* "Research for Forest Management" (J. J. Landsberg and W. Parsons, eds.), pp. 73–93. CSIRO, Melbourne, Australia.

Flint, H. L. (1972). Cold hardiness of twigs of *Quercus rubra* L. as a function of geographic origin. *Ecology* **53**, 1163–1170.

Flint, M. L., and van den Bosch, R. (1981). "Introduction to Integrated Pest Management." Plenum, New York and London.

Flore, J. A. (1981). Influence of light interception on cherry production and orchard design. *Proc. Mich. State Hortic. Soc.* **111**, 161–169.

Flückiger, W., Oertli, J. J., and Flückiger-Keller, H. (1978). The effect of wind gusts on leaf growth and foliar water relations of aspen. *Oecologia* **34**, 101–106.

Fluhr, R., Kuhlmeier, C., Nagy, F., and Chua, N.-H. (1986). Organ-specific and light-induced expression of plant genes. *Science* **232**, 1106–1112.

Foil, P. R., and Ralston, C. W. (1967). The establishment and growth of loblolly pine seedlings on compacted soil. *Soil Sci. Soc. Am. Proc.* **31**, 565–568.

Ford, E. D. (1992). The control of tree structure and productivity through the interaction of morphological development and physiological processes. *Int. J. Plant Sci.* **153**, 5147–5162.

Ford, E. D., Deans, J. D., and Milne, R. (1987a). Shoot extension in *Picea sitchensis*. I. Seasonal variation within a forest canopy. *Ann. Bot.* **60**, 531–542.

Ford, E. D., Milne, R., and Deans, J. D. (1987b). Shoot extension in *Picea sitchensis*. II. Analysis of weather influences on daily growth rate. *Ann. Bot.* **60**, 543–552.

Fordham, R. (1972). Observations on the growth of roots and shoots of tea (*Camellia sinensis* L.) in southern Malawi. *J. Hortic. Sci.* **47**, 221–229.

Forney, C. F., and Peterson, S. J. (1990). Chilling induced potassium leakage of cultured citrus cells. *Physiol. Plant.* **78**, 193–196.

Forshey, C. G. (1963). A comparison of soil nitrogen fertilization and urea sprays as sources of nitrogen for apple trees in sand culture. *Proc. Am. Soc. Hortic. Sci.* **83**, 32–45.

Forsyth, C., and Van Staden, J. (1986). The metabolism and cell division activity of adenine derivatives in soybean callus. *J. Plant Physiol.* **124**, 275–287.

Forward, D. F., and Nolan, N. J. (1961). Growth and morphogenesis in the Canadian forest species. IV. Further studies of wood growth in branches and main axis of *Pinus resinosa* Ait. under conditions of open growth, suppression, and release. *Can. J. Bot.* **39**, 411–436.

Foster, D. R. (1988). Species and stand responses to catastrophic wind in central New England, U.S.A. *J. Ecol.* **76**, 135–151.

Foster, D. R., and Boose, E. R. (1992). Patterns of forest damage resulting from catastrophic wind in central New England, USA. *J. Ecol.* **80**, 79–98.

Fowke, L. C., Attree, S. M., Binarova, P., Galway, M. E., and Wang, H. (1995).

Conifer somatic embryogenesis for studies in plant cell biology. *In Vitro Cell Dev. Biol.* **31P**, 1–7.

Fowler, D. P., and Dwight, T. W. (1964). Provenance differences in the stratification requirements of white pine. *Can. J. Bot.* **42**, 669–675.

Francois, L. E., and Clark, R. A. (1980). Salinity effects on yield and fruit quality of 'Valencia' orange. *J. Am. Soc. Hortic. Sci.* **105**, 199–202.

Frankie, G. W., Baker, H. G., and Opler, P. (1974). Tropical plant phenology: Applications for studies in community ecology. *In* "Phenology and Seasonality Modeling" (H. Lieth, ed.), pp. 287–296. Springer-Verlag, New York.

Fraser, D. A. (1952). Initiation of cambial activity in some forest trees in Ontario. *Ecology* **33**, 259–273.

Freer-Smith, P. H. (1984). Response of six broad-leaved trees during long-term exposure to SO_2 and NO_2. *New Phytol.* **97**, 49–61.

Friedland, A. J., Gregory, R. A., Kärenlampi, L., and Johnson, A. H. (1984). Winter damage to foliage as a factor in red spruce decline. *Can. J. For. Res.* **14**, 963–965.

Friend, A. D., and Woodward, F. I. (1990). Evolutionary and ecophysiological responses of mountain plants to the growing season environment. *Adv. Ecol. Res.* **20**, 59–124.

Friend, A. I., and Tomlinson, P. T. (1992). Mild ozone exposure alters ^{14}C dynamics of foliage of *Pinus taeda* L. *Tree Physiol.* **11**, 215–227.

Frisby, J. W., and Seeley, S. D. (1993). Chilling of endodormant peach propagules. II. Initial seedling growth. *J. Am. Soc. Hortic. Sci.* **118**, 253–257.

Fritts, H. C. (1959). The relations of radial growth to maximum and minimum temperatures in three tree species. *Ecology* **40**, 261–265.

Fritts, H. C., Smith, D. G., and Stokes, M. A. (1965). The biological model for the paleoclimatic interpretation of Mesa Verde tree-ring series. *Am. Antiq.* **31**, 101–121.

Froehlich, H. A., Miles, D. W. R., and Robbins, R. W. (1985). Soil bulk density recovery on compacted skid trails in central Idaho. *Soil Sci Soc. Am. J.* **49**, 1015–1017.

Froehlich, H. A., Miles, D. W. R., and Robbins, R. W. (1986). Growth of young *Pinus ponderosa* and *Pinus contorta* on compacted soil in central Washington. *For. Ecol. Manage.* **15**, 285–294.

Fuchs, Y., and Lieberman, M. (1968). Effects of kinetin, IAA, and gibberellin on ethylene production, and their interactions in growth of seedlings. *Plant Physiol.* **43**, 2029–2036.

Fukuda, H., and Komamine, A. (1985). Cytodifferentiation. *In* "Cell Culture and Somatic Cell Genetics of Plants" (I. K. Vasil, ed.), Vol. 2, pp. 149–212. Academic Press, New York and London.

Gadgil, R. L. (1971a). The nutritional role of *Lupinus arboreus* in coastal sand dune forestry. I. The potential influence of undamaged lupin plants on nitrogen uptake by *Pinus radiata*. *Plant Soil* **34**, 357–367.

Gadgil, R. L. (1971b). The nutritional role of *Lupinus arboreus* in coastal sand dune forestry. II. The potential influence of damaged lupin plants on nitrogen uptake by *Pinus radiata*. *Plant Soil* **34**, 575–593.

Gadgil, R. L. (1971c). The nutritional role of *Lupinus arboreus* in coastal sand dune forestry. III. Nitrogen distributions in the ecosystem before tree planting. *Plant Soil* **35**, 113–126.

Gamborg, O. L., Murashige, T., Thorpe, T. A., and Vasil, I. K. (1976). Plant tissue culture media. *In Vitro Cell Dev. Biol.* **12P**, 473–478.

Gara, R. I., Geiszler, D. R., and Littke, W. R. (1984). Primary attraction of the mountain pine beetle to lodgepole pine in Oregon. *Ann. Entomol. Soc. Am.* **77**, 333–334.

Garcia, R. L., Idso, S. B., Wall, G. W., and Kimball, B. A. (1994). Changes in net photosynthesis and growth of *Pinus eldarica* seedlings in response to atmospheric CO_2 enrichment. *Plant Cell Environ.* **17**, 971–978.

Garrett, P. W., and Zahner, R. (1973). Fascicle density and needle growth responses of red pine to water supply over two seasons. *Ecology* **54**, 1328–1334.

Gaspar, T., and Coumans, M. (1987). Root formation. *In* "Cell and Tissue Culture in Forestry" (J. M. Bonga and D. J. Durzan, eds.), Vol. 2, pp. 202–217. Martinus Nijhoff, Dordrecht, The Netherlands.

Gaspar, T., and Hofinger, M. (1988). Auxin metabolism during adventitious rooting. *In* "Adventitious Root Formation in Cuttings" (T. D. Davis, B. E. Haissig, and N. Sankhla, eds.), pp. 117–131. Dioscorides Press, Portland, Oregon.

Gaspar, T., Kevers, C., deBergh, P., Maene, M., Paques, M., and Boxus, P. (1987). Vitrification: Morphological, physiological, and ecological aspects. *In* "Cell and Tissue Culture in Forestry" (J. M. Bonga and D. J. Durzan, eds.), Vol. 1, pp. 152–166. Martinus Nijhoff, Dordrecht, The Netherlands.

Gates, D. M. (1993). "Climate Change and Its Biological Consequences." Sinauer, Sunderland, Massachusetts.

Gäumann, E. A. (1950). "Principles of Plant Infection" (English translation by W. B. Brierley), Crosby, Lockwood, London.

Gebhardt, K., and Goldbach, H. (1988). Establishment, graft union characteristics and growth of *Prunus* micrografts. *Physiol. Plant.* **72**, 153–159.

Geiger, D. R., and Fondy, B. R. (1980). Phloem loading and unloading: Pathways and mechanisms. *What's New Plant Physiology* **11**, 25–28.

Geiger, D. R., and Savonick, S. A. (1975). Effects of temperature, anoxia and other metabolic inhibitors on translocation. *Encycl. Plant Physiol.* **1**, 256–286.

Geisler, D., and Ferree, D. C. (1984). Response of plants to root pruning. *Hortic. Rev.* **6**, 155–188.

Geiszler, D. R., Gara, R. I., Driver, C. H., Gallucci, V. F., and Martin, R. E. (1980). Fire, fungi, and beetle influences on a lodgepole pine ecosystem in south-central Oregon. *Oecologia* **46**, 239–243.

Gertych, Z. (1953). Delaying blossoming in fruit trees by means of growth substances. *Rocz. Nauk Roln. Ser. A.* **66**, 115–128.

George, M. F., and Burke, M. J. (1977). Supercooling in overwintering azalea buds. *Plant Physiol.* **59**, 326–328.

Gianfagna, T. J. (1989). Chemical control with ethephon of bud dormancy, cold hardiness, and time of bloom in peach trees. *Plant Growth Regul. Soc. Am. Q.* **17**, 39–47.

Gianfagna, T. J., and Mehlenbacher, S. A. (1985). Importance of heat requirement for bud break and time of flowering in apple. *HortScience* **20**, 909–911.

Gianfagna, T. J., and Rachmiel, S. (1986). Changes in gibberellin-like substances of peach seed during stratification. *Physiol. Plant.* **66**, 154–158.

Giaquinta, R. T. (1980). Translocation of sucrose and oligosaccharides. *In* "The Biochemistry of Plants" (P. K. Stumpf and E. E. Conn, eds.), Vol. 3, pp. 271–320. Academic Press, New York.

Gibbs, R. D. (1940). Studies in tree physiology. II. Seasonal changes in the food reserves of field birch (*Betula populifolia* Marsh.). *Can. J. Res.* **C18**, 1–9.

Gibson, D. M., Ketchum, R. E. B., Vance, N. C., and Christen, A. A. (1993). Initiation and growth of cell lines of *Taxus brevifolia* (Pacific yew). *Plant Cell Rep.* **12**, 479–482.

Giertych, M. M., and Forward, D. F. (1966). Growth regulator changes in relation to growth and development of *Pinus resinosa* Ait. *Can. J. Bot.* **44**, 718–738.

Gifford, D. J., and Tolley, M. C. (1989). The seed proteins of white spruce and their mobilization following germination. *Physiol. Plant.* **77**, 254–261.

Gifford, D. J., Wenzel, K. A., and Lammer, D. L. (1989). Lodgepole pine seed germination. I. Changes in peptidase activity in the megagametophyte and embryonic axis. *Can. J. Bot.* **67**, 2539–2543.

Gifford, R. M., Thorne, J. H., Hitz, W. D., and Giaquinta, R. T. (1984). Crop productivity and photoassimilate partitioning. *Science* **225**, 801–808.

Gilbert, O. L. (1983). The growth of planted trees subject to fumes from brickworks. *Environ. Pollut., Ser. A* **31**, 301–310.

Gill, A. M., and Tomlinson, P. B. (1975). Aerial roots: An array of forms and functions. *In* "The Development and Function of Roots" (J. G. Torrey and D. T. Clarkson, eds.), pp. 237–260. Academic Press, London.

Gill, C. J. (1975). The ecological significance of adventitious rooting as a response to flooding in woody species, with special reference to *Alnus glutinosa* (L.) Gaertn. *Flora (Jena)* **164**, 85–97.

Ginter-Whitehouse, D. L., Hinckley, T. M., and Pallardy, S. G. (1983). Spatial and temporal aspects of water relations of three tree species with different vascular anatomy. *For. Sci.* **29**, 317–329.

Giovannoni, J. J., DellaPenna, D., Bennett, A. B., and Fischer, R. L. (1992). Polygalacturonase and tomato fruit ripening. *Hortic. Rev.* **13**, 87–103.

Girona, J., Mata, M., Goldhamer, D. A., Johnson, R. S., and DeJong, T. J. (1993). Patterns of soil and tree water status and leaf functioning during regulated deficit irrigation scheduling in peach. *J. Am. Soc. Hortic. Sci.* **118**, 580–586.

Givnish, T. J. (1987). Comparative studies of leaf form: Assessing the relative roles of selective pressures and phylogenetic constraints. *New Phytol.* **106**(Suppl.), 131–160.

Givnish, T. J. (1988). Adaptation to sun and shade: A whole-plant perspective. *Aust. J. Plant Physiol.* **15**, 63–92.

Glass, B. P., and McKenzie, H. (1989). Decay distribution in relation to pruning and growth stress in plantation-grown *Eucalyptus regnans* in New Zealand. *N. Z. J. For. Sci.* **19**, 210–222.

Gleed, J. A.. (1993). Development of plantlings and stecklings of radiata pine. *In* "Clonal Forestry" (M. R. Ahuja and W. J. Libby, eds.), Vol. 1, pp. 149–157. Springer-Verlag, Berlin.

Glerum, C. (1976). Frost hardiness of forest trees. *In* "Tree Physiology and Yield

Improvement" (M. G. R. Cannell and F. T. Last, eds.), pp. 403–420. Academic Press, New York.

Glerum, C. (1980). Food sinks and food reserves of trees in temperate climates. *N. Z. J. For. Sci.* **10**, 176–185.

Gloyne, R. W. (1976). Shelter in agriculture, forestry, and horticulture—A review. *Agric. Dev. Advisory Service Q. Rev.* **21**, 197–207.

Godbold, D. L., and Hüttermann, A. (1985). Effect of zinc, cadmium, and mercury on root elongation of *Picea abies* (Karst.) seedlings and the significance of these metals to forest die-back. *Environ. Pollut. Ser. A*, 375–382.

Goldhamer, D. A., and Snyder, R. S., eds. (1989). "Irrigation Scheduling." Univ. of California, Berkeley, Leaflet 2145A.

Goldschmidt, E. E., and Golomb, A. (1982). The carbohydrate balance of alternate-bearing citrus trees and the significance of reserves for flowering and fruiting. *J. Am. Soc. Hortic. Sci.* **107**, 206–208.

Goldwin, G. C. (1978). Improved fruit setting with plant hormones. *Acta Hortic.* **80**, 115–121.

Goldwin, G. K. (1985). The use of plant growth regulators to improve fruit setting. *Br. Plant Growth Regul. Group* **13**, 71–88.

Gonzalez, A., Rodriguez, R., and Sanchez Tomes, R. (1991). Ethylene and *in vitro* rooting of hazelnut (*Corylus avellana*) cotyledons. *Physiol. Plant.* **81**, 227–233.

Goo, M. (1952). When cell division begins in germinating seeds of *Pinus thunbergii*. *J. Jpn. For. Soc.* **34**, 3.

Goode, J. E., and Hyrycz, K. J. (1964). The response of Laxton's Superb apple trees to different soil moisture conditions. *J. Hortic. Sci.* **39**, 254–276.

Goode, J. E., and Ingram, J. (1971). The effect of irrigation on the growth, cropping, and nutrition of Cox's Orange Pippin apple trees. *J. Hortic. Sci.* **46**, 195–208.

Goode, J. E., Higgs, K. H., and Hyrycz, K. J. (1979). Effects of water stress control in apple trees by misting. *J. Hortic. Sci.* **54**, 1–11.

Goodwin, T. W., and Mercer, E. I. (1983). "Introduction to Plant Biochemistry." Pergamon, New York.

Gordon, A. G., and Gorham, E. (1963). Ecological aspects of air pollution from an iron sintering plant at Wawa, Ontario. *Can. J. Bot.* **41**, 1063–1078.

Gordon, J. C. (1968). Effect of IAA and sucrose concentration on cell walls of woody stem segments. *Plant Physiol.* **43**, 5–17.

Gordon, J. C. (1983). Silvicultural systems and biological nitrogen fixation. *In* "Biological Nitrogen Fixation in Forest Ecosystems: Foundations and Applications" (J. C. Gordon and C. T. Wheeler, eds.), pp. 1–6. Martinus Nijhoff/Junk, The Hague.

Gordon, J. C., and Dawson, J. O. (1979). Potential use of nitrogen-fixing trees and shrubs in commercial forestry. *Bot. Gaz.* **40**(Suppl), 588–590.

Gordon, J. C., and Larson, P. R. (1968). Seasonal course of photosynthesis, respiration, and distribution of ^{14}C in young *Pinus resinosa* trees as related to wood formation. *Plant Physiol.* **43**, 1617–1624.

Gordon, J. C., and Larson, P. R. (1970). Redistribution of ^{14}C- labelled reserve food in young red pines during shoot elongation. *For. Sci.* **16**, 14–20.

Gordon, J. C., and Wheeler, C. T., eds. (1983). "Biological Nitrogen Fixation in Forest Ecosystems." Martinus Nijhoff/W. Junk, The Hague.

Gorham, J., Tomar, O. S., and Wyn Jones, R. G. (1980). Salinity-induced changes in the chemical composition of *Leucaena leucocephala* and *Sesbania bispinosa J. Plant Physiol.* **132**, 678–682.

Gorski, P. M., and Creasy, C. L. (1977). Color development in 'Golden Delicious' apples. *J. Am. Soc. Hortic. Sci.* **102**, 73–75.

Gosling, P. D., and Ross, J. D. (1981). Peroxidase levels in the cotyledons of hazel seed (*Corylus avellana*). *Phytochemistry* **20**, 31–33.

Goulet, F. (1995). Frost heaving of forest tree seedlings: A review. *New Forests* **9**, 67–94.

Govind, S., and Prasad, A. (1982). Effect of nitrogen nutrition on fruit-set, fruit drop and yield in sweet orange. *Punjab Hortic. J.* **22(1/2)**, 15–20.

Gower, S. T., Vogt, K. A., and Grier, C. C. (1992). Carbon dynamics of Rocky Mountain Douglas-fir: Influence of water and nutrient availability. *Ecol. Monogr.* **62**, 43–65.

Gower, S. T., Gholz, H. L., Nakane, K., and Baldwin, V. C. (1994). Production and carbon allocation patterns of pine forests. *Ecol. Bull.* **43**, 115–133.

Grace, J. (1987). Climatic tolerance and distribution of plants. *New Phytol.* **106** (Suppl.), 113–130.

Grace, J. (1988). Temperature as a determinant of plant productivity. *In* "Plants and Temperature" (S. P. Long and F. I. Woodward, eds.), pp. 91–107. Dept. of Zoology, Univ. of Cambridge, Cambridge.

Grace, J. (1989). Tree lines. *Philos. Trans. R. Soc. London* **B324**, 233–245.

Grace, J., and Thompson, J. R. (1973). The after-effect of wind on photosynthesis and transpiration of *Festuca arundinacea*. *Physiol. Plant.* **28**, 541–547.

Grace, J., Allen, S., and Wilson, C. (1989). Climate and the meristem temperatures of the plant communities near the tree line. *Oecologia* **79**, 198–204.

Graebe, J. (1987). Gibberellin biosynthesis and control. *Annu. Rev. Plant Physiol.* **38**, 419–465.

Graham, R. L., Turner, M. G., and Dale, V. H. (1990). How increasing CO_2 and climate change affect forests. *BioScience* **40**, 575–587.

Grant, J. A., and Ryugo, K. (1984). Influence of within-canopy shading on fruit size, shoot growth, and return bloom in kiwifruit. *J. Am. Soc. Hortic. Sci.* **109**, 799–802.

Grant, M. C., and Mitton, J. B. (1977). Genetic differentiation among growth forms of Engelmann spruce and alpine fir at tree line. *Arct. Alp. Res.* **9**, 259–263.

Grant, S. A., and Hunter, R. F. (1962). Ecotypic differentiation in *Calluna vulgaris* (L.) in relation to altitude. *New Phytol.* **61**, 44–55.

Grasmanis, V. O., and Edwards, G. R. (1974). Promotion of flower initiation in apple trees by short exposure to the ammonium ion. *Aust. J. Plant Physiol.* **1**, 99–105.

Grattan, S. R., and Grieve, C. M. (1992). Mineral element acquisition and growth response of plants grown in saline environments. *Agric. Ecosyst. Environ.* **38**, 275–300.

Graybill, D. A. (1986). A network of high elevation conifers in the western U.S. for detection of tree-ring growth responses to increasing atmospheric carbon dioxide. *In* "Symposium on Ecological Aspects of Tree-Ring Analysis" (G. C. Jacoby and J. W. Hornbeck, eds.), pp. 463–474. U.S. Dept. of Energy, Washington, D. C.

Greacen, E. L., and Sands, R. (1980). Compaction of forest soils: A review. *Aust. J. Soil. Res.* **18**, 163–189.

Green, T. H., and Mitchell, R. J. (1992). Effects of nitrogen on the response of loblolly pine to water stress. I. Photosynthesis and stomatal conductance. *New Phytol.* **122**, 627–633.

Greenberg, B. M., and Glick, B. R. (1993). The use of recombinant DNA technology to produce genetically modified plants. *In* "Methods in Plant Molecular Biology and Biotechnology" (B. R. Glick and J. E. Thompson, eds.), pp. 1–10. CRC Press, Boca Raton, Florida.

Greene, D. W. (1989). Regulation of fruit set in tree fruits with plant growth regulators. *Acta Hortic.* **239**, 323–334.

Greene, D. W. (1991). Reduced rates and multiple sprays of paclobutrazol control growth and improve fruit quality of apple. *J. Am. Soc. Hortic. Sci.* **116**, 807–812.

Greenspan, M. D., Shackel, K. A., and Matthews, M. A. (1994). Developmental changes in the diurnal water budget of the grape berry exposed to water deficits. *Plant Cell Environ.* **17**, 811–820.

Greenwood, M. S., and Hutchison, K. W. (1993). Maturation as a developmental process. *In* "Clonal Forestry" (M. R. Ahuja and W. J. Libby, eds.), Vol. 1, pp. 14–33. Springer-Verlag, Berlin.

Greer, D. H., and Stanley, C. J. (1985). Regulation of the loss of frost hardiness in *Pinus radiata* by photoperiod and temperature. *Plant Cell Environ.* **8**, 111–116.

Greer, D. H., and Warrington, I. J. (1982). Effect of photoperiod, night temperature, and frost incidence on development of frost hardiness in *Pinus radiata*. *Aust. J. Plant Physiol.* **9**, 333–342.

Gregory, R. A., Williams, M. W., Jr., Wong, B. L., and Hawley, G. J. (1986). Proposed scenario for dieback and decline of *Acer saccharum* in northeastern U.S.A. and southeastern Canada. *IAWA Bull. N.S.* **7**, 357–369.

Greszta, J., Braniewski, S., and Nosek, A. (1982). The effect of dusts from different emitters on the height increment of the seedlings of selected tree species. *Fragmenta Floristica et Geobotanica* **28**, 67–75.

Grier, C. C., Vogt, K. A., Keyes, M. R., and Edmonds, R. L. (1981). Biomass distribution and above- and belowground production in young and mature *Abies amabilis* zone ecosystems of the Washington Cascades. *Can. J. For. Res.* **11**, 155–167.

Grierson, W., Soule, J., and Kawada, K. (1982). Beneficial aspects of physiological stress. *Hortic. Rev.* **4**, 247–271.

Griffin, A. R. (1982). Clonal variation in radiata pine seed orchards. I. Some flowering, cone and seed production traits. *Aust. For. Res.* **12**, 295–302.

Griffin, A. R. (1984). Clonal variation in radiata pine seed orchards. II. Flowering phenology. *Aust. For. Res.* **14**, 271–281.

Griffin, A. R., Crane, W. J. B., and Cromer, R. N. (1984). Irrigation and fertilizer effects on productivity of a *Pinus radiata* seed orchard: Response to treatment of an established orchard. *N. Z. J. For. Sci.* **14**, 289–302.

Griffin, A. R., Whiteman, P., Rudge, T., Burgess, I. P., and Moncur, M. (1993). Effect of paclobutrazol on flower-bud production and vegetative growth in two species of *Eucalyptus*. *Can. J. For. Res.* **23**, 640–647.

Griffin, D. H., Quinn, K., and McMillen, B. (1986). Regulation of hyphal growth rate of *Hypoxylon mammatum* by amino acids: Stimulation by proline. *Exp. Mycol.* **10**, 307–314.

Griffith, B. G. (1968). Phenology, growth, and flower cone production on 154

Douglas-fir trees on the University Research Forest as influenced by climates and fertilizer, 1957–1967. *Univ. B. C. Fac. For. Bull.* **6**, 1–70.

Grill, D., Liegl, E., and Windisch, E. (1979). Holzanatomische Untersuchungen an abgasbelesteten Bäumen. *Phytopathol. Z.* **94**, 335–342.

Grochowska, M. J., Karaszewska, A., Jankowska, B., and Mika, A. (1983). The pattern of hormones of intact apple shoots and its changes after spraying with growth regulators. *Acta Hortic.* **149**, 13–23.

Grochowska, M. J., Karaszewska, A., Jankowska, B., Maksymiuk, J., and Williams, M. W. (1984). Dormant pruning influences on auxin, gibberellin, and cytokinin levels in apple trees. *J. Am. Soc. Hortic. Sci.* **109**, 312–318.

Groome, M. C., Oxler, S. R., and Gifford, D. J. (1991). Hydrolysis of lipid and protein reserves in loblolly pine seeds in relation to protein electrophoretic patterns following imbibition. *Physiol. Plant.* **83**, 99–106.

Gross, D. C., Proebsting, E. L., Jr., and Zimmerman, H. (1988). Development, distribution, and characteristics of intrinsic, nonbacterial ice nuclei in *Prunus* wood. *Plant Physiol.* **88**, 915–922.

Grossmann, K. (1990). Plant growth retardants as tools in physiological research. *Physiol. Plant.* **78**, 640–648.

Guardiola, J. L., Monerri, C., and Augusti, M. (1982). The inhibitory effect of gibberellic acid on flowering in citrus. *Physiol. Plant.* **55**, 136–142.

Gucinski, H., Vance, E., and Reiners, W. A. (1995). Potential effects of global climate change. *In* "Ecophysiology of Coniferous Forests" (W. K. Smith and T. M. Hinckley, eds.), pp. 309–331. Academic Press, San Diego.

Guderian, R., and Kueppers, H. (1980). Responses of plant communities to air pollution. *U.S. For. Serv. Gen. Tech. Rep. PSW* **PSW-43**, 187–199.

Guinn, G., and Brummett, D. L. (1988). Changes in abscisic acid and indoleacetic acid before and after anthesis relative to change in abscission rates of cotton fruiting forms. *Plant Physiol.* **87**, 629–631.

Günthardt-Georg, M. S., Matyssek, R., Scheidegger, C., and Keller, T. (1993). Differentiation and structural decline in the leaves and bark of birch (*Betula pendula*) under low ozone concentrations. *Trees* **7**, 104–114.

Gupta, P. K., and Durzan, D. J. (1987). Biotechnology of somatic polyembryogenesis and plantlet regeneration in loblolly pine. *Bio/Technology* **5**, 147–151.

Gutterson, N. (1995). Anthocyanin biosynthetic genes and their application to flower color modification through sense suppression. *HortScience* **30**, 964–966.

Guy, C. L. (1990). Cold acclimation and freezing stress tolerance: Role of protein metabolism. *Annu. Rev. Plant Physiol.* **41**, 187–223.

Hacskaylo, E. (1971). Metabolic exchanges in ectomycorrhizae. *In* "Mycorrhizae" (E. Hacskaylo, ed.), pp. 175–196. U. S. Dept. Agric. Forest Service Misc. Publ. 1189. Washington, D. C.

Hacskaylo, E. (1973). Carbohydrate physiology of ectomycorrhizae. *In* "Ectomycorrhizae" (G. C. Marks and T. T. Kozlowski, eds.), pp. 207–230. Academic Press, New York.

Hacskaylo, E., Palmer, J. G., and Vozzo, J. A. (1965). Effect of temperature on growth and respiration of ectotrophic mycorrhizal fungi. *Mycologia* **57**, 748–756.

Hadas, A. (1982). Seed–soil contact and germination. *In* "The Physiology and

Biochemistry of Seed Development, Dormancy and Germination" (A. A. Khan, ed.), pp. 507–527. Elsevier, Amsterdam.

Haddon, L. E., and Northcote, D. H. (1975). Quantitative measurement of the course of bean callus differentiation. *J. Cell Sci.* **17**, 11–26.

Hadley, J. L., and Smith, W. K. (1983). Influence of wind exposure on needle desiccation and mortality for timberline conifers in Wyoming, U.S.A. *Arct. Alp. Res.* **15**, 127–133.

Hadley, J. L., and Smith, W. K. (1987). Influence of krummholz mat microclimate on needle physiology and survival. *Oecologia* **73**, 82–90.

Hahn, G. G., Hartley, C., and Rhoads, A. S. (1920). Hypertrophied lenticels on the roots of conifers and their relation to moisture and aeration. *J. Agric. Res.* **20**, 253–265.

Haines, L. H., Maki, T. E., and Sanderford, S. G. (1975). The effect of mechanical site preparation treatments on soil productivity and tree (*Pinus taeda* L. and *P. elliottii* var. *elliottii*) growth. *In* "Forest Soils and Forest Land Management" (B. Bernier and C. H. Winget, eds.), pp. 379–395. Laval Univ. Press, Quebec.

Haines, S. G., Haines, L. W., and White, G. (1978). Leguminous plants increase sycamore growth in northern Alabama. *Soil Sci. Soc. Am. J.* **42**, 130–132.

Haissig, B. E. (1972). Meristematic activity during adventitious root primordium development. Influences of endogenous auxin and applied gibberellic acid. *Plant Physiol.* **49**, 886–892.

Haissig, B. E. (1974a). Origins of adventitious roots. *N. Z. J. For. Sci.* **4**, 299–310.

Haissig, B. E. (1974b). Influence of auxins and auxin synergists on adventitious root primordium initiation and development. *N. Z. J. For. Sci.* **4**, 311–323.

Haissig, B. E. (1974c). Metabolism during adventitious root primordium initiation and development. *N. Z. J. For. Sci.* **4**, 324–337.

Haissig, B. E. (1979). Influence of aryl esters of indole-3-acetic acid and indole-3-butyric acids on adventitious root primordium initiation and development. *Physiol. Plant.* **47**, 29–33.

Haissig, B. E. (1982a). Carbohydrate and amino acid concentrations during adventitious root primordium development in *Pinus banksiana* Lamb. cuttings. *For. Sci.* **28**, 813–821.

Haissig, B. E. (1982b). Activity of some glycolytic and pentose phosphate pathway enzymes during development of adventitious roots. *Physiol. Plant.* **55**, 261–272.

Haissig, B. E. (1983). The rooting stimulus in pine cuttings. *Proc. Int. Plant Prop. Soc.* **32**, 625–638.

Haissig, B. E. (1986). Metabolic processes in adventitious rooting of cuttings. *In* "New Root Formation in Plants and Cuttings" (M. B. Jackson, ed.), pp. 141–189. Martinus Nijhoff, Dordrecht, Boston, and Lancaster.

Hakman, I., Rennie, P., and Fowke, L. (1987). A light and electron microscope study of *Picea glauca* (white spruce) somatic embryos. *Protoplasma* **140**, 100–109.

Hale, C. R., and Weaver, R. J. (1962). The effect of developmental stage on direction of translocation of photosynthate in *Vitis vinifera*. *Hilgardia* **33**, 89–131.

Halevy, A. H., Whitehead, C. S., and Kofranek, A. M. (1984). Does pollution induce corolla abscission of cyclamen flowers by promoting ethylene production? *Plant Physiol.* **75**, 1090–1093.

Hall, D. O., and de Groot, P. J. (1987). Introduction: The biomass framework. *In* "Biomass" (D. O. Holland and R. P. Overend, eds.), pp. 3–24. Wiley, New York.

Hall, J. B., and Swaine, M. D. (1980). Seed stocks in Ghanaian forest soils. *Biotropica* **12**, 256–263.

Hampp, R., Egger, B., Effenberger, S., and Einig, W. (1994). Carbon allocation in developing spruce needles. Enzymes and intermediates of sucrose metabolism. *Physiol. Plant.* **90**, 299–306.

Han, K. H., and Keathley, D. E. (1989). Regeneration of whole plants from seedling-derived callus of black locust. *In* "Nitrogen Fixing Tree Research Report," Vol. 7, pp. 112–114. Thailand Inst. Sci. Technol. Res., Bangkok.

Han, K. H., Keathley, D. E., and Gordon, M. P. (1993). Cambial tissue culture and subsequent shoot regeneration from mature black locust (*Robinia pseudoacacia* L.). *Plant Cell Rep.* **12**, 185–188.

Hance, B. A., and Bevington, J. M. (1992). Changes in protein synthesis during stratification and dormancy release in embryos of sugar maple (*Acer saccharum*). *Physiol. Plant.* **86**, 365–371.

Hanger, B. C. (1979). The movement of calcium in plants. *Commun. Soil Sci. Plant Anal.* **10**, 171–193.

Hansen, E. A. (1988). Irrigating short rotation intensive culture hybrid poplars. *Biomass* **16**, 237–250.

Hansen, E. A., and Dawson, J. O. (1982). Effect of *Alnus glutinosa* on hybrid *Populus* height growth in a short-rotation intensively cultured plantation. *For. Sci.* **28**, 49–59.

Hansen, J. (1988). Influence of gibberellins on adventitious root formation. *In* "Adventitious Root Formation in Cuttings" (T. D. Davis, B. E. Haissig, and N. Sankhla, eds.), pp. 162–173. Dioscorides Press, Portland, Oregon.

Hansen, P. (1967a). ^{14}C-studies on apple trees. I. The effect of the fruit on the translocation and distribution of photosynthates. *Physiol. Plant.* **20**, 382–391.

Hansen, P. (1967b). ^{14}C studies on apple trees. II. Distribution of photosynthates from top and base leaves from extension shoots. *Physiol. Plant.* **20**, 720–725.

Hansen, P. (1971). ^{14}C-studies on apple trees. VII. The early seasonal growth in leaves, flowers, and shoots as dependent upon current photosynthates and existing reserves. *Physiol. Plant.* **25**, 469–473.

Hansen, P. (1977). Carbohydrate allocation. *In* "Environmental Effects on Crop Physiology" (J. J. Landsberg and C. V. Cutting, eds.), pp. 247–255. Academic Press, London.

Hansen, P. (1987). Source–sink relations in fruits. I. Effects of pruning in apple. *Gartenbauwissenschaft* **52**, 193–195.

Hansen, P., and Grauslund, J. (1973). ^{14}C-studies on apple trees. VIII. The seasonal variation and nature of reserves. *Physiol. Plant.* **28**, 24–32.

Hansen, P., Ryugo, K., Ramos, D. F., and Fitch, I. (1982). Influence of cropping on Ca, K, Mg and carbohydrate status of 'French' prune trees grown on potassium-limited soils. *J. Am. Soc. Hortic. Sci.* **107**, 511–515.

Hanson, P. J., Isebrands, J. G., Dickson, R. E., and Dixon, R. K. (1988a). Ontogenetic patterns of CO_2 exchange of *Quercus rubra* L. leaves during three flushes of shoot growth. I. Median flush leaves. *For. Sci.* **34**, 55–68.

Hanson, P. J., Isebrands, J. G., Dickson, R. E., and Dixon, R. K. (1988b). Ontogenetic patterns of CO_2 exchange of *Quercus rubra* L. leaves during three flushes of shoot growth. II. Insertion gradients of leaf photosynthesis. *For. Sci.* **34**, 69–76.

Hanson, P. J., McLaughlin, S. B., and Edwards, N. T. (1988c). Net CO_2 exchange of *Pinus taeda* shoots exposed to variable ozone levels and rain chemistries in field and laboratory settings. *Physiol. Plant.* **74**, 635–642.

Harcharik, D. A. (1981). The timing and economics of irrigation in loblolly pine seed orchards. Ph.D. Dissertation, North Carolina State University, Raleigh.

Hardie, W. J., and Considine, J. A. (1976). Response of grapes to water deficits in particular stages of development. *Am. J. Enol. Vitic.* **27**, 55–61.

Hare, R. C. (1984). EL-500: An effective growth retardant for dwarfing southern pine seedlings. *Can. J. For. Res.* **14**, 123–127.

Harley, J. L., and Smith, S. E. (1983). "Mycorrhizal Symbiosis." Academic Press, New York.

Harlow, W. M., Coté, W. A., Jr., and Day, A. C. (1964). The opening mechanism of pine cone scales. *J. For.* **62**, 538–540.

Harrington, C. A., and Reukema, D. L. (1983). Initial shock and long-term stand development following thinning in a Douglas-fir plantation. *For. Sci.* **29**, 33–46.

Harris, J. M., Kriedemann, P. E., and Possingham, J. V. (1968). Anatomical aspects of grape berry development. *Vitis* **7**, 106–119.

Harris, R. W. (1992). "Arboriculture." Prentice-Hall, Englewood Cliffs, New Jersey.

Hart, J. W. (1988). "Light and Plant Growth." Unwin Hyman, London.

Hartmann, H. T., and Panetsos, C. (1961). Effect of soil moisture deficiency during floral development on fruitfulness of the olive. *Proc. Am. Soc. Hortic. Sci.* **78**, 209–217.

Hartmann, H. T., Kester, D. E., and Davies, F. T., Jr. (1990). "Plant Propagation: Principles and Practices." 5th Ed. Prentice-Hall, Englewood Cliffs, New Jersey.

Harty, A. R., and Van Staden, J. (1988). Mini review. The use of growth retardants in citriculture. *Isr. J. Bot.* **37**, 155–164.

Hatano, K., and Asakawa, S. (1964). Physiological processes in forest tree seeds during maturation, storage, and germination. *Int. Rev. For. Res.* **1**, 279–323.

Hatcher, E. S. J. (1959). Auxin relations of the woody shoot. *Ann. Bot. (London)* **23**, 409–423.

Haug, A. (1983). Molecular aspects of aluminum toxicity. *Crit. Rev. Plant Sci.* **1**, 345–373.

Havas, P., and Huttunen, S. (1972). The effects of air pollution on the radial growth of Scots pine (*Pinus sylvestris*). *Biol. Conserv.* **4**, 361–368.

Havis, J. R. (1971). Water movement in stems during freezing. *Cryobiology* **8**, 581–585.

Havranek, W. M., and Tranquillini, W. (1995). Physiological processes during winter dormancy and their ecological significance. *In* "Ecophysiology of Coniferous Forests" (W. K. Smith and T. M. Hinckley, eds.), pp. 95–124. Academic Press, San Diego.

Hawley, G. J., and DeHayes, D. H. (1994). Genetic diversity and population structure of red spruce (*Picea rubens*). *Can. J. Bot.* **72**, 1778–1786.

Haywood, J. D., and Burton, J. D. (1989). Loblolly pine plantation development is influenced by site preparation and soil in the west gulf coastal plain. *South. J. Appl. For.* **13**, 17–21.

Haywood, J. D., Tiarks, A. E., and Shoulders, E. (1990). Loblolly pine and slash pine height and diameter are related to soil drainage on poorly drained silt loams. *New Forests* **4**, 81–96.

Head, G. C. (1969). The effects of fruiting and defoliation on seasonal trends in new root production on apple trees. *J. Hortic. Sci.* **42**, 169–180.

Heagle, J. S., and Johnston, J. W. (1979). Variable responses of soybeans to mixtures of ozone and sulfur dioxide. *J. Air Pollut. Control Assoc.* **29**, 729–732.

Hedgcock, G. G. (1912). Winter-killing and smelter injury in the forests of Montana. *Torreya* **12**, 25–30.

Heide, O. M. (1974). Growth and dormancy in Norway spruce ecotypes (*Picea abies*). I. Interaction of photoperiod and temperature. *Physiol. Plant.* **30**, 1–12.

Heide, O. M. (1993a). Daylength and thermal time responses of budburst during dormancy release in some northern deciduous trees. *Physiol. Plant.* **88**, 531–540.

Heide, O. M. (1993b). Dormancy release in beech buds (*Fagus sylvatica*) requires both chilling and long days. *Physiol. Plant.* **89**, 187–191.

Heidmann, L. J. (1982). Effect of selected natural and synthetic growth regulators on the growth of ponderosa pine seedlings. *For. Sci.* **28**, 156–160.

Heilman, P. E. (1981). Root penetration of Douglas-fir seedlings into compacted soil. *For. Sci.* **27**, 660–666.

Heilman, P. E., and Stettler, R. F. (1985). Genetic variation and productivity of *Populus trichocarpa* and its hybrids. II. Biomass production in a 4-year plantation. *Can. J. For. Res.* **15**, 384–388.

Heinicke, D. R. (1963). The microclimate of fruit trees. II. Foliage and light distribution patterns in apple trees. *Proc. Am. Soc. Hortic. Sci.* **83**, 1–11.

Heinicke, D. R. (1964). The microclimate of fruit trees. III. The effect of tree size on light penetration and leaf area in Red Delicious apple trees. *Proc. Am. Soc. Hortic. Sci.* **85**, 33–41.

Heinicke, D. R. (1966). The effect of natural shade on photosynthesis and light intensity in Red Delicious apple trees. *Proc. Am. Soc. Hortic. Sci.* **88**, 1–8.

Hejnowicz, A., and Tomaszewski, M. (1969). Growth regulators and wood formation in *Pinus silvestris*. *Physiol. Plant.* **22**, 984–992.

Helgerson, O. T. (1990). Heat damage in tree seedlings and its prevention. *New Forests* **3**, 333–358.

Hellmers, H. (1962). Temperature effect upon optimum tree growth. *In* "Tree Growth" (T. T. Kozlowski, ed.), pp. 275–287. Ronald, New York.

Helmisaari, H. S. (1992). Nutrient retranslocation within the foliage of *Pinus sylvestris*. *Tree Physiol.* **10**, 45–58.

Hennerty, M. J., O'Kennedy, B. T., and Titus, J. S. (1980). Conservation and reutilization of bark proteins in apple trees. *In* "Mineral Nutrition of Fruit Trees" (D. Atkinson, J. E. Jackson, R. O. Sharples, and W. M. Waller, eds.), pp. 369–377. Butterworth, London.

Hennessey, T. C., Dougherty, P. M., Cregg, B. M., and Wittwer, R. F. (1992). Annual variation in needle fall of a loblolly pine stand in relation to climate and stand density. *For. Ecol. Manage.* **51**, 329–338.

Henry, P. H., and Blazich, F. A. (1990). Seed germination of Fraser fir: Timing of irradiation and involvement of phytochrome. *J. Am. Soc. Hortic. Sci.* **115**, 231–234.

Herrero, J. (1956). Incompatibilidad entre patron e injerto. III. Comparacion de sintomas producidas por incompatibilidad y por el anillado del tronco. *An. Aula Del* **4**, 262–265.

Heuser, W. (1987). Graft incompatability: Effect of cyanogenic glycosides on almond and plum callus growth. *Proc. Int. Plant Prop. Soc.* **36**, 91–97.

Hickman, G. W. (1993). Summer irrigation of established oak trees. *J. Arboric.* **19**, 35–37.

Hicks, G. S. (1994). Shoot induction and organogenesis in vitro: A developmental perspective. *In Vitro Cell Dev. Biol.* **30P**, 10–15.

Hidano, Y., and Niizeki, M. (1988). Protoplast culture of deciduous fruit trees. *Sci. Hortic. (Amsterdam)* **37**, 201–216.

Hidy, G. (1995). Acid rain. *In* "Encyclopedia of Environmental Biology" (W. A. Nierenberg, ed.), Vol. 1, pp. 1–17. Academic Press, San Diego.

Higgins, T. J. B. (1984). Synthesis and regulation of major proteins in seeds. *Annu. Rev. Plant Physiol.* **35**, 191–221.

Higgs, K. H., and Jones, H. G. (1990). Response of apple tree rootstocks to irrigation in south-east England. *J. Hortic. Sci.* **65**, 129–141.

Higuchi, T. (1985). Biosynthesis of lignin. *In* "Biosynthesis and Biodegradation of Wood Components" (T. Higuchi, ed.), pp. 141–160. Academic Press, Orlando, Florida.

Hilder, V. A., Gatehouse, A. M. R., Sheerman, S. E., Barker, R. F., and Boulter, D. (1987). A novel mechanism of insect resistance engineered into tobacco. *Nature (London)* **330**, 160–163.

Hilgeman, R. H., and Reuther, W. (1967). Evergreen tree fruits. *Am. Soc. Agron. Monogr.* **11**, 704–718.

Hill-Cottingham, D. G., and Williams, R. R. (1967). Effect of time of application of fertilizer nitrogen on the growth, flower development, and fruit set of maiden apple trees, var. Lord Lambourne, and on the distribution of total nitrogen within the trees. *J. Hortic. Sci.* **42**, 319–338.

Hill, M. O., and Stevens, P. A. (1981). The density of viable seed in soils of forest plantations in upland Britain. *J. Ecol.* **69**, 693–709.

Hinckley, T. M., Teskey, R. O., Duhme, F., and Richter, H. (1981). Temperate hardwood forests. *In* "Water Deficits and Plant Growth" (T. T. Kozlowski, ed.), Vol. 6, pp. 153–208. Academic Press, New York.

Hinesley, L. E., and Maki, T. E. (1980). Fall fertilization helps longleaf pine nursery stock. *South. J. Appl. For.* **4**, 132–135.

Ho, L. C. (1992). Fruit growth and sink strength. *In* "Fruit and Seed Production: Aspects of Development, Environmental Physiology and Ecology" (C. Marshall and J. Grace, eds.), pp. 101–124. Cambridge Univ. Press, Cambridge.

Hoad, C. V. (1984). Hormonal regulation of fruit-bud formation in fruit trees. *Acta Hortic.* **149**, 12–23.

Hochbichler, E., Krapfenbauer, A., and Mayrhofer, F. (1990). Ein Pflegemodell für Eichenjungbestände-Grünastung, eine wirtschaftliche Problemlösung der Wertholzerziehung. *Centralbl. Gesamte Forstwes.* **107**, 1–12.

Hocking, D., and Nyland, R. D. (1971). Cold storage of coniferous seedlings. A review. "AFRI Research Report Number 6." Applied Forestry Research Institute, College of Forestry at Syracuse, Syracuse, New York.

Hoekstra, P. E., and Mergen, F. (1957). Experimental induction of female flowers on young slash pine. *J. For.* **55**, 827–831.

Hoff, R. V. (1987). Dormancy in *Pinus monticola* seed related to stratification times, seed coat, and genetics. *Can. J. For. Res.* **17**, 294–298.

Högberg, P., and Johannison, C. (1993). ^{15}N abundance of forests is correlated with losses of nitrogen. *Plant Soil* **157**, 147–150.

Hogsett, W. E., Plocher, M., Wildman, V., Tingey, D. T., and Bennett, J. P. (1985). Growth response of two varieties of slash pine seedlings to chronic ozone exposures. *Can. J. Bot.* **63**, 2369–2376.

Höll, W. (1975). Radial transport in rays. *Encycl. Plant Physiol.* **N.S. 1**, 432–450.

Holmes, F. W. (1961). Salt injury to trees. *Phytopathology* **51**, 712–718.

Holmes, J. J., and Davidson, J. A. (1984). Integrated pest management for arborists: Implementation of a pilot program. *J. Arboric.* **10**, 65–70.

Holst, M. J. (1961). Experiments with flower promotion in *Picea glauca* (Moench) Voss and *Pinus resinosa*. *Rec. Adv. Bot.* **2**, 1654–1658.

Holzer, K. (1973). Die Vererbung von physiologischen und morphologischen Eigenschaften der Fichte. II. Mutterbaummerkmall. Unpublished manuscript.

Hook, D. D. (1984). Adaptations to flooding with fresh water. *In* "Flooding and Plant Growth" (T. T. Kozlowski, ed.), pp. 265–294. Academic Press, Orlando, Florida.

Hook, D. D., and Brown, C. L. (1973). Root adaptations and relative flood tolerance of five hardwood species. *For. Sci.* **19**, 225–229.

Hook, D. D., Brown, C. L., and Kormanik, P. O. (1970a). Lenticels and water root development of swamp tupelo under various flooding conditions. *Bot. Gaz.* **131**, 217–224.

Hook, D. D., Langdon, O. G., Stubbs, J., and Brown, C. L. (1970b). Effects of water regime on the survival, growth, and morphology of tupelo seedlings. *For. Sci.* **16**, 304–311.

Hopper, G. M., Parrish, D. J., and Smith, D. W. (1985). Adenylate energy charge during stratification and germination of *Quercus rubra*. *Can. J. For. Res.* **15**, 829–832.

Hori, H., and Elbein, A. D. (1985). The biosynthesis of plant cell walls. *In* "Biosynthesis and Biodegradation of Wood Components" (T. Higuchi, ed.), pp. 109–139. Academic Press, Orlando, Florida.

Horsley, S. B. (1977a). Allelopathic inhibition of black cherry by fern, grass, goldenrod, and aster. *Can. J. For. Res.* **7**, 205–216.

Horsley, S. B. (1977b). Allelopathic inhibition of black cherry. II. Inhibition by woodland grass, ferns, and club moss. *Can. J. For. Res.* **7**, 515–519.

Hosner, J. F. (1957). Effects of water upon the seed germination of bottomland trees. *For. Sci.* **3**, 67–70.

Hosner, J. F. (1958). The effects of complete inundation upon seedlings of six bottomland tree species. *Ecology* **39**, 371–373.

Hosner, J. F. (1960). Relative tolerance of complete inundation of fourteen bottomland tree species. *For. Sci.* **6**, 246–251.

Hosner, J. F. (1962). The southern bottomland region. *In* "Regional Silviculture of the United States" (J. W. Barrett, ed.), pp. 296–333. Wiley, New York.

Houck, L. G., Jenner, J. F., and Bianchi, J. (1990). Holding lemon fruit at 5°C or 15°C before cold treatment reduces chilling injury. *HortScience* **25**, 1174.

Houghton, R. A., and Woodwell, G. M. (1989). Global climatic change. *Sci. Am.* **260**, 36–44.

Houston, D. B. (1984). Stress related to diseases. *Arboric. J.* **8**, 137–149.

Houston, D. B., and Dochinger, L. S. (1977). Effects of ambient air pollution on cone, seed, and pollen characteristics in eastern white and red pine. *Environ. Pollut.* **12**, 1–5.

Howard, B. H. (1987). Propagation. *In* "Rootstocks for Fruit Crops" (R. C. Rom and R. F. Carlson, eds.), pp. 29–77. Wiley, New York and Chichester.

Howarth, K. J., and Ougham, H. J. (1993). Gene expression under temperature stress. *New Phytol.* **125**, 1–26.

Howe, G. T., Hackett, W. P., Furnier, G. R., and Klevorn, R. E. (1995). Photoperiodic responses of a northern and southern ecotype of black cottonwood. *Physiol. Plant.* **93**, 695–708.

Hsiao, T. C. (1973). Plant responses to water stress. *Annu. Rev. Plant Physiol.* **24**, 519–570.

Huang, H., and Villanueva, R. (1993). Amino acids, polyamines and proteins during seed germination of two species of Dipterocarpaceae. *Trees* **7**, 189–193.

Huang, Y., Diner, A. M., and Karnosky, D. F. (1991). *Agrobacterium rhizogenes*-mediated genetic transformation and regeneration of a conifer: *Larix decidua. In Vitro Cell Dev. Biol.* **27P**, 201–207.

Hubbard, M., Kelly, J., Rajapske, S., Abbott, A., and Ballard, R. (1992). Restriction fragment length polymorphisms in rose and their use for cultivar identification. *HortScience* **27**, 172–173.

Huber, D. M. (1978). Disturbed mineral nutrition. *In* "Plant Disease: An Advanced Treatise" (J. G. Horsfall and E. B. Cowling, eds.), pp. 163–181. Academic Press, New York.

Huddle, J. A., and Pallardy, S. G. (1996). Effects of soil and stem base heating on survival, resprouting and gas exchange of *Acer* and *Quercus* seedlings. *Tree Physiol.* **16**, 583–589.

Huettl, R. F. (1989). "New types" of forest damages in central Europe. *In* "Air Pollution's Toll on Forests and Crops" (J. A. MacKenzie and M. T. El-Ashry, eds.), pp. 22–74. Yale Univ. Press, New Haven, Connecticut.

Huffaker, C. B., ed. (1980). "New Technology of Pest Control." Wiley, New York.

Hunt, R. S., and O'Reilly, H. J. (1984). Evaluation of control of Lawson cypress root rot with resistant root stocks. *Can. J. Plant Pathol.* **6**, 172–174.

Hutchinson, B. A., and Matt, D. R. (1977). The distribution of solar radiation within a deciduous forest. *Ecol. Monogr.* **47**, 185–207.

Hutchinson, J. G., and Bramlett, D. L. (1964). Frost damage to shortleaf pine flowers. *J. For.* **62**, 343.

Hutchison, K. W., and Greenwood, M. S. (1991). Molecular approaches to gene expression during conifer development and maturation. *For. Ecol. Manage.* **43**, 273–286.

Hutton, M. J., and Van Staden, J. (1983). Transport and metabolism of 8-[^{14}C] zeatin applied to leaves of *Citrus sinensis*. *Z. Pflanzenphysiol.* **111**, 75–83.

Huxley, P. A., ed. (1983). "Plant Research and Agroforestry." Pillans and Wilson, Edinburgh.

Hwang, S. C., Chen, C. L., Lin, J. C., and Lin, H. L. (1984). Cultivation of banana using plantlets from meristem culture. *HortScience* **19**, 231–233.

Hytönen, J. (1992). Allelopathic potential of peatland plant species on germination and early growth of Scots pine, silver birch and downy birch. *Silva Fenn.* **26**, 63–73.

Idso, S. B. (1991). The aerial fertilization effect of CO_2 and its implications for global carbon dioxide cycling and maximum greenhouse warming. *Bull. Am. Meteorol. Soc.* **72**, 962–965.

Idso, S. B., and Kimball, B. A. (1992a). Seasonal fine-root biomass development of sour orange trees grown in atmospheres of ambient and elevated CO_2 concentration. *Plant Cell Environ.* **15**, 337–341.

Idso, S. B., and Kimball, B. A. (1992b). Effects of atmospheric CO_2 enrichment on photosynthesis, respiration, and growth of sour orange trees. *Plant Physiol.* **99**, 341–343.

Idso, S. B., Kimball, B. A., and Allen, S. G. (1991). CO_2 enrichment of sour orange trees: 2.5 years into a long-term experiment. *Plant Cell Environ.* **14**, 351–353.

Imaseki, H. (1985). Hormonal control of wound-induced responses. *Encycl. Plant Physiol. New Ser.* **11**, 485–512.

Inderjit, and Dakshini, K. M. M. (1995). On laboratory bioassays in allelopathy. *Bot. Rev.* **61**, 28–44.

Ingemarsson, B. S. M., Eklund, L., and Eliasson, L. (1991). Ethylene effects on cambial activity and cell wall formation in hypocotyls of *Picea abies* seedlings. *Physiol. Plant.* **82**, 219–221.

Ingestad, T. (1980). Growth, nutrition, and nitrogen fixation in grey alder at varied rate of nitrogen addition. *Physiol. Plant.* **50**, 353–364.

Ingestad, T. (1981). Nutrition and growth of birch and grey alder seedlings in low conductivity solutions and at varied relative rate of nutrient addition. *Physiol. Plant.* **52**, 454–456.

Ingestad, T. (1982). Relative addition rate and external concentration: Driving variables used in plant nutrition research. *Plant Cell Environ.* **5**, 443–453.

Ingestad, T. (1987). New concepts on soil fertility and plant nutrition as affected by research on forest trees and stands. *Geoderma* **40**, 237–252.

Ingestad, T. (1988). A fertilization model based on the concepts of nutrient flux density and nutrient productivity. *Scand. J. For. Res.* **3**, 157–173.

Ingestad, T., and Lund, A.-B. (1979). Nitrogen stress in birch seedlings. I. Growth technique and growth. *Physiol. Plant.* **45**, 137–148.

Ingestad, T., and Lund, A.-B. (1986). Theory and techniques for steady-state mineral nutrition and growth of plants. *Scand. J. For. Res.* **1**, 439–453.

Ingram, D. L. (1986). Root cell membrane heat tolerance of two dwarf hollies. *J. Am. Soc. Hortic. Sci.* **111**, 270–272.

International Rules for Seed Testing. (1993). The germination test. *Seed Sci. Tech.* **21**(Suppl.).

Isaia, A., and Bulard, C. (1978). Relative levels of some bound and free gibberellins in dormant and after-ripened embryos of *Pyrus malus* cv. Golden Delicious. *Z. Pflanzenphysiol.* **110**, 89–95.

Ishikawa, M., and Sakai, A. (1985). Extra organ freezing in wintering flower buds of *Cornus officinalis.* Sieb. et Zucc. *Plant Cell Environ.* **8**, 333–338.

Issarakraisila, M., and Considine, J. A. (1994). Effects of temperature on pollen viability in mango cv. 'Kensington.' *Ann. Bot.* **73**, 231–240.

Iwasa, Y., Sato, K., Kakita, M., and Kubo, T. (1993). Modelling biodiversity: Latitudinal gradient of forest species diversity. *In* "Biodiversity and Ecosystem Function" (E.-D. Schulze and H. A. Mooney, eds.), pp. 433–451. Springer-Verlag, Berlin and New York.

Iwasaki, K. (1980). Effects of bud scale removal, calcium cyanamide, GA$_3$, and ethephon on bud break of 'Muscat of Alexandria' grape (*Vitis vinifera* L.). *J. Jpn. Soc. Hortic. Sci.* **48**, 395–398.

Jackson, C. H. (1975). Date of blossoming of Bramley's seedling apple in relation to temperature. *Commonw. Agric. Bur. Res. Rev.* **5**, 47–50.

Jackson, D. I. (1969). Effects of water, light, and nutrition on flower-bud initiation in apricots. *Aust. J. Biol. Sci.* **22**, 69–75.

Jackson, D. I., and Coombe, B. G. (1966). Gibberellin-like substances in developing apricot fruit. *Science* **154**, 277–278.

Jackson, D. I., and Sweet, G. B. (1972). Flower initiation in temperate woody plants. *Hortic. Abstr.* **42**, 9–24.

Jackson, G. A. D., and Blundell, J. B. (1963). Germination in *Rosa. J. Hortic. Sci.* **38**, 310–320.

Jackson, J. E. (1970). Aspects of light climate within apple orchards. *J. Appl. Ecol.* **7**, 207–216.

Jackson, J. E. (1975). Patterns of distribution of foliage and light. *In* "Climate and the Orchard" (H. C. Pereira, ed.), pp. 31–40. Farnham Royal: Commonwealth Agricultural Bureaux.

Jackson, J. E. (1980). Light interception and utilization by orchard systems. *Hortic. Rev.* **2**, 208–267.

Jackson, J. E. (1981). Theory of light interception by orchards and a modelling approach to optimizing orchard design. *Acta Hortic.* **114**, 69–79.

Jackson, J. E. (1985). Future fruit orchard design: Economics and biology. *In* "Attributes of Trees as Crop Plants" (M. G. R. Cannell and J. E. Jackson, eds.), pp. 441–459. Institute of Terrestrial Ecology, Huntingdon, England.

Jackson, J. E. (1989). The manipulation of fruiting. *In* "Manipulation of Fruiting" (C. J. Wright, ed.), pp. 3–12. Butterworth, London.

Jackson, J. E., and Palmer, J. W. (1977a). Effects of shade on the growth and cropping of apple trees. I. Experimental details and effects on vegetative growth. *J. Hortic. Sci.* **52**, 245–252.

Jackson, J. E., and Palmer, J. W. (1977b). Effects of shade on the growth and cropping of apple trees. II. Effects on components of yield. *J. Hortic. Sci.* **52**, 253–266.

Jackson, J. E., Palmer, J. W., Penning, M. A., and Sharples, R. O. (1977). Effects of shade on the growth and cropping of apple trees. III. Effects on fruit growth, chemical composition and quality at harvest and after storage. *J. Hortic. Sci.* **52**, 267–282.

Jackson, M. B., Herman, B., and Goodenough, A. (1982). An examination of the importance of ethanol in causing injury to flooded plants. *Plant Cell Environ.* **5**, 163–172.

Jaffe, M. J. (1980). Morphogenetic responses of plants to mechanical stimuli or stress. *BioScience* **30**, 239–243.

James, R. L., Cobb, F. W., Jr., Miller, P. R., and Parmeter, J. R., Jr. (1980). Effects of oxidant air pollution on susceptibility of pine roots to *Fomes annosus*. *Phytopathology* **70**, 560–563.

Jancarik, V. (1961). Vyskyt drevokaznych hub v kourem poskozovani oblasti krusnych hor. *Lesnictvi* **7**, 667–692.

Janzen, D. H. (1972). *Jacquinia pungens*, a heliophile from the understory of tropical deciduous forest. *Biotropica* **2**, 112–119.

Jarrell, W. M., and Virginia, R. A. (1990). Response of mesquite to nitrate and salinity in a simulated phreatic environment: Water use, dry matter and mineral nutrient accumulation. *Plant Soil* **125**, 185–196.

Jarvis, B. C., Frankland, B., and Cherry, J. H. (1968). Increased nucleic acid synthesis in relation to the breaking of dormancy of hazel seed by gibberellic acid. *Planta* **83**, 257–266.

Jarvis, P. G., and Jarvis, M. S. (1963). The water relations of tree seedlings. I. Growth and water use in relation to soil water potential. *Physiol. Plant.* **16**, 215–235.

Jarvis, P. G., and Leverenz, J. W. (1983). Productivity of temperate, deciduous and evergreen forests. *Encycl. Plant Physiol. New Ser.* **12D**, 233–280.

Jayawickrama, K. J. S., Jett, J. B., and McKeand, S. E. (1991). Rootstock effects in grafted conifers: A review. *New Forests* **5**, 157–173.

Jefferies, C. J., Brain, P., Scott, K. G., and Belcher, A. R. (1982). Experimental systems and mathematical models for studying temperature effects on pollen-tube growth and fertilization in plum. *Plant Cell Environ.* **5**, 231–236.

Jeffords, M. R., and Endress, A. G. (1984). Possible role of ozone in tree defoliation by the gypsy moth (*Lepidoptera*: Lymantriidae). *Environ. Entomol.* **13**, 1249–1252.

Jemison, G. M. (1944). The effect of basal wounding by forest fires on the diameter growth of some southern Appalachian hardwoods. *Duke (University) School of Forestry Bulletin* **9**, 1–63.

Jenkins, M. A., and Pallardy, S. G. (1995). The influence of drought on red oak group species growth and mortality in the Missouri Ozarks. *Can. J. For. Res.* **25**, 1119–1127.

Jenkins, P. A. (1974). Influence of applied indoleacetic acid and abscisic acid on xylem cell dimensions in *Pinus radiata* D. Don. *In* "Mechanisms of Regulation of Plant Growth" (R. L. Bieleski, A. R. Ferguson, and M. M. Creswell, eds.), Bull. 12, pp. 737–742. Royal Society of New Zealand, Wellington.

Jenkins, P. A., and Shepherd, K. R. (1974). Seasonal changes in levels of indoleacetic acid and abscisic acid in stem tissues of *Pinus radiata*. *N. Z. J. For. Sci.* **4**, 511–519.

Jensen, K. F. (1981). Ozone fumigation decreased the root carbohydrate content and dry weight of green ash seedlings. *Environ. Pollut.* (Ser. A) **26**, 147–152.

Jensen, K. F. (1982). An analysis of the growth of silver maple and eastern cottonwood seedlings exposed to ozone. *Can. J. For. Res.* **12**, 420–424.

Jensen, K. F., and Kozlowski, T. T. (1975). Absorption and translocation of sulfur dioxide by seedlings of four forest tree species. *J. Environ. Qual.* **4**, 379–381.

Jeremias, K. (1964). Über die jahrperiodisch bedingten Veränderungen der Ablagerungs form der Kohlenhydrate in Vegetativen Pflanzenteilen. *Bot. Stud.* **15**, 1–96.

Jia, H., and Ingestad, T. (1984). Nutrient requirements and stress response of *Populus simonii* and *Paulownia tomentosa. Physiol. Plant.* **62**, 117–124.

Johansen, L. G., Oden, P.-C., and Junttila, O. (1986). Abscisic acid and cessation of apical growth in *Salix pentandra. Physiol. Plant.* **66**, 409–412.

Johnson, A. H., Friedland, A. J., and Dushoff, J. G. (1986). Recent and historic red spruce mortality: Evidence of climatic influence. *Water Air Soil Pollut.* **30**, 319–330.

Johnson, A. H., Cook, E. R., and Siccama, T. G. (1988). Climate and red spruce growth and decline in the northern Appalachians. *Proc. Natl. Acad. Sci. U.S.A.* **85**, 5369–5373.

Johnson, D. S., Stinchcombe, D. R., and Scott, K. G. (1983). Effect of soil management on mineral composition and storage quality of Cox's Orange Pippin apples. *J. Hortic. Sci.* **58**, 317–326.

Johnson, D. W., and Taylor, G. E., Jr. (1989). Role of air pollution in forest decline in eastern North America. *Water Air Soil Pollut.* **48**, 21–43.

Johnson, M. A., and Croteau, R. (1987). Biochemistry of conifer resistance to bark beetles and their fungal symbionts. *In* "Ecology and Metabolism of Plant Lipids" (G. Fuller and W. D. Nes, eds.), pp. 76–92. Series No. 325. Am. Chem. Soc. Symp., Washington, D.C.

Johnson, R. W., Tyree, M. T., and Dixon, M. A. (1987). A requirement for sucrose in xylem sap flow from dormant maple trees. *Plant Physiol.* **84**, 495–500.

Jones, C. H., and Bradlee, J. L. (1933). The carbohydrate contents of the maple tree. *Bull. Univ. Vt., Agric. Exp. Stn.* **358**.

Jones, H. C., and Curlin, J. W. (1968). The role of fertilzers in improving the hardwoods of the Tennessee Valley. *In* "Forest Fertilization: Theory and Practice," pp. 185–196. Tennessee Valley Authority, Muscle Shoals, Alabama.

Jones, H. G. (1985). Strategies for optimizing the yield of tree crops in suboptimal environments. *In* "Attributes of Trees as Crop Plants" (M. G. R. Cannell and J. E. Jackson, eds.), pp. 68–79. Institute of Terrestrial Ecology, Huntingdon, England.

Jones, H. G., and Higgs, K. H. (1982). Surface conductance and water balance of developing apple (*Malus pumila* Mill.) fruits. *J. Exp. Bot.* **33**, 67–77.

Jones, H. G., Lakso, A. N., and Syvertsen, J. P. (1985). Physiological control of water status in temperate and subtropical fruit trees. *Hortic. Rev.* **7**, 301–343.

Jones, L. (1961). Effect of light on germination of forest tree seeds. *Proc. Int. Seed Test. Assoc.* **26**, 437–452.

Jones, O. P. (1993). Propagation of apple *in vitro. In* "Micropropagation of Woody Plants" (M. R. Ahuja, ed.), pp. 169–186. Kluwer, Dordrecht, The Netherlands.

Jones, R. H., Sharitz, R. R., Dixon, P. M., Segal, D. S., and Schneider, R. L. (1994). Woody plant regeneration in four flood plain forests. *Ecol. Monogr.* **64**, 345–367.

Jones, R. L. (1983). The role of gibberellins in plant cell elongation. *Crit. Rev. Plant Sci.* **1**, 23–47.

Jongsma, M. A., Bakker, P. L., Bosch, D., and Stiekma, W. J. (1995). Adaptation of *Spodoptera exigua* larvae to plant proteinase inhibitors by induction of gut proteinase activity insensitive to inhibition. *Proc. Natl. Acad. Sci. U.S.A.* **92**, 8041–8045.

Jordan, C. F. (1985). "Nutrient Cycling in Tropical Forest Ecosystems." Wiley, New York.

Josza, L. A., and Powell, J. M. (1987). Some climatic aspects of biomass productivity of white spruce stem wood. *Can. J. For. Res.* **17**, 1075–1079.

Julin-Tegelman, Å., and Pinfield, N. J. (1982). Changes in the level of endogenous cytokinin-like substances in *Acer pseudoplatanus* embryos during stratification and germination. *Physiol. Plant.* **54**, 318–322.

Junttila, O. (1976). Apical growth cessation and shoot tip abscission in *Salix. Physiol. Plant.* **38**, 278–286.

Junttila, O. (1982). Gibberellin-like activity in shoots of *Salix pentandra* as related to the elongation growth. *Can. J. Bot.* **60**, 1231–1234.

Junttila, O. (1991). Gibberellins and the regulation of shoot elongation in woody plants. *In* "Gibberellins" (N. Takahashi, B. O. Phinney, and J. MacMillan, eds.), pp. 199–210. Springer-Verlag, New York and Berlin.

Junttila, O., and Jensen, E. (1988). Gibberellins and photoperiodic control of shoot elongation in *Salix. Physiol. Plant.* **74**, 371–376.

Kamaly, J. C., and Goldberg, R. B. (1980). Regulation of structural gene expression in tobacco. *Cell (Cambridge, Mass.)* **19**, 935–946.

Kangasjärvi, J., Talvinen, J., Utriainen, M., and Karjalainen, R. (1994). Plant defense systems induced by ozone. *Plant Cell Environ.* **17**, 783–794.

Kappel, F., and Flore, J. A. (1983). Effect of shade on photosynthesis, specific leaf weight, leaf chlorophyll content, and morphology of young peach trees. *J. Am. Soc. Hortic. Sci.* **108**, 541–544.

Kappel, F., Flore, A. J., and Layne, R. E. C. (1983). Characterization of the light microclimate in four peach hedgerow canopies. *J. Am. Soc. Hortic. Sci.* **108**, 102–105.

Kargiolaki, H., Osborne, D. J., and Thompson, F. B. (1991). Leaf abscission and stem lesions (intumescences) on poplar clones after SO_2 and O_3 fumigation: A link with ethylene release? *J. Exp. Bot.* **42**, 1189–1198.

Kärki, L., and Tigerstedt, P. M. A. (1985). Definition and exploitation of forest tree ideotypes in Finland. *In* "Attributes of Trees as Crop Plants" (M. G. R. Cannell and J. E. Jackson, eds.), pp. 102–109. Inst. of Terrestrial Ecology, Huntingdon, England.

Karlsson, I., and Carson, D. (1985). Survival and growth of *Abies amabilis* scions grafted on four species of understock. *The Plant Propagator* **31**(2), 6–8.

Karlsson, M. G., Heins, R. D., Gerberick, J. O., and Hackmann, M. E. (1991). Temperature driven leaf unfolding rate in *Hibiscus rosa-sinensis. Sci. Hortic. (Amsterdam)* **45**, 323–331.

Karnosky, D. F., and Stairs, G. R. (1974). The effects of SO_2 on *in vitro* forest tree pollen germination and tube elongation. *J. Environ. Qual.* **3**, 406–409.

Kasana, M. S., and Mansfield, T. A. (1986). Effects of air pollutants on the growth and functioning of roots. *Proc. Indian Acad. Sci. (Ser.) Plant Sci.* **96**, 429–441.

Katterman, F. R. (1991). Environmental injury to plants. *In* "Encyclopedia of Earth System Science" (W. A. Nierenberg, ed.), Vol. 2, pp. 153–162. Academic Press, San Diego.

Kaufmann, M. R. (1969). Effects of water potential on germination of lettuce, sunflower, and citrus seeds. *Can. J. Bot.* **47**, 1761–1764.

Kaufmann, M. R. (1972). Water deficits and reproductive growth. *In* "Water Deficits and Plant Growth" (T. T. Kozlowski, ed.), Vol. 3, pp. 99–124. Academic Press, New York.

Kaufmann, M. R., and Eckard, A. N. (1977). Water potential and temperature effects on germination of Engelmann spruce and lodgepole pine seeds. *For. Sci.* **23**, 27–33.

Kaurin, A., Stushnoff, C., and Junttila, O. (1982). Vegetative growth and frost hardiness of cloudberry (*Rubus chamaemorus*) as affected by temperature and photoperiod. *Physiol. Plant.* **55**, 76–81.

Kawada, K., Grierson, W., and Soule, J. (1979). Seasonal resistance to chilling injury of 'Marsh' grapefruit as related to winter field temperature. *Citrus Ind.* **60**(10), 5–9.

Kawase, M. (1971). Causes of centrifugal root promotion. *Physiol. Plant.* **25**, 64–70.

Kawase, M. (1972). Effect of flooding on ethylene concentration in horticultural plants. *J. Am. Soc. Hortic. Sci.* **97**, 584–588.

Keeling, C. D. (1986). "Atmospheric CO_2 Concentrations—Mauna Loa Observatory, Hawaii 1958–1986." NDP-001/R1. Carbon Dioxide Inf. Cent., Oak Ridge Natl. Lab., Oak Ridge, Tennessee.

Keller, T. (1980). The effect of a continuous springtime fumigation with SO_2 on CO_2 uptake and structure of the annual ring in spruce. *Can. J. For. Res.* **10**, 1–6.

Keller, T., and Beda, H. (1984). Effects of SO_2 on the germination of conifer pollen. *Environ. Pollut.* (*Ser. A*) **33**, 237–243.

Kelliher, F. M., and Tauer, C. G. (1980). Stomatal resistance and growth of drought stressed eastern cottonwood from a wet and dry site. *Silvae Genet.* **29**, 166–171.

Kellogg, R. M., and Barber, F. J. (1981). Stem eccentricity in coastal western hemlock. *Can. J. For. Res.* **11**, 714–718.

Kellomäki, S., Hänninen, H., and Kolström, M. (1995). Computations on frost damage to Scots pine under climatic warming in boreal conditions. *Ecol. Appl.* **5**, 42–52.

Kelly, J. D. (1995). Use of random amplified polymorphic DNA markers in breeding for major gene resistance of plant pathogens. *HortScience* **30**, 461–465.

Kelly, K. M., Van Staden, J., and Bell, W. E. (1992). Seed coat structure and dormancy. *Plant Growth Regul.* **11**, 201–209.

Kennard, W. C., Tukey, L. D., and White, D. G. (1951). Further studies with maleic hydrazide to delay blossoming of fruits. *Proc. Am. Soc. Hortic. Sci.* **58**, 26–32.

Kercher, J. R., Axelrod, M. C., and Bingham, G. E. (1980). Forecasting effects of SO_2 pollution on growth and succession in a western conifer forest. *U.S. For. Serv. Gen. Tech. Rep. PSW* **PSW-43**.

Kerr, R. A. (1989). Greenhouse skeptic out in the cold. *Science* **246**, 1118–1119.

Kerstiens, G., Townsend, J., Heath, J., and Mansfield, T. A. (1995). Effects of water and nutrient availability on physiological responses of woody species to elevated CO_2. *Forestry* **68**, 303–315.

Kester, D. E., and Grasselly, C. (1987). Almond rootstocks. *In* "Rootstocks for Fruit Crops" (R. C. Rom and R. F. Carlson, eds.), pp. 265–293. Wiley, New York and Chichester.

Ketchie, D. O., and Ballard, A. L. (1968). Environments which cause heat injury to Valencia oranges. *Proc. Am. Soc. Hortic. Sci.* **93**, 166–172.

Ketchie, D. O., and Burts, W. D. (1973). The relation of lipids to cold acclimation in 'Red Delicious' apple trees. *Cryobiology* **10**, 529.

Keyes, M. R., and Grier, C. C. (1981). Above- and below-ground net production in 40-year-old Douglas-fir stands on low and high productivity sites. *Can. J. For. Res.* **11**, 599–605.

Khan, A. A., and Samimy, C. (1982). Hormones in relation to primary and secondary seed dormancy. *In* "The Physiology and Biochemistry of Seed Development, Dormancy and Germination" (A. A. Khan, ed.), pp. 203–241. Elsevier, Amsterdam, and New York.

Kieliszewska-Rokicka, B. (1985). Activity of respiratory enzymes during dormancy breakage in *Acer platanoides* L. seeds. *Acta Univ. Agric. Fac. Agron.* (*Brno*) *A* **33**, 457–462.

Killingbeck, K. T., May, J. D., and Nyman, S. (1990). Foliar senescence in an aspen (*Populus tremuloides*) clone: The response of element resorption to interramet variation and timing of abscission. *Can. J. For. Res.* **20**, 1156–1164.

Kimmerer, T. W., and Kozlowski, T. T. (1981). Stomatal conductance and sulfur uptake of five clones of *Populus tremuloides* exposed to sulfur dioxide. *Plant Physiol.* **67**, 990–995.

Kimmerer, T. W., and Kozlowski, T. T. (1982). Ethylene, ethane, acetaldehyde, and ethanol production by plants under stress. *Plant Physiol.* **69**, 840–847.

King, D. A. (1994). Influence of light level on the growth and morphology of saplings in a Panamanian forest. *Am. J. Bot.* **81**, 948–957.

Kirby, E. G. (1988). Recent advances in protoplast culture of horticultural crops: Conifers. *Sci. Hortic.* (*Amsterdam*) **37**, 267–276.

Kitajima, K. (1992). Relationship between photosynthesis and thickness of cotyledons for tropical tree species. *Funct. Ecol.* **6**, 582–589.

Klawitter, R. A. (1964). Water tupelos like it wet. *South. Lumberman* **209**, 108–109.

Klein, I. (1983). Drip irrigation based on soil matric potential conserves water in peach and grape. *HortScience* **18**, 942–944.

Kleiner, K. W., Ellis, D. D., McCown, B. H., and Raffa, K. F. (1995). Field evaluation of transgenic poplar expressing a *Bacillus thuringiensis cryIA(a)* δ-endotoxin gene against the forest tent caterpillar (Lepidoptera: Lasiocampidae) and gypsy moth (Lepidoptera: Lymantriidae) following winter dormancy. *Environ. Entomol.* **24**, 1358–1364.

Klimaszewska, K., Ward, C., and Cheliak, W. M. (1992). Cryopreservation and plant regeneration from embryogenic cultures of larch (*Larix* × *eurolepis*) and black spruce (*Picea mariana*). *J. Exp. Bot.* **43**, 73–79.

Kling, G. J., Meyer, M. M., Jr., and Siegler, D. (1988). Rooting cofactors in five *Acer* species. *J. Am. Soc. Hortic. Sci.* **113**, 252–257.

Klinger, T., and Ellstrand, N. C. (1994). Engineered genes in wild populations: Fitness of weed–crop hybrids of *Raphanus sativus*. *Ecol. Appl.* **4**, 117–120.

Knee, J. (1972). Anthocyanin, carotenoid and chlorophyll changes in the peel of Cox's orange pippin apples during ripening on and off the tree. *J. Exp. Bot.* **23**, 184–196.

Kobe, R. K., Pacala, S. W., Silander, J. A., Jr., and Canham, C. D. (1995). Juvenile tree survivorship as a component of shade tolerance. *Ecol. Appl.* **5**, 517–532.

Koch, K. E., and Johnson, C. R. (1984). Photosynthate partitioning in split-root citrus seedlings with mycorrhizal and non-mycorrhizal root systems. *Plant Physiol.* **75**, 26–30.

Koller, D. (1972). Environmental control of seed germination. *In* "Seed Biology" (T. T. Kozlowski, ed.), Vol. 2, pp. 1–101. Academic Press, New York.

Kondo, S., and Takahashi, Y. (1987). Effects of high temperature in the night-time and shading in the daytime on the early drop of apple fruit 'Starking Delicious.' *J. Jpn. Soc. Hortic. Sci.* **56**, 142–150.

Königshofer, H. (1991). Distribution and seasonal variation of polyamines in shoot-axes of spruce [*Picea abies* (L.) Karst.]. *J. Plant Physiol.* **137**, 607–612.

Kopecky, F., Sebanek, J., and Blazkova, J. (1975). Time course of the changes in the level of endogenous growth regulator during the stratification of the seeds of the 'Panenske ceske' apple. *Biol. Plant.* **17**, 81–87.

Koppenaal, R. S., and Colombo, S. J. (1988). Heat tolerance of actively growing, bud-initiated, and dormant black spruce seedlings. *Can. J. For. Res.* **18**, 1103–1105.

Koppenaal, R. S., Colombo, S. J., and Blumwald, E. (1991). Acquired thermotolerance of jack pine, white spruce and black spruce seedlings. *Tree Physiol.* **8**, 83–91.

Körner, C., and Arnone, J. A. I. (1992). Responses to elevated carbon dioxide in artificial tropical ecosystems. *Science* **257**, 1672–1675.

Kosiyachinda, S., and Young, R. E. (1976). Chilling sensitivity of avocado fruit at different stages of respiratory climacteric. *J. Am. Soc. Hortic. Sci.* **101**, 665–667.

Kovac, M., and Kregar, I. (1989a). Protein metabolism in silver fir seeds during germination. *Plant Physiol. Biochem. (Paris)* **27**, 35–41.

Kovac, M., and Kregar, I. (1989b). Starch metabolism in silver fir seeds during germination. *Plant Physiol. Biochem. (Paris)* **27**, 873–880.

Kozlowski, T. T. (1949). Light and water in relation to growth and competition of Piedmont forest tree species. *Ecol. Monogr.* **19**, 207–231.

Kozlowski, T. T. (1958). Water relations and growth of trees. *J. For.* **56**, 498–502.

Kozlowski, T. T. (1960). Some problems in the use of herbicides in forestry. *Proc. North Cent. Weed Control Conf.* **17**, 1–10.

Kozlowski, T. T. (1962). Photosynthesis, climate, and tree growth. *In* "Tree Growth" (T. T. Kozlowski, ed.), pp. 149–170. Ronald, New York.

Kozlowski, T. T. (1965). Expansion and contraction of plants. *Adv. Front. Plant Sci.* **13**, 100–101.

Kozlowski, T. T. (1967a). Diurnal variation in stem diameters of small trees. *Bot. Gaz. (Chicago)* **128**, 60–68.

Kozlowski, T. T. (1967b). Growth and development of *Pinus resinosa* seedlings under controlled temperatures. *Adv. Front. Plant Sci.* **19**, 17–27.

Kozlowski, T. T. (1968a). Soil water and tree growth. *In* "The Ecology of Southern Forests" (N. E. Linnartz, ed.), pp. 30–57. Louisiana State Univ. Press, Baton Rouge.

Kozlowski, T. T. (1968b). Diurnal changes in diameters of fruits and tree stems of Montmorency cherry. *J. Hortic. Sci.* **43**, 1–15.

Kozlowski, T. T. (1968c). Water balance in shade trees. *Proc. 44th Int. Shade Tree Conf.*, 29–42.

Kozlowski, T. T. (1969). Tree physiology and forest pests. *J. For.* **69**, 118–122.

Kozlowski, T. T. (1971a). "Growth and Development of Trees. Volume 1: Seed Germination, Ontogeny, and Shoot Growth." Academic Press, New York.

Kozlowski, T. T. (1971b). "Growth and Development of Trees, Volume 2: Cambial Growth, Root Growth, and Reproductive Growth." Academic Press, New York.

Kozlowski, T. T. (1972a). Physiology of water stress. *U.S. For. Serv. Gen. Tech. Rep. INT* **INT-1**, 229–244.

Kozlowski, T. T. (1972b). Shrinking and swelling of plant tissues. *In* "Water Deficits and Plant Growth" (T. T. Kozlowski, ed.), Vol. 3, pp. 1–64. Academic Press, New York.

Kozlowski, T. T., ed. (1972c). "Seed Biology." Vols. 1–3. Academic Press, New York.

Kozlowski, T. T. (1973). Extent and significance of shedding of plant parts. *In* "Shedding of Plant Parts" (T. T. Kozlowski, ed.), pp. 1–44. Academic Press, New York.

Kozlowski, T. T. (1975). Effects of transplanting and site on water relations of trees. *Am. Nurseryman* **141**(9), 84–94.

Kozlowski, T. T. (1976a). Drought resistance and transplantability of shade trees. *U.S. For. Serv. Gen. Tech. Rep. NE* **NE-22**, 77–90.

Kozlowski, T. T. (1976b). Water supply and leaf shedding. *In* "Water Deficits and Plant Growth" (T. T. Kozlowski, ed.), Vol. 4, pp. 191–231. Academic Press, New York.

Kozlowski, T. T. (1977). How healthy plants grow. *In* "Plant Pathology—An Advanced Treatise" (J. Horsfall and E. Cowling, eds.), pp. 19–51. Academic Press, New York.

Kozlowski, T. T. (1979). "Tree Growth and Environmental Stresses." Univ. of Washington Press, Seattle.

Kozlowski, T. T. (1980a). Impacts of air pollution on forest ecosystems. *BioScience* **30**, 88–93.

Kozlowski, T. T. (1980b). Responses of shade trees to pollution. *J. Arboric.* **6**, 29–41.

Kozlowski, T. T. (1982a). Water supply and tree growth. Part I. Water deficits. *For. Abstr.* **43**, 57–95.

Kozlowski, T. T. (1982b). Water supply and tree growth. Part II. Flooding. *For. Abstr.* **43**, 145–161.

Kozlowski, T. T. (1983). Reduction in yield of forest and fruit trees by water and temperature stress. *In* "Crop Reactions to Water and Temperature Stresses in Humid, Temperate Climates" (C. D. Raper and P. J. Kramer, eds.), pp. 67–88. Westview, Boulder, Colorado.

Kozlowski, T. T. (1984a). Extent, causes, and impacts of flooding. *In* "Flooding and Plant Growth" (T. T. Kozlowski, ed.), pp. 1–7. Academic Press, New York.

Kozlowski, T. T. (1984b). Responses of woody plants to flooding. *In* "Flooding and Plant Growth" (T. T. Kozlowski, ed.), pp. 129–164. Academic Press, New York.

Kozlowski, T. T. (1984c). Plant responses to flooding of soil. *BioScience* **34**, 162–167.

Kozlowski, T. T., ed. (1984d). "Flooding and Plant Growth." Academic Press, New York.

Kozlowski, T. T. (1985a). Soil aeration, flooding, and tree growth. *J. Arboric.* **11**, 85–90.

Kozlowski, T. T. (1985b). Tree growth in response to environmental stresses. *J. Arboric.* **11**, 97–111.

Kozlowski, T. T. (1985c). Effects of SO_2 on plant community structure. *In* "Sulfur Dioxide and Vegetation: Physiology, Ecology, and Policy Issues" (W. E. Winner, H. A. Mooney, and R. Goldstein, eds.), pp. 431–451. Stanford Univ. Press, Stanford, California.

Kozlowski, T. T. (1985d). Effects of direct contact of *Pinus resinosa* seeds and young seedlings with *N*-dimethylaminosuccinamic acid, (2-chloroethyl) trimethylammonium chloride, or maleic hydrazide. *Can. J. For. Res.* **15**, 1000–1004.

Kozlowski, T. T. (1986a). Effects on seedling development of direct contact of *Pinus resinosa* seeds or young seedlings with captan. *Eur. J. For. Pathol.* **16**, 87–90.

Kozlowski, T. T. (1986b). Effects of 2,3,6-TBA on seed germination, early development, and mortality of *Pinus resinosa* seedlings. *Eur. J. For. Pathol.* **16**, 385–390.

Kozlowski, T. T. (1986c). Soil aeration and growth of forest trees. *Scand. J. For. Res.* **1**, 113–123.

Kozlowski, T. T. (1991). Effects of environmental stresses on deciduous trees. *In* "Response of Plants to Multiple Stresses" (H. A. Mooney, W. E. Winner, and E. J. Pell, eds.), pp. 391–411. Academic Press, San Diego.

Kozlowski, T. T. (1992). Carbohydrate sources and sinks in woody plants. *Bot. Rev.* **58**, 107–222.

Kozlowski, T. T. (1996). Reflections on research and editing. *In* "The Literature of Forestry and Agroforestry" (P. McDonald and J. Lassoie, eds.), pp. 198–235. Cornell Univ. Press, Ithaca, New York and London.

Kozlowski, T. T., and Borger, G. A. (1971). Effect of temperature and light intensity early in ontogeny on growth of *Pinus resinosa* seedlings. *Can. J. For. Res.* **1**, 57–65.

Kozlowski, T. T., and Clausen, J. J. (1965). Changes in moisture contents and dry weights of buds and leaves of forest trees. *Bot. Gaz. (Chicago)* **126**, 20–26.

Kozlowski, T. T., and Constantinidou, H. A. (1986a). Responses of woody plants to environmental pollution. Part I. Sources, types of pollutants, and plant responses. *For. Abstr.* **47**, 5–51.

Kozlowski, T. T., and Constantinidou, H. A. (1986b). Responses of woody plants to environmental pollution. Part II. Factors affecting responses to pollution. *For. Abstr.* **47**, 105–132.

Kozlowski, T. T., and Davies, W. J. (1975a). Control of water balance in transplanted trees. *J. Arboric.* **1**, 1–10.

Kozlowski, T. T., and Davies, W. J. (1975b). Control of water loss in shade trees. *J. Arboric.* **1**, 81–90.

Kozlowski, T. T., and Gentile, A. C. (1959). Influence of the seed coat on germination, water absorption, and oxygen uptake of eastern white pine seed. *For. Sci.* **5**, 389–395.

Kozlowski, T. T., and Greathouse, T. E. (1970). Shoot growth characteristics of tropical pines. *Unasylva* **24**, 1–10.

Kozlowski, T. T., and Huxley, P. A. (1983). The role of controlled environments in agroforestry research. *In* "Plant Research and Agroforestry" (P. A. Huxley, ed.), pp. 551–567. Pillans and Wilson, Edinburgh.

Kozlowski, T. T., and Kuntz, J. E. (1963). Effect of simazine, atrazine, propazine, and eptam on growth of pine seedlings. *Soil Sci.* **95**, 165–174.

Kozlowski, T. T., and Pallardy, S. G. (1979). Stomatal responses of *Fraxinus pennsylvanica* seedlings during and after flooding. *Physiol. Plant.* **46**, 155–158.

Kozlowski, T. T., and Pallardy, S. G. (1984). Effects of flooding on water, carbohydrate and mineral relations. *In* "Flooding and Plant Growth" (T. T. Kozlowski, ed.), pp. 165–193. Academic Press, New York.

Kozlowski, T. T., and Pallardy, S. G. (1997). "Physiology of Woody Plants," 2nd Ed. Academic Press, San Diego.

Kozlowski, T. T., and Peterson, T. A. (1962). Seasonal growth of dominant, intermediate, and suppressed red pine trees. *Bot. Gaz.* (*Chicago*) **124**, 146–154.

Kozlowski, T. T., and Sasaki, S. (1968a). Effects of direct contact of pine seeds or young seedlings with commercial formulations, active ingredients, or inert ingredients of triazine herbicides. *Can. J. Plant Sci.* **48**, 1–7.

Kozlowski, T. T., and Sasaki, S. (1968b). Germination and morphology of red pine seeds and seedlings in contact with EPTC, CDEC, CDAA, 2,4-D, and picloram. *Proc. Am. Soc. Hortic. Sci.* **93**, 655–662.

Kozlowski, T. T., and Sasaki, S. (1970). Effects of herbicides on seed germination and development of young pine seedlings. *In* "Proceedings of the International Symposium on Seed Physiology of Woody Plants," pp. 19–24. Poznan, Poland (1968).

Kozlowski, T. T., and Torrie, J. H. (1964). Effects of hydrogen peroxide on germination of eastern white pine seed. *Adv. Front. Plant Sci.* **9**, 131–144.

Kozlowski, T. T., and Torrie, J. H. (1965). Effect of soil incorporation of herbicides on seed germination and growth of pine seedlings. *Soil Sci.* **100**, 139–146.

Kozlowski, T. T., and Winget, C. H. (1964a). The role of reserves in leaves, branches, stems, and roots on shoot.growth of red pine. *Am. J. Bot.* **51**, 522–529.

Kozlowski, T. T., and Winget, C. H. (1964b). Diurnal and seasonal variation in radii of tree stems. *Ecology* **45**, 149–155.

Kozlowski, T. T., Winget, C. H., and Torrie, J. H. (1962). Daily radial growth of oak in relation to maximum and minimum temperature. *Bot. Gaz.* (*Chicago*) **124**, 9–17.

Kozlowski, T. T., Sasaki, S., and Torrie, J. H. (1967a). Influence of temperature on phytotoxicity of triazine herbicides to pine seedlings. *Am. J. Bot.* **54**, 790–796.

Kozlowski, T. T., Sasaki, S., and Torrie, J. H. (1967b). Effects of temperature on phytotoxicity of monuron, picloram, CDEC, EPTC, CDAA, and sesone to young pine seedlings. *Silva Fenn.* **32**, 13–28.

Kozlowski, T. T., Torrie, R. H., and Marshall, P. E. (1973). Predictability of shoot length from bud size in *Pinus resinosa*. *Can. J. For. Res.* **3**, 34–38.

Kozlowski, T. T., Kramer P. J., and Pallardy, S. G. (1991). "The Physiological Ecology of Woody Plants." Academic Press, San Diego.

Kramer, P. J. (1936). Effect of variation in length of day on growth and dormancy of trees. *Plant Physiol.* **11**, 127–137.

Kramer, P. J. (1937). Photoperiodic stimulation of growth by artificial light as a cause of winter killing. *Plant Physiol.* **12**, 881–883.

Kramer, P. J. (1957). Some effects of various combinations of day and night temperatures and photoperiod on height growth of loblolly pine seedlings. *For. Sci.* **3**, 45–55.

Kramer, P. J. (1969). "Plant and Soil Water Relationships: A Modern Synthesis." McGraw-Hill, New York.

Kramer, P. J. (1978). The use of controlled environments in research. *HortScience* **13**, 447–451.

Kramer, P. J. (1983). "Water Relations of Plants." Academic Press, San Diego.

Kramer, P. J., and Boyer, J. S. (1995). "Water Relations of Plants and Soils." Academic Press, San Diego.

Kramer, P. J., and Clark, W. S. (1947). A comparison of photosynthesis in individual pine needles and entire seedlings at various light intensities. *Plant Physiol.* **22**, 51–57.

Kramer, P. J., and Decker, J. P. (1944). Relation between light intensity and rate of photosynthesis of loblolly pine and certain hardwoods. *Plant Physiol.* **19**, 350–358.

Kramer, P. J., and Kozlowski, T. T. (1979). "Physiology of Woody Plants." Academic Press, New York.

Kramer, P. J., and Rose, R. W., Jr. (1986). Physiological characteristics of loblolly pine in relation to field performance. *In* "Nursery Management Practices for the Southern Pines" (D. B. South, ed.), pp. 416–440. Auburn University, Auburn, Alabama.

Krastanova, S., Perrin, M., Barbier, P., Demangeat, G., Cornuet, P., Bardonnet, N., Otten, L., Pinck, L., and Walter, B. (1995). Transformation of grapevine rootstocks with the coat protein gene of grapevine fanleaf nepovirus. *Plant Cell Rep.* **14**, 550–554.

Kriebel, H. B., and Wang, C. (1962). The interaction between provenance and degree of chilling in bud break of sugar maple. *Silvae Genet.* **11**, 125–130.

Kriedemann, P. E., and Barrs, H. (1981). Citrus orchards. *In* "Water Deficits and Plant Growth" (T. T. Kozlowski, ed.), Vol. 6, pp. 325–417. Academic Press, New York.

Kronenberg, H. G. (1985). Apple growing potentials in Europe. 2. Flowering dates. *Neth. J. Agric. Sci.* **3**, 45–52.

Kronenberg, H. G. (1994). Temperature influences on the flowering dates of *Syringa vulgaris* L. and *Sorbus aucuparia* L. *Sci. Hortic. (Amsterdam)* **57**, 59–71.

Kruger, B. M., and Manion, P. D. (1994). Antifungal compounds in plants: Effects of water stress. *Can. J. Bot.* **72**, 454–460.

Kubler, H. (1983). Mechanism of frost crack formation in trees—A review and synthesis. *For. Sci.* **29**, 559–568.

Kubler, H. (1987). Origin of frost cracks in stems of trees. *J. Arboric.* **13**, 93–97.

Kubler, H. (1988). Frost cracks in stems of trees. *Arboric. J.* **12**, 163–175.

Kuboi, T., and Yamada, Y. (1978). Regulation of the enzyme activities related to lignin synthesis in cell aggregates of tobacco cell cultures. *Biochim. Biophys. Acta* **542**, 181–190.

Kuc, J. A. (1976). Phytoalexins. *Encycl. Plant Physiol.* **4**, 632–652.

Kuhns, M. R., Garrett, H. E., Teskey, R. O., and Hinckley, T. M. (1985). Root growth of black walnut trees related to soil temperature, soil water potential and leaf water potential. *For. Sci.* **31**, 617–629.

Kulkarni, V. J. (1988). Chemical control of tree vigour and the promotion of flowering and fruiting in mango (*Mangifera indica* L.) using paclobutrazol. *J. Hortic. Sci.* **63**, 557–566.

Kulman, H. M. (1971). Effects of insect defoliation on growth and mortality of trees. *Annu. Rev. Entomol.* **16**, 289–324.

Kunik, T., Salomon, R., Zamir, D., Navot, N., Zeidan, M., Michelson, L., Gafni, Y., Czosnek, H. (1994). Transgenic tomato plants expressing the tomato leaf curl virus capsid protein are resistant to the virus. *Bio/Technology* **12**, 500–504.

Kuntz, J. E., and Kozlowski, T. T. (1963). Effect of herbicides and land preparation on replanting of heavy soils in Northern Wisconsin. *Univ. Wisc. For. Res.* **Note 89**.

Kuntz, J. E., Kozlowski, T. T., Wojahn, K. E., and Brener, W. H. (1964). Nursery weed control with Dacthal. *Tree Planters Notes* **61**, 8–10.

Küppers, M. (1994). Canopy gaps: Competitive light interception and economic space filling—A matter of whole-plant allocation. *In* "Exploitation of Environmental Heterogeneity by Plants: Ecological Processes Above and Below Ground" (M. M. Caldwell and R. W. Pearcy, eds.), pp. 111–144. Academic Press, San Diego.

Kurian, R. M., and Iyer, C. P. A. (1993a). Chemical regulation of tree size in mango (*Mangifera indica* L.) cv. Alphonso. I. Effects of growth retardants on vegetative growth and tree vigour. *J. Hortic. Sci.* **68**, 349–354.

Kurian, R. M., and Iyer, C. P. A. (1993b). Chemical regulation of tree size in mango (*Mangifera indica* L.) cv. Alphonso. II. Effects of growth retardants on flowering and fruit set. *J. Hortic. Sci.* **68**, 355–360.

Kurian, R. M., and Iyer, C. P. A. (1993c). Chemical regulation of tree size in mango (*Mangifera indica* L.) cv. Alphonso. III. Effects of growth retardants on yield and quality of fruits. *J. Hortic. Sci.* **68**, 361–364.

Kurian, R. M., Murti, G. S. R., and Iyer, C. P. A. (1991). Abscisic acid levels in shoot apices in relation to tree vigour in mango (*Mangifera indica* L.). *Gartenbauwissenschaft* **56**, 84–86.

Kurian, R. M., Murti, G. S. R., and Iyer, C. P. A. (1992). Changes in cytokinin level in mango leaf extracts following soil drenches with paclobutrazol. *Gartenbauwissenschaft* **57**, 84–87.

Kuroda, H., and Sagisaka, S. (1993). Ultrastructural changes in cortical cells of apple (*Malus pumila* Mill.) associated with cold hardiness. *Plant Cell Physiol.* **34**, 357–365.

Kwesiga, F. R., and Grace, J. (1986). The role of the red/far-red ratio in the response of tropical tree seedlings to shade. *Ann. Bot.* **57**, 283–290.

Labanauskas, C. K., and Handy, M. F. (1972). Nutrient removal by Valencia orange fruit from citrus orchards in California. *Calif. Agric.* **26**, 3–4.

Labyak, L. F., and Schumacher, F. X. (1954). The contribution of its branches to the main stem growth of loblolly pine. *J. For.* **52**, 333–337.

Lachaud, S. (1983). Xylogénèse chez les Dicotylédones arborescentes. IV. Influence des bourgeous, de l'acide β-indolylacétique et de l'acide gibberellique sur la réactivation cambial et la xylogénèse dans les jeunes tiges de Hêtre. *Can. J. Bot.* **61**, 1768–1774.

Lachaud, S. (1989). Participation of auxin and abscisic acid in the regulation of seasonal variations in cambial activity and xylogenesis. *Trees* **3**, 125–137.

Lachaud, S., and Bonnemain, J. L. (1982). Xylogenèse chez les Dicotylédones arborescentes. III. Transport de l'auxine et activité cambiale dans les jeunes tiges de Hêtre. *Can. J. Bot.* **60**, 869–876.

Lachaud, S., and Bonnemain, J. L. (1984). Seasonal variations in the polar transport pathways and retention sites of [^{3}H]indole-3-acetic acid in young branches of *Fagus sylvatica* L. *Planta* **161**, 207–215.

Ladiges, P. Y. (1977). Differential susceptibility of two populations of *Eucalyptus viminalis* Labill. to iron chlorosis. *Plant Soil* **48**, 581–597.

Lai, R., Wooley, D. J., and Lawes, G. S. (1989). Effect of leaf to fruit ratio on fruit growth of kiwifruit (*Actinidia deliciosa*). *Sci. Hortic. (Amsterdam)* **39**, 247–255.

Lakso, A. N. (1987). The importance of climate and microclimate to yield and quality in horticultural crops. *In* "International Conference on Agrometeorology" (F. Prodi, F. Rossi, and G. Cristoferi, eds.), pp. 287–298. Fondazione Cesena Agricultura Publications, Cesena, Italy.

Lakso, A. N., Robinson, T. L., and Pool, R. M. (1989). Canopy microclimate effects on patterns of fruiting and fruit development in apples and grapes. *In* "Manipulation of Fruiting" (C. J. Wright, ed.), pp. 263–274. Butterworth, London.

Lammer, D. L., and Gifford, D. J. (1989). Lodgepole pine seed germination. II. The seed proteins and their mobilization in the megagametophyte and embryonic axis. *Can. J. Bot.* **67**, 2544–2551.

Lampinen, B. D., Shackel, K. A., Southwick, S. M., Olson, B., and Yeager, J. T. (1995). Sensitivity of yield and fruit quality of French prune to water deprivation at different fruit growth stages. *J. Am. Soc. Hortic. Sci.* **120**, 139–147.

Lancaster, J. E. (1992). Regulation of skin color in apples. *Crit. Rev. Plant Sci.* **10**, 487–502.

Landis, T. D., Tinus, R. W., McDonald, S. E., and Barnett, J. P., eds. (1984). "The Container Tree Nursery Manual." U.S. Dept. of Agriculture, Forest Service Agriculture Handbook No. 674, Washington D.C.

Landmark, T., and Hällgren, J. E. (1987). Effects of frost on shaded and exposed spruce and pine seedlings planted in the field. *Can. J. For. Res.* **17**, 1197–1204.

Landry, B. (1993). DNA mapping in plants. *In* "Methods in Plant Molecular Biology and Biotechnology" (B. R. Glick and J. E. Thompson, eds.), pp. 269–285. CRC Press, Boca Raton, Florida.

Landsberg, J. J. (1986). "Physiological Ecology of Forest Production." Academic Press, New York.

Landsberg, J. J., and Jones, H. G. (1981). Apple orchards. *In* "Water Deficits and Plant Growth" (T. T. Kozlowski, ed.), Vol. 6, pp. 419–467. Academic Press, New York.

Lang, G. A. (1989). Dormancy—Models and manipulations of environmental/physiological regulation. *In* "Manipulation of Fruiting" (C. J. Wright, ed.), pp. 79–98. Butterworth, London.

Lange, E., and Fica, J. (1982). Storage of Spartan, Melrose and Idared apples in ultra-low oxygen controlled atmospheres. *Fruit Sci. Rep.* **9**, 123–131.

Langenfeld-Heyser, R. (1987). Distribution of leaf assimilates in the stem of *Picea abies* L. *Trees* **1**, 102–109.

Langenheim, J. H. (1984). The role of plant secondary chemicals in wet tropical ecosystems. *In* "Physiological Ecology of Plants of the Wet Tropics" (E. Medina, H. A. Mooney, and C. Vazquez-Yanes, eds.), pp. 189–208. Junk, The Hague.

Lanini, W. T., and Radosevich, S. R. (1986). Responses of three conifer species to site preparation and shrub control. *For. Sci.* **9**, 497–506.

Larocque, G. R., and Marshall, P. L. (1993). Evaluating the impact of competition using relative growth rate in red pine (*Pinus resinosa*) stands. *For. Ecol. Manage.* **58**, 65–83.

Larson, K. D., Davies, F. S., and Schaffer, B. (1991). Floodwater temperature and stem lenticel hypertrophy in *Mangifera indica* (Anacardiaceae). *Am. J. Bot.* **78**, 1397–1403.

Larson, P. R. (1960). A physiological consideration of the springwood–summerwood transition in red pine. *For. Sci.* **6**, 110–122.

Larson, P. R. (1962a). The indirect effect of photoperiod on tracheid diameter in red pine. *Am. J. Bot.* **49**, 132–137.

Larson, P. R. (1962b). Auxin gradients and the regulation of cambial activity. *In* "Tree Growth" (T. T. Kozlowski, ed.), pp. 97–117. Ronald, New York.

Larson, P. R. (1963a). Stem form and silviculture. *Proc. Soc. Am. For.*, 103–107.

Larson, P. R. (1963b). Stem form development in forest trees. *For. Sci. Monogr.* **5**, 1–42.

Larson, P. R. (1963c). The indirect effect of drought on tracheid diameter in red pine. *For. Sci.* **9**, 52–62.

Larson, P. R. (1964). Some indirect effects of environment on wood formation. *In* "The Formation of Wood in Forest Trees" (M. H. Zimmermann, ed.), pp. 345–365. Academic Press, New York.

Larson, P. R. (1965). Stem form of young *Larix* as influenced by wind and pruning. *For. Sci.* **11**, 412–424.

Larson, P. R. (1969). Wood formation and the concept of wood quality. *Bull. Yale Univ., Sch. For.* **74**.

Larson, P. R., and Dickson, R. E. (1986). ^{14}C translocation pathways in honey locust and green ash: Woody plants with complex leaf forms. *Physiol. Plant.* **66**, 21–30.

Larson, P. R., and Gordon, J. C. (1969). Leaf development, photosynthesis and ^{14}C distribution in *Populus deltoides* seedlings. *Am. J. Bot.* **56**, 1058–1066.

Larsson, S., Oren, R., Waring, R. H., and Barnett, J. W. (1983). Attacks of mountain pine beetle as related to tree vigor of ponderosa pine. *For. Sci.* **29**, 395–402.

Latimer, J. G., and Robitaille, H. A. (1981). Sources of variability in apple shoot selection and handling for bud rest determinations. *J. Am. Soc. Hortic. Sci.* **106**, 794–798.

Latsague, M., Acevedo, H., Fernandez, J., Romero, M., Cristi, R., and Alberdi, M. (1992). Frost-resistance and lipid composition of cold-hardened needles of Chilean conifers. *Phytochemistry* **31**, 3419–3426.

Lau, O. L., and Looney, N. E. (1982). Improvement of fruit firmness and acidity in controlled atmosphere-stored 'Golden Delicious' apples by a rapid O_2 reduction procedure. *J. Am. Soc. Hortic. Sci.* **107**, 531–534.

Lauri, P.-E. (1993). Long-term effects of (2*RS*,3*RS*)-paclobutrazol on vegetative and fruiting characteristics of sweet cherry spurs. *J. Hortic. Sci.* **68**, 149–159.

Lauriere, C. (1983). Enzymes and leaf senescence. *Physiol. Veg.* **21**, 1159–1177.

Lavee, S., and Nir, G. (1986). Grape. *In* "CRC Handbook of Fruit Set and Development" (S. P. Monselise, ed.), pp. 167–191. CRC Press, Boca Raton, Florida.

Lavee, S., Haskal, A., and Ben Tal, Y. (1983). Girdling olive trees, a partial solution to biennial bearing: I. Methods, timing and direct tree response. *J. Hortic. Sci.* **58**, 209–218.

Layne, R. E. C. (1987). Peach rootstocks. *In* "Rootstocks for Fruit Trees" (R. C. Rom and R. F. Carlson, eds.), pp. 185–216. Wiley, New York and Chichester.

Lazada, R., and Cardemil, L. (1990). Further characterization of sucrose uptake by cotyledons of *Araucaria araucana.* Energy requirements and specificity of the uptake system. *Plant Physiol. Biochem.* **28**, 773–778.

Leach, R. W. A., and Wareing, P. F. (1967). Distribution of auxin in horizontal woody stems in relation to gravimorphism. *Nature (London)* **214**, 1025–1027.

Leakey, R. R. B., and Coutts, M. P. (1989). The dynamics of rooting in *Triplochiton scleroxylon* cuttings: Their relation to leaf area, node position, dry weight accumulation, leaf water potential and carbohydrate composition. *Tree Physiol.* **5**, 135–146.

Lechowicz, M. (1984). Why do temperate deciduous trees leaf out at different times? Adaptations and ecology of forest communities. *Am. Nat.* **124**, 821–842.

Lee, E. H., Byun, J. K., and Wilding, S. J. (1985). A new gibberellin biosynthesis inhibitor paclobutrazol (PP333), confers increased SO_2 tolerance on snap bean plants. *Environ. Exp. Bot.* **25**, 265–275.

Lee, K. J. (1979). Factors affecting cone initiation in pines: A review. *Res. Rep. Inst. For. Genet. Korea* **15**, pp. 45–85.

Lee, S. S., Teng, S. Y., Lim, M. T., and Razali, A. K. (1988). Discolouration and heart rot of *Acacia mangium* Willd.—Some preliminary results. *J. Trop. For. Sci.* **1**, 170–177.

Lee, T. D. (1988). Patterns of fruit and seed production. *In* "Plant Reproductive Ecology—Patterns and Strategies" (J. L. Doust and L. L. Doust, eds.), pp. 179–202. Oxford, New York.

Legg, M. H., and Schneider, G. (1977). Soil deterioration in campsites: Northern forest types. *Soil Sci.* **41**, 437–441.

Leike, H. (1967). Wirkung von Gibberellinsäure und Kinetin auf ruhende Knospen verschiedener Gehölze. *Flora A* **158**, 351–362.

Leinhos, V., and Savidge, R. A. (1993). Isolation of protoplasts from developing xylem of *Pinus banksiana* and *Pinus strobus. Can. J. For. Res.* **23**, 343–348.

Leishman, M. R., and Westoby, M. (1994). The role of large seed size in shaded conditions: Experimental evidence. *Funct. Ecol.* **8**, 205–214.

Leopold, A. C., and Kriedemann, P. E. (1975). "Plant Growth and Development." McGraw-Hill, New York.

Leopold, A. C., and Nooden, L. D. (1984). Hormonal regulatory systems in plants. *Encycl. Plant Physiol. New Ser.* **10**, 4–22.

Leopold, A. C., Brown, K. M., and Emerson, F. H. (1972). Ethylene in the wood of stressed trees. *HortScience* **7**, 175.

Lerner, R. H., and Evenari, M. (1961). The nature of the germination inhibitor present in leaves of *Eucalyptus rostrata. Physiol. Plant.* **14**, 221–229.

Levitt, J. (1980a). "Responses of Plants to Environmental Stresses, Second Edition. Volume I: Chilling, Freezing, and High Temperature Stresses." Academic Press, New York.

Levitt, J. (1980b). "Responses of Plants to Environmental Stresses, Second Edition, Volume 2: Water, Radiation, Salt, and Other Stresses." Academic Press, New York.

Levy, Y., Bar-Akiva, A., and Vaadia, Y. (1978). Influence of irrigation and environmental factors on grapefruit acidity. *J. Am. Soc. Hortic. Sci.* **103**, 73–76.

Lev-Yadun, S., and Aloni, R. (1995). Differentiation of the ray system in woody plants. *Bot. Rev.* **61**, 45–84.

Lewak, S., and Rudnicki, R. M. (1977). After-ripening in cold requiring seeds. *In* "The Physiology and Biochemistry of Seed Dormancy and Germination" (A. A. Khan, ed.), pp. 193–217. North-Holland, Amsterdam.

Lewis, W. M. (1981). Precipitation chemistry and nutrient loading by precipitation in a tropical watershed. *Water Resour. Res.* **17**, 169–181.

Li, J.-R., Proctor, T. A., and Murr, D. P. (1985). Effects of cotyledon removal on apple seedling growth and distribution of ^{14}C-labelled photosynthates. *Can. J. Bot.* **63**, 1736–1739.

Li, L., and Ross, J. D. (1990). Lipid mobilization during dormancy breakage in oilseed of *Corylus avellana. Ann. Bot.* **66**, 501–505.

Li, P. H., and Weiser, C. J. (1967). Evaluation of extraction and assay methods for nucleic acids from red osier dogwood and RNA, DNA, and protein changes during cold acclimation. *Proc. Am. Soc. Hortic. Sci.* **91**, 716–727.

Li, S.-H., Huguet, J.-G., Schoch, P. G., and Orlando, P. (1989). Response of peach tree growth and cropping to soil water deficit at various phenological stages of fruit development. *J. Hortic. Sci.* **64**, 541–552.

Libby, W. J., and Ahuja, M. R. (1993). Micropropagation and clonal options in forestry. *In* "Micropropagation of Woody Plants" (M. R. Ahuja, ed.), pp. 425–442. Kluwer, Dordrecht, The Netherlands.

Lieffers, V. J., and Rothwell, R. L. (1987). Effects of drainage on substrate temperature and phenology of some trees and shrubs in an Alberta peatland. *Can. J. For. Res.* **17**, 97–104.

Likens, G. E., Bormann, F. H., Pierce, R. S., Eaton, J. S., and Johnson, N. M. (1979). "Biogeochemistry of a Forested Ecosystem." Springer-Verlag, New York.

Lill, R. E., O'Donaghue, E. M., and King, G. A. (1989). Postharvest physiology of peaches and nectarines. *Hortic. Rev.* **11**, 413–452.

Lin, C., and Powell, L. E. (1981). The effect of bud scales in dormancy of apple buds. *HortScience* **16**(Abstr.), 441.

Lin, C.-Y., Roberts, J. K., and Key, J. L. (1984). Acquisition of thermotolerance in soybean seedlings. *Plant Physiol.* **74**, 152–160.

Lin, C.-Y., Chen, Y.-M., and Key, J. L. (1985). Solute leakage in soybean seedlings under various heat shock regimes. *Plant Cell Physiol.* **26**, 1493–1498.

Lin, T.-P., Chen, M.-H., and Lin, C.-H. (1994). Dormancy in seeds of *Phellodendron wilsonii* is mediated in part by abscisic acid. *Plant Cell Physiol.* **35**, 115–119.

Lindow, S. E. (1983). The role of bacterial ice nucleation in frost injury to plants. *Annu. Rev. Phytopathol.* **21**, 363–384.

Linzon, S. N. (1966). Damage to eastern white pine by sulfur dioxide, semimature-tissue needle blight, and ozone. *J. Air Pollut. Control. Assoc.* **16**, 140–144.

Linzon, S. N. (1978). Effects of airborne sulfur pollutants on plants. *In* "Sulfur in the Environment" (J. A. Nriagu, ed.), pp. 109–162. Wiley, New York.

Linzon, S. N. (1986). Effects of gaseous pollutants on forests in eastern North America. *Water Air Soil Pollut.* **31**, 537–550.

Little, C. E. (1995). "The Dying of the Trees." Viking Penguin, New York.

Little, C. H. A. (1970). Apical dominance in long shoots of white pine (*Pinus strobus*). *Can. J. Bot.* **48**, 239–253.

Little, C. H. A. (1975). Inhibition of cambial activity in *Abies balsamea* by internal water stress: Role of abscisic acid. *Can. J. Bot.* **53**, 3041–3050.

Little, C. H. A. (1981). Effect of cambial dormancy state on the transport of [1-¹⁴C]indol-3-ylacetic acid in *Abies balsamea* shoots. *Can. J. Bot.* **59**, 342–348.

Little, C. H. A., and Bonga, J. M. (1974). Rest in the cambium of *Abies balsamea. Can. J. Bot.* **52**, 1723–1730.

Little, C. H. A., and Eidt, D. C. (1968). Effects of abscisic acid on budbreak and transpiration in woody species. *Nature (London)* **220**, 498–499.

Little, C. H. A., and Eidt, D. C. (1970). Relationship between transpiration and cambial activity in *Abies balsamea. Can. J. Bot.* **48**, 1027–1028.

Little, C. H. A., and Loach, K. (1975). Effect of gibberellic acid on growth and photosynthesis in *Abies balsamea. Can. J. Bot.* **53**, 1805–1810.

Little, C. H. A., and Savidge, R. A. (1987). The role of plant growth regulators in forest tree cambial growth. *Plant Growth Regul.* **6**, 137–169.

Little, C. H. A., and Wareing, P. F. (1981). Control of cambial activity and dormancy in *Picea sitchensis* by indole-3-acetic acid and abscisic acid. *Can. J. Bot.* **59**, 1480–1493.

Little, C. H. A., Sundberg, B., and Ericsson, A. (1990). Induction of acropetal ¹⁴C-photosynthate transport and radial growth by indole-3-acetic acid in *Pinus sylvestris* shoots. *Tree Physiol.* **6**, 177–189.

Litz, R. E., and Gray, D. J. (1992). Organogenesis and somatic embryogenesis. *Biotechnol. Agric.* **8**, 3–34.

Liu, J. R., Sink, K. C., and Dennis, F. G. (1983). Plant regeneration from apple seedling explants and callus cultures. *Plant Cell Tissue Org. Cult.* **2**, 293–304.

Liu, Z. J., and Dickmann, D. I. (1992a). Abscisic acid accumulation in leaves of two contrasting hybrid poplar clones affected by nitrogen fertilization plus cyclic flooding and soil drying. *Tree Physiol.* **11**, 109–122.

Liu, Z. J., and Dickmann, D. I. (1992b). Responses of two hybrid *Populus* clones to flooding, drought, and nitrogen availability. I. Morphology and growth. *Can. J. Bot.* **70**, 2265–2270.

Livingston, N. J., Von-Aderkas, P., Fuchs, E. E., and Reaney, M. J. T. (1992). Water relation parameters of embryogenic cultures and seedlings of larch. *Plant Physiol.* **100**, 1304–1309.

Livingston, R. B., and Allessio, M. L. (1968). Buried viable seed in successional field and forest stands, Harvard Forest, Massachusetts. *Bull. Torrey Bot. Club* **95**, 58–69.

Loach, K., and Little, C. H. A. (1973). Production, storage, and use of photosynthate during shoot elongation in balsam fir (*Abies balsamea*). *Can. J. Bot.* **51**, 1161–1168.

Lockhard, R. G., and Schneider, G. W. (1981). Stock and scion growth relationship and the dwarfing mechanisms in apple. *Hortic. Rev.* **3**, 315–375.

Lockhart, J. A. (1965). An analysis of irreversible plant cell elongation. *J. Theor. Biol.* **8**, 264–275.

Loehle, C. (1988a). Tree life history strategies. The role of defenses. *Can. J. For. Res.* **18**, 209–222.

Loehle, C. (1988b). Forest decline: Endogenous dynamics, tree defenses, and the elimination of spurious correlation. *Vegetatio* **77**, 65–78.

Logan, K. T. (1965). Growth of tree seedlings as affected by light intensity. I. White birch, yellow birch, sugar maple, and silver maple. *Can. For. Branch, Dep. Publ.* **1121**.

Logan, K. T. (1966a). Growth of tree seedlings as affected by light intensity. II. Red pine, white pine, jack pine, and eastern larch. *Can. For. Branch, Dep. Publ.* **1160**.

Logan, K. T. (1966b). Growth of tree seedlings as affected by light intensity. III. Basswood and white elm. *Can. For. Branch, Dep. Publ.* **1176**.

Lombard, P. B., and Westwood, M. N. (1987). Pear rootstocks. *In* "Rootstocks for Fruit Crops" (R. C. Rom and R. F. Carlson, eds.), pp. 145–183. Wiley, New York and Chichester.

Longman, K. A., and Jenik, J. (1974). "Tropical Forest and Its Environment." Longman, London.

Looney, N. E. (1983). Growth regulator usage in apple and pear production. *In* "Plant Growth Regulating Chemicals" (L. G. Nickell, ed.), Vol. 1, pp. 1–26. CRC Press, Boca Raton, Florida.

Looney, N. E., McGlasson, W. B., and Coombe, B. E. (1974). Control of fruit ripening in peach, *Prunus persica*: Action of succinic acid-2,2-dimethylhydrazide and (2-chloroethyl)-phosphonic acid. *Aust. J. Plant Physiol.* **1**, 77–86.

Looney, N. E., Pharis, R. P., and Noma, M. (1985). Promotion of flowering in apple trees with gibberellin A_4 and C-3 epi-gibberellin A_4. *Planta* **165**, 292–294.

Lord, E. M., and Eckard, K. (1987). Shoot development in *Citrus sinensis* L. (Washington Navel Orange). II. Alteration of developmental fate of flowering shoots after GA_3 treatment. *Bot. Gaz.* (*Chicago*) **148**, 17–22.

Lorio, P. L., Jr. (1980). Loblolly pine stocking levels affect potential for southern pine beetle infestations. *South. J. Appl. For.* **4**, 162–165.

Lotan, J. E., and Zahner, R. (1963). Shoot and needle responses of 20-year-old red pine to current soil moisture regimes. *For. Sci.* **9**, 497–506.

Lucier, A. A., and Haines, S. G. (1990). Overview and synthesis. *In* "Mechanisms of Forest Responses to Acidic Deposition" (A. A. Lucier and S. G. Haines, eds.), pp. 1–9. Springer-Verlag, New York and Berlin.

Luckwill, L. C. (1978). Meadow orchards and fruit walls. *Acta Hortic.* **65**, 237–243.

Lugo, A. E., Brown, S., and Chapman, J. (1988). An analytical review of production rates and stemwood biomass of tropical forest plantations. *For. Ecol. Manage.* **23**, 179–200.

Lurquin, P. F. (1989). Uptake and integration of exogenous DNA in plants. *In* "Plant Protoplasts and Genetic Engineering II" (Y. P. S. Bajaj, ed.), pp. 54–74. Springer-Verlag, Berlin.

Lürssen, K. (1987). The use of inhibitors of gibberellin and sterol biosynthesis to probe hormone action. *In* "Hormone Action in Plant Development—A Critical Appraisal" (G. V. Hoad, J. R. Lenton, M. B. Jackson, and R. K. Atkin, eds.), pp. 133–144. Butterworth, London.

Luxmoore, R. J., Grizzard, T., and Strand, R. H. (1981). Nutrient retranslocation in the outer canopy and understorey of an eastern deciduous forest. *For. Sci.* **27**, 505–518.

Luxmoore, R. J., Oren, R., Sheriff, D. W., and Thomas, R. B. (1995). Source–sink–storage relationships of conifers. *In* "Resource Physiology of Conifers: Acquisition, Allocation, and Utilization" (W. K. Smith and T. M. Hinckley, eds.), pp. 179–216. Academic Press, San Diego.

Luza, J. G., Polito, V. S., and Weinbaum, S. A. (1987). Staminate bloom date and temperature responses of pollen germination and tube growth in two walnut (*Juglans*) species. *Am. J. Bot.* **74**, 1898–1903.

Luza, J. G., van Gorsel, R., Polito, V. S., and Kader, A. A. (1992). Chilling injury in peaches: A cytochemical and ultrastructural cell wall study. *J. Am. Soc. Hortic. Sci.* **117**, 114–118.

Lynch, D. V. (1990). Chilling injury in plants: The relevance of membrane lipids. *In* "Environmental Injury to Plants" (F. Katterman, ed.), pp. 17–34. Academic Press, San Diego.

Lyons, J. M., and Breidenbach, R. W. (1987). Chilling injury. *In* "Postharvest Physiology of Vegetables" (J. Weichmann, ed.), pp. 305–326. Dekker, New York.

Lyr, H., and Garbe, V. (1995). The influence of root temperature on growth of *Pinus sylvestris, Fagus sylvatica, Tilia cordata,* and *Quercus robur. Trees* **9**, 220–223.

McBride, K. E., Svab, Z., Schaaf, D. J., Hogan, P. S., Stalker, D. M., and Maliga, P. (1995). Amplification of a chimeric *Bacillus* gene in chloroplasts leads to an extraordinary level of an insecticidal protein in tobacco. *Bio/Technology* **13**, 362–365.

McClenahen, J. R. (1983). The impact of an urban-industrial area on deciduous forest tree growth. *J. Environ. Qual.* **12**, 64–69.

McCown, B. H. (1985). From gene manipulation to forest establishment: Shoot cultures could be a central tool. *Tappi J.* **68**, 116–119.

McCown, B. H. (1988). Recent advances in protoplast culture of horticultural crops: Ornamental trees and shrubs. *Sci. Hortic. (Amsterdam)* **37**, 257–265.

McCown, B. H. (1989a). Birch. *In* "Biotechnology in Agriculture and Forestry, Volume 5: Trees II" (Y. P. S. Bajaj, ed.), pp. 324–341.

McCown, B. H. (1989b). The biotechnology of urban trees. *J. Arboric.* **15**, 77–83.

McCown, B. H., and Lloyd, G. B. (1983). A survey of the response of *Rhododendron* to *in vitro* culture. *Plant Cell Tissue Organ Cult.* **2**, 77–85.

McCown, B. H., and Russell, J. A. (1987). Protoplast culture in hardwoods. *In* "Cell and Tissue Culture in Forestry" (J. M. Bonga and D. J. Durzan, eds.), Vol. 2, pp. 16–30. Martinus Nijhoff, Dordrecht, The Netherlands.

McCown, B. H., and Sellmer, J. C. (1987). General media and vessels suitable for woody plant culture. *In* "Cell and Tissue Culture in Forestry" (J. M. Bonga and D. J. Durzan, eds.), Vol. 1, pp. 4–16. Martinus Nijhoff, Dordrecht, The Netherlands.

McCown, B., McCabe, D. E., Russell, D. R., Robison, D. J., Barton, K. A., and Raffa, K. F. (1991). Stable transformation of *Populus* and incorporation of pest resistance by electric discharge particle acceleration. *Plant Cell Rep.* **9**, 590–594.

McCracken, I. W., and Kozlowski, T. T. (1965). Thermal contraction in twigs. *Nature (London)* **208**, 910–912.

MacDonald, S. E., and Lieffers, V. J. (1990). Photosynthesis, water relations, and foliar nitrogen of *Picea mariana* and *Larix laricina* from drained and undrained peatlands. *Can. J. For. Res.* **20**, 995–1000.

MacDonald, J. E., and Owens, J. N. (1993a). Bud development in coastal Douglas-fir seedlings under controlled environment conditions. *Can. J. For. Res.* **23**, 1203–1212.

MacDonald, J. E., and Owens, J. N. (1993b). Bud development in coastal Douglas-fir seedlings in response to different dormancy-induction treatments. *Can. J. Bot.* **71**, 1280–1290.

McDonald, M. A., Hawkins, B. J., Prescott, C. E., and Kimmins, J. P. (1994). Growth and foliar nutrition of western red cedar fertilized with sewage sludge, pulp sludge, fish silage, and wood ash on northern Vancouver Island. *Can. J. For. Res.* **24**, 297–301.

MacDonald, S. E., and Lieffers, V. J. (1990). Photosynthesis, water relations, and foliar nitrogen of *Picea mariana* and *Larix laricina* from drained and undrained peatlands. *Can. J. For. Res.* **20**, 995–1000.

McGranahan, G. H., Leslie, C. A., Dandekar, A. M., Uratsu, S. L., and Yates, I. E. (1993). Transformation of pecan and regeneration of transgenic plants. *Plant Cell Rep.* **12**, 634–638.

McIntosh, R. (1984). Fertiliser experiments in established conifer stands. *Forest Record* **127**, 1–23. Forestry Commission, United Kingdom.

McKeand, S. E., and Frampton, L. J. (1984). Performance of tissue culture plantlets of loblolly pine *in vivo*. *In* "Proceedings of the International Symposium of Recent Advances in Forest Biotechnology," pp. 82–91. Michigan Biotechnology Institute, Michigan State University, East Lansing, Michigan.

McKee, W. H., McLeod, K. W., Davis, C. E., McKelvin, M. R., and Thomas, H. A. (1986). Growth response of loblolly pine to municipal and industrial sewage sludge applied at four ages on upper coastal plain sites. *In* "The Forest Alternative for Treatment and Utilization of Municipal and Industrial Wastewater and Sludge" (D. W. Cole, C. L. Henry, and W. L. Nutter, eds.), pp. 272–281. Univ. of Washington Press, Seattle.

McKersie, B. D., and Leshem, Y. Y. (1994). "Stress and Stress Coping in Cultivated Plants." Kluwer, Dordrecht, The Netherlands.

McLaughlin, D. L., Linzon, S. N., Dimma, D. E., and McIlveen, W. D. (1985). Sugar maple decline in Ontario. Report No. ARB-144-85-Phyto. Ministry of the Environment, Ontario, Canada.

McLaughlin, D. L., Downing, D. J., Blasing, T. J., Cook, E. R., and Adams, H. S. (1987). An analysis of climate and competition as contributors to decline of red spruce at high elevation Appalachian forests of the eastern United States. *Oecologia* **72**, 487–501.

McLaughlin, J. M., and Greene, D. W. (1991a). Fruit and hormones influence flowering of apple. I. Effect of cultivar. *J. Am. Soc. Hortic. Sci.* **116**, 446–449.

McLaughlin, J. M., and Greene, D. W. (1991b). Fruit and hormones influence flowering of apple. II. Effects of hormones. *J. Am. Soc. Hortic. Sci.* **116**, 450–453.

McLaughlin, S. B. (1985). Effects of air pollution on forests. *J. Air Pollut. Control Assoc.* **35**, 512–534.

McLaughlin, S. B., Andersen, C. P., Edwards, N. T., Roy, W. K., and Layton, P. A. (1990). Seasonal patterns of photosynthesis and respiration of red spruce saplings from two elevations in declining southern Appalachian stands. *Can. J. For. Res.* **20**, 485–495.

McLaughlin, S. B., Andersen, C. P., Hanson, P. J., Tjoelker, M. G., and Roy, W. K. (1991). Increased dark respiration and calcium deficiency of red spruce in relation to acidic deposition at high-elevation southern Appalachian Mountain sites. *Can. J. For. Res.* **21**, 1234–1244.

McLaughlin, S. B., Tjoelker, M. G., and Roy, W. K. (1993). Acid deposition alters red spruce physiology: Laboratory studies support field observations. *Can. J. For. Res.* **23**, 380–386.

McLaughlin, S. B., Layton, P. A., Adams, M. B., Edwards, N. T., Hanson, P. J., O'Neill, E. G., and Roy, W. K. (1994). Growth responses of 53 open-pollinated loblolly pine families to ozone and acid rain. *J. Environ. Qual.* **23**, 247–257.

McMinn, R. G. (1963). Characteristics of Douglas-fir root systems. *Can. J. Bot.* **41**, 105–122.

McMurtrie, R. E., Benson, M. L., Linden, S., Running, S. W., Talsma, T., Crane, W. J. B., and Myers, B. J. (1990). Water/nutrient interactions affecting the productivity of stands of *Pinus radiata*. *For. Ecol. Manage.* **30**, 415–423.

MacNair, M. R. (1993). The genetics of metal tolerance in vascular plants. *New Phytol.* **124**, 541–559.

McQueen-Mason, S. J., Durachko, D. M., and Cosgrove, D. J. (1992). Two endogenous proteins that induce cell wall extension in plants. *Plant Cell* **4**, 1425–1433.

McQueen-Mason, S. J., and Cosgrove, D. J. (1995). Expansion mode of action on cell walls. Analysis of wall hydrolysis, stress relaxation, and binding. *Plant Physiol.* **107**, 87–100.

Maestri, M., and Barros, R. M. (1977). Coffee. *In* "Ecophysiology of Tropical Crops" (P. de T. Alvim and T. T. Kozlowski, eds.), pp. 249–278. Academic Press, New York.

Magness, J. R., and Regeimbal, L. O. (1938). The nitrogen requirement of the apple. *Proc. Am. Soc. Hortic. Sci.* **36**, 51–55.

Magnussen, S., Smith, V. G., and Yeatman, C. W. (1986). Foliage and canopy characteristics in relation to aboveground dry matter increment of seven jack pine provenances. *Can. J. For. Res.* **16**, 464–470.

Mahall, B. E., and Callaway, R. M. (1992). Root communication mechanisms and intercommunity distributions of two Mojave desert shrubs. *Ecology* **73**, 2145–2151.

Maillard, P., Deléens, E., Daudet, F. A., Lacointe, A., and Frossard, J. S. (1994). Carbon economy in walnut seedlings during the acquisition of autotrophy studied by long-term labelling with $^{13}CO_2$. *Physiol. Plant.* **91**, 359–368.

Maisenhelder, L. C. (1969). Identifying juvenile seedlings in southern hardwood forests. *U.S.D.A. For. Serv. Res. Paper* **SO-47**.

Malone, S. R., and Ashworth, E. N. (1991). Freezing stress response in woody tissues observed using low-temperature scanning electron microscopy and freeze substitution techniques. *Plant Physiol.* **95**, 871–881.

Maltby, E. (1986). "Waterlogged Wealth." International Institute for Environment and Development, London and Washington, D.C.

Mamajev, S. A., and Shkarlet, O. D. (1972). Effects of air and soil pollution by industrial waste on the fructification of Scotch pine in the Urals. *Mitteil Forst. Bundes-Versuch., Wien* **97**, 443–450.

Manion, P. D. (1981). "Tree Disease Concepts." Prentice-Hall, Englewood Cliffs, New Jersey.

Manion, P. D., and Lachance, D. (1992). Forest decline concepts: An overview. *In* "Forest Decline Concepts" (P. D. Manion and D. Lachance, eds.), pp. 181–190. American Phytopathology Society, St. Paul, Minnesota.

Mante, S., Morgens, P. H., Scorza, R., Cordts, J. M., and Callahan A. M. (1991).

Agrobacterium-mediated transformation of plum (*Prunus domestica* L.) hypocotyl slices and regeneration of transgenic plants. *Bio/Technology* **9**, 853–857.

Margolis, H. A., and Waring, R. H. (1986). Carbon and nitrogen allocation patterns of Douglas-fir seedlings fertilized with nitrogen in autumn. II. Field performance. *Can. J. For. Res.* **16**, 903–909.

Margolis, H. A., Vezina, L. P., and Ouimet, R. (1988). Relation of light and nitrogen source to growth, nitrate reductase, and glutamine synthetase activity of jack pine seedlings. *Physiol. Plant.* **72**, 790–795.

Margolis, H., Oren, R., Whitehead, D., and Kaufmann, M. R. (1995). Leaf area dynamics of conifer forests. *In* "Ecophysiology of Coniferous Forests" (W. K. Smith and T. M. Hinckley, eds.), pp. 181–223. Academic Press, San Diego.

Marini, R. P., and Barden, J. A. (1987). Summer pruning of apple and peach trees. *Hortic. Rev.* **9**, 351–375.

Marks, G. C., Ditchburne, N., and Foster, R. C. (1968). Quantitative estimates of mycorrhiza population in radiata pine forests. *Aust. For.* **32**, 26–38.

Maronek, D., and Flint, H. L. (1974). Cold hardiness of needles of *Pinus strobus* L. as a function of geographic source. *For. Sci.* **20**, 135–141.

Marquard, R. D. (1985). Chemical growth regulation of pecan seedlings. *HortScience* **20**, 919–921.

Marquard, R. D., and Hanover, J. W. (1984). Relationship between gibberellin $GA_{4/7}$ concentration, time of treatment, and crown position on flowering of *Picea glauca*. *Can. J. For. Res.* **14**, 547–553.

Marschner, H. (1986). "The Mineral Nutrition of Higher Plants." Academic Press, London.

Marschner, H. (1995). "The Mineral Nutrition of Higher Plants," 2nd Ed. Academic Press, London and San Diego.

Marshall, P. E., and Kozlowski, T. T. (1974a). The role of cotyledons in growth and development of woody angiosperms. *Can. J. Bot.* **52**, 239–245.

Marshall, P. E., and Kozlowski, T. T. (1974b). Photosynthetic activity of cotyledons and foliage leaves of young angiosperm seedlings. *Can. J. Bot.* **52**, 2023–2032.

Marshall, P. E., and Kozlowski, T. T. (1975). Changes in mineral contents of cotyledons and young seedlings of woody angiosperms. *Can. J. Bot.* **53**, 2026–2031.

Marshall, P. E., and Kozlowski, T. T. (1976a). Importance of photosynthetic cotyledons for early growth of woody angiosperms. *Physiol. Plant.* **37**, 336–340.

Marshall, P. E., and Kozlowski, T. T. (1976b). Importance of endosperm for nutrition of *Fraxinus pennsylvanica* seedlings. *J. Exp. Bot.* **27**, 572–574.

Marshall, P. E., and Kozlowski, T. T. (1976c). Compositional changes in cotyledons of woody angiosperms. *Can. J. Bot.* **54**, 2473–2477.

Marshall, P. E., and Kozlowski, T. T. (1977). Changes in structure and function of epigeous cotyledons of woody angiosperms during seedling growth. *Can. J. Bot.* **55**, 208–215.

Marth, P. C., Havis, L., and Batjer, L. P. (1947). Further results with growth regulators in retarding flower opening of peaches. *Proc. Am. Soc. Hortic. Sci.* **49**, 49–54.

Martin, G. C. (1991). Bud dormancy in deciduous fruit trees. *In* "Plant Physiology: A Treatise" (F. C. Steward, ed.), Vol. 10, pp. 183–225. Academic Press, San Diego.

Marx, D. H. (1980). Ectomycorrhizal fungus inoculations: A tool for improving forestation practices. *In* "Tropical Mycorrhiza Research" (P. Mikola, ed.), pp. 13–71. Oxford Univ. Press, London.

Marx, D. H., and Artman, J. D. (1979). *Pisolithus tinctorius* ectomycorrhizae improve survival and growth of pine seedlings on acid coal spoils in Kentucky and Virginia. *Reclam. Rev.* **2**, 23–31.

Marx, D. H., Ruehle, J. L., Kenny, D. S., Cordell, C. E., Riffle, J. W., Molina, R. J., Pawuk, W. H., Navratil, S., Tinus, R. W., and Goodwin, O. C. (1982). Commercial vegetative inoculum of *Pisolithus tinctorius* and inoculation techniques for development of ectomycorrhizae on container-grown tree seedlings. *For. Sci.* **28**, 373–400.

Marx, D. H., Cordell, C. E., Kenney, J. G., Mexal, J. G., Artman, J. D., Riffle, J. W., and Molina, R. J. (1984). Commercial vegetative inoculum of *Pisolithus tinctorius* and inoculum techniques for development of ectomycorrhizae on bare-root tree seedlings. *For. Sci. Monogr.* **25**.

Marx, D. H., Cordell, C. E., Maul, S. B., and Ruehle, J. L. (1989). Ectomycorrhizal development on pine by *Pisolithus tinctorius* in bare-root and container seedling nurseries. *New Forests* **3**, 45–56.

Masaru, N., Syozo, F., and Saburo, K. (1976). Effects of exposure to various injurious gases on germination of lily pollen. *Environ. Pollut.* **11**, 181–187.

Matheson, N. K. (1984). The synthesis of reserve oligosaccharides and polysaccharides in seeds. *In* "Seed Physiology" (D. R. Murray, ed.), Vol. 1, pp. 167–208. Academic Press, Sydney.

Matson, P. A., Hain, F. P., and Mawby, W. (1987). Indices of tree susceptibility to bark beetles vary with silvicultural treatment in a loblolly pine plantation. *For. Ecol. Manage.* **22**, 107–118.

Mathews, H., Litz, R. E., Wilde, D. E., Merkle, S. A., Wetzstein, H. Y. (1992). Stable integration and expression of β-glucuronidase and NPT II genes in mango somatic embryos. *In Vitro Cell Dev. Biol.* **29P**, 172–178.

Matthews, J. D. (1963). Factors affecting the production of seed by forest trees. *For. Abstr.* **24**, 1–13.

Mattson, W. J., and Haack, R. A. (1987a). The role of drought in outbreaks of leaf-eating insects. *In* "Insect Outbreaks: Ecological and Evolutionary Perspectives" (P. Barbosa and J. C. Schultz, eds.), pp. 365–407. Academic Press, Orlando, Florida.

Mattson, W. J., and Haack, R. A. (1987b). Effects of drought on host plants, phytophagous insects, and their natural enemies to induce insect outbreaks. *BioScience* **37**, 110–118.

Matyssek, R., Günthardt-Goerg, M. S., Saurer, M., and Keller, T. (1992). Seasonal growth δ^{13} of leaves and stem and phloem structure in birch (*Betula pendula*) under low ozone concentrations. *Trees* **6**, 69–76.

Matyssek, R., Keller, T., and Koike, T. (1993). Branch growth and leaf gas exchange of *Populus tremula* exposed to low ozone concentrations throughout two growing seasons. *Environ. Pollut.* **79**, 1–7.

Matyssek, R., Reich, P., Oren, R., and Winner, W. E. (1995). Response mechanisms of conifers to air pollutants. *In* "Ecophysiology of Coniferous Forests" (W. K. Smith and T. M. Hinckley, eds.), pp. 255–308. Academic Press, San Diego.

May, F. E., and Ash, J. E. (1990). An assessment of the allelopathic potential of *Eucalyptus. Aust. J. Bot.* **38**, 245–254.

May, G. M., and Pritts, M. P. (1993). Phosphorus, zinc, and boron influence yield components in 'Earliglow' strawberry. *J. Am. Soc. Hortic. Sci.* **118**, 43–49.

Mayer, A. M., and Marbach, I. (1981). Biochemistry of the transition from resting to germinating state in seeds. *In* "Progress in Biochemistry" (L. Reinhold *et al.*, eds.), Vol. 7, pp. 95–136. Pergamon, Oxford.

Mayer, A. M., and Poljakoff-Mayber, A. (1963). "The Germination of Seeds." Mac-Millan, New York.

Mayer, A. M., and Poljakoff-Mayber, A. (1989). "The Germination of Seeds." 4th Ed. Pergamon, Oxford and New York.

Mazzoleni, S., and Dickmann, D. I. (1988). Differential physiological and morphological responses of two hybrid *Populus* clones to water stress. *Tree Physiol.* **4**, 61–78.

Mead, D. J. (1984). Diagnosis of nutrient deficiencies in plantations. *In* "Nutrition of Plantation Forests" (G. D. Bowen and E. K. S. Nambiar, eds.), pp. 259–291. Academic Press, London.

Medina, C., and Cardemil, L. (1993). *Prosopis chilensis* is a plant highly resistant to heat shock. *Plant Cell Environ.* **16**, 305–310.

Meharg, A. A. (1994). Integrated tolerance mechanisms: Constitutive and adaptive plant responses to elevated metal concentrations in the environment: Opinion. *Plant Cell Environ.* **17**, 989–993.

Mejnartowicz, L. E. (1984). Enzymatic investigations on tolerance in forest trees. *In* "Gaseous Air Pollutants and Plant Metabolism" (M. J. Koziol and F. R. Whatley, eds.), pp. 381–398. Butterworth, London.

Melillo, J. M., Steudler, P. A., Aber, J. D., and Bowden, R. D. (1989). Atmospheric deposition and nutrient cycling. *In* "Exchange of Trace Gases between Terrestrial Ecosystems and the Atmosphere" (M. O. Andreae and D. S. Schimel, eds.), pp. 263–280. Wiley, Chichester.

Melillo, J. M., McGuire, A. D., Kicklighter, D. W., Moore III, B., Vorosmarty, C. J., and Schloss, A. L. (1993). Global climate change and terrestrial net primary productivity. *Nature (London)* **363**, 234–240.

Meng, X., and Seymour, R. S. (1992). Influence of soil drainage on early development and biomass production of young herbicide-released fir-spruce stands in north central Maine. *Can. J. For. Res.* **22**, 955–967.

Menzel, C. M. (1983). The control of floral initiation in lychee: A review. *Sci. Hortic. (Amsterdam)* **21**, 201–215.

Menzies, M. I., Holden, D. G., Rook, D. A., and Hardacre, A. K. (1981). Seasonal frost tolerance of *Eucalyptus saligna, E. regnans* and *E. fastigiata. N. Z. J. For. Sci.* **11**, 254–261.

Merkle, S. A. (1991). Maturation of yellow-poplar somatic embryos. *In* "Woody Plant Biotechnology" (M. R. Ahuja, ed.), pp. 179–187. Plenum, New York.

Merkle, S. A., Wetzstein, H. Y., and Sommer, H. E. (1987). Somatic embryogenesis in tissue cultures of pecan. *HortScience* **22**, 128–130.

Metcalfe, J. C., Davies, W. J., and Pereira, J. S. (1990). Leaf growth of *Eucalyptus globulus* seedlings under water deficit. *Tree Physiol.* **6**, 221–228.

Meyer, O. (1993). Functional groups of microorganisms. *In* "Biodiversity and Ecosystem Functioning" (E.-D. Schulze and H. A. Mooney, eds.), pp. 67–96. Springer-Verlag, Berlin and New York.

Meynhardt, J. T., and Malan, A. H. (1963). Translocation of sugars in double-stem grape vines. *S. Afr. J. Agric. Sci.* **6**, 337–338.

Michael, J. L. (1980). Long-term impact of aerial application of 2,4,5-T to longleaf pine (*Pinus palustris*). *Weed Sci.* **28**, 255–257.

Michelena, V. A., and Boyer, J. S. (1982). Complete turgor maintenance at low water potentials in the elongating region of maize leaves. *Plant Physiol.* **69**, 1145–1149.

Michler, C. H. (1991). Biotechnology of woody environmental crops. *HortScience* **26**, 1042–1044.

Michler, C. H., and Haissig, B. E. (1988). Increased herbicide tolerance of *in vitro* selected hybrid poplar. *In* "Somatic Cell Genetics of Woody Plants" (M. R. Ahuja, ed.), pp. 183–189. Kluwer, Dordrecht, The Netherlands.

Mielke, E. A., and Dennis, F. G., Jr. (1978). Hormonal control of flower bud dormancy in sour cherry (*Prunus cerasus* L.). II. Effects of leaves, defoliation and temperature on levels of abscisic acid in flower primordia. *J. Am. Soc. Hortic. Sci.* **103**, 446–449.

Mies, E., and Zoettl, H. W. (1985). Zeitliche Aenderung der Chlorophyll- und Elementgehalte in den Nadeln eines gelbchlorotischen Fichtenbestandes. *Forstwiss. Centralbl.* **104**, 1–8.

Mika, A. (1986). Physiological responses of fruit trees to pruning. *Hortic. Rev.* **8**, 338–378.

Mikami, S., Asakawa, S., Tizuka, M., Yokoyama, T., Nagao, A., Takehana, S., and Kaneko, T. (1980). Flower induction in Japanese larch. *Biol. Abstr.* **70**, 36943.

Miki, B. L., Fobert, P. F., Charest, P. J., and Iyer, V. N. (1993). Procedures for introducing foreign DNA into plants. *In* "Methods in Plant Molecular Biology and Biotechnology" (B. R. Glick and J. E. Thompson, eds.), pp. 67–88. CRC Press, Boca Raton, Florida.

Mikola, P. (1989). Suot suomen metsätaloudessa. *Suo* **40**, 71–78.

Millard, P., and Proe, M. F. (1992). Storage and internal cycling of nitrogen in relation to seasonal growth of Sitka spruce. *Tree Physiol.* **10**, 33–43.

Miller, A. N., Walsh, C. S., and Cohen, J. D. (1987). Measurement of indole-3-acetic acid in peach fruits (*Prunus persica* L. Batch cv. Redhaven) during development. *Plant Physiol.* **84**, 491–494.

Miller, A. R., Crawford, D. L., and Roberts, L. W. (1985). Lignification and xylogenesis in *Lactuca* pith explants cultured *in vitro* in the presence of auxin and cytokinin. A role for endogenous ethylene. *J. Exp. Bot.* **36**, 110–118.

Miller, H. G. (1981). Forest fertilization: Some guiding concepts. *Forestry* **54**, 157–167.

Miller, J. H., Zutter, B. R., Zedaker, S. M., Edwards, M. B., Heywood, J. D., and Newbold, R. A. (1991). A regional study on the influence of woody and herbaceous competition on early loblolly pine growth. *South. J. Appl. For.* **15**, 169–178.

Miller, L. N. (1965). Changes in radiosensitivity of pine seedlings subjected to water stress during chronic gamma irradiation. *Health Phys.* **11**, 1653–1662.

Miller, P. R. (1989). Forest decline syndromes in the southeastern United States. *In* "Air Pollution Toll on Forests and Crops" (J. J. MacKenzie and M. T. El-Ashry, eds.), pp. 75–112. Yale Univ. Press, New Haven, Connecticut.

Miller, P. R., and McBride, J. R. (1975). Effects of air pollutants on forests. *In* "Responses of Plants to Air Pollution" (J. B. Mudd and T. T. Kozlowski, eds.), pp. 195–235. Academic Press, New York.

Miller, R. E., and Tarrant, R. F. (1983). Long-term growth response of Douglas-fir to ammonium nitrate fertilzer. *For. Sci.* **29**, 127–137.

Milne, R., Satlin, M., Deans, J. D., Jarvis, P. G., and Cannell, M. G. R. (1992). The biomass production of three poplar clones in relation to intercepted solar radiation. *For. Ecol. Manage.* **55**, 1–14.

Minore, D., and Weatherly, H. G. (1990). Effects of site preparation on Douglas-fir seedling growth and survival. *West. J. Appl. For.* **5**, 49–51.

Mircetich, S. M., Sanborn, R. R., and Ramos, D. E. (1980). Natural spread, graft transmission and possible etiology of walnut blackline disease. *Phytopathology* **70**, 962–968.

Mitchell, P. D., Jerie, P. H., and Chalmers, D. J. (1984). The effects of regulated water deficits on pear tree growth, flowering, fruit growth and yield. *J. Am. Soc. Hortic. Sci.* **109**, 604–606.

Mitchell, P. D., Chalmers, D. J., Jerie, P. H., and Burge, G. (1986). The use of initial withholding of irrigation and tree spacing to enhance the effect of regulated deficit irrigation in pear trees. *J. Am. Soc. Hortic. Sci.* **111**, 858–864.

Mitton, J. B. (1995). Genetics and the physiological ecology of conifers. *In* "Ecophysiology of Coniferous Forests" (W. K. Smith and T. M. Hinckley, eds.), pp. 1–36. Academic Press, San Diego.

Miyanishi, K., and Kellman, M. (1986). The role of fire in recruitment of two neotropical savanna shrubs, *Miconia albicans* and *Clidemia sericea. Biotropica* **18**, 224–230.

Mobayen, R. G., and Milthorpe, F. L. (1980). Response of seedlings of three citrus rootstock cultivars to salinity. *Aust. J. Agric. Res.* **31**, 117–121.

Modlibowska, I., and Ruxton, J. P. (1953). Preliminary studies of spring frost resistance of black currant varieties. *Rep. East Malling Res. Stn. (Maidstone, Engl.)* **1952**, 67–72.

Moffat, A. J., Mathews, R. W., and Hall, J. E. (1991). The effects of sewage sludge on growth and foliar and soil chemistry in pole-stage Corsican pine at Ringwood Forest, Dorset, U.K. *Can. J. For. Res.* **21**, 902–909.

Mohamed-Yasseen, Y., Barringer, S. A., Splittstoesser, W. E., and Costanza, S. (1994). The role of seed coats in seed viability. *Bot. Rev.* **60**, 426–439.

Mohammed, G. H., and Vidaver, W. E. (1990). The influence of acclimatization treatment and plantlet morphology on early greenhouse-performance of tissue-cultured Douglas-fir [*Pseudotsuga menziesii* (Mirb.) Franco]. *Plant Cell Tissue Organ Cult.* **21**, 111–117.

Moing, A., and Gaudillère, J. P. (1992). Carbon and nitrogen partitioning in peach/plum grafts. *Tree Physiol.* **10**, 81–92.

Moncusin, C., and Gaspar, T. (1983). Perioxidase as a marker for rooting improvement of *Cynara scolymus* L. cultured *in vitro. Biochem. Physiol. Pflanz.* **178**, 263–271.

Monselise, S. P. (1951). Light distribution in citrus trees. *Bull. Res. Council Isr.* **1**, 36–53.

Mooney, H. A., and Dunn, E. L. (1970). Photosynthetic systems of Mediterranean climate shrubs and trees of California and Chile. *Am. Midl. Nat.* **104**, 447–453.

Mooney, H. A., and Hays, R. I. (1973). Carbohydrate storage cycles in two California Mediterranean-climate trees. *Flora* **162**, 295–304.

Mooney, H. A., and Koch, G. W. (1994). The impact of rising CO_2 concentrations on the terrestrial biosphere. *Ambio* **23**, 74–76.

Mooney, H. A., Dunn, E. L., Shropshire, F., and Song, L., Jr. (1970). Vegetative comparisons between the Mediterranean climatic areas of California and Chile. *Flora* **159**, 480–496.

Mooney, H. A., Drake, B. G., Luxmoore, R. J., Oechel, W. C., and Pitelka, L. F. (1991a). Predicting ecosystem responses to elevated CO_2 concentrations. *BioScience* **41**, 96–104.

Mooney, H. A., Medina, E., Schindler, D. W., Schulze, E.-D., and Walker, B. H., eds. (1991b). "Ecosystem Experiments." Wiley, Chichester and New York.

Moore, R. (1984). A model for graft compatibility–incompatibility in higher plants. *Am. J. Bot.* **71**, 752–758.

Moore, R. (1986). Graft incompatibility between pear and quince: The influence of metabolites of *Cydonia oblonga* on suspension cultures of *Pyrus communis*. *Am. J. Bot.* **73**, 1–4.

Moore, R., and Walker, D. B. (1983). Studies on vegetative compatibility–incompatibility in higher plants. VI. Grafting of *Sedum* and *Solanum* callus tissue *in vitro*. *Protoplasma* **115**, 114–121.

Mopper, S., and Whitham, T. G. (1992). The plant stress paradox—Effects of pinyon sawfly sex ratios and fecundity. *Ecology* **73**, 515–525.

Moreau, R. A., and Huang, A. H. C. (1977). Gluconeogenesis from storage wax in the cotyledons of jojoba seedlings. *Plant Physiol.* **60**, 329–333.

Morel, G. (1960). Producing virus-free cymbidiums. *Am. Orchid Soc. Bull.* **29**, 495–497.

Moreno-Casasola, P., Grime, J. P., and Martinez, M. L. (1994). A comparative study of the effects of fluctuations in temperature and moisture supply on hard coat dormancy in seeds of coastal tropical legumes in Mexico. *J. Trop. Ecol.* **10**, 67–86.

Morey, P. R. (1973). "How Trees Grow." Arnold, London.

Morikawa, Y., Hattori, S., and Kiyono, Y. (1986). Transpiration of a 31-year-old *Chamaecyparis obtusa* Endl. stand before and after thinning. *Tree Physiol.* **2**, 105–114.

Morini, S., Guerriero, R., and Scalibrelli, G. (1976). Winter treatments with SADH on peach trees. Effects on bud opening. *Riv. Ortoflorofruttic. Ital.* **6**, 356–363.

Morris, L. A., and Lowery, R. F. (1988). Influence of site preparation on soil conditions affecting stand establishment and tree growth. *South. J. Appl. For.* **12**, 170–178.

Morris, L. A., Pritchett, W. L., and Swindel, B. F. (1983). Displacement of nutrients into windrows during site preparation of a flatwood forest. *Soil Sci. Soc. Am. J.* **47**, 591–594.

Morris, L. L. (1982). Chilling injury of horticultural crops: An overview. *HortScience* **17**, 161–164.

Morrison, D. A., Auld, T. D., Rish, S., Porter, C., and McClay, K. (1992). Patterns of testa-imposed seed dormancy in native Australian legumes. *Ann. Bot.* **70**, 157–163.

Moss, G. I. (1969). Influence of temperature and photoperiod on flower induction and inflorescence development in sweet orange [*Citrus sinensis* (L.) Osbeck]. *J. Hortic. Sci.* **44**, 311–320.

Mosse, B. (1962). Graft-incompatibility in fruit trees with particular reference to its underlying causes. *Tech. Commun. Bur. Hortic. Plant. Crops, East Malling* **28**.

Mudd, J. B., and Kozlowski, T. T., eds. (1975). "Responses of Plants to Air Pollution." Academic Press, New York.

Mudge, K. W. (1988). Effect of ethylene on rooting. *In* "Adventitious Root Formation in Cuttings" (T. D. Davis, B. E. Haissig, and N. Sankhla, eds.), pp. 150–161. Dioscorides Press, Portland, Oregon.

Muir, W. (1978). Pest control—A perspective. *In* "Pest Control Strategies" (E. H. Smith and D. Pimentel, eds.), pp. 3–7. Academic Press, New York.

Mulkey, S. S., Wright, S. J., and Smith, A. P. (1993). Comparative physiology and demography of three neotropical forest shrubs: Alternative shade-adaptive character syndromes. *Oecologia* **96**, 526–536.

Mullin, R. E. (1964). Reduction in growth of white spruce after outplanting. *For. Chron.* **40**, 488–502.

Munns, R. (1993). Physiological processes limiting plant growth in saline soils: Some dogmas and hypotheses. *Plant Cell Environ.* **16**, 15–24.

Murashige, T. (1974). Plant propagation through tissue cultures. *Annu. Rev. Plant Physiol.* **25**, 135–166.

Murashige, T. (1977). Plant cell culture as horticultural practices. *Acta Hortic.* **78**, 17–30.

Murdock, B. A., and Ferguson, N. H. (1990). Effects of fall ethephon and gibberellic acid applications on bloom delay, flowering, and fruiting of plum. *HortScience* **25**, 1110.

Murneek, A. E. (1942). Quantitative distribution of nitrogen and carbohydrates in apple trees. *Res. Bull.—Mo., Agric. Exp. Stn.* **348**.

Murphy, J. B., and Noland, T. L. (1980). Activity of L-phenylalanine ammonia-lyase during stratification and germination of sugar pine seeds. *Z. Pflanzenphysiol.* **97**, 161–169.

Murphy, J. B., and Noland, T. L. (1981a). Activity of L-phenylalanine ammonia-lyase during stratification. *Physiol. Plant.* **52**, 370–374.

Murphy, J. B., and Noland, T. L. (1981b). Changes in phenolic acids and abscisic acid in sugar pine embryos and megagametophytes during stratification. *Physiol. Plant.* **52**, 375–379.

Murphy, J. B., and Noland, T. L. (1982). Temperature effects on seed imbibition and leakage mediated by viscosity and membranes. *Plant Physiol.* **69**, 428–431.

Murphy, P. G., and Lyugo, A. E. (1986). Ecology of tropical dry forest. *Annu. Rev. Ecol. Syst.* **17**, 67–88.

Murray, D. R. (1984). "Seed Physiology," Vols. 1 and 2. Academic Press, Sydney.

Myers, C. A. (1963). Vertical distribution of annual increment in thinned ponderosa pine. *For. Sci.* **9**, 394–404.

Nadelhoffer, K. J., and Raich, J. W. (1992). Fine root production estimates and belowground carbon allocation in forest ecosystems. *Ecology* **73**, 1139–1147.

Nadelhoffer, K. J., Downs, M. R., Fry, B., Aber, J. D., Magill, A. H., and Melillo, J. (1995). The fate of ^{15}N-labelled nitrate additions to a northern hardwood forest in eastern Maine, U.S.A. *Oecologia* **103**, 292–301.

Nair, P. K. R. (1989). "Agroforestry Systems in the Tropics." Kluwer, Dordrecht, The Netherlands.

Nambiar, E. K. S., and Fife, D. N. (1987). Growth and nutrient retranslocation in needles of radiata pine in relation to nitrogen supply. *Ann. Bot.* **60**, 147–156.

Nambiar, E. K. S., and Fife, D. N. (1991). Nutrient retranslocation in temperate conifers. *Tree Physiol.* 9, 185–207.

Näsholm, T., and Ericsson, A. (1990). Seasonal changes in amino acids, protein and total nitrogen in needles of fertilized Scots pine trees. *Tree Physiol.* **6**, 267–281.

Näsholm, T., Edfast, A.-B., Ericsson, A., and Norden, L.-G. (1994). Accumulation of amino acids in some boreal forest plants in response to increased nitrogen availability. *New Phytol.* **126**, 137–143.

National Academy of Sciences. (1977a). Effects of nitrogen oxides on vegetation. *In* "Nitrogen Oxides" Natl. Acad. Sci., Washington, D.C.

National Academy of Sciences. (1977b). "Ozone and Other Photochemical Oxidants." Natl. Acad. Sci., Washington, D.C.

National Academy of Sciences. (1979). "Tropical Legumes: Resources for the Future." Natl. Acad. Sci., Washington, D.C.

Navratil, S., and McLaughlin, M. C. (1979). Field survey techniques can detect SO_2 pollution effects on white pine of up to 120 km. *Phytopathology* **69**, 918.

Naylor, A. W. (1984). Functions of hormones at the organ level of organisation. *Encycl. Plant Physiol. New Ser.* **10**, 192–204.

Neale, D. B., and Williams, C. G. (1991). Restriction fragment length polymorphism mapping in conifers and applications to forest genetics and tree improvement. *Can. J. For. Res.* **21**, 545–554.

Neave, I. A., Dawson, J. O., and DeLucia, E. H. (1989). Autumnal photosynthesis is extended in nitrogen-fixing European black alder compared with white basswood; possible adaptive significance. *Can. J. For. Res.* **19**, 12–17.

Nebeker, T. E. (1981). Manipulation of stand density for prevention of southern pine beetle infestation in loblolly pine plantations. *U.S.D.A. Combined Forest Pest Research and Development Program Final Report.*

Neely, D., and Himelick, E. B. (1987). Fertilizing and watering trees. *Circ.—Ill. Nat. Hist. Surv.* **56**, 1–24.

Neely, D., Himelick, E. B., and Crowley, W. R., Jr. (1970). Fertilization of established trees. *Circ.—Ill. Nat. Hist. Surv.* **30**, 235–266.

Negi, G. C. S., and Singh, S. P. (1993). Leaf nitrogen dynamics with particular retranslocation in evergreen and deciduous tree species of Kumaun Himalaya. *Can. J. For. Res.* **23**, 349–357.

Nelson, C. D. (1964). The production and translocation of photosynthate-C^{14} in conifers. *In* "The Formation of Wood in Forest Trees" (M. H. Zimmermann, ed.), pp. 243–257. Academic Press, New York.

Nelson, N. D., and Hillis, W. E. (1978). Ethylene and tension wood formation in *Eucalyptus gomphocephala*. *Wood Sci. Technol.* **12**, 309–315.

Nelson, N. D., Burk, T., and Isebrands, J. G. (1981). Crown architecture of short-rotation intensively cultured *Populus*. I. Effects of clone characteristics. *Can. J. For. Res.* **11**, 73–81.

Nelson, N. D., Dickmann, D. I., and Gottschalk, K. W. (1982). Autumnal photosynthesis in short-rotation intensively cultured *Populus* clones. *Photosynthetica* **16**, 321–333.

Nelson, R. S., McCormick, S. M., Delannay, X., Dube, P., Layton, J., Anderson, E. J., Kaniewska, M., Proksch, R. K., Horsch, R. B., Rogers, S. G., Fraley, R. T., and

Beachy, R. N. (1988). Virus tolerance, plant growth, and field performance of transgenic tomato plants expressing coat protein from tobacco mosaic virus. *Bio/Technology* **6**, 403–409.

Nemeth, G. (1979). Benzyladenine-stimulated rooting in fruit-tree rootstocks cultured *in vitro. Z. Pflanzenphysiol.* **95**, 389–396.

NeSmith, D. S., and Krewer, G. (1992). Flower bud stage and chill hours influence the activity of GA_3 applied to rabbiteye blueberry. *HortScience* **27**, 316–318.

Neumann, D., Lichtenberger, O., Günther, D., Tschiersh, K., and Nover, O. (1994). Heat-shock proteins induce heavy-metal tolerance in plants. *Planta* **194**, 360–367.

Newsome, R. D., Kozlowski, T. T., and Tang, Z. C. (1982). Responses of *Ulmus americana* seedlings to flooding of soil. *Can. J. Bot.* **60**, 1688–1695.

Newton, A. C., and Pigott, C. D. (1991a). Mineral nutrition and mycorrhizal infection of seedling oak and birch. I. Nutrient uptake and the development of mycorrhizal infection during seedling establishment. *New Phytol.* **117**, 37–44.

Newton, A. C., and Pigott, C. D. (1991b). Mineral nutrition and mycorrhizal infection of seedling oak and birch. II. The effect of fertilizers on growth, nutrient uptake and ectomycorrhizal infection. *New Phytol.* **117**, 45–52.

Newton, A. C., Muthoka, P. N., and Dick, J. McP. (1992). The influence of leaf area on the rooting physiology of leafy stem cuttings of *Terminalia spinosa. Trees* **6**, 210–215.

Ng, F. S. P. (1980). Germination ecology of Malaysian woody plants. *Malays. For.* **43**, 406–437.

Nguyen, P. V., Dickmann, D. I., Pregitzer, K. S., and Hendrick, R. (1990). Late-season changes in allocation of starch and sugar to shoots, coarse roots, and fine roots in two hybrid poplar clones. *Tree Physiol.* **7**, 95–105.

Nickell, L. G., ed. (1983a). "Plant Growth Regulating Chemicals," Vol. 1. CRC Press, Boca Raton, Florida.

Nickell, L. G., ed. (1983b). "Plant Growth Regulating Chemicals," Vol. 2. CRC Press, Boca Raton, Florida.

Nickell, L. G. (1991). Use of growth regulating chemicals. *In* "Physiology of Trees" (A. S. Raghavendra, ed.), pp. 467–487. Wiley(Interscience), New York.

Nieves, M., Cerdá, A., and Botella, M. (1991). Salt tolerance of two lemon scions measured by leaf chloride and sodium accumulation. *J. Plant Nutr.* **14**, 623–636.

Niki, T., and Sakai, A. (1981). Ultrastructural changes related to frost hardiness in the cortical parenchyma cells from mulberry twigs. *Plant Cell Physiol.* **22**, 171–183.

Niki, T., Yoshida, S., and Sakai, A. (1978). Studies on chilling injury in plant cells. I. Ultrastructural changes associated with chilling injury in callus tissues of *Cornus stolonifera. Plant Cell Physiol.* **19**, 139–148.

Nilsson, M.-C. (1994). Separation of allelopathy and resource competition by the boreal dwarf shrub *Empetrum hermaphroditum* Hagerup. *Oecologia* **98**, 1–7.

Nitsch, J. P. (1957). Photoperiodism in woody plants. *Proc. Am. Soc. Hortic. Sci.* **70**, 526–544.

Nitsch, J. P. (1962). Photoperiodic regulation of growth in woody plants. *Proc. 16th Int. Hortic. Congr.* **3**, 14–22.

Nix, L. E., and Wodzicki, T. J. (1974). The radial distribution and metabolism of IAA-^{14}C in *Pinus echinata* stems in relation to wood formation. *Can. J. Bot.* **52**, 1349–1355.

Noble, R. O., and Jensen, K. F. (1980). Effects of sulfur dioxide and ozone on growth of hybrid poplar leaves. *Am. J. Bot.* **67**, 1005–1009.

Noland, T. L., and Kozlowski, T. T. (1979). Influence of potassium nutrition on susceptibility of silver maple to ozone. *Can. J. For. Res.* **9**, 501–503.

Noland, T. L., and Murphy, J. B. (1984). Changes in isocitrate lyase and ATP content during stratification and germination of sugar pine seeds. *Seed Sci. Technol.* **12**, 777–787.

Noland, T. L., and Murphy, J. B. (1986). Protein synthesis and aminopeptidase activity in dormant sugar pine (*Pinus lambertiana*) seeds during stratification and warm incubation. *J. Plant Physiol.* **124**, 1–10.

Nonami, H., and Boyer, J. S. (1990). Primary events regulating stem growth at low water potentials. *Plant Physiol.* **93**, 1601–1609.

Norby, R. J. (1987). Nodulation and nitrogenase activity in nitrogen-fixing woody plants stimulated by CO_2 enrichment of the atmosphere. *Physiol. Plant.* **71**, 77–82.

Norby, R. J. (1989). Direct responses of forest trees to rising atmospheric carbon dioxide. *In* "Air Pollution Effects on Vegetation including Forest Ecosystems" (R. D. Noble, J. L. Martin and K. F. Jensen, eds.), Proc. 2nd U.S.–USSR Symposium, pp. 243–249. U.S. Dept. of Agriculture, U.S. Forest Service, Broomall, Pennsylvania.

Norby, R. J. (1994). Issues and perspectives for investigating root responses to elevated atmospheric carbon dioxide. *Plant Soil* **165**, 9–20.

Norby, R. J., and Kozlowski, T. T. (1980). Allelopathic potential of ground cover species on *Pinus resinosa* seedlings. *Plant Soil* **57**, 363–374.

Norby, R. J., and Kozlowski, T. T. (1981a). Response of SO_2-fumigated *Pinus resinosa* seedlings to post fumigation temperature. *Can. J. Bot.* **59**, 470–475.

Norby, R. J., and Kozlowski, T. T. (1981b). Relative sensitivity of three species of woody plants to SO_2 at high or low exposure temperature. *Oecologia* **51**, 33–36.

Norby, R. J., and Kozlowski, T. T. (1981c). Interactions of SO_2 concentration and post-fumigation temperature on growth of woody plants. *Environ. Pollut.* Ser. A **25**, 27–39.

Norby, R. J., and Kozlowski, T. T. (1982). The role of stomata in sensitivity of *Betula papyrifera* Marsh seedlings to SO_2 at different humidities. *Oecologia* **53**, 34–39.

Norby, R. J., and Kozlowski, T. T. (1983). Flooding and SO_2-stress interaction in *Betula papyrifera* and *B. nigra* seedlings. *For. Sci.* **29**, 739–750.

Norby, R. J., O'Neill, E. G., and Luxmoore, R. J. (1986a). Effects of atmospheric CO_2 enrichment on the growth and mineral nutrition of *Quercus alba* seedlings in nutrient-poor soil. *Plant Physiol.* **82**, 83–89.

Norby, R. J., Pastor, J., and Melillo, J. M. (1986b). Carbon–nitrogen interactions in CO_2-enriched white oak: Physiological and long-term perspectives. *Tree Physiol.* **2**, 233–241.

Norby, R. J., O'Neill, E. G., and Hood, W. G. (1987). Carbon allocation, root exudation and mycorrhizal colonization of *Pinus echinata* seedlings grown under CO_2 enrichment. *Tree Physiol.* **3**, 203–210.

Norby, R. J., Gunderson, C. A., Wullschleger, S. D., O'Neil, E. G., and McCracken, M. K. (1992). Productivity and compensatory responses of yellow-poplar trees to elevated CO_2. *Nature (London)* **357**, 322–324.

Norby, R. J., Wullschleger, S. D., Gunderson, C. A., and Neitsch, C. T. (1995). Increased growth efficiency of *Quercus alba* trees in a CO_2-enriched atmosphere. *New Phytol.* **131**, 91–97.

Novy, R. G., and Vorsa, N. (1995). Identification of intracultivar genetic heterogeneity in cranberry using silver-stained RAPDs. *HortScience* **30**, 600–604.

Nylund, J.-E. (1988). The regulation of mycorrhiza formation—Carbohydrate and hormone theories reviewed. *Scand. J. For. Res.* **3**, 465–479.

Nylund, J.-E., and Unestam, T. (1982). Structure and physiology of ectomycorrhizae. I. The process of mycorrhiza formation in Norway spruce *in vitro. New Phytol.* **91**, 63–79.

Nyman, B. (1961). Effect of red and far-red irradiation on the germination process in seeds of *Pinus sylvestris* L. *Nature (London)* **191**, 1219–1220.

O'Malley, D. M., Whetten, R., Bao, W., Chen, C.-L., and Sederoff, R. N. (1993). The role of laccase in lignification. *Plant J.* **4**, 751–757.

O'Neill, E. G., Luxmoore, R. J., and Norby, R. J. (1987a). Increases in mycorrhizal colonization and seedling growth in *Pinus echinata* and *Quercus alba* in an enriched CO_2 atmosphere. *Can. J. For. Res.* **17**, 878–887.

O'Neill, E. G., Luxmoore, R. J., and Norby, R. J. (1987b). Elevated atmospheric CO_2 effects on seedling growth, nutrient uptake, and rhizosphere bacterial populations of *Liriodendron tulipifera* L. *Plant Soil* **104**, 3–11.

O'Reilly, C., and Parker, W. H. (1982). Vegetative phenology in a clonal seed orchard of *Picea glauca* and *Picea mariana* in northwestern Ontario. *Can. J. For. Res.* **12**, 408–413.

Oberarzbacher, P. (1977). Beiträge zur physiologischen Analyse des Hochenzuwachses von der schiedenen Fichtenklonen entlang eines Hohenprofils im Wipptal (Tirol) und in Klimakammern. Dissertation, Univ. Innsbruck.

Ochatt, S. J., Cocking, E. C., and Power, J. B. (1987). Isolation, culture and plant regeneration of Colt cherry (*Prunus avium* × *pseudocerasus*) protoplasts. *Plant Sci.* **50**, 139–143.

Ogasanovic, D., Paunovic, S. A., and Plazinic, R. (1983). The effect of Alar 85, TIBA, TMBA, and GA_3 on the flowering time of the apricot cv. Hungarian Best. *Acta Hortic.* **121**, 111–114.

Ogigirigi, M., Kozlowski, T. T., and Sasaki, S. (1970). Effect of soil moisture depletion on stem shrinkage and photosynthesis of tree seedlings. *Plant Soil* **32**, 33–49.

Ohgawara, T., Kobayashi, S., Ohgawara, E., Uchimiya, H., and Ishii, S. (1985). Somatic hybrid plants obtained by protoplast fusion between *Citrus sinensis* and *Poncirus trifoliata. Theor. Appl. Genet.* **71**, 1–4.

Ohmart, C. P., and Williams, C. B. (1979). The effects of photochemical oxidants on radial growth increment for five species of conifers in San Bernardino National Forest. *Plant Dis. Rep.* **63**, 1038–1042.

Oker-Blom, P., and Kellomaki, S. (1983). Effect of grouping of foliage on the within-stand and within-crown light regime: A comparison of random and grouping models. *Agric. Meteorol.* **28**, 143–155.

Okie, W. R. (1987). Plum rootstocks. *In* "Rootstocks for Fruit Crops" (R. C. Rom and R. F. Carlson, eds.), pp. 321–360. Wiley, New York and Chichester.

Okoro, O. O., and Grace, J. (1976). The physiology of rooting of *Populus* cuttings. I. Carbohydrates and photosynthesis. *Physiol. Plant.* **36**, 133–138.

Oland, K. (1963). Changes in the content of dry matter and major nutrient elements of apple foliage during senescence and abscission. *Physiol. Plant.* **16**, 682–694.

Olien, W. C. (1987). Effect of seasonal soil waterlogging on vegetative growth and fruiting of apple trees. *J. Am. Soc. Hortic. Sci.* **112**, 209–214.

Oliver, C. D. (1981). Forest development in North America following major disturbances. *For. Ecol. Manage.* **3**, 153–168.

Oliver, C. D., O'Hara, K. L., McFadden, G., and Nagame, I. (1986). Concepts of thinning regimes. *In* "Douglas-Fir: Stand Management for the Future" (C. D. Oliver, D. Hawley, and J. Johnson, eds.), pp. 246–257. College of Forest Resources, Univ. of Washington, Seattle.

Olney, H. O., and Pollock, B. M. (1960). Studies of rest period. II. Nitrogen and phosphorus changes in embryonic organs of after-ripening cherry seed. *Plant Physiol.* **35**, 970–975.

Olofinboba, M. O. (1969). Seasonal variations in the carbohydrates in the xylem of *Antiaris africana. Ann. Bot.* **33**, 339–349.

Olofinboba, M. O. (1975). Studies on seedlings of *Theobroma cacao* L., variety F_3 Amazon. I. Role of cotyledons in seedling development. *Turrialba* **25**, 121–127.

Olofinboba, M. O., and Kozlowski, T. T. (1973). Accumulation and utilization of carbohydrate reserves in shoot growth of *Pinus resinosa. Can. J. For. Res.* **3**, 346–353.

Olofinboba, M. O., and Kozlowski, T. T. (1982). Effects of three systemic insecticides on seed germination and growth of *Pinus halepensis* seedlings. *Plant Soil* **64**, 255–258.

Olsen, J. A., and Huang, A. H. C. (1988). Glyoxysomal acyl-CoA synthetase and oxidase from germinating elm, rape, and maize seed. *Phytochemistry* **27**, 1601–1603.

Olsen, S. R. (1972). Micronutrient interactions. *In* "Micronutrients in Agriculture" (J. J. Mortvedt, P. M. Giordano, and W. L. Lindsay, eds.), pp. 559–589. Am. Soc. Agronomy, Madison, Wisconsin.

Olson, J. S., Stearns, F., and Nienstaedt, H. (1959). Eastern hemlock seeds and seedlings. Response to photoperiod and temperature. *Conn., Agric. Exp. Stn., New Haven Bull.* **620**.

Omi, S. K., Yoder, B., and Rose, S. (1991). Fall lifting and long-term freezer storage of ponderosa pine seedlings: Effects on post-storage leaf water potential, stomatal conductance, and root growth potential. *Tree Physiol.* **8**, 315–325.

Oo, C., and Stumpf, P. K. (1983a). Some enzymic activities in the germinating oil palm (*Elaeis guineensis*) seedling. *Plant Physiol.* **73**, 1028–1032.

Oo, C., and Stumpf, P. K. (1983b). The metabolism of the germinating oil palm (*Elaeis guineensis*) seedlings. *Plant Physiol.* **73**, 1033–1037.

Oosting, H. J., and Humphreys, M. E. (1940). Buried viable seeds in a successional series of old field and forest soils. *Bull. Torrey Bot. Club* **67**, 253–273.

Opik, H. (1980). "The Respiration of Higher Plants." Arnold, London.

Opler, P. A., Frankie, G. W., and Baker, H. G. (1976). Rainfall as a factor in the release, timing, and synchronization of anthesis by tropical trees and shrubs. *J. Biogeogr.* **3**, 231–236.

Oren, R., Werk, K. S., Schulze, E.-D., Meyer, J., Schneider, B. U., and Schramel, P. (1988). Performance of two *Picea abies* (L.) Karst. stands at different stages of decline. VI. Nutrient concentration. *Oecologia* **77**, 151–162.

Ormrod, D. P. (1978). "Pollution in Horticulture." Elsevier, Amsterdam.

Orozco-Segovia, A., Sanchez-Coronado, M. E., and Vazquez-Yanes, C. (1993). Light environment and phytochrome-controlled germination in *Piper auritum. Funct. Ecol.* **7**, 585–590.

Ortola, A. G., Monerri, A. G., and Guardiola, J. L. (1991). The use of naphthalene acetic acid as a fruit growth enhancer in Satsuma mandarin: A comparison with the fruit thinning effect. *Sci. Hortic. (Amsterdam)* **47**, 15–25.

Orunfleh, M. M. (1991). Studies on the hawthorn (*Crataegus azarollus* L.): II. changes in abscisic acid content during cold stratification in relation to seed germination. *J. Hortic. Sci.* **66**, 223–226.

Osborne, D. J. (1980). Senescence in seeds. *In* "Senescence in Plants" (K. V. Thimann, ed.), pp. 13–37. CRC Press, Boca Raton, Florida.

Osborne, D. J. (1982). Deoxyribonucleic acid integrity and repair in seed germination: The importance in viability and survival. *In* "The Physiology and Biochemistry of Seed Development, Dormancy and Germination" (A. A. Khan, ed.), pp. 435–463. Elsevier, Amsterdam.

Osborne, D. J., and Hallaway, M. (1960). The role of auxins in the control of leaf senescence. Some effects of local applications of 2,4-dichlorophenoxyacetic acid on carbon and nitrogen metabolism. *In* "Plant Growth Regulation" (W. Klein, ed.), pp. 329–340. Iowa State Univ. Press, Ames.

Ostman, N. L., and Weaver, G. T. (1982). Autumnal nutrient transfer by retranslocation, leaching and litterfall in a chestnut oak forest in southern Illinois. *Can. J. For. Res.* **12**, 40–51.

Owens, J. N. (1991). Measuring growth and development of reproductive structures. *In* "Techniques and Approaches to Forest Tree Ecophysiology" (J. P. Lassoie and T. M. Hinckley, eds.), pp. 423–452. CRC Press, Boca Raton, Florida.

Owens, J. N., and Colangeli, A. M. (1989). Promotion of flowering in western hemlock by gibberellin $A_{4/7}$ applied at different stages of bud and shoot development. *Can. J. For. Res.* **19**, 1051–1058.

Owens, J. N., Webber, J. E., Ross, S. D., and Pharis, R. P. (1985). Interaction between gibberellin $A_{4/7}$ and root pruning on the reproductive and vegetative processes in Douglas-fir. III. Effects on anatomy of shoot elongation and terminal bud development. *Can. J. For. Res.* **15**, 354–361.

Owens, J. N., Webber, J. E., Ross, S. D., and Pharis, R. P. (1986). Interaction between gibberellin $A_{4/7}$ and root pruning on the reproductive and vegetative processes in Douglas-fir. IV. Effects on lateral bud development. *Can. J. For. Res.* **16**, 211–221.

Pack, M. R. (1972). Response of strawberry fruiting to hydrogen fluoride fumigation. *J. Air Pollut. Control Assoc.* **22**, 714–717.

Paine, T. D., Millar, J. G., Bellows, T. S., Hanks, L. M., and Gould, J. R. (1993). Integrating classical biological control with plant health in the urban forest. *J. Arboric.* **19**, 125–130.

Painter, J. W., and Stembridge, G. E. (1972). Peach flowering response as related to time of gibberellin application. *HortScience* **7**, 389–390.

Paiva, E., and Robitaille, H. A. (1978a). Breaking bud rest on detached apple shoots: Effects of wounding and ethylene. *J. Am. Soc. Hortic. Sci.* **103**, 101–104.

Paiva, E., and Robitaille, H. (1978b). Breaking bud rest on detached apple shoots: Interaction of gibberellic acid with some rest-breaking chemicals. *HortScience* **13**, 57–58.

Pal, M., and Nanda, K. K. (1981). Rooting of etiolated stem segments of *Populus robusta*—Interaction of temperature, catechol, and sucrose in the presence of IAA. *Physiol. Plant.* **53**, 540–542.

Pallardy, S. G., and Kozlowski, T. T. (1979a). Stomatal response of *Populus* clones to light intensity and vapor pressure deficit. *Plant Physiol.* **64**, 112–114.

Pallardy, S. G., and Kozlowski, T. T. (1979b). Early root and shoot growth of *Populus* clones. *Silvae Genet.* **28**, 153–156.

Pallardy, S. G., and Kozlowski, T. T. (1979c). Relationship of leaf diffusion resistance of *Populus* clones to leaf water potential and environment. *Oecologia* **40**, 371–380.

Pallardy, S. G., and Rhoads, J. L. (1993). Morphological adaptations to drought in seedlings of deciduous angiosperms. *Can. J. For. Res.* **23**, 1766–1774.

Palmer, J. W. (1989). Canopy manipulation for optimum utilization of light. *In* "Manipulation of Fruiting" (C. J. Wright, ed.), pp. 245–262. Butterworth, London.

Pan, Y., and Raynal, D. J. (1995). Predicting growth of plantation conifers in the Adirondack Mountains in response to climate change. *Can. J. For. Res.* **25**, 48–56.

Paoletti, E. (1992). Effects of acidity and detergent on *in vitro* pollen germination and tube growth in forest tree species. *Tree Physiol.* **10**, 357–366.

Papadol, C. S. (1982). Some effects of water supply on the accumulation of poplar biomass and energy budget. North Amererican Poplar Council, 19th Annual Meeting, Rhinelander, Wisconsin.

Pâques, M. (1991). Vitrification and micropropagation: Causes, remedies and prospects. *Acta Hortic.* **289**, 283–290.

Pardossi, A., Pritchard, J., and Tomes, A. D. (1994). Leaf illumination and root cooling inhibit bean leaf expansion by decreasing turgor pressure. *J. Exp. Bot.* **45**, 415–422.

Parker, M. L. (1986). Recent abnormal increase in tree-ring widths: A possible effect of elevated atmospheric carbon dioxide. *In* "Symposium on Ecological Aspects of Tree-Ring Analysis" (G. C. Jacoby and J. W. Hornbeck, eds.), pp. 511–521. CONF-8608144 U.S. Dept. of Energy, Washington, D.C.

Parker, W. C., and Pallardy, S. G. (1985). Drought-induced leaf abscission and whole-plant drought tolerance of seedlings of seven black walnut families. *Can. J. For. Res.* **15**, 818–821.

Parsons, L. R., Combs, B. S., and Tucker, D. P. H. (1985). Citrus freeze protection with micro-sprinkler irrigation during advective freeze. *HortScience* **20**, 1078–1080.

Parsons, L. R., Wheaton, T. A., Faryna, N. D., and Jackson, J. L. (1991). Elevated microsprinklers improve protection of citrus trees in advective freeze. *HortScience* **26**, 1149–1151.

Parviainen, J. (1979). Einfluss des Verpflanzens und das Wurzelschnittes auf den Tagesverlauf des xylem Wasserpotentials bei Fichtenpflanzen. *Forstarchiv* **50**, 148–153.

Passioura, J. B. (1994). The physical chemistry of the primary cell wall: Implications for the control of expansion rate. *J. Exp. Bot.* **45**, 1675–1682.

Passioura, J. B., and Fry, S. C. (1992). Turgor and cell expansion—Beyond the Lockhart equation. *Aust. J. Plant Physiol.* **19**, 565–576.

Pastor, J., and Post, W. M. (1988). Response of northern forests to CO_2-induced climate change. *Nature (London)* **334**, 55–58.

Paton, D. M. (1982). A mechanism for frost resistance in *Eucalyptus*. *In* "Cold Hardiness and Freezing Stress" (P. H. Li and A. Sakai, eds.), pp. 77–92. Academic Press, London.

Patten, K. D., and Proebsting, E. L. (1986). Effect of different artificial shading times and natural light intensities on fruit quality of 'Bing' sweet cherry. *J. Am. Soc. Hortic. Sci.* **111**, 360–363.

Patterson, J. C. (1976). Soil compaction and its effects upon urban vegetation. *U.S. For. Serv. Gen. Tech. Rep. NE* **NE-22**, 91–101.

Pauley, S. S., and Perry, T. O. (1954). Ecotypic variation of the photoperiodic response in *Populus*. *J. Arnold Arbor., Harv. Univ.* **35**, 167–188.

Pearcy, R. W. (1977). Acclimation of photosynthetic and respiratory carbon dioxide exchange to growth temperature in *Atriplex lentiformis* (Torr.) Wats. *Plant Physiol.* **59**, 795–799.

Pearson, M., and Mansfield, T. A. (1993). Interacting effects of ozone and water stress on the stomatal resistance of beech (*Fagus sylvatica* L.). *New Phytol.* **123**, 351–358.

Pearson, M., and Mansfield, T. A. (1994). Effects of exposure to ozone and water stress on the following season's growth of beech (*Fagus sylvatica* L.). *New Phytol.* **126**, 511–515.

Peart, D. R., Poage, N. J., and Jones, M. B. (1992). Winter injury to subalpine red spruce: Influence of prior vigor and effects on subsequent growth. *Can. J. For. Res.* **22**, 888–892.

Pellett, H., Litzow, M., and Mainquist, L. (1980). Use of metal compounds as root pruning agents. *HortScience* **15**, 308–309.

Peña, L., Cervera, M., Juárez, J., Navarro, A., Pina, J., Durán-Vila, N., and Navarro, L. (1995). *Agrobacterium*-mediated transformation of sweet orange and regeneration of transgenic plants. *Plant Cell Rep.* **14**, 616–619.

Percy, K. E., and Riding, R. T. (1978). The epicuticular waxes of *Pinus strobus* subjected to air pollutants. *Can. J. For. Res.* **8**, 474–477.

Pereira, H. C. (1975). Climate and the orchard. *Commonw. Bur. Hortic. Plant. Crops (G.B.). Tech. Commun.* **5**.

Pereira, J. S., and Kozlowski, T. T. (1976a). Diurnal and seasonal changes in water balance of *Abies balsamea* and *Pinus resinosa*. *Oecol. Plant.* **11**, 397–412.

Pereira, J. S., and Kozlowski, T. T. (1976b). Influence of light intensity, temperature, and leaf area on stomatal aperture and water potential of woody plants. *Can. J. For. Res.* **7**, 145–153.

Pereira, J. S., and Kozlowski, T. T. (1977a). Water relations and drought resistance of young *Pinus banksiana* and *P. resinosa* plantation trees. *Can. J. For. Res.* **7**, 132–137.

Pereira, J. S., and Kozlowski, T. T. (1977b). Variations among woody angiosperms in response to flooding. *Physiol. Plant.* **41**, 184–192.

Pereira, J. S., and Kozlowski, T. T. (1978). Diurnal and seasonal changes in water balance of *Acer saccharum* and *Betula papyrifera*. *Physiol. Plant.* **43**, 19–30.

Perkins, T. D., Adams, G. T., and Klein, R. M. (1991). Desiccation or freezing? Mechanisms of winter injury to red spruce foliage. *Am. J. Bot.* **78**, 1207–1217.

Perlak, F. J., Fuchs, R. L., Dean, D. A., McPherson, S. L., and Fischoff, D. A. (1991).

Modification of the coding sequence enhances plant expression of insect control protein genes. *Proc. Natl. Acad. Sci. U.S.A.* **88**, 3324–3328.

Perry, D. A., Molina, R., and Amaranthus, A. P. (1987). Mycorrhizae, mycorrhizospheres, and reforestation: Current knowledge and research needs. *Can. J. For. Res.* **17**, 929–940.

Perry, R. L. (1987). Cherry rootstocks. *In* "Rootstocks for Fruit Crops" (R. C. Rom and R. F. Carlson, eds.), pp. 217–284. Wiley, New York and Chichester.

Perry, T. O., and Wang, C. W. (1960). Genetic variation in the winter chilling requirement for date of dormancy break for *Acer rubrum. Ecology* **41**, 785–790.

Pessi, Y. (1958). On the influence of bog draining upon thermal conditions in the soil and air near the ground. *Acta Agric. Scand.* **24**, 359–374.

Peters, R. L., and Darling, D. S. (1985). The greenhouse effect and nature reserves. *BioScience* **35**, 707–717.

Peterson, C. J., and Facelli, J. M. (1992). Contrasting germination and seedling growth of *Betula alleghaniensis* and *Rhus typhina* subjected to various amounts and types of plant litter. *Am. J. Bot.* **79**, 1209–1216.

Peterson, J. K. (1983). Mechanisms involved in delayed germination of *Quercus nigra* seeds. *Ann. Bot.* **52**, 81–92.

Petty, J. A., and Worrell, R. (1981). Stability of coniferous tree stems in relation to damage by snow. *Forestry* **54**, 115–128.

Pharis, R. P., and Kramer, P. J. (1964). The effects of nitrogen and drought on loblolly pine seedlings. I. Growth and composition. *For. Sci.* **10**, 143–150.

Pharis, R. P., Ruddat, M. D. E., Glenn, J. L., and Morf, W. (1970). A quantitative requirement for long day in the induction of staminate strobili by gibberellin in the conifer *Cupressus arizonica. Can. J. Bot.* **48**, 653–658.

Pharis, R. P., Jenkins, P. A., Aoki, H., and Sassa, T. (1981). Hormonal physiology of wood growth in *Pinus radiata* D. Don: Effects of gibberellin A_4 and the influence of abscisic acid upon [^{3}H]gibberellin A_4 metabolism. *Aust. J. Plant Physiol.* **8**, 559–570.

Pharis, R. P., Webber, J. E., and Ross, S. D. (1987). The promotion of flowering in forest trees by gibberellin $A_{4/7}$ and cultural treatments: A review of the possible mechanisms. *For. Ecol. Manage.* **19**, 65–84.

Phelps, J. E., McGinnes, E. A., Smolinski, M., Saniewski, M., and Pieniazek, J. (1977). A note on the formation of compression wood induced by morphactin IT 3456 in *Thuja* shoots. *Wood Fiber* **8**, 223–227.

Phelps, J. E., McGinnes, E. A., Jr., Saniewski, M., Pieniazek, J., and Smolinski, M. (1980). Some anatomical observations on the effect of morphactin IT 3456 and Ethrel on wood formation in *Salix fragilis* L. *IAWA Bull. New Ser.* **1–2**, 76–82.

Phelps, J. E., Isebrands, J. G., and Jowett, J. D. (1982). Raw material quality of short-rotation intensively cultured *Populus* clones. I. A comparison of stem and branch properties at three spacings. *IAWA Bull. New Ser.* **3**, 193–200.

Philipson, J. J. (1985). The promotion of flowering in large field-grown Sitka spruce by girdling and stem injections of gibberellin $A_{4/7}$. *Can. J. For. Res.* **15**, 166–170.

Philipson, J. J. (1988). Root growth in Sitka spruce and Douglas-fir transplants: Dependence on the shoot and stored carbohydrates. *Tree Physiol.* **4**, 101–108.

Philipson, J. J. (1992). Optimal conditions for inducing coning of container-grown *Picea sitchensis* grafts: Effects of applying different quantities of $GA_{4/7}$, timing and duration of heat and drought treatment, and girdling. *For. Ecol. Manage.* **53**, 39–52.

Philipson, J. J., and Coutts, M. P. (1977). The influence of mineral nutrition on the root development of trees. 2. The effect of specific nutrient elements on the growth of individual roots of Sitka spruce. *J. Exp. Bot.* **28**, 864–871.

Philipson, J. J., and Coutts, M. P. (1978). Tolerance of tree roots to waterlogging. III. Oxygen transport in lodgepole pine and Sitka spruce roots of primary structure. *New Phytol.* **80**, 341–349.

Philipson, J. J., and Coutts, M. P. (1980a). Effects of growth hormone application on the secondary growth of roots and stems in *Picea sitchensis* (Bong.) Carr. *Ann. Bot.* **46**, 747–755.

Philipson, J. J., and Coutts, M. P. (1980b). The tolerance of tree roots to waterlogging. *New Phytol.* **85**, 489–494.

Phillips, I. D. J., and Wareing, P. F. (1958). Studies in the dormancy of sycamore. I. Seasonal changes in growth substance content of the shoot. *J. Exp. Bot.* **9**, 350–364.

Phillips, I. D. J., Miners, J., and Roddick, J. G. (1980). Effects of light and photoperiodic conditions on abscisic acid in leaves and roots of *Acer pseudoplatanus*. *Planta* **149**, 118–122.

Phillips, J. H. H., and Weaver, G. M. (1975). A high density peach orchard. *Hort-Science* **10**, 580–582.

Piagnani, C., and Eccher, T. (1990). Somatic embryogenesis in chestnut. *Acta Hortic.* **280**, 159–161.

Pickett, S. T. A., and Kempf, J. S. (1980). Branching patterns in forest shrubs and understory trees in relation to habitat. *New Phytol.* **86**, 219–220.

Pieniazek, J., Smolinski, M., and Saniewski, M. (1970). Induced structural changes in anatomy of apple shoots after treament with morphactin IT 3456 and other growth regulators (NAA, GA, BA). *Acta Agrobot.* **23**, 387–396.

Pigott, C. D. (1992). Are the distributions of species determined by failure to set seed? *In* "Fruit and Seed Production: Aspects of Development, Environmental Physiology and Ecology" (C. Marshall and J. Grace, eds.), pp. 203–216. Cambridge Univ. Press, Cambridge.

Pinfield, N. J. (1968). The promotion of isocitrate lyase activity in hazel cotyledons by exogenous gibberellin. *Planta* **82**, 337–341.

Pinfield, N. J., and Davies, H. V. (1978). Hormonal changes during after-ripening of *Acer platanoides* (L.) seeds. *Z. Pflanzenphysiol.* **90**, 171–181.

Pinfield, N. J., and Dungey, N. O. (1985). Seed dormancy in *Acer:* An assessment of the role of the structures covering the embryo. *J. Plant Physiol.* **120**, 65–81.

Pinfield, N. J., and Stutchbury, P. A. (1990). Seed dormancy in *Acer:* The role of testa-imposed and embryo dormancy in *Acer velutinum*. *Ann. Bot.* **66**, 133–138.

Pinfield, N. J., Stutchbury, P. A., and Bazaid, S. A. (1987). Seed dormancy in *Acer:* Is there a common mechanism for all *Acer* species and what part is played in it by abscisic acid? *Physiol. Plant.* **71**, 365–371.

Pinney, K., and Polito, V. S. (1990). Flower initiation in 'Manzanillo' olive. *Acta Hortic.* **286**, 203–206.

Piringer, A. A., and Borthwick, H. A. (1955). Photoperiodic responses in coffee. *Turrialba* **5**, 72–77.

Piringer, A. A., and Downs, R. J. (1959). Responses of apple and peach trees to various photoperiods. *Proc. Am. Soc. Hortic. Sci.* **73**, 9–15.

Pitel, J. A., and Cheliak, W. M. (1986). Enzyme activities during imbibition and germination of seeds of tamarack (*Larix laricina*). *Physiol. Plant.* **67**, 562–569.

Pitel, J. A., Cheliak, W. M., and Wang, B. S. P. (1989). Some biochemical changes associated with stratification and germination of basswood seeds. *Seed Sci. Technol.* **17**, 57–71.

Place, I. C. M. (1955). The influence of seed-bed conditions on the regeneration of spruce and fir. *Can. For. Branch Dep. Publ.* **117**.

Plass, W. T. (1977). Growth and survival of hardwoods and pine interplanted with European alder. *U.S.D.A. For. Serv., Res. Pap. NE* **NE-376**.

Plochmann, R. (1984). Air pollution and the dying forests of Europe. *Am. For.* **90**(6), 17–21, 56.

Pollard, A. S., Parr, A. J., and Loughman, B. C. (1977). Boron in relation to membrane function in higher plants. *J. Exp. Bot.* **28**, 831–841.

Pollard, D. F. W., and Logan, K. T. (1977). The effects of light intensity, photoperiod, soil moisture potential, and temperature on bud morphogenesis in *Picea* species. *Can. J. For. Res.* **7**, 415–421.

Pollock, B. M., and Olney, H. O. (1959). Studies of the rest period. I. Growth, translocation, and respiratory changes in the embryonic organs of the afterripening cherry seed. *Plant Physiol.* **34**, 131–142.

Pomeroy, M. K., Siminovitch, D., and Wightman, F. (1970). Seasonal biochemical changes in living bark and needles of red pine (*Pinus resinosa*) in relation to adaptation to freezing. *Can. J. Bot.* **45**, 953–967.

Ponnamperuma, F. N. (1984). Effects of flooding on soils. *In* "Flooding and Plant Growth" (T. T. Kozlowski, ed.), pp. 9–45. Academic Press, New York.

Pook, E. W., Costin, A. B., and Moore, C. W. E. (1966). Water stress in native vegetation during the drought of 1965. *Aust. J. Bot.* **14**, 257–267.

Porandowski, J., Rakowski, K., and Wodzicki, T. J. (1982). Apical control of xylem formation in the pine stem. II. Responses of differentiating tracheids. *Acta Soc. Bot. Pol.* **51**, 203–214.

Porpiglia, P. J., and Barden, J. A. (1980). Seasonal trends in net photosynthetic potential, dark respiration, and specific leaf weight of apple leaves as affected by canopy position. *J. Am. Soc. Hortic. Sci.* **105**, 920–923.

Porsild, A. E., Harrington, C. R., and Mulligan, G. A. (1967). *Lupinus arcticus* Wats. grown from seeds of Pleistocene age. *Science* **158**, 114–115.

Portlock, C. C., Shea, S. R., Majer, J. D., and Bell, D. T. (1990). Stimulation of germination of *Acacia pulchella*: Laboratory basis for forest management options. *J. Appl. Ecol.* **27**, 319–324.

Potrykus, I., Paszkowski, J., Shillito, R. D., and Saul, M. W. (1987). Direct gene transfer to plants. *In* "Plant DNA Infectious Agents" (T. Hohn and J. Schell, eds.), pp. 229–247. Springer-Verlag, Vienna.

Potter, D. A., and Kimmerer, T. W. (1989). Evidence for the defensive role of saponins. *Oecologia* **78**, 322–329.

Powell, D. B. B. (1974). Some effects of water stress in late spring on apple trees. *J. Hortic. Sci.* **49**, 257–272.

Powell, D. B. B. (1976). Some effects of water stress on the growth and development of apple trees. *J. Hortic. Sci.* **51**, 75–90.

Powell, L. E. (1972). Naturally occurring plant growth regulators and their physiological roles in fruit trees. *Acta Hortic.* **34**, 33–39.

Powell, L. E. (1982). Shoot growth in woody plants and possible participation of abscisic acid. *In* "Plant Growth Substances" (P. F. Wareing, ed.), pp. 363–372. Academic Press, London.

Powell, L. E. (1987). Hormonal aspects of bud and seed dormancy in temperate-zone woody plants. *HortScience* **22**, 845–850.

Powell, L. E. (1988). The hormonal control of bud and seed dormancy in woody plants. *In* "Plant Hormones and Their Role in Plant Growth and Development" (P. J. Davies, ed.), pp. 539–552. Kluwer, Dordrecht, The Netherlands.

Powles, S. B., Berry, J. A., and Björkmann, O. (1983). Interaction between light and chilling temperature on the inhibition of photosynthesis in chilling-sensitive plants. *Plant Cell Environ.* **6**, 117–123.

Pregitzer, K. S., Dickmann, D. I., Hendrick, R., and Nguyen, P. V. (1990). Whole-tree carbon and nitrogen partitioning in young hybrid poplars. *Tree Physiol.* **7**, 79–93.

Pregitzer, K. S., Hendrick, R. L., and Fogel, R. (1993). The demography of fine roots in response to patches of water and nitrogen. *New Phytol.* **125**, 575–580.

Priestley, D. A. (1986). "Seed Aging." Cornell Univ. Press, Ithaca, New York.

Pritchard, J. (1994). Tansley review no. 68—The control of cell expansion in roots. *New Phytol.* **127**, 3–26.

Pritchard, J., and Tomos, A. D. (1993). Correlating biophysical and biochemical control of root cell expansion. *In* "Water Deficits: Plant Responses from Cell to Community" (J. A. C. Smith and H. Griffiths, eds.), pp. 53–72. BIOS, Oxford.

Pritchett, W. L. (1979). "Properties and Management of Forest Soils." Wiley, New York.

Proebsting, E. L., Jr., and Middleton, J. E. (1980). The behaviour of peach and pear trees under extreme drought stress. *J. Am. Soc. Hortic. Sci.* **105**, 380–385.

Pukacka, S., and Kuiper, P. J. C. (1988). Phospholipid composition and fatty acid peroxidation during ageing of *Acer platanoides* seeds. *Physiol. Plant.* **72**, 89–93.

Quamme, H. A. (1985). Avoidance of freezing injury in woody plants by deep supercooling. *Acta Hortic.* **168**, 11–30.

Quinlan, J. D. (1965). The pattern of distribution of ^{14}C in a potted rootstock following assimilation of $^{14}CO_2$ by a single leaf. *Annu. Rep. East Malling Res. Sta.*, 117–118.

Quinlan, J. D. (1981). New chemical approaches to the control of tree form and size. *Acta Hortic.* **120**, 95–106.

Rachie, K. O. (1983). Intercropping tree legumes with annual crops. *In* "Plant Research and Agroforestry" (P. A. Huxley, ed.), pp. 103–116. Pullans and Wilson, Edinburgh.

Rademacher, W. (1991). Biochemical effects of plant growth retardants. *In* "Plant Biochemical Regulators" (H. W. Gausman, ed.), pp. 169–200. Dekker, New York.

Radwan, M. A., Shumway, J. S., DeBell, D. S., and Kraft, J. M. (1991). Variance in response of pole-size trees and seedlings of Douglas-fir and western hemlock to nitrogen and phosphorus fertilizers. *Can. J. For. Res.* **21**, 1431–1438.

Radwanski, S. A., and Wickens, G. E. (1967). The ecology of *Acacia albida* on mantle soils in Zalingeri, Jebel Marsa, Sudan. *J. Appl. Ecol.* **4**, 569–579.

Raffa, K. F., and Berryman, A. A. (1983). The role of host plant resistance in the colonization behavior and ecology of bark beetles (Coleoptera: Scolytidae). *Ecol. Monogr.* **53**, 27–49.

Raffa, K. F., Berryman, A. A., Simasko, J., Teal, W., and Wang, B. L. (1985). Effects of grand fir monoterpenes on the fir engraver, *Scolytus ventralis* (Coleoptera: Scolytidae). *Environ. Entomol.* **14**, 552–556.

Rahozinski, J., Edward, G. R., and Hoskyns, P. (1986). Effects of brief exposure to nitrogenous compounds on floral initiation in apple. *Physiol. Veg.* **24**, 673–677.

Raison, R. J., Khanna, P. K., Benson, M. L., Myers, B. J., McMurtrie, R. E., and Lang, A. R. G. (1992). Dynamics of *Pinus radiata* foliage in relation to water and nitrogen stress. II. Needle loss and temporal changes in total foliage mass. *For. Ecol. Manage.* **52**, 159–178.

Rajasekaran, K., Vine, J., and Mullins, M. G. (1982). Dormancy in somatic embryos and seeds of *Vitis*: Changes in endogenous abscisic acid during embryogeny and germination. *Planta* **154**, 139–144.

Rakowski, K. J., Zwiazek, J. J., and Sumner, M. J. (1995). Hydrogen fluoride effects on plasma membrane composition, ATPase activity and cell structure in needles of eastern white pine (*Pinus strobus*) seedlings. *Trees* **9**, 190–194.

Rallo, L., and Martin, G. C. (1991). The role of chilling in releasing olive floral buds from dormancy. *J. Am. Soc. Hortic. Sci.* **116**, 1058–1062.

Randolph, W. S., and Wiest, C. (1981). Relative importance of tractable factors affecting the establishment of transplanted holly (*Ilex crenata*). *J. Am. Soc. Hortic. Sci.* **106**, 207–210.

Rangenekar, P. V., Forward, D. F., and Nolan, N. J. (1969). Foliar nutrition and wood growth in red pine. The distribution of radiocarbon photoassimilated by individual branches of young trees. *Can. J. Bot.* **47**, 1701–1711.

Rani, V., Parida, A., and Raina, S. N. (1995). Random amplified polymorphic DNA (RAPD) markers for genetic analysis in micropropagated plants of *Populus deltoides* Marsh. *Plant Cell Rep.* **14**, 459–462.

Ranney, T. G., and Peet, M. M. (1994). Heat tolerance of five taxa of birch (*Betula*): Physiological responses to supraoptimal leaf temperatures. *J. Am. Soc. Hortic. Sci.* **119**, 243–248.

Rao, C. D., Goh, C. J., and Kumar, P. P. (1993). High frequency plant regeneration from excised leaves of *Paulownia fortunei*. *In Vitro Cell Dev. Biol.* **29P**, 72–76.

Rashid, H. G. (1987). Effects of fire on soil carbon and nitrogen in a Mediterranean oak forest of Algeria. *Plant Soil* **103**, 89–93.

Rasmussen, G. K. (1975). Cellulase activity, endogenous abscisic acid, and ethylene in four citrus cultivars during maturation. *Plant Physiol.* **56**, 765–767.

Raupp, M. J., Koehler, C. S., and Davidson, J. A. (1992). Advances in implementing integrated pest management for woody landscape plants. *Annu. Rev. Entomol.* **37**, 561–585.

Rech, E. L., Ochatt, S. J., Chand, P. K., Power, J. B., and Davey, M. R. (1987). Electroenhancement of division of plant protoplast-derived cells. *Protoplasma* **141**, 169–176.

Reddy, D. V., and Jokela, J. J. (1982). Variation in specific gravity within trees and between clones in the cottonwood evaluation test in southern Illinois. *Proceedings of the North American Poplar Council Annual Meeting, Rhinelander, Wisconsin*, pp. 127–133.

Redenbaugh, K., Paasch, B. D., Nichol, J. W., Kossler, M. E., Viss, P. R., and Walker, K. A. (1986). Somatic seeds: Encapsulation of asexual plant embryos. *Bio/Technology* **4**, 797–801.

Redenbaugh, K., Slade, D., Viss, P. R., and Fujii, J. A. (1987). Encapsulation of somatic embryos in synthetic seed coats. *HortScience* **22**, 803–809.

Reich, P. B. (1995). Phenology of tropical forests: Patterns, causes, and consequences. *Can. J. Bot.* **73**, 164–174.

Reich, P. B., and Borchert, R. (1982). Phenology and ecophysiology of the tropical tree, *Tabebuia neochrysantha* (Bignoniaceae). *Ecology* **63**, 294–299.

Reich, P. B., and Borchert, R. (1984). Water stress and tree phenology in a tropical dry forest in the lowlands of Costa Rica. *J. Ecol.* **72**, 61–74.

Reich, P. B., and Lassoie, J. P. (1985). Influence of low concentrations of ozone on growth, biomass partitioning, and leaf senescence in young hybrid poplar plants. *Environ. Pollut. (Ser. A)* **39**, 39–51.

Reich, P. B., Teskey, R. O., Johnson, P. S., and Hinckley, T. M. (1980). Periodic root and shoot growth in oak. *For. Sci.* **26**, 590–598.

Reich, P. B., Schoettle, A. W., Stroo, H. F., Troiano, J., and Amundson, R. G. (1985). Effects of O_3, SO_2, and acidic rain on mycorrhizal infection in northern red oak seedlings. *Can. J. Bot.* **63**, 2049–2055.

Reich, P. B., Stroo, H. F., Schoettle, A. W., and Amundson, R. G. (1986). Acid rain and ozone influence mycorrhizal infection of tree seedlings. *J. Air Pollut. Control Assoc.* **36**, 724–726.

Reich, P. B., Schoettle A. W., Stroo, H. F., Troiano, J., and Amundson, R. G. (1987). Effects of O_3 and acid rain on white pine seedlings grown in five soils. I. Net photosynthesis and growth. *Can. J. Bot* **65**, 977–987.

Reid, D. M., and Bradford, K. J. (1984). Effects of flooding on hormone relations. *In* "Flooding and Plant Growth" (T. T. Kozlowski, ed.), pp. 195–219. Academic Press, New York.

Reinert, R. A., Heagle, A. S., and Heck, W. W. (1975). Plant responses to pollutant combinations. *In* "Responses of Plants to Air Pollution" (J. B. Mudd and T. T. Kozlowski, eds.), pp. 159–177. Academic Press, New York.

Remphrey, W. R. (1989). Shoot ontogeny in *Fraxinus pennsylvanica* (green ash). I. Seasonal cycle of terminal meristem activity. *Can. J. Bot.* **67**, 1624–1632.

Renwick, J. A. A., and Potter, J. (1981). Effects of sulfur dioxide on volatile terpene emission from balsam fir. *J. Air Pollut. Control Assoc.* **31**, 65–66.

Repo, T. (1992). Seasonal changes of frost hardiness in *Picea abies* and *Pinus sylvestris* in Finland. *Can. J. For. Res.* **22**, 1949–1957.

Reukema, D. L. (1959). Missing annual rings in branches of young Douglas-fir. *Ecology* **40**, 480–482.

Rey, M., Diaz-Sala, C., and Rodriguez, R. (1994). Comparison of endogenous polyamine content in hazel leaves and buds between the annual dormancy and flowering phases of growth. *Physiol. Plant.* **91**, 45–50.

Rhoades, D. F. (1983). Responses of alder and willow to attack by tent caterpillars and webworms: Evidence for pheromonal sensitivity of willows. *Am. Chem. Soc. Symp. Ser.* **208**, 55–68.

Rhodenbaugh, E. J., and Pallardy, S. G. (1993). Water stress, photosynthesis and early growth patterns of cuttings of 3 *Populus* clones. *Tree Physiol.* **13**, 213–226.

Rhodes, M. J. C. (1980). The maturation and ripening of fruits. *In* "Senescence in Plants" (K. V. Thimann, ed.), pp. 157–205. CRC Press, Boca Raton, Florida.

Rhodes, M. J. C., and Wooltorton, L. S. C. (1973). Formation of CoA esters of cinnamic acid derivatives by extracts of *Brassica napobrassica* root tissue. *Phytochemistry* **12**, 2381–2387.

Rice, E. L. (1974). "Allelopathy." Academic Press, New York.

Rice, E. L. (1984). "Allelopathy." 2nd Ed. Academic Press, Orlando, Florida.

Richards, D., and Cockcroft, B. (1974). Soil physical properties and root concentrations in an irrigated apple orchard. *Aust. J. Exp. Agric. Anim. Husb.* **14**, 103–107.

Richards, D., and Cockroft, B. (1975). The effect of soil water on root production of peach trees in summer. *Aust. J. Agric. Res.* **26**, 173–180.

Richards, D., and Rowe, R. N. (1977a). Effects of root restriction, root pruning and 6-benzylaminopurine on the growth of peach seedlings. *Ann. Bot.* **41**, 729–740.

Richards, D., and Rowe, R. N. (1977b). Root–shoot interactions in peach: The function of the root. *Ann. Bot.* **41**, 1211–1216.

Richards, L. A., and Wadleigh, C. H. (1952). Soil water and plant growth. *In* "Soil Physical Conditions and Plant Growth" (B. T. Shaw, ed.), pp. 73–251. Academic Press, New York.

Richards, P. W. (1966). "Tropical Rain Forest." Cambridge Univ. Press, London and New York.

Richardson, C. J. (1983). Pocosins: Vanishing wastelands or valuable wetlands? *BioScience* **33**, 626–633.

Richardson, P. J., Webster, A. D., and Quinlan, J. D. (1986). The effect of paclobutrazol sprays with or without the addition of surfactants on the shoot growth, yield, and fruit quality of the apple cultivars Cox and Suntan. *J. Hortic. Sci.* **61**, 439–446.

Richardson, S. D. (1953). A note on some differences in root-hair formation between seedlings of sycamore and American oak. *New Phytol.* **52**, 80–82.

Richardson, S. D. (1956a). Studies of root growth in *Acer saccharinum* L. III. The influence of seedling age on the short term relation between photosynthesis, and root growth. *Proc. K. Ned. Akad. Wet.* **C59**, 416–427.

Richardson, S. D. (1956b). Studies of root growth in *Acer saccharinum* L. V. The effect of a long-term limitation of photosynthesis on root growth rate in first-year seedlings. *Proc. Kon. Ned. Akad. Wet.* **C59**, 694–701.

Ride, J. P. (1978). The role of cell wall alterations in resistance to fungi. *Ann. Appl. Biol.* **89**, 302–306.

Rieger, M. (1989). Freeze protection for horticultural crops. *Hortic. Rev.* **11**, 45–109.

Rinne, P., Saarelainen, A., and Junttila, O. (1994). Growth cessation and bud dormancy in relation to ABA level in seedlings and coppice shoots of *Betula pubescens* as affected by a short photoperiod, water-stress and chilling. *Physiol. Plant.* **90**, 451–458.

Ristic, Z., and Ashworth, E. N. (1993). Ultrastructural evidence that intracellular ice formation and possibly cavitation are the sources of freezing injury in supercooling wood tissue of *Cornus florida* L. *Plant Physiol.* **103**, 753–761.

Ritchie, G. A. (1982). Carbohydrate reserves and root growth potential in Douglas fir seedlings before and after cold storage. *Can. J. For. Res.* **12**, 905–912.

Ritchie, G. A. (1984). Assessing seedling quality. *In* "Forest Nursery Manual" (M. L.

Duryea and T. D. Landis, eds.), pp. 243–259. Martinus Nijhoff and W. Junk, The Hague, Boston, and Lancaster.

Ritchie, G. A., and Dunlap, J. R. (1980). Root growth potential: Its development and expression in forest tree seedlings. *N. Z. J. For. Sci.* **10**, 218–248.

Ritchie, G. A., Roden, J. R., and Kleyn, N. (1985). Physiological quality of lodgepole pine and interior spruce seedlings: Effect of lift date and duration of freezer storage. *Can. J. For. Res.* **15**, 636–645.

Robards, A. W., Davidson, E., and Kidwai, P. (1969). Short-term effects of some chemicals on cambial activity. *J. Exp. Bot.* **20**, 912–920.

Roberts, B. R. (1989). Some physiological considerations on sulfur dioxide sensitivity in drought-stressed seedlings of sweetgum (*Liquidambar styraciflua* L.). *Sci. Hortic. (Amsterdam)* **41**, 141–149.

Roberts, B. R., Kramer, P. J., and Karl, C. M., Jr. (1963). Long-term effects of gibberellin on the growth of loblolly pine seedlings. *For. Sci.* **9**, 202–205.

Roberts, E. H. (1972). Storage environment and the control of viability. *In* "Viability of Seeds" (E. H. Roberts, ed.), pp. 14–58. Chapman & Hall, London.

Roberts, E. H. (1973). Predicting the storage life of seeds. *Seed Sci. Technol.* **1**, 499–514.

Roberts, E. H., King, M. W., and Ellis, R. H. (1984). Recalcitrant seeds, their recognition and storage. *In* "Crop Genetic Resources: Conservation and Evaluation" (J. H. W. Holden and J. T. Williams, eds.), pp. 38–52. Allen & Unwin, London.

Roberts, J. A., Schindler, B., and Tucker, G. A. (1984). Ethylene-promoted tomato flower abscission and the possible involvement of an inhibitor. *Planta* **160**, 159–163.

Roberts, J. K. M., Andrade, F. H., and Anderson, I. C. (1985). Further evidence that cytoplasmic acidosis is a determinant of flooding intolerance in plants. *Plant Physiol.* **77**, 492–494.

Roberts, L. W., and Miller, A. R. (1983). Is ethylene involved in xylem differentiation? *Vistas Plant Sci.* **6**, 1–24.

Robinson, L. W., and Wareing, P. F. (1969). Experiments on the juvenile–adult phase change in some woody plants. *New Phytol.* **68**, 67–78.

Robinson, N. J., Shirsat, A. H., and Gatehouse, J. A. (1993). Regulation of gene expression. *In* "Plant Biochemistry and Molecular Biology" (P. J. Lea and R. C. Leegood, eds.), pp. 221–240. Wiley, Chichester and New York.

Robinson, T. L., Seeley, E. J., and Barritt, B. H. (1983). Effect of light environment and spur age on 'Delicious' apple fruit size and quality. *J. Am. Soc. Hortic. Sci.* **108**, 855–861.

Robitaille, H. A., and Leopold, A. C. (1974). Ethylene and the regulation of apple stem growth under stress. *Physiol. Plant.* **32**, 301–304.

Rochange, F., Serrano, L., Marque, C., Teulières, C., and Boudet, A.-M. (1995). DNA delivery into *Eucalyptus globulus* zygotic embryos through biolistics: Optimization of the biological and physical parameters of bombardment for two different particle guns. *Plant Cell Rep.* **14**, 674–678.

Roden, J., Van Volkenburgh, E., and Hinckley, T. M. (1990). Cellular basis for limitation of poplar leaf growth by water deficit. *Tree Physiol.* **6**, 211–220.

Roe, E. I. (1941). Effect of temperature in seed germination. *J. For.* **39**, 413–414.

Rogers, D. L., Stettler, R. F., and Heilman, P. E. (1989). Genetic variation and productivity of *Populus trichocarpa* and its hybrids. III. Structure and pattern of variation in a 3-year field test. *Can. J. For. Res.* **19**, 372–377.

Rokach, A. (1953). Water transfer from fruits to leaves in the Shamouti orange tree and related topics. *Palest. J. Bot. Rehovot Ser.* **8**, 146–151.

Rolf, K. (1994). A review of preventative and loosening measures to alleviate soil compaction in tree planting areas. *Arboric. J.* **18**, 431–448.

Rom, C., and Brown, S. A. (1979). Water tolerance of apples on clonal rootstocks and peaches on seedling rootstocks. *Compact Fruit Tree* **12**, 30–33.

Rom, R. C., and Carlson, R. F., eds. (1987). "Rootstocks for Fruit Crops." Wiley, New York and Chichester.

Romberger, J. A. (1963). Meristems, growth and development in woody plants. *U.S. Dep. Agric. Tech. Bull.* **1293**.

Rook, D. A. (1969). Water relations of wrenched and unwrenched *Pinus radiata* seedlings on being transplanted into conditions of water stress. *N. Z. J. For.* **14**, 50–58.

Rook, D. A. (1971). Effect of undercutting and wrenching on growth of *Pinus radiata* D. Don seedlings. *J. Appl. Ecol.* **8**, 477–490.

Rook, D. A., and Whyte, A. G. D. (1976). Partial defoliation and growth of 5-year-old radiata pine. *N. Z. J. For. Sci.* **6**, 40–56.

Roos, E. E. (1982). Induced genetic changes in seed germplasm during storage. *In* "The Physiology and Biochemistry of Seed Development, Dormancy, and Germination" (A. A. Khan, ed.), pp. 409–434. Elsevier, Amsterdam.

Roper, T. R., and Kennedy, R. A. (1986). Photosynthetic characteristics during leaf development in 'Bing' sweet cherry. *J. Am. Soc. Hortic. Sci.* **111**, 352–356.

Roper, T. R., Loescher, W. H., Keller, J., and Rom, C. R. (1987). Sources of photosynthate for fruit growth in 'Bing' sweet cherry. *J. Am. Soc. Hortic. Sci.* **112**, 808–812.

Roper, T. R., Keller, J. D., Loescher, W. H., and Row, C. R. (1988). Photosynthesis and carbohydrate partitioning in sweet cherry: Fruiting effects. *Physiol. Plant.* **72**, 42–47.

Roques, A., Kerjean, M., and Auclair, D. (1980). Effets de la pollution atmosphérique par le fluor et le dioxyde de soufre sur l'appareil reproducteur femelle de *Pinus sylvestris* en Forêt de Roumare (Seine-Maritime, France). *Environ. Pollut. (Ser. A)* **29**, 191–201.

Rosen, C. J., and Carlson, R. M. (1984). Influence of root zone oxygen stress on potassium and ammonium absorption by myrobalan plum rootstock. *Plant Soil* **80**, 345–353.

Rosen, K., Gunderson, P., Tenghammar, L., Johansson, M., and Frogner, T. (1992). Nitrogen enrichment in nordic forest ecosystems. *Ambio* **21**, 364–368.

Rosenzweig, M. L. (1968). Net primary productivity of terrestrial communities: Prediction from climatological data. *Am. Nat.* **102**, 67–74.

Ross, J. D. (1984). Metabolic aspects of dormancy. *In* "Seed Physiology" (D. R. Murray, ed.), pp. 45–75. Academic Press, Sydney.

Ross, J. D., and Bradbeer, J. W. (1968). Concentrations of gibberellin in chilled hazel seeds. *Nature (London)* **220**, 85–86.

Ross, J. D., and Bradbeer, J. W. (1971). Studies in seed dormancy. V. The content of endogenous gibberellins in seeds of *Corylus avellana*. *Planta* **100**, 288–302.

Ross, S. D., and Bower, R. C. (1989). Cost-effective promotion of flowering in a

Douglas-fir seed orchard by girdling and pulsed stem injection of gibberellin $A_{4/7}$. *Silvae Genet.* **38**, 189–195.

Ross, S. D., and Pharis, R. P. (1985). Promotion of flowering in tree crops: Different mechanisms and techniques, with special reference to conifers. *In* "Attributes of Trees as Crop Plants" (M. G. R. Cannell and J. E. Jackson, eds.), pp. 383–397. Institute of Terrestrial Ecology, Huntingdon, England.

Ross, S. D., Pharis, R. P., and Binder, W. D. (1983). Growth regulators and conifers: Their physiology and potential uses in forestry. *In* "Plant Growth Regulating Chemicals" (L. G. Nickell, ed.), Vol. 2, pp. 35–78. CRC Press, Boca Raton, Florida.

Ross, S. D., Bollman, M. P., Pharis, R. P., and Sweet, G. B. (1984). Gibberellin $A_{4/7}$ and the promotion of flowering in *Pinus radiata*: Effects of partitioning of photoassimilate within the bud during primordia differentiation. *Plant Physiol.* **76**, 320–330.

Rottink, B. A. (1986). Comparison of gibberellins A_3 and $A_{4/7}$ for promoting strobili in western hemlock. *Can. J. For. Res.* **16**, 389–390.

Rowe, R. N., and Beardsell, D. V. (1973). Waterlogging of fruit trees. *Hortic. Abstr.* **43**, 533–548.

Rubery, H. (1988). Auxin transport. *In* "Plant Hormones and Their Role in Plant Growth and Development" (P. J. Davies, ed.), pp. 341–362. Kluwer, Dordrecht, The Netherlands.

Rudinsky, J. A. (1962). Ecology of Scolytidae. *Annu. Rev. Entomol.* **7**, 327–348.

Rupp, L. A., and Mudge, K. W. (1985). Ethephon and auxin induce mycorrhiza-like changes in the morphology of root organ cultures of mugo pine. *Physiol. Plant.* **64**, 316–322.

Rushton, B. S., and Toner, A. E. (1989). Wind damage to leaves of sycamore (*Acer pseudoplatanus* L.) in coastal and noncoastal stands. *Forestry* **62**, 67–88.

Russell, C. E., and Berryman, A. A. (1976). Host resistance to the fir engraver beetle. I. Monoterpene composition of *Abies grandis* pitch blisters and fungus infected wounds. *Can. J. Bot.* **54**, 14–18.

Russell, E. W. (1973). "Soil Conditions and Plant Growth." Longmans, London.

Russell, J. A. (1993). Advances in the protoplast culture of woody plants. *In* "Micropropagation of Woody Plants" (M. R. Ahuja, ed.). pp. 67–91. Kluwer, Dordrecht, The Netherlands.

Ryan, D. F., and Bormann, F. H. (1982). Nutrient resorption in northern hardwood forests. *BioScience* **32**, 29–32.

Ryugo, K. (1988). "Fruit Culture: Its Science and Art." Wiley, New York.

Ryugo, K., and Davis, L. D. (1959). The effect of the time of ripening on the starch content of bearing peach branches. *Proc. Am. Soc. Hortic. Sci.* **74**, 130–133.

Ryugo, K., Marangoni, B., and Ramos, D. E. (1980). Light intensity and fruiting effects on carbohydrate contents, spur development, and return bloom of 'Hartley' walnut. *J. Am. Soc. Hortic. Sci.* **105**, 223–227.

Ryugo, K., Bartolini, G., Carlson, R. M., and Ramos, D. E. (1985). Relationship between catkin development and cropping in the Persian walnut 'Serr.' *HortScience* **20**, 1094–1096.

Saab, I. N., Sharp, R. E., Pritchard, J., and Voetberg, G. S. (1990). Increased endogenous abscisic acid maintains primary root growth and inhibits shoot growth of maize seedlings at low water potentials. *Plant Physiol.* **93**, 1329–1336.

Saastamoinen, T., and Halopainen, T. (1989). Needle and root responses of small *Pinus sylvestris* seedlings exposed to sulphur dioxide and simulated acid rain. *Scand. J. For. Res.* **4**, 273–283.

Sachs, P. M., Ripperda, J., Forister, G., Miller, G., Kasemsap, P., Murphy, M., and Benyl, G. (1988). Maximum biomass yields on prime agricultural land. *Calif. Agric. November–December*, 23–24.

Sachs, R. M., and Hackett, W. (1977). Chemical control of flowering. *HortScience* **12**, 220–222.

Saieed, N. T., Douglas, G. C., and Fry, D. J. (1994a). Induction and stability of somaclonal variation in growth, leaf phenotype and gas exchange characteristics of poplar regenerated from callus culture. *Tree Physiol.* **14**, 1–16.

Saieed, N. T., Douglas, G. C., and Fry, D. J. (1994b). Somaclonal variation in growth, leaf phenotype and gas exchange characteristics of poplar—Utilization of leaf morphotype analysis as a basis for selection. *Tree Physiol.* **14**, 17–26.

Saiki, R. K., Scharf, S., Faloona, F. Mullis, K. B., Horn, G. T., Erlich, H. A., and Arnheim, N. (1985). Enzymatic amplification of β-globin genomic sequences and restriction site analysis for diagnosis of sickle cell anemia. *Science* **230**, 1350–1354.

Sakai, A., and Larcher, W. (1987). "Frost Survival of Plants." Springer-Verlag, Berlin and New York.

Sakai, A., and Weiser, C. J. (1973). Freezing resistance of trees in North America with reference to tree regions. *Ecology* **54**, 118–126.

Salisbury, F. B. (1982). Photoperiodism. *Hortic. Rev.* **4**, 66–105.

Salisbury, F. B., and Ross, C. W. (1985). "Plant Physiology," 3rd Ed. Wadsworth, Belmont, California.

Sanchez, E., Khemira, H., Sugar, D., and Righetti, T. L. (1995). Nitrogen management in orchards. *In* "Nitrogen Fertilization in the Environment" (P. E. Bacon, ed.), pp. 327–380. Dekker, New York.

Sandberg, G., Anderson, B., and Dunberg, A. (1981). Identification of 3-indoleacetic acid in *Pinus silvestris* L. by gas chromatography–mass spectrometry, and quantitative analysis by ion-pair on reversed-phase liquid chromatography with spectrofluorimetric detection. *J. Chromatogr.* **205**, 125–137.

Sanderson, P. L., and Armstrong, W. A. (1980a). The responses of conifers to some of the adverse factors associated with waterlogged soils. *New Phytol.* **85**, 351–362.

Sanderson, P. L., and Armstrong, W. A. (1980b). Phytotoxins in periodically waterlogged forest soils. *J. Soil Sci.* **31**, 643–653.

Santamour, F. S. (1988a). Graft incompatability related to cambial peroxidase isozymes in Chinese chestnut. *J. Environ. Hortic.* **6**, 33–39.

Santamour, F. S. (1988b). Cambial peroxidase enzymes related to graft incompatability in red oak. *J. Environ. Hortic.* **6**, 87–93.

Santamour, F. S., McArdle, A. J., and Jaynes, R. A. (1986). Cambial peroxidase patterns in *Castanea. J. Environ. Hortic.* **4**, 14–16.

Santantonio, D., and Santantonio, E. (1987). Effect of thinning on production and mortality of fine roots in a *Pinus radiata* plantation on a fertile site in New Zealand. *Can. J. For. Res.* **17**, 919–928.

Sarvas, R. (1962). Investigations on the flowering and seed crop of *Pinus silvestris*. *Commun. Inst. Forst. Fenn.* (see *Metsantutkimuslaitoksen Julk.*) **53**, 1–98.

Sasaki, S. (1980a). Storage and germination of some Malaysian legume seeds. *Malays. For.* **43**, 161–165.

Sasaki, S. (1980b). Storage and germination of dipterocarp seed. *Malays. For.* **43**, 290–308.

Sasaki, S., and Kozlowski, T. T. (1968a). The role of cotyledons in early development of pine seedlings. *Can. J. Bot.* **46**, 1173–1183.

Sasaki, S., and Kozlowski, T. T. (1968b). Effects of herbicides on respiration of red pine (*Pinus resinosa* Ait.) seedlings. I. *S*-Triazine and chlorophenoxy acid herbicides. *Adv. Front. Plant Sci.* **22**, 187–202.

Sasaki, S., and Kozlowski, T. T. (1968c). Effects of herbicides on seed germination and early seedling development of *Pinus resinosa. Bot. Gaz.* (*Chicago*) **129**, 238–246.

Sasaki, S., and Kozlowski, T. T. (1968d). Effects of herbicides on respiration of red pine (*Pinus resinosa* Ait.) seedlings. II. Monuron, diuron, DCPA, Dalapon, CDEC, CDAA, EPTC, and NPA. *Bot. Gaz.* (*Chicago*) **129**, 286–293.

Sasaki, S., and Kozlowski, T. T. (1969). Utilization of seed reserves and currently produced photosynthates of embryonic tissues of pine seedlings. *Ann. Bot.* (*London*) **33**, 472–482.

Sasaki, S., and Kozlowski, T. T. (1970). Effects of cotyledon and hypocotyl photosynthesis on growth of young pine seedlings. *New Phytol.* **69**, 493–500.

Sasaki, S., Kozlowski, T. T., and Torrie, J. H. (1968). Effect of pretreatment of pine seeds with herbicides on seed germination and growth of young seedlings. *Can. J. Bot.* **46**, 255–262.

Sastry, C. B. R., and Anderson, H. W. (1980). Clonal variation in gross heat combustion of juvenile *Populus* hybrids. *Can. J. For. Res.* **10**, 245–249.

Satoo, T. (1962). Wind, transpiration, and tree growth. *In* "Tree Growth" (T. T. Kozlowski, ed.), pp. 299–310. Ronald, New York.

Satoo, T. (1966). Variation in response of conifer seed germination to soil moisture conditions. *Tokyo Univ. For., Misc. Inf.* (*Tokyo Noko Daigaku Nogakubu Enshurin Hokoku*) **16**, 17–20.

Saure, M. C. (1985). Dormancy release in deciduous fruit trees. *Hortic. Rev.* **7**, 239–300.

Saure, M. C. (1990). External control of anthocyanin formation in apple. *Sci. Hortic.* (*Amsterdam*) **42**, 181–218.

Sauter, J. J. (1980). Seasonal variation of sucrose content in the xylem sap of *Salix. Z. Pflanzenphysiol.* **98**, 377–391.

Sauter, J. J. (1981). Sucrose uptake in the xylem of *Populus. Z. Pflanzenphysiol.* **103**, 165–168.

Sauter, J. J., and Ambrosius, T. (1986). Changes in the partitioning of carbohydrates in the wood during bud break in *Betula pendula* Roth. *J. Plant Physiol.* **124**, 31–43.

Sauter, J. J., and Kloth, S. (1986). Plasmodesmatal frequency and radial translocation rates in ray cells of poplar (*Populus* × *canadensis* Moench 'Robusta'). *Planta* **168**, 377–380.

Sauter, J. J., Kammerbauer, H., Pambor, L., and Hock, B. (1987). Evidence for the accelerated micromorphological degradation of epistomatal waxes in Norway spruce by motor vehicle emissions. *Eur. J. For. Pathol.* **17**, 444–448.

Savidge, R. A. (1988). Auxin and ethylene regulation of diameter growth in trees. *Tree Physiol.* **4**, 401–414.

Savidge, R. A., and Burnett, J. R. (1993). Protoplasmic changes in cambial cells induced by a tracheid-differentiation factor from pine needles. *J. Exp. Bot.* **44**, 395–405.

Savidge, R. A., and Wareing, P. F. (1981). Plant-growth regulators and the differentiation of vascular elements. *In* "Plant-Growth Regulators and the Differentiation of Vascular Elements" (J. R. Barnett, ed.), pp. 192–235. Castle House, Tunbridge Wells, England.

Savidge, R. A., and Wareing, P. F. (1982). Apparent auxin production and transport during winter in the nongrowing pine tree. *Can. J. Bot.* **60**, 681–691.

Savidge, R. A., and Wareing, P. F. (1984). Seasonal cambial activity and xylem development in *Pinus contorta* in relation to endogenous indol-3-yl-acetic acid and (*S*)-abscisic acid levels. *Can. J. For. Res.* **14**, 676–682.

Savidge, R. A., Mutumba, G. M. C., Heald, J. K., and Wareing, P. F. (1983). Gas chromatography–mass spectroscopy identification of 1-aminocyclopropane-1-carboxylic acid in compression-wood vascular cambium of *Pinus contorta* Dougl. *Plant Physiol.* **71**, 434–436.

Savidge, R., and Udagama-Randeniya, P. (1992). Cell wall-bound coniferyl alcohol oxidase associated with lignification in conifers. *Phytochemistry* **31**, 2959–2966.

Savill, P. S. (1983). Silviculture in windy climates. *For. Abstr.* **44**, 473–488.

Savin, K. W., Baudinette, S. C., Graham, M. W., Michael, M. Z., Nugent, G. D., Lu, C.-Y., Chandler, S. F., and Cornish, E. C. (1995). Antisense ACC oxidase RNA delays carnation petal senescence. *HortScience* **30**, 970–972.

Scarano, F. R., and Crawford, R. M. M. (1992). Ontogeny and concept of anoxia-tolerance: The case of the leguminous tree *Parkia pendula. J. Trop. Ecol.* **8**, 349–352.

Schäfer, H., Bossel, H., and Krieger, K. (1988). Modelling the responses of mature forest trees to air pollution. *GeoJournal* **17**, 279–287.

Schaff, D. A. (1991). Biotechnology–gene transfer—Terminology, techniques, and problems involved. *HortScience* **26**, 1021–1024.

Schechter, I., Proctor, J. T. A., and Elfving, D. C. (1994). Apple fruit removal and limb girdling affect fruit and leaf characteristics. *J. Am. Soc. Hortic. Sci.* **119**, 157–162.

Scheffer, T. C., and Hedgcock, G. G. (1955). Injury to northwestern forest trees by sulfur dioxide from smelters. *Tech. Bull., U.S.D.A.* **117**.

Schier, G. A. (1975). Effect of ethephon on adventitious root and shoot development in aspen. *Plant Physiol.* **56**(Suppl.), 71.

Schier, G. A. (1985). Seedling growth and nutrient relationships in a New Jersey Pine Barrens soil treated with 'acid rain.' *Can. J. For. Res.* **16**, 136–142.

Schlesinger, W. H. (1978). Community structure, dynamics and nutrient cycling in the Okefenokee cypress swamp forest. *Ecol. Monogr.* **44**, 43–65.

Schmidtling, R. C. (1983). Rootstock influences flowering, growth, and survival of loblolly pine grafts. *For. Sci.* **29**, 117–124.

Schmidtling, R. C. (1988). Influence of rootstock on flowering, growth, and foliar nutrients of slash pine grafts. *Proc. North Am. For. Biol. Workshop, 10th, Vancouver, B.C.,* 120–127.

Schneider, A., and Schmitz, K. (1989). Seasonal course of translocation and distribution of ^{14}C-labelled photoassimilate in young trees of *Larix decidua* Mill. *Trees* **4**, 185–191.

Schneider, G. W. (1978). Abscission mechanism studies with apple fruitlets. *J. Am. Soc. Hortic. Sci.* **103**, 455–458.

Schneider, S. H. (1990). The global warming debate heats up: An analysis and perspective. *Bull. Am. Meteorol. Soc.* **71**, 1292–1304.

Schneider, W. L., and Gifford, D. J. (1994). Loblolly pine seed dormancy. I. The relation between protein synthesis and the loss of dormancy. *Physiol. Plant.* **90**, 246–252.

Schoeneweiss, D. F. (1978a). Water stress as a predisposing factor in plant disease. *In* "Water Deficits and Plant Growth" (T. T. Kozlowski, ed.), Vol. 5, pp. 61–99. Academic Press, New York.

Schoeneweiss, D. F. (1978b). The influence of stress on diseases of nursery and landscape plants. *J. Arboric.* **4**, 217–225.

Schoeneweiss, D. F. (1986). Water stress predisposition to disease—an overview. *In* "Water, Fungi, and Plants" (P. G. Ayres, ed.), pp. 157–174. Cambridge, New York.

Schoettle, A. W., and Smith, W. K. (1991). Interrelation between shoot characteristics and solar irradiance in the crown of *Pinus contorta* spp. *latifolia. Tree Physiol.* **9**, 245–254.

Scholander, P. F. (1968). How mangroves desalinate seawater. *Physiol. Plant.* **21**, 251–261.

Scholander, P., van Dam, L., and Scholander, S. I. (1955). Gas exchange in the roots of mangroves. *Am. J. Bot.* **42**, 92–98.

Scholefield, P. B., Sedgley, M., and Alexander, M. (1985). Carbohydrate cycling in relation to shoot growth, floral initiation and development and yield in the avocado. *Sci. Hortic. (Amsterdam)* **25**, 99–110.

Scholz, R., Gregorius, H.-R., and Rudin, D., eds. (1989). "Genetic Effects of Air Pollutants in Forest Tree Populations." Springer-Verlag, New York.

Schopfer, P. (1977). Phytochrome control of enzymes. *Annu. Rev. Plant Physiol.* **28**, 223–252.

Schowalter, T. D., Hargrove, W. W., and Crossley, D. A., Jr. (1986). Herbivory in forested ecosystems. *Annu. Rev. Entomol.* **31**, 177–196.

Schramm, J. E. (1966). Plant colonization studies on black wastes from anthracite mining in Pennsylvania. *Trans. Am. Philos. Soc.* **56**, 1–194.

Schuett, P., and Cowling, E. B. (1985). Waldsterben, a general decline of forests in central Europe: Symptoms, development, and possible causes. *Plant Dis.* **69**, 548–558.

Schultz, H. R., and Matthews, M. A. (1993). Growth, osmotic adjustment, and cell-wall mechanics of expanding grape leaves during water deficits. *Crop Sci.* **33**, 287–294.

Schulze, E.-D., Lange, O. L., and Oren, R., eds. (1989). "Forest Decline and Air Pollution." Springer-Verlag, Berlin.

Schumacher, R., Frankhauser, F., and Stadler, W. (1978). Beeinflussung der Fruchtbarkeit und der Fruchtqualität durch den Wurzelschnitt. *Schweiz. Obst Weinbau* **114**, 56–61.

Scifres, C. J., and Brock, J. H. (1969). Moisture–temperature interrelations in germination and early seedling development of mesquite. *J. Range Manage.* **22**, 334–337.

Scurfield, G. (1973). Reaction wood: Its structure and function. *Science* **179**, 647–655.

Sedgley, M. (1977). The effect of temperature on floral behaviour, pollen tube growth and fruit set in the avocado. *J. Hortic. Sci.* **52**, 135–141.

Sedgley, M. (1990). Flowering of deciduous perennial fruit crops. *Hortic. Rev.* **12**, 223–264.

Sedgley, M., and Griffin, A. R. (1989). "Sexual Reproduction of Tree Crops." Academic Press, London and San Diego.

Seeley, E. J., Micke, W. C., and Kammereck, R. (1980). 'Delicious' apple fruit size and quality as influenced by radiant flux density in the immediate growing environment. *J. Am. Soc. Hortic. Sci.* **105**, 645–657.

Seeley, S. (1990). Hormonal transduction of environmental stresses. *HortScience* **25**, 1369–1376.

Seeley, S. D., and Powell, L. E. (1981). Seasonal changes of free and hydrolyzable abscisic acid in vegetative apple buds. *J. Am. Soc. Hortic. Sci.* **106**, 405–409.

Seidel, K. W. (1986). Tolerance of seedlings of ponderosa pine, Douglas-fir, grand fir, and Engelmann spruce for high temperatures. *Northwest Sci.* **60**, 1–7.

Seipp, D. (1974). CA-Lagerungsversuche mit Apfeln. 2. Einfluss des CO_2-Anteils der Lagerluft auf die Erhaltung der Fruchtqualität. *Erwerbsobstbau* **16**, 165–169.

Sekse, L. (1995). Fruit cracking in sweet cherries (*Prunus avium* L.). Some physiological aspects—A minireview. *Sci. Hortic. (Amsterdam)* **63**, 135–141.

Sena Gomes, A. R., and Kozlowski, T. T. (1980a). Growth responses and adaptations of *Fraxinus pennsylvanica* seedlings to flooding. *Plant Physiol.* **66**, 267–271.

Sena Gomes, A. R., and Kozlowski, T. T. (1980b). Responses of *Melaleuca quinquenervia* seedlings to flooding. *Physiol. Plant.* **49**, 373–377.

Sena Gomes, A. R., and Kozlowski, T. T. (1980c). Responses of *Pinus halepensis* seedlings to flooding. *Can. J. For. Res.* **10**, 308–311.

Sena Gomes, A. R., and Kozlowski, T. T. (1986). The effects of flooding on water relations and growth of *Theobroma cacao* var. *catongo* seedlings. *J. Hortic. Sci.* **61**, 265–276.

Sena Gomes, A. R., and Kozlowski, T. T. (1988). Physiological and growth responses to flooding of seedlings of *Hevea brasiliensis*. *Biotropica* **20**, 286–293.

Sena Gomes, A. R., and Kozlowski, T. T. (1989). Responses of seedlings of two varieties of *Theobroma cacao* to wind. *Trop. Agric.* **66**, 137–141.

Sena Gomes, A. R., Kozlowski, T. T., and Reich, P. B. (1987). Some physiological responses of *Theobroma cacao* var. *Catongo* seedlings to air humidity. *New Phytol.* **107**, 591–602.

Senser, M. (1982). Frost resistance in spruce [*Picea abies* (L.) Karst.]. III. Seasonal changes in the phospho- and galactolipids of spruce needles. *Z. Pflanzenphysiol.* **105**, 229–239.

Senser, M., and Beck, E. (1979). Kälteresistenz der Fichte. II. Einfluss von Photoperiode und Temperatur auf die Struktur und photochemischen Reaktionen von Chloroplasten. *Ber. Dtsch. Bot. Ges.* **92**, 243–259.

Serres, R., Ostry, M., McCown, B., and Skilling, D. (1991). Somaclonal variation in *Populus* hybrids regenerated from protoplast culture. *In* "Woody Plant Biotechnology" (M. R. Ahuja, ed.), pp. 59–61. Plenum, New York.

Serres, R., Stang, E., McCabe, D., Russell, D., Mahr, D., and McCown, B. H. (1992). Gene transfer using electric discharge particle bombardment and recovery of transformed cranberry plants. *J. Am. Soc. Hortic. Sci.* **117**, 174–180.

Sexton, R., Lewis, L. N., Trewavas, A. J., and Kelly, P. (1985). Ethylene and abscission. *In* "Ethylene and Plant Development" (J. A. Roberts and G. A. Tucker, eds.), pp. 173–196. Butterworth, London.

Sexton, R., Struthers, W. A., and Lewis, L. N. (1983). Some observations on the very rapid abscission of the petals of *Geranium robertianum* L. *Protoplasma* **116**, 179–186.

Shalhevet, J., and Levy, Y. (1990). Citrus trees. *In* "Irrigation of Agricultural Crops" (B. A. Stewart and D. R. Nielsen, eds.), pp. 951–986. Am. Soc. Agron., Madison, Wisconsin.

Shaltout, A. D., and Unrath, C. R. (1983). Rest completion prediction model for 'Starkrimson Delicious' apples. *J. Am. Soc. Hortic. Sci.* **108**, 957–961.

Shaltout, A. D., Salem, A. T., and Kilany, A. S. (1988). Effect of pre-bloom sprays and soil drenches of paclobutrazol on growth, yield, and fruit composition of 'Roumi Red' grapes. *J. Am. Soc. Hortic. Sci.* **113**, 13–17.

Shanklin, J. J., and Kozlowski, T. T. (1984). Effect of temperature preconditioning on response of *Fraxinus pennsylvanica* seedlings to SO_2. *Environ. Pollut. (Ser. A)* **36**, 311–326.

Shanklin, J., and Kozlowski, T. T. (1985). Effect of flooding of soil on growth and subsequent responses of *Taxodium distichum* seedlings to SO_2. *Environ. Pollut. (Ser. A)* **38**, 199–212.

Shanks, J. B. (1969). Some effects and potential uses of Ethrel on ornamental crops. *HortScience* **4**, 56–58.

Shannon, M. C., Grieve, C. M., and Francois, L. E. (1994). Whole-plant response to salinity. *In* "Plant–Environment Interactions" (R. E. Wilkinson, ed.), pp. 199–244. Dekker, New York.

Sharp, R. E., and Davies, W. J. (1979). Solute regulation and growth by roots and shoots of water-stressed maize plants. *Planta* **147**, 43–49.

Sharp, R. E., Wu, Y. J., Voetberg, G. S., Saab, I. N., and Lenoble, M. E. (1994). Confirmation that abscisic acid accumulation is required for maize primary root elongation at low water potentials. *J. Exp. Bot.* **45**, 1743–1751.

Sharples, R. O. (1980). The influence of orchard nutrition on the storage quality of apples and pears grown in the United Kingdom. *In* "Mineral Nutrition of Fruit Trees" (D. Atkinson, J. E. Jackson, R. O. Sharples, and W. M. Waller, eds.), pp. 17–28. Butterworth, London.

Shaw, C. G., and Taes, E. H. A. (1977). Impact of *Dothistroma* needle blight and *Armillaria* root rot on diameter growth of *Pinus radiata*. *Phytopathology* **66**, 1319–1323.

Shear, C. B. (1975). Calcium-related disorders of fruits and vegetables. *HortScience* **10**, 361–365.

Shekhawat, N. S., Rathore, T. S., Singh, R. P., Deora, N. S., and Rao, S. R. (1993). Factors affecting *in vitro* clonal propagation of *Prosopis cineraria*. *Plant Growth Regul.* **12**, 273–280.

Sheldrake, A. R. (1971). Auxin in the cambium and its differentiating derivatives. *J. Exp. Bot.* **22**, 735–740.

Sheppard, L. J. (1994). Causal mechanisms by which sulphate nitrate and acidity influence frost hardiness in spruce: Review and hypothesis. *New Phytol.* **127**, 69–82.

Sheriff, D. W. (1983). Control by indole-3-acetic acid of wood production in *Pinus radiata* D. Don segments in culture. *Aust. J. Plant Physiol.* **10**, 131–135.

Sheriff, D. W., and Whitehead, D. (1984). Photosynthesis and wood structure in *Pinus radiata* D. Don during dehydration and immediately after rewatering. *Plant Cell Environ.* **7**, 53–62.

Shigo, A. L. (1984). Compartmentalization: A conceptual framework for understanding how trees grow and defend themselves. *Annu. Rev. Phytopathol.* **22**, 189–214.

Shirazi, A. M., and Fuchigami, L. H. (1993). Recovery of plants from "Near-Lethal" stress. *Oecologia* **93**, 429–434.

Shiver, B. D., Knowe, S. A., and Kline, W. N. (1990). Comparison of chemical site preparation treatments in the Georgia piedmont. *South. J. Appl. For.* **14**, 24–32.

Sierpinski, Z., and Cholodny, J. (1977). Entofauna of forest plantations in the zone of disastrous industrial pollution. *In* "Relationship between Increase in Air Pollution Toxicity and Elevation Above Ground" (J. Woldk, ed.), pp. 81–150. Inst. Badawczego Lesnictwa, Warsaw.

Silim, S. N., and Lavender, D. P. (1994). Seasonal patterns and environmental regulation of frost hardiness in shoots of seedlings of *Thuja plicata, Chamaecyparis nootkatensis,* and *Picea glauca. Can. J. Bot.* **72**, 309–316.

Sill, W. H., Jr. (1982). "Plant Protection: An Integrated Interdisciplinary Approach." Iowa State Univ. Press, Ames.

Silverborg, S. B., and Ross, E. W. (1968). Ash dieback disease development in New York State. *Plant Dis. Rep.* **52**, 105–107.

Simmonds, J. A., and Dumbroff, E. B. (1974). High energy charge as a requirement for axis elongation in response to gibberellic acid and kinetin during stratification of *Acer saccharum* seeds. *Plant Physiol.* **53**, 91–95.

Simmonds, N. W. (1982). "Bananas." Longmans, London.

Simons, R. K. (1959). Anatomical and morphological responses of four varieties of apples to frost injury. *Proc. Am. Soc. Hortic. Sci.* **74**, 10–24.

Simons, R. K. (1970). Phloem tissue development response to freeze injury to trunks of apple trees. *J. Am. Soc. Hortic. Sci.* **95**, 182–190.

Simons, R. K., and Lott, R. V. (1963). The morphological and anatomical development of apple injured by late spring frost. *Proc. Am. Soc. Hortic. Sci.* **83**, 88–100.

Simpson, D. G. (1990). Frost hardiness, root growth capacity and field performance relationships in interior spruce, lodgepole pine, Douglas fir and western hemlock seedlings. *Can. J. For. Res.* **20**, 566–572.

Sinclair, W. A. (1969). Polluted air: Potent new selective force in forests. *J. For.* **67**, 305–309.

Singh, Z., and Browning, G. (1991). The role of ABA in the control of apple seed dormancy re-appraised by combined gas chromatography–mass spectrometry. *J. Exp. Bot.* **42**, 269–275.

Sionit, N., and Kramer, P. J. (1986). Woody plant reactions to carbon dioxide enrichment. *In* "Carbon Dioxide Enrichment of Greenhouse Crops" (H. Z. Enoch and B. A. Kimball, eds.), pp. 69–84. CRC Press, Boca Raton, Florida.

Sionit, N., Strain, B. R., Hellmers, H., Riechers, G. H., and Jaeger, C. H. (1985). Long-term atmospheric CO_2 enrichment affects the growth and development of *Liquidambar styraciflua* and *Pinus taeda* seedlings. *Can. J. For. Res.* **15**, 468–471.

Sipes, D. L., and Einset, J. W. (1982). Role of ethylene in stimulating stylar abscission in pistil explants of lemon. *Physiol. Plant.* **56**, 6–10.

Siwecki, R., and Kozlowski, T. T. (1973). Leaf anatomy and water relations of excised leaves of six *Populus* clones. *Arbor. Kornickie* **8**, 83–105.

Skeffington, R. A., and Roberts, T. M. (1985). The effects of ozone and acid mist on Scots pine saplings. *Oecologia* **65**, 201–206.

Skinner, P. W., and Matthews, M. A. (1989). Reproductive development in grape (*Vitis vinifera* L.) under phosphorus-limited conditions. *Sci. Hortic.* (*Amsterdam*) **38**, 49–60.

Skirvin, R. M., Abu-Qaoud, H., Sriskandarajah, S., and Harry, D. E. (1993). Genetics of micropropagated plants. *In* "Micropropagation of Woody Plants" (M. R. Ahuja, ed.). pp. 121–152. Kluwer, Dordrecht, The Netherlands.

Skoog, F., and Miller, C. O. (1957). Chemical regulation of growth and organ formation in plant tissue cultures *in vitro*. *Symp. Soc. Exp. Biol.* **11**, 118–130.

Slabaugh, P. E. (1957). Effects of live crown removal on the growth of red pine. *J. For.* **55**, 904–906.

Slankis, V. (1973). Hormonal relationships in mycorrhizal development. *In* "Ectomycorrhizae" (G. C. Marks and T. T. Kozlowski, eds.), pp. 231–298. Academic Press, New York.

Slankis, V. (1974). Soil factors influencing formation of mycorrhizae. *Annu. Rev. Phytopathol.* **12**, 437–457.

Slee, M. U. (1977). A model relating needleless shoots and dieback in *Pinus caribaea* to strobilus production and climatic conditions. *Silvae Genet.* **26**, 135–141.

Sleigh II, P. A., Collin, A., and Hardwick, K. (1984). Distribution of assimilate during the flush cycle of growth in *Theobroma cacao* L. *Plant Growth Regul.* **2**, 381–391.

Small, J. A., and Monk, C. D. (1959). Winter changes in tree radii and temperature. *For. Sci.* **5**, 229–233.

Smart, R. E., and Coombe, B. G. (1983). Water relations of grapevines. *In* "Water Deficits and Plant Growth" (T. T. Kozlowski, ed.), Vol. 7, pp. 137–196. Academic Press, New York.

Smit-Spinks, B., Swanson, B. T., and Markhart, A. H. (1985). The effect of photoperiod and thermoperiod on cold acclimation and growth of *Pinus sylvestris*. *Can. J. For. Res.* **15**, 453–460.

Smith, C., Watson, C., Ray, J., Bird, C., Morris, P., Schuch, W., and Grierson, D. (1988). Antisense RNA inhibition of polygalacturonase gene expression in transgenic tomatoes. *Nature* (*London*) **334**, 724–726.

Smith, D. M. (1986). "The Practice of Silviculture." Wiley, New York.

Smith, D. M., and Wilsie, M. C. (1961). Some anatomical responses of loblolly pine to soil-water deficiencies. *Tappi* **44**, 179–185.

Smith, D. R., and Thorpe, T. A. (1975a). Root initiation in cuttings of *Pinus radiata* seedlings. I. Developmental sequence. *J. Exp. Bot.* **26**, 184–192.

Smith, D. R., and Thorpe, T. A. (1975b). Root initiation in cuttings of *Pinus radiata* seedlings. II. Growth regulator interactions. *J. Exp. Bot.* **26**, 193–202.

Smith, D. R., and Wareing, P. F. (1972a). Rooting of hardwood cuttings in relation to bud dormancy and the auxin content of the excised stems. *New Phytol.* **71**, 63–80.

Smith, D. R., and Wareing, P. F. (1972b). The rooting of actively growing and dormant leafy cuttings in relation to the endogenous hormone levels and photoperiod. *New Phytol.* **71**, 483–500.

Smith, J. H. G., Walters, J., and Kozak, A. (1968). Influences of fertilizers on cone production and growth of young Douglas-fir, western hemlock, and western red cedar on U. B. C. research forest. *U. B. C. Fac. For. Bull.* **5.**

Smith, M. A. L., Palta, J. P., and McCown, B. H. (1986). Comparative anatomy and physiology of microcultured, seedling, and greenhouse-grown Asian white birch. *J. Am. Soc. Hortic. Sci.* **111,** 437–442.

Smith, R. C., and Fry, S. C. (1991). Endotransglycosylation of xyloglucans in plant cell suspension cultures. *Biochem. J.* **279,** 529–535.

Smith, R. H. (1966a). Resin quality as a factor in the resistance of pines to bark beetles. *In* "Breeding Pest Resistant Trees" (H. D. Gerhold, E. J. Schreiner, R. E. McDermott, and J. A. Wisnieski, eds.), pp. 189–196. Pergamon, New York.

Smith, R. H. (1995). Air pollution and forests. *In* "Encyclopedia of Environmental Biology" (W. A. Nierenberg, ed.), Vol. 1, pp. 37–51. Academic Press, San Diego.

Smith, R. H. (1966b). The monoterpene composition of *Pinus ponderosa* xylem resin and of *Dendroctonus brevicomis* pitch tubes. *For. Sci.* **12,** 63–68.

Smith, S. E., and Gianinazzi-Pearson, V. (1988). Physiological interactions between symbionts in vesicular–arbuscular mycorrhizal plants. *Annu. Rev. Plant Physiol. Mol. Biol.* **39,** 221–244.

Smith, W. H. (1981). "Air Pollution and Forests." Springer-Verlag, New York.

Smith, W. H. (1990). "Air Pollution and Forests." 2nd Ed. Springer-Verlag, New York.

Smith, W. H. (1995). Air pollution and forests. *In* "Encyclopedia of Environmental Biology" (W. A. Nierenberg, ed.), Vol. 1, pp. 37–51. Academic Press, San Diego.

Smith, W. K., and Brewer, C. A. (1994). The adaptive importance of shoot and crown architecture in conifer trees. *Am. Nat.* **143,** 528–532.

Smith, W. K., Knapp, A. K., and Reiners, W. A. (1989). Pnenumbral effects on sunlight penetration in plant communities. *Ecology* **70,** 1603–1609.

Smock, R. M. (1979). Controlled atmosphere storage of fruits. *Hortic. Rev.* **1,** 301–336.

Sneed, R. E., and Patterson, R. P. (1983). The future role of irrigation in a humid climate. *In* "Crop Reactions to Water and Temperature Stresses in Humid Temperate Climates" (C. D. Raper, Jr., and P. J. Kramer, eds.), pp. 187–199. Westview Press, Boulder, Colorado.

Snellgrove, R. C., Splittstoesser, W. E., Stribley, D. P., and Tinker, P. B. (1982). The carbon distribution and the demand of the fungal symbiont in leek plants with vesicular–arbuscular mycorrhizas. *New Phytol.* **92,** 75–81.

Sniezko, R. A. (1986). Influence of *Pinus taeda* rootstock on growth and cone production of *P. kesiya* and *P. elliottii* clones. *In* "Proceedings of the IUFRO Conference, A Joint Meeting of Working Parties on Breeding Theory, Progeny Testing and Seed Orchards," pp. 429- 439. Williamsburg, Virginia.

Snowball, A. M., Halligan, E. A., Warrington, I. J., and Mullins, M. G. (1994a). Phase change in citrus: Growth and flowering of citrus seedlings from thirteen genetically diverse seedling families. *J. Hortic. Sci.* **69,** 141–148.

Snowball, A. M., Warrington, I. J., Halligan, E. A., and Mullins, M. G. (1994b). Phase change in citrus: The effects of main stem node number, branch habit and paclobutrazol application on flowering in citrus seedlings. *J. Hortic. Sci.* **69,** 149–160.

Soikkeli, S., and Kärenlampi, S. (1984). Cellular and ultrastructure effects. *In* "Air Pollution and Plant Life" (M. Treshow, ed.), pp. 159–186. Wiley, Chichester.

Solomon, M. G. (1987). Fruit and hops. *In* "Integrated Pest Management" (A. J. Burn, T. H. Coaker, and P. C. Jepson, eds.), pp. 329–360. Academic Press, London.

Sondheimer, E., Tzou, D. S., and Galson, E. C. (1968). Abscisic acid levels and seed dormancy. *Plant Physiol.* **43**, 1443–1447.

Songstad, D. D., Somers, D. A., and Griesbach, R. J. (1995). Advances in alternative DNA delivery techniques. *Plant Cell, Tissue Organ Cult.* **40**, 1–15.

Sotak, R. J., Sommer, H. E., and Merkle, S. A. (1991). Relation of the developmental stage of zygotic embryos of yellow-poplar to their somatic embryogenic potential. *Plant Cell Rep.* **10**, 175–178.

South, D. B., Boyer, J. N., and Bosch, L. (1985). Survival and growth of loblolly pine as influenced by seedling grade: 13-Year results. *South. J. Appl. For.* **9**, 76–81.

Southern, E. M. (1975). Detection of specific sequences among DNA fragments separated by gel electrophoresis. *J. Mol. Biol.* **98**, 503–517.

Southwick, S. M., and Davenport, T. L. (1986). Characterization of water stress and low temperature effects on flower induction in citrus. *Plant Physiol.* **81**, 26–29.

Spalding, D. H., and Reeder, W. F. (1983). Conditioning 'Tahiti' limes to reduce chilling injury. *Proc. Fla. State Hortic. Soc.* **96**, 231–232.

Sparks, D. (1989). Drought stress induces fruit abortion in pecan. *HortScience* **24**, 78–79.

Sparks, D. (1992). Abnormal flowering in pecan associated with freezing temperature. *HortScience* **27**, 801–802.

Spiegel-Roy, P., and Kochba, J. (1980). Embryogenesis in *Citrus* tissue cultures. *Adv. Biochem. Eng.* **18**, 27–48.

Spollen, W. G., and Sharp, R. E. (1991). Spatial distribution of turgor and root growth at low water potentials. *Plant Physiol.* **96**, 438–443.

Spollen, W. G., Sharp, R. E., Saab, I. N., and Wu, Y. (1993). Regulation of cell expansion in roots and shoots at low water potentials. *In* "Water Deficits: Plant Responses from Cell to Community" (J. A. C. Smith and H. Griffiths, eds.), pp. 37–52. BIOS, Oxford.

Sprugel, D. G. (1976). Dynamic structure of wave regenerated *Abies balsamea* forests in the northeastern United States. *J. Ecol.* **64**, 889–910.

Spurr, S. H., and Barnes, B. V. (1980). "Forest Ecology," 3rd Ed., Wiley, New York.

St. Clair, J. B. (1994a). Genetic variation in tree structure and its relation to size in Douglas-fir. I. Biomass partitioning, foliage efficiency, stem form, and wood density. *Can. J. For. Res.* **24**, 1226–1235.

St. Clair, J. B. (1994b). Genetic variation in tree structure and its relation to size in Douglas-fir. II. Crown form, branch characters, and foliage characters. *Can. J. For. Res.* **24**, 1236–1247.

St. Clair, J. B., Kleinschmit, J., and Svolba, J. (1985). Juvenility and serial vegetative propagation of Norway spruce clones (*Picea abies* Karst.). *Silvae Genet.* **34**, 42–48.

Staaf, H., and Stjernquist, I. (1986). Seasonal dynamics, especially autumnal retranslocation, of nitrogen and phosphorus in foliage of dominant and suppressed trees of beech, *Fagus sylvatica*. *Scand. J. For. Res.* **1**, 333–342.

Stachurski, A., and Zimka, J. R. (1975). Methods of studying forest ecosystems: Leaf area, leaf production and withdrawal of nutrients from leaves of trees. *Ekol. Pol.* **23**, 637–648.

Stanek, W., Hopkins, J. C., and Simmons, C. S. (1986). Effects of spacing in loblolly pine stands on incidence of *Atropellis* canker. *For. Chron.* **62**, 91–95.

Staniforth, R. J., and Sidhu, S. S. (1984). Effects of atmospheric fluorides on foliage, flower, fruit and seed production in wild raspberry and blueberry. *Can. J. Bot.* **62**, 2827–2834.

Stanley, R. G. (1958). Gross respiratory and water uptake patterns in germinating sugar pine seed. *Physiol. Plant.* **11**, 503–515.

Stassen, P. J. C., Strydom, D. K., and Stindt, H. W. (1981). Seasonal changes in carbohydrates of young 'Kakamas' peach trees. *Agroplantae* **13**, 47–53.

Stead, A. D., and Moore, K. G. (1983). Studies on flower longevity in *Digitalis*. *Planta* **157**, 15–21.

Steane, D. A., West, A. K., Potts, B. M., Ovenden, J. R., and Reid, J. B. (1991). Restriction fragment length polymorphisms in chloroplast DNA from six species of *Eucalyptus*. *Aust. J. Bot.* **39**, 399–414.

Stein, L. A., McEachern, G. R., and Storey, J. B. (1989). Summer and fall moisture stress and irrigation scheduling influence pecan growth and production. *Hort-Science* **24**, 607–611.

Steingraber, D. A. (1982). Phenotypic plasticity of branching pattern in sugar maple (*Acer saccharum*). *Am. J. Bot.* **69**, 638–640.

Steingraber, D. A., Kascht, L. J., and Franck, D. H. (1979). Variation of shoot morphology and bifurcation ratio in sugar maple (*Acer saccharum*) saplings. *Am. J. Bot.* **66**, 441–445.

Stephens, G. R., Jr. (1964). Stimulation of flowering in eastern white pine. *For. Sci.* **10**, 28–34.

Stephenson, A. G. (1980). Fruit set, herbivory, and the fruiting strategy of *Catalpa speciosa* (Bignoniaceae). *Ecology* **61**, 57–64.

Stephenson, A. G. (1981). Flower and fruit abortion: Proximate causes and ultimate functions. *Annu. Rev. Ecol. Syst.* **12**, 253–279.

Stephenson, R. A., and Gallagher, E. C. (1986). Effects of night temperature on floral initiation and raceme development in macadamia. *Sci. Hortic.* (*Amsterdam*) **30**, 213–218.

Steponkus, P. L. (1990). Cold acclimation and freezing injury from a perspective of the plasma membrane. *In* "Enviromental Injury to Plants" (F. Katterman, ed.), pp. 1–16. Academic Press, San Diego.

Steponkus, P. L., and Lynch, D. V. (1989). Freeze–thaw induced destabilization of the plasma membrane and the effects of cold acclimation. *J. Bioenerg. Biomembr.* **21**, 21–41.

Steponkus, P. L., and Webb, M. S. (1992). Freeze-induced dehydration and membrane destabilization in plants. *In* "Water and Life" (G. N. Somero, C. B. Osmond, and C. L. Bolis, eds.), pp. 338–362. Springer-Verlag, Berlin, Heidelberg, and New York.

Steponkus, P. L., Uemura, M., Balsamo, R. A., Arvinte, T., and Lynch, D. V. (1988). Transformation of the cryobehavior of rye protoplasts by modification of the plasma membrane lipid compositon. *Proc. Natl. Acad. Sci. U.S.A.* **85**, 9026–9030.

Steponkus, P. L., Lynch, D. V., and Uemura, M. (1990). The influence of cold acclimation on the lipid composition and cryobehavior of the plasma membrane of isolated rye protoplasts. *Philos. Trans. R. Soc. London* **B326**, 571–588.

Sterrett, J. P. (1985). Paclobutrazol: A promising growth inhibitor for injection into woody plants. *J. Am. Soc. Hortic. Sci.* **110**, 4–8.

Sterrett, J. P. (1990). Translocation and degradation of injected uniconazole in apple during a 4-month growing period. *J. Plant Growth Regul.* **9**, 147–150.

Sterrett, J. P., and Tworkoski, T. J. (1987). Flurprimidol: Plant response, translocation and metabolism. *J. Am. Soc. Hortic. Sci.* **112**, 341–345.

Sterrett, J. P., Tworkoski, T. J., and Kujawski, P. T. (1989). Physiological responses of deciduous tree root collar drenched with flurprimidol. *J. Arboric.* **15**, 120–124.

Stettler, R. F., Fenn, R. C., Heilman, P. E., and Stanton, B. J. (1988). *Populus trichocarpa* × *Populus deltoides* hybrids for short-rotation culture: Variation patterns and 4-year field performance. *Can. J. For. Res.* **18**, 745–783.

Stevens, G. A., and Westwood, M. N. (1984). Fruit set and cytokinin-like activity in the xylem sap of sweet cherry (*Prunus avium*) as affected by rootstock. *Physiol. Plant.* **61**, 464–468.

Stewart, C. M., Thom, S. H., and Rolfe, D. L. (1973). Diurnal variations of water in developing secondary stem tissues of eucalypt trees. *Nature (London)* **242**, 479–480.

Stewart, H. T. L., and Salmon, G. R. (1986). Irrigation of tree plantations with recycled water. Some economic analyses. *Aust. For.* **49**, 89–96.

Stinchcombe, G. R., Copas, E., Williams, R. R., and Arnold, G. (1984). The effects of paclobutrazol and daminozide on the growth and yield of cider apple trees. *J. Hortic. Sci.* **59**, 323–327.

Stolzy, L. H., and Sojka, R. E. (1984). Effects of flooding in plant disease. *In* "Flooding and Plant Growth" (T. T. Kozlowski, ed.), pp. 221–264. Academic Press, New York.

Stomp, A. M., Weissinger, A., and Sederoff, R. R. (1991). Transient expression from microprojectile-mediated DNA transfer in *Pinus taeda*. *Plant Cell Rep.* **10**, 187–190.

Stone, E. C., and Jenkinson, J. L. (1970). Influence of soil water on root growth capacity of ponderosa pine transplants. *For. Sci.* **16**, 230–239.

Stone, E. C., Jenkinson, J. L., and Krugman, S. L. (1962). Root-regenerating potential of Douglas-fir seedlings lifted at different times of the year. *For. Sci.* **8**, 288–297.

Stone, E. L. (1968). Microelement nutrition of forest trees: A review. *In* "Forest Fertilization: Theory and Practice," pp. 132–175. Tennessee Valley Authority, Muscle Shoals, Alabama.

Stone, E. L. (1973). Biological objectives in forest fertilization. *Proceedings of the Forest Fertilzation Symposium. U.S. For. Serv. Gen. Tech. Rep. NE* **NE-3**.

Stout, P. R., and Hoagland, D. R. (1939). Upward and lateral movement of salt in certain plants as indicated by radioactive isotopes of potassium, sodium and phosphorus absorbed by roots. *Am. J. Bot.* **26**, 320–324.

Stow, T. K., Allen, H. L., and Kress, L. W. (1992). Ozone impacts on seasonal foliage dynamics of young loblolly pine. *For. Sci.* **38**, 102–119.

Strain, B. R. (1987). Direct effects of increasing atmospheric CO_2 on plants and ecosystems. *Trends Ecol. Evol.* **2**, 18–21.

Strain, B. R., and Cure, J. D. (1985). "Direct Effects of Increasing Carbon Dioxide on Vegetation." U.S. Dep. of Energy, NTIS, Springfield, Virginia.

Stransky, J. J., Roese, J. H., and Watterston, K. G. (1985). Soil properties and pine growth affected by site preparation after clear cutting. *South. J. Appl. For.* **9**, 40–43.

Strauss, S. H., Howe, G. T., and Goldfarb, B. (1991). Prospects for genetic engineering of insect resistance in forest trees. *For. Ecol. Manage.* **43**, 181–209.

Strauss, S. H., Rottmann, W. H., Brunner, A. M., and Sheppard, L. A. (1995). Genetic engineering of reproductive sterility in forest trees. *Mol. Breed.* **1**, 5–26.

Strauss-Debenedetti, S. I. (1989). Responses to light in tropical Moraceae of different successional stages. Ph.D. Dissertation. Yale University, New Haven, Connecticut.

Striem, M. J., Ben-Haym, G., and Spiegel-Roy, P. (1994). Developing molecular markers for grape breeding, using polymerase chain reaction procedures. *Vitis* **33**, 53–54.

Stroo, H. F., Reich, P. B., Schoettle, A. W., and Amundson, R. G. (1988). Effects of ozone and acid rain on white pine (*Pinus strobus*) seedlings grown in five soils. II. Mycorrhizal infection. *Can. J. Bot.* **66**, 1510–1516.

Subbaiah, T. K., and Powell, L. E. (1992). Abscisic acid relationships in the chill-related dormancy mechanism in apple seed. *Plant Growth Regul.* **11**, 115–123.

Sullivan, D. T., and Widmoyer, F. B. (1968). Effect of Alar on bloom date of Richared apples. *Fruit. Var. Hortic. Dig.* **22**, 70–71.

Sullivan, D. T., and Widmoyer, F. B. (1970). Effects of succinic acid 2,2-dimethylhydrazide (Alar) on bloom delay and fruit development of Delicious apples. *HortScience* **5**, 91–92.

Sumimoto, M., Shiraga, M., and Kondo, T. (1975). Ethane in pine needles preventing the feeding of the beetle, *Monochamus alternatus*. *J. Insect Physiol.* **21**, 713–722.

Sun, D., and Dickinson, D. (1993). Responses to salt stress of 16 *Eucalyptus* species, *Grevillea robusta, Lophostemon confertus,* and *Pinus caribaea* var. *hondurensis. For. Ecol. Manage.* **60**, 1–4.

Sundberg, B., and Little, C. H. A. (1987). Effect of defoliation on tracheid production and the level of indole-3-acetic acid in *Abies balsamea* shoots. *Physiol. Plant.* **71**, 430–435.

Sundberg, B., and Little, C. H. A. (1991). Tracheid production in response to changes in the internal level of indole-3-acetic acid in 1-year-old shoots of Scots pine. *Plant Physiol.* **94**, 1721–1727.

Sundberg, B., Little, C. H. A., Riding, R. T., and Sandberg, G. (1987). Levels of endogenous indole-3-acetic acid in the vascular cambium region of *Abies balsamea* trees during the activity-rest-quiescence transition. *Physiol. Plant.* **71**, 163–170.

Sutinen, S., Skärby, L., Wallin, G., and Sellden, G. (1990). Long-term exposure of Norway spruce, *Picea abies* (L.) Karst., to ozone in open-top chambers. II. Effects on the ultrastructure of needles. *New Phytol.* **115**, 345–355.

Suwannapinunt, W., and Kozlowski, T. T. (1980). Effect of SO_2 on transpiration, chlorophyll content, growth, and injury in young seedlings of woody angiosperms. *Can. J. For. Res.* **10**, 78–81.

Suzuki, T., and Kaneko, M. (1970). The effect of suction pressure in the soil solution during summer on growth and fruiting of young satsuma orange trees. *J. Jpn. Soc. Hortic. Sci.* **39**, 99–106.

Svihra, P. (1994). Principles of eradicative pruning. *J. Arboric.* **20**, 262–271.

Sweet, G. B., and Hong, S. O. (1978). The role of nitrogen in relation to cone production in *Pinus radiata. N. Z. J. For. Sci.* **8**, 225–238.

Sweet, G. B., and Will, G. M. (1965). Precocious male cone production associated with low nutrient status in clones of *Pinus radiata. Nature (London)* **206**, 739.

Swietlik, D. (1986). Effect of gibberellin inhibitors on growth and mineral nutrition of sour orange seedlings. *Sci. Hortic. (Amsterdam)* **29**, 325–333.

Swietlik, D., and Faust, M. (1984). Foliar nutrition of fruit crops. *Hortic. Rev.* **6**, 287–355.

Swietlik, D., and Slowik, K. (1981). The uptake of ^{15}N-labelled urea by tart cherry and apple trees and the distribution of absorbed nitrogen in tart cherry trees. *Fruit Sci. Rep.* **8**, 49–59.

Switzer, G. L., and Nelson, L. E. (1972). Nutrient accumulation and cycling in loblolly pine plantation ecosystems; the first twenty years. *Soil Sci. Soc. Am. Proc.* **36**, 143–147.

Sydes, C., and Grime, J. P. (1981a). Effect of tree leaf litter on herbaceous vegetation in the deciduous woodlands. I. Field investigations. *J. Ecol.* **69**, 237–248.

Sydes, C., and Grime, J. P. (1981b). Effects of tree leaf litter on herbaceous vegetation in the deciduous woodlands. II. An experimental investigation. *J. Ecol.* **69**, 249–262.

Tai, E. A. (1977). Banana. *In* "Ecophysiology of Tropical Crops" (P. de T. Alvim and T. T. Kozlowski, eds.), pp. 441–460. Academic Press, New York.

Tainter, F. H., Fraedrich, S. W., and Benson, D. M. (1984). The effect of climate on growth, decline, and death of northern red oaks in the western North Carolina Nantahala Mountains. *Castanea* **49**, 127–137.

Tal, M. (1985). Genetics of salt tolerance in higher plants: Theoretical and practical considerations. *Plant Soil* **89**, 199–226.

Tal, M. (1994). *In vitro* selection for salt tolerance in crop plants: Theoretical and practical considerations. *In Vitro Cell Dev. Biol.* **30P**, 175–180.

Talbert, C. B., Ritchie, G. A., and Gupta, P. (1993). Conifer vegetative propagation: An overview from a commercial perspective. *In* "Clonal Forestry" (M. R. Ahuja and W. J. Libby, eds.), Vol. 1, pp. 145–181. Springer-Verlag, Berlin.

Talboys, P. W. (1978). Dysfunction of the water system. *In* "Plant Disease: An Advanced Treatise" (J. G. Horsfall and E. B. Cowling, eds.), pp. 141–162. Academic Press, New York.

Tang, Z. C., and Kozlowski, T. T. (1982a). Some physiological and morphological responses of *Quercus macrocarpa* seedlings to flooding. *Can. J. For. Res.* **10**, 308–311.

Tang, Z. C., and Kozlowski, T. T. (1982b). Physiological, morphological, and growth responses of *Platanus occidentalis* seedlings to flooding. *Plant Soil* **66**, 243–255.

Tang, Z. C., and Kozlowski, T. T. (1982c). Some physiological and growth responses of *Betula papyrifera* seedlings to flooding. *Physiol. Plant.* **55**, 415–420.

Tang, Z. C., and Kozlowski, T. T. (1983). Responses of *Pinus banksiana* and *Pinus resinosa* seedlings to flooding. *Can. J. For. Res.* **13**, 633–639.

Tang, Z. C., and Kozlowski, T. T. (1984a). Water relations, ethylene production, and morphological adaptations of *Fraxinus pennsylvanica* seedlings to flooding. *Plant Soil* **77**, 183–192.

Tang, Z. C., and Kozlowski, T. T. (1984b). Ethylene production and morphological adaptation of woody plants to flooding. *Can. J. Bot.* **62**, 1659–1664.

Tao, D. L., Xu, Z. B., and Li, X. (1987). Effect of litter layer on natural regeneration of companion tree species in the Korean pine forest. *Environ. Exp. Bot.* **27**, 53–66.

Tattar, T. A. (1978). "Diseases of Shade Trees." Academic Press, New York.

Tautorus, T. E., Lulsdorf, M. M., Kikcio, S. I., and Dunstan, D. I. (1994). Nutrient utilization during bioreactor culture, and maturation of somatic embryo cultures of *Picea mariana* and *Picea glauca-engelmannii*. *In Vitro Cell Dev. Biol.* **30P**, 58–63.

Taylor, B. H., and Ferree, D. C. (1981). The influence of summer pruning on photosynthesis, transpiration, leaf abscission, and dry weight accumulation of young apple trees. *J. Am. Soc. Hortic. Sci.* **106**, 389–393.

Taylor, G. E., Jr., McLaughlin, S. B., and Shriner, D. S. (1982). Effective pollutant dose. *In* "Effects of Gaseous Air Pollution in Agriculture and Horticulture" (M. H. Unsworth and D. P. Ormrod, eds.), pp. 458–460. Butterworth, London.

Taylor, G. E., Jr., Johnson, D. W., and Andersen, C. P. (1994). Air pollution and forest ecosystems: A regional to global perspective. *Ecol. Appl.* **4**, 662–689.

Taylor, J. S., and Dumbroff, E. B. (1975). Bud, root, and growth-regulator activity in *Acer saccharum* during the dormant season. *Can. J. Bot.* **53**, 321–331.

Taylor, J. S., and Wareing, P. F. (1979). The effect of stratification on the endogenous levels of gibberellins and cytokinins in seeds of Douglas-fir [*Pseudotsuga menziesii* (Mirb.) Franco] and sugar pine (*Pinus lambertiana* Dougl.). *Plant Cell Environ.* **2**, 165–171.

Taylorson, R. B., and Hendricks, S. B. (1976). Aspects of dormancy in vascular plants. *BioScience* **26**, 95–101.

Taylorson, R. B., and Hendricks, S. B. (1977). Dormancy in seeds. *Annu. Rev. Plant Physiol.* **28**, 331–354.

Telewski, F. W. (1995). Wind-induced physiological and developmental responses in trees. *In* "Wind and Trees" (M. P. Coutts and J. Grace, eds.), pp. 237–263. Cambridge Univ. Press, Cambridge.

Telewski, F. W., and Jaffe, M. J. (1981). Thigmomorphogenesis: Changes in the morphology and chemical composition induced by mechanical perturbation in 6-month-old *Pinus taeda* seedlings. *Can. J. For. Res.* **11**, 380–387.

Telewski, F. W., and Jaffe, M. J. (1986a). Thigmomorphogenesis: Field and laboratory studies of *Abies fraseri* in response to wind or mechanical perturbation. *Physiol. Plant.* **66**, 211–218.

Telewski, F. W., and Jaffe, M. J. (1986b). Thigmomorphogenesis: The role of ethylene in the response of *Pinus taeda* and *Abies fraseri* to mechanical perturbation. *Physiol. Plant.* **66**, 227–233.

Tepper, H. B., Yang, C. S., and Schaedle, M. (1989). Effect of aluminum on growth of root tips on honeylocust and loblolly pine. *Environ. Exp. Bot.* **29**, 165–173.

Terazawa, K., and Kikuzawa, K. (1994). Effects of flooding on leaf dynamics and other seedling responses in flood-tolerant *Alnus japonica* and flood-intolerant *Betula platyphylla* var. *japonica*. *Tree Physiol.* **14**, 251–261.

Terborgh, J. (1985). The vertical component of plant species diversity in temperate and tropical forests. *Am. Nat.* **126**, 761–776.

Terry, M. E., McGraw, D., and Jones, R. L. (1982). Effect of IAA on growth and soluble cell wall polysaccharides centrifuged from pine hypocotyl sections. *Plant Physiol.* **69**, 323–326.

Terry, T. A., and Hughes, J. H. (1975). The effects of intensive management on planted loblolly pine (*Pinus taeda* L.) grown on poorly drained soils of the Atlantic coastal plain. *In* "Forest Soils and Forest Land Mangement" (B. Bernier and C. H. Winget, eds.), pp. 351–377. Laval Univ. Press, Quebec.

Teskey, B. J. E., and Shoemaker, J. S. (1978). "Tree Fruit Production." AVI, Westport, Connecticut.

Teskey, R. O., and Hinckley, T. M. (1981). Influence of temperature and water potential on root growth of white oak. *Physiol. Plant.* **52**, 363–369.

Theodorou, C., and Bowen, G. D. (1971). Influence of temperature on the mycorrhizal associations of *Pinus radiata* D. Don. *Aust. J. Bot.* **19**, 13–20.

Thimann, K. V. (1965). Toward an endocrinology of higher plants. *Recent Prog. Horm. Res.* **21**, 579–596.

Thimann, K. V. (1980). "Senescence in Plants." CRC Press, Boca Raton, Florida.

Thompson, C. R., and Taylor, O. C. (1969). Effects of air pollutants on growth, leaf-drop, fruit-drop, and yield of citrus trees. *Environ. Sci. Technol.* **3**, 934–940.

Thompson, C. R., Hensel, E. G., Kats, D., and Taylor, O. C. (1970). Effects of continuous exposure of navel oranges to NO_2. *Atmos. Environ.* **4**, 349–355.

Thompson, R. G., Fensom, D. S., Anderson, R. R., Drouin, R., and Leiper, W. (1979). Translocation of ^{14}C from leaves of *Helianthus, Heracleum, Nymphoides, Ipomoea, Tropaeolum, Zea, Fraxinus, Ulmus, Picea,* and *Pinus.* Comparative shapes and some fine structure profiles. *Can. J. Bot.* **57**, 845–863.

Thomson, A. J., and Moncrief, S. M. (1982). Prediction of bud burst in Douglas-fir by degree-day accumulation. *Can. J. For. Res.* **12**, 448–452.

Thomson, B. D., Grove, T. S., Malajczuk, N., and Hardy, G. E. St. (1994). The effectiveness of ectomycorrhizal fungi in increasing the growth of *Eucalyptus globulus* Labill. in relation to root colonization and hyphal development in soil. *New Phytol.* **126**, 517–524.

Thorne, D. W., and Thorne, M. D. (1979). "Soil Water and Crop Production." AVI, Westport, Connecticut.

Thorne, J. H., and Giaquinta, R. T. (1984). Pathways and mechanisms associated with carbohydrate translocation in plants. *In* "Storage Carbohydrates in Vascular Plants. Distribution, Physiology and Metabolism" (D. H. Lewis, ed.). pp. 75–96. Cambridge, Univ. Press, Cambridge.

Tiedje, J. M., Colwell, R. K., Grossman, Y. L., Hodson, R. E., Lenski, R. R., Mack, R. N., and Regal, P. J. (1989). The planned introduction of genetically engineered organisms: Ecological considerations and recommendations. *Ecology* **70**, 298–315.

Tillberg, E. (1984). Levels of endogenous indole-3-acetic acid in achenes of *Rosa rugosa* during dormancy release and germination. *Plant Physiol.* **76**, 84–87.

Tillberg, E., and Pinfield, N. J. (1981). The dynamics of indole-3-acetic acid in *Acer platanoides* seeds during stratification and germination. *Physiol. Plant.* **53**, 34–38.

Tillman-Sutela, E., and Kauppi, A. (1995a). The morphological background to imbibition in seeds of *Pinus sylvestris* L. of different provenances. *Trees* **9**, 123–133.

Tillman-Sutela, E., and Kauppi, A. (1995b). The significance of structure for imbibition in seeds of the Norway spruce, *Picea abies* (L.) Karst. *Trees* **9**, 269–278.

Timell, T. E. (1973). Ultrastructure of the dormant and active cambial zones and the dormant phloem associated with formation of normal and compression woods in *Picea abies* (L.) Karst. *State Univ. Coll. Tech. Publ.* **96**, *Environ. Sci. For. Syracuse Univ., New York.*

Timell, T. E. (1986). "Compression Wood." Vols. 1–3. Springer-Verlag, Berlin and New York.

Timmer, V. R., and Armstrong, G. (1987). Growth and nutrition of containerized *Pinus resinosa* at exponentially increasing nutrient additions. *Can. J. For. Res.* **17**, 644–647.

Tinus, R. W., and McDonald, S. E. (1979). How to grow tree seedlings in containers in greenhouses. *U.S.D.A. For. Serv. Gen. Tech. Rep. RM* **RM-60**.

Toda, R. (1974). Vegetative propagation in relation to Japanese forest tree improvement. *N. Z. J. For. Sci.* **4**, 410–417.

Todd, G. W., Chadwick, D. L., and Tosi, S.-D. (1972). Effect of wind on plant respiration. *Physiol. Plant.* **27**, 342–346.

Tolley, L. C., and Strain, B. R. (1984a). Effects of CO_2 enrichment on growth of *Liquidambar styraciflua* and *Pinus taeda* seedlings under different irradiance levels. *Can. J. For. Res.* **14**, 343–355.

Tolley, L. C., and Strain, B. R. (1984b). Effects of atmospheric CO_2 enrichment and water stress on growth of *Liquidambar straciflua* and *Pinus taeda* seedlings. *Can. J. Bot.* **62**, 2135–2139.

Tomer, E. (1984). Inhibition of flowering in mango by gibberellic acid. *Sci. Hortic. (Amsterdam)* **24**, 299–303.

Tomiczek, C. (1992). Schäden durch Grünastung on Fichte. *Centralbl. Gesamte Forstwes.* **109**, 185–192.

Tomos, D., and Pritchard, J. (1994). Biophysical and biochemical control of cell expansion in roots and leaves. *J. Exp. Bot.* **45**, 1721–1731.

Toole, V. K., Toole, E. H., Hendricks, S. B., Borthwick, H. S., and Snow, A. G., Jr. (1961). Responses of seeds of *Pinus virginiana* to light. *Plant Physiol.* **36**, 285–290.

Topa, M. A., and McLeod, K. W. (1986). Aerenchyma and lenticel formation in pine seedlings: A possible avoidance mechanism in anaerobic growth conditions. *Physiol. Plant.* **68**, 540–550.

Torrey, J. G., Fosket, D. E., and Hepler, P. K. (1971). Xylem formation: A paradigm of cytodifferentiation in higher plants. *Am. Sci.* **59**, 338–352.

Tranquillini, W. (1969). Photosynthese und Transpiration einiger Holzarten bei verschieden starkem Wind. *Centralbl. Gesamte Forstwes.* **86**, 35–48.

Tranquillini, W., and Unterholzner, R. (1968). Das Wachstum zweijähriger Lärchen einheitlicher Herkunft in verschiedener Seehöhe. *Centralbl. Gesamte Forstwes.* **85**, 43–49.

Trapp, E. (1938). Untersuchung über die Verteilung der Helligkeit in einem Buchenstand. *Bioklimatologie Ser. B* **5**, 153–158.

Trappe, J. M. (1983). Effects of herbicides bifenox, DCPA, and napropamide on mycorrhiza development of ponderosa pine and Douglas-fir seedlings in six western nurseries. *For. Sci.* **29**, 464–468.

Treharne, K. J., Quinlan, J. N., Knight, J. D., and Ward, D. A. (1985). Hormonal regulation of fruit development in apple: A minireview. *Plant Growth Regul.* **3**, 125–132.

Trelease, R. N., and Doman, D. C. (1984). Mobilization of oil and wax reserves. *In* "Seed Physiology" (Murray, D. R., ed.), Vol. 2, pp. 202–245. Academic Press, Sydney.

Treshow, M., and Anderson, F. K. (1982). Ecological assessment of potential fluoride effects on plants. *In* "Fluoride Emissions: Their Monitoring and Effects on Vegetation and Ecosystems" (F. Murray, ed.), pp. 177–189. Academic Press, Sydney.

Treshow, M., and Pack, M. R. (1970). Fluoride. *In* "Recognition of Air Pollution Injury to Vegetation. A Pictorial Atlas" (J. S. Jacobson and A. C. Hill, eds.), pp. D1–D7. Air Pollution Control Association, Pittsburgh, Pennsylvania.

Tricoli, D. M., Maynard, C. A., and Drew, A. P. (1985). Tissue culture propagation of mature trees of *Prunus serotina* Ehrh. I. Establishment, multiplication, and rooting *in vitro. For. Sci.* **31**, 201–208.

Tripepi, R. R., and Mitchell, C. A. (1984). Stem hypoxia and root respiration of flooded maple and birch seedlings. *Physiol. Plant.* **60**, 567–571.

Tromp, J. (1984a). Diurnal fruit shrinkage in apple as affected by leaf water potential and vapour pressure deficit of the air. *Sci. Hortic. (Amsterdam)* **22**, 81–87.

Tromp, J. (1984b). Flower-bud formation in apple as affected by air and root temperature, air humidity, light intensity and day length. *Acta Hortic.* **149**, 39–47.

Tromp, J. (1987). Growth and flower-bud formation in apple as affected by paclobutrazol, daminozide, and tree orientation in combination with various gibberellins. *J. Hortic. Sci.* **62**, 433–440.

Tschaplinski, T. J., Johnson, D. W., Norby, R. J., and Todd, D. E. (1991). Optimum nitrogen nutrition in short rotation sycamore plantations. *Soil Sci. Soc. Am. J.* **55**, 841–847.

Tschaplinski, T. J., Stewart, D. B., Hanson, P. J., and Norby, R. J. (1995). Interactions between drought and elevated CO_2 on growth and gas exchange of seedlings of three deciduous tree species. *New Phytol.* **129**, 63–71.

Tsukahara, H., and Kozlowski, T. T. (1984). Effect of flooding on growth of *Larix leptolepis* seedlings. *J. Jpn. For. Soc.* **66**, 33–66.

Tsukahara, H., and Kozlowski, T. T. (1985). Importance of adventitious roots to growth of flooded *Platanus occidentalis* seedlings. *Plant Soil* **88**, 123–132.

Tsukahara, H., and Kozlowski, T. T. (1986). Effect of flooding and temperature regime on growth and stomatal resistance of *Betula platyphylla* var. *japonica* seedlings. *Plant Soil* **92**, 103–112.

Tsukahara, H., Kozlowski, T. T., and Shanklin, J. (1985). Tolerance of *Pinus densiflora, Pinus thunbergii,* and *Larix leptolepis* seedlings to SO_2. *Plant Soil* **88**, 385–397.

Tsukahara, H., Kozlowski, T. T., and Shanklin, J. (1986). Effects of SO_2 on two age classes of *Chamaecyparis obtusa* seedlings. *J. Jpn. For. Soc.* **68**, 349–353.

Tubbs, F. R. (1973a). Research fields in the interaction of rootstock and scions in woody perennials. Part I. *Hortic. Abstr.* **43**, 247–253.

Tubbs, F. R. (1973b). Research fields in the interaction of rootstock and scions in woody perennials. Part II. *Hortic. Abstr.* **43**, 325–335.

Tucker, C. M., and Evert, R. F. (1969). Seasonal development of the secondary phloem in *Acer negundo. Am. J. Bot.* **56**, 275–284.

Tucker, G. A. (1993). Introduction. *In* "Biochemistry of Fruit Ripening" (G. B.

Seymour, J. E. Taylor, and G. A. Tucker, eds.), pp. 1–37. Chapman & Hall, London, Glasgow, and New York.

Tucker, G. A., and Grierson, D. (1987). Fruit Ripening. *In* "The Biochemistry of Plants" (D. D. Davies, ed.), Vol. 12, pp. 265–318. Academic Press, New York.

Tukey, L. D. (1983). Vegetative control and fruiting on mature apple trees treated with PP-333. *Acta Hortic.* **137**, 103–109.

Tukey, L. D. (1986). Cropping characteristics of bearing apple trees annually sprayed with paclobutrazol (PP-333). *Acta Hortic.* **179**, 481–488.

Tukey, L. D. (1989). Growth factors and plant regulants in the manipulation of plant development and cropping in tree fruits. *In* "Manipulation of Fruiting" (C. J. Wright, ed.), pp. 343–361. Butterworth, London.

Tulecke, W. (1987). Somatic embryogenesis in woody perennials. *In* "Cell and Tissue Culture in Forestry" (J. M. Bonga and D. J. Durzan, eds.), Vol. 2, pp. 61–91. Martinus Nijhoff, Dordrecht, The Netherlands.

Tulecke, W., and McGranahan, G. H. (1985). Somatic embryogenesis and plant regeneration from cotyledons of walnut, *Juglans regia* L. *Plant Sci.* **40**, 57–63.

Turvey, N. D., and Smethurst, P. J. (1983). Nitrogen fixing plants in forest plantation management. *In* "Biological Nitrogen Fixation in Forest Ecosystems" (J. C. Gordon and C. T. Wheeler, eds.), pp. 233–259. Martinus Nijhoff and W. Junk. The Hague.

Tyree, M. T., Cochard, H., Cruiziat, P., Sinclair, B., and Ameglio, T. (1993). Drought-induced leaf shedding in walnut: Evidence for vulnerability segmentation. *Plant Cell Environ.* **16**, 879–882.

Ulrich, B. (1986). Die Rolle der Bodenversauerung beim Waldsterben: Langfristige Konsequenzen und forstliche Moeglichkeiten. *Forstwiss. Centralbl.* **105**, 421–435.

Ursino, D. J., Nelson, C. D., and Krotkov, G. (1968). Seasonal changes in the distribution of photoassimilated ^{14}C in young pine plants. *Plant Physiol.* **43**, 845–852.

Usher, R. W., and Williams, W. T. (1982). Air pollution toxicity to eastern white pine in Indiana and Wisconsin. *Plant Dis.* **66**, 199–204.

Vaartaja, O. (1959). Evidence of photoperiodic ecotypes in trees. *Ecol. Monogr.* **29**, 91–111.

Van Bel, A. J. E. (1990). Xylem–phloem exchange via the rays: The undervalued route of transport. *J. Exp. Bot.* **41**, 631–644.

van Camp, W., Willekens, H., Bowler, C., van Montagu, M., Inze, D., Reupold-Popp, P., Sandermann, H., Jr., and Langbartels, C. (1994). Elevated levels of superoxide dismutase protect transgenic plants against ozone. *Bio/Technology* **12**, 165–168.

van den Driessche, R. (1972). Different effects of nitrate and ammonium forms of nitrogen on growth and photosynthesis of slash pine seedlings. *Aust. For.* **36**, 125–137.

van den Driessche, R. (1984). Relationship between spacing and nitrogen fertilization of seedlings in the nursery, seedling mineral nutrition, and outplanting performance. *Can. J. For. Res.* **14**, 431–436.

van den Driessche, R. (1985). Late-season fertilization, mineral nutrient reserves, and retranslocation in planted Douglas-fir [*Pseudotsuga menziesii* (Mirb.) Franco] seedlings. *For. Sci.* **31**, 485–496.

van den Driessche, R. (1988). Nursery growth of conifer seedlings using fertilizers of different solubilities and application time and their forest growth. *Can. J. For. Res.* **18**, 172–180.

van den Driessche, R. (1991a). Effects of nutrients on stock performance in the forest. *In* "Mineral Nutrition of Conifer Seedlings" (R. van den Driessche, ed.), pp. 229–260. CRC Press, Boca Raton, Florida.

van den Driessche, R., ed. (1991b). "Mineral Nutrition of Conifer Seedlings." CRC Press, Boca Raton, Florida.

van Die, J., and Willemse, P. C. M. (1975). Mineral and organic nutrients in sieve tube exudate and xylem vessel sap of *Quercus rubra* L. *Acta Bot. Neerl.* **24**, 237–239.

Van der Kamp, B. J. (1991). Pathogens as agents of landscapes. *For. Chron.* **67**, 353–354.

van der Krol, A. R., Lenting, P. E., Veenstra, J., van der Meer, I. M., Koes, R. E., Gerats, A. G. M., Mol, J. N. M., and Stuitje, A. R. (1988). An anti-sense chalcone synthase gene in transgenic plants inhibits flower pigmentation. *Nature (London)* **333**, 866–869.

Van Dijk, H. F. G., and Bienfait, H. F. (1993). Iron-deficiency chlorosis in Scots pine growing on acid soils. *Plant Soil* **153**, 255–263.

Van Eerden, E., and Gates, J. W. (1990). Seedling production and processing: Container. *In* "Regenerating British Columbia's Forests" (D. P. Lavender, R. Parish, C. Johnson, G. Montgomery, A. Vyse, R. A. Willes, and D. Winston, eds.), pp. 226–234. Univ. of British Columbia Press, Vancouver, B.C.

van Loon, L. C., and Bruinsma, J. (1992). The new plant physiology—Molecular approaches to studying hormonal regulation of plant development. *Acta Bot. Neerl.* **41**, 1–23.

Van Ryn, D. M., Jacobson, J. S., and Lassoie, J. P. (1986). Effects of acidity on *in vitro* pollen germination and tube elongation in four hardwood species. *Can. J. For. Res.* **16**, 397–400.

Van Ryn, D. M., Lassoie, J. P., and Jacobson, J. S. (1988). Effects of acid mist on *in vitro* pollen tube growth in red maple. *Can. J. For. Res.* **18**, 1049–1052.

Van Staden, J., and Davey, J. E. (1979). The synthesis, transport and metabolism of endogenous cytokinins. *Plant Cell Environ.* **2**, 93–106.

Van Staden, J., and Harty, A. R. (1988). Cytokinins and adventitious root formation. *In* "Adventitious Root Formation in Cuttings" (T. D. Davis, B. E. Haissig, and N. Sankhla, eds.), pp. 185–201. Dioscorides Press, Portland, Oregon.

Van Volkenburgh, E., and Cleland, R. E. (1986). Wall yield threshold and effective turgor in growing bean leaves. *Planta* **167**, 37–43.

Van Volkenburgh, E., and Davies, W. J. (1983). Inhibition of light-stimulated leaf expansion by abscisic acid. *J. Exp. Bot.* **34**, 835–845.

Vardi, A., and Galun, E. (1988). Recent advances in protoplast culture of horticultural crops: Citrus. *Sci. Hortic. (Amsterdam)* **37**, 217–230.

Vardi, A., Spiegel-Roy, P., and Galun, E. (1982). Plant regeneration from *Citrus* protoplasts: Variability in methodological requirements among cultivars and species. *Theor. Appl. Genet.* **62**, 171–176.

Vasil, I. K. (1994). Automation of plant propagation. *Plant Cell, Tissue Organ Cult.* **39**, 105–108.

Vaughan, A. K. F., and Bate, G. C. (1977). Changes in the levels of ethylene, abscisic-acid-like substances and total non-structural carbohydrate in young cotton bolls in relation to abscission induced by a dark period. *Rhod. J. Agric. Res.* **15**, 51–63.

Vazquez-Yanes, C., and Orozco-Segovia, A. (1994). Signals for seeds to sense and respond to gaps. *In* "Exploitation of Environmental Heterogeneity by Plants: Ecophysiological Processes Above and Below Ground" (M. M. Caldwell and R. W. Pearcy, eds.), pp. 209–236. Academic Press, San Diego.

Vazquez-Yanes, C. A., Orozco-Segovia, A., Rincon, E., Sanchez-Coronado, M. E., Haunte, P., Toledo, J. R., and Barradas, V. L. (1990). Light beneath the litter in a tropical forest: Effect on seed germination. *Ecology* **71**, 1952–1953.

Veierskov, B. (1988). Relations between carbohydrates and adventitious root formation. *In* "Adventitious Root Formation in Cuttings" (T. D. Davis, B. E. Haissig, and N. Sankhla, eds.), pp. 70–78. Dioscorides Press, Portland, Oregon.

Veihmeyer, F. J., and Hendrickson, A. H. (1952). The effects of soil moisture on deciduous fruit trees. *Proc. Int. Hortic. Congr., 13th,* **1**, 306–319.

Vendrell, M. (1970). Acceleration and delay of ripening in banana fruit tissue by gibberellic acid. *Aust. J. Biol. Sci.* **23**, 553–559.

Veres, J. S., and Pickett, S. T. A. (1982). Branching patterns of *Lindera benzoin* beneath gaps and closed canopies. *New Phytol.* **91**, 767–772.

Vergara, N. T., and Nair, P. K. R. (1985). Agroforestry in the South Pacific region— An overview. *Agroforestry Systems* **3**, 363–379.

Vidal, M. T., Azcon-Aguilar, C., and Barea, J. M. (1992). Mycorrhizal inoculation enhances growth and development of micropropagated plants of avocado. *HortScience* **27**, 785–787.

Vieitez, A. M., Ballester, A., Garcia, M. T., and Vieitez, E. (1980). Starch depletion and anatomical changes during the rooting of *Castanea sativa* Mill. cuttings. *Sci. Hortic. (Amsterdam)* **13**, 261–266.

Vierling, E. (1991). The roles of heat shock proteins in plants. *Annu. Rev. Plant Physiol. Plant Mol. Biol.* **42**, 579–620.

Villiers, T. A. (1972). Seed dormancy. *In* "Seed Biology" (T. T. Kozlowski, ed.), Vol. 2, pp. 220–281. Academic Press, New York.

Villiers, T. A. (1975). "Dormancy and the Survival of Plants." Arnold, London.

Vince-Prue, D. (1975). "Photoperiodism and Plants." McGraw-Hill, Maidenhead, England.

Visser, T. (1954). After-ripening and germination of apple seeds in relation to the seed coats. *Proc. K. Ned. Akad. Wet. Ser. C* **57**, 175–185.

Vité, J. P. (1961). The influence of water supply on oleoresin exudation pressure and resistance to bark beetle attack in *Pinus ponderosa. Contrib. Boyce Thompson Inst.* **21**, 37–66.

Vitousek, P. M. (1982). Nutrient cycling and nutrient use efficiency. *Am. Nat.* **119**, 553–572.

Vogt, A. R. (1970). Effect of gibberellic acid on germination and initial seedling growth of northern red oak. *For. Sci.* **16**, 453–459.

Vogt, A. R. (1974). Physiological importance of changes in endogenous hormones during red oak stratification. *For. Sci.* **20**, 187–191.

Vogt, K. A., Grier, C. C., Meier, C. E., and Edmonds, R. L. (1982). Mycorrhizal role in net primary production and nutrient cycling in *Abies amabilis* ecosystems in western Washington. *Ecology* **63**, 370–380.

Von Aderkas, P., and Bonga, J. M. (1993). Plants from haploid tissue culture of *Larix decidua. Theor. Appl. Genet.* **87**, 225–228.

Vose, J. M. (1988). Patterns of leaf area distribution within crowns of nitrogen- and phosphorus-fertilized loblolly pine trees. *For. Sci.* **34**, 564–573.

Vose, J. M., and Allen, H. L. (1988). Leaf area, stemwood growth and nutrient relationships in loblolly pine. *For. Sci.* **34**, 547–563.

Vozzo, J. A., and Young, R. W. (1975). Carbohydrate, lipid, and protein distribution in dormant, stratified and germinated *Quercus nigra* embryos. *Bot. Gaz.* (*Chicago*) **136**, 306–311.

Wade, N. L., and Brady, C. J. (1971). Effect of kinetin on respiration, ethylene production, and ripening of banana fruit slices. *Aust. J. Biol. Sci.* **24**, 165–167.

Wadleigh, C. H., Gauch, H. G., and Magistad, O. C. (1946). Growth and rubber accumulation in guayule as conditioned by soil salinity and irrigation regimes. *Tech. Bull., U.S. Dep. Agric.* **925**.

Wagenmakers, P. S. (1991). Planting systems for fruit trees in temperate climates. *Crit. Rev. Plant Sci.* **10**, 369–385.

Wainhouse, D. (1987). Forests. *In* "Integrated Pest Management" (A. J. Burn, T. H. Coaker, and P. C. Jepson, eds.), pp. 361–401. Academic Press, London.

Waisel, Y. (1991). Adaptation to salinity. *In* "Physiology of Trees" (A. S. Raghavendra, ed.), pp. 359–383. Wiley, New York.

Waisel, Y., and Fahn, A. (1965). The effect of environment on wood formation and cambial activity in *Robinia pseudoacacia* L. *New Phytol.* **64**, 436–442.

Waisel, Y., Eshel, A., and Agami, M. (1986). Salt balance of leaves of the mangrove *Avicennia marina. Physiol. Plant.* **67**, 67–72.

Waister, P. D. (1972). Wind damage in horticultural crops. *Hortic. Abstr.* **42**, 609–615.

Walden, R. (1993). Cell culture, transformation and gene technology. *In* "Plant Biochemistry and Molecular Biology" (P. J. Lea and R. C. Leegood, eds.), pp. 275–295. Wiley, Chichester and New York.

Waldman, M., Rikin, A., Dovrat, A., and Richmond, A. E. (1975). Hormonal regulation of morphogenesis and cold-resistance. *J. Exp. Bot.* **26**, 853–859.

Walker, R. B. (1991). Measuring mineral nutrient utilization. *In* "Techniques and Approaches in Forest Tree Ecophysiology" (J. P. Lassoie and T. M. Hinckley, eds.), pp. 183–206. CRC Press, Boca Raton, Florida.

Walker, R. B., and Douglas, T. J. (1982). Effect of salinity on uptake and distribution of chloride, sodium, and potassium ions in citrus plants. *Aust. J. Agric. Res.* **34**, 145–153.

Walser, R. H., and Davis, T. D. (1989). Growth, reproductive development, and dormancy characteristics of paclobutrazol-treated tart cherry trees. *J. Hortic. Sci.* **64**, 435–441.

Walser, R. H., Walker, D. R., and Seeley, S. D. (1981). Effect of temperature, fall defoliation, and gibberellic acid on the rest period of peach leaf buds. *J. Am. Soc. Hortic. Sci.* **106**, 91–94.

Walter, M. H. (1992). Regulation of lignification in defense. *In* "Plant Gene Research. Genes Involved in Plant Defense" (T. Boller and F. Meins, eds.), pp. 327–352. Springer-Verlag, Vienna.

Walters, M. B., and Reich, P. B. (1989). Response of *Ulmus americana* seedlings to varying nitrogen and water status. I. Photosynthesis and growth. *Tree Physiol.* **5**, 159–172.

Wang, C. Y. (1993). Approaches to reducing chilling injury of fruits and vegetables. *Hortic. Rev.* **15**, 63–95.

Wang, C. Y., ed. (1990). "Chilling Injury of Horticultural Crops." CRC Press, Boca Raton, Florida.

Wang, J. R., Simard, S. W., and Kimmins, J. P. (1995). Physiological responses of paper birch to thinning in British Columbia. *For. Ecol. Manage.* **73**, 177–184.

Wang, K., Drayton, P., Frame, B., Dunwell, J., and Thompson, J. (1995). Whisker-mediated plant transformation: An alternative technology. *In Vitro Cell Dev. Biol. Plant.* **31**, 101–104.

Wang, Q., Little, C. H. A., Sheng, C., Oden, P. C., and Pharis, R. P. (1992). Effects of exogenous gibberellins $A_{4/7}$ on tracheid production, longitudinal growth and levels of indole-3-acetic acid and gibberellins A_4, A_7, and A_9 in the terminal shoot of *Pinus sylvestris* seedlings. *Physiol. Plant.* **86**, 202–208.

Wang, S. Y., Faust, M., and Steffens, G. L. (1985). Metabolic changes in cherry flower buds associated with breaking of dormancy in early and late blooming cultivars. *Physiol. Plant.* **65**, 89–94.

Wang, S. Y., Sun, T., and Faust, M. (1986). Translocation of paclobutrazol, a gibberellin biosynthesis inhibitor in apple seedlings. *Plant Physiol.* **82**, 11–14.

Wang, T., and Zhu, Z. (1981). The prevention of black heart disorder in Chinese Duck pears. *Food Sci. (Beijing)* [*Shipin Kexue (Beijing)*] **10**, 39–43.

Wang, Z.-Y., Patterson, K. J., Gould, K. S., and Lowe, R. G. (1994a). Rootstock effects on budburst and flowering in kiwifruit. *Sci. Hortic. (Amsterdam)* **57**, 187–199.

Wang, Z.-Y., Gould, K. S., and Patterson, K. J. (1994b). Comparative root anatomy of five *Actinidia* species in relation to rootstock effects on kiwifruit flowering. *Ann. Bot.* **73**, 403–413.

Wardle, P. (1981). Winter desiccation of conifer needles simulated by artificial freezing. *Arct. Alp. Res.* **13**, 419–423.

Wardrop, A. B. (1961). The structure and organization of thickened cell walls. *Rep. For. Prod. Austr.* **476**.

Wardrop, A. B. (1964). The reaction anatomy of arborescent gymnosperms. *In* "The Formation of Wood in Forest Trees" (M. H. Zimmermann, ed.), pp. 405–456. Academic Press, New York.

Wardrop, A. B. (1981). Lignification and xylogenesis. *In* "Xylem Cell Development" (J. R. Barnett, ed.), pp. 115–152. Castle House, Tunbridge Wells, England.

Wardrop, A. B., and Dadswell, H. E. (1955). The nature of reaction wood. IV. Variations in cell wall organization of tension wood fibers. *Aust. J. Bot.* **3**, 177–189.

Wareing, P. F. (1951). Growth studies in woody species. III. Further photoperiodic effects in *Pinus sylvestris*. *Physiol. Plant.* **4**, 41–56.

Wareing, P. F. (1953). Growth studies in woody species. V. Photoperiodism in dormant buds of *Fagus sylvatica* L. *Physiol. Plant.* **6**, 692–706.

Wareing, P. F. (1954). Growth studies in woody species VI. The locus of photoperiodic perception in relation to dormancy. *Physiol. Plant.* **7**, 261–277.

Wareing, P. F. (1956). Photoperiodism in woody plants. *Annu. Rev. Plant Physiol.* **7**, 191–214.

Wareing, P. F. (1958). Interaction between indole-acetic acid and gibberellic acid in cambial activity. *Nature (London)* **181**, 1745–1746.

Wareing, P. F. (1963). The germination of seeds. *Vistas Bot.* **3**, 195–227.

Wareing, P. F., and Roberts, D. L. (1956). Photoperiodic control of cambial activity in *Robinia pseudoacacia* L. *New Phytol.* **55**, 356–366.

Wareing, P. F., and Robinson, L. W. (1963). Juvenility problems in woody plants. *Rep. For. Res.*, 125–127.

Wareing, P. F., Hanney, C. E. A., and Digby, J. (1964). The role of endogenous hormones in cambial activity and xylem differentiation. *In* "The Formation of Wood in Forest Trees" (M. H. Zimmermann, ed.), pp. 323–344. Academic Press, New York.

Wargo, P. M. (1979). Starch storage and radial growth in woody roots of sugar maple. *Can. J. For. Res.* **9**, 49–56.

Wargo, P. M., and Montgomery, M. E. (1983). Colonization by *Armillaria mellea* and *Agrilus bilineatus* of oaks injected with ethanol. *For. Sci.* **29**, 848–857.

Waring, R. H. (1983). Estimating forest growth and efficiency in relation to canopy leaf area. *Adv. Ecol. Res.* **13**, 327–354.

Waring, R. H. (1987). Characteristics of trees predisposed to die. *BioScience* **37**, 569–574.

Waring, R. H., Newman, K., and Bell, J. (1981). Efficiency of tree crowns and stemwood production at different canopy leaf densities. *Forestry* **54**, 129–137.

Warrington, I. J., Rook, D. A., Morgan D. C., and Turnbull, H. L. (1988). The influence of simulated shadelight and daylight on growth, development and photosynthesis of *Pinus radiata*, *Agathis australis* and *Dacrydium cupressinum*. *Plant Cell Environ.* **12**, 343–356.

Waters, E. R., and Schaal, B. A. (1991). No variation is detected in the chloroplast genome of *Pinus torreyana*. *Can. J. For. Res.* **21**, 1832–1835.

Watson, G. W. (1987). The relationship of root growth and tree vigour following transplanting. *Arboric. J.* **11**, 97–104.

Watson, G. W., and Sydnor, T. D. (1987). The effect of root pruning on the root system of nursery trees. *J. Arboric.* **13**, 126–130.

Watson, M. D., and Murphy, D. J. (1993). Genome organization, protein synthesis and processing in plants. *In* "Plant Biochemistry and Molecular Biology" (P. J. Lea and R. C. Leegood, eds.), pp. 197–219. Wiley, Chichester and New York.

Watts, S., Rodriguez, J. L., Evans, S., and Davies, W. J. (1981). Root and shoot growth in plants treated with abscisic acid. *Ann. Bot.* **47**, 595–602.

Watzig, H., and Fischer, B. (1987). Untersuchungen über den Jahrringausfall on Fichtenbeständen. *Wiss. Z. Tech. Univ. Dresden* **36**, 273–275.

Weaver, H. (1974). Effects of fire on temperate forests: Western United States. *In* "Fire and Ecosystems" (T. T. Kozlowski and C. E. Ahlgren, eds.), pp. 279–319. Academic Press, New York.

Webb, D. P., Van Staden, J., and Wareing, P. F. (1973a). Seed dormancy in *Acer*. Changes in endogenous cytokinins, gibberellins and germination inhibitors during the breaking of dormancy in *Acer saccharum* Marsh. *J. Exp. Bot.* **24**, 105–116.

Webb, D. P., Van Staden, J., and Wareing, P. F. (1973b). Seed dormancy in *Acer*. Changes in endogenous germination inhibitors, cytokinins, and gibberellins during the breaking of dormancy in *Acer pseudoplatanus* L. *J. Exp. Bot.* **24**, 741–750.

Weber, J. C., and Stettler, R. F. (1981). Isoenzyme variation among ten populations of *Populus trichocarpa* Torr. et Gray in the Pacific Northwest. *Silvae Genet.* **30**, 82–87.

Weber, J. C., Stettler, R. F., and Heilman, P. E. (1985). Genetic variation and productivity in *Populus trichocarpa* and its hybrids. I. Morphology and phenology of 50 native clones. *Can. J. For. Res.* **15**, 376–383.

Webster, A. D. (1980). Flower and fruitlet thinning of the plum (*Prunus domestica* L.) cv. Victoria. *J. Hortic. Sci.* **55**, 19–26.

Webster, A. D. (1989). Opportunities for high density plantings of European plum and sweet cherry. *Acta Hortic.* **243**, 309–317.

Webster, A. D. (1990). The influence of paclobutrazol on the growth and cropping of sweet cherry cultivars. II. Rootstock effects on scion sensitivity and the longevity of residual effects. *J. Hortic. Sci.* **65**, 519–527.

Webster, A. D., and Wertheim, S. J. (1993). Comparisons of species and hybrid rootstocks for European plum cultivars. *J. Hortic. Sci.* **68**, 861–869.

Webster, A. D., Quinlan, J. D., and Richardson, P. J. (1986). The influence of paclobutrazol on the growth and cropping of sweet cherry cultivars. I. The effect of cv. Early Rivers. *J. Hortic. Sci.* **61**, 471–478.

Weichmann, J. (1986). The effect of controlled atmosphere storage on the sensory and nutritional quality of fruits and vegetables. *Hortic. Rev.* **8**, 101–127.

Weigel, D., and Nilsson, O. (1995). A developmental switch sufficient for flower initiation in diverse plants. *Nature (London)* **377**, 495–500.

Weinbaum, S. A., Parfitt, D. E., and Polito, V. S. (1984). Differential cold sensitivity of pollen grain germination in two *Prunus* species. *Euphytica* **3**, 419–426.

Weinstein, L. H. (1977). Fluoride and plant life. *J. Occup. Med.* **19**, 49–78.

Wenger, K. F. (1954). The stimulation of loblolly pine seed trees by preharvest release. *J. For.* **52**, 115–118.

Wertheim, S. J. (1990). Results of plum rootstock trials in the Netherlands. *Acta Hortic.* 211–227.

West, D. W., and Taylor, J. A. (1984). Response of six grape cultivars to the combined effects of high salinity and rootzone waterlogging. *J. Am. Soc. Hortic. Sci.* **109**, 844–851.

Weste, G. (1986). Vegetation changes associated with invasion of *Phytophthora cinnamomi* of defined plots in the Brisbane Ranges, Victoria, 1975–1985. *Aust. J. Bot.* **34**, 633–648.

Westgate, M. E., and Boyer, J. S. (1985). Osmotic adjustments and the inhibition of leaf, root, stem, and silk growth at low water potentials in maize. *Planta* **164**, 540–549.

Westing, A. H. (1965). Formation and function of compression wood in gymnosperms. II. *Bot. Rev.* **31**, 381–480.

Westing, A. H. (1968). Formation and function of compression wood in gymnosperms. *Bot. Rev.* **34**, 51–78.

Westwood, M. N. (1993). "Temperate Zone Pomology: Physiology and Culture." Timber Press, Portland, Oregon.

Westwood, M. N., and Stevens, G. (1979). Factors influencing cherry and prune set. *Proc. Oreg. Hortic. Soc.* **70**, 175–179.

Wetmore, R. H., and Rier, J. P. (1963). Experimental induction of vascular tissue in callus of angiosperms. *Am. J. Bot.* **50**, 418–423.

Wheeler, N. C. (1987). Effect of paclobutrazol on Douglas-fir and loblolly pine. *J. Hortic. Sci.* **62**, 101–106.

Wheeler, N. C., and Bramlett, D. L. (1991). Flower stimulation treatments in a loblolly pine seed orchard. *South. J. Appl. For.* **15**, 44–49.

Wheeler, N. C., Masters, C. J., Cade, S. C., Ross, S. D., and Hsin, L. Y. (1985). Girdling: An effective and practical treatment for enhancing seed yields in Douglas-fir seed orchards. *Can. J. For. Res.* **15**, 505–510.

White, M. J., and Lott, J. N. A. (1983). Protein body inclusions in seeds of *Eucalyptus maculata* and *Eucalyptus erythrocorys*. *Can. J. Bot.* **61**, 1911–1918.

Whitmore, F. W., and Zahner, R. (1966). Development of the xylem ring in stems of young red pine trees. *For. Sci.* **12**, 198–210.

Whitmore, F. W., and Zahner, R. (1967). Evidence for a direct effect of water stress in the metabolism of cell walls in *Pinus*. *For. Sci.* **13**, 397–400.

Whitmore, T. C. (1983). Secondary succession from seed in tropical rain forests. *For. Abstr.* **44**, 769–779.

Whittaker, R. H. (1975). "Communities and Ecosystems." Macmillan, New York.

Wibbe, M. L., Blanke, M. M., and Lenz, F. (1993). Effect of fruiting on carbon budgets of apple tree canopies. *Trees* **8**, 56–60.

Wightman, F., Schneider, E. A., and Thimann, K. V. (1980). Hormonal factors controlling the initiation and development of lateral roots. II. Effects of exogenous growth factors on lateral root formation in pea roots. *Physiol. Plant.* **49**, 304–314.

Wilcox, J. R. (1968). Sweetgum seed stratification requirements related to winter climate at seed source. *For. Sci.* **14**, 16–19.

Wilde, H. D., Meagher, R. B., and Merkle, S. A. (1992). Expression of foreign genes in transgenic yellow-poplar plants. *Plant Physiol.* **98**, 114–120.

Wilde, S. A. (1958). "Forest Soils." Ronald, New York.

Wilhite, L. P., and McKee, W. H., Jr. (1985). Site prepartation and phosphorus application alter early growth of loblolly pine. *South. J. Appl. For.* **9**, 103–109.

Wilkinson, R. C. (1990). The effects of winter injury on basal area and height growth of 30-year-old red spruce from 12 provenances growing in northern New Hampshire. *Can. J. For. Res.* **20**, 1616–1622.

Will, G. M. (1985). Nutrient deficiencies and fertilizer use in New Zealand exotic forests. *FRI Bulletin, Forest Research Institute, Rota Rua, New Zealand.*

Williams, C. B., Jr. (1967). Spruce budworm damage symptoms related to radial growth of grand fir, Douglas-fir, and Engelmann spruce. *For. Sci.* **13**, 274–285.

Williams, J. G., Kubelik, A. R., Livak, K., Rafalski, J. A., and Tingey, S. V. (1990). DNA polymorphisms amplified by arbitrary primers are useful as genetic markers. *Nucleic Acids Res.* **18**, 6531–6535.

Williams, M. W. (1979). Chemical thinning of apples. *Hortic. Rev.* **1**, 270–300.

Williams, M. W., and Edgerton, L. J. (1981). Fruit thinning of apples and pears with chemicals. *U.S.D.A. Agric. Info. Bull.* **289**.

Williams, M. W., and Edgerton, L. J. (1983). Vegetative growth control of apple and pear trees with ICI PP333 (paclobutrazol), a chemical analog of Boyleton. *Acta Hortic.* **137**, 111–116.

Williams, M. W., and Stahly, E. A. (1969). Effect of cytokinins and gibberellins on shape of 'Delicious' apple fruits. *J. Am. Soc. Hortic. Sci.* **94**, 17–18.

Williams, R. A., Hoffman, B. F., and Seymour, R. S. (1991). Comparison of site index and biomass production of spruce–fir stands by soil drainage class in Maine. *For. Ecol. Manage.* **41**, 279–290.

Williams, R. F. (1955). Redistribution of mineral elements during development. *Annu. Rev. Plant Physiol.* **6**, 25–40.

Williamson, M. J. (1966). Premature abscissions and white oak acorn crops. *For. Sci.* **12**, 19–21.

Wilson, B. F. (1968). Effect of girdling on cambial activity in white pine. *Can. J. Bot.* **46**, 141–146.

Wilson, B. F. (1986). Apical control of compression wood action in white pine branches. *Wood Sci. Technol.* **20**, 111–117.

Wilson, B. F., and Bachelard, E. P. (1975). Effects of girdling and defoliation on root activity and survival of *Eucalyptus regnans* and *E. viminalis* seedlings. *Aust. J. Plant Physiol.* **2**, 197–206.

Wilson, B. F., and Fischer, B. C. (1977). Striped maple: Shoot growth and bud formation related to light intensity. *Can. J. For.* **7**, 1–7.

Wilson, C., Grace, J., Allen, S., and Slack, F. (1987). Temperature and stature: A study of temperatures in montane vegetation. *Funct. Ecol.* **1**, 405–413.

Wilson, J. (1980). Macroscopic featues of wind damage to leaves of *Acer pseudoplatanus* L. and its relationship with season, leaf age, and windspeed. *Ann. Bot. (London)* **46**, 303–311.

Wilson, J. (1984). Microscopic features of wind damage to leaves of *Acer pseudoplatanus* L. *Ann. Bot. (New Ser.)* **53**, 73–82.

Wilson, J. B. (1989). A review of evidence on the control of shoot:root ratio, in relation to models. *Ann. Bot.* **61**, 433–449.

Wilson, S. A., Rahe, T. A., and Webber, W. B., Jr., eds. (1985). Municipal wastewater sludge as a soil amendment for revegetating landfill cover. *J. Soil Water Conserv.* **40**, 296–299.

Winget, C. H., and Kozlowski, T. T. (1965a). Yellow birch germination and seedling growth. *For. Sci.* **11**, 386–392.

Winget, C. H., and Kozlowski, T. T. (1965b). Seasonal basal growth area as an expression of competition in northern hardwoods. *Ecology* **46**, 786–793.

Winkler, A. J. (1948). Maturity tests for table grapes—The relation of heat summation to time of maturing and palatability. *Proc. Am. Soc. Hortic. Sci.* **51**, 295–298.

Winner, W. E., Koch, G. W., and Mooney, H. A. (1982). Ecology of SO_2 resistance. IV. Predicting metabolic responses of fumigated trees and shrubs. *Oecologia* **52**, 16–21.

Witkosky, J. J., Schowalter, T. D., and Hansen, E. M. (1986). The influence of time of precommercial thinning on the colonization of Douglas-fir by three species of root-colonizing insects. *Can. J. For. Res.* **16**, 745–749.

Witt, H. J., Allison, J. R., and Daniell, J. W. (1989). Economic analyses of space management practices in high-density pecan groves. *J. Am. Soc. Hortic. Sci.* **114**, 61–64.

Witt, W., and Sauter, J. J. (1994). Starch metabolism in poplar wood ray cells during spring mobilization and summer deposition. *Physiol. Plant.* **92**, 9–16.

Wodzicki, T. J. (1965). Annual ring of wood formation and seasonal changes of natural growth-inhibitors in larch. *Acta Soc. Bot. Pol.* **34**, 117–151.

Wodzicki, T. J. (1978). Seasonal variation of auxin in stem cambial region of *Pinus silvestris* L. *Acta Soc. Bot. Pol.* **47**, 225–231.

Wodzicki, T. J., and Wodzicki, A. B. (1973). Auxin stimulation of cambial activity in *Pinus sylvestris.* II. Dependence on basipetal transport. *Physiol. Plant.* **29**, 288–297.

Wodzicki, T. J., and Wodzicki, A. B. (1980). Seasonal abscisic acid accumulation in the stem cambial region of *Pinus silvestris,* and its contribution to the hypothesis of a latewood control system in conifers. *Physiol. Plant.* **48**, 443–447.

Wodzicki, T. J., and Wodzicki, A. B. (1981). Modulation of the oscillatory system involved in polar transport of auxin by other phytohormones. *Physiol. Plant.* **53**, 176–180.

Wodzicki, T. J., and Zajaczkowski, S. (1974). Effect of auxin on xylem tracheid differentiation in decapitated stems of *Pinus silvestris* L. and its interaction with some vitamins and growth regulators. *Acta Soc. Bot. Pol.* **43**, 129–148.

Wodzicki, T. J., Rakowski, K., Starck, Z., Porandowski, J., and Zajaczkowski, S. (1982). Apical control of xylem formation in the pine stem. I. Auxin effects and distribution of assimilates. *Acta Soc. Bot. Pol.* **51**, 187–201.

Wolk, W. D., and Herner, R. C. (1982). Chilling injury of germinating seeds and seedlings. *HortScience* **17**, 169–173.

Wolter, K. E., and Zinkel, D. F. (1984). Observations on the physiological mechanisms and chemical constituents of induced oleoresin synthesis in *Pinus resinosa. Can. J. For. Res.* **14**, 452–458.

Wolters, J. H. B., and Martens, M. J. M. (1987). Effects of air pollutants on pollen. *Bot. Rev.* **53**, 372–414.

Wong, T. L., Harris, R. W., and Fissell, R. E. (1971). Influence of high soil temperatures on five woody plant species. *J. Am. Soc. Hortic. Sci.* **96**, 80–83.

Wood, B. W. (1983). Changes in indoleacetic acid, abscisic acid, gibberellins, and cytokinins during bud break of pecan. *J. Am. Soc. Hortic. Sci.* **108**, 333–338.

Wood, B. W. (1986). Cold injury susceptibility of pecan as influenced by cultivar, carbohydrates, and crop load. *HortScience* **21**, 285–286.

Wood, B. W. (1988a). Paclobutrazol suppresses shoot growth and influences nut quality and yield of young pecan trees. *J. Am. Soc. Hortic. Sci.* **113**, 374–377.

Wood, B. W. (1988b). Paclobutrazol, uniconazole, and flurprimidol influence shoot growth and nut yield of young pecan trees. *HortScience* **23**, 1026–1028.

Wood, B. W., and McMeans, J. L. (1981). Carbohydrate changes in various organs of bearing and nonbearing pecan trees. *J. Am. Soc. Hortic. Sci.* **106**, 758–761.

Woodcock, D. W. (1989). Climate sensitivity of wood-anatomical features in a ring-porous oak (*Quercus macrocarpa*). *Can. J. For. Res.* **19**, 639–644.

Woodman, J. N. (1987). Pollution-induced injury in North American forests: Facts and suspicions. *Tree Physiol.* **3**, 1–15.

Woodmansee, R. G., and Wallach, L. S. (1981). Effects of fire regimes on biogeochemical cycles. *U.S.D.A. For. Serv. Gen. Tech. Rep. WO* **WO-26**, 379–400.

Woods, D. B., and Turner, N. C. (1971). Stomatal response to changing light by four tree species of varying shade tolerance. *New Phytol.* **70**, 77–84.

Woodson, W. (1991). Biotechnology of floricultural crops. *HortScience* **26**, 1029–1033.

Woodward, F. I. (1987). "Climate and Plant Distribution." Cambridge Univ. Press, Cambridge.

Woodward, F. I. (1995). Ecophysiological controls of conifer distributions. *In* "Ecophysiology of Coniferous Forests" (W. K. Smith and T. M. Hinckley, eds.), pp. 79–94. Academic Press, San Diego.

Worley, J. F., and Hacskaylo, E. (1959). The effect of available moisture on the mycorrhizal association of Virginia pine. *For. Sci.* **5**, 267–268.

Worley, R. E. (1982). Tree yield and nut characteristics of pecans with drip irrigation under humid conditions. *J. Am. Soc. Hortic. Sci.* **107**, 30–34.

Worrall, J., Draper, D. A., and Anderson, S. A. (1985). Periphysis in stagnant lodgepole pine: An hypothesis demolished. *In* "Crop Physiology of Forest Trees" (P. M. A. Tigerstedt, P. Puttonen, and V. Koski, eds.), pp. 65–70. Department of Plant Breeding, Univ. of Helsinki, Finland.

Wright, E. (1931). The effect of high temperatures on seed germination. *J. For.* **29**, 679–687.

Wright, H. A., and Bailey, A. W. (1982). "Fire Ecology." Wiley, New York.

Wright, H. A., Churchill, F. M., and Stevens, W. C. (1982). Soil loss and runoff on seeded vs. nonseeded watersheds following prescribed burning. *J. Range Manage.* **35**, 382–385.

Wright, J. W. (1976). "Introduction to Forest Genetics." Academic Press, New York.

Wright, L. C., Berryman, A. A., and Gurusiddaiah, S. (1979). Host resistance to the fir engraver beetle, *Scolytus ventralis* (Coleoptera: Scolytidae). 4. Effect of defoliation on wound monoterpene and inner bark carbohydrate concentrations. *Can. Entomol.* **111**, 1255–1262.

Wright, L. C., Berryman, A. A., and Wickman, B. E. (1984). Abundance of the fir engraver, *Scolytus ventralis*, and the Douglas-fir beetle, *Dendroctonus pseudotsugae*, following tree defoliation by the Douglas-fir tussock moth, *Orygia pseudotsuga*. *Can. Entomol.* **116**, 293–305.

Wright, R. D., and Niemiera, A. X. (1987). Nutrition of container-grown woody nursery crops. *Hortic. Rev.* **9**, 75–101.

Wright, S. J., and van Schaik, C. P. (1994). Light and the phenology of tropical trees. *Am. Nat.* **143**, 192–199.

Wu, C. C., and Kozlowski, T. T. (1972). Some histological effects of direct contact on *Pinus resinosa* seeds and young seedlings with 2,4,5-T. *Weed Res.* **12**, 229–233.

Wu, C. C., Kozlowski, T. T., Evert, R. F., and Sasaki, S. (1971). Effects of direct contact on *Pinus resinosa* seeds and young seedlings with 2,4,-D or picloram on seedling development. *Can. J. Bot.* **49**, 1737–1742.

Wu, F. T., and Barnes, M. F. (1981). The hormone levels in stem cuttings of difficult-to-root and easy-to-root rhododendrons. *Biochem. Physiol. Pflanz.* **176**, 13–22.

Wu, Y., Spollen, W. G., Sharp, R. E., Hetherington, P. R., and Fry, S. C. (1994). Root growth maintenance at low water potentials. Increased activity of xyloglucan endotransglycosylase and its possible regulation by abscisic acid. *Plant Physiol.* **106**, 607–615.

Wuenscher, J. E., and Kozlowski, T. T. (1971). The relationship of gas exchange resistance to tree seedling ecology. *Ecology* **52**, 1016–1023.

Wutscher, H. K., and Dube, D. (1977). Performance of young nucellar grapefruit on 20 rootstocks. *J. Am. Soc. Hortic. Sci.* **102**, 267–270.

Wyman, D. (1950). Order of bloom. *Arnoldia (Boston)* **10**, 41–56.

Yamaguchi, K., Shimaji, K., and Itoh, T. (1983). Simultaneous inhibition and induction of compression wood formation by morphactin in artificially inclined stems of Japanese larch (*Larix leptolepis* Gordon). *Wood Sci. Technol.* **17**, 81–89.

Yamamoto, F., and Kozlowski, T. T. (1987a). Effect of Ethrel on stem anatomy of *Ulmus americana* seedlings. *IAWA Bull. New Ser.* **8**, 3–9.

Yamamoto, F., and Kozlowski, T. T. (1987b). Effect of ethrel on growth and stem anatomy of *Pinus halepensis* seedlings. *IAWA Bull. New Ser.* **8**, 11–19.

Yamamoto, F., and Kozlowski, T. T. (1987c). Effects of flooding, tilting of stems, and Ethrel application on growth, stem anatomy and ethylene production of *Pinus densiflora* seedlings. *J. Exp. Bot.* **38**, 293–310.

Yamamoto, F., and Kozlowski, T. T. (1987d). Effects of flooding of soil on growth, stem anatomy, and ethylene production of *Cryptomeria japonica* seedlings. *Scand. J. For. Res.* **2**, 45–58.

Yamamoto, F., and Kozlowski, T. T. (1987e). Effects of flooding of soil and application of NPA and NAA to stems on growth and stem anatomy of *Acer negundo* seedlings. *Environ. Exp. Bot.* **27**, 329–340.

Yamamoto, F., and Kozlowski, T. T. (1987f). Effects of flooding, tilting of stems, and Ethrel application on growth, stem anatomy, and ethylene production of *Acer platanoides* seedlings. *Scand. J. For. Res.* **2**, 141–156.

Yamamoto, F., Angeles, G., and Kozlowski, T. T. (1987a). Effect of Ethrel on stem anatomy of *Ulmus americana* seedlings. *IAWA Bull. New Ser.* **8**, 11–19.

Yamamoto, F., Kozlowski, T. T., and Wolter, K. E. (1987b). Effect of flooding on growth, stem anatomy and ethylene production of *Pinus halepensis* seedlings. *Can. J. For. Res.* **17**, 69–79.

Yamamoto, F., Sakata, T., and Terazawa, K. (1995). Physiological, morphological, and anatomical responses of *Fraxinus mandshurica* seedlings to flooding. *Tree Physiol.* **15**, 713–719.

Yang, S. F. (1987). The role of ethylene and ethylene synthesis in fruit ripening. *In* "Plant Senescence: Its Biochemistry and Physiology" (W. W. Thomson, E. A. Nothnagel, and R. C. Huffaker, eds.), pp. 156–166. Am. Soc. Plant Physiologists, Rockville, Maryland.

Yao, J.-L., Cohen, D., Atkinson, R., Richardson, K., and Morris, B. (1995). Regeneration of transgenic plants from the apple cultivar Royal Gala. *Plant Cell Rep.* **14**, 407–412.

Yeiser, J. L., and Barnett, J. P. (1991). Growth and physiological response of four shortleaf pine families to herbicidal control of herbaceous competition. *South. J. Appl. For.* **15**, 199–204.

Yelenosky, G. (1976). Cold hardening young 'Valencia' orange trees on 'Swingle' citrumelo (CPB-4475) and other rootstocks. *Proc. Fla. State Hortic. Soc.* **89**, 9–10.

Yelenosky, G. (1979). Water-stress induced cold hardening of young citrus trees. *J. Am. Soc. Hortic. Sci.* **104**, 270–273.

Ying, C. C., and Bagley, W. T. (1977). Variation in rooting capability of *Populus deltoides*. *Silvae Genet.* **26**, 204–206.

Yogaratnam, N., Allen, M., and Greenham, D. W. P. (1981). The phosphorus concentration in apple leaves as affected by foliar application of its compounds. *J. Hortic. Sci.* **56**, 255–260.

Yoshida, S. (1974). Studies on lipid changes associated with frost hardiness in cortex in woody plants. *Contrib. Inst. Low Temp. Sci., Ser. B* (*Hokkaido Univ.*) 1–43.

Yoshikawa, M., Yamaoka, N., and Takeuchi, Y. (1993). Elicitors: Their significance and primary modes of action in the induction of plant defense reactions. *Plant Cell Physiol.* **34**, 1163–1173.

Young, A. G., Warwick, S. I., and Merriam, H. G. (1993). Genetic variation and structure at three spatial scales for *Acer saccharum* (sugar maple) in Canada and the implications for conservation. *Can. J. For. Res.* **23**, 2569–2578.

Young, E. (1989). Cytokinin and soluble carbohydrate concentration in xylem sap of apple during dormancy and budbreak. *J. Am. Soc. Hortic. Sci.* **114**, 297–300.

Young, E. (1992). Timing of high temperature influences chilling negation in dormant apple shoots. *J. Am. Soc. Hortic. Sci.* **117**, 271–272.

Young, H. E., and Kramer, P. J. (1952). The effect of pruning on the height and diameter growth of loblolly pine. *J. For.* **50**, 474–479.

Young, V. A., and Young, C. G. (1992). "Seeds of Woody Plants in North America." Dioscorides Press, Portland, Oregon.

Zachrisson, A., and Bornman, C. H. (1984). Application of electric field fusion in plant tissue culture. *Physiol. Plant.* **61**, 314–320.

Zackrisson, O., and Nilsson, M. C. (1992). Allelopathic effects by *Empetrum hermaphroditum* on seed germination of two boreal tree species. *Can. J. For. Res.* **22**, 1310–1319.

Zahner, R. (1958). September rains bring growth gains. *U.S.D.A. For. Serv., South. For. Exp. Stn. For.* Note 113.

Zahner, R. (1962). Terminal growth and wood formation by juvenile loblolly pine under two soil moisture regimes. *For. Sci.* **8**, 345–352.

Zahner, R. (1968). Water deficits and growth of trees. *In* "Water Deficits and Plant Growth" (T. T. Kozlowski, ed.), Vol. 2, pp. 191–254. Academic Press, New York.

Zahner, R., and Oliver, W. W. (1962). The influence of thinning and pruning on the date of summerwood initiation in red and jack pines. *For. Sci.* **8**, 51–63.

Zahner, R., and Whitmore, F. W. (1960). Early growth of radically thinned loblolly pine. *J. For.* **58**, 628–634.

Zajaczkowski, S. (1973). Auxin stimulation of cambial activity in *Pinus silvestris* L. I. The differential cambial response. *Physiol. Plant.* **29**, 231–287.

Zakrzewski, J. (1983). Hormonal control of cambial activity and vessel differentiation in *Quercus robur. Physiol. Plant.* **57**, 537–542.

Zanchin, A., Marcato, C., Trainotti, L., Casadoro, G., and Rascio, N. (1995). Characterization of abscission zones in the flowers and fruits of peach [*Prunus persica* (L.) Batsch]. *New Phytol.* **129**, 345–354.

Zandvoort, E. A., and Holdgate, D. P. (1991). Mechanisation in tissue culture systems. *Acta Hortic.* **289**, 203–212.

Zarska-Maciejewska, B., and Lewak, S. (1983). The role of proteolytic enzymes in the release from dormancy of apple seeds. *Z. Pflanzenphysiol.* **110**, 409–417.

Zavitkovski, J. (1979). Energy production in irrigated, intensively cultured plantations of *Populus* 'Tristis #1' and jack pine. *For. Sci.* **25**, 383–392.

Zekri, M. (1993). Salinity and calcium effects on emergence, growth and mineral composition of seedlings of eight citrus rootstocks. *J. Hortic. Sci.* **68**, 53–62.

Zhang, J., and Davies, W. J. (1990). Changes in the concentration of ABA in xylem sap as a function of changing soil water status can account for changes in leaf conductance and growth. *Plant Cell Environ.* **13**, 277–286.

Zhu, Q., Maher, E. A., Masoud, S., Dixon, R. A., and Lamb, C. J. (1994). Enhanced protection against fungal attack by constitutive co-expression of chitinase and glucanase genes in transgenic tobacco. *Bio/Technology* **12**, 807–812.

Zimmerman, R. H. (1971). Flowering in crabapple seedlings. Methods of shortening the juvenile phase. *J. Am. Soc. Hortic. Sci.* **96**, 404–411.

Zimmermann, M. H., and McDonough, J. (1978). Dysfunction in the flow of food. *In* "Plant Disease: An Advanced Treatise" (J. G. Horsfall and E. B. Cowling, eds.), pp. 117–140. Academic Press, New York.

Zimmermann, M. H., and Potter, D. (1982). Vessel-length distribution in branches, stems, and roots of *Acer rubrum. IAWA Bull. New Ser.* **3**, 103–109.

Zimmermann, U., and Scheurich, P. (1981). High frequency fusion of plant protoplasts by electric fields. *Planta* **151**, 26–32.

Zisa, R. P., Halverson, H. G., and Stout, B. J. (1980). Establishment and early growth of conifers on compost soils in urban areas. *U.S.D.A. For. Serv. Res. Pap. NE* **NE-451**.

Zobel, B. J., and Talbert, J. (1984). "Applied Forest Tree Improvement." Wiley, New York.

Zobel, B., and van Buijtenen, J. P. (1989). "Wood Variation." Springer-Verlag, Berlin and New York.

Zsuffa, L., Sennerby-Forsse, L., Weisgerber, H., and Hall, R. B. (1993). Strategies for clonal forestry with poplars, aspens and willows. *In* "Clonal Forestry" (M. R. Ahuja and W. J. Libby, eds.), Vol. 2, pp. 91–119. Springer-Verlag, Berlin.

Index

Physiological Ecology
A Series of Monographs, Texts, and Treatises

Series Editor
Harold A. Mooney
Stanford University, Stanford, California

Editorial Board
Fakhri A. Bazzaz   F. Stuart Chapin   James R. Ehleringer
Robert W. Pearcy   Martyn M. Caldwell   E.-D. Schulze

T. D. SHARKEY, E. A. HOLLAND, and H. A. MOONEY (Eds.). Trace Gas Emissions by Plants, 1991

U. SEELIGER (Ed.). Coastal Plant Communities of Latin America, 1992

JAMES R. EHLERINGER and CHRISTOPHER B. FIELD (Eds.). Scaling Physiological Processes: Leaf to Globe, 1993

JAMES R. EHLERINGER, ANTHONY E. HALL, and GRAHAM D. FARQUHAR (Eds.). Stable Isotopes and Plant Carbon–Water Relations, 1993

E.-D. SCHULZE (Ed.). Flux Control in Biological Systems, 1993

MARTYN M. CALDWELL and ROBERT W. PEARCY (Eds.). Exploitation of Environmental Heterogeneity by Plants: Ecophysiological Processes Above- and Belowground, 1994

WILLIAM K. SMITH and THOMAS M. HINCKLEY (Eds.). Resource Physiology of Conifers: Acquisition, Allocation, and Utilization, 1995

WILLIAM K. SMITH and THOMAS M. HINCKLEY (Eds.). Ecophysiology of Coniferous Forests, 1995

MARGARET D. LOWMAN and NALINI M. NADKARNI (Eds.). Forest Canopies, 1995

BARBARA L. GARTNER (Ed.). Plant Stems: Physiology and Functional Morphology, 1995

GEORGE W. KOCH and HAROLD A. MOONEY (Eds.). Carbon Dioxide and Terrestrial Ecosystems, 1996

CHRISTIAN KÖRNER and FAKHRI A. BAZZAZ (Eds.). Carbon Dioxide, Populations, and Communities, 1996

THEODORE T. KOZLOWSKI and STEPHEN G. PALLARDY. Growth Control in Woody Plants, 1997

J. J. LANDSBERG and S. T. GOWER. Applications of Physiological Ecology to Forest Management, 1997

Printed and bound by CPI Group (UK) Ltd, Croydon, CR0 4YY

08/05/2025

01864897-0001